Immunocytochemistry

Immunocytochemistry
Modern Methods and Applications

Edited by
Julia M. Polak and
Susan Van Noorden

Second Edition

Bristol
1986

Published by
John Wright & Sons Ltd, Techno House, Redcliffe
Way, Bristol BS1 6NX, England

First edition, 1983
Second edition, 1986
Reprinted, 1987

British Library Cataloguing in Publication Data

Immunocytochemistry: modern methods and
applications.——2nd ed.
  1. Immunochemistry   2. Cytochemistry
  I. Polak, Julia M.   II. Van Noorden, Susan
  574.2′9   QR183.6

ISBN 0 7236 0870 9

Typeset by Activity Limited, Salisbury, Wilts

Printed in Great Britain by J. W. Arrowsmith Ltd, Bristol

Preface to the Second Edition

In the three years since the First Edition of this book was published, immunocytochemistry has become firmly established in routine diagnostic laboratories and its applications have multiplied. Greater familiarity with the techniques has spawned further modifications and exploitation of new antigens, to the extent that the book required up-dating. The need now was not so much for instruction in the techniques that have become standard but for explanation of the latest localization methods and their uses. The result is largely a new book which complements, but does not supersede, the first edition. Of its forty chapters, twenty-eight are completely new and the remaining twelve are fully revised versions of the originals.

We have kept the same general format as before with a section on techniques leading to a section on applications in the widest sense, followed by discussions of current diagnostic uses of immunocytochemistry in histopathology and consideration of how a diagnostic service can best be provided.

The 'techniques' section has been enlarged by separate chapters devoted to raising monoclonal antibodies, the use of colloidal gold for light microscopical immunocytochemistry, radiolabelling, avidin–biotin methods and a survey of immunostaining methods for electron microscopy. Localization other than strictly immunocytochemical is covered in chapters dealing with lectin histochemistry, localization of receptor sites and the latest means of showing cell products in the making rather than at their storage site — hybridization histochemistry. Some methods of comparative quantification are described and the section includes an important but often forgotten aspect of microscopy: how best to record results by photomicrography.

In the 'applications' section new subjects are uses of cell lines, immunocytochemistry of oncogenes, applications to evolutionary studies, and greatly expanded accounts of intermediate filaments and studies on the nervous system. There is an additional chapter on autoimmunity with regard to non-organ-specific autoantibodies and a discussion of organ transplant studies.

New subjects in the section most directly concerned with diagnosis are thyroid disorders, cytology, haematology, micro-organisms and skeletal muscle disorders.

Technical instructions are incorporated in the text or included in appendices to the chapters as appropriate. We have tried to avoid repetition while ensuring that methods are presented in sufficient detail to be followed by the reader. Source lists of immune reagents used are provided.

The strength of immunocytochemistry lies in its visual impact, as emphasized in this book and elsewhere by numerous references to contrast, intensity of stain and low background. We have been able to provide colour plates so that the reader may experience the same 'lift' as the experimenter on looking down the microscope at a successfully immunostained preparation. We acknowledge with thanks the contributions of the companies, listed on p. xix, to the cost of colour reproduction.

Authors were asked to give general accounts of their fields and to point out the pitfalls and problems that may be encountered, as well as to describe their own expert contributions to immunocytochemical studies. Once again, they have responded magnificently and we are more than grateful to them for making this book possible.

Julia M. Polak
Susan Van Noorden

Preface to the First Edition

In 1941 the genius of Albert H. Coons planted the seed of a new way of localizing tissue constituents. His revolutionary concept, known variously as 'immunohistochemistry' or 'immunocytochemistry', took root so firmly that its offshoots now reach to the distant corners of all the fields of biology and diagnostic pathology from cell membranes to tumours.

The subject's literature has burgeoned likewise, and is scattered as widely as its many applications. Thus scientists seeking the best method of solving a particular problem and wondering whether immunocytochemistry might be the answer have had a hard struggle to find reliable information on the latest methods, all propounded eagerly by their inventors but not necessarily right for the job in hand.

It is therefore opportune to present a compact and comprehensive book giving in one volume an account of how immunocytochemical methods may be applied to many problems of pathology and biology, showing the variety of methods in use today and providing enough practical detail to inspire the nervous or uninitiated reader with confidence to try them.

We do not attempt to discuss the principles of antigen–antibody interactions as applied to immunocytochemistry because this has recently been admirably done by Ludwig Sternberger, who has himself graciously agreed to write the Introduction to this book. Following the introduction of enzymes as antibody markers by Paul Nakane, Sternberger created the unlabelled antibody–enzyme bridge procedures and the peroxidase–anti-peroxidase complex which enormously increased the sensitivity of the method. Under his influence, immunocytochemistry has grown from an interesting but not always reliable technique into an essential tool to help us not only in diagnosis but also in understanding more about basic inter- and intracellular processes. In fact, the entire discipline of histochemistry, which used to consist largely of the localization of enzymes by their reactions in situ, has been altered by the expansion of immunocytochemical methods, as a glance at the journals dealing with histochemistry will show. Even enzymes are now localized as often as not by their immunocytochemical reactivity, rather than by their biochemical actions.

Sternberger's own recent work on the relationship between the different cell types of the nervous system uses the newest immunocytochemical techniques made possible by Cesar Milstein's concept of monoclonal antibodies of absolutely uniform specificity and is fully described in his introductory chapter.

Some pathologists argue that immunocytochemistry, though of great value as a research tool, currently has rather few day-to-day practical applications in diagnosis, and that retrospective confirmation by immunocytochemistry of a diagnosis made on clinical, morphological and biochemical evidence is not vital to the patient. This viewpoint is put forward in Chapter 18, together with the author's acknowledgement of the very real use for immunocytochemical techniques in diagnosis of renal disease, lymphomas, skin diseases and the identification of organisms. Other chapters (11, 14, 15, 17, 19, 20) make abundantly clear the view that immunocytochemistry has a wide application in these and other branches of diagnostic histopathology. Indeed, the advances in technique that have come about during the past decade will surely soon be outdated by further improvements in the sensitivity and specificity of methods of localizing substances in tissues, defining diseases and contributing to their cure.

The authors of this book have been chosen particularly because they are all active exponents of the technique. They daily use up-to-date immunocytochemical methods in their routine diagnostic and research laboratories; they innovate, adapt and experiment with modifications and have themselves made large contributions to the available methodology and applications. A certain degree of overlap between the chapters was inevitable but we hope there is not too much. Each chapter stands by itself but references to other chapters are included. Some of the chapters deal mainly with techniques, and a technical appendix follows many of the others. Although several authors may use the same technique (e.g. the PAP method), there are so many variations in practice that we thought it useful for authors to describe their preferred variants, partly to show that there are more ways than one of going about this very flexible method and partly to suggest the sort of modification that might be tried. We hope that people will read the whole book, enjoy the reviews and get an idea of what is worth doing, how to do it and where to look for further information.

The whole point of immunocytochemistry lies in its visual impact and to this end we have aimed at high quality photographic reproduction. We have been able to produce colour prints as well, where they are essential for the appreciation of results, for example in double staining methods. We are most grateful to the many commercial firms whose generous financial help has made the colour reproduction possible.

The idea for this book arose as a result of our annual practical course in immunocytochemistry at the Royal Postgraduate Medical School. The number of applicants far exceeds the available places and it seemed to us that there was an urgent demand for information on the usage of immunocytochemistry from practitioners of every branch of biological science. Needless to say, the idea could not have grown into the present reality without the dedicated cooperation of all our authors, many of whom have also been kind enough to participate in the courses. We have indeed been fortunate in their enthusiastic response to our requests that they write chapters for this book. We owe them our heartfelt thanks for their toleration of our exigent instructions that they keep to the deadline, and above all for their hard work and splendid achievements.

Julia M. Polak
Susan Van Noorden

References

1. Coons A. H., Creech H. J. and Jones R. N. Immunological properties of an antibody containing a fluorescent group. *Proc. Soc. Exp. Biol. Med.* 1941, **47**, 200–202.
2. Sternberger L. A. *Immunocytochemistry,* 2nd ed. 1979. New York, John Wiley & Sons.
3. Nakane P. K. and Pierce G. B. Jr. Enzyme-labelled antibodies: preparation and application for the localization of antigen. *J. Histochem. Cytochem.* 1966, **14**, 929–931.
4. Sternberger L. A., Hardy P. H. Jr, Cuculis J. J. and Meyer H. G. The unlabelled antibody–enzyme method of immunohistochemistry. Preparation of soluble antigen–antibody complex (horseradish peroxidase–antihorseradish peroxidase) and its use in the identification of spirochetes. *J. Histochem. Cytochem.* 1970, **18**, 315–333.

List of Contributors

L. F. Agnati

Department of Human Physiology, University of Modena, Modena, Italy. *Chapter 13*

J. Allanson

Wolfson College, Oxford, UK. *Chapter 7*

Y. S. Allen

Department of Neuropathology, Institute of Pathology, De Crespigny Park, Denmark Hill, London SE5, UK. *Chapter 21*

S. T. Appleyard

Jerry Lewis Muscle Research Center, Royal Postgraduate Medical School, Du Cane Road, London W12 0HS, UK. *Chapter 39*

P. A. Berg

Department of Internal Medicine II, University Hospital of Tübingen, D-7400 Tübingen, Germany. *Chapter 29*

A. Bignami

Spinal Cord Injury Research Laboratory, Veterans Administration Medical Center, 1400 VFW Parkway, West Roxbury, Massachusetts 02132, USA. *Chapter 24*

M. Blaschek

Department of Internal Medicine II, University Hospital of Tübingen, D-7400 Tübingen, Germany. *Chapter 29*

S. R. Bloom

Department of Medicine, Royal Postgraduate Medical School, Hammersmith Hospital, Du Cane Road, London W12 0HS, UK. *Chapter 20*

G. F. Bottazzo

Department of Immunology, The Middlesex Hospital Medical School, 40–50 Tottenham St, London W1P 9PG, UK. *Chapters 28, 29*

S. Bradbury

Department of Human Anatomy, University of Oxford, South Parks Road, Oxford OX1 3QX, UK. *Chapter 14*

J. Broers

Department of Pathology, University of Nijmegen, Nijmegen, The Netherlands. *Chapter 23*

M. J. Capaldi

Advanced Drug Delivery Research Unit, Ciba Geigy Pharmaceuticals, Wimblehurst Road, Horsham, West Sussex RH12 4AB, UK. *Chapter 39*

A. C. Chu

Department of Dermatology, Royal Postgraduate Medical School, Hammersmith Hospital, Du Cane Road, London W12 0HS, UK. *Chapter 36*

A. R. M. Coates

Department of Medical Microbiology, The London Hospital Medical College, University of London, Turner Building, Turner St., London E1 2AD, UK. *Chapter 38*

G. Coggi

University of Milan Medical School, IV Department of Pathology, Istituto di Scienze Biomediche 'S. Paolo', via A di Rudini 8, 20142 Milano, Italy. *Chapter 4*

D. Dahl

Spinal Cord Injury Research Laboratory, Veterans Administration Medical Center, 1400 VFW Parkway, West Roxbury, Massachusetts 02132, USA. *Chapter 24*

J. De Mey

Laboratory of Biochemical Cytology, Division of Cellular Biology and Chemotherapy, Janssen Pharmaceutica Research Laboratories, B-2340 Beerse, Belgium. *Chapters 1, 5, 8*

M. De Waele

Department of Haematology, University Hospital of the Free University of Brussels, B-1090 Brussels, Belgium. *Chapter 5.*

B. M. Dean

Department of Immunology, The Middlesex Hospital Medical School, 40–50 Tottenham St, London W1P 9PG, UK. *Chapter 28.*

P. Dell'Orto

University of Milan Medical School, IV Department of Pathology, Istituto di Scienze Biomediche 'S. Paolo', via A di Rudini 8, 20142 Milano, Italy. *Chapter 4.*

D. Doniach   Department of Immunology, The Mid-
                              dlesex Hospital Medical School, 40–50
                              Tottenham Street, London W1P 9PG, UK.
                              Chapter 29.

M. J. Dunn   Jerry Lewis Muscle Research Centre, Royal
                              Postgraduate Medical School, Hammers-
                              smith Hospital, Du Cane Road, London
                              W12 0HS, UK. *Chapter 39.*

D. J. Evans   Department of Histopathology, Royal Post-
                              graduate Medical School, Hammersmith
                              Hospital, Du Cane Road, London W12
                              0HS, UK. *Chapter 37.*

K. Fuxe   Department of Histology, Karolinska Insti-
                              tutet, Solnavagen 1, S-704-01 Stockholm 60,
                              Sweden. *Chapter 13.*

G. Gastl   Department of Internal Medicine, Universi-
                              ty of Innsbruck, Austria. *Chapter 17.*

S. J. Gibson   Department of Histochemistry, Royal Post-
                              graduate Medical School, Hammersmith
                              Hospital, Du Cane Road, London W12
                              0HS, UK. *Chapter 22.*

R. B. Goudie   Department of Pathology, Royal Infirmary,
                              Glasgow G4 0SF, UK. *Chapter 40.*

B. A. Gusterson   Ludwig Institute for Cancer Research,
                              Royal Marsden Hospital, Sutton, Surrey
                              SM2 5PX, UK. *Chapter 16.*

G. Hacker   Department of Histochemistry, Royal Post-
                              graduate Medical School, Hammersmith
                              Hospital, Du Cane Road, London W12
                              0HS, UK.
                              and
                              University of Salzburg, Department of
                              Zoology, Akademiestrasse 26, A-5020 Salz-
                              burg, Austria. *Chapter 5.*

A. Härfstrand   Department of Histology, Karolinska Insti-
                              tutet, Solnavagen 1, S-704-01 Stockholm 60,
                              Sweden. *Chapter 13.*

E. Heyderman   Department of Histopathology, 2nd Floor
                              North Wing, St Thomas' Hospital Medical
                              School, London SE1 7EH, UK. *Chapter 31.*

S. P. Hunt

MRC Neurochemistry and Pharmacology Unit, Medical School, Hills Road, Cambridge CB2 2QB, UK. *Chapter 7.*

P. G. Isaacson

Department of Histopathology, School of Medicine, University College London, University Street, London WC1E 6JJ, UK. *Chapter 34.*

G. Janossy

Academic Department of Immunology, Royal Free Hospital School of Medicine, Pond Street, London NW3 2QG, UK. *Chapter 27.*

A. M. Janson

Department of Histology, Karolinska Institutet, Solnavagen 1, S-704-01 Stockholm 60, Sweden. *Chapter 13.*

K. R. Jessen

Department of Anatomy and Embryology, University College, London University, Gower Street, London WC1E 6BT, UK. *Chapter 15.*

A. Leathem

Bland Sutton Institute, The Middlesex Hospital, Mortimer Street, London W1, UK. *Chapter 10.*

P. W. Mantyh

CURE, VA Center, Wadsworth Building 115, Room 217, Los Angeles, California 90073, USA. *Chapter 7.*

E. Matutes

Leukaemia Unit, Haematology Department, Royal Postgraduate Medical School, Hammersmith Hospital, Du Cane Road, London W12 0HS, UK. *Chapter 35.*

R. Mirakian

Department of Immunology, The Middlesex Hospital Medical School, 40–50 Tottenham St, London W1P 9PG, UK. *Chapter 28.*

M. Moeremans

Laboratory of Biochemical Cytology, Division of Cellular Biology and Chemotherapy, Janssen Pharmaceutica Research Laboratories, B-2340 Beerse, Belgium. *Chapter 1.*

P. Monaghan

Ludwig Institute for Cancer Research, Royal Marsden Hospital, Sutton, Surrey SM2 5PX, UK. *Chapter 16.*

P. Ordronneau	Department of Anatomy, University of North Carolina, 108 Swing Building 217H, Chapel Hill, North Carolina 27514, USA. *Chapter 26.*
P. Petrusz	Department of Anatomy, University of North Carolina, 108 Swing Building 217H, Chapel Hill, North Carolina 27514, USA. *Chapter 26.*
J. M. Polak	Department of Histochemistry, Royal Postgraduate Medical School, Hammersmith Hospital, Du Cane Road, London W12 0HS, UK. *Chapters 9, 20, 22.*
R. Pujol-Borrell	Department of Immunology, The Middlesex Hospital Medical School, 40–50 Tottenham St, London W1P 9PG, UK. *Chapter 28.*
U. R. Rapp	Laboratory of Viral Carcinogenesis, National Cancer Institute, Frederick Cancer Research Facility, Frederick, Maryland 21701, USA. *Chapter 17.*
F. Ramaekers	Department of Pathology, University of Nijmegen, Nijmegen, The Netherlands. *Chapter 23.*
R. C. Richards	Department of Medical and Cell Biology, University of Liverpool, PO Box 147, Liverpool L69 38X, UK. *Chapter 6.*
M. A. Ritter	Department of Immunology, Royal Postgraduate Medical School, Hammersmith Hospital, Du Cane Road, London W12 0HS, UK. *Chapter 2.*
G. W. Roberts	CRC Division of Psychiatry, Northwick Park Hospital, Harrow, Middlesex, UK. *Chapter 21.*
J. L. Roberts	Department of Biochemistry, Columbia University, 630 West 168th Street New York, New York 100032, USA. *Chapter 12*
M. Rose	Immunology Department, Harefield Hospital, Uxbridge, Middlesex, UK. *Chapter 30*

W. Scherbaum — Department of Internal Medicine I, University Hospital of Ulm, D-7900 Ulm/ Danube, Germany (FRG). *Chapters 28, 29.*

S. Semoff — Department of Parasitology, Liverpool School of Tropical Medicine, Liverpool, UK. *Chapter 6.*

C. Sewry — Jerry Lewis Muscle Research Centre, Royal Postgraduate Medical School, Hammersmith Hospital, Du Cane Road, London W12 0HS, UK. *Chapter 39.*

D. R. Springall — Cell Pathology and Cytogenetics Unit, Wandle Valley Hospital, Mitcham Junction, Surrey CR4 4XL, UK. and Department of Histochemistry, Royal Postgraduate Medical School, Hammersmith Hospital, Du Cane Road, London W12 0HS, UK. *Chapters 5, 33.*

H. W. M. Steinbusch — Department of Pharmacology, Free University of Amersterdam, 1081 BT Amsterdam, The Netherlands. *Chapter 18.*

A. Suitters — Immunology Department, Harefield Hospital, Uxbridge, Middlesex, UK. *Chapter 30.*

M. C. Thorndyke — Department of Zoology, Royal Holloway College, Egham Hill, Egham, Surrey TW20 0EX, UK. *Chapter 19.*

K. R. Trenholm — Department of Parasitology, Liverpool School of Tropical Medicine, Liverpool, UK. *Chapter 6.*

J. Q. Trojanowski — Division of Neuropathology, Department of Pathology and Laboratory Medicine, University of Pennsylvania School of Medicine, Philadelphia, Pennsylvania 19104, USA. *Chapter 25.*

S. Van Noorden — Department of Histopathology, Royal Postgraduate Medical School, Hammersmith Hospital, Du Cane Road, London W12 0HS, UK. *Chapter 3.*

I. M. Varndell

Department of Histochemistry, Royal Post-graduate Medical School, Hammersmith Hospital, Du Cane Road, London W12 0HS, UK. *Chapter 9.*

A. A. J. Verhofstad

Dept of Anatomy and Embryology, University of Nijmegen, PO Box 9101, 6500 HB Nijmegen, The Netherlands. *Chapter 18*

G. Viale

University of Milan Medical School, IV Department of Pathology, Istituto di Scienze Biomediche 'S. Paolo', via A di Rudini 8, 20142 Milano, Italy. *Chapter 4.*

G. P. Vooijs

Department of Pathology, University of Nijmegen, Nijmegen, The Netherlands. *Chapter 23.*

R. A. Walker

Department of Pathology, Clinical Sciences Building, Leicester Royal Infirmary, PO Box 65, Leicester LE2 7EX, UK. *Chapter 11.*

J. M. Ward

Laboratory of Comparative Carcinogenesis, National Cancer Institute, Bethesda, Maryland 21701, USA. *Chapter 17.*

E. D. Williams

Department of Pathology, The Welsh National School of Medicine, Heath Park, Cardiff CF4 4XN, UK. *Chapter 32.*

J. N. Wilcox

Dept of Biochemistry, Columbia University, 630 West 168th Street, New York 10032, USA. *Chapter 12.*

D. H. Wright

Department of Pathology, University of Southampton, South Laboratory/Pathology Block, Southampton General Hospital, Tremona Road, Southampton SO9 4XY, UK. *Chapter 34.*

M. Zoli

Department of Human Physiology, University of Modena, Modena, Italy. *Chapter 13.*

Acknowledgements

We acknowledge with gratitude the generous contribution to the cost of colour reproductions for this book made by the firms listed below.

Amersham International plc,
White Lion Road,
Amersham,
Buckinghamshire HP7 9LL, UK

BDH Chemicals Ltd,
Broom Road,
Poole BH12 4NN, UK

N. V. Becton Dickinson Benelux SA,
Postbus 13,
Denderstraat 24,
9440 Erembodegem-Aalst,
Belgium

Biogen SA,
46 Route des Acacias,
1227 Geneva,
Switzerland

Bio-Nuclear Services Ltd,
24 Westleigh Drive,
Sonning Common,
Reading RG4 9LB, UK

Cambridge Research Biochemicals Ltd,
Button End Industrial Estate,
Harston,
Cambridgeshire CB2 5NX, UK

Dako Ltd,
22 The Arcade,
The Octagon,
High Wycombe,
Buckinghamshire HP11 2HT, UK

Euro-Diagnostics bv,
Wilmersdorf 24,
7327 AC Apeldoorn,
The Netherlands

Milab,
Malmö Immunolaboratorium AB,
Box 20047,
S–20074 Malmö 20,
Sweden

Peninsula Laboratories Inc.,
611 Taylor Way,
Belmont,
California 94002, USA

Triton Biosciences Inc.,
6900 Fannin Street,
Suite 270,
Houston,
Texas 77030,
USA

Contents

List of plates   xxv

Foreword to First Edition   xxix

Plates 1–44   facing page 546

TECHNIQUES

Antibody raising and testing
1   Raising and testing polyclonal antibodies for immunocytochemistry   3
   J. De Mey and M. Moeremans
2   Raising and testing monoclonal antibodies for immunocytochemistry   13
   M. A. Ritter

Immunostaining for light microscopy
3   Tissue preparation and immunostaining techniques for light microscopy   26
   S. Van Noorden
4   Avidin–biotin methods   54
   G. Coggi, P. Dell'Orto and G. Viale
5   Gold probes in light microscopy   71
   J. De Mey, G. Hacker, M. De Waele and D. R. Springall
6   The use of semithin frozen sections in immunocytochemistry   89
   R. C. Richards, K. R. Trenholm and S. Semoff
7   Radioimmunocytochemistry   99
   S. P. Hunt, J. Allanson and P. W. Mantyh

Immunostaining for electron microscopy
8   The preparation and use of gold probes   115
   J. De Mey
9   Electron microscopical immunocytochemistry   146
   I. M. Varndell and J. M. Polak

Newer types of localization

10 Lectin histochemistry 167
A. Leathem

11 The localization of receptors and binding sites with reference to steroids 188
R. A. Walker

12 Hybridization histochemistry: identification of specific mRNAs
in individual cells 198
J. L. Roberts and J. N. Wilcox

Ancillary techniques

13 Quantitative analysis: computer-assisted morphometry and microdensitometry applied to immunostained neurones 205
L. F. Agnati, K. Fuxe, A. M. Janson, M. Zoli and A. Härfstrand

14 Photomicrography 225
S. Bradbury

APPLICATIONS

15 Immunocytochemistry of cell and tissue cultures 245
K. R. Jessen

16 Immunocytochemistry of cell lines 261
B. A. Gusterson and P. Monaghan

17 Immunocytochemistry of oncogenes 273
G. Gastl, J. M. Ward and U. R. Rapp

18 Immunocytochemical localization of intrinsic amines 284
A. A. J. Verhofstad and H. M. W. Steinbusch

19 Immunocytochemistry and evolutionary studies with particular
reference to peptides 308
M. C. Thorndyke

20 Immunocytochemistry of the diffuse neuroendocrine system 328
J. M. Polak and S. R. Bloom

21 Immunocytochemistry of brain neuropeptides 349
G. W. Roberts and Y. S. Allen

22 Neurochemistry of the spinal cord 360
S. J. Gibson and J. M. Polak

23 Immunocytochemistry of intermediate filament proteins 390
F. Ramaekers, J. Broers and G. P. Vooijs

24 Intermediate filaments and differentiation in the central nervous system 401
D. Dahl and A. Bignami

25 Neurofilaments and glial filaments in neuropathology 413
J. Q. Trojanowski

26 Non-hormonal markers in the pituitary 425
P. Ordronneau and P. Petrusz

27 Two-colour immunofluorescence: analysis of the lymphoid system with
monoclonal antibodies 438
G. Janossy, M. Bofill and L. W. Poulter

28 Immunocytochemistry in the study and diagnosis of organ-specific
autoimmune diseases 456
W. A. Scherbaum, R. Mirakian, R. Pujol-Borrell, B. M. Dean and G. F.
Bottazzo

29 Spectrum and profiles of non-organ-specific autoantibodies in auto-
immune diseases 477
W. A. Scherbaum, M. Blaschek, P. A. Berg, D. Doniach and G. F.
Bottazzo

30 Immunocytochemistry in cardiac and renal transplantation 492
M. L. Rose and A. Suitters

31 Tumour markers 502
E. Heyderman

32 Immunocytochemistry in diagnosis of thryoid diseases 533
E. D. Williams

33 Immunocytochemistry in diagnostic cytology 547
D. R. Springall

34 Immunocytochemistry of lymphoreticular tumours 568
P. G. Isaacson and D. H. Wright

35 Cell markers in diagnostic haematology 599
E. Matutes

36 Immunocytochemistry in dermatology 618
A. C. Chu

37 Immunohistology in the diagnosis of glomerular disease 638
D. J. Evans

38 Immunocytochemistry of micro-organisms 650
A. R. M. Coates

39 Immunocytochemistry of human skeletal muscle diseases 664
C. A. Sewry, S. T. Appleyard, M. J. Dunn and M. J. Capaldi

40 Some implications of immunocytochemistry for routine diagnostic histopathology 674
R. B. Goudie

Index 683

List of Plates

1 Detection of myosin by an immunoblotting technique using colloidal gold-labelled antibody with silver enhancement. (*Chapter 1*)

2 Human pituitary stained for four hormones by a multiple immunoenzymatic method. (*Chapter 3*)

3 Chorionic gonadotrophin in human placenta: avidin–biotin immunofluorescence compared with indirect immunofluorescence. (*Chapter 4*)

4 Human peripheral blood mononuclear leucocytes immunostained for OKT4 antigen by an indirect immunogold method and for endogenous peroxidase by cytochemistry: a comparison of bright field and epipolarization microscopy. (*Chapter 5*)

5 Indirect immunogold staining of tubulin in an endosperm cell of *Hemanthus* in metaphase. (*Chapter 5*)

6 Double staining by immunogold–silver for neurofilament protein and peroxidase–anti-peroxidase for S-100. (*Chapter 5*)

7 T-suppressor cells in a cryostat section of human tonsil stained by OKT8 antibody and immunogold–silver. (*Chapter 5*)

8 Human peripheral blood mononuclear leucocytes stained for OKT4 antigen by immunogold–silver, viewed with combined bright field and epipolarization microscopy. (*Chapter 5*)

9 Cells of arcuate nucleus labelled with Fast Blue and showing pro-opiomelanocortin localized by hybridization histochemistry. (*Chapter 12*)

10 Substance P-like immunoreactivity in substantia nigra of rat brain: absorbance map and comparison of lateral and medial areas by immunocytochemistry and radioimmunoassay. (*Chapter 13*)

11 Azan-stained blood vessel: comparison of photography at correct and reduced colour temperature. (*Chapter 14*)

12 Cultured astrocytes from rat cerebellum. Surface immunostaining for a glial cell antigen. (*Chapter 15*)

13 Double immunofluorescence with two monoclonal antibodies on cultured MCF7 cells. (*Chapter 16*)

14   Focal non-specific staining of a colonic carcinoma with antibody to a peptide in a transforming gene protein. (*Chapter 17*)

15   Non-specific membrane staining of a colonic carcinoma with antibody to a peptide in a transforming gene protein. (*Chapter 17*)

16   Mouse sarcoma induced by BALB/c 3T3 cells transfected by *Ras*^{Ha} oncogene; membrane staining by antibody to a peptide in *Ras*^{Ha} p21. (*Chapter 17*)

17   Mouse lymphoblastic lymphoma induced by *raf/myc* recombinant virus showing intranuclear staining for *myc* protein and some non-specific cytoplasmic staining. (*Chapter 17*)

18   *Ras*^{Ha} p21 staining in virus-induced sarcoma. (*Chapter 17*)

19   *Ras*^{Ha} p21 staining of splenic erythroblast cell membrane in virus-induced erythroblastosis. (*Chapter 17*)

20   *Ras*^{Ha} p21 on membrane of reticular cells in lymph node adjacent to virus-induced sarcoma. (*Chapter 17*)

21   Double staining for Rauscher leukaemia virus p30 in a megakaryocyte and *Ras*^{Ha} p21 in an erythroblast in the spleen of a Harvey sarcoma/Moloney leukaemia virus-injected mouse. (*Chapter 17*)

22   V-*raf* oncogene protein in lymphoblasts induced by *raf/myc* recombinant virus shown by antibody to a peptide in v-*raf* transforming protein. (*Chapter 17*)

23   Cell membranes of a chemically induced hepatocellular adenoma stained by antibody to SP-63, a peptide in v-*raf* transforming protein. (*Chapter 17*)

24   Gastrin/CCK-like immunoreactivity in the neural ganglion of *Ciona intestinalis*. (*Chapter 19*)

25   Temporal cortex from case of Alzheimer's dementia. Silver-impregnated tangles are present in neurones immunoreactive for somatostatin. (*Chapter 21*)

26   Caudate nucleus in Huntington's chorea showing survival of neuropeptide Y-immunoreactive neurones. (*Chapter 21*)

27   Immunostaining of intermediate filaments in three tumours leading to revised diagnosis. (*Chapter 23*)

28   Human antrum. Double indirect immunofluorescence demonstrating autoantibodies to G cells, identified on the same section by antibodies to gastrin. (*Chapter 28*)

29   Double indirect immunofluorescence demonstrating that the autoantibody-stained cells shown in *Plate* 28 are not somatostatin immunoreactive. (*Chapter 28*)

30   Different immunofluorescent staining patterns of anti-nuclear autoantibodies on cells in tissue culture. (*Chapter 29*)

31   Immunofluorescence staining patterns of autoantibodies to various nuclear and cytoplasmic components. (*Chapter 29*)

32   Normal heart showing binding of monoclonal antibody W6/32. (*Chapter 30*)

33   Binding of monoclonal antibody on a cardiac transplant biopsy showing mild rejection. (*Chapter 30*)

34   Multinucleate cell in malignant teratoma immunoreactive for human chorionic gonadotrophin. (*Chapter 31*)

35   Papillary carcinoma of thyroid immunostained for thyroglobulin. (*Chapter 31*)

36   Metastasis from testicular tumour showing alpha-fetoprotein-positive immunoreactivity. (*Chapter 31*)

37   Malignant cells in ascitic fluid from ovarian carcinoma immunostained for human milk fat globule antigen (HMFG2). (*Chapter 33*)

38   Tumour cells from small cell carcinoma of lung in smear of pleural effusion immunostained for neuron-specific enolase. (*Chapter 33*)

39   Pleural effusion from case of T-cell lymphoblastic leukaemia immunostained for terminal deoxynucleotidyl transferase. (*Chapter 35*)

40   Skin from case of junctional epidermolysis bullosa stained with bullous pemphigoid serum. (*Chapter 36*)

41   Skin from case of junctional epidermolysis bullosa stained for type IV collagen. (*Chapter 36*)

42   *Leishmania tropica major* promastigotes immunostained with monoclonal antibody WIC 79.3. (*Chapter 38*)

43   Purkinje cells in rat cerebellar cortex immunostained with monoclonal antibody CE5. (*Chapter 38*)

44   Skeletal muscle in polymyositis immunostained for T-lymphocytes. (*Chapter 39*)

Foreword to the First Edition

by Professor A. G. E. Pearse
Department of Histopathology,
Royal Postgraduate Medical School,
Hammersmith Hospital, London.

This book is written to good purpose and its publication is timely. Suddenly the pace of advance in the twin fields of immuno- and affinity cytochemistry has accelerated to such effect that the great majority of newcomers have been left floundering. So here gathered together are some twenty chapters, by something more than twenty authors, all of whom are currently engaged in the effective use of one or other of the main technologies encompassed in the title.

Sternberger's 'Introduction' sets the tone and purpose and the chapters which follow can be divided, loosely, into three main groups. A single overview chapter covers the field of immunocytochemistry, the succeeding five chapters are devoted to more or less pure technology. These are backed up by six which are more purely cytochemical, albeit applied to scientific problems as opposed to disease processes. Finally there are seven chapters dealing with different pathological topics, most of them relatively restricted.

It is clear that multiple author works of this kind depend to no mean extent on the judgement of the editors and on their skill in persuading their chosen authors to write both swiftly and cogently on their selected topics. In this book is reflected that skill and I forsee that it will be a trailblazer and the forerunner of many similar productions. But the field is wide enough and competition is the essence of advance. I wish the editors, and their galaxy of authors, every success.

Techniques

1

Raising and Testing Polyclonal Antibodies for Immunocytochemistry

J. De Mey and M. Moeremans

Immunocytochemical techniques have given us the potential to localize antigens in cells and tissues with strongly improved diversity, selectivity and specificity of staining reactions, as compared to classical staining methods for the demonstration of tissue components.[1-6] They completely rely on the availability of antibodies that will react in a specific way with the antigen in situ (which may be very different from *in vitro* conditions). Obtaining such antibodies often presents difficulties, sometimes because immunohistochemistry laboratories lack the necessary equipment for raising and testing antibodies. Nevertheless, it is important that raising and testing of antibodies for immunocytochemistry occur in a coordinated fashion, in order to select those antisera that perform best in a particular localization problem.

We will consider in this chapter general problems relating to the preparation of antigens, suitable for immunization, raising and detecting antibodies, and problems related to the antibody specificity, and how one can test and improve this.

1. THE ANTIGEN

An antigen can take many different forms. It can be whole micro-organisms (viruses, bacteria, fungi) or parts thereof,[7,8] whole eukaryotic cells or subfrac-

3

tions (e.g. membrane preparations). These kinds of antigen have a complex composition and antisera against them will mostly be used for diagnostic purposes to distinguish presence or absence of the group of antigens they are directed to. It is very difficult to obtain monospecific antibodies from this class of antigen, unless extensive absorptions are done. Hybridoma technology[9,10] clearly has opened new possibilities in this area, and it is recommended to use this approach whenever one wants to obtain monospecific antibodies from mixtures of antigens.

Often, however, the immunocytochemist wants to raise a specific antibody to one single antigen. Here too, it may be worth considering the hybridoma approach, but it is still very useful to raise serum antibodies as well. In order to obtain monospecific antibodies, it is essential to use highly purified antigens. Often, it will be necessary to use techniques with high resolving power such as preparative isoelectric focusing or affinity chromatography where possible. If the antigen is unsuitable for these techniques, it has often been helpful first to purify it as far as possible and then to separate the mixture by high resolution sodium dodecylsulphate (SDS)–polyacrylamide gel electrophoresis, cutting out the bands containing the antigen of interest, and using this material to immunize the animal.[11,12] Using this approach, antibodies have been prepared against proteins that were otherwise very difficult to purify. For testing the purity of protein antigens the recently introduced, very sensitive silver-staining methods can be used.[13] They are very simple and 20–50 times more sensitive than the normal staining procedures.

Low-molecular-weight antigens such as the peptide hormones are mostly conjugated to a larger carrier protein, for example bovine serum albumin, bovine thyroglobulin or keyhole limpet haemocyanin.[14,15] The possible presence of impurities in the carrier proteins has to be considered. The purity of the peptides must be verified by highly sensitive techniques such as high performance liquid chromatography.

2. RAISING ANTISERA

Rabbits are most often used, but other animals such as goats, sheep or guinea-pigs, or even turkeys can be excellent alternatives. Availability of antibodies made in different species offers the possibility of performing indirect double-labelling experiments without the need to elute antibodies from sections (*see* Chapters 3, 10).

It is difficult to make firm statements, based on hard facts, concerning the best possible immunization scheme (e.g. quantity and form of antigen, route, time-schedule). We will, therefore, give as one example the procedure that we have used in our laboratory for the past few years, for antigens such as conjugated peptides and purified soluble proteins.

In general, at least two animals are injected per antigen. When enough antigens and animals are available, it is recommended to immunize as many animals as possible. For rabbits and goats, about 1 mg of antigen is solubilized in phosphate-buffered saline (PBS) and homogenized with 1 ml of complete Freund's adjuvant. It does not matter whether the antigen is readily soluble or not. Often, peptide protein conjugates do not dissolve at all.

The homogenate is intradermally and subcutaneously injected at multiple sites along the spinal cord. Usually, 50–200 µl per injection site is given. Four and eight weeks later, identical booster injections are given, followed by monthly boosters of antigen made with incomplete Freund's adjuvant.

The first bleeding is taken one week after the second booster, followed by a second bleeding one week later. This is repeated after each booster. For rabbits, blood is taken from the central ear artery with a cannulated needle and a 50-ml plastic syringe. Up to 60 ml blood per bleeding is taken, without damage to the animal. For goats, blood is taken from the jugular vein. Up to 350 ml per bleeding is taken.

Serum is made by incubating the blood (in 12-ml tubes) for 1 h at 37 °C. Serum is centrifuged at low (2500 rev./min) and high speed (3500 rev./min) to remove cells and debris.

Aliquots are stored frozen at −25 °C, or can be lyophilized.

3. DETECTING ANTIBODIES

For most bleedings, it is highly recommended to screen for the presence of antibodies, using an immunocytochemical procedure. This approach presents important advantages.

a. The same principles are used for detection as will be used in future applications. Radioimmunoassay, for example, uses binding of labelled antigen under *in vitro* conditions. Radioimmunoassay will, indeed, be able to detect the presence of antibodies, but it will not give information about their usefulness in immunocytochemistry.

b. Conditions for tissue or cell preparation which are most appropriate for the localization problem can be selected. The first localization results will give very useful information about the general quality (background and titre) of the antiserum, and its usefulness under staining conditions which should be used to obtain meaningful results. Therefore, direct selection of the animal that gives good quality antiserum, and elimination of the unsatisfactory ones is possible.

c. The particular staining pattern obtained after the first trials will already give an idea of possible specific and non-specific reactions since, for many antigens, the localization in the test tissue is known (e.g. restricted to only one cell type or associated with one type of organelle).

For the screening of antibodies, a simple technique, such as immunofluorescence is recommended. A second screening can be done with more complicated but more efficient techniques such as immunoperoxidase.

4. TESTING ANTIBODIES: ANTIBODY SPECIFICITY IN IMMUNOCYTOCHEMISTRY

The above described screening will have yielded an antiserum that gives a particular staining pattern.

Now, the specificity of the staining reaction has to be firmly established. The immunocytochemical specificity of an antiserum has different aspects. We will assume that the method specificity of the immunocytochemical procedure used is satisfactory (after performing the usual tests; *see* Vandesande[3]).

4.1. Method Specificity of the Primary Antibody

Grube[16] has shown that antibodies in the primary antiserum can bind to structures such as secretory granules by ionic forces. To avoid the possibility of false positives, he has recommended the use of only high titre antisera at high dilutions (>1 : 1500) and/or raising the salt (NaCl) content of the buffer used as diluent or as rinsing solution to 0·5 M.

4.2. Antibody Specificity

Even a highly purified antigen may contain traces of immunogenic impurities that will induce unwanted antibodies. Such unwanted antibodies may produce a totally normal and specific-looking immunocytochemical staining pattern. This kind of 'specificity', however, is not the desired one. A valuable test for controlling the specificity is the antigen-absorption test. This is often done by adding excess antigen to the diluted antiserum. A safer way is to use solid phase absorption by which the majority of the antibodies are removed.[3] A negative immunostaining result of such an absorption test indicates that the immunoreactive substances previously shown contain the same antigenic determinants as the ones introduced to the animal by the antigen.

4.3. Affinity Purification of Serum Antibodies

When the antigen absorption test is satisfactorily negative and the serum can be used at high dilutions, giving low background, it can be used as such. The only problems remaining are cross-reactions (*see* Section 4.4) and problems related to the procedure used, such as false negative results because of the use of an inappropriate dilution with the peroxidase–anti-peroxidase (PAP) technique,[3] or differential modification of the immunoreactivity of the antigen, localized at different sites. When the overall staining quality is satisfactory, for example when there is a strong specific component, using the most satisfactory tissue preparation, but there is also a non-specific non-absorbed staining reaction, it will be worth isolating the specific antibodies with antigen-specific affinity chromatography. In our experience, this has practically always been necessary for antibodies against cytoskeletal proteins (actin, myosin, α-actinin) and other structural proteins. Fortunately, the yield of such antibodies is often satisfactory. Antisera against peptide hormones, on the other hand, often prove to be of sufficient quality to be used as such. The yield of specific antibodies after affinity purification from such antisera is often also very low.

   The following describes the rather simple procedure that is used in our laboratory for purifying antibodies from serum in one step, yielding specific antibodies with >95 per cent immunoglobulin content as judged by analytical SDS–acrylamide electrophoresis.

The antigen is coupled to Sepharose-4B-CNBr (Pharmacia) according to the instructions of the manufacturer. Other carriers and linking groups may be used. Usually, 5–10 mg of the highly purified antigen is coupled per gramme of dry gel to give a total gel volume of 3·5 ml. This amount of gel is added to ±15 ml antiserum and incubated for 2–3 h at room temperature with gentle shaking or rolling. The serum–gel mixture is poured into a column and washed with 10 mM Tris-buffered saline (TBS) at pH 7·6 until the absorbance of the effluent is equal to zero. Then the gel is washed with TBS + 1 M NaCl to elute a small peak of non-specifically absorbed material. The salt is removed from the gel by washing with TBS. To elute the specifically bound antibody, the column is eluted with 0·1 M glycine-HCl buffer pH 2·8 until the baseline is almost reached. The entire trailing peak is collected since the antibodies that are released more slowly are those with the highest affinity. The acid eluant is neutralized with ± 10 per cent of 1 M Tris-HCl buffer pH 8·5. This is done during the elution step. Aliquots are stored at −20 °C or −70 °C without further treatment. The gel is removed from the column, washed with 10^{-2} M HCl (pH 2) on a sintered glass filter and re-equilibrated with TBS + 0·02 M NaN_3, for storage at 4° C. It can be reused several times. The antibodies thus purified have to be retested to determine the antibody concentration (μg/ml) giving optimal staining reaction and for further characterization of the specificity. An alternative elution method consists of using 4 M $MgCl_2$ in H_2O pH 6·5–7. The eluted antibody is dialysed against 20 mM Tris-buffered saline pH 8·2. Affinity-purified antibodies give mostly very specific staining patterns with extremely low background levels, comparable to those of monoclonal antibodies. An interesting alternative to the above procedure, especially when only very limited quantities of purified antigens are available, has recently been introduced. It involves electrophoretic transfer of electrophoretically separated proteins (see Section 4.5), reaction of antisera with the blots and eluting the antibodies bound to different bands. These antibodies can be retested and used for immunocytochemistry.[17]

4.4. Cross-reactions and Site Specificity

Many peptides and proteins have homologies in their amino-acid sequences and therefore exhibit immunological cross-reactivity. These homologies occur very frequently in secretory peptides found in neurones, and endocrine or paracrine cells, but are also common in related proteins.

An antigenic determinant mostly comprises a sequence of about three to eight amino acid residues. It has therefore been claimed that antibodies should be regarded as site or region-specific detection reagents rather than antigen-specific reagents.[5] With serum antibodies, this important aspect of antibody specificity may be largely hidden by the heteroclonality of the antibodies: various antigenic determinants of the same antigen may be recognized by an even greater variety of antibody molecules. Such antibodies can be made more site-specific by using differential absorption, using either a cross-reacting molecule or parts of the whole antigen. These will remove only certain, cross-reacting antibody populations.

Immunization with smaller fragments of a protein or peptide greatly increases the chances of obtaining region-specific animal sera, and thus helps in avoiding predictable cross-reactions. Such antisera have been very important for specifically localizing different molecular forms belonging to the same family of peptides and proteins.[5,6] Monoclonal antibodies are, by definition, specific for only one antigenic determinant and offer the advantage that their specificity can be precisely determined. By selecting for specificity towards defined sequences, certain predictable cross-reactivities can also be avoided.

The region or site specificity of both animal antisera and monoclonal antibodies must be carefully characterized. This has mostly been achieved through absorption of the antisera or purified antibodies (including monoclonals) against various possible cross-reacting substances such as related proteins or natural and synthetic peptides and peptide fragments. Elegant quantitative cross-reaction studies have been done directly on tissue sections.[18] The test consisted of adding serial dilutions of various peptides and peptide–albumin conjugates to the diluted antisera. The effect of added peptides on labelling intensity was qualitatively observed or quantitatively measured by fluorimetry (depending on the type of section). This approach has the important advantage of using the in situ material as substrate for competition reactions. It can give information on the probable degree of identity or similarity of the immunoreactive substances, in different localizations. This study has also given a clear example of the fact that radioimmunological tests do not ensure the specificity of immunohistochemical labelling (see also Vandesande[3]). Results of radioimmunological tests reflect the binding properties of a tracer (e.g. a pure preparation of ^{125}I-labelled antigen), to an antiserum, while immunocytochemical procedures reveal the binding properties of an antiserum to an unknown population of immunoreactive tissue substances, as modified by fixation and embedding procedures.[18] The recently introduced gold-labelled antigen detection technique[19] also lends itself very well to this type of antibody characterization. In addition to the above used in situ method, immunocytochemical model systems are useful for defining antisera specificity. Such systems include antigens, covalently linked to a solid phase such as Sepharose-4B-CNBr beads.[20] A more elegant model system has recently been introduced by Larsson.[21] It uses submicrogramme quantities of antigens (mostly small peptides) and allows for simultaneous and identical testing of virtually unlimited numbers of peptides under standard immunocytochemical staining conditions.

The test involves immobilization (by formaldehyde vapour) of nanogramme amounts of peptides on filter paper and staining them with the PAP procedure.

A similarly elegant method uses nitrocellulose or diazobenzyloxymethyl (DBM) paper on which various antigens (soluble proteins) have been spotted.[22]

4.5. Positive Identification of Immunoreactive Substances

All the above methods are quite useful for characterizing antisera and antibodies against available molecular forms with more or less predictable cross-reactions. They are, however, dependent on availability of pure test antigens and do not account for unpredictable cross-reactions. Such cross-reactions do occur, certainly when instead of a homologous tissue, a heterologous system is investigated.

The only way to overcome this limitation is to positively identify the immunoreactive substances in the tissue or cells under study. One serious limitation here is that the conditions for testing (especially the state of the antigen that will not be the same as after fixation and tissue embedding) cannot be the same as those of the immunocytochemical procedure.

Three major approaches exist for positively identifying the antigen: biochemical identification, immunoprecipitation and immunoreplica techniques.

4.5.1. Biochemical Identification

The tissue containing the immunoreactive material is extracted and the components are separated by high resolution liquid chromatographic techniques such as high performance liquid chromatography. The fractions are screened by a competitive radioimmunoassay and the positive ones are further analysed to identify the immunoreactive substances. The chromatographic characteristics of the immunoreactive substances are compared with those of known immunoreactive standards, mostly the antigen itself or some probable cross-reacting substances. In this way many of the precursor molecules of peptide hormones have been identified, but cross-reactions have also been clarified.

4.5.2. Immunoprecipitation

The tissues or cells are extracted. Denaturing conditions, including urea- and detergent-containing buffers are often used, to ensure solubilization of as many components as possible. For cultured cells, metabolic labelling is often used. Tissue extracts are often iodinated.

The extracts are reacted with antibodies and then the antibody–antigen complexes are reacted with either solid phase immunoadsorbent, such as protein A–Sepharose-4B or secondary antibody–Sepharose-4B, or formalin-fixed *Staphylococcus aureus* which has active protein A at its surface.[23] The complexes thus formed are washed, dissolved and analysed by one or two-dimensional electrophoresis techniques, followed by fluorography to detect the tissue components that reacted with the antibodies. Suitable controls, involving pre-immune antiserum or unrelated purified antibodies, have to be included to provide a standard for non-specific absorptions. These techniques are very useful for identification of the antigen. Coprecipitation of components, showing close association with the antigen, may give false positives. On the other hand, such coprecipitation may reveal interesting associations of the antigen with other cell components.

4.5.3. Immunoblots

The development of techniques for transferring proteins separated by polyacrylamide gel electrophoresis to nitrocellulose,[24,25] diazobenzyloxymethyl paper[8] and, more recently, positively charged nylon membranes[26] has led to a very elegant and efficient antibody characterization method: the immunoblot technique.

Initial attempts to characterize antibody binding to separated proteins in acrylamide gels[11,27] were successful but laborious and time consuming. The transfer techniques offer a means for making exact protein blots which are subsequently reacted with antisera. These techniques are extremely sensitive and easy to perform. With two-dimensional separation techniques, it is possible to identify the immunoreactive products unequivocally.[28]

Indirect techniques, using either iodinated protein A or secondary antibodies followed by fluorography[29] are still widely used. The known disadvantages of radioactive tracers have led to the increasing use of immunoenzymatic detection techniques derived from immunocytochemical methods such as the indirect peroxidase, PAP[25,30] and avidin–biotin complex methods.[31] Their sensitivity is reportedly as high as that of autoradiography. Unfortunately, the reaction products of these methods bleach relatively quickly and the substrates used are often noxious to health.

Recently, we have introduced the use of immunogold staining (IGS) and immunogold–silver (IGSS) staining[32] as a non-radioactive, non-enzymatic alternative.[33] With the IGS method, involving the use of gold-labelled secondary antibodies, a clearly visible red signal is generated at the sites of immune reaction. No additional development is needed and the sensitivity equals that of indirect immunoperoxidase. With the IGSS method very high sensitivity is obtained by silver enhancement. The signal becomes jet-black against a white background. As in light microscopy (see Chapter 5), it appears that this method is the most sensitive available today. In addition, we have introduced AuroDye® (available from Janssen Life Sciences Products, Beerse, Belgium), a protein stain for protein transfers on nitrocellulose membranes. It consists of stabilized colloidal gold, buffered at pH 3·0. At this low pH, the gold particles bind very strongly and selectively to the transferred proteins to stain them dark red-purple. The sensitivity is somewhat better than silver staining of proteins in acrylamide gels and matches that of the IGSS immunoblot technique. AuroDye® staining of protein blots is of great help for controlling the quality of transferring proteins to nitrocellulose membranes and for correlating, in duplicate blots, the bands or spots (2D-gels) detected with the IGS or IGSS methods with the total electrophoretogram. It is also possible to combine IGSS and AuroDye® on one single blot[35] (Plate 1), when Tween 20 is used (a) to quench the protein-binding sites and (b) in the diluent for immune reagents and the washing buffer. The immunodetection is performed first and is followed by AuroDye® staining. For more details refer to the original papers and to the instruction leaflets of the manufacturer, Janssen Life Sciences Products, Beerse, Belgium.

It should be mentioned that immunoblot techniques can give false negatives, since it is not know whether all possible antigenic sites can still react with the antibody after SDS denaturation and adsorption to the nicrocellulose paper.

5. CONCLUSION

Immunocytochemical results can only be validated by the use of high quality carefully characterized antibodies. Recent developments, briefly described in this chapter, have greatly improved the possibilities for performing this characterization. Nevertheless, a number of pitfalls such as false positives or negatives remain. Specificity of immunocytochemical reactions therefore cannot always be 100 per cent proven and this has to be considered when results are interpreted. Nevertheless, with the techniques for antibody characterization at hand today, and the possibilities offered by newly developed procedures for tissue preparation and immunocytochemistry, one can obtain more reliable results than a few years ago.

REFERENCES

1. Coons A. H. Fluorescent antibody methods. In: Danielli, J. F., ed., *General Cytochemical Methods*. New York, Academic Press, 1978: 399–422
2. Sternberger L. A. *Immunocytochemistry*, 2nd ed. New York, Wiley, 1979.
3. Vandesande F. A critical review of immunocytochemical methods for light microscopy. *J. Neurosci. Methods* 1979, **1**, 3–23.
4. Forsmann W. G., Pickel V., Reinecke M., Hack D. and Hetz J. Immunohistochemistry and immunocytochemistry of nervous tissue. In: Heym C. and Forsmann W. G., ed., *Techniques in Neuranatomical Research*. Berlin, Springer Verlag, 1981: 171–205.
5. Larsson L. I. Peptide immunocytochemistry. *Prog. Histochem. Cytochem.* 1981, **13**, 1–83.
6. Polak J. M., Buchan A. M. J., Probert L., Tapia F., De Mey J. and Bloom S. R. Regulatory peptides in endocrine cells and autonomic nerves. Electron immunocytochemistry. *Scand. J. Gastroenterol.* 1981, **16**, Suppl. 70, 11–23.
7. Legocki R. and Verma D. Multiple immunoreplica technique: screening for specific proteins with a series of different antibodies using one polyacrylamide gel. *Anal. Biochem.* 1981, **111**, 385–392.
8. Symington J., Green M. and Brackmann K. Immunoautoradiographic detection of proteins after electrophoretic transfer from gels to diazo-paper: analysis of adenovirus encoded proteins. *Proc. Natl Acad. Sci. USA* 1981, **78**, 177–181.
9. Köhler G. and Milstein C. Continuous cultures of fused cells secreting antibodies of predefined specificity. *Nature* 1975, **256**, 495–497.
10. Kennett R. H. Monoclonal antibodies. *Hybridomas: a New Dimension in Biological Analyses.* London, Plenum Press, 1980.
11. Willingham M., Yamada S., Bechtel P., Rutherford A. and Pastan I. Ultrastructural immunocytochemical localization of myosin in cultured fibroblastic cells. *J. Histochem. Cytochem.* 1981, **29**, 1289–1301.
12. Cozzani C. and Hartmann B. Preparation of antibodies specific to choline acetyl transferase from bovine caudate nucleus and immunohistochemical localization of the enzyme. *Proc. Natl Acad. Sci. USA* 1980, **77**, 7453–7457.
13. Merril C., Goldman D., Sedman S. and Ebert M. Ultrasensitive stain for proteins in polyacrylamide gels shows regional variation in cerebrospinal fluid proteins. *Science* 1981, **211**, 1437–1438.
14. Tateishi K., Hamaoka T., Soguira W., Yanaihara C. and Yanaihara N. A novel immunization procedure for production of anti-cholecystokinin-specific antiserum of low cross-reactivity. *J. Immunol. Methods* 1981, **47**, 249–258.
15. Vandesande F. and Dierickx K. Identification of the vasopressin producing and of the oxytocin producing neurons of the hypothalamic magnocellular neurosecretory system of the rat. *Cell Tissue Res.* 1975, **164**, 153–162.
16. Grube D. Immunoreactivities of gastrin (G-) cells. II. Non-specific binding of immunoglobulins to G-cells by ionic interactions. *Histochemistry* 1980, **66**, 149–167.
17. Olmsted J. B. Affinity purification of antibodies from diazotized paper blots of heterogeneous protein samples. *J. Biol. Chem.* 1981, **256**, 11, 955–957.
18. Garaud J. C., Eloy R., Moody A. J., Stock C. and Grenier J. F. Glucagon and glicentin-immunoreactive cells in the human digestive tract. *Cell Tissue Res.* 1980, **213**, 121–136.
19. Larsson L.-I. Simultaneous ultrastructural demonstration of multiple peptides in endocrine cells by a novel immunocytochemical method. *Nature* 1979, **282**, 743–746.
20. Streefkerk J., Deelder A., Kors N. and Kornelius D. Antigen-coupled beads adherent to slides: a simplified method for immunological studies. *J. Immunol. Methods* 1975, **8**, 251.
21. Larsson L.-I. A novel immunocytochemical model system for specificity and sensitivity screening of antisera against multiple antigens. *J. Histochem. Cytochem.* 1981, **29**, 408–410.

22. Herbrink P., Van Bussel F. and Warmaar S. The antigen spot test (AST): a highly sensitive assay for the detection of antibodies. *J. Immunol Methods.* 1982, **48**, 293–298.
23. Oshima R. Identification and immunoprecipitation of cytoskeletal proteins from murine extraembryonic endodermal cells. *J. Biol. Chem.* 1981, **236**, 8124–8133.
24. Towbin H., Staekelin T. and Gordon J. Electrophoretic transfer of proteins from polyacrylamide gels to nitrocellulose sheets: procedure and some applications. *Proc. Natl. Acad. Sci. USA* 1979, **76**, 4350–4354.
25. Peferoen M., Huybrechts R. and De Loof A. Vacuum-blotting: a new simple and efficient transfer of proteins from sodium dodecyl sulfate–polyacrylamide gels to nitrocellulose. *FEBS Lett.* 1982, **145**, 369–372.
26. Gershoni J. M. and Palade G. E. Electrophoretic transfer of proteins from sodium dodecyl sulfate–polyacrylamide gels to a positively charged membrane filter. *Anal. Biochem.* 1982, **124**, 396–465.
27. Adair W., Jurnick D. and Goodinough V. Localization of cellular antigens in sodium dodecylsulfate–polyacrylamide gels. *J. Cell Biol.* 1978, **79**, 281–285.
28. Gershoni J. M. and Palade G. E. Protein blotting: principles and applications. *Anal. Biochem.* 1983, **131**, 1–15.
29. Laskey R. and Mills A. Enhanced autoradiography detection of ^{32}P and ^{125}I using intensifying screens and hypersensitized film. *FEBS Lett.* 1977, **82**, 314–316.
30. De Mey J. Raising and testing antibodies for immunocytochemistry. In: Polak J. M. and Van Noorden S., eds., *Immunocytochemistry: Practical Applications in Pathology and Biology.* Wright PSG, Bristol, 1983: 43–52.
31. Hsu S., Raine L. and Fanger H. Use of avidin–biotin–peroxidase complex (ABC) in immunoperoxidase techniques: a comparison between ABC and unlabeled antibody (PAP) procedures. *J. Histochem. Cytochem.* 1981, **29**, 577–580.
32. Holgate C., Jackson P., Cowen P. and Bird C. Immunogold–silver staining: new method of immunostaining with enhanced sensitivity. *J. Histochem. Cytochem.* 1983, **31**, 938–944.
33. Moeremans M., Daneels G., Van Dijck A., Langanger G. and De Mey J. Sensitive visualization in dot and blot immune overlay assays with immunogold and immunogold/silver staining. *J. Immunol. Methods* 1984, **74**, 353–360.
34. Moeremans M., Daneels G. and De Mey J. Sensitive colloidal metal (gold or silver) staining of protein blots on nitrocellulose membranes. *Anal. Biochem.* 1985, **145**, 315–321.
35. Daneels G., Moeremans M. and De Mey J. Combined immunogold silver antigen detection and AuroDye® or FerriDye® protein staining on one single nitrocellulose blot of one or two-dimensional polyacrylamide gels. Submitted for publication.

2

Raising and Testing of Monoclonal Antibodies for Immunocytochemistry

M. A. Ritter

Antibodies are produced as part of an individual's immune response to foreign antigen, e.g. invading pathogen/toxin produced by pathogen. These molecules bind to and are highly specific for distinct sites (antigenic determinant/epitope) on an antigenic molecule — a property that has led to the use of antibodies as specific probes for membrane-bound and soluble molecules in many areas of research, medicine and industry.

Structurally, antibodies are immunoglobulins, with a basic unit of four polypeptide chains (two heavy and two light, with molecular weights of 50 kdal and 25 kdal, respectively) as shown in *Fig.* 2.1. Such immunoglobulin molecules are bifunctional: there are two antigen-binding sites at the N-terminal end of the molecule, each formed by the combination of a heavy and a light chain variable domain; the C-terminal portion of the molecule is relatively constant, carries binding sites for complement (C1q, C4) and cellular Fc receptors, and is thus responsible for the effector functions of immunoglobulin. Small amino acid sequence differences distinguish between the various classes (IgM, IgG, IgA, IgD and IgE) and subclasses (mouse: IgG1, IgG2a, IgG2b, IgG3; man: IgG1, IgG2, IgG3, IgG4). IgA and IgM form dimers and pentamers, respectively, giving increased valency for antigen binding (IgA — 4 binding sites/molecule; IgM — 10 binding sites/molecule, although only 5 of these are available at any given time). Every individual has the capacity to produce all classes/subclasses of immunoglobulin; however, the actual isotype of antibody produced during an immune response will depend upon many factors, in particular the route of immunization and the immunization schedule. This will be discussed later.

The production of antibodies to an invading pathogen involves the stimulation of B-lymphocytes whose cell surface immunoglobulin receptors for antigen are

13

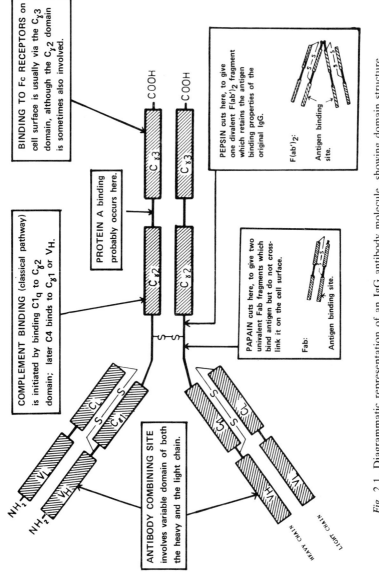

Fig. 2.1. Diagrammatic representation of an IgG antibody molecule, showing domain structure, binding sites for antigen and effector functions, and papain and pepsin cleavage points. Different subclasses of IgG differ slightly in their structure and in their ability to perform different effector functions. (Reproduced with permission from Morris, R. J. [1982] In: Pfeiffer, S. E., ed., *Neuroscience approached through Cell Culture*, p.9. Copyright, Boca Raton, Florida, CRC Press.)

BINDING TO Fc RECEPTORS on cell surface is usually via the $C_{\gamma}3$ domain, although the $C_{\gamma}2$ domain is sometimes also involved.

COMPLEMENT BINDING (classical pathway) is initiated by binding $C1_q$ to $C_{\gamma}2$ domain; later C4 binds to $C_{\gamma}1$ or V_H.

PROTEIN A binding probably occurs here.

PEPSIN cuts here, to give one divalent $F(ab')_2$ fragment which retains the antigen binding properties of the original IgG.

F(ab')₂:

Antigen binding site.

PAPAIN cuts here, to give two univalent Fab fragments which bind antigen but do not cross-link it on the cell surface.

Fab:

Antigen binding site.

ANTIBODY COMBINING SITE involves variable domain of both the heavy and the light chain.

NH_2

NH_2

V_L

V_H

C_L

$C_{\gamma}1$

$C_{\gamma}2$

$C_{\gamma}3$

COOH

V_H

$C_{\delta}1$

$C_{\delta}2$

$C_{\delta}3$

COOH

V_L

C_L

HEAVY CHAIN

LIGHT CHAIN

specific for, i.e. will bind to, antigenic determinants on the relevant antigen. This will also involve many other different processes, including the processing of antigen by antigen-presenting cells, e.g. macrophages and dendritic cells, and the production by T-lymphocytes of soluble inducers (lymphokines) for B-lymphocyte proliferation and maturation. The result of this process is that the specifically antigen-stimulated B-lymphocytes proliferate and some will mature to give plasma cells that secrete large amounts of immunoglobulin whose specificity is identical to that of the original B-lymphocyte's surface receptor. Others will become 'memory' B-cells ready for the next encounter with antigen. Hence each antigen-activated B-cell will give rise to a clone of daughter cells (some memory, some plasma cells), each member of which produces membrane-bound/secreted immunoglobulin of the same antigen-binding specificity. However, the situation is more complicated during a normal *in vivo* immune response since many different B-cells will be stimulated (to either the same or to different antigenic determinants, according to the specificity of their cell surface immunoglobulin receptors) and these will ultimately give rise to many different clones of memory and antibody-secreting plasma cells. Thus, many different antibody molecules will be secreted (each typical of the clone from which it has come) and the immune response is said to be 'polyclonal'.

1. WHAT IS A MONOCLONAL ANTIBODY?

Conventional antibodies are prepared from the sera of animals who have been polyclonally activated, as described above. Such reagents can be purified and absorbed to be operationally specific for the antigen of interest. Their main disadvantage lies in the polyclonality of the antibodies whereby there will be a mixture of specificities for different determinants on the antigen and each will have a different binding affinity and avidity. In addition, individual animals will respond in different ways to the same immunogen, thus making it almost impossible to produce a consistent reagent on a large scale. These problems have largely been overcome by the development of the technique of monoclonal antibody production by Kohler and Milstein in 1975.[1] A 'monoclonal' antibody is a reagent that is the product of a single clone of immortalized B-cells and, as such, is uniform in molecular structure, specificity and affinity/avidity, and can be produced in large amounts. These reagents have revolutionized many areas of experimental, clinical and industrial work — a fact recognized by the shared award of the 1984 Nobel Prize for Medicine to the above two authors.

The technique currently used for raising monoclonal antibodies is basically the same as that originally described. B-lymphocytes can produce antibody of a desired specificity but are very short lived *in vitro*. In contrast, there are myeloma cell lines (derived from plasma cell tumours) that have lost their ability to secrete immunoglobulin, but which can grow indefinitely in tissue culture. These two types of cell can be fused together to give hybrids that possess both the myeloma parent's property of immortal growth and the B-lymphocyte parent's ability to synthesize and secrete antibody. These secreted antibodies are then tested to find those that are specific for the immunogen. The relevant hybridomas can then be cloned and raised in large numbers. The supernatant medium from such cultures will yield large amounts (approximately 10 μg/ml antibody). As this is produced by a single clone, it is by definition monoclonal and is homogeneous. Alternatively, the hybridomas can be grown as ascitic tumours which produce ascites fluid containing approximately 10 mg/ml of specific antibody.

The methods of fusion of immunized spleen cells with myeloma cells, HAT (hypoxanthine, aminopterin, thymidine) selection to separate hybrids from unfused spleen and myeloma cells, screening of antibodies produced by the hybrids and cloning of hybridomas of interest are discussed below. The routine procedures will be only briefly referred to since these are fully described elsewhere,[2,3] whilst those aspects that pertain more specifically to the production/use of monoclonal antibodies in immunocytochemistry will be dealt with in greater detail.

2. MONOCLONAL ANTIBODY METHODOLOGY

2.1. Immunization

Immunization protocols tend to be determined empirically and, once successful, are adhered to with great conservatism (for the obvious reason that if a method works well it is a waste of time trying to modify it). However, certain generalizations can be made.

2.1.1. Choice of Animal to be Immunized

This will be determined by the source of antigen; unless an immunogen is an alloantigen the immunized animal must be from another species. It will also be affected by the choice of myeloma cell line, since intraspecies hybrids are more stable than interspecies hybrids. For monoclonal antibodies that are going to be used in immunocytochemistry these species are most likely to be mouse or rat (*see* Section 2.2).

2.1.2. The Route of Immunization

This will depend upon the nature of the antigen: whether soluble (intravenous route) or particulate (intraperitoneal, intradermal or intramuscular routes). Wherever possible the final immunization prior to fusion should be an intravenous one as this will drive the immune response to the spleen — the lymphoid organ normally used to obtain B-cells for producing hybridomas. If the intravenous route is not possible, sufficient activated B-cells are present in the murine spleen for this still to provide an adequate routine source for fusion. In contrast, few activated B-cells appear to be present in the rat spleen after, for example, intraperitoneal immunization (Dr H. Waldmann, 1985, personal communication) and the regional lymph nodes provide a better source of specifically activated B-lymphocytes. An alternative approach is to immunize directly into the spleen.[4]

2.1.3. Immunogenicity of the Antigen

This can be enhanced by the addition of an adjuvant to the preparation used for immunization, providing the site of injection is not intravascular. Adjuvants are of two main types: colloidal, e.g. aluminium hydroxide, or oils in which the antigen can be dispersed as an emulsion, e.g. Freund's incomplete adjuvant. Adjuvant–antigen mixtures work by providing a local 'depot' from which antigen can be shed slowly over a long period of time. This increases the length of time that antigen is available to the immune system, since free antigen in the absence of adjuvant is rapidly cleared by phagocytic cells. Complete Freund's adjuvant is frequently used for primary immunization since the killed mycobacteria that form the additional ingredient in this preparation stimulate both general immune responsiveness and the formation of a granuloma at the site of injection — creating a temporary local lymphoid organ and so increasing the efficiency of immunogen–lymphocyte interactions. For small highly conserved antigen molecules, immunogenicity may be enhanced by coupling them to a larger heterogeneous 'carrier' protein, e.g. bovine serum albumin.[5] Such a carrier has the non-specific effect of reducing the rapid loss of antigen via the reticuloendothelial system, thus increasing the chances of antigen–B-cell interactions. More importantly, the larger molecule will act as an immunological 'carrier', providing additional antigenic determinants for T-helper lymphocytes and so augmenting the T-dependent antibody response to the small immunogen.

2.1.4. Factors affecting Class and Affinity of Antibodies produced

Many factors beyond the control of the experimenter, such as the nature of the antigen (carbohydrate antigens typically generate IgM antibodies) and individual variability in the animals that are immunized, will be involved in determining the class of antibody produced in an immune response. However, certain precautions can be taken to optimize the chances of obtaining the desired class of antibody. Primary immunization yields an IgM response, whilst secondary immunization is associated with progressive class switching — mainly to the various subclasses of IgG (unless skin or secretory tissues are involved). Thus, hyperimmunization will tend to give IgG antibodies whereas primary immunization followed by only a single boost will yield predominantly IgM antibodies. In general, IgG antibodies are of more use in immunocytochemistry since they are smaller (better penetration in tissue sections) and can be digested to $F(ab')_2$ fragments (to avoid background staining due to Fc binding, see below). However, an IgM monoclonal antibody can be useful in combination with an IgG antibody in dual immunostaining since class-specific secondary reagents can be used. These may not give such a strong signal as that obtained when the first layers are from different species (see Section 2.2), but it is considerably stronger than that obtained when the primary antibodies differ at only the subclass level.

Finally, several immunizations (more than 3) with relatively long gaps between (4–6 weeks), will increase the chance of producing high affinity antibodies, as these are typically activated at low antigen concentrations and are produced late in an immune response.

2.2. Choice of Myeloma Cell Line

2.2.1. Range

A wide range of myeloma cell lines are now available. Of these, those most frequently used are: P3X63–Ag8.653 (Ag8) and P3/NS–1/1–Ag4–1 (NS–1) mouse myeloma lines; Y3–Ag1.2.3 (Y3) and YB2/3.0Ag20 (Y0) rat myeloma lines; SKO–007 and LICR/LON/HMy2 human myeloma cell lines.[6-9]

2.2.2. Non-secretor Myeloma Cell Lines

The most suitable myelomas for the production of monoclonal antibodies for use in immunocytochemistry are the mouse Ag8 and the rat Y0 cell lines. These have the advantage that they have been selected for the inability to synthesize their own immunoglobulin heavy or light chains — thus ensuring that all immunoglobulin produced and secreted by a myeloma–lymphocyte hybrid (hybridoma) is of the specificity coded for by the B-cell parent's genome. In contrast, NS–1 and Y3 myeloma lines do produce light chains. Although these are not secreted on their own by the myeloma, during immunoglobulin biosynthesis in a hybridoma cell they can be assembled with the B-cell's heavy chains to form complete immunoglobulin molecules of irrelevant specificity that are subsequently secreted by the hybrid cell in conjunction with the relevant, i.e. selected for, monoclonal antibody. The effect of this will be two-fold: to reduce the amount of relevant monoclonal antibody that is produced, and to increase the risk of background labelling due to non-specific binding of antibody.

2.2.3. What Species?

Both mouse and rat monoclonal antibodies are suitable for immunocytochemistry. The fact that the majority of such reagents are of mouse origin is mainly historical, in that the mouse system was the first to be developed, and hence has been used in a greater number of studies. One obvious factor determining the choice of myeloma is the choice of animal to be immunized, since intraspecies hybrids are more stable and give a higher percentage of antibody-secreting hybrids;[9] this, in turn, is determined by the source of antigen (see Section 2.1). However, hybrids between closely related species (rat and mouse) have been successful[10] (our recent unpublished data). Where possible, there are good reasons for trying both mouse and rat systems. Firstly, B-cells from rat and mouse may 'see' different antigenic determinants on an immunogen; the use of both species will, therefore, increase the range of antibodies produced. Secondly, a combination of mouse and rat monoclonal antibodies is very useful for dual labelling in immunocytochemistry, e.g. two colour immunofluorescence/immuno-enzyme staining, since species-specific anti-immunoglobulin secondary (label-led) reagents can then be used. These give stronger labelling than the class/subclass-specific secondary reagents that are necessary when both first layer antibodies are of mouse origin.

2.3. Method of Fusion

Hybridomas are produced by the fusion of an activated B-lymphocyte (last boost with antigen should be 3–4 days before the fusion) with a myeloma cell. The easiest and largest source of B-cells is the spleen. Lymph nodes can also be used, e.g. rat after intraperitoneal immunization (*see* Section 2.1.2). Fusion was initially performed using Sendai virus. However, a much easier system is to use polyethylene glycol (PEG) as the fusogen. PEG reduces surface tension and so permits closer contact between cells, resulting in the fusion of their plasma membranes. Nuclear fusion occurs during mitosis resulting in hybrid cells that carry a complete diploid set of chromosomes from each parent cell. Proliferation of such a hybrid cell will give rise to a clone of identical cells that secrete immunoglobulin of the specificity and class/subclass coded for by the B-cell's genome and are immortal in culture (hybridoma). Interspecies hybrids are less stable and tend to lose chromosomes — obviously a disaster if this loss involves those chromosomes that carry the B-cell parent's immunoglobulin genes.

2.4. Plating out of Hybrids/Feeder Cells

2.4.1. Culture Dishes

After fusion the cells are plated out into multiwell tissue culture dishes. There is a choice of 96, 48 or 24 well plates. Ninety-six well plates have the advantage of containing only one or two clones per well, thus reducing the risk of losing interesting but relatively slow-growing clones by overgrowth with a faster one; however, cell growth *in toto* is less good in small wells. Alternatively, the large wells of a 24 well plate support better overall cell growth, but do carry a higher risk of overgrowth. The 48 well plates that have recently been introduced on the market may provide a successful compromise.

2.4.2. Feeder Cells

These are required in all the early stages of hybridoma culture, until cloning is complete. Although not routinely used after this stage, the addition of feeder cells to 'unhealthy' hybridoma cells at any stage of their culture, even when in flasks, has a very beneficial effect. The cells most frequently used as feeders are peritoneal macrophages; although normally from the same species as the myeloma cells, this is not obligatory — for example mouse macrophages are very successful feeders in a rat myeloma system. The role of macrophages is probably two-fold: they 'condition' the medium, presumably via the secretion of various hormones/growth factors, e.g. prostaglandins; they also act as the 'vacuum cleaners' of tissue culture by phagocytosing debris left by dying cells — especially important during HAT selection. However, some non-phagocytic cell types, e.g. rat thymocytes, have been used successfully as feeders; hence the ability to phagocytose, although useful, is not essential.

2.5. HAT Selection

Immediately after fusion and plating out of cells, two problems have to be dealt with. Firstly, only hybrid cells are of interest; unfused spleen and myeloma cells must therefore be removed from the cultures. Spleen cells present no problem as they die within a few days of culture. Unfused myeloma cells are obviously a more serious problem as they, like the hybrids they give rise to, are immortal. Elegant selection systems have been used to overcome this.[11] The method developed for all the most commonly used myeloma lines is that of HAT selection, where HAT stands for hypoxanthine, aminopterin and thymidine. Aminopterin blocks the normal pathway for DNA synthesis from nucleoside precursors; however, if provided with exogenous hypoxanthine and thymidine a cell can use the enzyme hypoxanthine phosphoribosyl transferase (HPRT) for producing DNA via a salvage pathway. Myeloma cell lines lack this enzyme (HPRT variants were selected for by growth in the toxic guanine analogue 8-azoguanine). Thus, after fusion HAT is added to the tissue culture medium. Under these conditions unfused HPRT myeloma cells will die, unfused spleen cells will die anyway, and the only surviving cells will be the hybrids in which the myeloma parent has contributed the property of immortality and the spleen parent the enzyme HPRT. Cultures are maintained in HAT for approximately two weeks. After this they should be weaned gently off the salvage pathway using medium containing HT before reverting to normal medium. The hybridoma growth rate increases rapidly as soon as aminopterin is removed from the cultures, and immunoglobulin will be secreted into the tissue culture supernatant. As soon as the supernatant has turned yellow, it should be tested for the specificity of this immunoglobulin (*Fig.* 2.2).

2.6. Screening Tissue Culture Supernates

There are two main requirements for a good screening assay for use in monoclonal antibody production. Firstly, it should be rapid. If the interval between testing and cloning is too long, cell death will occur and a valuable hybrid may be lost. Ideally the cells in wells whose supernatant contains antibody of interest should be cloned/split within 24 hours of their supernatant medium turning yellow. Secondly, the screening method should be identical, or as close as possible, to the system in which the monoclonal antibody is finally to be used. Antibodies of different classes and subclasses do not perform equally well in all assay systems and the antigenic determinants to which they bind are not all equally preserved in different antigen preparations. For example, there is no point selecting a good complement-binding IgM antibody by cytotoxicity assay if what is required is a monoclonal that can be used as a $F(ab')_2$ preparation in immunocytochemistry. Similarly, if an antibody is wanted for use on routine formaldehyde-fixed tissue sections, the supernatants should be screened on such fixed tissue rather than on frozen tissue where more antigenic determinants are preserved. Fortunately, immunocytochemistry is very suitable for screening a large number of superna-tants in a short time. Although not so rapid as an enzyme-linked immunosorbent assay (ELISA), since screening is by human eye rather than multiscan spectrophotometer, the data obtained are more informative, being both qualitative and semiquantitative. We have routinely used immunocytochemical

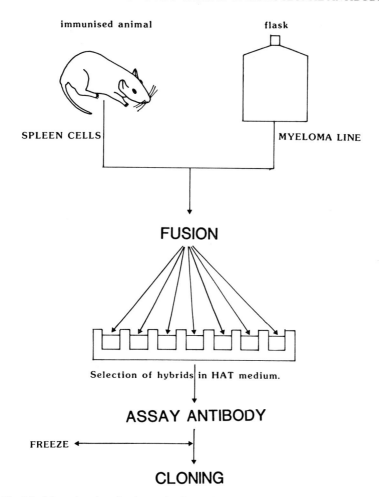

immunised animal

flask

SPLEEN CELLS

MYELOMA LINE

FUSION

Selection of hybrids in HAT medium.

ASSAY ANTIBODY

FREEZE

CLONING

Fig. 2.2. Monoclonal antibody production: scheme for fusion and HAT selection.

methods to screen for monoclonal antibodies against unknown molecules in or on epithelial cells in the human thymus, as part of a project to analyse the molecular basis of microenvironmental influences on T-lymphocyte differentiation in man. We have raised rat, mouse and rat/mouse (rat B-cells/mouse myeloma) monoclonal antibodies using fresh homogenized tissue as immunogen and frozen tissue sections for screening by indirect immunofluorescence. We use 10 sections per multispot slide (Teflon coating between spots prevents antibody spreading from one section to another), incubate the sections with tissue culture supernatant at 4 °C overnight and with the fluorescein isothiocyanate (FITC)-conjugated second layer reagent next morning. Sections can then be read (at a rate of approximately 100 sections per hour) later the same morning. In this way, yellow supernatants can be tested and cells in the interesting wells cloned within 24 hours. This type of immunocytochemical screening provides dual information, showing not only which wells contain secreted immunoglobulin that binds to antigenic molecules in the screening tissue, but also showing where within the tissue these molecules are localized.

2.7. Cloning Hybridomas

When the fused cells are first plated out, more than one hybrid may be present in each well of the tissue culture dish; hence the antibody that is subsequently secreted and tested for in the supernatant may be of more than one type, and staining patterns seen in immunocytochemistry may be a composite of the individual specificities present. It is therefore essential to ensure that subsequent cultures contain only the progeny of a single hybridoma, and hence that the antibody secreted is truly monoclonal. There are three main cloning methods: limiting dilution, soft agar cloning and single cell manipulation. Of these, cloning by limiting dilution is the simplest and fastest, and hence is the method most frequently used. We clone three times by limiting dilution. For the first cloning, the cells from a single positive well are diluted such that when plated out into a 96 well plate there will be, on average, 6 cells per well. For the second and third clonings the dilution is greater, yielding an average of 0·3 cells per well (*Fig.* 2.3). Feeders are particularly important when dealing with such small numbers of cells. The first cloning very often produces a single clone in each well and provides a safety factor for the two subsequent clonings where, at lower dilution, cell survival is less secure. Cloning in soft agar relies upon the fact that in a semisolid medium all the members of a clone will remain grouped together. Single clones can therefore be picked out, washed free of agar and grown up in culture. However, hybridomas do not always grow well under these conditions and washing out all the agar can present problems. The third method, cloning by single cell manipulation is the most rigorous method for ensuring monclonality. However, it takes great patience and a considerable amount of time, and is thus impracticable at the stage where (with a successful fusion) there are several wells whose contents require cloning on the same day. Possibly its main use is after two or three limiting dilutions have been performed and a final crucial cloning is needed for a small number of important 'clones'. In practice, however, we find limiting dilution to be perfectly adequate.

2.8. Class/Subclass Testing of Monoclonal Antibodies

The determination of the class/subclass of your monoclonal antibodies is not simply of academic interest. The knowledge is useful in several ways: prediction of the behaviour of a reagent, ability to choose combinations of monoclonals for dual immunocytochemistry using subclass-specific second layer antibodies, and monitoring the clonality of the hybridoma. This testing can be performed directly on a smear of hybridoma cells using subclass specific-labelled antibodies (available commercially) or indirectly with the same reagents following application of the monoclonal antibodies to sections of appropriate tissue.

Cloning

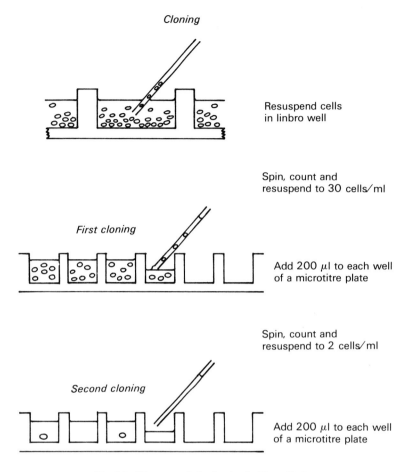

Resuspend cells
in linbro well

Spin, count and
resuspend to 30 cells/ml

First cloning

Add 200 μl to each well
of a microtitre plate

Spin, count and
resuspend to 2 cells/ml

Second cloning

Add 200 μl to each well
of a microtitre plate

Fig. 2.3. Diagram of cloning by limiting dilution.

2.9. Bulk Production of Monoclonal Antibodies

Immunoglobulin secreted by hybridoma cells into their tissue culture medium is present at a concentration of approximately 10 μg/ml, and is more than sufficient for use in immunocytochemistry. Large amounts of immunoglobulin, e.g. for purification, digestion, conjugation, can be produced either from bulk tissue culture or from ascitic fluid of animals in which the hybridoma cells have been grown as an ascites tumour (*Fig.* 2.4).[12]

2.10. Purification, Digestion and Conjugation of Monoclonal Antibodies

These methods are beyond the scope of this article, and are well covered in the literature.[13] However, it is worth stressing the point that each monoclonal antibody is unique and, as such, will have its own set of unique requirements for

POSITIVE WELLS

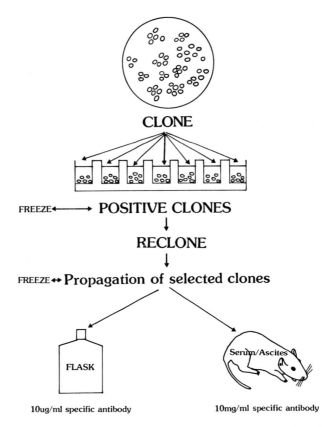

CLONE

FREEZE ←——→ **POSITIVE CLONES**

RECLONE

FREEZE ↔ **Propagation of selected clones**

FLASK

Serum/Ascites

10ug/ml specific antibody   10mg/ml specific antibody

Fig. 2.4. Monoclonal antibody production: scheme for cloning and propagation of selected clones.

purification, digestion and conjugation. Most methods are based upon the average behaviour of a particular class or subclass of immunoglobulin; your monoclonal may be the exception! For example, although the IgG2a subclass of mouse immunoglobulin can be purified from serum by protein A affinity chromatography, there are individual IgG2a monoclonal antibodies that will not bind to this ligand and hence cannot be purified by this method.

2.11. Storage

Monoclonal antibodies are best stored in small aliquots at −70 °C. Once thawed they should be kept at 4 °C as repeated freezing and thawing will destroy antibody activity. Storage is better at either high antibody concentrations (as in ascites) or in the presence of a carrier protein (as provided by the fetal calf serum present in the tissue culture medium of supernatants). Where antibodies are diluted out to low concentrations, 5 per cent fetal calf serum or 1 per cent bovine serum albumin can be used as carrier.

Finally, be paranoid! Freeze hybridoma cells down at all stages of their production; you will then have an insurance against both contamination in tissue culture and mechanical faults in incubators or gassing facilities. It is also wise to store frozen duplicates in separate liquid nitrogen tanks; it is easy to forget to fill a tank, but very difficult to repeat a year's work.

Acknowledgements

I should like to thank the following members of the Department of Immunology, Royal Postgraduate Medical School, whose expertise in monoclonal antibody production has made a major contribution to this chapter: Wilma MacKenzie, Kevin Price, Mark Larché.

REFERENCES

1. Köhler G. and Milstein C. Continuous cultures of fused cells producing antibodies of predefined specificity. *Nature* 1975, **256**, 495–497.
2. Galfre G. and Milstein C. Preparation of monoclonal antibodies: strategies and procedures. *Methods Enzymol.* 1981, **73B**, 3–46.
3. Bastin J., Kirkley J. and McMichael A. J. Production of monoclonal antibodies: a practical guide. In: McMichael A. J. and Fabre J. W., eds., *Monoclonal Antibodies in Clinical Medicine.* Academic Press, London, 1982: 503–517.
4. Gearing A. J. H., Thorpe R., Spitz L. and Spitz M. Use of single shot intra splenic immunization for production of monoclonal antibodies specific for IgM. *J. Immunol. Methods* 1985, **76**, 337–343.
5. Dadi H. K., Morris R. J., Hulme E. C. and Birdsell N. J. M. Antibodies to a covalent antagonist used to isolate the muscarinic cholinergic receptor from rat brain. In: Reid E., Cook G. M. W. and Morre D. J., eds., *Investigation of Membrane Located Receptors.* New York, Plenum Press, 1984: 425–428.
6. Reading C. Theory and method for immunization in culture and monoclonal antibody production. *J. Immunol. Methods* 1983, **53**, 261–291.
7. Edwards P. A. W., Smith C. M., Neville A. M. and O'Hare M. J. A human–human hybridoma system based on a fast growing mutant of the ARH–77 plasma cell leukaemia-derived line. *Eur. J. Immunol.* 1982, **12**, 641–648.
8. Kaplan H. S., Olsson L. and Raubitschek A. Monoclonal human antibodies: a recent development with wide-ranging clinical potential. In: McMichael A. J. and Fabre J. W., eds., *Monoclonal Antibodies in Clinical Medicine.* London, Academic Press, 1982: 17–35.
9. Milstein C. Monoclonal antibodies from hybrid myelomas: theoretical aspects and some general comments. In: McMichael A. J. and Fabre J. W., eds., *Monclonal Antibodies in Clinical Medicine.* London, Academic Press, 1982: 3–16.
10. Van Vliet E., Melis M. and Van Ewijk W. Monoclonal antibodies to stromal cell types of the mouse thymus. *Eur. J. Immunol.* 1984, **14**, 524–529.
11. Szybalski W., Szybalska E. H. and Ragin G. Genetic studies with human cell lines. *Natl Cancer Inst. Monogr.* 1962, **7**, 75–89.
12. Hoogenraad N., Helman T. and Hoogenraad J. The effects of pre-injection of mice with pristane on ascites tumour formation and monoclonal antibody production. *J. Immunol. Methods* 1983, **61**, 317–320.
13. Johnstone A. P. and Thorpe R. *Immunochemistry in Practice.* Oxford, Blackwell Scientific Press, 1982.

3 Tissue Preparation and Immunostaining Techniques for Light Microscopy

S. Van Noorden

Immunocytochemical techniques, invented by A. H. Coons[1,2] have gradually, with modifications and improvements, come into use since the sixties. Lately, with the added fillip of monoclonal antibodies,[3] they have been widely adopted in all fields of biological science. In this chapter an introduction to currently standard methods and common problems will be given with a practical guide in the Appendix. For greater detail, the reader is referred to several books on the subject.[4–11]

1. DEFINITION

Immunocytochemistry is the identification of a tissue constituent in situ by means of a specific antigen–antibody reaction, tagged by a microscopically visible label. Provided that a suitable antibody can be produced and the antigen preserved, there is no limit to the substances that may be localized in this way.

2. METHODS

2.1. Direct Method

A labelled primary antibody is applied directly to the tissue preparation.

2.2. Indirect Method

The primary antibody is unlabelled and is identified by a labelled secondary antibody raised to the immunoglobulin of the species providing the primary antibody. Because at least two secondary antibody molecules can bind to each primary antibody molecule, this method is more sensitive than the direct method, i.e. smaller amounts of antigen can be detected, or a stronger signal is given for the same number of bound primary antibody molecules. An additional advantage is that a labelled secondary antibody to the immunoglobulin of one species can be used to identify any number of primary antibodies raised in that species.

2.3. Unlabelled Antibody–Enzyme Methods

An unconjugated bridging secondary antibody is used between the primary antibody and the labelled detecting reagent, which is now usually an enzyme–anti-enzyme complex or an avidin–biotin-enzyme complex (*see* Chapter 4).

2.3.1. Peroxidase–Anti-peroxidase

Sternberger[5] combined peroxidase with anti-peroxidase to form a stable cyclic complex with three peroxidase molecules to two antibody molecules. The peroxidase–anti-peroxidase (PAP) complex is used as a third layer antigen, the first layer being (rabbit) anti-tissue antigen, the second layer being unconjugated (goat) anti-rabbit IgG, in excess, so that one of the two identical binding sites (Fab portion) is free to combine with the third layer, the rabbit PAP complex. There is thus a high ratio of peroxidase label to primary antigen (*Fig.* 3.1). The rabbit PAP molecules will react only with the anti-rabbit IgG of the second layer, so that provided there is no unwanted binding to the tissue of the second layer this is a

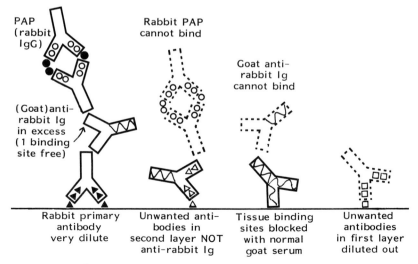

| Rabbit primary antibody very dilute | Unwanted anti-bodies in second layer NOT anti-rabbit Ig | Tissue binding sites blocked with normal goat serum | Unwanted antibodies in first layer diluted out |

Fig. 3.1. The peroxidase–anti-peroxidase (PAP) method.

very specific and background-free technique. The increased amount of label allows the primary antibody to be highly diluted which has the advantage of reducing unwanted background staining (*see* Section 6), of allowing preferential attachment of high affinity antibodies and of decreasing the possibility of dissociation of the antibody from the tissue antigen because both combining sites of each antibody molecule are bound. Because the second antibody is unconjugated it has full activity and because the peroxidase is bound by an antigen–antibody reaction rather than a chemical conjugation, it too is fully active.

Although in fact the end-results of the PAP method are not always better than those of the indirect method (*see* Chapter 31), the high degree of purification and specific binding of this final layer provides a very clean and sensitive method which is still the most popular one in use (*see* Appendix, Section I.C.).

2.3.2. *Alkaline Phosphatase–Anti-alkaline Phosphatase*

This method works on the same principle as the PAP method and, like it, can be extremely sensitive. To date only monoclonal mouse anti-alkaline phosphatase has been produced, so the alkaline phosphatase–anti-alkaline phosphatase (APAAP) method is used with mouse monoclonal primary antibodies.[12]

3. ESSENTIAL CONDITIONS

The necessary conditions for accurate localization of an antigen by a labelled antibody are: preservation of the tissue and antigen; production of a specific antibody; efficient labelling.

3.1. Preservation of Tissue and Antigen

It is conventional and necessary to treat fresh tissue with chemical fixatives, such as formalin, to preserve the structure and prevent tissue components from disintegrating during histological staining procedures. Soluble antigens must also be 'fixed' in the tissue before being immunostained. Unfortunately, the standard fixation process with formaldehyde or glutaraldehyde can sometimes cross-link proteins so strongly that antigenic sites become obscured and the immunoreaction cannot take place. A compromise between good tissue preservation and antigen availability has to be reached for each antigen.

3.1.1. *Fixation*

The method of fixation depends not only on the antigen to be localized, but also on the tissue structure in which it is present. *Fig.* 3.2 illustrates the main preparative processes. Fixation is fully discussed by Pearse[13] and Brandtzaeg.[14] The use of various fixatives is outlined in *Table* 3.1.

Table 3.1. **Fixation methods**

Method	Useful for
Fresh frozen tissue Cryostat sections, unfixed or postfixed in alcohol or acetone Freeze-dried, paraffin sections, postfixed in acetone Smears, impressions, unfixed or fixed in alcohol or acetone	Extracellular antigens, e.g. immunoglobulin in glomerular basement membrane Cell surface antigens; tissue antigens for diagnosis of autoimmune disease
Cryostat sections of tissue prefixed by perfusion or immersion in buffered formaldehyde or buffered p-benzoquinone (NB Alteration of temperature or pH may improve fixation)	Peptides in endocrine cells and nerves Amines, enzymes (as antigens) etc.
Freeze-drying followed by vapour-fixation in formaldehyde, p-benzoquinone, diethyl pyrocarbonate and embedding in paraffin	Intracellular water-soluble antigens; peptides in endocrine cells (not suitable for membrane antigens)
Rinse in saline, alcohol fixation, paraffin embedding	Immune deposits and intracellular immunoglobulins
Buffered formalin, formol saline, formol mercury, Bouin's fixative etc. Routine surgical material, paraffin sections (dried at 37 °C) May be treated with a protease before use	Histopathological diagnosis; peptide hormones, immunoglobulins etc.
Periodate–lysine–paraformaldehyde	Glycoproteins
Glutaraldehyde Glutaraldehyde–formaldehyde, periodate–lysine–paraformaldehyde	Electron microscopical immunocytochemistry

Conventional formalin fixation in one of its many forms and paraffin embedding is often quite satisfactory for immunocytochemical techniques with the sensitive antibodies and detection systems that are now available. A significant advance in this area was made by Huang and coworkers[15] with their introduction of protease digestion to reveal 'overfixed' antigenic sites. They treated fixed material with trypsin and showed an improvement in the intensity of the immunostain and number of stained sites, probably due to the breaking of the protein cross-linkages caused by formalin fixation. In practice, it is advisable to use both enzyme-treated and non-enzyme-treated sections since, in some cases, the enzyme can digest antigens that are still available at the same time as revealing others that were hidden.[16] The best fixative and protease to use will depend on the antigen to be uncovered.[17,18] We have found, in common with others, that protease VII (Sigma, P–5255) is best for revealing extracellular immunoglobulin deposits in glomerular basement membrane (*see* Chapter 37), but we use trypsin as a general purpose protease and find it useful (*see* Appendix, Section II).

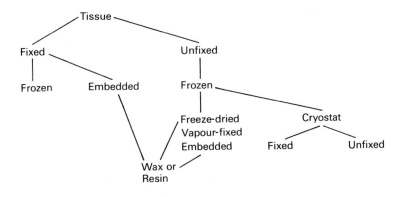

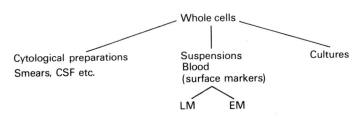

Fig. 3.2. Preparation of material for immunocytochemistry.

Alcohol has been used as a fixative, followed by paraffin embedding, where intracellular immunoglobulins are to be detected, the tissue sample being presoaked in saline for 48 hours.[14] This washes out intercellular immunoglobulin deposits, giving the immunoglobulin-containing cells a better contrast with the background in immunofluorescence. As can be imagined, the pre-treatment does not do the structure of the tissue much good, and it is thus not suitable for transmitted light microscopical immunoenzyme methods.

Adding mercuric chloride to formalin fixative, which is still done in some histopathology departments, does not seem to affect immunostains adversely and may even enhance them. Mason et al.[19] state that protease digestion is unnecessary for revealing immunoglobulin antigens if formol–mercury is used as the primary fixative. The preparations must be treated with iodine to remove pigment deposits caused by the mercury fixation (not necessary if immunofluorescence is to be used). This is usually done before the immunostain but can be carried out at the end of the (peroxidase) immunoreaction if necessary. Eneström[20] has suggested a more complicated procedure to remove divalent as well as monovalent mercury ions, incubating sections in a mixture of ethylenediaminetetraacetic acid and dithiothreitol. He used this to improve the immunofluorescence of IgG and complement components in kidney. We have found it helpful in immunoperoxidase staining of immunoglobulins in formol-–mercury-fixed paraffin sections of kidney (*see* Chapter 37).

3.1.2. Tissue Preparation

3.1.2.1. LYMPHOID TISSUE. If cell surface antigens of lymphoid tissue are to be studied, the tissue must be fresh frozen, and well-dried cryostat sections must be used with short postfixation in, for example, acetone (*see* Chapter 34). Smears or imprints should be air dried and then fixed in acetone or other fixative for a short time before staining (some antigens are best preserved in preparations stored in buffer — *see* Chapter 33 for cytological preparations). If the antigens are to be examined in blood cells, the cell suspension is exposed to the primary antibody before being fixed for electron microscopy (*see* Chapter 35). Smears and cryostat sections may be stored at −20 °C after drying or freeze drying. They should be wrapped in foil or cling-film and sealed in a bag with some dessicant such as silica gel. The preparations may be fixed after removal from the deep-freeze. The whole bag is allowed to come to room temperature before being opened, thus preventing the deleterious effects of atmospheric water condensing on the preparations.

We stated in the first edition of this book that freeze drying with vapour fixation was unsuitable for membrane antigens. A method has recently been published[21] for the preservation of lymphoid cell surface antigens by freeze drying a tissue block and embedding it in paraffin without fixation. Sections are subsequently dewaxed in xylene and briefly fixed in acetone before proceeding to the immunostain. The method has the advantage that it is more convenient to store wax blocks than frozen material and the results are said to be comparable to those achieved with freeze-dried cryostat sections. Whether substances such as basement membrane proteins also survive this process has yet to be determined.

3.1.2.2. PERMEABILIZATION. Whole-cell preparations such as smears, imprints or cell cultures, may need to be made permeable to the antibodies as well as being fixed, if acetone is not the fixative, because lipid membranes can exclude large molecules and need to be broken down (*see* Chapter 15). A similar consideration applies to whole-mount preparations, for instance of layers of the wall of the intestine,[22] the iris membrane[23] or thick (> 10 μm) cryostat sections.[24] One way of achieving permeabilization is to fix the preparations, then bring them through graded alcohols to xylene and back to water, provided that the antigen will survive this treatment.

3.1.2.3. NERVE FIBRES. Fine nerve fibres in sections are best seen in fairly thick sections (> 15 μm), whether neuropeptides, neurofilaments or other constituents are being immunostained. Thick cryostat or Vibratome sections of prefixed tissue are suitable (*see* Chapter 22). Immunofluorescence is sometimes more effective than immunoperoxidase because the thickness of the section does not always allow sufficient penetration of transmitted light to give a clear background. This is particularly so when *p*-benzoquinone is the fixative. This mildly cross-linking fixative, very effective for preserving neuro-peptides,[25–28] provides tissue with a brown colour that gives an excellent dark background for fluorescence but can be confusing if peroxidase stains are used. In thinner sections, the background colour is not intrusive, but the structural preservation is in any case not very good (*see* Appendix, Section IV for method). Formalin in many different fixatives, e.g. Zamboni's,[29] or phosphate-buffered paraformaldehyde,[30] is also useful for neuropeptide immunocytochemistry and does not give the brown background. It provides good structural preservation and allows the use of conventional histological counterstains. However, it has the disadvantage for immunofluorescence techniques that autofluorescence may be disturbing and that catecholamines may be induced by it to fluoresce.[31] In practice, most induced catecholamine fluorescence would be quenched by the many aqueous solutions through which the immunostained preparations must pass before reaching the fluorescence microscope.

The problem with thick Vibratome sections is that of penetration of large antibody molecules. Permeabilization of the sections by the method described above or by brief freezing and thawing, or inclusion of a detergent in the buffer solutions may be useful.

3.1.2.4. FREEZE DRYING FOLLOWED BY VAPOUR FIXATION. This is an excellent preparative method for many substances, including peptides in endocrine cells.[28] The proviso about induced fluorescence of catecholamines should be borne in mind if formaldehyde vapour is used as the fixative. If benzoquinone vapour is used the problem does not arise.

3.1.3. Sectioning Media

Reference has already been made to conventional paraffin sections. Cryostat sections of fresh-frozen tissue are useful for cell surface and other antigens that are labile to routine fixation and processing methods. Thick cryostat sections of prefixed material may be used as free-floating preparations,[24] as may prefixed unfrozen sections cut on a Vibratome or tissue chopper or unsectioned whole-mount preparations.[22,23] Epoxy resins are suitable for semithin sections as well as ultrathin sections. Sections ($0.5-1$ μm) are mounted on glass slides on a drop of water or 10 per cent acetone and dried down on a warm ($< 45\,°C$) surface. The sections are then thoroughly dried at 37 °C and the resin is removed completely, for example by sodium hydroxide in alcohol,[32] before immunostaining is carried out. Glycol methacrylate is unfortunately not very suitable as an embedding medium for immunostaining of semithin sections. There have been a few papers describing successful immunostaining for various antigens, using a variety of proteases and types of methacrylate,[33,34] but in our hands little success has been achieved. A newly marketed type of glycol methacrylate, 'Immunobed' from Polysciences, promised to be an improvement because of its increased mesh size, but probably requires very careful low-temperature fixation and processing if it is to be successful. Historesin (Kulzer/LKB) is another variant which may prove useful.

3.1.3.1. ADHERENCE OF SECTIONS TO SLIDES. To prevent sections from becoming detached from the glass slide during immunostaining, particularly when protease digestion is used, or fragile cryostat sections are involved, it is useful to coat the slide with poly(L-lysine)[35] before mounting the sections (see Appendix, Section III). Paraffin sections should not be baked on a hot-plate, which reduces the immunoreactivity of some antigens, but dried at 37 °C overnight. Cryostat sections should be air dried for 1–3 h before being stained and may be stored at −20 °C before use (*see* Section 3.1.1).

3.2. Specific Antibody Production

Antibodies are still produced by immunizing animals, and even the monoclonal antibodies begin by this procedure. Immunostaining schedules are of wide variety and there is no standard procedure guaranteed to produce an antibody that is of high specificity and affinity. Antibody production and testing is described in Chapters 1 and 2. Antibodies must be rigorously tested for specificity and the possibility of cross-reactivity must always be borne in mind (*see* Sections 4.3 and 6.9).

3.2.1. Working Dilution of Antibodies

Every antiserum must be tested on a known positive control to find the 'correct' working dilution for the staining method to be used. The dilution will obviously vary according to whether incubation is to be for half-an-hour at room temperature or overnight at 4 °C. For overnight incubation and the PAP method, for example, an unknown antibody should be tested at doubling dilutions from 1/100 to 1/32 000 to find the initial range, and then further tested if necessary. Even if a dilution of 1/5000 is satisfactory on the positive test tissue, it may have to be altered for another tissue or another fixation method, but the initial test provides a guideline. More concentrated does not usually mean better; for reasons already given, a dilute antibody provides less non-specific background staining and better contrast and, for the PAP method, too concentrated a primary antibody may prevent the PAP layer from attaching.[36] Linking or conjugated antisera and PAP, used for half-an-hour at room temperature, can usually be diluted in the range 1/50 to 1/300. Antisera may also be tested by radioimmunoassay and by the enzyme-linked immunosorbent assay.[37] However, conditions for these *in vitro* techniques are not the same as for antibody exposed to tissue-bound antigen, together with all the other tissue constituents that may be present, and the golden rule for assessing antibodies for immunocytochemistry is to test them by immunocytochemistry (*see* Chapters 1 and 2). The other methods have more value, as far as immunocytochemistry is concerned, in assessing potential regional specificity and cross-reactivity of the antibody, since they may expose it to a variety of known antigens.

3.2.2. Storage of Antibodies

Unconjugated primary antisera may be stored frozen, undiluted or at a low dilution (1/10–1/50) in buffer. They should not be repeatedly thawed and refrozen so the solution should be divided into quantities sufficient for a once or twice only dilution. The vials of antibody should be snap-frozen in liquid nitrogen, before storing at −20 to −40 °C.

Alternatively, conjugated or unconjugated antisera may be stored neat at 4 °C with the addition of sodium azide (0·01–0·1 per cent) or merthiolate (0·01 per cent) to prevent bacterial growth. Azide inhibits peroxidase activity but the amount added by commercial suppliers to the neat antibody–peroxidase conjugate or to PAP does not prevent the activity in the diluted working solution. Azide should not be included in the buffers used to dilute peroxidase-conjugated antibodies or PAP. Unconjugated primary antiserum may be kept at its working dilution in buffered saline containing sodium azide or merthiolate and 1 per cent normal serum or albumin to provide a high protein concentration. The inert protein competes with the antibody which is present in very small quantities, and prevents it from sticking to the walls of the vessel. Antibody solutions may be kept for many weeks under these conditions. Conjugated antibodies should not be kept diluted as dilution increases the chance of the bond between the label and the antibody becoming dissociated. For the same reason, they should not be repeatedly frozen and thawed. Similar considerations apply to PAP.

3.3. Efficient Labelling

The label or tracer must be attached to the antibody in such proportion that there are few or no unlabelled antibody molecules in the solution which would compete with the labelled molecules for the available antigens. As much label as possible must be attached in order to increase the sensitivity of the method, but the antibody must not be so overburdened with label that its reactivity is impaired. Similarly, the label must not be too harshly treated, particularly in the case of an enzyme, so that its reactivity is reduced. The unlabelled antibody–enzyme method,[38] and the peroxidase–anti-peroxidase method[5,39] avoid these difficulties. Commercial supplies of immune reagents are now generally available, but for information on how to label antibodies *see* the literature.[40]

3.3.1. Labels

3.3.1.1. FLUORESCENT. *Fluorescein* and *rhodamine* isothiocyanates (FITC, RITC) are well-known fluorescent labels, the former fluorescing green at an excitation wavelength of 490 nm, the latter red at 530 nm. Newer fluorescent labels are Texas Red[41] which fluoresces red at the same wavelength as rhodamine but is more stable under the irradiation, and phycoerythrin[42] which also gives a red fluorescence and has the advantage that its excitation optimum for fluorescence is near that of fluorescein. Thus with a double fluorescence immunostain both labels can be seen together without the need for switching filters.

3.3.1.2. ENZYMES. Enzymes are still the most popular labels, particularly *peroxidase* which can be developed with the diaminobenzidine (DAB)–hydrogen peroxide reaction of Graham and Karnovsky[43] to give an insoluble, intensely coloured, brown end-product. DAB was formerly considered to be potentially carcinogenic but is now thought to be less dangerous,[44] while one of the substitutes, 3-amino-9-ethylcarbazole,[45] has become known as a possible carcinogen.[46] It is probably advisable to treat all reagents with care. The brown end-product can be made more intensely coloured (*see* Section 7). Alternatives to DAB for development of peroxidase are the above-mentioned 3-amino-9-ethylcarbazole which gives a red end-product or 4-chloro-1-naphthol which gives a blue end-product and can be useful in multiple staining techniques[47] (*see* Section 8). Both these end-products are alcohol soluble and the preparations must be mounted in an aqueous medium. Endogenous peroxidase, which might otherwise react with the development reagents and be confused with the applied peroxidase, may be blocked, if necessary, by a variety of procedures, e.g. 0·3 per cent H_2O_2 in water, buffer or methanol, or nitroferricyanide[48] (*see also* Appendix, Chapter 31).

Alkaline phosphatase is increasingly popular as a label. It is used in the indirect method conjugated to a second antibody or in a variant of the unlabelled antibody–enzyme method using an anti-alkaline phosphatase combined with alkaline phosphatase, either as two steps or as a single step, the alkaline phosphatase being pre-incubated with the antibody to form an alkaline phosphatase–anti-alkaline phosphatase (APAAP) complex. The enzyme used comes from calf intestine and one of the advantages is that endogenous alkaline phosphatase, except that of the intestine, can easily be inhibited by levamisole (Sigma) added to the final incubating medium to a 10 mM concentration.[49] This blocking method is not suitable if the intestine is being stained and if endogenous intestinal alkaline phosphatase survives the tissue preparation it must be blocked by some other means, such as 20 per cent acetic acid or 0·3 per cent H_2O_2 and 2·5 per cent periodic acid before use (*see* Chapter 31). Monoclonal APAAP is commercially available (Dako) and alkaline phosphatase can also be bought (Sigma). It may be necessary to use a highly purified enzyme (Sigma) if cryostat sections are to be stained as traces of other enzymes in the impure preparation can destroy this delicate material (D. Y. Mason, 1985, personal communication). Alkaline phosphatase can be developed to give blue or red products that are alcohol soluble and must be mounted in an aqueous medium, or a red product that survives rapid dehydration and mounting in permanent medium. These methods are also useful in double staining[50,51] (*see* Section 8) and are described in the Appendix, Sections I.*D, E.*

Glucose oxidase has the advantage as a label that it is not present in animal tissue and therefore there are no problems with endogenous enzyme activity in this material.[52] The end-product is an intense navy-blue which provides a good contrast.

β-*D-Galactosidase* is also a popular enzyme label. The enzyme is derived from bacteria and its pH optimum is different from that of the mammalian enzyme. Thus endogenous enzyme is not a problem, unless there happen to be bacteria in the preparation. The end-product is a bright turquoise-blue with the indigogenic method of development and can be dehydrated and permanently mounted.[53] We have found it particularly useful in multiple staining[54] (*see* Section 6).

3.3.1.3. HAPTENS. Haptens such as dinitrophenol or arsanilic acid have been used as labels for primary antibodies in bridging techniques, the hapten being localized by an unlabelled anti-hapten, following by a haptenized PAP complex.[55] One such method is described in Chapter 32.

3.3.1.4. COLLOIDAL METALS. Colloidal gold as a label can be seen in the light microscope as a deep pink, provided the antibody is used at high concentration, and has its uses in double staining.[56] It can also be seen under dark field illumination or epipolarization (*see* Chapter 5). However, colloidal gold as a label for light microscopical immunocytochemistry has really come into its own with the introduction of labelling by immunogold with silver intensification.[57] This extremely sensitive method, producing an intensely black end-product, is fully described in Chapter 5. Colloidal silver may be used in the same way as colloidal gold and gives a yellow colour.[58]

3.3.1.5. BIOTIN. This is extensively used as a bridging label for attachment to avidin and for attaching enzymes or gold to avidin (*see* Chapter 4).

3.3.1.6. RADIOACTIVE LABELS. These are described in Chapter 7.

4. SPECIFICITY CONTROLS

4.1. Positive Controls

A positive control preparation, known to contain the antigen in question, should be carried through with every batch of immunostain in order to confirm that all the reagents are in working order. Ideally, a specimen that stains weakly for the antigen in question should be used, as a preparation containing a lot of the antigen may be stained adequately by suboptimal reagents so that a fall-off in staining power is not noticed.

4.2. Negative Controls

The specific antibody should not stain inappropriate tissues, and on appropriate preparations no reaction should be seen if any of the steps of the reaction are omitted, or if an inappropriate antibody or non-immune serum (at the same dilution) is substituted for the primary antibody. If staining occurs under these circumstances a cause for the 'non-specific background' should be sought (*see* Section 6).

4.3. Absorption Controls

The ultimate test of specificity is to show that the immunoreactivity of the antibody is abolished by prior absorption with excess of its specific antigen, but not by absorption with similar quantitities of related substances. Preferably, a synthetic preparation of antigen should be used in case impurities in the original antigen cause antibody formation and tissue binding; this would be absorbed out if the original antigen were used for absorption, giving the impression of true staining.

A further hazard is that of cross-reactivity, the original antigen containing immunogenic sequences shared by related molecules. As an example, an antibody raised to cholecystokinin which shares the C-terminal pentapeptide with gastrin, may be treated with pure cholecystokinin and with gastrin. If the activity is removed by cholecystokinin but not by gastrin, the implication is that the antibody is wholly or mainly reactive with a portion of the cholecystokinin molecule not shared by gastrin. If the immunoreactivity is removed by addition of gastrin, the antibody must be directed to the common C terminal and is therefore of no use for discriminating between the two peptides in an area where both are present, such as the mammalian small intestine. Where cholecystokinin is known to be absent but gastrin present, e.g. the pyloric antrum, the antibody could be used to detect gastrin.

An absorption test should be carried out whenever a new antibody or previously unexamined tissue is used. It is convenient to measure the quantity of antigen added in nanomoles per ml of optimally diluted antibody, i.e. the dilution which gives strong staining and low background. The antigen should be added in a series of decreasing concentrations from 10 nmol/ml to 0·001 nmol/ml. Usually immunoreactivity will be abolished by addition of 10–0·1 nmol/ml and will be incompletely removed by 0·01–0·001 nmol/ml. Measurement of antigen concentration in molar terms allows comparison between several antigens and one antibody, or one antigen and several antibodies. One proviso for this absorption test is that the absorbed antibody should not be used to stain tissue that may contain receptors for the antigen. Sternberger[5] found that addition of luteinizing hormone-releasing hormone (LHRH) to anti-LHRH resulted in increased staining of pituitary gonadotrophs. In such a case, a solid-phase absorption on antigen-coated sepharose beads to remove the antibody from the solution would be a better test, although it is more complicated to calculate the amount of antigen added when it is used to coat beads.

We have found it convenient to divide an antigen solution into vials containing one nanomole of antigen. In general, this is a large enough amount to react completely with the antibody contained in 100 µl of antibody solution at its highest usable dilution for the PAP method with 24 hours of incubation at 4 °C. The nanomole quantities also provide a convenient means of comparing the amounts of different antigens required to saturate an antibody, and avoid the problems of dealing with antigens of different molecular weights. The nanomole aliquots are lyophilized and sealed under vacuum and may then be stored indefinitely in a freezer or at room temperature. An antigenically inert substance, such as lactose or albumin, may be added to the antigen solution to provide a visual marker for the lyophilized antigen in the sealed vial. When antibody is added to the vial, it should be vigorously shaken to distribute the antibody over the entire surface to make sure that all the antigen is taken up. The antigen is incubated with the antibody for 12–24 h at 4 °C before the solution is used for staining. As a control, the antibody should be 'absorbed' with an inappropriate antigen under the same conditions.

5. PURIFICATION OF ANTIBODIES

5.1. Absorption on Tissue Powders

Absorption on tissue powders or homogenates from tissue not containing the antigen in question, has been used to get rid of unwanted tissue reactions. In some cases this may be successful, but random non-specific absorption of protein does reduce the amount of antibody available, which is a serious consideration.

5.2. Affinity Purification

Purification of a mixed antiserum may be achieved by affinity absorption. The pure antigen is absorbed onto a solid phase such as cyanogen bromide-activated sepharose beads; the coated beads are then mixed with the antibody, which reacts with the antigen. Non-attached proteins are washed off and the specific antibody is eluted from its antigen by washing with a low or high pH buffer. For the details of the method *see* Chapter 1 and the literature.[59] The disadvantage of this method is that antibodies of high affinity and avidity that are most useful for immunocytochemistry are, by their very nature, difficult to dissociate when bound to specific antigen on the beads. The antibody that is eluted is therefore likely to be of low affinity, although this may be compensated for by its purity.

Absorption of an antibody onto antigen-coated beads may also be used as a negative control for staining (*see* Section 4.3.).

Another way in which affinity absorption may be useful is in removing a known contaminant from an antibody. For example, polyvalent antiserum to follicle-stimulating hormone (FSH) will contain antibodies both to the β subunit, specific to the FSH molecule, and to the α subunit, shared by FSH, luteinizing hormone, thyroid-stimulating hormone and human chorionic gonadotrophin (HCG). Absorption of this antibody on, for example, HCG-coated beads will remove the confusing antibodies to the α subunit, leaving only the antibodies specific to the β subunit of the molecule that is not shared with the other three peptides. The unwanted tissue antibodies will of course remain after this treatment.

6. BACKGROUND STAINING: CAUSES AND REMEDIES

The antigen–antibody reaction is by nature highly specific. However, immunocytochemists are dogged by problems put down to 'non-specific' or 'background' staining. The 'background' may be diffuse and all-over, or it may take the form of 'non-specific' clear staining of structures other than the expected one. There are several causes for a high background which reduces the signal-to-noise ratio and confuses the observer. Many of them can be remedied, the first step being to identify the reagent causing the problem by omitting each in turn from the staining sequence.

6.1. Binding of Immunoglobulin to Fc Receptors

Some tissues, particularly in the form of cryostat sections or whole-cell preparations, have receptors for the Fc portion of the immunoglobulins, resulting in antibody becoming attached to non-antigenic sites. Usually this can be avoided by a preliminary blocking of the receptors with normal serum from the species donating the second antibody. Fc-binding sites will be covered by immunoglobulin from the normal serum and will not be available for attachment of the primary antibody. The second antibody will not identify these immunoglobulins as they will be from the same species. An alternative method is to use the $F(ab)_2$ portion of the antibodies. When the primary antibody is labelled with protein A, which attaches to the Fc portion of immunoglobulins, albumin is better than serum for non-specific blocking. Protein A, which has been extensively used in electron microscopical immunocytochemistry, adsorbed to colloidal gold (*see* Chapters 5 and 9) is extracted from the cell coat of certain bacteria and can be labelled with fluorescent or enzyme markers for light microscopy (*see*[60] for review).

6.2. Ionic Binding and Physical Forces

Hydrophobic and electrostatic forces can bind immunoglobulin to tissue, particularly when some of the tissue components are positively charged. Complement may also be bound in this way, causing further attachment of immunoglobulins.[61] These types of truly non-specific binding can be prevented by diluting the primary (and secondary) antisera as far as possible, by decomplementing all sera (heat to 56 °C for 30 min), or by increasing the ionic strength of the washing buffers or the antibody diluent. Grube[62] recommends a high salt concentration (2·5 per cent NaCl) in the buffers. Addition of a detergent such as Triton X-100 to the buffer can also help,[63] as can use of Tris buffer which is of higher ionic strength than phosphate buffer.

6.3. Reaction between Secondary Antibody and Tissue Immunoglobulin

Antibodies to immunoglobulins of one species may have some reactivity towards immunoglobulins of another species, and thus the secondary antibody may bind to the tissue substrate in a specific but unwanted fashion. This problem can be avoided easily by adding to the second antibody solution 5 per cent of normal serum from the species providing the tissue to be stained.

6.4. Binding due to Active Groups in the Tissue

After fixation in formaldehyde or glutaraldehyde in particular, if the tissue is not thoroughly washed after fixation and before further processing, free aldehyde groups may be present which can attract immunoglobulins from the applied antibodies. This is avoided by reducing the aldehyde groups with sodium or potassium borohydride before immunostaining (*see* Chapter 31).

6.5. Endogenous Peroxidase

This has been dealt with in Section 3.3.1.2. It is difficult to remove endogenous peroxidase from cryostat sections without also removing antigenicity. One answer may be to block the enzyme after binding of the primary antibody. However, endogenous peroxidase activity of macrophages, for example, is usually very obvious and blocking may be unnecessary. Acid haematin in paraffin sections can be bleached with a strong (6 per cent) solution of hydrogen peroxidase. This has no apparent effect on antigens in the tissue (*see* Chapter 31).

6.6. Intrinsic Fluorescence

As well as above-mentioned causes for binding of antibody to tissue, endogenous tissue components may be intrinsically fluorescent, e.g. elastic fibres, or have fluorescence induced by formaldehyde (catecholamines). Intrinsic autofluorescence may sometimes be altered with a counterstain such as Evans Blue,[64] or Pontamine sky blue[65] which fluoresce red. Catecholamine fluorescence is removed by water and is not usually a problem by the end of the immunostaining sequence.

6.7. Unwanted but Specific Staining

Since polyclonal antibodies are heterogeneous, antisera may contain antibodies to tissue constituents, to the coupler used in immunization, and to natural hazards previously encountered by the immunized animal. In general, if the antibody is of high titre and a sensitive method is used, these can be removed by using high dilutions of the antibody. An alternative is to use a monoclonal antibody to the antigen under investigation.

6.8. Immunoreactivity of the Antibody with Molecules Structurally Similar to the Antigen

This is the most intractable of background problems because it is due to genuine antigen–antibody reaction. Many naturally occurring substances are members of 'families', probably derived in evolution from the same precursor molecule by gene duplication and mutation[66] and, though separate substances, still retaining some common structural sequences and also actions. Peptides are well known for this, but newcomers to the minefield are the intermediate filaments, several of which are now known to share antigenic sites.[67]

If possible, the detecting antibody should be raised to a portion of the molecule not shared with structurally related substances. Region-specific antibodies have been valuable in peptide immunocytochemistry[68] and monoclonal antibodies again are useful in this respect, provided that they react with unshared portions of the molecules.

Continual suspicion and awareness of the possibilities of cross-reactions, both suspected and unsuspected, are advised in all branches of immunocytochemistry. An excellent account of specificity and related problems is given by Pool et al.[69]

7. METHODS OF INCREASING SENSITIVITY OF THE IMMUNOSTAIN

Normally, standard methods of localization suffice, but if a very small amount of antigen is present in the preparation, a highly sensitive technique is required. Many methods have been adopted to this end. One way of increasing sensitivity is to increase the amount of label over the antigenic site. This can be done by using the PAP or avidin–biotinylated peroxidase complex methods (*see* Chapter 4) in preference to an indirect method, or by repeating the layers of antibody and labelled detection reagent.[70] The colour of the DAB reaction product may be intensified by addition of imidazole,[71] or a heavy metal such as cobalt,[72] or by osmication or treatment with gold chloride after the reaction. An ultrasensitive technique is the immunogold stain with silver enhancement (*see* Chapter 5). Protease digestion (Section 3.1.1.) is also a method of enhancing the reaction.

8. MULTIPLE STAINING TECHNIQUES

It is often of interest to identify more than one antigen in the same preparation and it is increasingly apparent that several products, even of the same type, may be made by a single cell. Thus a conventional neurotransmitter may be found with a peptide in a single neurone or two peptides may be present together (*see* Chapter 22), either related to the same precursor, e.g. vasoactive intestinal polypeptide and peptide histidine isoleucine, or completely separate, e.g. calcitonin gene-related peptide and substance P.[74]

8.1. Serial Sections

The simplest and probably the safest way of studying co-localization is to immunostain serial sections for different antigens and compare the structures that contain them. The sections must be very thin (1–2 μm) or there is a risk that the same cells will not be present in both sections. An alternative, particularly useful for studying nerve fibres, where thick prefixed cryostat sections are the optimal medium (*see* Chapter 22) is to use 'mirror-image' sections. In the case of cryostat sections, the first section of a pair is picked up on a slide from the knife in the usual way so that the cut surface will be stained. The second, serial section is flipped over onto a cold surface and picked up from that so that the side that was originally apposed to the cut surface of the first section is uppermost. Thus the two surfaces to be stained are those which were separated by the knife. For easier comparison of photographs of the two immunostained sections, one of the pair is printed in reverse so that the mirror-image is cancelled and the two sections appear to be serial.

These methods are suitable where it is easy to identify particular structures by tissue landmarks, e.g. for a comparison of neuron-specific enolase or chromogranin with peptides in intestinal endocrine cells.[75,76] However, in homogeneous tissue, such as many tumours or lymphoid tissue, cell-by-cell comparisons can present a problem. If it is desired to see the relation of one antigen-containing structure with another in the same preparation, other strategies must be adopted.

8.2. Double Direct Method

A method using two primary antibodies labelled with different fluorophores, such as fluorescein and rhodamine, has been successfully used, e.g. for labelling κ and λ light immunoglobulin chains,[14] and this is a very satisfactory way of double labelling. Each fluorescent label is examined with a different filter so excess of one antigen does not obscure or confuse localization of the other, even if both are present in the same cell (*see* Chapter 27). Photographs of the two labels on the same frame will show areas of co-localization in a mixed colour (orange). If phycoerythrin is used instead of rhodamine, both labels can be detected with the same filter system and double-stained structures will be immediately obvious. However, if one antigen is present in such a structure in excess of the other, the presence of the latter may be obscured.

8.3. Double Indirect Method

The primary antibodies may be unlabelled, detected by two differently labelled second antibodies. In this case the primary antibodies should be raised in different species, and the detecting antibodies should be free of species cross-reactivity. Alternatively, the two primary antibodies could be monoclonals of different immunoglobulin subclass or heavy chain type, with non-cross-reacting second antibodies. In both these methods, enzymes or other markers could be substituted for the fluorescent labels (*see* Section 6.6).

8.4. Double or Multiple Immunoperoxidase with Elution

Nakane[47] achieved an immunostain for three pituitary hormones in the same preparation with rabbit antibodies, by carrying out an indirect immunoperoxidase reaction for the first antigen, growth hormone, resulting in an insoluble brown $DAB–H_2O_2$ peroxidase reaction product. He then eluted the entire antigen–antibody complex from the tissue with a glycine-HCl buffer, leaving the brown product, and carried out a second stain, for thyrotrophin. This was developed with α-naphthol and pyronin to give a red end-product. The procedure was repeated for luteinizing hormone which was developed with 4-chloro-1-napthol to give a blue end-product. In practice, even if only two antigens are stained for, this method requires a large number of controls to ensure that each entire antigen–antibody complex has been removed before the next sequence is carried out, or there is the danger of cross-reaction between the layers of the second sequence with the remaining antibodies from the first sequence, and residual peroxidase from the first sequence might continue to react with the second developing solution. There is also the problem that high affinity antibodies are hard to detach from their antigens.

Tramu and his colleagues[77] suggested using acidified potassium permanganate to remove the antibodies of the first reaction. We have found it difficult to establish the correct concentration of eluting solution for each antigen.

8.5. Double Staining without Elution

Problems of cross-reactivity between two sequential immunoperoxidase stains were overcome by Sternberger and Joseph[78,79] who stated that very strong development of the peroxidase in the first reaction masked reactive sites on the whole system, and prevented binding by components of the second reaction, which could then be developed with a chromogen, such as 4-chloro-1-naphthol, giving a contrasting colour. This strategy works well in some circumstances but we have found that an easier way is to develop the first antibody peroxidase with a heavy metal enhancement technique such as the DAB–cobalt chloride system.[72] This results in such an intense blue-black colour that any further build-up of second antibody and subsequent peroxidase development with the DAB alone to give a contrasting brown colour is not noticeable. Similar considerations apply to the use of the immunogold–silver stain with peroxidase (see Chapter 5).

These two methods are suitable only when it is known that the two antigens to be localized are present in separate cells, because the strong stain of the first reaction in both cases would mask any staining for a second antigen on the same site.

An ingenious method for sequential multiple immunofluorescence or immuno-gold staining using primary antibodies raised in the same species has recently been published.[80] Potential immunoglobulin-binding sites remaining after the first set of reactions are destroyed by hot formaldehyde vapour before application of the next primary antibody.

8.6. Double or Multiple Immunoenzymatic Stain

A double immunoenzymatic staining method which avoids the problems of elution and can allow two antigens to be demonstrated in the same location was described by Mason and Sammons.[50] The technique has been fully discussed in previous publications.[81,82] Briefly, it consists of simultaneous application of two primary antibodies, raised in different species, say rabbit and mouse, to two antigens. Two monoclonal antibodies from the same species but of different Ig subclass would also be suitable. The second layer consists of two species- or Ig-class-specific detecting antibodies to immunoglobulin, say goat anti-rabbit Ig and goat anti-mouse Ig. Either these antibodies are conjugated with different enzymes, e.g. peroxidase and alkaline phosphatase, or they are unconjugated and a third layer of reagents, say rabbit PAP and mouse APAAP, is applied. The two enzymes are developed separately to give differently coloured end-products, say brown for peroxidase and blue for alkaline phosphatase, with the naphthol AS-MX phosphate and Fast Blue BB reaction. The two antigens are thus revealed in blue and brown in the same preparation. If the two antigens are present in the same cell, a purple-grey colour is obtained. However, if one antigen is present in greater quantity than the other, the reaction product for the least concentrated antigen may be obscured by that for the other. It is necessary that thin sections or smears are used for this method so that cells stained for different antigens do not overlap and give a spurious impression of double staining.

The use of different development methods with different enzymes can result in a variety of single and mixed colours. Using specific rabbit polyclonal and mouse monoclonal antibodies to various pituitary hormones[54] we have demonstrated cells containing both growth hormone and prolactin in a pituitary adenoma producing both hormones, and the dual presence of follicle-stimulating hormone and luteinizing hormone in gonadotrophs. We have also achieved a triple immunostain by preceding the double immunoenzymatic method with detection of a hormone known to be present in a different cell-type, developed with the DAB–cobalt chloride technique.

A quadruple stain on sections of normal human postmortem pituitary was achieved using the DAB–cobalt chloride development for the first hormone in black, followed by application of a mixture of a rabbit antibody to a second hormone plus a mouse antibody to a third hormone and both rabbit and mouse antibodies to a fourth. The rabbit antibodies were detected by a β-D-galactosidase system and the mouse antibodies by a non-cross-reacting peroxidase system, developed to give a brown colour. The pure turquoise β-D-galactosidase end-product marked the second hormone, the pure brown peroxidase end-product marked the third hormone and a mixture of turquoise and brown gave a green colour to the cells containing the fourth hormone (*Plate 2*).

9. CONCLUSION

It is impossible in a short chapter to give a full account of all the immunocytochemical methods that have been and are being used to achieve highly specific localization of cell products and other substances in situ. This brief guide should serve to direct the reader to more detailed literature, and the chapters that follow to show how the methods are being adapted and applied to solving problems of biology and pathology.

Appendix

I. SAMPLE SCHEDULES FOR IMMUNOSTAINING

These methods are those used by the author. Slight variants are given in the appendices to Chapters 31 and 34, and others may be found in the literature.

All primary antibodies are diluted in 0·01 M phosphate-buffered normal saline pH 7·0–7·2 (PBS) containing 0·1 per cent sodium azide and 0·1 per cent bovine serum albumin. For peroxidase methods the second layer antisera and PAP are diluted in PBS alone. Tris-buffered saline (TBS), 0·05 M Tris-HCl buffer pH 7·6, containing 0·85 per cent NaCl, may be used throughout instead of PBS and is *essential* for dilution of alkaline-phosphatase conjugates or APAAP, and for rinsing preparations after application of these solutions.

A. Indirect Method (Immunofluorescence)

1. (*a*) *Wax sections*: Remove wax in xylene or other suitable solvent and bring sections to water through graded alcohols *Protease digestion* is carried out at this stage if required.

   (*b*) *Prefixed cryostat sections*: Mount on poly(L-lysine)-coated slides (*see* Appendix, Section III) or on chrome–gelatine- or formol–gelatine-coated slides. Allow to dry at room temperature for 1–3 h.

   (*c*) *Fresh cryostat sections*: allow to dry thoroughly (*see* Section 3.1.1.), then postfix as required.

2. *Background blocking* (not usually necessary for monoclonal antibodies): Apply normal serum from the species supplying the second antibody at a dilution of 1:5 to 1:20 in PBS, 10–30 min at room temperature. Do not rinse; draw off serum with a pipette or tissue.

3. First layer antibody at established dilution overnight at 4 °C (recommended) or at a lower dilution for 1 h at room temperature.

4. Rinse three times in PBS, 5 min each rinse. Triton X-100 or other detergents may be added to the buffer rinse to reduce background staining. If this is done, slide-mounted preparations should be rinsed again in buffer without detergent before application of the next antibody to ensure that the next drop of antibody solution remains on the preparation and does not spread over the entire surface of the slide.

   Dry slide except for the area of the section which should remain moist with buffer.

   It is particularly important that sections, particularly cryostat sections, are not allowed to become dry at any stage after application of the first antibody (or blocking serum), as drying results in a very poor final preparation.

5. Apply second layer (anti-immunoglobulin of the species providing the primary antibody, conjugated with fluorescein isothiocyanate or other fluorescent label) at the established dilution in PBS, 30 min at room temperature.

6. Rinse three times in PBS, 5 min each rinse.

7. Dry slides except for the area of the section and mount in a non-fluorescent aqueous mountant.* Examine with fluorescence microscope.

B. Indirect Immunoperoxidase

1. As (*A*).

2. Blocking of endogenous peroxidase activity: immerse preparations for 30 min in a solution of 0·3 per cent hydrogen peroxide in PBS, water or methanol, or use the nitroferricyanide solution of Straus.[48] Rinse in PBS.

---

*The traditional mountant is buffered glycerine, one part of PBS to one or nine parts of glycerine. Some fluorescent labels fade rapidly under ultraviolet irradiation and modified mountants have been developed which are said to retard the fading. Citifluor (*see* Chapter 6) is one; another has recently been reported by Balaton et al.[83]

3.–5. As (2) to (4) in Section I.A.
6. Apply peroxidase-conjugated anti-immunoglobulin at the established dilution in PBS (*no* azide), for 30 min at room temperature.
7. Rinse three times in PBS, 5 min each rinse.
8. Develop peroxidase in the following solution:

Incubating solution for peroxidase development: Prepare the solution just before use (*see* (ii) *below*). Dissolve 25–50 mg of diaminobenzidine tetrahydrochloride dihydrate (DAB) (Aldrich Chemical Co.) in 100 ml PBS. Add 50–100 μl of 100-vol. (30 per cent) H_2O_2 (0·015–0·03 per cent final concentration); the solution should be almost colourless but some samples of DAB may produce a brown solution. Filtering this should result in a colourless solution, giving less background staining. Immerse the sections. The development time is usually 1–5 min. Control by microscopical examination. When development is complete, rinse the preparations well in water, counterstain *lightly* with haematoxylin (the nuclear stain can be differentiated in acid alcohol which will not affect the DAB reaction product), dehydrate, clear and mount in a permanent medium.

Notes:
i. The incubating medium may be made up in smaller quantities and applied in drops on the preparations. This is economical of DAB, which is expensive in some countries, but makes it complicated to incubate a batch of sections for a standard time.
ii. The DAB solution will remain effective for several hours, particularly if it is kept in the dark, but may become brown on standing and give more background staining.
iii. A convenient and safe method of dealing with DAB, avoiding the necessity of continually weighing out small quantities of this possibly hazardous substance, is to make a large batch of aqueous concentrate containing 25 mg/ml and dispense it into vials containing 1 or 2 ml. The vials are sealed and kept at −20 °C until used.[84]
iv. If Tris buffer at pH 7·6 is used instead of PBS at pH 7·0–7·2 the peroxidase reaction may be weaker (though the background staining may be less). Addition of 10 mM imidazole (Sigma, I-0250) at pH 7·0 to the incubating medium increases the strength of the reaction.[71]
v. The preparations may be treated with osmium tetroxide after the development of the peroxidase to increase the intensity of the stain. A dilute (0·05–0·1 per cent) solution of gold chloride may also be used (personal experience of the author).

C. Peroxidase–Anti-peroxidase

1.–3. As for (1) to (3) in Section I.B.
4. The primary antiserum is highly diluted in PBS and applied overnight at 4 °C (recommended) or at a lower dilution for 1 h at room temperature.
5. Rinse three times in PBS, 5 min each rinse.
6. The second antibody is unconjugated antibody to the immunoglobulin of the

species providing the primary antibody, at the established dilution in PBS for 30 min at room temperature.

7. Rinse three times in PBS, 5 min each rinse.
8. The third layer is PAP complex (raised from the same species as the primary antibody) at the established dilution in PBS (*no* azide), 30 min at room temperature.
9. Rinse three times in PBS and develop peroxidase in DAB as for (*B*).

Note: For alternative methods of developing peroxidase *see* the literature.[11]

D. Alkaline Phosphatase–Anti-alkaline Phosphatase

This method is essentially the same as the PAP method with the substitution of APAAP for PAP, except that Tris buffer or TBS is used instead of PBS throughout or at least for the rinses preceding and following the APAAP step (Step 8). If extra intensification of the reaction is desired, the second, unconjugated antibody may be reapplied after the APAAP, followed by rinsing and another application of APAAP. (This build-up of layers could also be applied to the PAP system.)

Development of the alkaline phosphatase is outlined below. The development methods are also applicable to an indirect alkaline phosphatase method.

For preparation of APAAP *see* Section 3.3.1.2.

E. Development of Alkaline Phosphatase

  i. *To give a blue end-product*:
  Dissolve 2 mg naphthol AS-MX phosphate in 0·2 ml dimethyl formamide (use a glass vessel).
  Immediately add 9·8 ml 0·1 M Tris-HCl buffer pH 8·2.
  Check pH.
  (This substrate solution may be stored at 4 °C for several weeks).
  Just before use add to a suitable quantity of substrate solution 1 mg Fast Blue BB per ml.
  Filter onto the sections. Incubate at room temperature or 37 °C for 5–10 min. When a satisfactory blue colour is obtained wash the sections well in water and mount in an aqueous mountant, e.g. Apathy's, glycerine jelly, Hydromount (National Diagnostics). Carmalum is a useful red nuclear counterstain.
 ii. *Red end-product (alcohol-soluble)*
  Use Fast Red TR instead of Fast Blue BB.
iii. *Red end-product for permanent preparations*[51]
  Dissolve 1 mg naphthol AS-TR phosphate, in 0·02 ml dimethyl formamide (use a glass vessel).
  Immediately add 4 ml 0·2 M Tris-HCl buffer pH 9·0.
  Just before use, mix. 25 µl 4 per cent New Fuchsin in 2 M HCl with 25 µl 4 per cent sodium nitrite.
  Allow to stand for 1 min (bubble formation — the solution should become a pale-straw colour) then add the 50 µl hexazotized New Fuchsin to the substrate solution, check the pH and filter onto the sections. Develop at room

temperature for 10 min. When a satisfactory red colour is achieved, rinse the sections well in running water, counterstain lightly with haematoxylin (rapid differentiation if required), dehydrate rapidly, clear and mount in a permanent medium.

Note: If cryostat sections are being immunostained with an alkaline phosphatase method, it may be necessary to inhibit endogenous alkaline phosphatase. This is simply done by adding 1 mM levamisole (Sigma, L-9756)[49] to the final incubating medium. This will inhibit all alkaline phosphatases except the intestinal enzyme. Addition of levamisole does not alter the pH of the incubating medium.

Chemicals obtainable from Sigma

Naphthol AS-MX phosphate, sodium salt, N-5000
                          (or free acid,   N-4875)
Naphthol AS-TR phosphate, sodium salt, N-6125
                          (or free acid,   N-6000)
Fast Blue BB, F-3378
Fast Red TR, F-1500
Levamisole,   F-9756

II. TRYPSIN DIGESTION

This can be a useful procedure for revealing overfixed antigenic sites and is frequently used in staining for immunoglobulins in routinely formaldehyde-fixed tissue. If tissue is fixed in formol sublimate as primary fixative, trypsin digestion is not necessary.

Make a solution of 0·1 per cent trypsin in Tris buffer pH 7·8, or 0·1 per cent calcium chloride, brought to pH 7·8 with 0·1 N NaOH. Warm the solution to 37°C and check the pH again. After removing wax and mercury pigment, if necessary, immerse the sections in the trypsin solution at 37 °C for 0·5 h. Rinse well in cold water and PBS (TBS) before proceeding with the immunostaining in the usual way.

Note: It is advisable to test a range of times and temperatures for the trypsin digestion to find the optimum for each tissue/antigen. Other proteases may be used in a similar way, with due regard to their optimal pH. Solutions may be applied in drops on the sections.

III. POLY(L-LYSINE)-COATED SLIDES

Sections tend to be detached from the slide during trypsin digestion. Make sure they are mounted on slides coated with poly(L-lysine) (from Sigma, P1399 mol. wt > 150 000) made up as a 0·1 per cent solution in water and applied to the slide in the same way as a blood film by smearing a small drop over the surface with the end of another slide. The film dries very quickly and the slide is ready to use. A batch of slides may be prepared in advance but should not be kept for more than a week, or they lose their 'stickiness'. It is essential to mark the coated side, because the dried film is invisible. Store the poly(L-lysine) solution frozen in aliquots and refreeze after use.

IV. FIXATION IN BENZOQUINONE SOLUTION

1. Prepare a 0.4 per cent solution of recrystallized *p*-benzoquinone* in 0·01 M phosphate-buffered saline (PBS) pH 7·1–7·4. Allow a tissue : solution volume ratio of approximately 1 : 10, using specimens no larger than 2 × 2 cm.
2. Add PBS to *p*-benzoquinone crystals and agitate until all crystals are in solution.
   Note: If solution is brown, discard it. The solution must be bright yellow when it is used. Prepare the solution a maximum of 20 min before it is to be used. Store the crystals in a cool dry dark place, otherwise they will turn brown and no longer be of use.
3. Fix the tissue by immersing it in the solution for the appropriate time (*see below*).
4. Following fixation, transfer the tissue into PBS containing 7–15 per cent sucrose and 0·01–0·1 per cent sodium azide. The solution need not be prepared freshly each time. It can be prepared in bulk and stored at 4 °C.
5. Rinse in PBS-sucrose at 4 °C for several hours.
6. Freeze a block of the tissue for cryostat sectioning.
7. Spare fixed tissue may be stored in PBS–sucrose solution at 4 °C. Check the tissue frequently for microbial growth.

   Fixation times (ordinary histological block size)

Small endoscopic biopsies	30 min
Thin-walled gut (child or animal)	1 h
Thick-walled gut	2 h
Pancreas	2 h
Tumour	2 h

REFERENCES

1. Coons A. H., Creech H. J. and Jones R. N. Immunological properties of an antibody containing a fluorescent group. *Proc. Soc. Exp. Biol. Med.* 1941, **47**, 200–202.
2. Coons A. H., Leduc E. H. and Connolly J. M. Studies on antibody production. I. A method for the histochemical demonstration of specific antibody and its application to a study of the hyperimmune rabbit. *J. Exp. Med.* 1955, **102**, 49–60.
3. Köhler G. and Milstein C. Continuous cultures of fused cells producing antibodies of predefined specificity. *Nature* 1975, **256**, 495–497.
4. Nairn R. C. *Fluorescent Protein Tracing*, 4th ed. Edinburgh, Churchill Livingstone, 1976.
5. Sternberger L. A. *Immunocytochemistry*, 2nd ed. New York, John Wiley & Sons Inc., 1979.
6. Bullock G. R. and Petrusz P. (Eds.) *Techniques in Immunocytochemistry*, Vol. 1. New York, Academic Press, 1982.
7. Bullock G. R. and Petrusz P. (Eds.) *Techniques in Immunocytochemistry*, Vol. 2. New York, Academic Press, 1983.

*Suitable grade available from Koch-Light (Cat no. 4939·6).

8. Bullock G. R. and Petrusz P. (Eds.) *Techniques in Immunocytochemistry*, Vol. 3. New York, Academic Press, 1985.

9. Wick G., Traill K. N. and Schauenstein, K. (Eds.) *Immunofluorescence Technology, Selected Theoretical and Clinical Aspects*. Amsterdam, Elsevier Biomedical Press, 1982.

10. Cuello A. C. (Ed.) *Immunohistochemistry*, Chichester, John Wiley & Sons, 1983.

11. Polak J. M. and Van Noorden S. (Eds.) *Immunocytochemistry, Practical Applications in Pathology and Biology*. Bristol, Wright PSG, 1983.

12. Cordell J. L., Falini B., Erber W. N., Ghosh A. K., Abdulaziz Z., Macdonald S., Pulford K. A. F., Stein H. and Mason D. Y. Immunoenzymatic labeling of monoclonal antibodies using immune complexes of alkaline phosphatase and monoclonal anti-alkaline phosphatase (APAAP complexes). *J. Histochem. Cytochem.* 1984, **32**, 219–229.

13. Pearse A. G. E. *Histochemistry, Theoretical and Applied, Vol. 2, Analytical Technology,* 4th ed. Edinburgh, Churchill Livingstone, 1985.

14. Brandtzaeg P. Tissue preparation methods for immunocytochemistry. In: Bullock G. R. and Petrusz P., eds., *Techniques in Immunocytochemistry*, Vol. 1. New York, Academic Press, 1983: 1–75.

15. Huang S., Minassian H. and More J. D. Application of immunofluorescent staining in paraffin sections improved by trypsin digestion. *Lab. Invest.* 1976, **35**, 383–391.

16. Finley J. W. C. and Petrusz P. The use of proteolytic enzymes for improved localization of tissue antigens with immunocytochemistry. In: Bullock G. R. and Petrusz P., eds., *Techniques in Immunocytochemistry*, Vol. 1. New York, Academic Press, 1983: 239–249.

17. Jacobsen M. and Jacobsen G. K. The influence of various fixatives on the immunohistochemical demonstration of a number of plasma proteins and on oncofetal proteins in paraffin embedded material. *Acta Pathol. Microbiol. Scand. Sect. A* 1984, **82**, 461–468.

18. Poston R. N., Sidhu Y. S., Makin C. A. and Mason D. Y. Specific enzyme treatment is required for individual monoclonal antibodies in immunohistochemistry with formalin fixed sections. In: Peeters H., ed., *Protides of the Biological Fluids*, Vol. **32**, Oxford, Pergamon Press, 1984, 557–559.

19. Mason D. Y., Bell J. I., Christensson B. and Biberfeld P. An immunohistological study of human lymphoma. *Clin. Exp. Immunol.* 1980, **40**, 235–248.

20. Eneström S. Immunohistology on formol-sublimate fixed and paraplast-embedded kidney tissue with comparison to formalin fixation and to pre-embedding immunofluorescent staining. *Stain Technology* 1983, **58**, 259–271.

21. Stein H., Gatter K., Asbahr H. and Mason D. Y. Use of freeze-dried paraffin embedded sections for immunohistologic staining with monoclonal antibodies. *Lab. Invest.* 1985, **52**, 676–683.

22. Costa M., Buffa R., Furness J. B. and Solcia E. L. Immunohistochemical localization of polypeptides in peripheral autonomic nerves using whole mount preparations. *Histochemistry* 1980, **65**, 157–165.

23. Terenghi G. L., Polak J. M., Probert L., McGregor G. P., Ferri G. -L., Blank M. A., Butler J. M., Unger W. G., Zhang S. Q., Cole D. F. and Bloom S. R. Mapping, quantitative distribution and origin of substance P- and VIP-containing nerves in the uvea of guinea pig eye. *Histochemistry* 1982, **75**, 399–417.

24. Rodrigo J., Polak J. M., Fernandez L., Ghatei M. A., Mulderry P. K. and Bloom S. R. Calcitonin gene-related peptide (CGRP)-immunoreactive sensory and motor nerves of the rat, cat and monkey oesophagus. *Gastroenterology* 1985, **88**, 444–451.

25. Pearse A. G. E. and Polak J. M. Bifunctional reagents as vapour and liquid phase fixatives for immunohistochemistry. *Histochem. J.* 1975, **7**, 179–186.

26. Bishop A. E., Polak J. M., Bloom S. R. and Pearse A. G. E. A new universal technique for immunocytochemical localisation of peptidergic innervation. *J. Endocrinol.* 1978, **77**, 25–26.

27. Bu'Lock A. J., Vaillant C., Dockray G. J. and Bu'Lock J. D. A rational approach to the fixation of peptidergic nerve cell bodies in the gut using parabenzoquinone. *Histochemistry* 1982, **74**, 49–55.

28. Van Noorden S. and Polak J. M. Immunocytochemistry of regulatory peptides. In: Bullock G. R. and Petrusz P., eds., *Techniques in Immunocytochemistry*, Vol. 3. New York, Academic Press, 1985: 115–154.

29. Stefanini M., De Martino C. and Zamboni L. Fixation of ejaculated spermatozoa for electron microscopy. *Nature* 1967, **216**, 173–174.

30. Schultzberg M., Hökfelt T., Nilsson G., Terenius L., Rehfeld J. F., Brown M., Elde R., Goldstein M. and Said S. Distribution of peptide and catecholamine containing nerves in the gastrointestinal tract of rat and guinea pig: immunohistochemical studies with antisera to

substance P, vasoactive intestinal polypeptide, enkephalins, somatostatin, gastrin/cholecy-stokinin, neurotensin and dopamine ß-hydroxylase. *Neuroscience* 1980, **5**, 689–744.

31. Falck B. and Owman C. A detailed methodological description of the fluorescence method for the cellular demonstration of biogenic monoamines. *Acta Univ. Lund, II* 1965, **7**, 1–23.

32. Lane B. P. and Europa D. L. Differential staining of ultrathin sections of epon-embedded tissues for light microscopy. *J. Histochem. Cytochem.* 1965, **13**, 579–582.

33. Hermanns W., Liebig K. and Schultz L. -C. Postembedding immunohistochemical demonstration of antigen in experimental polyarthritis using plastic embedded whole joints. *Histochemistry* 1981, **73**, 439–466.

34. Franklin R. M. Immunohistochemistry on semi-thin sections of hydroxypropyl methacrylate embedded tissues. *J. Immunol. Methods* 1984, **68**, 61–72.

35. Huang W. M., Gibson S. J., Facer P., Gu J. and Polak J. M. Improved section adhesion for immunocytochemistry using high molecular weight polymers of L-lysine as a slide coating. *Histochemistry* 1983, **77**, 275–279.

36. Bigbee J. W., Kosek J. C. and Eng L. E. The effects of primary antiserum dilution on staining of 'antigen-rich' tissue with the peroxidase antiperoxidase technique. *J. Histochem. Cytochem.* 1977, **25**, 443–447.

37. Engvall E. and Perlman P. Enzyme-linked immunosorbent assay (ELISA). Quantitative assay of immunoglobulin G. *Immunochemistry* 1971, **8**, 871–874.

38. Mason T. E., Phifer R. F., Spicer S. S., Swallow R. S. and Dreskin R. D. An immunoglobulin–enzyme bridge method for localizing tissue antigens. *J. Histochem. Cytochem.* 1969, **17**, 563–569.

39. Sternberger L. A., Hardy P. H. Jr, Cuculis J. J. and Meyer H. G. The unlabeled antibody–enzyme method of immunohistochemistry. Preparation and properties of soluble antigen–antibody complex (horseradish peroxidase–antihorseradish peroxidase) and its use in identification of spirochetes. *J. Histochem. Cytochem.* 1970, **18**, 315–333.

40. Johnstone A. and Thorpe R. *Immunochemistry in Practice*. Oxford, Blackwell Scientific Publications, 1982.

41. Titus J. A., Haugland R., Sharrow S. O. and Segal D. M. Texas Red, a hydrophilic, red-emitting fluorophore for use with fluorescein in dual parameter flow microfluorimetric and fluorescence microscopic studies. *J. Immunol. Methods* 1982, **50**, 193–204.

42. Oi V. T., Glazer A. N. and Stryer L. Fluorescent phycobiliprotein conjugates for analyses of cells and molecules. *J. Cell Biol.* 1982, **93**, 981–986.

43. Graham R. C. and Karnovsky M. J. The early stages of absorption of injected horseradish peroxidase in the proximal tubules of mouse kidney: ultrastructural cytochemistry by a new technique. *J. Histochem. Cytochem.* 1966, **14**, 291–302.

44. Weisburger E. K., Russfield A. B., Homburger F., Weisburger J. H., Boger E., Van Dongen C. G. and Chu K. C. Testing of twenty-one environmental aromatic amines or derivatives for long-term toxicity or carcinogenicity. *J. Environ. Pathol. Toxicol.* 1978, **2**, 325–356.

45. Graham R. C., Jr, Ludholm U. and Karnovsky M. J. Cytochemical demonstration of peroxidase activity with 3-amino-9-ethylcarbazole. *J. Histochem. Cytochem.* 1965, **13**, 150–152.

46. Tubbs R. R. and Sheibani K. Chromogens for immunohistochemistry. *Arch. Pathol. Lab. Med.* 1982, **106**, 205.

47. Nakane P. K. Simultaneous localization of multiple tissue antigens using the peroxidase-labeled antibody method: a study in pituitary glands of the rat. *J. Histochem. Cytochem.* 1968, **16**, 557–560.

48. Straus W. Inhibition of peroxidase by methanol and methanol nitroferricyanide for use in immunoperoxidase procedures. *J. Histochem. Cytochem.* 1971, **19**, 682–688.

49. Ponder B. A. and Wilkinson M. M. Inhibition of endogenous tissue alkaline phosphatase with the use of alkaline phosphatase conjugates in immunohistochemistry. *J. Histochem. Cytochem.* 1981, **29**, 981–984.

50. Mason D. Y. and Sammons R. E. Alkaline phosphatase and peroxidase for double immunoenzymatic labelling of cellular constituents. *J. Clin. Pathol.* 1978, **31**, 454–462.

51. Malik D. Y. and Damon M. E. Improved double immunoenzymatic labelling using alkaline phosphatase and horseradish peroxidase. *J. Clin. Pathol.* 1978, **35**, 1092–1094.

52. Suffin S. C., Muck K. B., Young J. C., Lewin K. and Porter D. D. Improvement of the glucose oxidase immunoenzyme technic. *Am. J. Clin. Pathol.* 1979, **71**, 492–496.

53. Bondi A., Chieregatti G., Eusebi V., Fulcheri E. and Bussolati G. The use of ß-galactosidase as a tracer in immunohistochemistry. *Histochemistry* 1982, **76**, 153–158.

54. Van Noorden S., Stuart M. C., Cheung A., Adams E. F. and Polak J. M. Localization of pituitary

hormones by multiple immunoenzyme staining procedures using monoclonal and polyclonal antibodies. *J. Histochem. Cytochem.* 1985, in press.

55. Cammisuli S. and Wofsy L. Hapten sandwich labelling. III. Bifunctional reagents for immunospecific labelling of cell surface antigens. *J. Immunol.* 1976, **117**, 1695–1704.

56. Gu J., De Mey J., Moeremans M. and Polak J. M. Sequential use of the PAP and immunogold staining methods for the light microscopical double staining of tissue antigens. *Reg. Peptides* 1981, **1**, 365–374.

57. Holgate C. S., Jackson P., Cowen P. N. and Bird C. C. Immunogold–silver staining — new method of immunostaining with enhanced sensitivity. *J. Histochem. Cytochem.* 1983, **31**, 938–944.

58. Roth J. Applications of immunocolloids in light microscopy. Preparation of protein A–silver and protein A–gold complexes and their application for the localization of single and multiple antigens in paraffin sections. *J. Histochem. Cytochem.* 1982, **30**, 691–696.

59. *Affinity Chromatography.* Publication of Pharmacia Fine Chemicals, 1982.

60. Roth J. The protein A–gold (pAg) technique — a qualitative and quantitative approach for antigen localization on thin sections. In: Bullock G. R. and Petrusz P. V., eds., *Techniques in Immunocytochemistry*, Vol. 1. New York, Academic Press, 1982: 107–133.

61. Buffa R., Solcia E., Fiocca R., Crivelli O. and Pera A. Complement-mediated binding of immunoglobulins to some endocrine cells of the pancreas and gut. *J. Histochem. Cytochem.* 1979, **27**, 1279–1280.

62. Grube D. Immunoreactivities of gastrin (G) cells. II. Nonspecific binding of immunoglobulins to G-cells by ionic interactions. *Histochemistry* 1980, **66**, 149–167.

63. Hartman B. K. Immunofluorescence of dopamine ß-hydroxylase. Application of improved methodology to the localization of the peripheral and central noradrenergic nervous system. *J. Histochem. Cytochem.* 1973, **21**, 312–332.

64. Goldman M. *Fluorescent Antibody Methods.* New York, Academic Press, 1968: 117.

65. Cowen T., Haven A. J. and Burnstock G. Pontamine sky blue: a counterstain for background autofluorescence in fluorescence and immunofluorescence histochemistry. *Histochemistry* 1985, **82**, 205–208.

66. Dockray G. J. Molecular evolution of gut hormones: application of comparative studies on the regulation of digestion. *Gastroenterology* 1977, **72**, 344–358.

67. Pruss R. M., Mirsky R., Raff M. C., Thorpe R., Dowding A. J. and Anderton B. H. All classes of intermediate filaments share a common antigenic determinant defined by a monoclonal antibody. *Cell* 1981, **27**, 419–428.

68. Polak J. M., Pearse A. G. E., Szelke M., Bloom S. R., Hudson D., Facer P., Buchan A. M. J., Bryant M. G., Christophides N. and MacIntyre I. Specific immunostaining of CCK cells by use of synthetic fragment antisera. *Experientia* 1977, **33**, 762–763.

69. Pool C. W., Buijs R. M., Swaab D. F., Boer G. J. and van Leeuwen F. W. On the way to a specific immunocytochemical localization. In: Cuello A. C., ed., *Immunohistochemistry. IBRO Handbook Series, Methods in the Neurosciences*, Vol. 3. Chichester, Wiley, 1983: 1–46.

70. Vacca L. L. "Double bridge" techniques of immunocytochemistry. In: Bullock G. R. and Petrusz P., eds., *Techniques in Immunocytochemistry*, Vol. 1. New York, Academic Press, 1982: 155–182.

71. Straus W. Imidazole increases the sensitivity of the cytochemical reaction for peroxidase with diaminobenzidine at a neutral pH. *J. Histochem. Cytochem.* 1982, **30**, 491–493.

72. Hsu S.-M. and Soban E. Colour modification of diaminobenzidine (DAB) precipitation by metallic ions and its application to double immunohistochemistry. *J. Histochem. Cytochem.* 1982, **30**, 1079–1082.

73. Bloom S. R., Christofides N. D., Delamarter J., Buell G., Kawashima E. and Polak J. M. Tumour co-production of VIP and PHI explained by a single coding gene. *Lancet* 1983, **ii**, 1163–1165.

74. Gibson S. J., Polak J. M., Bloom S. R., Sabate I. M., Mulderry P. K., Ghatei M. A., Morrison J. F. B., Kelly J. S. and Rosenfeld M. G. Calcitonin gene-related peptide (CGRP) immunoreactivity in the spinal cord of man and of eight species. *J. Neurosci.* 1984, **4**, 3101–3111.

75. Bishop A. E., Polak J. M., Facer P., Ferri G. -L., Marangos P. J. and Pearse A. G. E. Neuron-specific enolase: a common marker for the endocrine cells and innervation of the gut and pancreas. *Gastroenterology* 1982, **83**, 902–915.

76. Facer P., Bishop A. E., Lloyd R. V., Wilson B. S., Hennessy R. J. and Polak J. M. Chromogranin: a newly recognised marker for endocrine cells of the human gastrointestinal tract. *Gastroenterology* 1985, in the press.

77. Tramu G., Pillez A. and Leonardelli J. An efficient method of antibody elution for the successive or simultaneous localization of two antigens by immunocytochemistry. *J. Histochem. Cytochem.* 1978, **26**, 322–324.
78. Sternberger L. A. and Joseph F. A. The unlabeled antibody method. Contrasting color staining of paired pituitary hormones without antibody removal. *J. Histochem. Cytochem.* 1979, **29**, 1424–1429.
79. Joseph F. A. and Sternberger L. A. The unlabeled antibody method. Contrasting color staining of ß-lipotropin and ACTH-associated hypothalamic peptides without antibody removal. *J. Histochem. Cytochem.* 1979, **27**, 1430–1437.
80. Wang B.-L. and Larsson L.-I. Simultaneous demonstration of multiple antigens by indirect immunofluorescence or immunogold staining. *Histochemistry* 1985, **83**, 47–56.
81. Mason D. Y. and Woolston R. -E. Double immunoenzymatic labelling. In: Bullock G. R. and Petrusz P., eds., *Techniques in Immunocytochemistry*, Vol. 1. New York, Academic Press, 1982: 135–152.
82. Mason D. Y., Abdulaziz Z., Falini B. and Stein H. Double immunoenzymatic labelling. In: Polak J. M. and Van Noorden S. eds., *Immunocytochemistry, Practical Applications in Pathology and Biology*, Bristol, Wright PSG, 1983: 113–128.
83. Balaton A. J., Dalix A. M. and Oriol R. An improved mounting medium for immunofluorescence microscopy. *Arch. Pathol. Lab. Med.* 1985, **109**, 108.
84. Pelliniemi L. J., Dym M. and Karnovsky J. Peroxidase histochemistry using diaminobenzidine tetrahydrochloride stored as a frozen solution. *J. Histochem. Cytochem.* 1980, **28**, 191–192.

4　Avidin–Biotin Methods

G. Coggi, P. Dell'Orto and G. Viale

The continuous expansion of immunocytochemistry in recent years has been accompanied by the development of a variety of techniques, based upon different theoretical approaches, aimed at ensuring high sensitivity and specificity. Among these techniques, those based upon the interaction between *avidin* and *biotin*, and therefore upon the use of biotinylated antibodies, have gained increasing popularity, to the point that they have been introduced by some manufacturers in their immunocytochemical kits.

The aim of this chapter is to describe the avidin–biotin methods, with special reference to their suitability in immunocytochemical practice.

1. HISTORICAL NOTES

In contrast to other immunocytochemical techniques, those based upon the avidin–biotin system were not originally conceived for immunocytochemistry. Indeed, these methods had been routinely used by biochemists in several extraction and chromatographic procedures, based on the same principle, i.e. the high affinity between avidin and biotin.[1-4] They were originally introduced as long as 40 years ago, and are still in use because of their reliability.

The application of the avidin–biotin system to immunocytochemistry was proposed by Heggeness and Ash[5] in 1977 for fluorescence microscopy. In 1979, Guesdon et al.[6] described the direct and bridge techniques using enzymes as markers and, in 1980, Warnke and Levy[7] stressed the peculiar suitability of the avidin–biotin system for immunocytochemistry with monoclonal antibodies.

Finally, in 1981 Hsu et al.[8] described an interesting development of the technique, the avidin-biotin complex (ABC) method, by which avidin–biotin immunocytochemistry gained widespread acceptance and popularity.

2. CHEMICAL BACKGROUND

2.1. General Principles

All avidin–biotin methods are based upon four main principles:

a. The extraordinary affinity between avidin and biotin molecules, which links one to the other in a practically permanent complex.
b. The possibility of coupling biotin to larger molecules, such as enzymes or antibodies, through a single biochemical reaction.
c. The possibility of labelling avidin with a variety of markers, such as enzymes, heavy metals and fluorochromes.
d. The possibility of using avidin as a bridge between two different biotinylated molecules, such as an antibody and peroxidase.

2.2. Avidin

Avidin is a basic glycoprotein (mol. wt, 67 kdal) present in large amounts in egg white, from which it is extracted. It consists of four subunits, each represented by a single polypeptide chain of 128 amino acid residues — among which there are four tryptophans in positions 10, 70, 97 and 110.[9,10] The sugar moiety of each subunit includes an oligosaccharide, containing mannose and glucosamine.

The most peculiar characteristic of the tertiary structure of the avidin molecule is the presence on its surface of four hydrophobic pockets (one for each subunit) which behave as specific binding sites for four biotin residues.[11]

2.3. Streptavidin

Although the suitability and reliability of egg white avidin for immunocytochemical purposes is widely confirmed, it has been recently observed that the presence of oligosaccharide residues[12] and its very high isoelectric point[10] may be minor disadvantages.

Indeed, the oligosaccharide residues, which have no effect on the biotin-binding capability of avidin, may (a) react with lectin-like endogenous proteins, and (b) stick to other tissue components.[13] Both these possibilities may occasionally result in reduction of the specificity of the immunoreaction, by increasing the background.

At the usually neutral pH of immunocytochemical procedures, the isoelectric point of avidin (10·5) can cause the protein to react with negatively charged tissue components, such as cell membranes[12] and nuclei, so producing unwanted staining.

In order to overcome these limitations, avidin can be replaced by *streptavidin*. This is a protein (mol. wt, 60 kdal) extracted from the culture broth of *Streptomyces avidinii*; like avidin, it consists of four subunits, each containing a single high affinity biotin-binding site; the lack of oligosaccharide residues, and the neutral isoelectric point[14] prevent streptavidin from binding non-specifically.

Since there are no differences in the practical use of streptavidin, as compared to avidin, in this chapter the two molecules will be referred to as 'avidin'.

2.4. Biotin

Biotin is a water-soluble vitamin (vitamin H) of very low molecular weight (244 dal) present in egg yolk, and in a variety of other tissues, both animal and vegetable.[15] Its very simple structure (*Fig.* 4.1.) enables it, via the ureido group, to fill one of the four pockets present on the surface of the avidin molecule.[16] The carboxylic group is responsible for its binding to NH_2 residues of different proteins[17] (*see* Section 3.1).

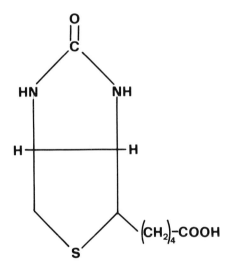

Fig. 4.1. Chemical structure of biotin.

2.5. The Avidin–Biotin Reaction

The reaction between avidin and biotin is due to non-covalent bonds; it has been shown[18] that the reaction is extremely rapid, and is characterized by a $K_A = 10^{-15}$ M^{-1}, which is considered to be one of the strongest bonds known in nature. The two molecules can be separated only by submitting them to extreme conditions, such as very low pH (approximately 1·5).[10,19]

3. BIOTINYLATION

3.1. Procedure

Biotinylation is the chemical procedure by which biotin is conjugated to a variety of molecules, such as enzymes, lectins, nucleic acids etc.[17] As already mentioned, the small size of the biotin molecule prevents the biotinylation procedure from modifying the chemical, immunological, or physical properties of the molecules to which biotin is bound. Moreover, multiple biotinylation of the same molecule can be performed without any adverse effect. This offers the possibility of 'coating' an antibody, or an enzyme, with a large number of biotin molecules, which eventually behave as very specific binding sites for avidin.

The number of biotin molecules which can be bound to an antibody has been estimated to be as high as 150.[13]

The biotinylation procedure requires that biotin be first transformed into its activated form, through a process of esterification; the carboxylic group of activated biotin will then react with the NH_2 residues of the protein to be biotinylated.[6,17,20,21]

Precise knowledge of the process of biotinylation has now become less important for the laboratory worker than in the past, since a variety of biotinylated molecules are now commercially available.

It is nevertheless worthwhile underlining that the biotinylation procedure can be applied to probes to be used in fields other than immunocytochemistry, for instance lectins[22] for affinity cytochemistry, or nucleic acids[23–25] for in situ hybridization.

3.2. The Spacer Arm

Following biotinylation, there is a danger that the reaction between the biotin residues present on the labelled macromolecule and avidin could be prevented by steric hindrance, with reduction in the number of bound avidins and subsequent loss of sensitivity.

To overcome such a possible limitation, the insertion of a spacer arm between biotin and the macromolecule was suggested by Costello et al.[26] The spacer arm, a 7-, 11-, or 16-atom long molecule, acts as a bridge linking the amino terminal group of biotin to the carboxylic group of the macromolecule, and widening the distance between the two.[13,27,28] Because of its effect on the sensitivity of the immunocytochemical reaction, the spacer arm procedure is now adopted by several manufacturers.

4. LABELLING OF AVIDIN

Avidin can be labelled with a marker, such as a fluorochrome (rhodamine, fluorescein),[5] an enzyme (peroxidase, phosphatases, β-galactosidase etc.),[6] ferritin,[29] or gold particles.[30]

Fluorochromes can be bound to avidin via an isothiocyanate derivative; enzymes and ferritin are coupled via a double-armed reagent such as glutaraldehyde. Finally, colloidal gold particles of different sizes (5–40 nm) can adsorb avidin by non-covalent electrostatic forces.

Labelled avidin is commercially available from various manufacturers; the commercial products show consistent and comparable reliability.

5. IMMUNOCYTOCHEMICAL METHODS

It should be first underlined that the immunocytochemical methods employing avidin–biotin do not differ, in their essential requirements and in the general

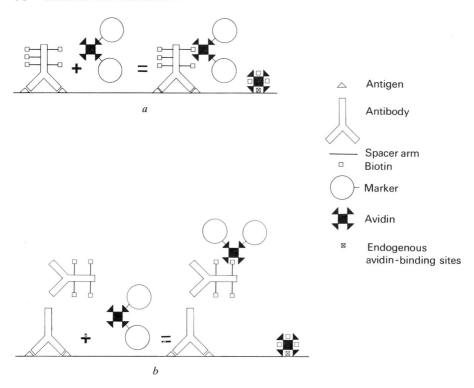

△ Antigen

Y Antibody

—— Spacer arm
□ Biotin

◯ ⊢ Marker

✦ Avidin

⊠ Endogenous
   avidin-binding sites

Fig. 4.2. Labelled avidin–biotin methods: (*a*) direct and (*b*) indirect. Note that the marker can be a fluorochrome, an enzyme or a metal.

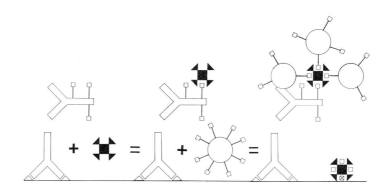

Fig. 4.3. Indirect bridge avidin–biotin method. Enzymes only can be used as markers.

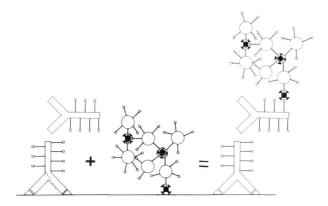

Fig. 4.4. Indirect ABC method, for enzymatic markers only.

outlines of the procedures, from more traditional ones. This holds particularly true for tissue processing, section incubation, washing procedures, controls etc. (*see* Appendix).

The main feature of the immunocytochemical methods employing avidin––biotin is that the biotinylated antibody, used either as a primary (in direct methods), or as a secondary (in indirect methods), can be visualized in three different ways:

    a. With labelled avidin (*Fig.* 4.2).
    b. With avidin acting as a bridge with biotinylated enzymes (*Fig.* 4.3).
    c. With the avidin–biotin–peroxidase complex (ABC) (*Fig.* 4.4).

5.1. Labelled Avidin

Direct methods, introduced in the early years of avidin–biotin immunocyto-chemistry, [5,6] have now lost much of their original interest. Nevertheless, recently a direct method using peroxidase-labelled avidin has been proposed in a commercially available kit: it is claimed to yield dependable results on formalin-fixed, paraffin-embedded material, and we ourselves have tested it in our laboratory with satisfactory results (*see Fig.* 4.5*a*).

Indirect methods,[7,31] in which the primary antibody is unlabelled, are still widely used, namely when fluorochromes or colloidal gold particles (*Fig.* 4.5*b*, *c*) are employed as markers (*see* Appendix).

5.2. Bridge Avidin

Use of avidin as a bridge is limited to immunoenzymatic procedures, since neither fluorochromes nor colloidal gold can be biotinylated. In this method, the biotinylated antibody, which can be either primary or secondary, reacts first with unlabelled avidin; the procedure is then completed by addition of biotinylated peroxidase, which reacts with the remaining free binding sites on the unlabelled avidin molecule.

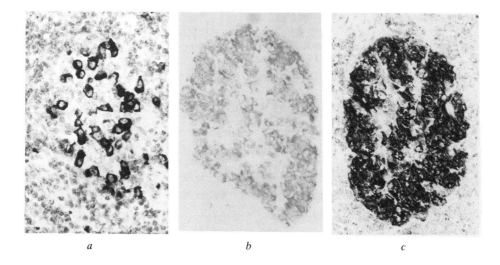

Fig. 4.5. (a) Direct peroxidase-labelled avidin–biotin method for the localization of κ chain of human immunoglobulins in formalin-fixed, paraffin-embedded lymph node. Protease digestion, 3-amino-9-ethylcarbazole, haematoxylin. (b) Indirect gold-labelled avidin–biotin technique for localization of insulin in human pancreas in formalin-fixed, paraffin-embedded tissue. (c) As in (b), with silver enhancement.

Both the direct bridge avidin–biotin method (BRAB)[6] and the indirect one (IBRAB)[8] have practically been abandoned in favour of the following more advanced procedure.

5.3. Avidin–Biotin–Peroxidase Complex

Methods based upon the use of an avidin–biotin–peroxidase complex[8] (ABC) have gained the highest popularity, and indeed ensure maximal sensitivity. They are commonly used in indirect procedures and actually represent a development of the already mentioned IBRAB technique.

The ABC method requires the preparation of an avidin–biotin–peroxidase complex in which at least one of the four reacting sites on the surface of the avidin molecule is left free to bind a biotin residue on the antibody molecule; the complex is then added to the biotinylated antibody (*see* Appendix).

5.4. Other Techniques

Other avidin–biotin techniques have been proposed, but their use in routine immunocytochemistry is purely theoretical; they include use of antibodies to biotin [10,24,28] or avidin,[32,33] and biotinylated protein A.[34]

6. ADVANTAGES OF THE AVIDIN–BIOTIN METHODS

The increasing use of the avidin–biotin methods, as shown by the immunocytochemical literature, definitely confirms the reliability of these procedures, up to the point of implying their possible superiority over conventional techniques.

The most relevant advantages of the avidin–biotin methods can be identified as (*a*) sensitivity and (*b*) versatility.

6.1. Sensitivity

The chemistry of biotinylation is responsible for the high sensitivity of these procedures. Indeed, as already mentioned, the possibility of coating an antibody molecule with up to 150 biotin residues,[13] each one capable of subsequently binding labelled avidin or the avidin–biotin–peroxidase complex, ensures even in the simplest methods a very high marker/antigen ratio.

Plate 3 shows the difference in sensitivity between traditional indirect immunofluorescence and FITC–avidin–biotin techniques for the localization of the β subunit of human chorionic gonadotrophin at the same dilution of the specific antiserum.

An even higher sensitivity is achieved by ABC techniques, in which both the antibody and the marker, namely the peroxidase, are biotinylated. In this case, the preformed avidin–biotin–peroxidase complex is actually a multimolecular network, in which a large number of peroxidase molecules are linked together via the interactions between biotin residues exposed on the enzyme and free binding sites of avidin. Addition of such a multimolecular complex, in which several biotin-binding sites of avidin are still unsaturated, to the biotinylated antibody results in an enzyme/antigen ratio not achievable even in the most sensitive techniques, such as the peroxidase–anti-peroxidase (PAP) complex staining.[8,35]

The different intensity in immunostaining obtained with the use of the PAP complex and ABC methods for the localization of cytokeratins, at the same dilution of the specific antiserum, is shown in *Fig.* 4.6.

6.2. Versatility

Besides being a dependable method for light microscopical immunocytochemis-

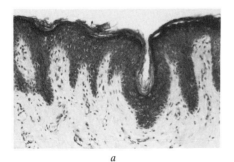
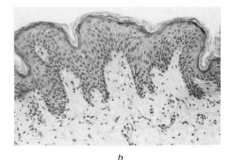

a   *b*

Fig. 4.6. Localization of cytokeratins in formalin-fixed, paraffin-embedded human epidermis. (*a*) ABC and (*b*) PAP complex methods. DAB, haematoxylin.

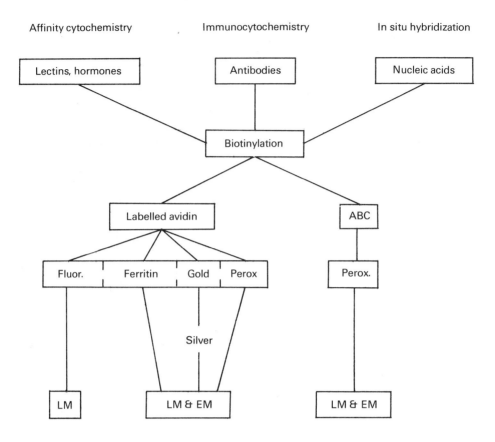

Fig. 4.7. Spectrum of biotinylated probes in cytochemistry.

try, the avidin–biotin techniques provide research and diagnostic laboratories with a multifaceted tool for the localization of tissue components.

In *Fig.* 4.7 is summarized the spectrum of cytochemical techniques based on the same theoretical and practical approach, all deriving from the unique biotinylation procedure. Addition of labelled avidin or ABC to the biotinylated probe, allows a great variety of technical possibilities and applications, at both light and electron microscopical levels.

7. PRACTICAL APPLICATIONS

The availability of such highly sensitive immunocytochemical methods is of particular relevance when localizing antigens present in very low amounts in the tissue, either naturally or as a consequence of the effects of fixatives and embedding media on their immunoreactivity. Such is the case, for instance, with epoxy-embedded semithin sections, in which the residual antigenicity is satisfactorily detected through the amplifying effects of avidin–biotin techniques (*Fig.* 4.8).

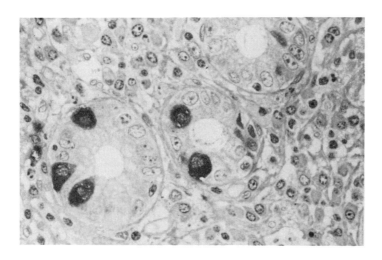

Fig. 4.8. Immunolocalization of gastrin in human gastric mucosa. Glutaraldehyde-fixation, OsO$_4$ postfixation, Epon-embedding, sodium ethoxide etching. DAB, haematoxylin.

A second practical advantage, again related to the sensitivity of the method, is the possibility of lowering the background staining by increasing the dilution of the primary antiserum and/or reducing the incubation time; this will result in a higher signal to noise ratio.

Moreover, the sensitivity of avidin–biotin methods makes them particularly suitable for immunolocalization of antigens by means of monoclonal antibodies. Indeed, the ABC and the indirect FITC-labelled avidin–biotin methods can be considered as first choice techniques, since they provide adequate staining, and are definitely superior to the traditional indirect immunoperoxidase or immuno-fluorescence procedures, and yield results very often comparable to those obtainable with PAP complexes raised in mice.

Electron microscopical immunocytochemistry is easily and advantageously performed with the ABC and gold-labelled avidin–biotin methods, both in pre-[36,37] and post-embedding[38,39] staining procedures (*Fig.* 4.9). It has been shown that, even at the electron microscopical level, the ABC technique confirms its higher sensitivity over the PAP complex staining method.[38]

Finally, in affinity cytochemistry, labelled avidin or ABC can be used for the detection of biotinylated lectins,[40-43] hormones,[44] and nucleic acid probes.[23-25,28]

8. PITFALLS OF AVIDIN–BIOTIN METHODS

As already mentioned, limitations in the use of avidin due to non-specific binding to tissue components have been, to a large extent, overcome by the introduction of streptavidin (*see* Section 2.3). In applying avidin–biotin techniques, one must be aware of two other minor problems: the possibility of unwanted binding between avidin and endogenous biotin, and the non-specific staining of mast cells.

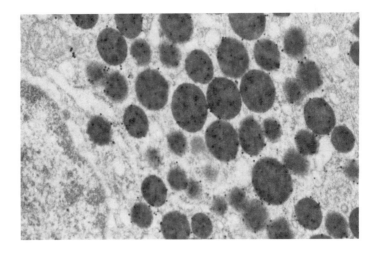

Fig. 4.9. Electron microscopical localization of growth hormone in human anterior pituitary cells. Glutaraldehyde fixation, OsO_4 postfixation, Epon embedding. Indirect gold-labelled avidin–biotin, uranyl acetate and lead citrate.

8.1. Endogenous Biotin

It has been observed that endogenous biotin is naturally present in several normal human tissues and organs, such as liver, breast, adipose tissue and kidney.[45] This can result in binding of avidin, leading to false-positive results.

It has therefore been proposed that endogenous avidin-binding sites can be blocked by preliminary treatment of tissue sections with free avidin, followed by biotin[45] (*see* Appendix). In our experience, however, this problem has not occurred when using the ABC method.

8.2.   Mast Cell Staining

Mast cell granules bind avidin, labelled either with fluorochromes or enzymes,[46] or ABC.[47] Although such an artefact is of minor relevance and can be easily tolerated, it can be avoided by using streptavidin or by increasing the pH of the reaction to 9·4; under these conditions, the immunoreaction will not be affected, and the mast cell granules will not bind avidin.[47]

9. CONCLUSIONS

Avidin–biotin methods appear to provide rather than an interesting curiosity, a real and effective alternative to conventional immunocytochemical methods, due to the additive effect of biotinylation on the marker/antigen ratio.

They are of particular relevance in diagnostic immunocytochemistry, where routine fixation, often performed under incompletely controlled conditions, may significantly reduce the antigenicity of tissue components; the same applies to retrospective studies on material subjected to very long or inadequate fixation.

Moreover, the general availability of monoclonal antibodies for diagnostic purposes has further underlined the usefulness of avidin–biotin methods.

In addition, the possibility of relying on the same method for a wide range of cytochemical techniques, from affinity cytochemistry to immunocytochemistry, and in situ hybridization, in both light and electron microscopy, qualifies avidin–biotin techniques for placement amongst the most versatile and effective tools in morphological research.

Appendix

I. PRE-TREATMENT OF TISSUES

A. Tissue Processing

The choice of fixatives and the procedures of tissue handling for avidin–biotin immunocytochemistry follow the same criteria adopted in conventional immunocytochemistry (*see* Chapter 3).

B. Buffers

Antibodies, biotinylated or not, and fluorochrome-labelled avidin are diluted in 0·01 M phosphate-buffered saline (PBS) pH 7·6, or 0·05 M Tris-buffered saline (TBS) pH 7·6, containing 0·2 per cent bovine serum albumin and 0·01 per cent sodium azide.

Peroxidase-labelled avidin and the avidin–biotin–peroxidase complex are diluted in the same buffers, without azide.

C. Section Handling Prior to Immunocytochemical Staining

  i. Cryostat sections (fresh or prefixed):
     Allow to dry for 30 min, at room temperature.
     Postfix, as required.
     Dip in distilled water.
     Block endogenous peroxidase (when needed).
 ii. Paraffin-embedded sections:
     Dewax in xylene or toluene.
     Rehydrate in graded alcohols and bring to distilled water.
     Remove mercury pigments (if necessary).
     Block endogenous peroxidase (when needed).
iii. Epoxy-embedded sections:
     Deresinate with a saturated solution of alcoholic sodium hydroxide for 30–60 min.

Wash in three changes of absolute ethanol, 5 min each.
Dip in distilled water.
Block endogenous peroxidase (when needed).

D. Blocking of Endogenous Peroxidase

Treat with 10 per cent H_2O_2, for 5 min.
Wash in tap water, 5 min.
Wash in distilled water, 3 min.
Transfer to the buffer (PBS or TBS) solution, 5 min.
(For cryostat sections: use 0·1 per cent phenylhydrazine in buffer for 60 min, at 37 °C instead of 10 per cent H_2O_2.)

II. INDIRECT LABELLED AVIDIN–BIOTIN STAINING TECHNIQUE

A. Fluorescence and Peroxidase

1. PBS or TBS, 5 min.
2. Normal serum (from the same species as the biotinylated antibody), 15 min, at room temperature.
3. Highly diluted primary antiserum, 30 min to 2 h at room temperature, or overnight at 4 °C.
4. PBS or TBS, three changes, 5 min each.
5. Biotinylated second antibody, 30 min, at room temperature.
6. Rinse as in Step 4.

Either

.Fluorescence:

7. Fluorescein isothiocyanate (FITC) or tetramethylrhodamine isothiocyanate (TRITC)-labelled avidin, 30 min, at room temperature.
8. Rinse as in Step 4.
9. Mount in buffered glycerol.

or

Peroxidase:

7. Peroxidase-labelled avidin, 30 min, at room temperature.
8. Rinse as in Step 4.
9. Chromogen substrate.
10. Mount in suitable medium.

B. Gold

(*See also* Chapter 5)

1. PBS or TBS containing 2·5 per cent sodium chloride and 0·5 per cent Tween 80, 3 changes, 5 min each.
2. Normal serum (from the same species as the biotinylated antibody), 15 min at room temperature.
3. Highly diluted primary antiserum, 30 min to 2 h, at room temperature or overnight at 4 °C.

4. Rinse as in Step 1.
5. Biotinylated second antibody, 30 min at room temperature.
6. Rinse as in Step 1.
7. PBS or TBS 0·02M, pH 8·2, containing 1 per cent bovine serum albumin, 5 min.
8. Streptavidin–gold, diluted in the same buffer as in step 7, 30 min, at room temperature.
9. Rinse as in Step 7.
10. Rinse as in Step 6.
11. Rinse in distilled water, three changes, 5 min each.
12. Counterstain, if desired, dehydrate and mount in permanent medium.

Silver enhancement technique[48,49]
Following Step 11:
12. Silver development solution, 5 min, in the dark; mix, just before use, 850 mg hydroquinone in 85 ml citrate buffer* pH 3·5–3·9 with 110 mg silver lactate dissolved in 15 ml glass-distilled water.
13. Photographic fixer 1 : 4, 2 min.
14. Tap water, 5 min.
15. Counterstain as desired, dehydrate and mount.

III. INDIRECT AVIDIN–BIOTIN–PEROXIDASE COMPLEX STAINING TECHNIQUE

As for the labelled avidin method (for peroxidase) up to Step 6, then as follows:
7. Preformed avidin–biotin–peroxidase complex, for 30 min, at room temperature: mix 10 µl biotinylated peroxidase in 1 ml buffer (Vector Vectastain) or use streptavidin-biotinylated peroxidase complex 1 : 200 (Amersham).†
8. PBS or TBS, three changes, 5 min each.
9. Chromogen substrate.
10. Mount in suitable medium.

IV. DIRECT LABELLED AVIDIN–BIOTIN OR ABC STAINING TECHNIQUES

Start at Step 4 of indirect methods and use in Step 5 highly diluted primary biotinylated antibody.
Continue through Step 6 to the end.

V. BLOCKING OF ENDOGENOUS AVIDIN–BINDING SITES (SO-CALLED 'ENDOGENOUS BIOTIN')[45]

1. Dewax.
2. Block endogenous peroxidase (if necessary).

---

*23·5 g trisodium citrate dihydrate and 25·5 g citric acid in 850 ml glass-distilled water.

†Alternatively, mix 1·3 ml of a 1 mg/ml avidin solution and 1 ml of a 33 mg/ml biotinylated peroxidase solution; let stand for 30 min and use unfiltered and diluted 1 : 20 in buffer.[50]

3. PBS or TBS, 5 min.
4. Avidin 1 mg/ml, 20 min.
5. Rinse as in Step 3.
6. Biotin 0·1 mg/ml, 20 min.
7. PBS or TBS, 5 min.
8. Continue with the immunocytochemical staining.

VI. SOURCES OF IMMUNE REAGENTS USED FOR ILLUSTRATED PREPARATIONS

Primary Antibodies

Kappa light chain	Biomeda Corp.
Gastrin	Biodata SpA
Insulin	Sclavo SpA
Cytokeratins	Sclavo SpA
Human growth hormone	Sclavo SpA
Human chorionic gonadotrophin	Sclavo SpA

Detecting Reagents

Biotinylated second antibodies	Vector
Avidin–peroxidase	Vector
Bridging antibodies and PAP complex	Dako
Streptavidin–gold	Gift from Dr J. De Mey, Janssen Pharmaceutica

REFERENCES

1. Swack J. A., Zander G. L. and Utter M. F. Use of avidin–Sepharose to isolate and identify biotin polypeptides from crude extracts. *Anal. Biochem.* 1978, **87**, 114–126.
2. Henrikson K. P., Allen S. H. G. and Maloy W. L. An avidin monomer affinity column for the purification of biotin-containing enzymes. *Anal. Biochem.* 1979, **94**, 366–370.
3. Haeuptle M. T., Aubert M. L., Djiane J. and Kraehenbuhl J. P. Binding sites for lactogenic and somatogenic hormones from rabbit mammary gland and liver. Their purification by affinity chromatography and their identification by immunoprecipitation and photoaffinity labeling. *J. Biol. Chem.* 1983, **258**, 305–314.
4. Finn F. M., Titus G., Horstman D. and Hofmann K. Avidin–biotin affinity chromatography: application to the isolation of human placental insulin receptor. *Proc. Natl Acad. Sci. USA*, 1984, **81**, 7328–7332.
5. Heggeness M. H. and Ash J. F. Use of the avidin–biotin complex for the localization of actin and myosin with fluorescence microscopy. *J. Cell Biol.* 1977, **73**, 783–788.
6. Guesdon J. L., Ternynck T. and Avrameas S. The use of avidin–biotin interaction in immunoenzymatic techniques. *J. Histochem. Cytochem.* 1979, **27**, 1131–1139.
7. Warnke R. and Levy R. Detection of T and B cell antigens with hybridoma monoclonal antibodies: a biotin–avidin–horseradish peroxidase method. *J. Histochem. Cytochem.* 1980, **28**, 771–776.

8. Hsu S. M., Raine L. and Fanger H. Use of avidin–biotin–peroxidase complex (ABC) in immunoperoxidase techniques: a comparison between ABC and unlabeled antibody (PAP) procedures. *J. Histochem. Cytochem.* 1981, **29**, 577–580.

9. Melamed M. D. and Green N. M. Avidin. 2. Purification and composition. *Biochem. J.* 1963, **89**, 591–599.

10. Green N. M. Avidin. *Adv. Prot. Chem.* 1975, **29**, 85–133.

11. Green N. M. Avidin. 3. The nature of the biotin-binding site. *Biochem. J.* 1963, **89**, 599–609.

12. Hofmann K., Wood S. W., Brinton C. C., Montibeller J. A. and Finn F. M. Iminobiotin affinity columns and their application to retrieval of streptavidin. *Proc. Natl Acad. Sci. USA,* 1980, **77**, 4666–4668.

13. Bonnard C., Papermaster D. S. and Kraehenbuhl J. P. The streptavidin–biotin bridge technique: application in light and electron microscope immunocytochemistry. In: Polak J. M. and Varndell I. M., eds., *Immunolabelling for Electron Microscopy.* Amsterdam, Elsevier Science Publishers, 1984: 95–111.

14. Chaiet L. and Wolf F. J. The properties of streptavidin, a biotin-binding protein produced by streptomycetes. *Arch. Biochem. Biophys.* 1964, **106**, 1–5.

15. Moss J. and Lane M. D. The biotin-dependent enzymes. *Adv. Enzymol.* 1971, **35**, 321–442.

16. Liu F. T. and Leonard N. J. Avidin–biotin interaction. Synthesis, oxidation, and spectroscopic properties of linked models. *J. Am. Chem. Soc.* 1979, **101**, 996–1005.

17. Bayer E. A. and Wilchek M. The use of the avidin–biotin complex as a tool in molecular biology. In: Glick D., ed., *Methods of Biochemical Analysis,* Vol. 26. New York, John Wiley & Sons, 1980: 1–45.

18. Green N. M. Avidin. 1. the use of [^{14}C] biotin for kinetic studies and for assay. *Biochem. J.* 1963, **89**, 585–591.

19. Green N. M. Avidin. 4. Stability at extremes of pH and dissociation into sub-units by guanidine hydrochloride. *Biochem. J.* 1963, **89**, 609–620.

20. Jasiewicz M. L., Schoenberg D. R. and Mueller G. C. Selective retrieval of biotin-labeled cells using immobilized avidin. *Exp. Cell Res.* 1976, **100**, 213–217.

21. Bayer E. A., Wilchek M. and Skutelsky E. Affinity cytochemistry: the localization of lectin and antibody receptors on erythrocytes via the avidin–biotin complex. *FEBS Lett.* 1976, **68**, 240–244.

22. Bayer E. A., Skutelsky E. and Wilchek M. The avidin–biotin complex in affinity cytochemistry. *Methods Enzymol.* 1979, **62**, 308–315.

23. Langer P. R., Waldrop A. A. and Ward D. C. Enzymatic synthesis of biotin-labeled polynucleotides: novel nucleic acid affinity probes. *Proc. Natl Acad. Sci. USA,* 1981, **78**, 6633–6637.

24. Manuelidis L., Langer-Safer P. R. and Ward D. C. High-resolution mapping of satellite DNA using biotin-labeled DNA probes. *J. Cell Biol.* 1982, **95**, 619–625.

25. Varndell I. M., Polak J. M., Sikri K. L., Minth C. D., Bloom S. R. and Dixon J. E. Visualisation of messenger RNA directing peptide synthesis by in situ hybridisation using a novel single-stranded cDNA probe. Potential for the investigation of gene expression and endocrine cell activity. *Histochemistry,* 1984, **81**, 597–601.

26. Costello S. M., Felix R. T. and Giese R. W. Enhancement of immune cellular agglutination by use of an avidin–biotin system. *Clin. Chem.* 1979, **25**, 1572–1580.

27. Mouton C. A., Pang D., Natraj C. V. and Shafer J. A. A reagent for covalently attaching biotin to proteins via a cleavable connector arm. *Arch. Biochem. Biophys.* 1982, **218**, 101–108.

28. Hutchison N. J. Hybridisation histochemistry: in situ hybridisation at the electron microscope level. In: Polak J. M. and Varndell I. M., eds., *Immunolabelling for Electron Microscopy.* Amsterdam, Elsevier Science Publishers, 1984: 341–351.

29. Heitzmann H. and Richards F. M. Use of the avidin–biotin complex for specific staining of biological membranes in electron microscopy. *Proc. Natl Acad. Sci. USA,* 1974, **71**, 3537–3541.

30. Morris, R. E. and Saelinger C. B. Visualization of intracellular trafficking: use of biotinylated ligands in conjunction with avidin–gold colloids. *J. Histochem. Cytochem.* 1984, **32**, 124–128.

31. Ghandour M. S., Langley O. K. and Keller A. A comparative immunohistological study of cerebellar enolases. Double labelling technique and immunoelectron-microscopy. *Exp. Brain Res.* 1981, **41**, 271–279.

32. Berman J. W. and Basch R. S. Amplification of the biotin–avidin immunofluorescence technique. *J. Immunol. Methods,* 1980, **36**, 335–338.

33. Hsu S. M., Raine L. and Fanger H. The use of antiavidin antibody and avidin–biotin–peroxidase complex in immunoperoxidase techniques. *Am. J. Clin. Pathol.* 1981, **75**, 816–821.

34. Hsu S. M. and Raine L. Protein A, avidin, and biotin in immunohistochemistry. *J. Histochem. Cytochem.* 1981, **29**, 1349–1353.
35. Childs G. and Unabia G. Application of the avidin–biotin–peroxidase complex (ABC) method to the light microscopic localization of pituitary hormones. *J. Histochem. Cytochem.* 1982, **30**, 713–716.
36. Löning Th., Schmitt D., Becker W. M., Weib J. and Jänner M. Application of the biotin–avidin system for ultrastructural identification of suppressor/cytotoxic lymphocytes in oral lichen planus. *Arch. Dermatol. Res.* 1982, **272**, 177–180.
37. Berti E., Monti M., Cavicchini S. and Caputo R. The avidin–biotin–peroxidase complex (ABCPx) in skin immunoelectron microscopy. *Arch. Dermatol. Res.* 1983, **275**, 134–138.
38. Childs G. and Unabia G. Application of a rapid avidin–biotin–peroxidase complex (ABC) technique to the localization of pituitary hormones at the electron microscopic level. *J. Histochem. Cytochem.* 1982, **30**, 1320–1324.
39. Viale G., Dell'Orto P., Braidotti P. and Coggi G. Ultrastructural localization of intracellular immunoglobulins in Epon-embedded human lymph nodes. An immunoelectron microscopic investigation using the immunogold staining (IGS) and the avidin–biotin–peroxidase complex (ABC) methods. *J. Histochem. Cytochem.*, 1985, **33**, 400–406.
40. Hsu S. M. and Raine L. Versatility of biotin-labelled lectins and avidin–biotin–peroxidase complex for localization of carbohydrates in tissue sections. *J. Histochem. Cytochem.* 1982, **30**, 157–161.
41. Coggi G., Dell'Orto P., Bonoldi E., Doi P. and Viale G. Lectins in diagnostic pathology. In: Bøg-Hansen T. C. and Spengler G. A., eds., *Lectins*, Vol. III. Berlin, Walter de Gruyter & Co., 1983: 87–103.
42. Viale G., Dell'Orto P., Colombi R., De Gennaro V., Comi A. and Coggi G. Lectin binding sites on semithin sections of epoxy-embedded tissues. In: Bøg-Hansen T. C. and Spengler G. A., eds., *Lectins*, Vol. III. Berlin, Walter de Gruyter & Co., 1983: 199–204.
43. Coggi G., Bonelli M., Bonoldi E., Viale G., Dell'Orto A., Clerici M., Pagani C., Cattaneo M., Betti R., Pavone E. and Marmini A. Lectin histochemistry in psoriasis. *Acta Derm. Venereol.* (Suppl.) 1984, **113**, 80–84.
44. Childs G. V., Naor Z., Hazum E., Tibolt R., Westlund K. N. and Hancock M. B. Localization of biotinylated gonadotropin releasing hormone on pituitary monolayer cells with avidin–biotin–peroxidase complex. *J. Histochem. Cytochem.* 1983, **31**, 1422–1425.
45. Wood G. S. and Warnke R. Suppression of endogenous avidin-binding activity in tissues and its relevance to biotin–avidin detection systems. *J. Histochem. Cytochem.* 1981, **29**, 1196–1204.
46. Bussolati G. and Gugliotta P. Non-specific staining of mast cells by avidin–biotin–peroxidase complexes (ABC). *J. Histochem. Cytochem.* 1983, **31**, 1419–1421.
47. Tharp M. D., Seelig L. L., Tigelaar R. E. and Bergstresser P. R. Conjugated avidin binds to mast cell granules. *J. Histochem. Cytochem.* 1985, **33**, 27–32.
48. Holgate C. S., Jackson P., Cowen P. N. and Bird C. C. Immunogold–silver staining: new method of immunostaining with enhanced sensitivity. *J. Histochem. Cytochem.* 1983, **31**, 938–944.
49. Springall D. R., Hacker G. W., Grimelius L. and Polak J. M. The potential of the immunogold–silver staining method for paraffin sections. *Histochemistry*, 1984, **81**, 603–608.
50. Lin C. W., Fujime M., Kirley S. D. and Prout G. R. Jr Visualization of urothelial blood group isoantigens A and B using direct biotin-labelled antibodies and avidin–biotin–peroxidase complex. *J. Histochem. Cytochem.* 1984, **32**, 1339–1343.

5

Gold Probes in Light Microscopy

J. De Mey, G. W. Hacker, M. De Waele
and D. R. Springall

The colloidal gold marker system for immunocytochemistry, introduced in 1971 by Faulk and Taylor[1] was originally limited to electron microscopy and has only recently emerged as an important tool in light microscopical immunocytochemistry. Colloidal gold sols display a natural red colour which is visible in bright-field light microscopy when gold particles accumulate at target sites (*see*[2] for review). Geoghegan and colleagues[3] were the first to report on an application of this principle which has now been expanded in a number of ways. The recent introduction of signal enhancement by microscopical, chemical or electronic means has shown that the colloidal gold marker system has far greater potential for light microscopical immuno- and affinity cytochemistry than originally believed. The aim of this chapter is to review these developments and to try to relate them to the state of the art in light microscopical immunocytochemistry.

1. APPLICATIONS USING THE RED COLOUR OF COLLOIDAL GOLD

1.1. Localization of Cell Surface Components

In 1978, Geoghegan and colleagues[3] found that B-lymphocytes, stained with antibodies to surface immunoglobulins and gold-labelled secondary antibodies, became readily visible in the bright-field light microscope. Positive lymphocytes were recognized because they were surrounded by a dark rim. Before incubation with immune reagents, the cells have to be fixed in order to prevent redistribution and internalization of the marker. This can also be achieved by carrying out the incubations at 4 °C, if prefixation interferes with antibody reactivity. In many

71

cases, however, the density of cell surface antigen is so low that a positive reaction cannot be detected, e.g. markers of T-lymphocytes. The use of special incubation conditions which permit patch formation, but not capping or internalization, has solved this problem.[4] Unlike capping and internalization, patching is an ATP-independent ligand-induced redistribution. ATP formation in cells can easily be blocked by addition of $0 \cdot 2$ per cent NaN_3 to the incubation media. Marking is then performed at room temperature and positive cells become readily visible, the patched cell surface marker being seen as numerous dark 'granules'. This approach has been used to enumerate lymphocytes and their subclasses with monoclonal antibodies which were either directly bound to colloidal gold[4] or were visualized with gold-labelled secondary antibodies (*Plate* 4a).[5] Optimal contrast was obtained with 30–40-nm diameter gold particles. Cytospin or smear preparations of marked cell suspensions were counterstained with methyl green and non-lymphocytic cells were reacted for endogenous peroxidase activity. In this way, cell identification and enumeration was greatly facilitated. For a detailed technical evaluation, *see* the literature.[6] The method can be adapted to examining a few drops of blood in which red blood cells have been eliminated by lysis.

Immunogold staining (IGS) of cell surface antigens has been combined with a variety of enzyme cytochemical reactions,[7–11] allowing the simultaneous determination of antigenic and cytochemical profiles of cells. Recently, an immunogold method in Terasaki plates has been described which allows the visualization of monoclonal antibodies binding to cell surface antigens and is suitable for large-scale screening.[12]

In conclusion, immunogold staining proves to be a very simple and reliable means of identifying and enumerating cells bearing surface membrane antigens without the need for special equipment. It produces preparations that can be re-examined at will. The marker lends itself particularly well to combination with enzyme cytochemistry. It can be applied to small volumes of cell suspensions prepared from any body source or from cell cultures and is also suitable for large-scale screening for the production of monoclonal antibodies.

1.2. Localization of Antigens in Sections of Paraffin-embedded Tissue and Semithin Sections of Resin-embedded Tissue

The natural colour of colloidal gold can also be used to localize antigens in tissue sections, particularly in sections from paraffin-embedded[2,13–16] and resin-embedded material (after removal of the resin).[2,17–19] The red colour appears during incubation with the gold probe (direct or indirect) and does not necessitate a revealing reaction. The best results are given by 20-nm gold particles. Indirect procedures using gold-labelled secondary antibodies[13] or protein-A–gold[14] have been described. Signal amplification was obtained by using a layer of gold-labelled secondary antibodies followed by a layer of gold-labelled IgG of the same species as the primary antibody.[13] The best results were obtained with paraffin-embedded material. Gu and colleagues[13] used immunogold staining in combination with the peroxidase–anti-peroxidase (PAP) method for double staining, while Roth[14] introduced its combination with another colloidal metal: colloidal silver. This novel colloid yielded a yellow staining of positive tissue structures. The IGS method was not successful in localizing nerve fibres, but was satisfactory for staining cells producing regulatory peptides, serotonin, exopancreatic enzymes and cytosolic, vitamin D-dependent, calcium-binding protein. The failure to stain nerve fibres efficiently is probably due to penetration problems; the 20-nm gold probes do not diffuse very well into tissue sections. Besides antigens, cellular and extracellular glycoconjugates were visualized by means of lectins directly adsorbed onto gold particles or by glycoprotein–gold complexes in an indirect technique.[5,18,19]

The IGS method for staining antigens in tissue sections is not widely used, and this will probably remain the case. The reason is that the gold probes have to be used at a relatively high concentration (absorbance at 520 nm, $A_{520} = 2$) and also that a much more sensitive technique requiring low concentrations of probe has been developed, the immunogold–silver staining method (*see below*). The IGS method is certainly not the most sensitive one, but in cases where a high sensitivity is not required or is even a handicap, it is attractive due to its ease of use and the higher resolution in comparison to fluorochromes or enzyme methods.[15,18] One application is described below in Section 1.3.

1.3. Localization of Microtubules in Cultured Cells and Plant Endosperm Cells

Microtubules are cell organelles (diameter 25 nm) composed of tubulin molecules. They play an important role in cell structure and functions, such as mitosis, intracellular transport and organelle translocation, signal transduction, directional cell locomotion, and cell shape.[20] Their distribution in whole cells can be illustrated in normal and experimental conditions, using a variety of immunocytochemical procedures with antibodies to tubulin. Colloidal gold can be used to visualize individual microtubules in the form of red strands.[21] Compared with PAP-stained microtubules, the picture obtained was clearer, resulting in improved visualization of microtubules in mitotic cells. The gold marker is more delicate and does not tend to overstain the large number of microtubules in the mitotic spindle. This characteristic is very useful for studying the distribution of microtubules in anastral spindles of plant endosperm cells (*Plate* 5).[22–24]

2. MICROSCOPICAL ENHANCEMENT OF THE GOLD MARKER

As discussed above, the gold marker used in direct and indirect immunogold staining of cell surface antigens is sufficiently visible, especially when high magnification ($100 \times$) oil immersion objectives are used. In some cases, however, the 'granules' associated with the plasmalemma are rather small, or become partially masked by subsequent cytochemical reactions on the same cells. De Waele and colleagues[5] have reported that dark-field illumination enhances the detectability of colloidal gold, but this is less practical and necessitates sequential use of bright-field illumination to appreciate the morphological characteristics of the cells. In 1983, epipolarization microscopy combined with epifluorescence optics was introduced as a means of enhancing the detectability of colloidal gold particles without the disadvantages of dark-field illumination.[17] The principle is illustrated in *Fig.* 5.1. Small clusters of gold particles[5] are now detectable as brightly shining gold-yellow dots, sharply contrasting with the dark background.

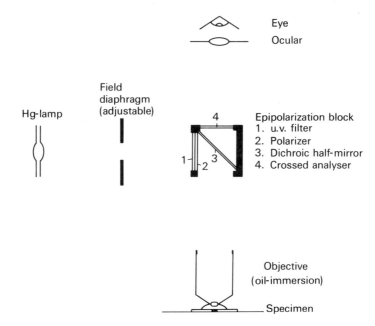

Eye

Ocular

Hg-lamp

Field
diaphragm
(adjustable)

4

Epipolarization block
1. u.v. filter
2. Polarizer
3. Dichroic half-mirror
4. Crossed analyser

Objective
(oil-immersion)

Specimen

Fig. 5.1. Schematic representation of the principle of epipolarization microscopy. The light, emitted by the mercury vapour lamp, is polarized and deflected into the objective via, respectively, a polarizer and a dichroic half mirror at 45°. The objective functions as the condenser. The light reflected by the preparation is extinguished by a crossed analyser, situated above the level of the half mirror. Clusters of colloidal gold, however, strongly backscatter the incident light. This signal is not extinguished and can be seen through the ocular against a dark background.

It has been shown that even individual gold particles as small as 20 nm can be detected with this method.[44] The appearance of this microscopic enhancement is shown in *Plate* 4. Mononuclear human blood cells, stained with OKT 4 (Ortho-Diagnostic Systems Inc.) and the immunogold staining procedure using 30-nm gold particles, are shown in bright-field illumination (*Plate* 4*a*). *Plate* 4*b* shows the same field in epipolarization illumination. The detectability of the gold particles associated with T-helper lymphocytes is strongly enhanced. The oxidized DAB product in a monocyte is not visible. If the epipolarization is combined with bright-field illumination, morphology of the cells is visible at the same time (*Plate* 4*c*). In this combined illumination mode, immunogold staining of surface antigens displays all its advantages:

a. The marker emits light, which is easily perceived by the eye.
b. The signal is stable and the preparations can be re-examined at will.
c. Counterstains and cytochemical reaction patterns can be appreciated simultaneously.
d. The increased visibility ensures that even very small clusters will be easily detected.

Decrease in size of the clusters is the first phenomenon observed when the bioactivity of the gold probe diminishes or when it is used in a more diluted form (unpublished observations). This can present problems when only bright-field illumination is used.

Epipolarization microscopy can be combined with (immuno)fluorescence microscopy. It has also been used to study the regulation of the expression of the interleukin 2 receptor on human T-lymphocytes.[25,26] In general, epipolarization microscopy has proved useful for observing small clusters or thin layers of colloidal gold and silver, at cell surfaces or in semithin sections, using high magnification objectives. Thicker layers absorb the light so that this technique cannot be used to enhance the signal produced by gold in thick histological sections. An epipolarization block is produced by Nikon for use with their Optiphot or Labophot microscopes (IGS block) equipped with epi-illumination.

3. ENHANCEMENT OF THE GOLD MARKER BY A PHYSICAL DEVELOPER

In 1983, Holgate and colleagues[27,28] reported that the silver precipitation reaction of Danscher, used for the detection of metal sulphides,[29] gold[30] or silver[31] in tissues, could be employed for strongly amplifying the visibility of colloidal gold probes used on paraffin sections. Metallic gold catalyses the reduction of silver ions to metallic silver in the presence of a reducing substance. During this 'physical' development, gold particles are encapsulated in growing shells of metallic silver, which gradually become more and more visible at light microscopical level, showing a brown-black to black colour. The physical developer is characterized by the use of silver lactate (as opposed to silver nitrate) as ion supplier and hydroquinone as reducing molecule, at a pH of around 3·5. It also contains a protective colloid in the form of gum arabic. The rationale of the choice of these components is explained in the literature;[29] the aim was to obtain a relatively slow formation of metallic silver with some tolerance for diffuse daylight.

Danscher and Nörgaard[32] used RNAase–gold complexes applied to semithin sections of Epon-embedded liver and pancreatic tissue for the localization of the substrate according to Bendayan.[33] The sections were coated with 0·5 per cent gelatine, which prevented catalytic precipitates from the developer and other non-specific staining from reaching the surface of the sections. Development took place in the dark, in jars placed in a waterbath at 26 °C. After 20–60 min, the slides were rinsed gently with running tap water at 40 °C for at least 40 min in order to remove the protecting gelatine coat, and then rinsed in distilled water and mounted. Silver precipitates seen with normal bright-field light microscopy were located similarly to the colloidal gold particles seen by electron microscopy of thin sections.

The combination of immunogold staining with the silver precipitation reaction proposed by Holgate and colleagues[27] was called 'immunogold–silver staining' (IGSS). These authors used antisera to human immunoglobulins, IgA, IgG, IgM, IgD, κ and λ chains on sections of reactive human tonsils, fixed in formol sublimate and embedded in paraffin wax. The sections were treated with Lugol's iodine and sodium thiosulphate after dewaxing and rehydration. Interestingly, it was noted that omission of iodine treatment prevented all immunostaining when the IGSS method was used. This was not the case with the peroxidase–anti-peroxidase (PAP) method of Sternberger.[34] After the iodine treatment, the sections were, as usual for these antigens, trypsinized with 0·1 per cent trypsin (a four times shorter time than with the PAP method was reported to be sufficient). This was followed by incubation with primary antibodies. A swine anti-rabbit immunoglobulin antiserum adsorbed to 20-nm gold particles was used as second layer (the colloid concentration [A_{520}] was not indicated). The sections were then thoroughly washed with distilled water and immersed in the physical developer. No protective gelatine coat was used. The development was performed in a dark room using an Ilford safe light S 902 or F 904 (placing the sections in a dark cupboard in an open laboratory is also satisfactory). Non-specific staining was reduced by incubation with neat normal swine serum before application of the primary and of the secondary antibodies, and by prolonged washing of sections after treatment with immune reagents. For the antibodies used, the authors clearly demonstrated the superior sensitivity of the IGSS method over the PAP method.

The potential of the IGSS method for staining a multitude of antigens in routinely formalin- or Bouin's-fixed and paraffin-embedded tissues has recently been investigated.[35,36] The antigens demonstrated included a variety of regulatory peptides, intermediate filaments, tumour antigens, cell surface antigens and micro-organisms. Various modifications of the IGSS method[27] were attempted in order to minimize background staining and to simplify the technique. These included the use of detergents, gelatine, high salt concentrations in buffer to reduce non-specific binding of antibodies, adjustment of buffer pH for the immunogold incubation, postfixation of the sections after incubation with the second layer to prevent elution of antibodies in the low pH buffer of the silver developer, and the use of high quality commercial immunogold reagents of various colloid sizes.

Of the tested modifications[35] to the original method,[27] both detergents and higher salt concentrations in buffers reduced the background, but not the specific staining; omission of Lugol's iodine pre-treatment usually prevented all immunostaining. Pre-treatment with trypsin had little or no effect on the demonstration of small peptides, but improved immunostaining for some larger molecules. Dilution of the immunogold reagent in buffer at pH 8·2–8·4 containing 0·8–1 per cent bovine serum albumin also resulted in lower background. Gold particles of 5–40 nm diameter were tested and the 5-nm size was found to be the most satisfactory. Incubation was optimal at 60 min using dilution of 1 : 50 to 1 : 500 ($A_{520} = 0·05$–0·005). The use of neat normal serum before incubation with the primary and the secondary antibodies was retained.

A modified silver precipitation reaction[37] involving omission of gum arabic from the developer was compared with Danscher's method[29] and was found to give equally good results. With the modified developer, optimum staining intensity is achieved after only 4–10 min, depending on temperature. Development may be checked microscopically after fixing the preparation for 1–2 min in photographic fixer or 5 per cent aqueous sodium thiosulphate and rinsing in water. If staining is found to be too weak, the silver precipitation step may be repeated after thoroughly washing the preparations in distilled water (G. W. Hacker and colleagues, 1985, unpublished results).

A modified photographic cutting reducer[35] was found to be useful for decreasing the intensity of staining in sections which had been overdeveloped or

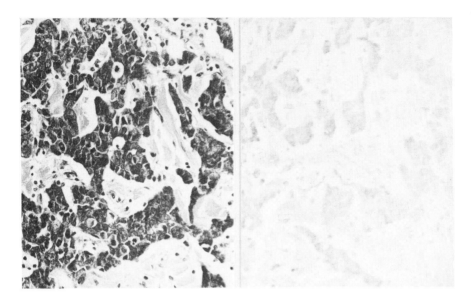

Fig. 5.2. Immunostaining for neuron-specific enolase in a surgical specimen of a Merkel cell tumour. Intense staining is obtained by use of the immunogold–silver method (left), whereas peroxidase– anti-peroxidase staining gives a rather weak reaction (right). Bouin's fluid-fixed near adjacent 5-μm wax sections.

which had a high degree of background staining.

The great potential of the IGSS method in immunostaining routinely processed tissues has become obvious.[35,36] The IGSS method was found to give superior or at least equal results to those obtained using the standard PAP technique for all antibodies tested. In some cases, staining was obtained with IGSS when no, or insufficient, immunoreactivity was visible with the PAP method (*Fig.* 5.2). This applied principally to regulatory peptides in nerves, such as calcitonin gene-related peptide, substance P and somatostatin, which have been clearly demonstrated in Bouin's- and formalin-fixed, wax-embedded tissue (*Fig.* 5.3), or to the demonstration of very thin terminal nerve fibres using antibodies to neurofilament proteins (*Fig.* 5.4). The technique allowed easier visualization of endocrine cells, largely due to the very high contrast of the immunostaining (*Fig.* 5.5). This intense black reaction product facilitated fast screening and rapid diagnosis in histopathological cases (*Fig.* 5.2). Positive reactions could easily be identified at low magnifications. In addition, the high contrast allowed counterstaining with conventional histological stains, such as haematoxylin and eosin, as an aid to assessment of the morphology. As a further advantage, in many cases the IGSS method allowed the primary antibodies to be used at a considerably higher dilution than for the PAP method and for an incubation time of only 90 min compared with overnight incubation.

The technique currently used by the authors is detailed in the Appendix. In this form, it was found to be suitable for wax-embedded tissues fixed in Zamboni's or Bouin's fixatives, formalin and formol mercury, paraformaldehyde and glutaraldehyde.

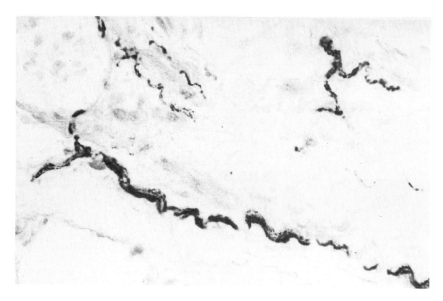

Fig. 5.3. Calcitonin gene-related peptide-immunoreactive nerves in rat lung. The immunogold–silver-stained fibres contrast sharply with other tissues in this preparation. Formalin-fixed, 5-μm wax section, counterstained with eosin.

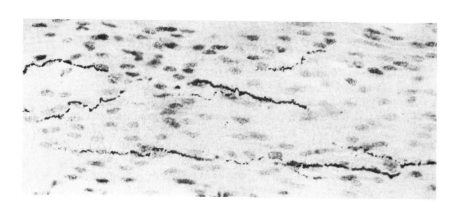

Fig. 5.4. Very thin terminal nerve fibres in smooth muscle of pig urinary bladder, demonstrated by immunogold–silver staining using rat monoclonal antibodies to bovine neurofilament proteins 150 and 200 kdal. Bouin's fluid-fixed, 5-μm wax section, counterstained with haematoxylin and eosin.

In a study aimed at a better understanding of the IGSS method, Lackie and colleagues[38] found that the silver shell size increases with time and the rate of deposition with temperature. Using electron microscopy, it could be confirmed that a colloid size of 5-nm gold particles produces a higher labelling density than particles of larger diameters, without affecting the ultimate silver shell size obtainable. This publication also describes the use of the IGSS method on semithin resin sections.

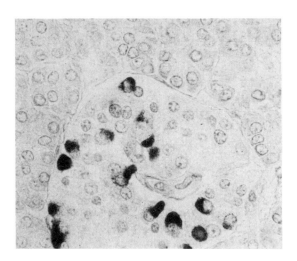

Fig. 5.5   Glucagon-immunoreactive endocrine cells in a Langerhans' islet of human pancreas. Formalin-fixed 5-μm wax section. Immunogold–silver staining method, counterstained with haematoxylin and eosin.

The possibility of using the IGSS method for the simultaneous demonstration of several antigens in different compartments of the same tissue section was first shown for double staining using the IGSS method followed by the PAP technique.[27] Double and triple staining is also possible using combinations of the IGSS method with one or more of the different immunoenzyme methods such as PAP, ß-galactosidase and alkaline phosphatase (*Plate* 6).[39] For multiple staining it is preferable to use primary antibodies produced in different species with non-cross-reacting secondary antibodies. For double immunostaining, the intense black staining product of the IGSS method allowed the sequential use of primary antibodies produced in the same species, due to a 'masking' of antigens of the first set of reactions by the immunogold–silver deposit. For the same reason, a combination of the IGSS method with another immunocytochemical method does not easily permit two substances to be detected in a single cell or fibre. However, there are many possible uses of this combination, for example to determine whether certain peptide-containing nerve fibres stained by the IGSS method for the peptide are surrounded by Schwann cells, stained with antibodies to S-100 using the PAP method (G. W. Hacker and colleagues, 1985, unpublished results).

Recently, it was shown that the IGSS method may be of great value for immunostaining cytological preparations[40] (Chapter 33). Here, it is even possible to detect bacterial or viral infections at the light microscopical level, and this in combination with conventional counterstaining to assess morphology.

The IGSS method has also been adapted to the localization of membrane antigens in cryostat sections of fresh frozen lymphoid tissues and found to be of great use (*Plate* 7).[41] As in previous studies,[35,38] a colloid size of 5 nm gave optimal results. The stained membranes were visible with sharp contrast. The IGSS method has been further applied to the visualization of cell surface antigens in cell suspensions.[42] These preparations were incubated with immune reagents in suspension. Smears or cytospins of reacted cells were then air dried and postfixed with formaldehyde–ethanol. The silver development is performed on the slide with modified developer,[37] and the cells are counterstained with conventional May–Grünewald, Giemsa or Wright staining.

The black gold–silver grains at the surface of positive cells are easily seen, and the cells are identified by conventional morphological criteria. Epipolarization microscopy can enhance further the detectability of the silver-coated gold particles. These emit a very intense white (silver) light that can be combined with almost normal levels of transmitted bright-field illumination (*Plate* 8). In this mode, staining of cell surface antigens acquires a level of convenience and reliability not previously available. Finally, the IGSS method is very well suited to staining cytoskeletal antigens in cultured cell monolayers. All types of filament (microfilaments, intermediate filaments and microtubules) and their associated proteins can be visualized with strong contrast and high resolution. *Fig.* 5.6 shows staining of microtubules, and *Fig* 8.11, Chapter 8, illustrates that myosin is stained identically with indirect immunofluorescence and IGSS. Cells were grown on coverslips, immunostained, fixed in glutaraldehyde, washed in distilled water and treated with the physical developer of Moeremans et al.[37]

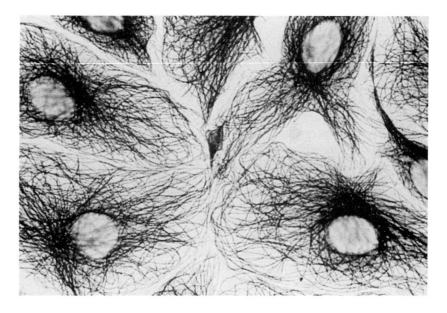

Fig. 5.6 Cytoplasmic microtubules in PTK2 cells are visualized by use of rabbit antibodies to tubulin and the immunogold–silver staining method. The jet-black microtubules are seen in sharp contrast.

Various modifications of the IGSS procedure have been attempted in order to further increase the sensitivity of the method. Bridge methods, involving the use of unconjugated bridging antibodies between the primary antibodies and the final gold–antibodies may result in a build-up of gold–antibody binding sites (D. R.

Springall and colleagues, 1985, unpublished results). The IGSS method was also combined with a streptavidin–biotin system by Hacker and colleagues (1985, unpublished results), using colloidal gold adsorbed to streptavidin molecules (*see also* Chapter 4). Colloidal gold in combination with silver enhancement may also become useful for in situ hybridization, the specific visualization of nucleic acid sequences in tissue sections and cultured cells.[43]

4. ENHANCEMENT OF THE GOLD MARKER WITH NANOMETER PARTICLE VIDEO MICROSCOPY (NANOVID ULTRAMICROSCOPY)

In normal optical microscopy, individual gold particles which are smaller than the theoretical limit of resolution are invisible to the eye or photographic plate. Recently, however, De Brabander and colleagues[44] have introduced the use of microinjected colloidal gold particles of 10–40 nm, stabilized with polyethylene glycol (PEG) or anti-tubulin, as light microscopic probes for studying intracellular transport.

Initial attempts with rhodamine-labelled BSA–gold were not successful: the rhodamine fluorescence was quenched. With epipolarization, however, aggregates and individual particles could be seen as shining dots and were distinguishable from organelles. Optimal visualization was obtained using transmitted light and high resolution optics at the full numerical aperture. The image was projected directly onto the face plate of a video camera which performed electronic subtraction of the background light and amplification of the remaining signal.[45] With illumination increased to completely saturate the camera, the shining dots, previously detected with epipolarization, appeared on the monitor screen as clearly defined dark spots on an entirely white background. By switching condenser settings or filter units, the same cells could be observed within less than 30 s with the sequence: transmitted light, differential interference contrast (DIC), and epipolarization (*Fig.* 5.7).

This new technique has been called 'nanometer particle video ultramicroscopy' or 'Nanovid ultramicroscopy'. It allows the visualization by light microscopy of individual gold probes, containing any of the various possible effector molecules, and can be used to study or measure their interaction with other molecules or cell organelles.

Nanovid ultramicroscopy will be widely applicable to the study of the fate and behaviour of microinjected or (receptor-mediated) endocytosed gold probes, increasing our means of understanding the molecular biology of the living cell.

In addition, it is clear that Nanovid ultramicroscopy has potential in light microscopical (immuno)cytochemistry, since it may be regarded as bridging the gap between light and electron microscopy. Indeed, it has already been shown that cell structures covered with small gold particles via antibodies can be visualized on the monitor screen with high resolution and variable contrast.

Nanovid ultramicroscopy has also potential for qualitative and quantitative assays *in vitro* since the number of gold particles bound to a transparent solid phase provides a measure of the molecular interaction on which the binding was based.

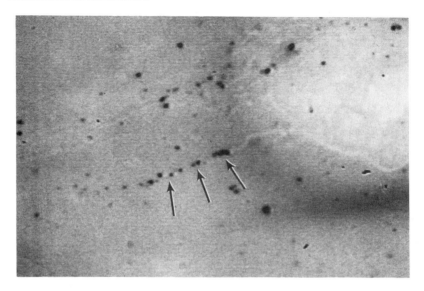

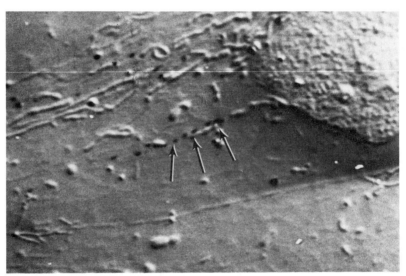

Fig. 5.7   Gold particles (40 nm diameter) were coupled to a monoclonal antibody to tubulin and microinjected into PTK2 cells. The top panel shows the gold particles in the bright-field mode. The lower panel shows the same cell with differential interference contrast. In contrast to 'non-specific' gold particles, coated with polyethylene glycol and albumin, which migrate constantly through the cytoplasm, the anti-tubulin-coated particles remain fixed for several hours. They define linear tracks (microtubules, arrows) in the cytoplasm converging on the perinuclear area. Along these tracks, endogenous organelles move in a typical saltatory fashion, demonstrating the functional integrity of the underlying microtubules. The gold particles are most probably fixed by the antibody to tubulin subunits in the microtubules, showing that subunit flux ('treadmilling') probably does not occur and is not needed for saltatory organelle motion. Note the clear visibility of the gold particles and the ease with which they can be distinguished unambiguously by Nanovid ultramicroscopy from cellular organelles.[44]

5. CONCLUSIONS

In this chapter, we have attempted to review the development of the gold marker system as applied to light microscopy, including the latest techniques. There is no doubt that the colloidal gold marker has an enormous potential for light microscopy. In a number of applications, the red colour generated by accumulations of gold particles will remain useful, especially when the tissue target is not too sparse, and is easily accessible. In these cases, it has been demonstrated that colloidal gold gives a very fine resolution and pictures of high quality.

The technique of epipolarization is an easy way of increasing the detectability of very small amounts of gold, and may become of even greater interest in combination with the silver-enhancement techniques.

Silver enhancement, which increases the visibility of gold particles dramatically, can indeed be regarded as a real departure from all existing marking techniques. The wide applicability of the principle has now been demonstrated, although there is still room for improvement. Smaller gold probes (e.g. 2–3 nm) are now available, containing more particles per unit volume at equal absorbance (*see* Chapter 8). These may allow better penetration making the IGSS method useful for thick frozen or Vibratome sections.

It is possible that the amplifying potential of the silver deposit and/or Nanovid ultramicroscopy will allow us to visualize single antibody–antigen or ligand binding-site reactions at light microscopical level. It is further likely that this high sensitivity will prove useful in the developing techniques of in situ hybridization, or in new types of solid phase (immuno)assays *in vitro*.

In conclusion, the recent developments reviewed in this chapter show that the colloidal gold marker system may be much more widely used in marking cytochemistry in the near future. The increased sensitivity of epipolarization, the IGSS method and Nanovid ultramicroscopy could provide us at a very timely moment with detection techniques for much smaller amounts of cell targets than hitherto possible. For routine applications, it may result in more reliable and complete localization, in combination either with conventional or specially adapted counterstaining or with other cytochemical techniques.

Acknowledgements

We gratefully acknowledge the contributions to this work of our colleagues M. Moeremans, G. Langanger and M. De Brabander. This work was supported by grants from the IWONL (Brussels) to J. De Mey and from the Wellcome Trust (London) to G. Hacker.

Appendix

I. IMMUNOGOLD–SILVER STAINING

1. Dewax sections in xylene and rehydrate through graded alcohols.
2. Wash in tap water for 10 min.

3. Immerse in Lugol's iodine (1 per cent iodine in 2 per cent aqueous potassium iodide solution) for 5 min.
4. Rinse briefly in tap water.
5. Place in 2·5 per cent aqueous sodium thiosulphate solution until sections are colourless.
6. Wash for 10 min in running tap water.
7. Wash in 2 × 5 min changes of IGSS buffer I (*see below*).
8. Apply neat normal goat serum for 10 min and drain off, but do not wash sections.
9. Incubate for 90 min in appropriately diluted primary antibody. This dilution should be determined by titration. Antibody diluent is 0·05 M Tris-buffered saline pH 7·4 containing 0·1 per cent bovine serum albumin and 0·01 per cent sodium azide.
10. Wash in 2 × 10 min changes of IGSS buffer I.
11. Immerse for 10 min in IGSS buffer II (*see below*).
12. Apply neat normal goat serum as before.
13. Incubate with gold-labelled second layer antibodies for 60 min. Optimal dilution is between 1/25 and 1/500 and should be determined by titration. Antibody diluent is IGSS buffer II containing 0·8 per cent bovine serum albumin.
14. Wash in IGSS buffer II, 3 × 10 min changes.
15. Rinse briefly several times in deionized or distilled water followed by 4 washes (5 min each) in the same. This washing must be very thorough to ensure that all halide is removed prior to silver development. Water purity is critical; since this is often variable, the use of glass-distilled deionized water may be necessary.
16. Prepare the physical silver development solution shortly before use (*see below*) and incubate the slides in the dark, or under photographic safelight illumination, for 4–8 min at 20 °C.
17. Wash in photographic fixer diluted 1 : 4 for 2 min or, alternatively, in a 5 per cent aqueous solution of sodium thiosulphate.
18. Wash in running tap water for 10 min. Staining intensity may now be checked microscopically; if staining is too weak, replace sections in water starting again at Step 15. Development will now have to be very short, as there is already some precipitate from the first devlopment.
19. Counterstain as desired (haematoxylin and eosin is suitable and does not affect the silver staining), dehydrate, clear and mount in a permanent medium, e.g. DPX or Styrolite.

A. IGSS Buffer I

0·05 M Tris-HCl buffer pH 7·4, containing 2·5 per cent sodium chloride and 0·5 per cent Tween 80 or 0·2 per cent Triton X-100.

B. IGSS Buffer II

0·05 M Tris-HCl buffer pH 8·2, containing 0·9 per cent sodium chloride.

C. Physical Silver Development Solution

This is prepared shortly before use from 85 ml citrate buffer pH 3·5–3·9 (2·35 g of trisodium citrate dihydrate and 2·55 g of citric acid in 85 ml distilled water) in which is freshly dissolved 850 mg hydroquinone. Add just before use in the dark or under photographic safelight illumination, 15 ml distilled water containing 110 mg silver lactate. Glassware should be scrupulously clean. Because silver lactate and hydroquinone are photosensitive, solutions should be protected from light and used immediately.

II. REAGENTS

Immunogold reagents were obtained from Janssen Life Sciences Products, Beerse, Belgium:

GAR G5 or LM grade = goat anti-rabbit IgG, 5 nm gold-labelled
GAM G5 or LM grade = goat anti-mouse IgG, 5 nm gold-labelled
GARa G5 or LM grade = goat anti-rat IgG, 5 nm gold-labelled
GAGP G5 or LM grade = goat anti-guinea-pig IgG, 5 nm gold-labelled

Immunogold–silver staining kits containing all the reagents necessary, as well as a separate silver enhancement kit, are available from the same company.

Bovine serum albumin (A-7906) and citric acid (C-7129) from Sigma Chemical Company; trisodium citrate (10242), hydroquinone (30011) and Tween 80 (56023) from BDH Chemicals; silver lactate (85210) from Fluka AG; photographic fixer (Amfix) from May and Baker Ltd.

Primary antisera:

Mouse monoclonal antibodies OKT 4 and OKT 8 from Ortho Diagnostic System, Raritan, NJ, USA
Rabbit antiserum to tubulin from J. De Mey, Beerse, Belgium
Mouse monoclonal antibodies to tubulin from J. V. Kilmartin, MRC, Cambridge, England
Rabbit antiserum to human neuron-specific enolase from P. Marangos, NIMH, Bethesda, Maryland, USA
Rat monoclonal antibody to bovine neurofilament proteins 150 and 200 kdal from J. Q. Trojanowski, Pennsylvania School of Medicine, Philadelphia, USA
Rabbit antiserum to chicken neurofilament protein triplet from D. Dahl, Harvard Medical School, Boston, Mass., USA
Rabbit antiserum to glucagon and to calcitonin gene-related peptide from the Departments of Histochemistry and Medicine, Royal Postgraduate Medical School, London, UK
Rabbit antiserum to bovine S-100 from D. Cocchia, Department of Anatomy, Università degli Studi di Roma, Italy.

REFERENCES

1. Faulk W. and Taylor G. An immunocolloid method for the electron microscope. *Immunochemistry*, 1971, **8**, 1081–1083.
2. Roth J. The colloidal gold marker system for light and electron microscopic cytochemistry. In: Bullock G. R. and Petrusz P., eds., *Immunocytochemistry*, Vol. 2. London, Academic Press, 1983: 217–284.
3. Geoghegan W. D., Scillian J. J. and Ackerman G. A. The detection of human B-lymphocytes by both light and electron microscopy utilizing colloidal gold-labelled anti-immunoglobulin. *Immunol. Commun.* 1978, **7**, 1–12.
4. De Mey J., Moeremans M., De Waele M., Geuens G. and De Brabander M. The IGS (Immuno-Gold-Staining) method used with monoclonal antibodies. In: Peeters, M., ed., *Proceedings of Colloquium on Protides of the Biological Fluids.* Oxford, Pergamon Press, 1981: 943–947.
5. De Waele M., De Mey J., Moeremans M., De Brabander M and Van Camp B. Immunogold staining method for the light microscopic detection of leukocyte cell surface antigens with monoclonal antibodies: its application to the enumeration of lymphocyte subpopulations. *J. Histochem. Cytochem.* 1983, **31**, 376–381.
6. De Waele M., De Mey J., Moeremans M., De Brabander M. and Van Camp B. Immunogold staining method for the detection of cell surface antigens with monoclonal antibodies. In: Bullock G. R. and Petrusz P., eds., *Immunocytochemistry*. London, Academic Press, 1983: 1–23.
7. Bergroth V., Konttinen Y. and Reitamo S. A method for the identification of human peripheral blood T lymphocytes by sequential immunogold and esterase double staining. *J. Histochem. Cytochem.* 1983, **31**, 837–839.
8. Crockard A. and Catovsky D. Cytochemistry of normal lymphocyte subsets defined by monoclonal antibodies and immunocolloidal gold. *Scand. J. Haematol.* 1983, **30**, 433–443.
9. De Waele M., De Mey J., Moeremans M., Smet L., Broodtaerts L. and Van Camp B. Cytochemical profile of immunoregulatory T-lymphocyte subsets defined by monoclonal antibodies. *J. Histochem. Cytochem.* 1983, **31**, 471–478.
10. Tavares De Castro J., San Miguel J., Soler J. and Catovsky D. Method for the simultaneous labelling of terminal deoxynucleotidyl transferase (TDT) and membrane antigens. *J. Clin. Pathol.* 1984, **1**, 628–632.
11. Crockard A. Cytochemistry of lymphoid cells, a review of findings in the normal and leukaemic state. *Histochem. J.* 1984, **16**, 1027–1050.
12. Coppe P., Letesson J. J., Saint-Guillain M. and Leloup R. The use of colloidal gold for screening monoclonal antibodies to cell surface antigens. *J. Immunol. Methods* 1985, **76**, 211–222.
13. Gu J., De Mey, J., Moeremans M. and Polak J. M. Sequential use of the PAP and immunogold staining methods for the light microscopical double staining of tissue antigens. Its application to the study of regulatory peptides in the gut. *Reg. Peptides*, 1981, **1**, 365–374.
14. Roth J. Applications of immunocolloids in light microscopy. Preparation of protein A–silver and protein A–gold complexes and their application for the localization of single and multiple antigens in paraffin sections. *J. Histochem. Cytochem.* 1982, **30**, 691–696.
15. Roth J. Applications of immunocolloids in light microscopy. II. Demonstration of lectin-binding sites in paraffin sections by the use of lectin–gold or glycoprotein–gold complexes. *J. Histochem. Cytochem.* 1983, **31**, 547–552.
16. Stein B., Buchan A., Morris J. and Polak J. M. The ontogeny of regulatory peptide-containing cells in the human fetal stomach: an immunocytochemical study. *J. Histochem. Cytochem.* 1983, **31**, 1117–1125.
17. De Mey J. Colloidal gold probes in immunocytochemistry. In: Polak J. M. and Van Noorden S., eds., *Immunocytochemistry, Practical Applications in Pathology and Biology*. Bristol, Wright-PSG, 1983: 82–112.
18. Lucocq J. and Roth J. Applications of immunocolloids in light microscopy. III. Demonstration of antigenic and lectin-binding sites in semithin resin sections. *J. Histochem. Cytochem.* 1984, **32**, 1075–1083.
19. Kunz A., Brown D. and Orci L. Appearance of *Helix pomatia* lectin-binding sites on podocyte plasma membrane during glomerular differentiation. *Lab. Invest.* 1984, **51**, 317–324.
20. De Brabander M. Microtubules, central elements of cellular organisation. *Endeavour*, 1982, **6**, 124–134.

21. De Mey J., Moeremans M., Geuens G., Nuydens R. and De Brabander M. High resolution light and electron microscopic localization of tubulin with the IGS (immuno-gold staining) method. *Cell Biol. Int. Rep.* 1981, **5**, 889–899.

22. De Mey J., Lambert A., Bajer A., Moeremans M. and De Brabander M. Visualization of microtubules in interphase and mitotic plant cells of *Haemanthus* endosperm with the immunogold-staining method. *Proc. Natl Acad. Sci. USA*, 1982, **79**, 1898–1902.

23. Bajer A. and Molè-Bajer J. Asters, poles and transport properties within spindle-like microtubule arrays. *Cold Spring Harbor Symp. Quant. Biol.* 1982, **XLVI**, 263–283.

24. Bajer A., Cypher C., Molè-Bajer J. and Howard H. Taxol-induced anaphase reversal: evidence that elongating microtubules can exert a pushing force in living cells. *Proc. Natl Acad. Sci. USA*, 1982, **79**, 6569–6573.

25. Van Wauwe J. P., Goossens J. G. and Beverly P. C. Human T-lymphocyte activation by monoclonal antibodies; OKT 3, but not UCHT 1, triggers mitogenesis via an interleukin 2-dependent mechanism. *J. Immunol.* 1984, **133**, 129–132.

26. Van Wauwe J. P., Goossens J. and Van Nyen G. Inhibition of lymphocyte proliferation by monoclonal antibody directed against the T3 antigen on human T cells. *Cell. Immunol.* 1984, **86**, 525–534.

27. Holgate C., Jackson P., Cowen P. and Bird C. Immunogold-silver staining: new method of immunostaining with enhanced sensitivity. *J. Histochem. Cytochem.* 1983, **31**, 938–944.

28. Holgate C., Jackson P., Lauder I., Cowen P. and Bird C. Surface membrane staining of immunoglobulins in paraffin sections of non-Hodgkin's lymphomas using immunogold-silver staining technique. *J. Clin. Pathol.* 1983, **36**, 742–746.

29. Danscher G. Histochemical demonstration of heavy metals. A revised version of the sulphide silver method suitable for both light and electron microscopy. *Histochemistry*, 1981, **71**, 1–16.

30. Danscher G. Localization of gold in biological tissue. A photochemical method for light and electron microscopy. *Histochemistry*, 1981, **71**, 81–88.

31. Danscher G. Light and electron microscopic localisation of silver in biological tissue. *Histochemistry*, 1981, **71**, 177–186.

32. Danscher G. and Nörgaard J. Light microscopic visualization of colloidal gold on resin-embedded tissue. *J. Histochem. Cytochem.* 1983, **31**, 1394–1398.

33. Bendayan M. Ultrastructural localization of nucleic acids by the use of enzyme–gold complexes. *J. Histochem. Cytochem.* 1981, **29** 531–541.

34. Sternberger L. A., Hardy P. H. Jr, Cuculis J. J. and Meyer H. G. The unlabeled antibody–enzyme method of immunohistochemistry. Preparation and properties of soluble antigen–antibody complex (horseradish peroxidase–antihorseradish peroxidase) and its use in identification of spirochetes. *J. Histochem. Cytochem.* 1970, **18**, 315–333.

35. Springall D. R., Hacker G. W., Grimelius L. and Polak J. M. The potential of the immunogold–silver staining method for paraffin sections. *Histochemistry*, 1984, **81**, 603–608.

36. Hacker G. W., Springall D. R., Van Noorden S., Bishop A. E., Grimelius L. and Polak J. M. The immunogold–silver staining method — a powerful tool in histopathology. *Virchows Archiv. Abt. A Pathol. Anat.* 1985, **406**, 449–461.

37. Moeremans M., Daneels G., Van Dijk A., Langanger G. and De Mey J. Sensitive visualisation of antigen antibody reaction in dot and blot immune-overlay assays with immunogold and immunogold/silver staining. *J. Immunol. Methods*, 1984, **74**, 353-360.

38. Lackie P. M., Hennessy R., Hacker G. W. and Polak J. M. Investigation of immunogold–silver staining by electron microscopy. *Histochemistry*, in press.

39. Hacker G. W., Springall D. R., Cheung A., Van Noorden S. and Polak J. M. Immunogold–silver staining and its potential use in multiple staining and histopathology. *J. Histochem. Cytochem.* 1985, in press.

40. Springall D. R., Tang S. -K., Hacker G. W., Levene M. M., Van Noorden S. and Polak J. M. Applications in diagnostic cytology of a new sensitive staining method — immunogold–silver staining. *J. Pathol.* in press.

41. De Waele M., De Mey J., Reynaert Ph., Dehou M. F., Gepts W. and Van Camp B. Detection of cell surface antigens in cryostat sections with immunogold–silver staining. *J. Histochem. Cytochem.* in press.

42. De Waele M., De Mey J., Jochmans K., Labeur C., Renmans W. and Van Camp B. Enumeration of lymphocyte subsets with immunogold–silver staining. *J. Histochem. Cytochem.* in press.

43. Varndell I. M., Polak J. M., Sikri K. L., Minth C. D., Bloom S. R. and Dixon J. E. Visualisation of messenger RNA directing peptide synthesis by *in situ* hybridisation using a novel

single-stranded cDNA probe. Potential for the investigation of gene expression and endocrine cell activity. *Histochemistry*, 1984, **81**, 597–601.

44. De Brabander M., Geuens G., Nuydens R., Moeremans M. and De Mey J. Probing microtubule-dependent intracellular motility with nanometer particle video ultramicroscopy (Nanovid ultramicroscopy). *Cytobios*, 1985, in press.

45. Allen R. D., Allen N. S. and Trans J. L. Video-enhanced contrast, differential interference contrast (AVEC-DIC) microscopy: A new method capable of analyzing microtubule-related motility in the reticulopodial network of *Allogromia laticollaris*. *Cell Motility*, 1981, **1**, 291–302.

6

The Use of Semithin Frozen Sections in Immunocytochemistry

R. C. Richards, K. R. Trenholm and
S. Semoff

For many years, frozen sections have been used to localize molecules by immunocytochemical techniques. Routinely, 5-μm thick sections are used and have helped to contribute to our knowledge in many fields of biology. However, with the advent of new technology, it is now possible to purchase attachments for most ultramicrotomes that adapt them to cut ultrathin frozen sections. Such sections are used in immunocytochemistry at the electron microscope level (*see* Chapter 9) or as unfixed material in straightforward structural studies (for a review *see* Tokuyasu[1]). In addition to cutting ultrathin frozen sections, such adapted ultramicrotomes are also able to cut sections in the semithin range (0·1−1 μm thick). Frozen sections of this thickness show increased clarity and resolution when viewed, either under the phase contrast microscope or in stained preparations.[2–4] In addition, when immunocytochemistry is performed on these sections, there is increased sensitivity in detecting antigens and Tokuyasu[1] has pointed out that 'use of these sections allows one to attain the highest resolution available with the light microscope, resulting in a new and powerful tool for immunocytochemical studies of cells and tissues.'

In order to illustrate the marked enhancement in immunocytochemical staining obtainable, we have chosen to take as our example the screening of monoclonal antibodies raised against the erythrocyte stage of the malarial parasite. The basic technique involved in this study is essentially that reported by Tokuyasu[1,5,6] and is described in the Appendix.

Rodent malarial parasites, such as *Plasmodium chabaudi*, are essentially typical mammalian malarial parasites and as such can be used as model systems when investigating human malaria. During part of the life cycle of this parasite, the merozoite stage penetrates the red blood cells (RBCs) of its mammalian host and during this time will express certain stage-specific antigens. As part of a study looking at the changing antigen repertoire during the life cycle of this parasite, monoclonal antibodies have been raised in Balb/C mice by challenging them with parasitized erythrocytes.

The usual method used for screening these antibodies, the standard indirect fluorescent antibody test, involves the lysis of whole RBCs to allow the antibodies to enter the RBC and react with antigens on the surface of the parasite. Lysis of the RBCs causes collapse of the cells and this makes it very difficult to say where in the parasite or RBC the fluorescence is localized. This method, therefore, has severe limitations in the study of parasite antigens. Furthermore, while being useful under some circumstances, the standard technique is relatively harsh and causes quite considerable ultrastructural damage.[7] Direct accessibility to antigens in intact specimens is often required before immunocytochemical labelling and this has been made possible by cutting frozen sections.

Frozen sections for light microscopy are usually cut at a thickness of aproximately 5 μm, but this creates problems with the RBC stages of the malarial parasite as sections of this thickness are considerably thicker than the parasite itself. However, improvements made in the technique by Tokuyasu[1] have allowed semithin frozen sections of 0·5 μm to be cut easily and in relatively large numbers.

In the study to be described here, blood was removed by cardiac puncture from mice infected with the AJ strain of *P.chabaudi*. Culture *in vitro* of these RBCs, which were infected with trophozoites, was used to obtain the schizont stage (*Fig. 6.1*)

I. STANDARD INDIRECT FLUORESCENT ANTIBODY SCREENING

Infected RBCs were washed in phosphate-buffered saline (PBS), spun at 1500 rev./min for 5 min, the supernatant removed and the process repeated. Cells were then resuspended in PBS and a 20-μl drop of cell suspension was placed in each well of a 15 well immunofluorescence slide (Flow Laboratories). The slides were then dried using a hair dryer and finally placed in a slide box and stored at −20 °C until required. The prepared slides were removed from the freezer, placed in a dessicator containing silica gel and left to dry for 30 min. Ten μl of the diluted monoclonal antibody was added to each of 2 wells and a control was also included in which PBS replaced the monoclonal. The slides were incubated overnight in a moist chamber at 4 °C and then washed in PBS for 10 min at room temperature. Excess moisture was carefully removed and 10 μl of fluorescein isothiocyanate-conjugated rabbit anti-mouse Ig (1 : 25) (Miles Laboratories) plus 10 μl of 1 per cent Evans Blue was added to each well. This incubation continued for 30 min at 37 °C in a moist chamber and the preparations were then washed in PBS for 30 min.

2. SEMITHIN FROZEN SECTION SCREENING

The RBCs were washed with PBS before fixing in 0·75 per cent paraformaldehyde at 4 °C for 24 h. This fixative was removed and replaced by 4 per cent

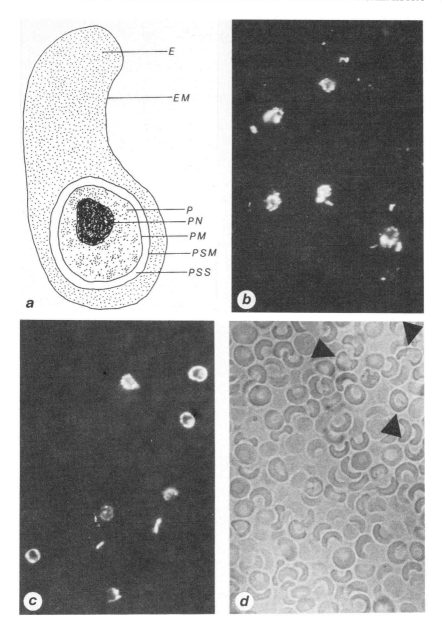

Fig. 6.1 (*a*) Representation showing the relationship between the malarial parasite *P. chabaudi* and the host erythrocyte (E, erythrocyte; EM, erythrocyte membrane; P, parasite; PN, parasite nucleus; PM, parasite membrane; PSM, parasitophorous membrane; PSS, parasitophorous space). (*b*) Pattern of fluorescence shown when using whole cells dried on a slide in the standard indirect fluorescent antibody test. (*c*) Pattern of fluorescence shown when using the same antibody as in (*b*) but on semithin frozen sections. (*d*) Bright-field illumination of the same section seen in (*c*) showing the parasitized erythrocytes (arrows). The increased resolution seen in the semithin frozen sections shows the staining to be localized in the region of the parasitophorous space.

paraformaldehyde at 4 °C for 24 h. The procedure was repeated with the 4 per cent fixative being replaced by 10 per cent paraformaldehyde. The cells were then stored in the 10 per cent fixative for at least 4 days or until required. This procedure was necessary as the RBCs lysed when placed directly in 10 per cent paraformaldehyde.

The RBCs were washed in PBS before being resuspended in a buffered 1 M sucrose solution for a minimum of 2 h at 4 °C. Thereafter, the cells were pelleted in a microfuge, transferred to an aluminium block and frozen by plunging into liquid Freon. The block was then stored in liquid nitrogen until required. Semithin frozen sections were cut at −30 °C on a Sorval MT600 ultramicrotome fitted with an FS1000 cryokit.

Sections, once cut, were picked up on a droplet of sucrose suspended on a fine platinum wire loop, gently warmed to flatten the section and transferred to a clean 15 well immunofluorescence slide (Flow Laboratories). Sections on the multiwell slide were washed in PBS for 20 min to remove sucrose and excess buffer was removed using strips of filter paper. A drop of conditioner (5 per cent IgG-free BSA + 0·25 per cent sodium azide) was placed on each section and the slides were then held in a moist chamber for 10 min at room temperature. At this point the monclonal antibodies being tested were prepared (in PBS) to give a range of dilutions down to 1 : 10 000. The drops of conditioner were removed using strips of filter paper and a drop of the diluted monclonal antibody was applied to each section. Three sections were used at each dilution and control sections incubated with PBS were also included. Sections were incubated overnight in a moist chamber at 4 °C and then washed in PBS for 30 min. Excess PBS was removed with strips of filter paper. The sections were then covered with a drop of fluorescein isothiocyanate-conjugated rabbit anti-mouse Ig diluted 1 : 25 in PBS. Slides were incubated at 37 °C for 30 min in a moist chamber, washed in PBS for 30 min and excess PBS removed by blotting. The sections were mounted in Citifluor mountant* in order to retard photofading and a coverslip applied. Sections were examined under a × 100 oil immersion lens using a Nikon Fluophot microscope with epifluorescence excitation.

During this investigation, improved resolution was seen when the immuno-fluorescence in semithin frozen sections was compared with that obtained in the standard screening. The general structure of an infected RBC is shown in *Fig. 6.1a*. The pattern of fluorescence observed when monoclonal antibody E8 was used in a standard indirect fluorescent antibody test is shown in *Fig. 6.1b*, while *Fig. 6.1c* shows the fluorescence obtained when E8 was used on a semithin frozen section. The improved resolution obtained when semithin frozen sections are used is readily apparent. A comparison of *Fig. 6.1c*, and *Fig. 6.1d* (which shows infected RBCs under bright-field illumination) shows that it is possible to localize the fluorescence to the outer edge of the parasite, possibly in the region of the parasitophorous space. In contrast, examination of preparations stained by the standard indirect fluorescent antibody test (*Fig. 6.1b*) does not allow the precise localization of the fluorescence. The resolution achieved by this semithin frozen section technique when compared to the standard technique allows one to distinguish between antigens associated with the RBC and those associated with

---

*Citifluor mountant is obtainable from Citifluor Ltd, Connaught Building (room 303), The City University, Northampton Square, London EC1V OHB.

the malarial parasite and thus allows a more precise screening of the monoclonal antibodies.

Appendix

I. THE PRODUCTION OF SEMITHIN FROZEN SECTIONS

The following technique for producing semithin frozen sections (0·1–1 μm thick) is essentially that described by Tokuyasu. [1,5,6] Such sections have been used in immunofluorescence microscopy[2,3,8,9,10] and have proved to be useful as serial sections in both light and electron microscopy. Indeed, Tokuyasu's technique also provides a system for carrying out immunocytochemistry at both the light microscope and electron microscope level on the same sample with almost identical procedures.[12] This is extremely useful as it enables many parameters for electron microscopy to be worked out at the light microscope level.

A. Fixation

The conditions of fixation to be used should be selected with a view to optimizing the following four variables: (1) preservation of structure, (2) preservation of antigenicity, (3) retention of accessibility of the antigen to the antibody molecules and (4) immobilization of the antigen. In general, there will be better preservation of antigenicity and accessibility of the antigen when cross-linking by the fixative is low. The best conditions of fixation for an immunocytochemical study will vary with different tissues/cells, different parts of these tissues/cells or with the different proteins under investigation and need to be worked out separately for each antigen/tissue to be studied.

The need for good structural preservation is obviously much less important for light microscopy than for electron microscopy. Conditions of fixation adopted for immunofluorescence microscopy are, therefore, more easily satisfied than for immunoelectron microscopy. Correspondingly, the cutting of 0·1–1 μm sections is much easier than the cutting of sections below 0·1 μm thick.

Formaldehyde is a good fixative for immunocytochemistry as it is a much milder chemical fixative than glutaraldehyde and allows greater retention of antigenicity of sensitive protein/peptide molecules.[13,14] Concentrations of formaldehyde up to 10 per cent have been used in our laboratory to good effect. A characteristic feature of formaldehyde fixation which should be kept in mind and which distinguishes it from glutaraldehyde fixation is its apparent reversibility.[6] In relation to this, formaldehyde, which penetrates tissues very quickly, has been used as a temporary stabilizer before a more permanent fixation by glutaraldehyde is achieved.[6]

Concentrations of glutaraldehyde from 0·1 to 2 per cent have been used successfully in immunocytochemistry.[10,15,17] However, if glutaraldehyde-fixed tissue is to be used then any free aldehyde groups need to be quenched. This can be accomplished by treating either tissue blocks or sections with a 1 per cent solution of sodium borohydride for 5–10 min.

B. Sucrose Infusion

In conventional ultramicrotomy, the sectioning quality is determined mainly by the resin embedding medium. However, in cryo-ultramicrotomy the variability in the physicochemical properties of cells or tissues in the frozen state is the decisive factor, there being no embedding medium present. These properties of biological specimens need therefore to be modified in order to obtain the required plasticity which will result in the best sectioning characteristics of the frozen material.

When tissues/cells are frozen, water is thought to be the most brittle component of the block, with any proteins providing a certain amount of plasticity. Experimental protein solutions sectioned in this way show no changes in their plasticity even after fixation with glutaraldehyde.[5] This suggests that areas of tissues/cells with high water content would be more difficult to cut frozen sections from than those areas having a high protein content.

By varying the cutting temperature, it is not possible to compensate completely for this variability in the composition of tissues. In order to control the plasticity of frozen tissue in a more direct manner, Tokuyasu[5] suggested that, before freezing, fixed pieces of tissue/cells be infused with chemically inert hydrophilic substances of low molecular weight. In addition, this treatment also reduces ice crystal damage, most probably by reducing the size of ice crystals formed. Tokuyasu chose sucrose for this purpose and found it to be superior to glucose, dimethylsulphoxide or glycerol.[5] The range of sucrose concentrations for optimum sectioning appears to be 0·6–1·0 M, depending on the tissue under investigation. The hardness of the tissue is inversely related to the sugar concentration within a certain range.[1] We routinely use buffered sucrose (1–2·3 M) for a variety of cells and tissues.

The infusion is carried out at 4 °C, for at least 30 min depending on the size of the tissue pieces to be used. It has been shown that there is no difference between using graded concentrations of sucrose and the one-step infusion technique.[5]

When cell suspensions are to be sectioned and the cells within the suspension are closely packed then they can be processed as for tissue pieces. However, if the cells are in a relatively dispersed state, the medium surrounding the cells becomes the major factor in influencing sectioning quality. It may thus become necessary to include macromolecular components in the medium in addition to the sucrose. These macromolecules will remain in the medium and will effectively equalize the physical properties of the medium and the cells. Addition of 1–2 per cent gelatin, agarose or dextran to the medium is suggested by Tokuyasu[1] to remedy the situation.

C. Freezing

Prior to freezing, blocks of tissue may be pretrimmed under a dissecting microscope and mounted on small copper or aluminium stubs with a small amount of infusion medium. Freezing of the material is accomplished by plunging the stub, held with a pair of forceps or an artery forceps, into liquid nitrogen and holding it there until the nitrogen ceases to boil. The stub together with the frozen tissue is then quickly transferred to the already cooled head of the microtome and secured firmly with a small screw.

Although liquid nitrogen is notoriously slow at freezing material and slow freezing results in the formation of large ice crystals which disrupt the tissue, tissue sections examined under the light and electron microscope show little evidence of ice-crystal damage when compared with material frozen in isopentane or Freon. As mentioned previously, the sucrose solution is most probably reponsible for this protective effect.

Frozen tissue can either be sectioned immediately or may be stored on the metal stubs under liquid nitrogen in a cell-storage unit until required. However, we have become aware of the possibility of the spontaneous occurrence of staining artefacts in tissue incorrectly stored in such units (*see also* Hühn and Nairn[17]).

D. Trimming

Trimming of frozen tissue after its transfer into the cryobowl of the ultramicrotome can be accomplished using a hand-held precooled scalpel but this risks damaging the specimen. An alternative method, which is similar to that used on the Reichert freezing ultramicrotome, is to use a piece of stainless steel in the position in the cryobowl normally occupied by the diamond knife. This piece of steel after cooling down to the temperature of the cryobowl can be advanced towards the specimen, during its sectioning cycle, enabling it to trim off peripheral areas of the tissue so forming a reasonably shaped block face which will result in greater ease of sectioning and the likelihood of producing a ribbon of sections. Trimming of frozen blocks is not absolutely essential for cutting semithin sections but it is critical for cutting ultrathin frozen sections.

E. Sectioning

The Sorvall MT2-B ultramicrotome and its LTC-2 cryokit attachment (originally developed by Christensen[18,19]) are used routinely in our laboratory, although other manufacturers produce similar equipment.

Glass knives of good quality perform better in ultracryotomy than in normal ultramicrotomy, but for our studies we routinely use a Dupont diamond knife. It is our experience that 'old' diamond knives, discarded from use in cutting resin sections, give excellent results when cutting semithin and ultrathin sections. This is probably due to the fact that frozen blocks of tissue tend to be softer than resin blocks and in addition any knife marks in the frozen sections seem to disappear when the sections melt on the sucrose pick-up droplet.

After transferring the frozen specimen to the ultramicrotome, a short time (10 min) is allowed for it to equilibrate with the temperature in the cryobowl (usually set at -30 to $-70\ °C$). The sections are cut dry at the knife edge and sometimes show small areas of interference colours which approximately correspond to the colours of plastic sections.[5] However, normally the average thickness of the sections produced is estimated from the thickness settings dialled on the controls of the microtome. Although sections of 0·1-2 µm can be used for light microscopy we routinely cut sections of 0·5-1 µm thickness.

In order to section tissue blocks or cell pellets smoothly, the correct combination of sucrose infusion concentration and cryobowl temperature is necessary. One normally starts cutting sections by dialling the required temperature and thickness settings on the microtome. However, it is important to remember that if the sections crumble as they come off the knife edge, then one needs to decrease the thickness setting or increase the temperature in the cryobowl. Because of the absence of a fluid on which to float the sections, a very slow speed of sectioning is possible. Although the automatic mode of operation can be used, we normally section extremely slowly by hand. This results in the smooth cutting of thin and flat sections and helps to prevent local heat production at the block face due to the friction of cutting. Faster sectioning also results in the generation of static electricity on the section. It is the production of static electricity which causes a ribbon of sections inconveniently to move back towards the specimen as it approaches the knife edge. At times the interior of the cryobowl becomes so highly charged that, as one attempts to move sections off the knife edge with an eyelash probe, they suddenly jump and disappear. We have found that the use of an antistatic pistol (Zerostat ASP21, Zerostat Instruments Ltd, Huntington) will at times help to reduce static,[20] as does the use of a humidifier in the cutting room. However, this latter procedure may give rise to an additional serious problem encountered in cryo-ultramicrotomy, namely, frosting at the knife edge. This is mainly due to condensation of moisture in the air. It dulls the knife edge and can also prevent the smooth movement of sections away from the knife edge as they are cut. When the knife edge is obviously frosting up one can attempt to clean off the frost using a piece of elder pith. It is also advisable to cover the plastic observation window, which closes off the top of the cryobowl, with a piece of polystyrene, whenever the cryobowl is left idle for any length of time.

A special feature of cryo-ultramicrotomy is that thermal instability is the main source of mechanical error, but this does not play such an important role when cutting semithin sections.

Sections, once cut, may be flattened using an eyelash probe, but it is more important to ensure that the sections are cut smooth and flat in the first place.

One major difficulty in dry cryosectioning is curling of the sections. We have found that this may be remedied by moving to a fresh part of the knife edge, or by cleaning the knife edge with elder pith. However, the sections may be curling because they are being cut too thick at that particular temperature. In this case, the remedy is to either reduce the thickness setting on the microtome or increase the temperature in the cryobowl.

F. Transfer of the Frozen Sections

Sections, once cut, are moved off the knife edge with the aid of an eyelash probe and are parked down in the boat of the diamond knife. When transfer of these sections onto glass slides is required, a small droplet of a 2·3 M sucrose solution, suspended on a fine platinum loop (1 mm diameter) at the end of a stick, is introduced into the cryobowl. This droplet remains in a fluid state for sufficient time to collect the frozen sections. The lower surface of the droplet is touched gently to the frozen sections, which stick to the sucrose. The correct timing, from insertion of the loop into the cryobowl to contact with the sections, will be apparent after several trials. If the sucrose droplet is too warm it can stick to the plastic of the diamond knife, while if it is too cold then the sections are disrupted in the pick-up process. A sucrose droplet of 2·3 M has a high surface tension and is suitable for structurally stable sections. However, in order to preserve sections of lightly fixed material a sucrose gelatin droplet (2M sucrose + 0·75 per cent gelatin) is sometimes used. Here the gelatin serves to reduce the surface tension acting on the section when it melts on the droplet. It should be noted, however, that sections do not flatten as easily on the surface of such a droplet.

When a droplet with sections on its lower surface is brought out of the cryobowl, it immediately becomes opaque as it is frosted by air moisture and then becomes clear again as it warms up to room temperature. The droplet plus attached sections is left for approximately 1–2 min at room temperature to ensure that the sections are completely flattened. The sections are then attached to alcohol-cleaned slides by touching the lower surface of the droplet to the slide. The sections attach firmly to the glass surface and it is only when sections of > 2 μm in thickness are used that a subbing solution for the slides, such as chrome–alum gelatin, is required to ensure adhesion of the sections.

G. Subsequent Treatment of Sections

After transfer to the surface of glass slides (multiwell immunofluorescence slides may be used), the sections remain covered by a small drop of sucrose from the pick-up loop. The position of the sections on the slide is marked by 'ringing' them using a diamond marker. A further droplet of 2·3 M sucrose is added to the existing droplet covering the sections and a coverslip is placed over them. They are then examined in a phase contrast microscope to check that they have retained their morphology and are properly flattened on the glass surface. Slides are then stored in a moist chamber until required. Prior to staining, the coverslips are soaked off in phosphate-buffered saline.

REFERENCES

1. Tokuyasu K. T. Immunochemistry on ultrathin frozen sections. *Histochem. J.*1980, **12**, 381–403.
2. Bourguignon L. Y. W., Tokuyasu K. T. and Singer S. J. The capping of lymphocytes and other cells, studied by an improved method of immunofluorescence staining of frozen sections. *J. Cell. Physiol.* 1978, **95**, 239–257.
3. Beyer E. C., Tokuyasu K. T. and Barondes S. H. Localization of an endogenous lectin in chicken liver, intestine and pancreas. *J. Cell Biol.* 1979, **82**, 565–571.
4. Franke W. W., Schmid E., Grund C., Müller H., Engelbrecht I., Moll R., Stadler J. and Jarasch E. D. Antibodies to high molecular weight polypeptides of desmosomes: specific localization of a class of junctional proteins in cells and tissues. *Differentiation* 1981, **20**, 217–241.
5. Tokuyasu K. T. A technique for ultracryotomy of cell suspensions and tissues. *J. Cell Biol.* 1973. **57**, 551–565.
6. Tokuyasu K. T. and Singer S. J. Improved procedures for immunoferritin labeling of ultrathin frozen sections, *J. Cell Biol.* 1976, **71**, 894–906.

7. Wartiovaara J., Linder E., Ruoslahti E. and Vaheri A. Distribution of fibroblast surface antigen. Association with fibrillar structures of normal cells and losses upon viral transformation. *J. Exptl Med.* 1974, **140**, 1522–1533.

8. Richards R. C. The use of trypsin to improve the localisation of immunoglobulins in semi-thin frozen sections of the mammary gland. *Histochem. J.* 1984, **16**, 565–572.

9. Baumgartner B., Tokuyasu K. T. and Chrispeels M. J. Localization of vicilin peptohydrolase in the cotyledons of mung bean seedlings by immunofluorescence microscopy. *J. Cell Biol.* 1978, **79**, 10–19.

10. Geiger B., Tokuyasu K. T. and Singer S. J. The immunocytochemical localization of α-actinin in intestinal epithelial cells. *Proc. Natl Acad. Sci. USA* 1979, **76**, 2833–2837.

11. Tokuyasu K. T., Slot J. W. and Singer S. J. Simultaneous observations of immunolabeled frozen sections in LM and EM. *Ninth International Congress of Electron Microscopy* 1978, **2**, 164–165.

12. Semoff S. Light and electron microscopic immunocytochemistry combined through cryomicrotomy. *Opt. Electron Microscopy* 1984, **10**, 4–7.

13. Kraehenbuhl J. P. and Jamieson J. D. Solid-phase conjugation of ferritin to Fab-fragments of immunoglobulin G for use in antigen localization on thin sections. *Proc. Natl Acad. Sci. USA* 1972, **69**, 1771–1775.

14. Kyte. J. Immunoferritin determination of the distribution of (Na^+, K^+) ATPase over the plasma membranes of renal convoluted tubules, I. Distal segment. *J. Cell Biol.* 1976, **68**, 287–303.

15. Painter R. G., Tokuyasu K. T. and Singer S. J. Immunoferritin localization of intracellular antigens: the use of ultracryotomy to obtain ultrathin sections suitable for direct immunoferritin staining. *Proc. Natl Acad. Sci. USA* 1973, **70**, 1649–1653.

16. Geuze H. J., Slot J. W. and Tokuyasu K. T. Immunocytochemical localization of amylase and chymotrypsinogen in the exocrine pancreatic cell with special attention to the Golgi complex. *J. Cell Biol.* 1979, **82**, 697–707.

17. Hühn A. and Nairn R. C. A nuclear staining artifact in immunofluorescence. *Clin. Exp. Immunol.* 1967, **2**, 697–700.

18. Christensen A. K. Frozen thin sections of fresh tissue for electron microscopy with a description of pancreas and liver. *J. Cell Biol.* 1971, **51**, 772–804.

19. Bernhard W. and Viron A. Improved techniques for the preparation of ultrathin frozen sections. *J. Cell Biol.* 1971, **49**, 731–746.

20. Frazer T. W. The anti-static pistol as an aid to ultrathin sectioning. *J. Microsc.* 1975, **106**, 97–99.

7   *Radioimmunocyto-chemistry*

S. P. Hunt, J. Allanson and P. W. Mantyh

Interest in autoradiographic approaches to immunocytochemistry has developed in response to several demands. Foremost was the need for a means of quantifying data, but equally important was the possibility of having a second type of immunological label with which to investigate the coexistence of antigens or multiple interactions at the ultrastructural level.

Two immunocytochemical techniques are widely used to localize specific antigens within nervous tissue. These are the indirect fluorescence technique of Coons[1], which consists essentially of identifying the primary antibody, with a fluorescent second antibody and the peroxidase–anti-peroxidase (PAP) modification of the immunoperoxidase method introduced by Sternberger.[2] The PAP technique involves joining the primary antibody to a large peroxidase–anti-peroxidase complex with a second linking antibody. The peroxidase marker can be visualized by reacting with hydrogen peroxide and a number of chromogens, most commonly diaminobenzidine, which produces a reaction product that can be viewed with both light and electron microscopy. Radioactive ligands have received some attention,[3] but have not generally been taken up, primarily because of their poor resolution at the light and electron microscopical levels, and because of the lack of commercially available labelled markers. However, the production of an internally labelled monoclonal antibody to substance P has recently been reported[4] and its use demonstrated in double-labelling experiments at both light and electron microscopical levels. Unfortunately, the limited availability of monoclonal antibodies and the additional problem of obtaining internally labelled antibodies of high specific activity limits the use of this technique at present.

More recently, tritiated[5] or iodinated[6] anti-rabbit immunoglobulins have been used but these suffer again from either lack of availability or a limited shelf-life (the half-life of ^{125}I is approximately 60 days).

We investigated the possibility of using radioactive ligands to identify tissue antigens[7] because of the ease with which a photographic record can be quantified. Using either biotinylated immunoglobulins or biotinylated protein A coupled sequentially to a tritiated biotin–avidin complex, it was possible to label a variety of antibodies in animal and human nervous tissue and to detect them autoradiographically at both light- and electron-microscopical levels. When coupled with tritium-sensitive film, a semiquantitative approach also proved possible. Further investigation revealed that, with suitable blocking techniques, tritiated biotin and avidin–peroxidase can be used to identify different antigens within the same tissue section. In the experiments described below we used these techniques to locate and quantitate peptide levels in the brain and spinal cord.

1. TISSUE PREPARATION

1.1. Light Microscopy

Rats (Sprague-Dawley 250 g) were anaesthetized with chloral hydrate and perfused intracardially with 400 ml of a modified[8,9] 4 per cent paraformaldehyde solution in 0·1 M sodium phosphate buffer (PB) pH 7·4 containing lysine (3·4 g/l) and sodium periodate (0·55 g/l) (PLP) at room temperature. Tissue was removed, postfixed for 2–4 h, washed overnight in PB with 30 per cent sucrose and sectioned the following day on a sliding microtome at 30 μm. Blocks of human tissue were obtained 1–3 days postmortem and immersion fixed for 7 days in PLP at 4 °C. Following a wash in 30 per cent sucrose in PB for 1–3 days tissue was either cut on a freezing microtome or embedded in paraffin wax and sectioned at 20 μm on a rotary microtome. All human tissue sections were passed through 1 per cent hydrogen peroxide in PB for 10 min before processing, as previously described.[10]

1.2. Electron Microscopy

Rats were perfused with 300 ml of 4 per cent paraformaldehyde in 0·1 M PB plus 0·1 per cent glutaraldehyde (EM grade) followed by 400 ml of PLP at room temperature. Tissue was removed, postfixed for 2 h in PLP, washed overnight in PB plus 8 per cent D-glucose and cut the following day on an MSE freezing microtome at 20–30 μm as previously described.[8]

1.3. Other Treatments

Some rats had received 30–150 μg of colchicine in 5 μl of sterile saline 24 h before perfusion. Injections were made either directly into the hindbrain, trigeminal ganglion or into the cerebrospinal fluid.

Others received topical applications (30 min) of 49 mM capsaicin or vinblastin (1–0·1 per cent) to the sciatic nerve 1–4 months before sacrifice.

2. IMMUNOCYTOCHEMICAL TECHNIQUES

The primary antibody was identified with a biotinylated anti-immunoglobulin, followed by an avidin–[^{3}H]biotin complex or by biotinylated protein A and the same complex (*see Fig. 7.1*). This was followed by development of the autoradiograph. At electron microscopical level the first method was used. Double labelling with immunoperoxidase for one antigen followed by radiolabelling for a second antigen was also used. The methods are described fully in the Appendix.

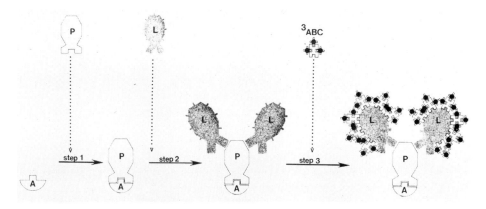

*Fig.*7.1. The basic [^{3}H]biotin technique described in the text. Step 1 represents the incubation of tissue containing antigen (A) with the primary antibody (P). Step 2 involves incubating with either biotinylated protein A or biotinylated anti-rabbit 1gG as the linking agent (L) between Step 1 and Step 3. Step 3 is the incubation with the [^{3}H]biotin–avidin complex (^{3}ABC). The central H-shaped structure represents avidin, the filled circles [^{3}H]biotin. Alternatively Step 3 could have been incubation with avidin–HRP for an electron-dense reaction product when finally incubated with DAB and hydrogen peroxide.

3. ANALYSIS OF DATA

Radiolabelled tissue was examined with light- and dark-field microscopy. LKB Ultrofilm film was printed to produce a direct but reversed image, or analysed either with a computer-controlled densitometry system previously described,[11] or manually using the photocell voltage output of a Leitz Vario Orthomat camera system as a measure of relative absorbance.[12]

4. RESULTS

4.1. Light Microscopy

The distribution of silver grains reflecting the binding of [^{3}H]biotin to a number of antibodies, closely matched the distribution previously reported for these antigens using either indirect fluorescence or immunoperoxidase techniques (*Figs.* 7.2 and 7.3.). However, the relative sensitivities of the two methods were

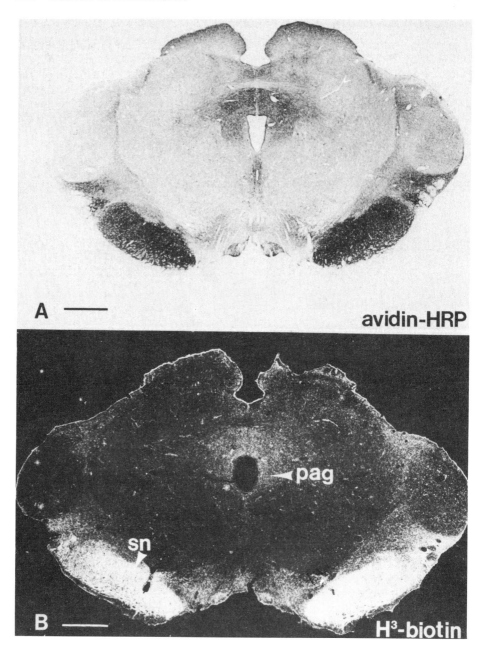

Fig. 7.2. Immunocytochemical detection of substance P (SP) in sections of the rat midbrain: (A) using avidin–HRP complex, (B) using [³H]biotin–avidin complex. The second antibody in each case was biotinylated anti-rabbit IgG. In both cases the substantia nigra (sn) and the periaqueductal grey (pag) are labelled. (A) Bright-field micrograph, scale bar = 850 μm; (B) dark-field micrograph, scale bar = 850 μm. (Reproduced from Hunt and Mantyh[7] with permission of authors and publishers.)

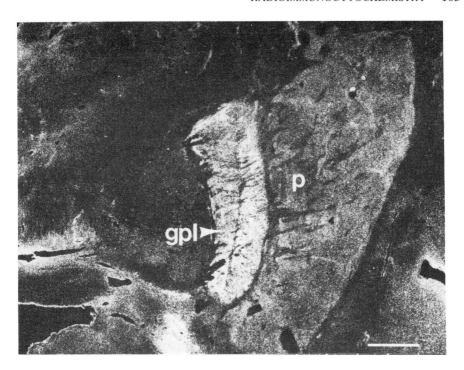

Fig. 7.3. [³H]Biotin used to detect the presence of methionine-enkephalin within the basal ganglia of the human brain. The section was apposed to LKB tritium-sensitive film and is a contact print taken from the LKB film. Notice the islands of immunoreactivity within the putamen (p) of the human brain and the heavy staining of the globus pallidus (gpl). Scale bar = 3 mm. (Reproduced from Hunt and Mantyh[7] with permission.)

difficult to compare directly. Because of the low energy of the β radiation emitted from a tritium source, only the radioactivity from approximately the superficial 2 μm of tissue register on the photographic emulsion[13]. Therefore, individually stained fibres were rarely outlined in the Golgi-like fashion seen with the PAP technique which stains throughout the section, although individually labelled perikarya could be seen (*see Fig.* 7.7.). Background levels varied, but were higher with biotinylated anti-rabbit IgG than with biotinylated protein A. Careful control of the initial concentration of the primary antibody minimized background problems, and antibody dilutions were in the range generally used for the PAP technique, or greater.

4.2. Semiquantitative Analysis

Use of LKB tritium-sensitive Ultrofilm[14] greatly simplified the analysis of large areas of tissue, such as those from postmortem human brain (*Fig.* 7.3.). In a series of animals in which the sciatic nerve was cut and ligated or had been exposed for 30 min to either of the drugs capsaicin (49 mM) or vinblastin (Velbe-Lilly) 0·1–1 per cent and allowed to survive 1–4 months, we were able to detect a unilateral drop in substance P-like immunoreactivity within the dorsal horn which depended upon both survival time and treatment (*Table* 7.1, *Figs.* 7.4 and 7.5).

Densitometric measurements were initially made with an automated system previously described. More recently we have modified the Leitz Vario Orthomat camera system to give measurements of relative absorbance (defined as the reciprocal of transmittance[15]). Readings were only taken over a predetermined range of film densities where the response of the emulsion is related in a linear fashion to the amount of radioactivity present in the section.[12]

Table 7.1. **Changes following sciatic nerve manipulation**

| | SPLI | | FRAP |
Reagent	4 week percentage drop	16 week percentage drop	(after 16 weeks)
Vinblastin			
1 per cent	81·2 ± 0·5	38·3 ± 4·6	Total loss
0·1 per cent	72·0 ± 2·9		Total loss
0·01 per cent	16·0 ± 3·1	n.c.	Partial loss
Capsaicin 46 mM	39·6 ± 2·7	34·0 ± 3·6	Total loss
Nerve section		46·5 ± 3·1	Total loss

The percentage loss (with ± s.e.m.) of substance P-like immunoreactivity (SPLI) following topical sciatic nerve treatment with vinblastin, capsaicin or nerve section and short (4 weeks) or long (16 weeks) survival times estimated from measurements of absorbance of LKB Film exposed for 1 week, i.e. *Fig.* 7.5, 3 sections from at least 3 animals were used for each value. These figures fit well with published data using radioimmunoassay.[33,34] The long-term loss of a second primary afferent marker fluoride-resistant acid phosphatase (FRAP) is also indicated (from J. Allanson and S. P. Hunt, 1983, unpublished observations).
n.c. = no change.

4.3. Electron Microscopy

Material taken from the rat substantia nigra was stained for substance P (*Fig.* 7.6) and, from the globus pallidus and spinal trigeminal nucleus, stained for enkephalin-like immunoreactivity using the [³H]biotin technique. Serial 1-μm sections dipped and examined by light microscopy indicated a rather restricted penetration of antibodies and/or tritiated marker complex limited to the superficial 2–3 μm. Thin sections (gold) taken from this superficial region and exposed for 70 days were lightly labelled with a higher concentration of silver grains on the most superficial sections. Background levels of silver grains over regions of the sections outside the tissue were extremely low and usually not more than one or two silver grains per 50 μm² grid square. The internal structure of the labelled profile was not obscured by the reaction product and could be observed easily.

4.4. Double Labelling

Tissue sections were stained for a number of antigens in an analysis of possible coexistence. These included tyrosine hydroxylase (TH), avian pancreatic polypep-

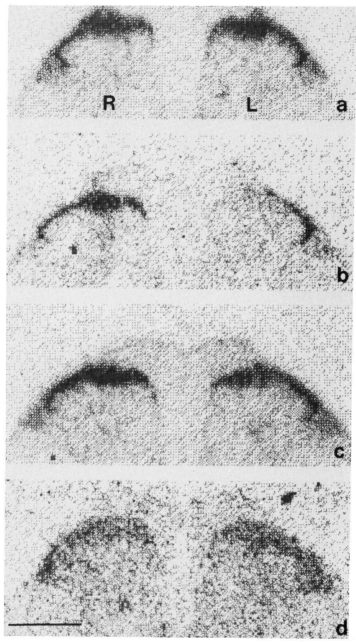

Fig. 7.4. Computer printout of LKB film images of [³H]biotin-labelled rat lumbar spinal cord sections. (*a*) Control rat, distribution of substance P; (*b*) ipsilateral loss of substance P one month after topical application of 1 per cent vinblastin to the sciatic nerve; (*c*) less pronounced ipsilateral loss of substance P, 1 month after topical application of 49 mM capsaicin to the sciatic nerve; (*d*) same rat as (*b*), but demonstrating the unchanged distribution of methionine-enkephalin-like immunoreactivity within the dorsal horn. L = drug-treated side; R = vehicle-treated side. Scale bar = 0·5 mm. (J. Allanson and S. P. Hunt, 1983, unpublished observations.)

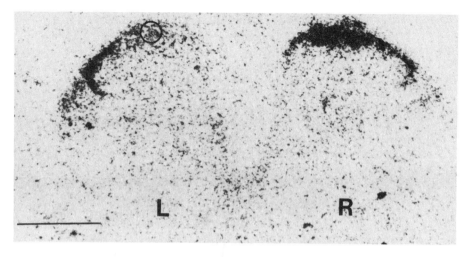

Fig. 7.5. A positive print of LKB film showing loss of substance P immunoreactivity from the lumbar dorsal horn of a rat following 1 month topical application of 1 per cent vinblastin. Relative absorbance measurements were made from the area indicated by the circle using a Leitz Vario Orthomat camera system.

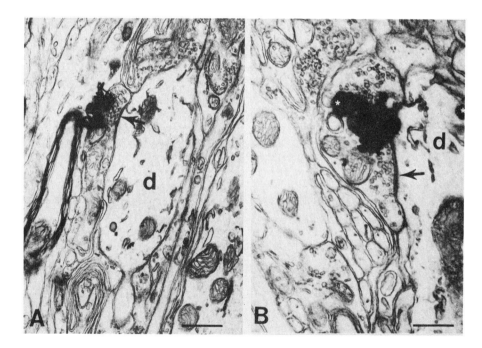

Fig. 7.6. Electron microscopical analysis of substance P-like immunoreactivity labelled with [³H]biotin. The second antibody was biotinylated anti-rabbit IgG. (*A*) and (*B*) are both taken from the rat substantia nigra. Presumed synaptic contacts indicated by arrows. d = dendrite, white asterisks = non-specific precipitate. Scale bar in (*A*) = 0·3 μm. Scale bar in (*B*) = 0·3 μm. (Reproduced from Hunt and Mantyh[7] with permission of authors and publishers.)

tide (APP),[16] now recognized as neuropeptide Y(NPY), substance P and somatostatin.[17] It was essential that the peroxidase reaction was performed first and the second antibody identified with [³H]biotin (*see* Section 5). Reversal of this order resulted in complete cross-over, so that the peroxidase reaction product was associated with both tissue antigens. It was possible to demonstrate minimal overlap between substance P-positive axons and the TH-positive neurones of the pars compacta of the substantia nigra and we were able to confirm the existence of APP (NPY) and TH within the A1–A3 medullary cell group (*Fig .7.7*) and of somatostatin and APP in neurones of the rat cerebral cortex (*Fig. 7.7*). It was also noted that neurones containing APP and somatostatin were found within the striatum and, to a lesser extent, within the olfactory tubercle.[18] We were unable to find coexistence of [Met⁵]enkephalin and APP within striatal neurones or between somatostatin and substance P in colchicine-treated rat trigeminal ganglion neurones (*Fig. 7.7*). Control material in which the second primary antibody had been omitted did not contain neurones heavily labelled with [³H]biotin although, in some cases, a slightly above background silver grain count was observed over peroxidase-positive neurones (*Fig.7.7*).

5. DISCUSSION

We have described a number of radioimmunocytochemical techniques based around the biotin–avidin system previously introduced as a more sensitive successor to the PAP technique.[19–21] Coupling a biotinylated second antibody or protein A to the primary antibody has allowed us to substitute a [³H]biotin complex for the usual avidin–peroxidase stage. We chose to pre-incubate the [³H]biotin with avidin to make the so-called, 'ABC complex' as a shortcut to incubating avidin at the third step and [³H]biotin as a fourth and final stage. In preliminary experiments we found that the four-stage procedure was equally effective as the ABC three-stage method, although 10 times lower concentrations of [³H]biotin were required in the four-stage procedure.

The mechanisms underlying the success of this approach are only partly understood. Avidin, a 68-kdal glycoprotein, has an affinity of 10^{-15} M^{-1} for the vitamin biotin. Each avidin molecule has four binding sites for biotin and it must be assumed that the formation of the ABC complex leaves the avidin molecule with some biotin-binding sites free to bind to the biotinylated second antibody or to biotinylated protein A. Protein A is a protein capable of binding to the Fc portion of a variety of immunoglobulins from a number of different species.[22–24]

The use of double-labelling procedures to localize two antigens within the same neurone has been hampered by the lack of sufficiently distinct immunological markers. Successive staining of sections with chemical elution[25] of the first antibody series has been widely used, but is limited by the necessity for a photographic image as the only remaining record of the first antibody binding site, by the small area of tissue which can be sampled and by the sensitivity of some antigens to the chemicals used for elution. Staining of adjacent tissue sections with different antibodies avoids most of these problems but necessitates the use of thin cryostat sections with the resulting reduction in antigenic sites.[26] Two-colour antigen labelling techniques have received some attention, but clear distinctions

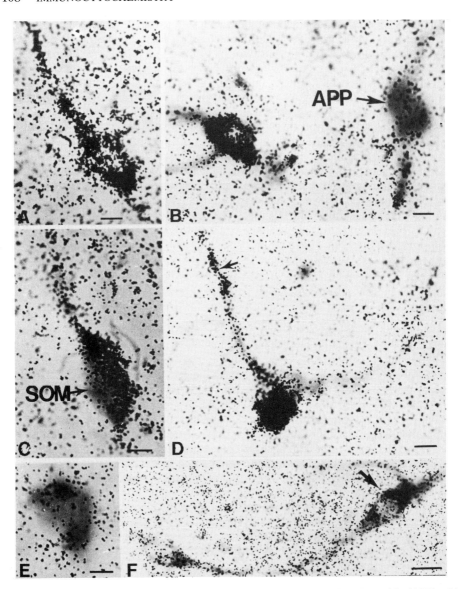

Fig. 7.7. (*A*) Neurone within the striatum labelled for avian pancreatic polypeptide (APP) with [³H]biotin using biotinylated protein A as the link antibody. Scale bar = 8 μm. (*B*) A single (arrow) and double-labelled neurone within the A1–A3 cell group of the rat medulla. APP was localized with avidin peroxidase, tyrosine hydroxylase with [³H]biotin. Scale bar = 12 μm. (*C*) Cortical neurone labelled for somatostatin (SOM) using avidin–peroxidase and APP using [³H]biotin. Scale bar = 7 μm. (*D*) Double-labelled neurone stained for APP using avidin–peroxidase and tyrosine hydroxylase using [³H]biotin. The cell body and a portion of the distal dendrite (arrow) are labelled. These are the regions of the neurone closest to the surface of the section. Scale bar = 12 μm. (*E*) Cortical neurone stained for somatostatin with avidin–peroxidase but with the second primary antibody omitted. Scale bar = 8 μm. (*F*) APP-labelled neurones (arrow) within the A1–A3 cell group labelled with avidin–peroxidase but with the second primary antibody omitted. Scale bar = 12 μm. (Reproduced from Hunt and Mantyh[7] with permission of authors and publishers.)

between chromogens within a single neurone can be difficult to make,[27–29] and in the case of fluorescent markers demand that each antibody is raised in a different animal species.[30] This severely limits the number of antigens which can be investigated as the majority of antibodies are raised in rabbits.

[³H]Biotin and avidin–HRP are two immunological markers which can be easily distinguished at the light- (and electron-) microscopical level. In the design of this protocol, we found it essential to use a biotinylated protein A link rather than biotinylated anti-rabbit IgG. After suitable blocking (with unlabelled protein A and paraformaldehyde fixation) there appeared to be little crossover between the two antigen detection systems provided that the peroxidase reaction was performed first. As has been suggested from two-colour peroxidase labelling studies, the reaction with diaminobenzidine appears to effectively neutralize binding sites related to the localization of the first antigen.

5.1. Semiquantitative Methods

Combination of [³H]biotin labelling of antigens and tritium-sensitive Ultrofilm film[5] provides a rapid and reproducible method for semiquantitative immunocytochemical analysis.[31] At present it is not possible to give absolute levels of antigen, as this demands a knowledge about the relationship between the immunohistochemical detection of fixed tissue antigen and antigen levels measured by radioimmunoassay. Nevertheless, it is possible, as in the case of the spinal cord given here, to measure ratios of a particular antigen in different regions of the brain. The use of [³H]biotin in combination with Ultrofilm has indicated levels of peptide loss in the spinal cord with low variability between animals and which are comparable to the results obtained from radioimmunoassay.[32,33]

5.2. Limitations of the Method

In general use, the PAP technique is superior to the tritiated biotin immunocytochemical procedure for detecting a single antigen. This would be expected as tritium radioactivity penetrates, at most, some 2 μm from the source. Thus the photographic emulsion records only the most superficial immunoreactive profiles. The PAP technique or the indirect fluorescence techniques are capable of staining tissue sections throughout and thus provide a more detailed staining of neuronal processes. At the electron microscopical level, [³H]biotin labelling lacks the resolution of peroxidase-labelling techniques, but does have the advantage that labelled profiles can be studied without the presence of peroxidase reaction product which obscures many internal details of the labelled profile. However, the interpretation of autoradiographic results, particularly at the electron microscopical level, largely depends on statistical analysis, and the significance of silver grains must be assessed in terms of both tissue and photographic background levels of silver grains.[13] Such problems do not arise when analysing immunoperoxidase techniques at the electron microscopical level. Nevertheless, it seems that the use of [³H]biotin, particularly as a second probe, at the electron microscopical level may be of some importance. The potential of such a probe has been indicated by the introduction of internally tritium-labelled monoclonal antibodies against substance P,[4] but the present protocol has the advantage that it applies equally to a wide variety of currently available antibodies.

Acknowledgements

It is a pleasure to acknowledge the excellent technical assistance of Annette Bond and David Chapman, and Mary Wynn for typing the manuscript. We are grateful to Dr Tony Crowther and Mrs J. Smith for their assistance with computerized densitometry.

Appendix

I. IMMUNOCYTOCHEMICAL TECHNIQUES

The immunocytochemical techniques detailed below give, in our hands, the most reliable and consistent results. Numerous steps were varied to obtain the optimum conditions described here.

A. Light Microscopy with Biotinylated Anti-rabbit IgG and Avidin–[³H]Biotin Complex

All incubations were caried out in PB with 1 per cent sheep serum, 0·3 per cent Triton X-100 and 0·95 per cent sodium chloride.
1. Incubate in primary antibody for 1–3 days at 4 °C with continuous agitation.
2. Wash in PB 30 min at room temperature.
3. Incubate tissue in biotinylated anti-rabbit IgG(IgG-B) at 20 µg/ml of incubation mixture for 30 min at 37 °C.
4. Wash in PB for 30 min at room temperature.
5. Incubate in a solution containing 10 µg avidin and 10 µCi [³H]biotin in 1 ml of 0·1 M Tris buffer pH 7·4 previously mixed for 20 min. Incubate for 30 min at 37 °C.
6. Wash in 3 changes of PB 20 min each.
7. Fix in PLP for 10 min.
8. Wash in PB for 30 min.
9. Mount sections on subbed slides.
10. Dehydrate in alcohols, clear in xylene, rehydrate through alcohols to water and dry.
11. Dip in Ilford K-5 nuclear emulsion diluted 2 : 1 in distilled water.
12. Expose in the dark at 4 °C for 1–28 days.
13. Develop in Kodak D–19 for 3 min at 17 °C. Rinse.
14. Fix in Kodafix 5 min.
15. Wash for 30 min, counterstain with cresyl violet, dehydrate and differentiate in alcohols, clear in xylene and mount in DePeX.

B. Light Microscopy with Biotinylated Protein A and Avidin–[³H]Biotin Complex

All incubations carried out in PB with 1 per cent bovine serum albumin and 0·3 per cent Triton X-100.

1. Incubate in primary antibody for 1–3 days at 4 °C with continuous agitation.
2. Wash in PB for 30 min.
3. Incubate in 1 μg biotinylated protein A/ml of incubation mixture for 60 min at 37 °C.
4. Wash in PB for 30 min.
5. Incubate in a solution containing 10 μg of avidin and 10 μCi of [³H]biotin in 1 ml 0·1 M Tris buffer pH 7·4 premixed for 20 min beforehand. Incubate for 60 min at 37 °C.
6. Wash in PB and prepare for autoradiography as in Steps A 7–15 above.

C. Autoradiography with LKB Ultrofilm

Instead of dipping in nuclear emulsion, slides were placed in an X-ray cassette with LKB Ultrofilm. Cassettes were stored at −20 °C for up to 8 weeks and the film was developed in D-19 for 5 min at 19 °C, washed and fixed in Kodafix for 5 min. The film was washed, dried and contact printed for easy reference.

D. Electron Microscopy with Biotinylated Anti-rabbit IgG and Avidin–[³H]Biotin Complex

The incubation mixture was always 0·1 M PB with 0·02 per cent saponin, 0·95 per cent sodium chloride and 1 per cent normal sheep serum added.

1. Sections were incubated overnight in primary antiserum.
2. Wash in PB for 30 min at room temperature.
3. Follow Steps A 3–6 above.
4. Fix in 0·5 per cent glutaraldehyde in PB for 30 min.
5. Wash in PB, 3 changes of 10 min each, and cut out area of interest.
6. Osmicate in 1 per cent osmic acid with 1·5 per cent potassium ferrocyanide added for 1 h at room temperature.
7. Rinse, dehydrate in ethyl alcohols, clear in propylene oxide and flat-embed in Araldite.
8. Cut ultrathin sections from the surface of the block without cutting thick sections.
9. Coat with carbon. Cover with loop of Ilford L4 emulsion as previously described.[34,35] Expose in dark for 70 days at 4 °C. Develop in D-19, fix and rinse and post-stain in lead citrate.
10. View with electron microscope.

E. Peroxidase Immunocytochemistry with Avidin–Horseradish Peroxidase

This procedure is used as a partial check for the success of [³H]biotin–avidin labelling procedures with either biotinylated immunoglobulins or biotinylated protein A. The steps outlined below follow incubation in the biotinylated second antibody (Step A 3) or biotinylated protein A (Step B 3) which bind to the primary antibody.

1. Wash in PB for 30 min at room temperature.
2. Incubate in 10 μg of avidin–HRP per ml of incubation mixture for 1 h at 37 °C. (Alternatively, commercially available avidin–biotin–HRP complexes may be used.)
3. Wash in PB for 30 min at room temperature.
4. Incubate in 50 mg diaminobenzidine in 80 ml of PB with 12 μl hydrogen peroxide 100 vol. (30 per cent) added for 5 min at room temperature.
5. Wash in PB three times, 10 min each.
6. Mount sections on subbed slides and process for light microscopy by dehydrating, clearing and mounting in DePeX.

F. Double Labelling with Peroxidase and [³H]Biotin

In this procedure the peroxidase reaction to identify the first primary antibody is always done first, while the second primary antibody is visualized by using the radioactive biotin complex. Biotinylated protein A but *not* biotinylated immunoglobulin is used in this procedure. All incubations are made in PB with 1 per cent BSA, 0·95 per cent sodium chloride and 0·3 per cent Triton X-100.

1. Localize the first antibody using avidin–peroxidase immunohistochemistry. The first incubation in primary antibody was generally for 2–3 days.
2. Wash in PB 30 min.
3. Incubate in biotinylated protein A at 1 μg/ml incubation mixture for 1 h at 37 °C.
4. Wash in PB, 30 min.
5. Incubate in avidin–HRP at 10 μg/ml incubation mixture for 1 h at 37 °C.
6. Wash in PB 30 min.
7. React with DAB (80 mg/100 ml) and 30 per cent hydrogen peroxide, 5 μl/100 ml for 5 min.
8. Wash in PB, 3 changes 10 min each.
9. Incubate in 100 μg/ml of protein A in PB at 37 °C for 10 min.
10. Rinse in PB.
11. Fix in PLP for 20 min.
12. Wash and incubate in the second primary antibody for 2–3 days.
13. Steps B2–6 are subsequently followed to tag the second primary antibody with [³H]biotin. Sections are dipped in nuclear emulsion and exposed for 2 weeks.

G. Controls

Controls include omission of the primary antibody or the use of pre-absorbed serum, generally with 50 μg of native peptide per ml of diluted antiserum. In double-labelling experiments, the second antibody was either omitted or absorbed to check for cross-over between the two primary antibodies.

II. REAGENTS

[^{3}H]Biotin (d-(8,9-[^{3}H](N)-) (41·4 Ci/mmol) was obtained from NEN Chemicals, biotinylated protein A, biotinylated anti-rabbit immunoglobulin, avidin-D and avidin-D–HRP were all obtained from Vector through their UK suppliers, Sera Labs Ltd. Protein A was obtained from Sigma; Ultrofilm was obtained from LKB (Sweden).

The substance P antibody recognized the C-terminal portion of the peptide and was a gift from P. C. Emson. The antibody against [Met^5]enkephalin was a gift from V. Clement Jones, the antibody against avian pancreatic polypeptide (APP) was a gift from J. Kimmel and the antibody against tyrosine hydroxylase (TH) obtained from M. Goldstein. Somatostatin antibody was obtained from Immunonuclear, Stillwater, Minn., USA.

REFERENCES

1. Coons A. H. Fluorescent antibody methods. In: Danielli, J. F., ed., *General Cytochemical Methods*. New York, Academic Press, 1958: 399–422.
2. Sternberger L. A. *Immunocytochemistry*, 2nd ed. New York, Wiley & Sons, 1979.
3. Larsson L-I. and Schwartz T. W. Radioimmunocytochemistry — a novel immunocytochemical principle. *J. Histochem. Cytochem.* 1977, **25**, 1140–1146.
4. Cuello A. C., Priestley J. V. and Milstein C. Immunocytochemistry with internally labelled monoclonal antibodies. *Proc. Natl Acad. Sci. USA* 1982, **79**, 665–669.
5. Glaser E. J., Ramachandran J. and Basbaum A. Radioimmunocytochemistry using a tritiated goat anti-rabbit second antibody. *J. Histochem. Cytochem.* 1984, **32**, 778–782.
6. McLean S., Skirboll L. R. and Pert C. B. Co-distribution of leu-enkephalin and substance P in the brain: mapping with radioimmunohistochemistry. *Soc. Neurosci. Abs.* 1983, **9**, 129.13.
7. Hunt S. P. and Mantyh P. W. Radioimmunocytochemistry with [^{3}H]biotin. *Brain Res.* 1984, **291**, 203–217.
8. Hunt S. P., Kelly J. S. and Emson P. C. The electron microscopic localisation of methionine-enkephalin within the superficial layers I and II, of the spinal cord. *Neuroscience* 1980, **5**, 1871–1890.
9. McLean I. W. and Nakane P. K. Periodate–lysine–paraformaldehyde fixative. A new fixative for immunoelectron microscopy. *J. Histochem. Cytochem.* 1974, **22**, 1077–1083.
10. Hunt S. P., Rossor M. N., Emson P. C. and Clement-Jones V. Substance P and enkephalins in spinal cord after limb amputation. *Lancet* 1982, **i**, 1023.
11. Arndt U. W., Barrington Leigh J., Mallet J. F. W. and Twinn K. E. A mechanical microdensitometer. *J. Phys. [E]* Series 2, Vol. 2, 1969, 385–387.
12. Mantyh P. W., Hunt S. P. and Maggio J. E. Substance P receptors: localization by light microscopic autoradiography in rat brain using [^{3}H]SP as the radioligand. *Brain Res.* 1984, **307**, 147–165.
13. Rogers A. W. *Techniques of Autoradiography*. Elsevier, North Holland, Biomedical Press, 1979.
14. Ehn E. and Larsson B. Properties of an antiscratch-layer-free X-ray film for autoradiographic registration of tritium. *Sci. Tools* 1979, **26**, 24–29.
15. Kuhar M. J. and Unnerstall J. R. Quantitative receptor mapping by autoradiography: some current technical problems. *Trends Neurosci.* 1985, **8**, 49–53.
16. Hunt S. P., Emson P., Gilbert R., Goldstein M. and Kimmel J. R. Presence of avian pancreatic polypeptide-like immunoreactivity in catecholamine and methionine-enkephalin containing neurons within the central nervous system. *Neurosci. Lett.* 1981, **21**, 125–130.
17. Hökfelt T., Elde R., Johansson O., Luft R., Nilsson G. and Arimura A. Immunohistochemical evidence for separate populations of somatostatin and substance P-containing primary afferent neurons. *Neuroscience* 1976, **1**, 131–136.
18. Vincent S. R., Skirboll L., Hökfelt T., Johansson O., Lundberg J. M., Elde R. P. Terenius L. and Kimmel J. Coexistence of somatostatin and avian pancreatic polypeptide-like immunoreactivity in some forebrain neurons. *Neuroscience* 1982, **7**, 439–446.

19. Hsu S-M., Raine L. and Fanger H. Use of avidin–biotin–peroxidase complex (ABC), in immunoperoxidase techniques: a comparison between ABC and unlabelled antibody (PAP), procedures. *J. Histochem. Cytochem.* 1981, **29**, 577–580.
20. Hsu S-M. and Raine L. Protein A, avidin and biotin in immunohistochemistry. *J. Histochem. Cytochem.* 1981, **29**, 1349–1353.
21. Guesdon J. L. Ternynck T. and Avrameas S. The use of avidin–biotin interaction in immunoenzymatic techniques. *J. Histochem. Cytochem.* 1979, **27**, 1131–1139.
22. Forsgren A. and Sjöquist J. 'Protein A' from *S. aureus*. I. Pseudoimmune reaction with human γ-globulin. *J. Immunol* 1966, **97**, 822–827.
23. Forsgren A. and Sjöquist J. Protein A from *S. aureus*. III. Reaction with rabbit γ-globulin. *J. Immunol*. 1967, **99**, 19–24.
24. Notani G. W., Parsons J. A. and Erlandsen S. L. Versatility of *Staphylococcus aureus* protein A in immunocytochemistry. *J. Histochem. Cytochem.* 1979, **27**, 1438–1444.
25. Tramu G., Pillez A. and Leonardelli J. An effective method of antibody elution for the successive or simultaneous localization of two antigens by immunocytochemistry. *J. Histochem. Cytochem.* 1978, **26**, 322–324.
26. Lundberg J. M., Hökfelt T., Änggärd A., Kimmel J., Goldstein M. and Markey K. Coexistence of an avian pancreatic polypeptide (APP), immunoreactive substances and catecholamines in some peripheral and central neurons. *Acta Physiol. Scand.* 1980, **110**, 107–109.
27. Hunt S. P. and Lovick T. A. The distribution of 5-HT, Met-enkephalin and β-lipotropin like immunoreactivity in neuronal perikarya in the cat brain stem. *Neurosci. Lett.* 1982, **30**, 139–140.
28. Joseph S. A. and Sternberger L. A. The unlabeled antibody method. Contrasting colour staining of β-lipotropin and ACTH-associated hypothalamic peptides without antibody removal. *J. Histochem. Cytochem.* 1979, **27**, 1430–1437.
29. Sternberger L. A. and Joseph S. A. The unlabeled antibody method. Contrasting colour staining of paired pituitary hormones without antibody removal. *J. Histochem. Cytochem.* 1979, **27**, 1424–1429.
30. Erichsen J. T., Reiner A. and Karten H. J. Co-occurrence of substance P-like and Leu-enkephalin-like immunoreactivities in neurones and fibres of avian nervous system. *Nature* 1982, **295**, 407–410.
31. Palacios J. M., Niehoff D. L. and Kuhar M. J. Receptor autoradiography with tritium-sensitive film: potential for computerized densitometry. *Neurosci. Lett.* 1981, **25**, 101–105.
32. Gamse R., Petsche U., Lembeck F. and Jansco G. Capsaicin applied to peripheral nerve inhibits axoplasmic transport of substance P and somatostatin. *Brain Res.* 1982, **239**, 447–462.
33. Jessell T., Tsunoo A., Kanazawa I. and Otsuka M. Substance P depletion in the dorsal horn of rat spinal cord after section of peripheral processes of primary sensory neurons. *Brain Res.* 1979, **168**, 247–259.
34. Caro L. G. and Van Tubergen R. P. High resolution autoradiography. I Methods. *J. Cell Biol.* 1962, **15**, 173–188.
35. Hunt S. P. and Schmidt J. The electron microscopic autoradiographic localisation of α-bungarotoxin binding sites within the central nervous system of the rat. *Brain Res.* 1978, **142**, 152–159.

8 The Preparation and Use of Gold Probes

J. De Mey

Immunocytochemical marking comprises the visualization and localization of antigens in cells and tissues.[1-11] The antigens are tagged by antibodies which can be visualized with appropriate *markers* attached directly or indirectly to them (through *linkers*). Examples of linkers for antigen-bound antibodies are secondary antibodies, protein A, antigens[11] or (strept)avidin (for biotinylated primary or secondary antibodies).

A variety of markers exist for light microscopy, transmission electron microscopy (TEM) and scanning electron microscopy (SEM). This chapter aims to focus on the colloidal gold marker system[12] and its use in electron microscopy. It is not intended to be a complete bibliography of the colloidal gold marker system because this has been provided in recent reviews,[13-22] and is being provided on a semi-annual basis by Janssen Life Sciences Products.[23] The essential steps in the establishment of the colloidal gold marker system are also covered in the above-mentioned review papers. The reader is further referred to reviews by Horisberger[13] and Roth[16] in which the chemistry and physics of metallic colloids are discussed.

Labelling will be used here exclusively to mean attaching the marker to the antibody or linker. *Probes* will be used for the antibody or linker/marker complexes. Terms such as 'marking', 'staining', 'localizing', 'visualizing', will be used for the binding to antigens and visualization of the probes.

1. THE COLLOIDAL GOLD MARKER SYSTEM

The spectacular growth of interest in the colloidal gold marker system is fully warranted by the following advantages:

a. Gold *sols* can be reproducibly and easily prepared in a range of sizes, making the system extremely flexible.

b. Gold *probes* are relatively easily prepared and purified, retain most of the binding activity of the unlabelled antibody or linker and are stable for a long time, particularly when stored frozen at $-20\,°C$ in the presence of at least 20 per cent glycerol.

c. Gold probes are very electron dense and have a characteristic appearance, which makes it virtually impossible to confuse them with biological structures.

d. Gold probes are capable of strong emission of secondary[24] and back-scattered[25] electrons. Thus, they are now considered as very useful in SEM marking techniques.

e. Gold probes absorb[16,26–29] or reflect[15] light or can be used for subsequent silver precipitation. Thus, they are applicable to a variety of light microscopical techniques[30] (*see* Chapter 5) and even immunoblotting.[31] (Immuno)gold–silver staining is now reportedly the most sensitive (immuno)cytochemical method available. In addition, it allows for the light microscopical visualization of gold particles, used in ultrastructural procedures. This makes it possible to correlate and control ultrastructural findings with light microscopy.

f. As recently reported, individual gold probes as small as 10 nm can be visualized by 'nanometer particle video-ultramicroscopy' (Nanovid microscopy) (*see* Chapter 5[44]).

All these new possibilities give the colloidal gold marker system an even greater flexibility and multivalency than enzymatic methods.

Some *disadvantages* of the gold marker system however, prevent it from becoming completely general; gold probes diffuse relatively badly and this restricts their use in pre-embedding localization procedures, although some success with cultured cells has been reported[29,32] and recently also in Vibratome sections of brain[33]. In addition, it is not yet possible to link any protein to gold particles. Proteins with low solubility in low ionic strength solutions, such as IgMs, have not yet been adsorbed successfully. Finally, our knowledge of the physical aspects of gold probes is still very limited.

In this chapter, I will also draw attention to some generally overlooked technical aspects of gold probes which have important implications for their use and applicability in cytochemistry.

2. THE PREPARATION OF COLLOIDAL GOLD SOLS

There are various methods of producing isodisperse gold sols of predefined sizes.[16] All methods are based on the controlled reduction of an aqueous solution of tetrachloroauric acid using a number of reducing agents under varying conditions (*see below*). According to Frens[34] all the gold ends up in the reduced form. This fact can be used to calculate for a known particle size the number of particles per unit volume and per unit absorbance at the peak of absorption (λ_{max}) or at 520 nm. Strong reducing agents, like white phosphorus, citrate–tannic acid or sodium borohydride induce far greater numbers of nuclei than the weaker agent, citrate. The available chloroauric acid will be incorporated in smaller, but far more

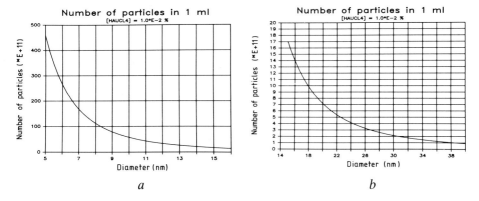

Fig. 8.1. The relationship between the number of particles per ml gold sol prepared from a 0·01 per cent HAuCl₄ solution and the diameter size for (*a*) the 5–15 nm and (*b*) 15–40 nm range.

numerous gold particles. *Fig.* 8.1 shows the relationship between the number of gold particles per unit volume (in the ordinate) and the mean diameter of the gold particles, for sols made from 0·01 per cent chloroauric acid. *Table* 8.1 gives the same relation calculated for the same sols, but expressed at an absorbance at 520 nm, $A_{520} = 1$. *For sols and probes having the same colour density, the number of particles decreases by a factor of approximately 8, for each doubling in size!* This relationship has important implications for both the amount of protein that is needed to protect the gold colloid against electrolyte-induced flocculation and the relative efficiency of smaller and larger gold probes in marking experiments. These implications will be further discussed below (*see* Sections 3.3, and 6.1).

Detailed instructions for making gold sols are given in the Appendix, Section I.

Table 8.1. The relationship between the number of particles of different sizes in 1 ml of gold sol prepared from 0·01 per cent HAuCl₄ and their absorbance

Diameter (nm)	No. particles/ml	A_{520}	No. particles/ml at $A_{520} = 1$
4·5	6.3×10^{13}	0·84	7.5×10^{13}
10·0	5.7×10^{12}	0·97	5.9×10^{12}
19·0	7.7×10^{11}	1·11	7.0×10^{11}
30·5	2.0×10^{11}	1·18	1.7×10^{11}
41·5	8.0×10^{10}	1·12	7.2×10^{10}

The correction only marginally influences the basic relationship in *Fig.* 8.1a,b.

3. THE PRODUCTION, PURIFICATION AND STORAGE OF GOLD PROBES

3.1. The Adsorption of Proteins to Colloidal Gold

The use of colloidal gold as a marker in cytochemical procedures is based on its physical characteristics and the fact that colloidal gold is stabilized against electrolyte-induced flocculation by macromolecules. Stabilization is due to the non-covalent adsorption of the macromolecules to the surface of the gold particles. The adsorption of proteins is influenced by particle size, ionic concentration (which is kept as low as possible and generally below 10 mM), the amount of protein added and its molecular weight. Above all, however, it is pH dependent.[35] In most cases, strong adsorption of macromolecules, resulting in stable probes, occurs at pH values close to or slightly basic to the isoelectric point of a given protein. At this pH interval, the zwitterion form of the protein is dominant and the interfacial tension maximal. This probably results in a stronger, more stable adsorption to the hydrophobic surface of gold particles. Adsorption of macromolecules is best under conditions allowing for maximal multipoint contact between the macromolecule and the edge surface of a good particle. These edges are thought to contain increased van der Waals' forces. Multipoint attachment of the macromolecule will result in irreversible adsorption. This can in our opinion be best achieved by using the minimal protecting amount of macromolecule, during the adsorption step, as opposed to an excess. Under optimal conditions of pH and ionic strength, each macromolecule will then be able to form a maximum number of contacts.

IgGs, such as affinity-purified polyclonal antibodies or chromotographically purified serum IgG fractions, present a special problem because their components display a broad range of isoelectric points (pI values) and have a variable tendency to form aggregates at low ionic strength. We have found empirically that for affinity purified polyclonal antibodies, pH 9–9·5 is optimal. Our unpublished results have shown that, at this pH, protection is obtained with the same amount as at neutral pH. Indeed, at pH 9, less than 60 per cent of the antibody added was adsorbed as opposed to more than 90 per cent at neutral pH. Thus, at pH 9·0, protection is obtained with fewer molecules per gold particle and this may lead to more stable binding.

Data for pH adjustments of colloidal gold for adsorption of various proteins, as published in the review of Roth,[16] are shown in *Table* 8.2.

3.2. Preparation of Protein Solutions and Colloidal Gold before the Adsorption Step

In general, proteins are dissolved in very dilute salt solutions or buffers or, if possible, in distilled water. They are centrifuged at 100 000 × g for 1 h just before use and the upper two-thirds of the supernatant are used. For most proteins, the pH is adjusted to ± 0·5 pH unit above their pI, provided they remain soluble.

Purified monoclonal and polyclonal IgGs present particular problems because they tend to form di- and oligomers. Their concentration must be kept below 1 mg/ml and they should not be lyophilized. If necessary, gel chromatography can be used to purify monomeric IgGs. The antibodies are dialysed against 2 mM borax buffer, pH 9·0 or higher[15,29] in which they have an increased solubility.

The pH of the colloidal gold is adjusted to ± 0·5 unit above the pI of the protein, or to pH ⩾ 9 in the case of affinity-purified polyclonal antibodies with 0·2 N K_2CO_3 or 0·2 M H_3PO_4. A gel-filled combination electrode (e.g. GX series, model 91–05, ORION Research, 380 Putnam Avenue, Cambridge, MA 02139, USA) is used to monitor the pH of the gold sol. Such an electrode has a

Table 8.2. *Data for pH adjustment of colloidal gold and proteins for complex formation*

Proteins	pH
Immunoglobulins (affinity-purified antibodies)	9·0
F(ab')$_2$	7·2
Protein A	5·9–6·2
Ricinus communis lectin I	8·0
Ricinus communis lectin II	8·0
Peanut lectin	6·3
Helix pomatia lectin	7·4
Soybean lectin	6·1
Lens culinaris lectin	6·9
Lotus tetragonolabus lectin	6·3
Ulex europaeus lectin I	6·3
Bandeiraea simplicifolia lectin	6·2
Mannan from *Candida utilis* or *Saccharomyces cerevisiae*	7
Horseradish peroxidase	7·2–8·0
Ovomucoid	4·8
Ceruloplasmin	7·0
Asialofetuin	6·0–6·5
Galactosyl bovine serum albumin	6·0–6·5
Bovine serum albumin	5·2–5·5
Peptide–bovine serum albumin conjugates	4·0–4·5
Insulin–bovine serum albumin conjugates	5·3
Cholera toxin	6·9
Tetanus toxin	6·9
DNAase	6·0
RNAase	9·0–9·2
Low density lipoprotein	5·5
α_2-Macroglobulin	6·0
Avidin (unmodified, egg white)	~ 10·0–10·6
Streptavidin	6·4–6·6
Wheat germ agglutinin	9·9

Modified from J. Roth.[16]

low electrolyte flow and, hence, does not induce gold flocculation at its surface, provided the gold sol is stirred.

3.3. Determination of the Minimal Protecting Amount of Protein

The minimal protecting amount is determined by constructing a concentration variable isotherm (CVAI) (*see* Appendix, Section II),[35] of which various types can be found in the literature.[14,36,37] If the optimal pH is unknown, one can determine the p*I* of the protein by analytical isoelectric focusing under non-denaturing conditions, and/or by constructing CVAIs at various pH values. *Fig*. 8.2 shows the determination of the minimal protecting amount of a goat secondary antibody at pH 9·0 for 6 nm and 15 nm gold made from the same amount of HAuCl$_4$. The larger amount of protein needed to protect the smaller

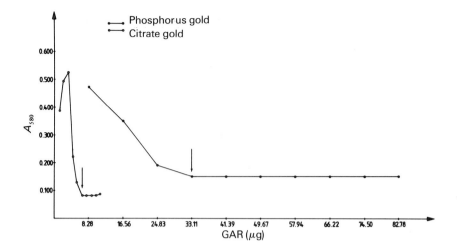

Fig. 8.2. Concentration variable isotherms for a goat secondary antibody, at pH 9·0 for 6–nm and 15–nm colloidal gold prepared from 0·01 per cent HAuCl₄. Note the larger amount of protein needed to protect the smaller gold particles.

gold particles can be explained by the fact that the higher number of gold particles (*see above*) presents a larger surface area.

3.4. Preparing, Purifying and Storing a Gold Probe

For a detailed review of different recipes, *see* the review of J. Roth.[16] The procedures used in the author's laboratory are outlined in the Appendix, Sections III–VI. For the preparation of a gold probe, the minimal protecting amount is simply added to the appropriate volume of suitably prepared gold.

The glassware used must be cleaned in the same way as for the preparation of a gold sol. The order of addition of the reagents has been claimed to be of importance. It is recommended to try both ways for each new application. Rapid stirring during mixing is important.

After a couple of minutes, a secondary stabilizer is added. This stabilizer is said to minimize possible aggregation of gold particles and to block remaining adsorption sites. Carbowax 20 M[24] is widely used. Polyvinyl pyrrolidone (mol. wt 10 000) in combination with Carbowax 20 M[35] and Tween 20[38] have also been satisfactory. Carbowax 20 M must be microfiltered (0·2 μm) before use. Bovine serum albumin (BSA), added to the stabilized gold sol after adjusting its pH to 9·0, can also be used for polyclonal and monoclonal IgGs,[15,29] streptavidin and protein A. BSA, however, is not an efficient stabilizer for particles larger than 25 nm.

Before a gold probe can be used, it must be purified and concentrated. Non-adsorbed protein, incompletely stabilized gold particles and aggregates of gold particles must be removed, particularly when the probe has to be used for electron microscopy. This is generally achieved by centrifugation. The first centrifugation (*see* Appendix for conditions) yields a loose, dark-red coloured sediment. Any tightly pelleted material at this stage should not be recovered: it contains too many aggregates. If most of the material is in this form, there will be a serious problem in obtaining a useful probe.

The supernatant, containing free protein, is carefully aspirated and discarded. The loose part of the sediment is decanted into a centrifuge tube and diluted with an appropriate buffer containing a suitable stabilizer(s). A suitable buffer is one compatible with (*a*) the solubility and binding activity of the adsorbed protein, (*b*) the storage conditions and (*c*) the use in marking experiments.

In order to remove essentially all unbound proteins, it is necessary to repeat the centrifugation twice. We have observed, however, that these additional centrifugations induce the formation of small aggregates, especially with immunoglobulin G–gold probes. These gold clusters are very disturbing and prevent the production of optimal staining results. It is therefore highly recommended to use the elegant and easy method recently introduced by Slot and Geuze for protein A[39] and IgGs.[18] After the initial concentration step of the stabilized gold probe, the decanted loose sediment is layered on top of a density gradient made with the final storage buffer and fractionated with velocity sedimentation. The important advantages of this method are: (*a*) it efficiently removes small clusters of aggregated gold particles; (*b*) it produces populations of gold particles with very homogeneous size distribution; (*c*) it leaves any remaining unadsorbed proteins in the supernatant; (*d*) it can give very good yields when the initial gold sol was already very homogeneous, as is the case for sols prepared by the citrate and the citrate–tannic acid methods.

It is recommended that the probes be standardized by diluting them with storage buffer to a predefined absorbance at 520 nm. Absorbances of 2·5–10 yield probes that can be diluted from 4 to 40 times in a number of applications.

Most often, the final probe is stored at 4–8 °C with the addition of 0·02 M NaN$_3$ in an airtight vial. The addition of glycerol to 20 per cent allows it to be frozen as small drops (25–50 µl) in liquid nitrogen and kept at −20 °C.[39] This increases the shelf-life of the probe to the same extent as keeping unlabelled proteins in the frozen state. After thawing, glycerol can be largely removed by dilution or by dialysis.

4. QUALITY CONTROL AND ANALYSIS OF GOLD SOLS AND PROBES

Each preparation of gold sol, as well as the gold probe, should be controlled for quality in a number of ways.

Absorption spectra of gold sols display a single peak of absorption (λ_{max}) in the visible wavelength spectrum ranging between 510 and 550 nm (*see* Fig. 8.3). With increasing mean particle diameter, the λ_{max} shifts to a longer wavelength. This causes a change in the colour from pale orange to dark purplish, and even brown.

Increasing heterogeneity in marker shape and size and the beginning of coagulation result in a broadening of the peak. Although absorption spectra only give approximate data, they provide a baseline from which deviations in the system can be assessed.[14,37]

Measuring the absorbance at λ_{max} or 520 nm can be used to calculate the yield of colloidal gold at different steps of the preparation of a gold probe.

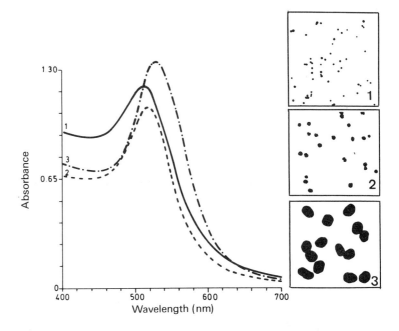

Fig. 8.3. Visible wavelength absorption spectra of (1) 5·5, (2) 17 and (3) 40–nm colloidal gold sols.

Measurement of the particle size from transmission electron micrographs of particles attached to Formvar-coated grids is an accurate way of determining the size distribution. Negatively stained catalase crystals can be used to calibrate the final magnification of the prints. Important parameters are not only the mean size, but also the coefficient of variance, and 95 per cent range.

The measurements can be performed manually or with an electronic image analyser (*see Fig.* 8.4). The data can be used to determine the degree of overlap between two different gold batches.

Artefactual clustering of gold particles (for example by air-drying) can be prevented by using Formvar–poly(L-lysine)-coated grids and a particle concentration that gives a relatively low particle density per unit surface. Such optimal preparations can be used to quantitatively evaluate the occurrence of gold clusters in a gold probe with electronic image analysis. Gold probes used for ultrastructural research should ideally consist of single, narrowly sized gold particles. In practice, probes containing more than 75 per cent singlets and less than 5 per cent triplets or larger aggregates are very well acceptable.

5. STABILITY OF GOLD PROBES

The stability of gold probes is in general very good (*see* Roth[16] for a complete review), although systematic data are scarce and often controversial. The bioactivity of the probes can be assessed by agglutination tests or indirect

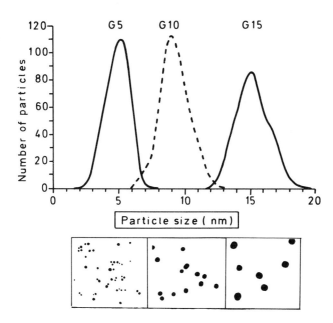

Fig. 8.4. Electron micrographs of 5, 10 and 15 nm gold sols and their respective size distributions determined with a video-image analyser (Cambridge Instr., Quantimet 900). Means and standard deviations are respectively 4·8 ± 0·9; 9·2 ± 1·1; 15·3 ± 1·8 nm.

radioassays.[14] The recently introduced application of gold probes in blot overlay assays[31] can also be used to measure their bioactivity. In such a test, 0·5 ml of a 1 : 25 dilution of the probe is reacted at room temperature until saturation with a nitrocellulose strip onto which spots, containing a dilution series of the corresponding antigen (25–0·1 ng per spot), have been blotted. Measurements of the red colour developed after reaction give interesting quantitative data. This system has the important advantage that it closely mimics conditions used in immunocytochemical experiments. Deviations in bioactivity can be monitored as a function of time and storage conditions. Using this approach, the working titre of a probe can also be monitored by reacting a dilution series of the probe with spots of antigen at a fixed concentration. Systematic work using this assay system is being conducted.

The bonds formed between the macromolecule and the gold marker are non-covalent in nature. It is therefore important to address the question of the stability of these bonds. From sparse data in the literature[37,40–43] and our own unpublished observations, it is clear that gold probes are not absolutely stable. Leakage of protein from the gold particles occurs to variable extents. There are no systematic data on this aspect and variations in the literature may be related to the characteristics of the protein and the conditions used for its adsorption to colloidal gold and subsequent storage. It is, however, not known yet how relevant leakage is for routine work. Experience has shown that gold probes perform well in many different marking experiments even when they have been mailed to distant places without cooling. There is also the bioactivity of the adsorbed molecule itself that, under normal storage conditions of gold probes (4–8 °C), will deteriorate with time. Storing the probes, supplemented with glycerol up to 20 per cent, in frozen aliquots is reported to increase their shelf-life.

Gold probes seem to be very 'stable' in experiments in which they are applied in excess over exposed immunoreactive sites, e.g. most on-grid stainings. Stability problems thus far seem to occur in applications in which such an excess is difficult to attain, e.g. most pre-embedding staining, where the number of immunoreactive sites exposed is much larger and the probes need to be used at high concentrations. With larger probes, an excess is also more difficult to obtain since, at the same colour density, they contain far fewer gold particles per unit volume. It is clear that in such situations relative loss of activity will show up much sooner. Relative freshness, e.g. not older than 5–6 months, and cold storage of the probe is then of importance.

6. CRITICAL EVALUATION OF THE USE OF GOLD PROBES IN SELECTED MARKING TECHNIQUES

6.1. Influence of the Size of the Gold Particles on Marking Efficiency

From experiments with iodinated IgGs, M. Moeremans and J. De Mey (1981, unpublished work) have determined that a 5–nm gold probe adjusted to an $A_{520} = 2 \cdot 5$ contains about 40 µg/ml IgG, while a 15–nm gold probe with the same colour density contains only 5 µg/ml. The approximate number of antibody molecules per gold particle was calculated to be 2 per particle for 6–nm gold and 6 per particle for 15–nm gold. These data, combined with the facts about the differences in the number of gold particles per unit volume at 520 nm (see Sections 2 and 5) can explain why gold probes used at the same A_{520} will progressively become less efficient in terms of marking density with increasing particle size.[18,39]

Smaller particles (3–10 nm) have the disadvantage that they are not as easily visible as slightly larger ones (10–19 nm). Two small sizes, e.g. 4 and 8 nm, are nevertheless the best for double marking experiments because their steric properties are still compatible with high resolution marking.

Which particle size is chosen by the investigator will depend on the particular needs of a localization problem. After all, one of the important advantages of the gold marker system is that it can be manufactured in different sizes. We believe, however, that for most applications particles smaller than 15 nm will be the best choice.

6.2. Gold Marking of Surface Components of Cells in Suspension and Monolayers

Cell surface marking was the first successful use of colloidal gold for both light and electron microscopy. Incubation conditions and controls are very similar to those used with other markers. The cells can be incubated in the living state, under conditions that inhibit capping and internalization, or after prefixation.

The material can be processed for thin sectioning, surface replicas[44] and negative staining[45,46] using standard techniques. For SEM, Horisberger et al.[13] recommend the use of non-coating methods to render the material conductive. Gold particles

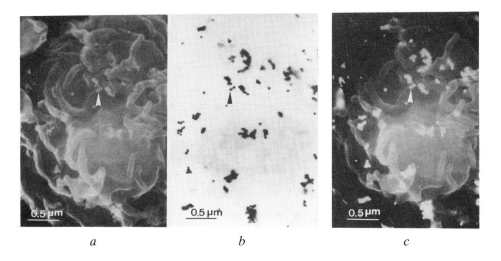

a   *b*   *c*

Fig. 8.5(*a*) A portion of the surface of an uncoated lymphocyte marked with goat anti-mouse IgG–40–nm gold particles bound to a monoclonal anti-T cell antibody, viewed in secondary electron image (SEI) mode. (*b*) The same cell as in (*a*), but viewed in the backscattered electron image (BEI) mode. The surface features of the cells have vanished but the gold particles (arrow) are seen in high contrast and their number is significantly higher. (*c*) The same cell as in (*a*) and (*b*), but after mixing the SEI and BEI signals. The normal polarity of the BEI signal makes the gold particles appear as white spots (arrow). Their localizations can now be precisely correlated with the surface structures of the cell. (Courtesy of Dr E. de Harven, Toronto, Canada.)

can be detected by back-scattered electron imaging.[25] Modern instruments have a good resolution and can produce separated back-scattered and secondary electron signals which can be mixed on the screen.[47] When adapted to colloidal gold marking, an optimal correlation between the distribution of stained sites and the cell surface structures is obtained (*Fig.* 8.5).[48] This approach can even be used to identify 10–15–nm gold particles, provided that the introduction of other materials with high back-scattering coefficients is carefully avoided (specimen support, some chemical fixatives, coating layers): *see* the literature[49] for a detailed description.

A novel technique called 'label-fracture'[50] was recently introduced (*see Fig.* 8.6). Cell surfaces marked with a colloidal gold probe are freeze-fractured and the fracture faces are replicated by platinum/carbon evaporation. Contrary to conventional freeze fracture, the Pt/C-replicated specimens are not digested with acids or bases. Instead, they are thawed and repeatedly washed in distilled water, which removes unfractioned cells. The marked outer half (exoplasmic 'E' face) of the membrane, and cells with fractured 'P' faces remain attached to the Pt/C replica. The replicas are then mounted on electron microscopic grids, dried and observed. This method reveals the high resolution ($\leqslant 15$ nm) surface distribution of the label, coincident with the Pt/C replicas of the 'E' fracture face, while preserving cell shape and relating this distribution to the freeze-fracture morphology of the 'E' face. All these novel approaches for the imaging of colloidal gold on cell surfaces will be quite useful.

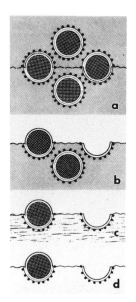

Fig. 8.6. Label-fracture. (*a*) Cells in suspension are labelled and frozen. (*b*) Freeze-fracture splits the plasma membranes of labelled cells into exoplasmic halves (with attached surface label; *right*) and protoplasmic halves (which remain attached to the cell body; *left*). Pt/C evaporation produces a high resolution cast of the fractured cells (only label at the interface of fracture is exposed and shadowed). (*c*) Fractured, shadowed specimens are thawed and washed with distilled water which removes unfractured cells. Exoplasmic membrane halves remain attached to the replica. Coincident images of the Pt/C replica of the E face and the surface label are produced. Cells with fractured P faces (left) remain attached to the replicas (here, electron density of the cell body prevents observation of the P face). (*d*) The replicas are then mounted on electron microscope grids, dried, and observed. (Reproduced from Pinto da Silva and Kan[50] with permission from the authors and publishers.)

6.3. Electron Microscopical Localization of Targets in Tissues

6.3.1. General Remarks

The gold marker system has not yet produced significant results with the pre-embedding approach. This approach is especially important in brain research because it allows selection at light microscopical level of the area that will be further studied at electron microscopical level.[20,51] The major reason for failure with gold markers is poor penetration of the probes in the tissue and the fact that the smallest gold probes do not generate a strong signal in the light microscope.

The finding that small gold probes (< 5 nm) can be made visible in high contrast by silver enhancement,[52] has eliminated the latter disadvantage. In addition, gold/silver particles are much larger and more easily detected in electron microscopical thin sections. It is therefore the poor penetration that today forms the major obstacle. It will be interesting to follow developments with still smaller gold probes (< 3 nm) and adapted fixation/permeabilization methods. (*See also* Van Den Pol.[33])

Two approaches now widely used are discussed here and in more detail in Chapter 9. They are commonly called 'on-grid' marking techniques: (*a*) post-embedding marking of thin, resin-embedded tissue sections and (*b*) marking of ultrathin frozen sections.

In addition, work is in progress to further develop and improve the marking of ultrathin sections of polythylene glycol(PEG)-embedded tissue after removal of the PEG,[53] and the marking of ultrathin sections of tissues (especially muscle) embedded in polyvinyl alcohols.[54]

Recent developments in freeze-fracture cytochemistry using particulate markers, especially colloidal gold, form another and potentially very interesting approach.

The 'fracture-label' techniques allow direct, in situ marking of freeze-fractured, thawed plasma and intracellular membranes as well as cross-fractured cytoplasm. Two main methods of fracture label for tissues are available: thin-section and critical point-drying fracture label.

Details and discussions on the limitations, possibilities and interpretation of results produced by these methods can be found in the literature.[55–57]

6.3.2. *Immunomarking of Ultrathin Sections of Resin-embedded Tissues or Cells*

In this approach, tissues are fixed, dehydrated and embedded in a resin. Ultrathin sections are mounted on coated or uncoated electron microscopic grids and these are incubated with a series of immune reagents, washed, contrasted, dried and viewed.

The fixation method and embedding material used will depend on the kind of antigen. These points are discussed in Chapter 9.

It is important to realize that resin embedding renders most potentially reactive targets inaccessible to antibodies and probes. Therefore, post-embedding 'on-grid' marking is incompatible with many localization studies of cytoskeletal proteins except in special situations where the elements are highly ordered and cut in a precise orientation (e.g. transversely or longitudinally). The filaments and their associated proteins will be largely hidden in the resin and, in most instances, marking will not be representative of their three-dimensional distribution (*see*[58] for a discussion of these problems). Post-embedding marking, however, is very suitable for reactive membrane-associated,[59,60] and membrane-lined components, e.g. endoplasmic reticulum, Golgi, contents of secretory granules, nuclear envelope. This approach yields the most conventional delineation of fine structure. It can be used for: determining the shape and size of marked secretory granules;[61] studying the anatomical relationships of target-defined entities;[62] assessing the co-localization of various compounds;[63] studying compartmentalization, for example during biosynthesis, intracellular transport and exo- or endocytosis (*Fig.* 8.7)[64,65] and vascular permeability.[66]

6.3.3. *Immunomarking of Thawed, Ultrathin Frozen Sections of Tissues or Cells*

In this approach,[58] tissues are fixed, infused with 2·3 M sucrose and frozen. Fixed cell suspensions are embedded in 10 per cent gelatin that is fixed itself. Ultrathin sections are mounted on carbon-coated copper grids and incubated on a series of immune reagents. After the immunostaining the sections must be contrasted and dried. Technically more difficult, thawed ultrathin frozen sections have the advantage of presenting the targets to the probes in a more accessible form. Therefore, this approach can be used for the marking of cytoskeletal elements[17,58,67] (*see Fig.* 8.8). Penetration problems are encountered even here, because of extensive cross-linking of the cytoplasmic ground substance by the fixative.

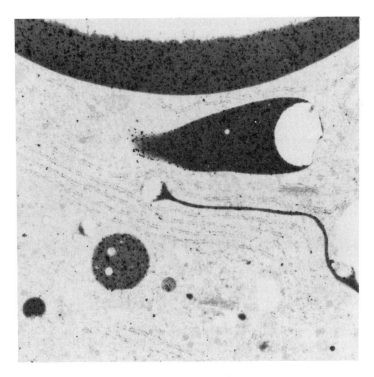

Fig. 8.7. Ultrathin section of a seed of the garden pea. The tissue has been fixed in glutaraldehyde and OsO_4, dehydrated through ethanol and embedded in Spurr's resin. The sections were treated with sodium metaperiodate and HCl before immunomarking. They were stained with antibodies against vicilin, one of the two storage proteins, followed by GAR G20 (Janssen Life Sciences Products). (Courtesy of Dr S. Craig, CSIRO, Canberra, Australia.)

After immunomarking, the sections are usually uranyl-stained and embedded in methyl cellulose after Tokuyasu[68] as modified by Griffiths et al.[69] This procedure produces good delineation of membranous structures but cytoplasmic filaments and other structures are often not well seen. A new approach introduced by Keller and colleagues[70] consists of plastic embedding (LR White resin) and subsequent contrasting of the section and looks promising also for filamentous elements.

6.3.4. *Protein A–Gold or Secondary Antibody–Gold Probes?*

So far, no extensive and systematic comparison has been made between protein A–gold and secondary antibody–gold probes for use in indirect immunocytochemistry. The techniques are very similar because the probe binds to the same identifier.

A protein A–gold probe will bind to one site on the Fc part of IgGs. With small probes (< 10 nm), a virtually one-to-one detection of a target-bound primary antibody can be achieved. Protein A is a pure product and therefore more standardized than affinity purified secondary antibodies. Its excellent solubility in

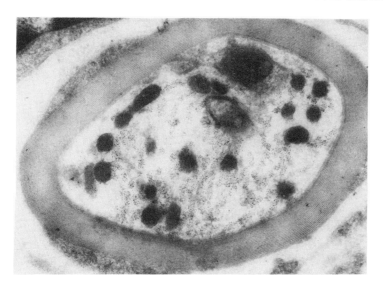

Fig. 8.8. Spinal ganglion, rat, ultrathin cryosection. Immunomarking with anti-tubulin and protein A–5 nm gold. Cross-section of a myelinated nerve. (Courtesy of Dr J. W. Slot, Utrecht, The Netherlands.)

water and its homogeneity facilitate the production of protein A–gold. In addition, small protein A–gold probes can be used for double marking experiments[71,72] (*see* Section 6.3.5) using primary antibodies made in the same animal species. Secondary antibody–gold probes will bind on various antigenic sites. With small probes, this can be exploited to generate some signal amplification seen as small clusters (2 to 4 particles). It is somewhat more difficult to produce high quality unclustered secondary antibody–gold probes, but this can now be accomplished with the procedure of Slot and Geuze.[18] With two primary antibodies made in different animal species, the probes are used in a very simple procedure for double marking (*see* Section 6.3.5). In conclusion, small secondary antibody–gold probes will often give a higher yield of marker than protein A–gold of similar size. Protein A–gold, however, gives a more correct impression of the antigen–primary antibody distribution. (Strept)avidin–gold probes in combination with biotinylated primary or secondary identifiers or linkers[73] can also be used for indirect (immuno)cytochemistry. It can also be expected that direct marking techniques will become more widely used, especially with monoclonal antibody–gold probes,[28] and when high resolution is required.

6.3.5. Double 'On-grid' Marking

Using gold probes of different sizes, double 'on-grid' marking becomes possible by both direct and indirect means. *Fig.* 8.9 shows a diagram of three currently used indirect procedures. With ultrathin frozen sections and resin sections mounted on

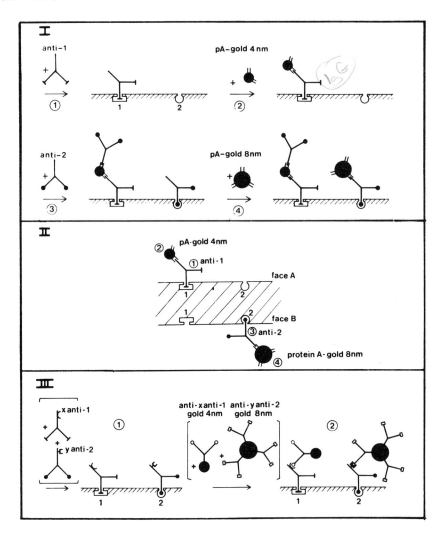

Fig. 8.9. Three currently used procedures for double 'on-grid' immunocytochemistry. (I) The double protein A–gold procedure of Geuze et al.;[71] (II) the two face double staining technique of Bendayan;[75] (III) the double two species, secondary antibody–gold technique of Tapia et al.[74]

support films, the double staining must be performed on the accessible face. The double protein A–gold procedure of Geuze et al.[71] or the double secondary antibody–gold procedure of Tapia et al.[74] can be used. Resin sections mounted on uncoated grids can be used in the two-face method.[75] The first two steps take place by floating the sections on drops, face A down. After rinsing and drying, the whole procedure is repeated with the third and fourth steps, face B down. A possible pitfall of this two-face method results in false negative results: when a structure, e.g. a storage granule, coated vesicle etc. is small or cut tangentially, its contents may be exposed only at one side of the section[22] (*see also* Chapter 9).

6.4. Pre-embedding Marking of Intracellular Targets in Cultured Cell Monolayers

Cultured cell monolayers are widely used to investigate the distribution of structural and regulatory proteins.[76] For cytoskeletal proteins, light microscopical techniques are well established, but the electron microscopical localization of these antigens has been far less successful. For reasons explained above, cytoskeletal antigens can rarely be adequately localized in resin-embedded thin sections.

Ultrathin frozen sections, showing the precise orientation needed to correlate the staining pattern with recognizable structures and with light microscopical data from whole cells, cannot yet be obtained easily.[67]

For these reasons, and despite many disadvantages, the pre-embedding approach remains the method of choice.[29,77] The cells are prefixed, made permeable to the probes and then further processed for electron microscopical observation (see Fig. 8.10 and the Appendix, Section VII). A major drawback of the pre-embedding approach is that it tends to damage the membrane systems, so that it is unsuitable for some studies, for example of cytoskeleton–membrane interactions. More gentle permeabilization techniques using saponin or digitonin in general do not render the cells permeable enough to gold probes.[78] In many instances, it is necessary to extract the cells to varying degrees with a detergent in a suitable buffer before fixation. Glutaraldehyde is practically the only fixative that gives sufficient preservation of fine structural details. In some instances, however, fixatives such as paraformaldehyde or even cold methanol or ethanol, followed by aldehyde fixation before and/or after the immunolabelling procedure,[79,80] can yield the necessary information.

The smallest gold probes will obviously penetrate better and give the best spatial resolution. In cultures, the amount of the antigen that needs to be marked with gold probes can be fairly high as compared to that in a thin section. Therefore, it is often necessary to use gold probes at $A_{520} \geqslant 1$.

The marking of cells may not be complete at all sites, especially in places where the filaments are arranged in dense arrays.

Specially adapted postfixation and contrasting methods make the correlation of the probe with recognizable structures easier.[77] The researcher will have to consider the kind of information he wants to obtain, and work out conditions that can produce meaningful results. Often, electron microscopical work is done in conjunction with light microscopical techniques. Immunogold–silver staining (IGSS)[30] allows the distribution of the gold probes to be visualized at the light microscopical level, and compared with the results obtained by indirect immunofluorescence (see Fig. 8.11 and Chapter 5).

6. CONCLUSIONS

The colloidal gold marking system is rapidly expanding. (Immuno)cytochemistry has recently evolved from a collection of descriptive target localization techniques to a range of potent and highly sensitive detection methods.

The availability of a flexible and multipurpose marking system, in addition to the other widely used systems (ferritin, Imposil, enzymes) will certainly contribute to the increasing use of electron microscopical immunocytochemistry.

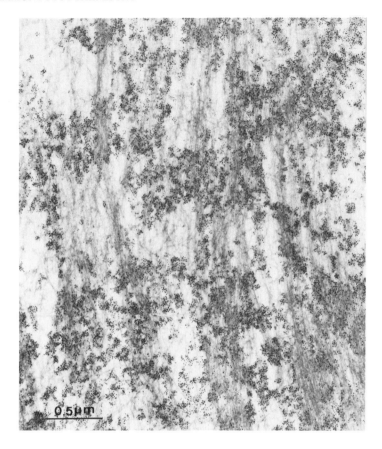

Fig. 8.10 Ultrathin section of an embryonic chicken fibroblast marked with anti-myosin and GAR G5 EM grade (Janssen Life Sciences Products, Beerse, Belgium) with the pre-embedding procedure of Langanger et al.[77] The label is distributed along the microfilament bundles of stress-fibres in a periodic pattern (*see also* Fig. 8.11). (Courtesy of Dr Gabriele Langanger, Janssen Pharmaceutica, Beerse, Belgium.)

Acknowledgements

My colleagues G. Daneels, M. De Raeymaecker, G. Geuens, G. Langanger, M. Moeremans, R. Nuydens, A. Van Dijck and M. De Brabander are gratefully acknowledged for their contributions. Mrs C. Verellen is thanked for typing the manuscript and Mr L Leyssen for preparing the micrographs. This work was supported by a grant from the IWONL (Instituut voor aanmoediging van Wetenschappelijk Onderzoek in Nijverheid en Landbouw), Brussels.

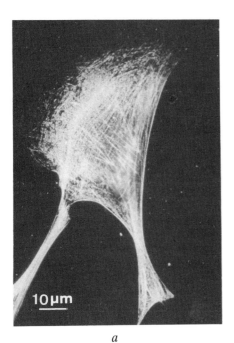

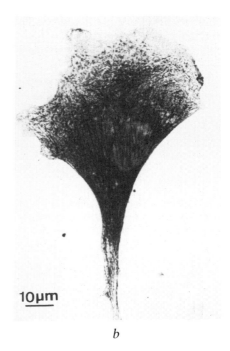

a   *b*

Fig. 8.11. Light microscopical localization of myosin in embryonic chicken fibroblasts: (*a*) indirect immunofluorescence; (*b*) indirect immunogold–silver staining. In (*b*), the silver development of the gold marker has revealed its distribution. It is identical to that obtained with indirect immunofluorescence. (Courtesy of Dr Gabriele Langanger, Janssen Pharmaceutica, Beerse, Belgium.)

Appendix

I. THE PREPARATION AND STORAGE OF COLLOIDAL GOLD SOLS

Reproducible production of gold sols is dependent on the use of scrupulously cleaned glassware, using freshly prepared, twice deionized, glass-distilled water (hereafter denoted as H_2O^*) for both final rinsing and making the solutions. Rapid stirring at all times is important, unless otherwise stated.

A. Citrate Gold (15 nm)

Reducing agent = sodium citrate.

i. Stock Solutions

Dissolve 5 g $HAuCl_4 \cdot 4H_2O$ (Merck 1582) in exactly 100 ml H_2O^* in a measuring flask. This yields a $HAuCl_4$ concentration of 4·12 per cent. Store this solution in the dark at 4 °C.

To prepare a 1 per cent solution, dilute 6·06 ml $HAuCl_4$ (4·12 per cent) to 25 ml with H_2O^*.

Dissolve 1·14 g trisodium citrate·$2H_2O$ (Merck 6448) in 100 ml H_2O^* in a measuring flask. This yields a citrate concentration of 1 per cent. Filter this solution through a 0·2–μm (Millipore) filter. Discard the first few millilitres. Store at room temperature and prepare daily.

ii. Procedure

A clean 500 ml Erlenmeyer flask, a Teflon-coated stirring bar and a water-cooled reflux tube are washed extensively with H_2O^*.

Boil about 100 ml H_2O^* in the Erlenmeyer flask with the stirring bar. This is the final cleaning of the flask.

Discard the wash water.

Pour exactly 240·0 ml H_2O^* into the Erlenmeyer flask.

Heat up to boiling and start refluxing.

Add 2·5 ml of the 1 per cent $HAuCl_4$ solution.

Increase the speed of stirring.

Take exactly 7·5 ml citrate (1 per cent) with a clean glass pipette and a pro-pipette.

Add the citrate to the boiling $HAuCl_4$ solution, as quickly as possible.

Reflux for an additional 30 min.

Pour the freshly prepared gold sol, while still hot, into cleaned, autoclaved, screw-capped, brown glass bottles. It can then be stored for longer than six months at 4–8 °C without deterioration.

Size control with electron microscope.

iii. Variation of the Size by using Different Citrate Concentrations

As mentioned in the introduction, the reaction rate determines the particle size. The citrate method has the advantage that a wide range of different particle diameters can be obtained by changing the volume of citrate added.

The procedure and stock solutions are exactly the same as in (A.i) and (A.ii). Adjust the volume of 0·01 per cent $HAuCl_4$ and citrate as follows:

Amount of 0·01 per cent $HAuCl_4$	Amount of citrate added (ml/250 ml total volume)	Average diameter (in nm)
242·5	7·5	± 15
246·25	3·75	± 30
247·5	2·50	± 50
248·10	1·90	± 60

These volumes are approximate. They may vary according to laboratory conditions and should be determined in a series of pilot experiments.

iv. A Modified Citrate Method for Producing 8–10 nm Gold

From M. Moeremans, F. Cornelissen, A. Van Dijck and J. De. Mey, 1982, unpublished data.

To decrease the particle diameter to below 12·5 nm, the 1 per cent $HAuCl_4$ stock solution is added to a boiling citrate solution. This simple modification produces very homogeneous 8–10 nm gold particles.

The stock solutions are exactly the same as in (A.i). The procedure is as follows:

Clean the glassware as in (A.ii).

Pour exactly 232·50 ml H_2O^* into a 500 ml Erlenmeyer flask.

Heat up to boiling and start refluxing.

Add 15 ml of the 1 per cent citrate stock solution.

Reflux for 5 min.

Increase the stirring speed.

Take 2·5 ml of the 1 per cent $HAuCl_4$ solution with a clean glass pipette and a pro-pipette.

Add this solution to the boiling citrate solution, as quickly as possible.

Reflux for an additional 15 min.

Continue as in (A.ii).

B. Phosphorus Gold

Reducing agent = white phosphorus.

i. Introduction

This method has long been the only way to obtain gold particles of 4–7 nm. It has the reputation of being dangerous because it uses white phosphorus, a substance that self-ignites in contact with air and that can even explode. White phosphorus must be kept under water in an air-tight container. A 1 per cent $CuSO_4$ solution should always be at hand to wash material that has been in contact with phosphorus. When using this method, proper safety precautions must be taken and it is advisable to discuss these with the local safety authorities. In practice, the method is very easy and reproducible, but the newer method with citrate–tannic acid (*see* Section E) should be tested as a non-hazardous alternative.

ii Stock Solutions

For preparing a 1 per cent $HAuCl_4$ solution, *see* (A.i).

Prepare a 0·2 N K_2CO_3 solution: dissolve 1·38 g K_2CO_3 in 100 ml H_2O^*. Filter this solution through a 0·2 μm (Millipore) filter. Discard the first few millilitres. Store at room temperature.

Prepare a 1 per cent $CuSO_4$ solution (1 litre).

Prepare a white phosphorus-saturated ether solution: Under water, cut 3 to 5 small pieces from phosphorus sticks (Merck 7271). Work under a hood. With a tweezer, quickly transfer them to about 20 ml diethyl ether in a small screw-capped bottle. Swirl from time to time to partly dissolve the phosphorus for two hours. Carefully decant the supernatant into two 15 ml sturdy glass centrifuge tubes with polyethylene caps. Fill the bottle with 1 per cent $CuSO_4$ solution. Pellet the remaining solids at $4000 \times g$ for 20 min, at room temperature. Decant the supernatant into a glass bottle and dilute it with diethyl ether (1 part phosphorus-saturated ether plus 4 parts diethyl ether). Fill the centrifuge tubes with 1 per cent $CuSO_4$ solution. Leave all material that has been in contact with the phosphorus immersed in 1 per cent $CuSO_4$ for 24 hours.

iii. Procedure

Clean the glassware as in (A.i).
Pour 120 ml H_2O^* into a 500 ml Erlenmeyer flask, add 1·5 ml of the 1 per cent $HAuCl_4$ solution and neutralize with 2·7 ml of the 0·2 N K_2CO_3 solution.
Stir this solution slowly and add 1 ml of the diluted phosphorus ether, immersing the tip of the pipette in the $HAuCl_4$ solution to minimize contact with air. Rinse the pipette at once with the 1 per cent $CuSO_4$ solution and destroy the remaining phosphorus ether (unless more than 1 batch is made at a time) by pouring it into a 1 per cent $CuSO_4$ solution.
Continue to stir slowly for 15 min at room temperature.
Heat to boiling and reflux for 30 min.
Continue as in (A.ii).

C. The Alternative Citrate–Tannic Acid Procedure[81]

i. Stock Solutions

A 1 per cent $HAuCl_4$ solution (see A.i).
A 1 per cent sodium citrate solution (see A.i).
A 1 per cent tannic acid (TA) solution (in H_2O^*) – store at room temperature and prepare fresh daily.
A 25 mM K_2CO_3 solution.

ii. Procedure

Clean the glassware as in (A.ii).
Prepare solution A: Add 1 ml 1 per cent $HAuCl_4$ to 79 ml H_2O^*. This solution is poured into an Erlenmeyer flask with reflux tube.
Prepare solution B: Mix 4 ml 1 per cent trisodium citrate with a variable volume of 1 per cent TA and a volume of 25 mM K_2CO_3 equal to the TA volume. Add H_2O^* to make the volume 20 ml.

A and B are heated to 60 °C and then solution B is quickly added to solution A while stirring. Red sols form within a second when high concentrations of TA are used and after minutes with low concentrations. In all cases the mixture is heated to 100 °C, while refluxing. When, in this procedure, the volume of 1 per cent TA (and that of 25 mM K_2CO_3) is increased from 0·1 to 5 ml, the diameter of the gold particles gradually decreases from 17 to nearly 3 nm. For example, when 0·02, 0·1, 0·5 or 3 ml 1 per cent TA is added, the particle size is about 15, 10, 6 and 3·5 nm, respectively. The sols are very homogeneous and very suitable for multiple staining procedures after complexing with various identifier or linker molecules.

II. DETERMINATION OF THE MINIMAL PROTECTING AMOUNT OF PROTEIN

After high-speed centrifugation of the dialysed protein (*see* Section 3.2), a small part of it is diluted with its solvent to a concentration of 0·2 mg/ml for 20-nm and 0·75 mg/ml for 5-nm gold.

With this solution, a linear dilution series in 5·0 ml plastic tubes is made:

100 μl protein + 0 μl solvent	50 μl protein + 50 μl solvent
90 μl protein + 10 μl solvent	40 μl protein + 60 μl solvent
80 μl protein + 20 μl solvent	30 μl protein + 70 μl solvent
70 μl protein + 30 μl solvent	20 μl protein + 80 μl solvent
60 μl protein + 40 μl solvent	10 μl protein + 90 μl solvent

Add 1·0 ml pH-adjusted (*see* Section 3.2) gold sol to each tube, vortex and let stand for 2 min.

Add 100 μl of a 10 per cent NaCl solution in H_2O^* and vortex.

Let stand 5 min and measure the absorbance at 580 nm (A_{580}) with the following blank:

    1 ml colloidal gold sol

    + 100 μl protein solvent

    + 100 μl H_2O^*

Construct a concentration variable absorption isotherm (CVAI) with the A_{580} in the ordinate and the amount of protein added in the abscissa. The amount of protein is expressed in μg/ml gold sol.

III. THE PREPARATION OF A POLYCLONAL ANTIBODY–GOLD PROBE FOR USE IN ELECTRON MICROSCOPICAL IMMUNOCYTOCHEMISTRY

(*See also* Section 3.4.

Prepare a 10 per cent bovine serum albumin (BSA) (Sigma, type V) solution in H_2O^* and adjust its pH to 9·0 with 1 N NaOH.

Avoid excessive foaming and microfilter (0·2 μm) just before use.

This solution can be stored frozen at −20 °C, in suitable aliquots.

Prepare 1 per cent BSA–Tris buffer: 20 mM Tris, 1 per cent BSA, 150 mM NaCl, 20 mM NaN_3 pH 8·2.

Adjust the pH of the gold sol (at room temperature).

Slowly add the minimal protecting amount of antibody, now undiluted, while continuously and rapidly stirring the colloidal gold.

After 2 min, add the BSA solution to a final concentration of 1 per cent BSA. Centrifuge:

60 000×g, 4 °C, 1 h for 5–nm gold

15 000×g, 4 °C, 1 h for 15–20 nm gold

Carefully aspirate and discard the supernatant, leaving ± 10 per cent of the initial volume.

Decant the loose part of the sediment, together with the remaining supernatant.

This is layered over a 10–30 per cent continuous sucrose or glycerol gradient (volume 10·5 ml, length 8 cm) in 20 mM TBS containing 0·1 per cent BSA pH 8·2.

The gradient is centrifuged in a SW41 rotor (Beckman Instruments) for 45 min at 41 000 rev./min (5–nm gold) or for 30 min at 20 000 rev./min (12–nm gold). The conditions must be adapted to the gold size.

The upper half of the gradient is collected in successive fractions of about 1 ml. These are analysed for the occurrence of clusters and for size distribution. Fractions containing few or no clusters can be pooled.

Measure the A_{520} of the pooled fractions (1 per cent BSA–Tris as blank) and dilute with 1 per cent BSA–Tris to an A_{520} of 2·5 for 5 and 10–nm gold, 3·5 for 15–nm, 5·0 for 20–nm and 7·0 for 40–nm gold.

IV. THE PREPARATION OF A MONOCLONAL ANTIBODY–GOLD PROBE FOR USE IN ELECTRON MICROSCOPICAL IMMUNOCYTOCHEMISTRY

Purified monclonal antibodies show a narrow pI profile. Determine it with agarose or polyacrylamide isoelectric focusing (native conditions!).

Prepare the monclonal antibody as if it were a polyclonal one (*see* Section 3.2). Adjust the pH of the gold sol to the pI of the antibody (take the pI of the most basic band).

If at this pH no satisfactory probe can be obtained, construct several CVAIs at various pH values above the pI.

Preparation of a monoclonal antibody–gold probe is very similar to the procedure for polyclonal antibodies. The latter procedure is modified as follows: after the coupling at pH x (x is lower than 9·0), the pH of the protected gold sol is raised to 9·0 with 1 N NaOH, *before* adding the BSA, and the procedure continues as in Section III.

V. PREPARATION OF STREPTAVIDIN–GOLD AND PROTEIN A–GOLD PROBES FOR USE IN ELECTRON MICROSCOPY

For gold probes smaller than 25 nm, BSA can be used as secondary stabilizer, in the same way as with monoclonal antibodies.

Streptavidin and protein A are dialysed against H_2O^* and centrifuged at 100 000 × g for 1 h at 4 °C.

Streptavidin is coupled at pH 6·5 and protein A at pH 6·0.

After the adsorption step, the pH of the protected gold sol is raised to 9·0 with 1 N NaOH, *before* adding the BSA, and the procedure continues as in Section III.

VI. USE OF CARBOWAX 20 M AS STABILIZER

Carbowax 20 M can be added to the protected gold sol at any pH from a microfiltered 1 per cent stock solution in H_2O, to a final concentration of 0·05 per cent. The storage buffer also contains 0·05 per cent Carbowax 20 M and can be supplemented with proteins such as BSA and ovalbumin.

It is advisable to test the stabilizing capacity of the Carbowax 20 M used by constructing a CVAI with it.[24]

VII. PRE-EMBEDDING MARKING OF INTRACELLULAR TARGETS IN CULTURED CELL MONOLAYERS

A. Cell Culture

Cells may be grown under standard conditions on round or 18 × 18 mm coverslips in plastic Petri dishes or directly in plastic Petri dishes. Treating the surface of the coverslip with a release agent, e.g. MS–123 Release Agent, dry lubricant (Miller–Stephenson Chemical Company, Inc., Danbury, CT 06810, USA), will help in separating the glass coverslip from the epoxy resin after flat embedding. The release agent is sprayed on the cell side of the coverslip which is directly afterwards wiped dry with a paper tissue. *Caution: avoid skin contact with and breathing vapour of the release agent. Consult the manufacturer's safety instructions.* For cells grown on Petri dishes, Permanox dishes (Lux, Lab-Tek Division, Miles Laboratories, Inc. Naperville, IL 60540, USA) are recommended, because they are resistant to acetone and epoxy resins. The cells should be used when they have reached about 50 per cent confluence.

B. Handling

When the cells are grown on larger coverslips or on plastic, most of the steps, e.g. fixation, permeabilization, incubation with immune reagents, washing, are done in the Petri dish in which the cells were grown. Removal of the liquids is done with a Pasteur pipette, fixed in a stand and connected to a vacuum line. This set-up forms a 'third hand' and greatly facilitates the handling. When the cells are grown on the bottom of the Petri dishes, a central area is marked with a sharp object and the plastic outside the area is dried with the 'third hand'. The immune reagents are applied within the marked area. For cells grown on coverslips, the immune reagents are applied on the coverslip (70 μl for a 18 × 18 mm slip) after removal with the 'third hand' of liquid remaining around the coverslip. The Petri dishes of one experiment are placed on rectangular inox trays, so that they can be transported together. For washing, the trays are put on a rocking table, such as the Heidolph Reax III.

C. Fixation–Permeabilization

i. Use of Glutaraldehyde Fixation

The glutaraldehyde needs to be of high purity, free of polymers. Glutaraldehyde can be used as such on living cells at 0·3–1 per cent, or mixed with variable concentrations of a detergent (e.g. Triton X–100, 0·1–1 per cent), for about 10 min. Glutaraldehyde fixation can be preceded by extraction with a detergent, such as Triton X–100, 0·1–0·5 per cent for 20 s or longer. Any suitable buffer system can be used. We have obtained good results for studying the microfilament system with the following buffer:[82] Hanks' basal balanced salts (HBBS): NaCl, 137 mM; KCl, 5 mM; Na_2HPO_4, 1·1 mM; KH_2PO_4, 0·4 mM; $NaHCO_3$, 4 mM; glucose, 5·5 mM; containing 5 mM PIPES, 2 mM $MgCl_2$, 2 mM EGTA. Adjust the pH to 6·1 with 1 N NaOH. When extracting cells with a detergent before fixation, the EGTA is left out. Another buffer that is useful when the microtubular system is investigated is: 65 mM PIPES, 25 mM HEPES, 10 mM EGTA and 2 mM $MgCl_2$ pH 6·9.[83]

Fixation and/or extraction are done in the Petri dish after briefly washing the cells with PBS without Mg^{2+} and Ca^{2+}. The washing fluid is removed and fixation fluid, optionally preceded by extraction fluid, is added with a Pasteur pipette. After the cells are fixed, they are rinsed with several changes of fixation buffer and washed 3 times for 10 min. For permeabilizing the cells, 0·2–0·5 per cent Triton X–100 for 30 min in fixation buffer is used. Do not include this step when surface labelling is the purpose. Other permeabilization methods exist and can be tried.[78] This step is followed by a treatment with $NaBH_4$, 1 mg/ml for 10 min, twice, in fixation buffer. $NaBH_4$ is very hydrophilic and should be stored in a desiccator. It should be dissolved just prior to use. The cells are washed with fixation buffer (3 × 5 min), and then left in 0·1 per cent BSA–Tris for immunocytochemistry: 0·1 per cent BSA–Tris is 20 mM Tris-buffered saline (0·9 per cent NaCl), supplemented with 1 mg/ml BSA type V and adjusted with HCl to pH 8·2.

D. Incubation with Immune Reagents (Indirect Method)

The cells are subsequently incubated with the following solutions at room temperature, in a water saturated atmosphere.

5 per cent normal serum (same species as secondary antibody) in 0·1 per cent BSA–Tris, 20 min.

Replace, without washing, by appropriately diluted antiserum, antibody or hybridoma culture supernatant in 1 per cent normal serum in 0·1 per cent BSA–Tris for 2 h.

Wash 3 × 10 min in 0·5 per cent BSA–Tris.

Appropriately diluted secondary antibody labelled with colloidal gold (preferably 5 nm) in 0·1 per cent BSA–Tris ($A_{520} = 1$) for 4 h to overnight.

Wash 3 × 10 min in 0·1 per cent BSA–Tris.

Wash 2 × 5 min in postfixation buffer (see Section E below).

E. Postfixation and Embedding Procedure

The cells can be postfixed with glutaraldehyde and osmium tetroxide and processed for embedding using standard methods. In some instances, especially for microfilaments, the following procedure produced good preservation and contrast.[74]

The washed cells are further washed twice in Sörensen's phosphate buffer pH 7·2 (Pi buffer : 36 ml 0·2 M Na_2HPO_4 + 14 ml 0·2 M NaH_2PO_4 diluted to 100 ml with H_2O). They are postfixed in 1 per cent glutaraldehyde + 0·2 per cent tannic acid in Pi buffer, for 30–60 min and then washed thoroughly in Pi buffer. They are postfixed in 0·5 per cent osmium tetroxide in Pi buffer, 10 min on ice and washed in Pi buffer for 3 × 5 min. The cells are partially dehydrated in 70 per cent ethanol (several changes), impregnated with 0·5 per cent uranyl acetate + 1 per cent phosphotungstic acid in 70 per cent ethanol for 30 minutes, further dehydrated in alcohol and embedded in Epon. Cells grown on coverslips are flat embedded on microscope slides treated with the same release agent as used for the coverslips. Spacers are formed by pieces of coverslips or any other appropriate material. A drop of Epon is put on the slide and the coverslip, cells down, positioned on the spacers. The Epon is hardened for 3 days at 50 °C. It is separated from the glass slide and coverslip by dipping alternately in boiling water and liquid nitrogen.

Suitable cells are selected under a phase microscope, cells facing lens, and circled with a diamond. They are cut out and re-embedded, the cell side up for horizontal sectioning, on edge for vertical sectioning. For the study of microfilament systems, closely apposed to the substrate, it is necessary to obtain flat sections from the surface of the block.

REFERENCES

1. Sternberger L. A. *Immunocytochemistry*. Wiley, Chichester, 1979.
2. Polak J. M. and Van Noorden S. (eds.) *Immunocytochemistry. Practical Applications in Pathology and Biology*. Bristol, Wright· PSG, 1983.
3. Polak J. M. and Varndell I. M. (eds.) *Immunolabelling for Electron Microscopy*. Amsterdam, Elsevier, 1984.
4. Bullock G. R. and Petrusz P. (eds.) *Techniques in Immunocytochemistry*, Vol. 1. New York, Academic Press, 1982.
5. Bullock G. R. and Petrusz P. (eds.) *Techniques in Immunocytochemistry*, Vol. 2. New York, Academic Press, 1983.
6. Coons A. H. Fluorescent antibody methods. In: Danielli J. F., ed., *General Cytochemical methods*. New York, Academic Press, 1978: 399.
7. Cuello A. C. (ed.) *Immunohistochemistry*. Chichester, Wiley, 1983.
8. De Lellis R. A. *Diagnostic Immunocytochemistry*. New York, Masson Publishing USA Inc., 1981.
9. Heym Ch. and Forsmann W. G. *Techniques in Neuroanatomical Research*. Berlin, Springer, 1981.
10. Molday R. and Moher P. A review of cell surface markers and labelling techniques for scanning electron microscopy. *Histochem. J.* 1980, **12**, 273–315.
11. Larsson L.–I. Peptide immunocytochemistry. *Proc. Histochem Cytochem.* 1981, **13**, 1–85.
12. Faulk W. and Taylor G. An immunocolloid method for the electron microscope. *Immunochemistry*. 1971, **8**, 1081–1083.

13. Horisberger M. Colloidal gold: cytochemical marker for light and fluorescent microscopy and for transmission and scanning electron microscopy. In: Johari O., ed., *Scanning Electron Microscopy II*. Illinois, S.E.M. Inc., AMF O'Hare, 1981: 9–31.

14. Goodman S. L., Hodges G. M. and Livingston D. C. A review of the colloidal gold marker system. In: Johari O., ed., *Scanning Electron Microscopy II*. Illinois, S.E.M. Inc., AMF O'Hare, 1981: 133–145.

15. De Mey J. Colloidal gold probes in immunocytochemistry. In: Polak J. M. and Van Noorden S., eds., *Immunocytochemistry. Practical Applications in Pathology and Biology*. Bristol, Wright PSG, 1983: 82.

16. Roth J. The colloidal gold marker system for light and electron microscopic cytochemistry. In: Bullock G. R. and Petrusz P., eds., *Techniques in Immunocytochemistry*, Vol 2. New York, Academic Press, 1983: 217.

17. Slot J. W. and Geuze H. J. The use of protein A–colloidal gold (PAG) complexes as immunolabels in ultra-thin frozen sections. In: Cuello A. C., ed., *Immunohistochemistry*. Chichester, Wiley, 1983: 323.

18. Slot J. W. and Geuze H. J. Gold markers for single and double immunolabelling of ultra-thin cryosections. In: Polak J. M. and Varndell I. M., eds., *Immunolabelling for Electron Microscopy*. Amsterdam, Elsevier, 1984: 129.

19. De Waele M. Haematological electron immunocytochemistry. In: Polak J. M. and Varndell I. M., eds., *Immunolabelling for Electron Microscopy*. Amsterdam, Elsevier, 1984: 267.

20. Van Den Pol A. Colloidal gold and biotin–avidin conjugates as ultrastructural markers for neural antigens. *Q. J. Exp. Phys.* 1984, **69**, 1–33.

21. Varndell I. M., Tapia F. J., Probert L., Buchan A. M. J., Gu J., De Mey J., Bloom S. R. and Polak J. M. Immunogold staining method for the localization of regulatory peptides. *Peptides* 1982, **3**, 259–272.

22. Bendayan M. Protein A–gold electron microscopic immunocytochemistry: methods, applications and limitations. *J. Electr. Microsc. Techn.* 1984, **1**, 243–270.

23. Konings F. *Colloidal Metal Marking Reference Book*, Volumes 1–3. Janssen Life Sciences Products, Beerse, Belgium. (Available on request from the author of this chapter.)

24. Horisberger M., Rosset J. and Bauer H. Colloidal gold granules as markers for cell surface receptors in the scanning electron microscope. *Experientia*. 1975, **31**, 1147–1151.

25. Sieber-Blum M., Sieber F. and Yamada K. Cellular fibronectin promotes adrenergic differentiation of quail neural crest cells *in vitro*. *Exp. Cell Res.* 1981, **193**, 285–295.

26. Geoghegan W. D., Scillian J. J. and Ackerman G. A. The detection of human B-lymphocytes by both light and electron microscopy utilizing colloidal gold labelled anti-immunoglobulin. *Immunol. Commun.* 1978, **7**, 1–12.

27. Gu J., De Mey J., Moeremans M. and Polak J. M. Sequential use of the PAP and immunogold staining methods for the light microscopical double staining of tissue antigens. Its application to the study of regulatory peptides in the gut. *Regul. Pept.* 1981, **1**, 365–374.

28. De Mey, J., Moeremans M., De Waele M., Geuens G. and De Brabander M. The IGS (immunogold staining) method used with monoclonal antibodies. In: Peeters M., ed., *Proceedings of Colloquium on the Protides of Biological Fluids*. Oxford, Pergamon Press, 1981: 943–947.

29. De Mey J., Moeremans M., Geuens G., Nuydens R. and De Brabander M. High resolution light and electron microscopic localization of tubulin with the IGS (Immunogold staining) method. *Cell. Biol. Int. Rep.* 1981, **5**, 889–899.

30. Holgate C., Jackson P., Lauder I., Cowen P. and Bird C. Immunogold–silver staining of immunoglobulins in paraffin sections of non-Hodgkin's lymphomas using immunogold–silver staining technique. *J. Clin. Pathol.* 1983, **36**, 742–746.

31. Moeremans M., Daneels G., Van Dijck A., Langanger G. and De Mey J. Sensitive visualization of antigen–antibody reactions in dot and blot immune overlay assays with immunogold and immunogold/silver staining. *J. Immunol. Methods.* 1984, **74**, 353–360.

32. Langanger G., De Mey J., Moeremans M., Daneels G., De Brabander M. and Small J. V. Ultrastructural localization of α-actinin and filamin in cultured cells with the immunogold staining (IGS) method. *J. Cell. Biol.* 1984, **99**, 1324–1334.

33. Van Den Pol A. N. Silver intensified gold and peroxidase on dual ultrastructural immunolabels for pre- and postsynaptic neurotransmitters. *Science* 1985, **228**, 332–335.

34. Frens G. Controlled nucleation for the regulation of the particle size in monodisperse gold suspensions. *Nature Phys. Sci.* 1973, **241**, 20–22.

35. Geoghegan W. D. and Ackerman G. A. Adsorption of horseradish peroxidase, ovomucoid and anti-immunoglobulin to colloidal gold for the indirect detection of Concanavalin A, wheat germ agglutinin and goat anti-human immunoglobulin G on cell surfaces at the electron microscopic level: a new method, theory and application. *J. Histochem. Cytochem.* 1977, **25**, 1187–1200.

36. Horisberger M. and Rosset J. Colloidal gold, a useful marker for transmission and scanning electron microscopy. *J. Histochem. Cytochem.* 1977, **25**, 295–305.

37. Goodman S. L., Hodges G. M., Trejdosiewicz L. and Livingston D. C. Colloidal gold markers and probes for routine application in microscopy. *J. Microsc.* 1981, **123**, 201–213.

38. Garaud J. C., Eloy R., Moody A. J., Stock C. and Grenier J. F. Glucagon- and glicentin-immunoreactive cells in the human digestive tract. *Cell Tissue Res.* 1980, **213**, 121–136.

39. Slot J. W. and Geuze H. J. Sizing of protein A–colloidal gold probes for immunoelectron microscopy. *J. Cell Biol.* 1981, **90**, 533–536.

40. Schwab M. and Thoenen M. Selective binding, uptake and retrograde transport of tetanus toxin by nerve terminals in the rat iris. An electron microscope study using colloidal gold as a tracer. *J. Cell Biol.* 1978, **77**, 1–13.

41. Warchol J., Brelinska R. and Herbert D. Analysis of colloidal gold methods for labelling proteins. *Histochemistry.* 1982, **76**, 567–575.

42. Beesley J., Orpin A. and Adlam C. An evaluation of the conditions necessary for optimal protein A–gold labelling of capsular antigen in ultrathin methacrylate sections of the bacterium *Pasteurella haemolytica. Histochem. J.* 1984, **16**, 151–163.

43. Horisberger M. and Vauthey M. Labelling of colloidal gold with protein. A quantitative study using beta-lactoglobulin. *Histochemistry*, 1984, **80**, 13–18.

44. Robenek H., Rassat J., Hesz A. and Grunwald J. A correlative study on the topographical distribution of the receptors for low density lipoprotein (LDL) conjugated to colloidal gold in cultured human skin fibroblasts employing thin sections, freeze-fracture, deep-etching and surface replication techniques. *Eur. J. Cell Biol.* 1982, **27**, 242–250.

45. Müller G. and Baigent C. Antigen controlled immunodiagnosis "acid test". *J. Immunol. Methods*, 1980, **37**, 185–190.

46. Beesley J. E. Recent advances in microbiological immunocytochemistry. In: Polak J. M. and Varndell I. M, eds, *Immunolabelling for Electron Microscopy*. Amsterdam, Elsevier, 1984, 289.

47. Becker R. P. and Sogard M. Visualization of subsurface structures in cells and tissues by backscattered electron imaging. In: Johari O., ed., *An International Review of Advances in Techniques and Applications of the Scanning Electron Microscope*. Chicago, S.E.M. Inc, 1979: 835–870.

48. de Harven E., Leung R. and Christensen H. A novel approach for scanning electron microscopy of colloidal gold-labeled cell surfaces. *J. Cell Biol.* 1984, **99**, 53–57.

49. Walther P., Kriz S., Müller M., Ariano B. H., Brodbeck U., Ott. P. and Schweingruber M. E. Detection of protein A gold 15 nm marked surface antigens by backscattered electrons. In: Makita T. and Roomans G. M., eds., *Scanning Electron Microscopy*. Vol. 3. Chicago, S.E.M. Inc., 1984: 1257–1266.

50. Pinto da Silva P. and Kan F. W. K. Label fracture: a method for high resolution labelling of cell surfaces. *J. Cell Biol.* 1984, **99** 1156–1161.

51. De Mey J. A critical review of light and electron microscopic immunocytochemical techniques used in neurobiology. *J. Neurosc. Methods*. 1983, **7**, 1–18.

52. Springall D. R., Hacker G. W., Grimelius L. and Polak J. M. The potential of the immunogold–silver staining method for paraffin sections. *Histochemistry*. 1984, **81**, 603–608.

53. Wolosewick J., De Mey J. and Meininger V. Ultrastructural localization of tubulin and actin in polyethylene glycol-embedded rat seminiferous epithelium by immunogold staining. *Biol. Cell* 1983, **49**, 219–226.

54. Small J. V. Polyvinylalcohol, a water soluble resin suitable for electron microscope immunocytochemistry. In: Csanady A., Röhlick P. and Szabo D., eds., *Proceedings of the 8th European Congress on Electron Microscopy*, Vol. 3, Budapest, 1984: 1799.

55. Pinto da Silva P. Freeze-fracture cytochemistry. In: Polak J. M. and Varndell I. M., eds., *Immunolabelling for Electron Microscopy*. Amsterdam, Elsevier, 1984:179.

56. Rash J. E., Johnson T. J. A., Hudson C. S., Giddins F. D., Graham W. F. and Eldefrani M. Labelled-replica techniques: post-shadow labelling of intramembrane particles in freeze-fracture replicas. *J. Microsc.* 1982, **128**, 121–138.

57. Severs N. and Robenek H. Detection of micro domains in biomembranes, an appraisal of recent developments in freeze-fracture cytochemistry. *Biochim. Biophys. Acta* 1983, **737**, 373–408.
58. Tokuyasu K. T. Present state of immunocryoultramicrotomy. *J. Histochem. Cytochem.* 1983, **31**, 164–167.
59. Roth J. and Berger E. Immunocytochemical localization of galactosyltransferase in Hela cells: codistribution with thiamine pyrophosphate in trans-Golgi cisternae. *J. Cell Biol.* 1982, **92**, 223–229.
60. Bendayan M. and Shore G. Immunocytochemical localization of mitochondrial proteins in the rat hepatocyte. *J. Histochem. Cytochem.* 1982, **30**, 139–147.
61. Probert L., De Mey J. and Polak J. M. Distinct subpopulations of enteric P-type neurones contain substance P and vasoactive intestinal polypeptide. *Nature* 1981, **294**, 470–471.
62. Doerr-Schott J. and Garaud J. Ultrastructural identification of gastrin-like immunoreactive nerve fibers in the brain of *Xenopus laevis* by means of colloidal gold or ferritin immunocytochemical methods. *Cell Tiss. Res.* 1981, **216**, 581–589.
63. Ravazzola M., Perrelet A., Unger R. and Orci L. Immunocytochemical characterization of secretory granule maturation in pancreatic A-cells. *Endocrinology*. 1984, **114**, 481–485.
64. Bendayan M., Roth J., Perrelet A. and Orci L. Quantitative immunocytochemical localization of pancreatic secretory proteins in subcellular compartments of the rat acinar cell. *J. Histochem. Cytochem.* 1980, **28**, 149–160.
65. Craig S. and Miller C. LR White resin and improved on-grid immunogold detection of vicilin, a pea seed storage protein. *Cell. Biol. Int. Rep.* 1984, **8**, 879–886.
66. Bendayan M. Use of the protein A–gold technique for the morphological study of vascular permeability. *J. Histochem. Cytochem.* 1980, **28**, 1251–1254.
67. Chen W. T. and Singer S. T. Immunoelectron microscopic studies of the sites of cell-substratum and cell-cell contacts in cultured fibroblasts. *J. Cell Biol.* 1982, **95**, 205–222.
68. Tokuyasu K. T. A study of positive staining of ultrathin frozen sections. *J. Ultrastr. Res.* 1978, **63**, 287–307.
69. Griffiths G., Brands R., Burke B., Louvard D. and Warren G. Viral membrane proteins acquire galactose in trans Golgi cisternae during intracellular transport. *J. Cell Biol.* 1982, **95**, 781–792.
70. Keller G. A., Tokuyasu K. T., Dutton A. H. and Singer S. J. An improved procedure for immunoelectron microscopy: ultrathin plastic embedding of immunolabeled ultrathin frozen sections. *Proc. Natl Acad. Sci. USA* 1984, **81**, 5744–5747.
71. Geuze H. J., Slot J. W., Van der Ley P. A. and Scheffer R. C. T. Use of colloidal gold particles in double labeling immunoelectron microscopy of ultrathin frozen tissue sections. *J. Cell Biol.* 1981, **89**, 653–665.
72. Roth J. The preparation of protein A–gold complexes with 3 nm and 15 nm gold particles and their use in labelling multiple antigens on ultrathin sections. *Histochem. J.* 1982, **14**, 791–801.
73. Bonnard C., Papermaster D. S. and Kraehenbuhl J.-P. The streptavidin–biotin bridge technique: application in light and electron microscope immunocytochemistry. In: Polak J. M. and Varndell I. M., eds., *Immunolabelling for Electron Microscopy*. Amsterdam, Elsevier, 1984: 95.
74. Tapia F., Varndell I. M., Probert L., De Mey J. and Polak J. M. Double immunogold staining method for the simultaneous ultrastructural localization of regulatory peptides. *J. Histochem. Cytochem.* 1983, **31**, 977–981.
75. Bendayan M. Double immunocytochemical labeling applying the protein A–gold technique. *J. Histochem. Cytochem.* 1982, **30**, 81–85.
76. Gröschel-Stewart U. Immunocytochemistry of cytoplasmic contractile proteins. *Int. Rev. Cytol.* 1980, **65**, 193–254.
77. Langanger G., De Mey J., Moeremans M., Daneels G., De Brabander M. and Small J. V. Ultrastructural localization of α-actinin and filamin in cultured non-muscle cells with the immunogold staining (IGS) method. *J. Cell Biol.* 1984, **99**, 1324–1334.
78. Willingham M. An alternative fixation-processing method for preembedding ultrastructural immunocytochemistry of cytoplasmic antigens: the GBS (glutaraldehyde–borohydride–saponin) procedure. *J. Histochem. Cytochem.* 1983, **31**, 791–798.
79. Herman I. M. and Pollard T. D. Electron microscopic localization of cytoplasmic myosin with ferritin-labelled antibodies. *J. Cell Biol.* 1979, **86**, 212–234.
80. Geuens G., De Brabander M., Nuydens R. and De Mey J. The interaction between microtubules and intermediate filaments in cultured cells treated with taxol and nocodazole. *Cell Biol. Int. Rep.* 1983, **7**, 35–47.

81. Slot J. W. and Geuze H. J. A new method of preparing gold probes for multiple labeling cytochemistry. *Eur. J. Cell Biol.* 1985, **38**, 87–93.
82. Small J. V. and Celis. Filament arrangements in negatively stained cultured cells: the organization of actin. *Eur. J. Cell Biol.* 1978, **16**, 308–325.
83. Schliwa M., Euteneuer U., Bulinski J. and Izant J. Calcium lability of cytoplasmic microtubules and its modulation by microtubule-associated proteins. *Proc. Natl Acad. Sci. USA*. 1981, **78**, 1037–1041.

9 *Electron Microscopical Immunocytochemistry*

I. M. Varndell and J. M. Polak

In the preceding chapter, De Mey introduced colloidal gold and described the procedures involved in the production of protein–gold complexes. There can be little doubt that the current awareness of electron microscopical immunocytochemical techniques is due in no small part to the impact of colloidal gold-labelled probes. In this chapter we detail the use of gold-labelled immunoreagents in a variety of different applications. However, it would be imprudent to fail to record the considerable contribution of other marker systems to electron microscopical immunocytochemistry.

1. MARKERS FOR ELECTRON MICROSCOPICAL IMMUNOCYTO-CHEMISTRY

1.1. Peroxidase

Of all immunomarkers peroxidase-labelled reagents are probably the most widely used. Peroxidase is stable, small (mol. wt. = about 40 kdal) and thus is able to penetrate tissue effectively, and its activity is not significantly impaired when conjugated to larger proteins, e.g. immunoglobulins. Several chromogens have been used to localize peroxidase activity, the most commonly encountered being 3,3'-diaminobenzidine (DAB). DAB is osmiophilic and can thus be rendered electron dense for visualization at the electron microscopical level. Indeed, gold chloride can also be used to enhance the ultrastructural appearance of oxidized DAB reaction product.[1] One advantage of the peroxidase–DAB system at the light microscopical level is the diffusion of reaction product from the true antigenic site. Such diffusion, when carefully controlled, may aid the identifica-

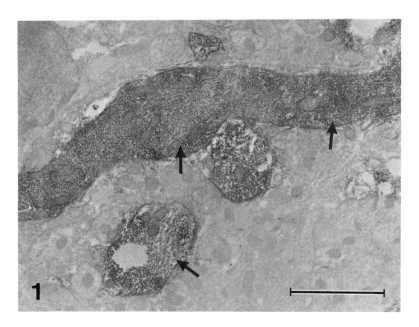

Fig. 9.1. Electron micrograph of a neurophysin-immunoreactive nerve fibre in the mouse supraoptic nucleus. Pre-embedding immunoperoxidase method. Oxidized diaminobenzidine reaction product (arrowed) is diffusely distributed throughout the fibre. Scale bar = 1 μm. (Figure provided by courtesy of Drs M. Castel, J. Morris and F. Shaw.)

tion of immunoreactive sites[2,3] by allowing rapid low magnification screening of the tissue. In favourable circumstances the diffusion of oxidized DAB reaction product may also be useful in electron microscopical immunocytochemistry. This is particularly true in pre-embedding immunocytochemical procedures (for review *see* the literature;[4-7] *Fig.* 9.1). However, it is acknowledged that peroxidase–DAB reaction product diffusion limits the usefulness of this technique in post-embedding (on-grid) procedures. In contrast, recent improvements in immunostaining technology, such as the dinitrophenol-labelled hapten sandwich staining technique,[8] have increased the sensitivity of the enzyme detection system with a consequent reduction in the probability of reaction product diffusion. Notwithstanding, oxidized DAB deposits tend to obliterate the fine structure of the marked organelle. This is strikingly evident at high magnification when the low electron opacity of the reaction product is also emphasized.

It should also be mentioned that 4-chloro-l-naphthol[9] has been used in peroxidase development at the electron microscopical level.[10-12]

1.2. Ferritin and Imposil

Horse spleen ferritin was the first reported electron immunocytochemical marker.[13,14] Ferritin has an iron-containing core (about 7 nm diameter) within a protein shell and is thus a particulate label of reasonable electron density. Imposil, an iron–dextran particle,[15] has also been used as an immunomarker. Both particles are particularly compatible with immunolabelling on ultrathin frozen sections (for review see[16]).

1.3. Radiolabels

The use of monoclonal antibodies internally labelled with tritium has already found applications in immunocytochemical procedures at both light and electron microscopical levels.[17–19] Of particular interest is the combination of auto-radiography with immunocytochemistry.[7,20–23] Chapter 7 deals with radioimmuno-cytochemical techniques in considerable detail.

1.4. Metals

Although they have not found widespread application, uranium,[24] uranyl–osmium bridges,[25–27] iron[28] and mercury[29,30] have been used to label antibody proteins. Colloidal particle–protein complexes are dealt with in Chapter 8.

1.5. Polyethyleneimine

Schurer and coworkers[31] introduced polyethyleneimine as a marker for use in electron microscopy. This ethylene derivative (mol. wt of 30–40 kdal) has been shown to complex with gold and silver to produce particles of up to 10 nm diameter.

2. METHODS

Some of the major procedures employed are given below, together with an indication of the markers used and particular applications.

2.1. Transmission Electron Microscopy

2.1.1. Non-embedding Methods

One of the most recent introductions to electron immunocytochemistry is that of cryo-ultramicrotomy (for review see[16]). Cells or tissues are lightly fixed, cryoprotected by infusion with 2·3 M sucrose, and snap-frozen. Ultrathin sections are then cut and mounted on carbon-stabilized coated copper grids. Immunostaining may then be performed (see Appendix, Section II). The technique is of particular importance for antigens which may be solubilized by solvent dehydration, heat-labile antigens, structural components etc. Peroxidase-,[32] gold-,[33–35] ferritin-[16,36,37] and Imposil-[15,16] labelled antibodies have all been used for immunocytochemistry on cryo-ultramicrotome sections. Receptors have also been localized on ultrathin frozen sections.[38]

2.1.2. Pre-embedding Methods

This is the method of choice for cell surface antigen and receptor immunolocalization[39–42] and for the detection of antigens prone to solubilization or denaturation by dehydrating agents and resin components. It is also the most suitable technique for electron microscopical studies of large and highly heterogeneous tissues, such as mammalian brain.[7,43]

Briefly, thick slices (from 20 to several hundred micrometres) of fresh or fixed tissue are cut on a vibrating knife microtome or tissue chopper. The slices are then immunostained (*see* Appendix, Section III, for technical details) using a modified indirect immunoperoxidase technique, usually in the presence of Triton X-100, saponin or a similar detergent which will aid penetration. The slices are then postfixed and contrasted with osmium tetroxide to enhance the DAB reaction product, dehydrated and flat-embedded in resin. Considerable care must be exercised during sectioning as the optimal reaction deposits are found in a narrow band some 2–4 μm below the cut surface of the slice. This penetrable depth appears to be independent of the total slice thickness.

Colloidal particle-labelled immunoreagents do not penetrate tissue slices but these markers have been used to localize cytoskeletal components in monolayers of cultured cells[44] at both light and electron microscopical levels. Indeed, a three-dimensional high resolution image of cytoskeletal proteins can be obtained using high voltage electron microscopy.

Recent developments in freeze-fracture cytochemistry using particulate markers should also be mentioned at this point. Fracture-label techniques permit direct marking of freeze-fractured, thawed plasma and intracellular membranes as well as cross-fractured cytoplasm (*see*[45] for review). In addition, work is in progress to establish the optimal conditions for immunolabelling of ultrathin sections of polyethylene glycol (PEG)-embedded[46] or polyvinylalcohol-embedded[47] tissue.

2.1.3. Post-embedding Methods

Two major categories should be considered.

2.1.3.1. SEMITHIN–THIN PROCEDURE. Alternate semithin (0·1–2 μm) and thin (60–100 nm) sections are cut from resin-embedded tissue blocks. The semithin sections are mounted on glass slides and immunostained using any of the immunohistochemical procedures described throughout this book. The serial thin sections are contrasted and viewed by transmission electron microscopy. Consequently immunohistochemical and conventional electron microscopical images can be correlated. Although immunoreactive cells identified at the light microscopical level can be correlated with their electron microscopical appearance the major disadvantage is that the actual subcellular site of immunoreactivity cannot be visualized.

2.1.3.2. ON-GRID IMMUNOSTAINING. The immunocytochemical reaction is performed directly on ultrathin, gridmounted tissue sections. Enzyme and colloidal particle-linked markers can be used in on-grid procedures. For reviews of the techniques employed *see* the literature.[3,8,44,48–51] Procedures for on-grid immunostaining are given in the Appendix, Section IV. On-grid immunostaining allows high resolution immunolocalization studies to be performed and elegant modifications, including serial section immunostaining and multiple immunolabelling procedures, have been reported (*see* Section 4).

3. TISSUE PREPARATION

Each author routinely practising electron microscopial immunocytochemistry advocates a personal preference for every stage in the preparation of tissue, from collection to staining. It would be impracticable to discuss each approach in this chapter. Specialized volumes[52–54] should be consulted for individual requirements.

Generally, tissue for immunocytochemical examination should be obtained as fresh as possible and should be processed to an inert state, e.g. resin block, without undue delay. High temperatures (greater than 55–60 °C) should be avoided (*see* Section 3.3.) and all solutions, wherever practicable, should be neutral buffered. Representative tissue samples should be taken to minimize the risk of errors induced by uneven distribution of antigens. If possible, a range of fixatives and processing protocols should be evaluated for each specimen. Whenever possible, and if deemed relevant, complementary data should be obtained from morphology (conventional light and electron microscopy and cytochemistry) and biochemical (radioimmunoassay, chromatography), physiological or pharmacological techniques.

3.1. Fixation

It is paradoxical that good morphological preservation often precludes maximal retention of antigen immunoreactivity; a compromise must be achieved to provide optimal reaction product deposition with acceptable ultrastructure. Formaldehyde and glutaraldehyde, either alone or in mixtures, have been used routinely in the majority of electron immunocytochemical procedures. This contrasts markedly with findings obtained at the light microscopical level which seem to indicate that glutaraldehyde is a poor fixative for immunohistochemistry in general. One explanation may be that glutaraldehyde is such an efficient protein cross-linking agent that immunoreagents are physically prevented from penetrating the tissue section. Immunolabelling at the electron microscopical level is a surface phenomenon and thus immunoreagent penetration is not critical. It should be emphasized, however, that many other reasons could be advanced. Acrolein,[55] periodate–lysine–paraformaldehyde,[56] ethylacetimidate[36] and picric acid-containing fixatives[57] have also been used for electron immunocytochemistry. Osmolarity and pH of the fixatives and washing buffers should be tailored to 'physiological' levels.

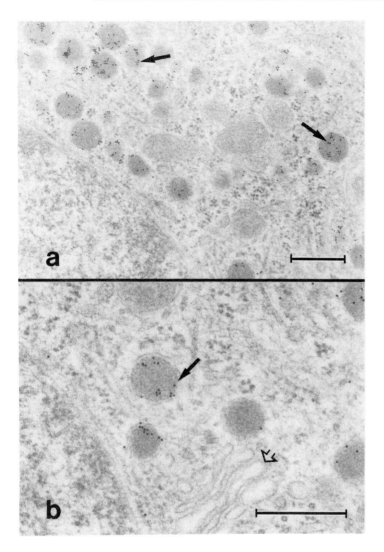

Fig. 9.2. Rat pancreatic D cell, fixed in glutaraldehyde and contrasted with osmium tetroxide. Immunogold staining procedure for somatostatin using 10-nm gold particles (solid arrows). Golgi cisternae are marked with an open arrowhead. Uranyl acetate and lead citrate counterstains. Scale bars = 300 nm.

3.1.1. Osmium Tetroxide

A second paradox in electron immunocytochemistry is that osmium tetroxide acts as an excellent membrane stabilizer and contrasting agent but also efficiently masks many protein antigens. There are many published exceptions to this dictum[58,59] and there is also evidence to suggest that the action of osmium tetroxide is reversible[60] (*Fig.* 9.2). Details for the treatment of osmicated tissue are given in the Appendix, Section IV.A.

3.2. Treatment after Fixation

Prolonged exposure to aqueous media, dehydrating agents and liquid phase resin components will elute small and soluble antigens totally or with such efficiency that the amount remaining is undetectable within the limits of immunocytochemical detection.[61] Indeed, in an experiment conducted in our laboratory, it was found that significant morphological deterioration was observed in ultrathin resin-embedded tissue sections exposed to immunoreagents for 24–48 h when compared to sections similarly treated but incubated for only 1 h.

Cryogenic procedures obviate this deterioration to a considerable degree, either by the use of cryo-ultramicrotome sections or embedding media which are hydrophilic (thus a degree of tissue hydration is tolerable) and/or may be used at low temperature (4 to −20 °C or lower).

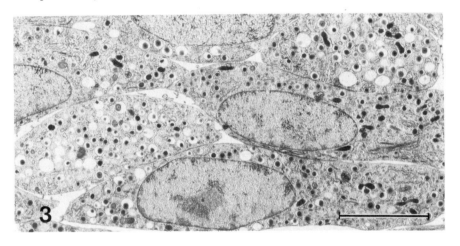

Fig. 9.3. Quick-frozen, freeze-dried, osmium vapour-fixed rat pancreatic B (insulin) cells from isolated islets. Note the distinct appearance of the secretory granules (arrowed). Uranyl acetate and lead citrate counterstains. Scale bar = 3 μm.

Recently, freeze substitution and freeze drying[62,63] techniques have been used successfully. In the latter procedure the tissue is snap-frozen, dried *in vacuo* for 1–2 days (initially at very low temperature), osmium vapour-fixed and embedded in Araldite resin — thus the only contact with liquid medium after freezing from the living state is the infiltrating resin (*Fig.* 9.3). Immunoreagent dilutions 25–100 times greater than those calculated to be optimal for on-grid immunostaining of conventionally treated (liquid phase fixed and alcohol-dehydrated) tissue can be used[63] (*Fig.* 9.4).

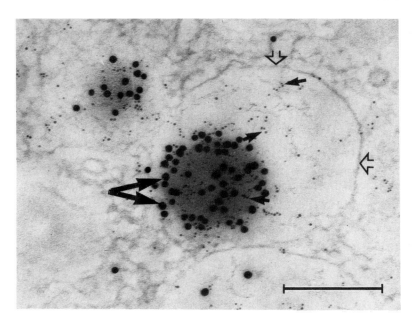

Fig. 9.4. Secretory granule from quick-frozen, freeze-dried, osmium vapour-fixed rat insulin-containing cell double immunostained (*see* Sections 4.3.3. and Appendix, Section VI) for insulin (large solid arrows) with 20-nm gold particles and C-peptide (pro-insulin; small solid arrows) with 5-nm gold particles. The insulin immunoreactivity is largely restricted to the core of the granule whilst the C-peptide (pro-insulin) immunoreactivity is distributed throughout the core and halo. The granule membrane is indicated by the open arrowheads. Uranyl acetate counterstain. Scale bar = 225 nm.

3.3. Embedding Media

Various types of resin are currently available and the relative advantages of each for on-grid electron immunocytochemistry are reviewed elsewhere.[64] There is a current trend away from the traditional hydrophobic epoxy resins, such as Araldite, Epon and Spurr in favour of the hydrophilic cross-linked acrylics (LR White, Lowicryl K4M) and new low acid glycol methacrylates. In general, the hydrophilic resins give better morphology and retention of antigenicity,[59,65–68] although they require punctilious attention to technical procedure to obtain high quality results. There is no doubt that some resins are not suitable for particular purposes and thus a range of embedding media should be tested for any given system to determine the optimal approach.

3.3.1. Etching

Much has been written in support and criticism of the oxidative 'etching' of resin sections prior to on-grid immunostaining. Causton[64] advocates the use of sodium alkoxide as a means of etching epoxy resin without denaturing the antigens under investigation. He pointed out that the oxidation of alkane chains in the polymerized resin by hydrogen peroxide, periodic acid or potassium permanganate, whilst increasing the hydrophilicity of the resin, will oxidize pendant alkane groups on the antigen with equal efficiency.

4. DOUBLE AND MULTIPLE IMMUNOLABELLING PROCEDURES

One logical extension of the capacity to localize a single antigen or groups of similar antigens at the electron microscopical level is the development of a reliable procedure which allows discrimination between two or more distinct, but coexisting or neighbouring, antigens. Various combinations of immunoenzyme, immunoferritin and immunogold procedures have been described for the ultrastructural demonstration of multiple antigens. In the overwhelming majority of cases one of the markers employed is the immunoperoxidase–DAB complex reaction product.[69] The immunoperoxidase system has been combined with radiolabelled antibodies[7,17,20] and ferritin-labelled antibodies,[70] in addition to techniques such as anterograde degeneration,[71,72] anterograde autoradiography,[73] Golgi impregnation[74] and retrograde peroxidase tracing.[6,75]

The quest for a reliable multiple immunostaining technique was made considerably easier by the introduction of colloidal gold. It is worth considering the basic techniques at this point.

4.1. Direct Immunolabelling

Direct marking of antigens with (gold-)labelled antibodies is the simplest, though not the most sensitive means of immunolocalization (*Fig. 9.5a*). This method has been revolutionized by the introduction of monoclonal antibodies. Simultaneous or sequential application of two or more monoclonal antibodies linked to different gold particle sizes is used frequently in the investigation of cell surface antigens.[40]

4.2. Indirect Immunolabelling — GLAD Method

The *g*old *l*abelled *a*ntigen *d*etection (GLAD) method (*Fig. 9.5b*) was initially described as a multiple immunolabelling procedure.[76] Each antigen (purified natural or synthetic) is conjugated to a different size of gold particles. Two (or more) antisera are applied to a tissue section and the sites of binding are marked using the differently gold-labelled antigens. This is an attractive technique, based on the bivalency of antibody molecules, as the antibody should recognize the same determinants on the section and on the gold-labelled reagent. In practice every antigen to be localized requires its own gold-labelled reagent. These are laborious and expensive to prepare. For further details see the literature.[49,50,76]

4.3. Indirect Immunolabelling — Protein A–Gold and Immunoglobulin–Gold

Protein A is a cell wall constituent produced by most strains of *Staphylococcus*

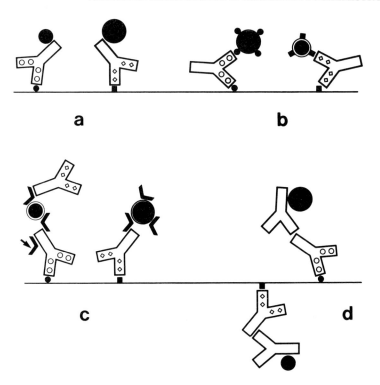

Fig. 9.5. Diagrammatic representations of double immunostaining procedures. (*a*) Antibodies directly labelled with different markers (a combination of PAP, ferritin, Imposil, radiolabel, gold etc. or two gold particle sizes). (*b*) GLAD method (*see* Section 4.2). (*c*) Double protein A–gold method. Antigen *circle* is localized with a specific antibody. Protein A–gold (3–5 nm) is used to localize this antibody. Free protein A (small arrow) is then added to block potential reactive sites remaining on the first antibody. Excess anti-*square* serum is added. This binds to antigen *square* and to free protein A sites on the protein A– small gold complex. Antigen *square* can then be detected using protein A–large gold particles. (*d*) Selected surface immunolabelling. Modification of Bendayan's[78] technique. Two antisera from a common donor species may be localized on a single electron microscope grid (*see* Section 4.3.2. and Appendix, Section V).

aureus. Protein A has a very high affinity for immunoglobulin molecules (particularly IgG) from many mammalian species. The interaction is pseudo-immune and takes place mainly at the C_{H_2} and C_{H_3} domains of the Fc region of immunoglobulin G. Protein A bound to gold particles is the most widely used gold-labelled reagent in electron immunocytochemistry. Two techniques have been described to date in which multiple antisera from the same, or from more than one, donor species can be visualized simultaneously.

4.3.1. Double Protein A–Gold Immunolabelling

This involves the sequential application of two separate antibody–protein A–gold procedures with intermediate blocking of potential immunoreactive sites with free protein A to prevent cross-contamination (*Fig.* 9.5c).[33,35,77]

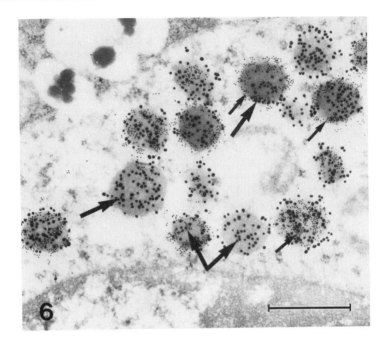

Fig. 9.6. Human pancreatic D cell. Secretory granules are double immunostained for rat somatostatin cryptic peptide with 20-nm gold particles (large arrows) and somatostatin-28-(1–12) with 10-nm gold particles (small arrows). Apparently, there is some intergranular segregation of immunoreactants. This may reflect a technical problem in that only antigens exposed at the surface of the section will be available to react.[85] Uranyl acetate and lead citrate counterstains. Scale bar = 300 nm.

4.3.2. Selected Surface Immunolabelling

By careful manipulation of electron microscope grids, one surface of the mounted section may be immunostained without contamination of the other face. Bendayan[78] used this technique to immunostain both faces of a single grid by the sequential application of different antisera which were visualized using two protein A–gold reagents with non-overlapping particle size ranges. We have modified this procedure by replacing the protein A–gold immunoreagents with immunoglobulin–gold complexes (*Figs.* 9.5d and 9.6). This elegant technique requires great care to ensure that the immunoreagents are not transferred from one face of the section to the other. Short incubation times and droplet washing procedures are recommended (*see* Appendix, Section V) to obtain optimal results.

4.3.3. Double Immunogold Staining Procedure

Occasionally, either by convention, design or convenience, some primary antisera are raised in species other than the rabbit. We have recently described a double

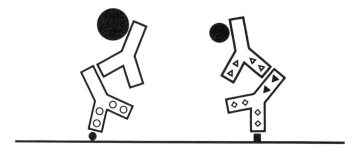

Fig. 9.7. Diagrammatic representation of the double immunogold staining procedure in which two antigens are localized simultaneously using antisera from two different donor species. Second layer antisera adsorbed with gold particles of different sizes are used to visualize the primary immunoglobulins.

immunolabelling technique based on the availability of primary antisera raised in different species.[79] This technique is described in detail in the literature.[69,80,81] Modifications to the published technique are given in the Appendix, Section VI. (*See also Figs*. 9.4 and 9.7.)

5. CONCLUSION

In this chapter we have attempted to provide an introduction to the rapidly expanding field of electron immunocytochemistry. It is beyond the means of a single chapter to cover all the techniques and immunomarkers currently available to electron microscopists and the reader is thus directed to more detailed texts (*see* the literature[52–54,82]). There can be little doubt that future developments with monoclonal antibodies will lead to an increased use of direct labelling techniques at the electron microscopical level. Moreover, the application of in situ hybridization histochemical procedures, in combination with immunocytochemistry at the electron microscopical level, will open still more cellular compartments to the scrutiny of morphologists.

Appendix

I. WASHING OF ELECTRON MICROSCOPE GRIDS

In previous accounts[48] we advocated the 'jet-washing' system for electron microscope grids. Recently we have evaluated a much more effective and less section destructive procedure for washing grids. Briefly, all immunostaining steps are performed by floating the grids on, or immersing the grids in, droplets of immunoreagent held within microtest (Terasaki) plates (60 × 15 µl well disposable plastic plates; Falcon Plastics, Becton Dickinson, Oxford, UK). After staining the grids are transferred, using anti-capillary forceps if possible, to larger

(100–200 μl) droplets of washing buffer on Parafilm® or dental wax sheets. Alternatively, larger volume microtest plates can be used. The plates are then placed on the top-plate of a magnetic stirrer. The stirrer is switched on and the spin speed increased until the grids are induced to rotate on or in the droplets of washing buffer. The speed of rotation should be slow (e.g. 1 rotation per second). We routinely use 6 washes of 1 min each per grid. We have encountered no significant induced magnetism of the grids but prolonged manipulation can result in the forceps becoming magnetized. Anti-magnetic forceps or a commercial loop demagnetizer can be used to alleviate this problem. One note of caution — some magnetic stirrers heat up due to friction at their bearings. This should be avoided at all costs as buffer will evaporate, antigens may be denatured etc. Other agitation platforms could be devised.

II. IMMUNOSTAINING CRYO-ULTRAMICROTOME SECTIONS

A. Infusion of Cryoprotectant and Freezing

Glycerol, polyethylene glycol (PEG) and sucrose are excellent cryoprotectants but only PEG and sucrose convey sufficient plasticity for cryosectioning to the frozen tissue.

Tissues are fixed in formaldehyde and/or glutaraldehyde, divided into 0·5-mm cubes, and then infused with 0·6–2·3 M sucrose. Infusion rates are tissue and concentration dependent. The tissue blocks are frozen by plunging into Arcton or isopentane slush, liquid nitrogen or by bounce-free contact against an ultra-cold polished metal plate.

B. Sectioning and Section Collection

A discussion of sectioning techniques is beyond the scope of this chapter. Full details may be found in the literature.[16,36,37]

C. Immunostaining

(Adapted from Tokuyasu.)[16] Cryo-ultramicrotome sections are thawed onto coated mesh grids.

i. Conditioning

Grids with sections are placed section down on a wet plate of 0·3 per cent agarose and 1 per cent gelatin to remove the sucrose by diffusion. PBS containing 0·01–0·02 M glycine (PBS-g) is used to float the sections off the plate. They are then removed from the plate and placed on droplets of 2 per cent gelatin (in PBS-g) and left for 10 min. Wire loops may be used to transfer the grids.

ii. Immunostaining

After conditioning the grids are treated as follows:

1. Float, section-side down, on PBS-g for 10 min, or 1 per cent freshly prepared sodium borohydride in PBS for 10 min.
2. Wash on PBS.
3. Primary antiserum. Calculate optimal time, temperature and dilution for each antiserum and tissue sample. The antiserum diluent (see Section IV, Step 4) may also be substantiated with gelatin and/or ovalbumin (up to 0·4 per cent w/v).
4. Wash on PBS.
5. Gold-labelled probe. Calculate optimal time, temperature and dilution for each new application.
6. Wash on PBS.
7. 2 per cent glutaraldehyde in distilled water — 5 min.
8. Wash in distilled water.

iii. Staining

Contrasting and positive staining procedures are described in detail elsewhere.[16,83,84]

III. PRE-EMBEDDING PEROXIDASE–ANTI-PEROXIDASE IMMUNOCYTOCHEMISTRY

Experimental details in pre-embedding immunocytochemistry vary widely and depend upon the particular antigen, primary antibody and biological system under study. (After Priestley.)[7]

Day 1

If practicable perfuse animal with 4 per cent paraformaldehyde and 0·2 per cent glutaraldehyde in 0·1 M sodium phosphate buffer pH 7·4. A suitable volume for a rat is exsanguination with 20–50 ml sodium phosphate buffer followed by 200–500 ml fixative. Dissect out the tissue of interest and immersion-fix for a further 2 h. Tissue that cannot be perfused should be immersion-fixed for 2–4 h at 4 °C.

1. Cut tissue into blocks, transfer to phosphate buffer containing 2·3 M sucrose and leave until fully infiltrated (6–8 h). Rapidly freeze blocks by plunging them into liquid nitrogen; thaw blocks by returning them to 2·3 M sucrose and then transfer to Tris-buffered saline or PBS. The purpose of the freezing/thawing is to disrupt the tissue sufficiently to facilitate immunoreagent penetration but minimize ultrastructural deterioration. Other protocols use dehydration/rehydration and low concentrations of detergents to attempt to produce similar results.

2. Cut 20–50 μm thick slices on a vibrating knife microtome (Vibratome).
3. Incubate slices in 10 per cent normal serum, e.g. goat or swine, for 30–60 min.
4. Wash in TBS/PBS.
5. Incubate in primary antibody, e.g. rabbit, overnight.

Day 2

1. Wash slices in TBS/PBS, incubate in link antibody for 60 min (e.g. goat anti-rabbit IgG 1 : 10–1 : 100).
2. Wash in TBS/PBS.
3. Incubate in PAP reagent (e.g. 1 : 50 rabbit IgG) for 90–120 min.
4. Wash in TBS/PBS.
5. Stain slices using 3,3′-diaminobenzidine (10 min in 0·06 per cent DAB in 0·05 M Tris-HCl pH 7·6, room temperature, add H_2O_2 to give 0·01 per cent). Terminate reaction (after 5–10 min) by replacing DAB solution with TBS/PBS.
6. Wash slices thoroughly, stain for 2 h with 1 per cent osmium tetroxide.
7. Wash well with TBS/PBS.
8. Place slices between two glass or plastic coverslips, dehydrate and bring to resin. For flat embedding, Durcupan is recommended because the polymerized resin will easily lift off glass microscope slides. Alternatively, silicone-based releasing agents such as dimethyldichlorosilane can be coated onto glass slides to prevent resin binding.

Day 3

Place slices in a drop of Durcupan (or other resin) on a warmed microscope slide. Cover with a plastic coverslip. Polymerize resin.

The slide-mounted slices can be examined in a light microscope. Subsequently the slide is warmed on a hot plate and areas of interest cut from the resin-embedded tissue. Such areas can be mounted in an electron-microscopical embedding capsule using fresh resin or can simply be glued onto a polymerized blank capsule using a rapid setting epoxy adhesive. Thin sections are cut for electron microscopical observation using standard procedures.

IV. ON-GRID IMMUNOSTAINING USING PROTEIN A–GOLD OR IMMUNOGLOBULIN–GOLD PROBES

1. Ultrathin sections exhibiting silver to silver-grey interference colours should be collected on coated or uncoated nickel or gold grids ('etching' agents will attack copper). Allow to dry for 2–24 h at 20–35 °C.
2. 'Etching' (*see* Section 3.3.1). An etching agent may be used at this stage if required. Grids should be floated section-side down onto the oxidizing agent. Typical etching regimes may be: 10 per cent H_2O_2 for 10 min; 1 per cent w/v periodic acid for 1 min; 1 per cent sodium (m)ethoxide in absolute

(m)ethanol for 10–30 min. Rinse well with appropriate solvent and bring to PBS or TBS pH 7·2–7·6.

3. Aldehyde block. To block reactive −CHO groups induced by fixation and etching, the grids should be immersed or floated onto a blocking solution. Typically, this would be either Tris–glycine buffer (0·1 M, pH 7·2); 1 per cent sodium borohydride freshly prepared; 1–10 per cent w/v bovine serum albumin (globulin free), ovalbumin or normal serum from second antiserum donor species etc. for 10–60 min at room temperature. NB Globulin-containing blockers should not be used in the protein A–gold technique as a 'non-specific' pseudo-immune protein A–globulin reaction may occur. Drain grids on fibre-free paper but do not dry.

4. Primary antiserum. Assess optimal time, temperature and dilution for each antiserum and tissue sample. The antiserum diluent used routinely is TBS or PBS containing 0·1 per cent bovine serum albumin (globulin-free) and 0·01 per cent sodium azide pH 7·2.

5. Rinse well in TBS or PBS.

[For protein A–gold go next to Step 8.]

6. Rinse in TBS containing 0·2 per cent BSA.

7. Stand grids in TBS containing 1 per cent BSA (globulin-free) pH 8·4 for 5–10 min.

8. Gold-labelled immunoreagent. Optimal dilution (*see* Chapter 8), time and temperature.

9. Rinse well in TBS or PBS.

10. Rinse well in distilled water.

11. Counterstain as desired. (M)ethanolic heavy metal salts may be used.

A. Osmicated Tissue

Bendayan and Zollinger[60] suggested the use of a saturated aqueous solution of sodium metaperiodate to reverse the antigen-masking effect of osmication. A saturated aqueous solution of sodium metaperiodate should be prepared 2–3 days before use. Store in a dark bottle in the refrigerator. Allow to warm to room temperature before use. Grids should be floated or immersed in this solution for 10–60 min at room temperature. The sodium metaperiodate is normally used in place of an etching agent. The mechanism of action is obscure but it is possible that low concentrations of periodic acid etch the section and create a hydrophilic surface gel from which osmium may be eluted by oxidative action. Essentially, the surface of the section, at which immunostaining normally occurs, is de-osmicated although the bulk of the section retains osmium and hence appears to be osmicated. It should be mentioned that periodic acid will cleave carbohydrate and peptidic bonds and may prove to be highly destructive to the antigen under investigation.

V. SELECTED SURFACE IMMUNOLABELLING

This modification, employing two sequential immunoglobulin–gold, or protein A–gold and immunoglobulin–gold procedures is adapted from Bendayan.[78]

1. Ultrathin sections must be mounted on uncoated 200 or 300-mesh nickel or gold grids. The grids must be floated throughout the procedure.
2. An on-grid technique is performed (*see* Section IV, Steps 2–10 inclusive); the grids are dried at the end of the final distilled water rinse.
3. A second immunostaining procedure is performed on the grid-side of the sections. The washing steps should be extended to ensure that immunoreagents are washed entirely from the recesses of the support grid. Great care must be exercised to ensure that reagents are not transferred from one face of the grid to the other.
4. Following completion of both immunostains the sections may be counterstained.

VI. DOUBLE IMMUNOGOLD STAINING PROCEDURE

This technique is described in considerable detail in previous publications.[69,80,81]

One modification we should like to advocate is the adoption of the magnetic stirrer washing procedure described in Section I.

VII. SOURCES OF IMMUNE REAGENTS USED FOR ILLUSTRATED PREPARATIONS

Antibody to	*Source*
Neurophysin (OT-NP and VP-NP cross-reactive)	Courtesy of Dr M. Castel, Hebrew University of Jerusalem, Jerusalem, Israel
Clono-PAP (monoclonal mouse PAP)	Sternberger-Meyer Immunochemicals Inc., Jarretsville, MD, USA
Somatostatin-14	Departments of Histochemistry and Medicine, Hammersmith Hospital
Somatostatin-28	Courtesy of Dr R. Benoit, General Hospital, Montreal, Canada
Rat somatostatin cryptic peptide (RSCP)	Courtesy of Dr R. H. Goodman, Tufts Medical Center, Boston, MA, USA
Insulin	Wellcome
C-Peptide	Courtesy of Professor N. Yanaihara, Shizuoka, Japan

OT = oxytocin; VP = vasopressin

REFERENCES

1. Newman G. R., Jasani B. and Williams E. D. Metal compound intensification of the electron density of diaminobenzidine. *J. Histochem.* 1983, **31**, 1430–1434.
2. Bretton R. Comparison of peroxidase and ferritin labelling for localization of specific cell surface antigens. *Proceedings of the Seventh International Congress of Electron Microscopy*, Grenoble, 1970: 527–528.

3. Romano E. L. and Romano M. Historical aspects. In: Polak J. M. and Varndell I. M., eds., *Immuno!abelling for Electron Microscopy*. Amsterdam, Elsevier Science Publishers, 1984: 3–15.

4. Pickel V. M. Immunocytochemical methods. In: Heimer L. and Robards M. J., eds., *Neuroanatomical Tract Tracing Methods*. New York, Plenum Press, 1983: 483–509.

5. Vaughn J. E., Barber R. P., Ribak C. E. and Houser C. R. Methods for the immunocytochemical localization of proteins and peptides involved in neurotransmission. In: Johnson J. E., ed., *Current Trends in Morphological Techniques, Vol. III*. Florida, CRC Press, 1981: 33–70.

6. Priestley J. V. and Cuello A. C. Electron microscopic immunocytochemistry for CNS transmitters and transmitter markers. In: Cuello A. C., ed., *Immunohistochemistry*. Chichester, John Wiley & Sons, 1983: 273–321.

7. Priestley J. V. Pre-embedding ultrastructural immunocytochemistry: Immunoenzyme techniques. In: Polak J. M. and Varndell I. M., eds., *Immunolabelling for Electron Microscopy*. Amsterdam, Elsevier Science Publishers, 1984: 37–52.

8. Newman G. R. and Jasani B. Post-embedding immunoenzyme techniques. In: Polak J. M. and Varndell I. M., eds., *Immunolabelling for Electron Microscopy*. Amsterdam, Elsevier Science Publishers, 1984: 53–70.

9. Nakane P. Simultaneous localization of multiple tissue antigens using the peroxidase-labeled antibody method: A study in pituitary glands of the rat. *J. Histochem. Cytochem*. 1968, **16**, 557–560.

10. Li J. Y., Dubois M. P. and Dubois P. M. Somatotrophs in the human fetal anterior pituitary. An electron microscopic immunocytochemical study. *Cell Tissue Res*. 1977, **181**, 545–552.

11. Li J. Y., Dubois M. P. and Dubois P. M. Ultrastructural localization of immunoreactive corticotropin, β-lipotropin, α- and β-endorphin in cells of the human fetal anterior pituitary. *Cell Tissue Res*. 1979, **204**, 37–43.

12. Baskin D. G., Mar H., Gorray K. C. and Fujimoto W. Y. Electron microscopic immunoperoxidase staining of insulin using 4-chloro-1-naphthol after osmium fixation. *J. Histochem. Cytochem*. 1982, **30**, 710–712.

13. Singer S. J. Preparation of an electron-dense antibody conjugate. *Nature* 1959, **183**, 1523–1524.

14. Singer S. J. and Schick A. I. The properties of specific stains for electron microscopy prepared by conjugation of antibody with ferritin. *J. Biophys. Biochem. Cytol*. 1961, **9**, 519–537.

15. Dutton A., Tokuyasu K. T. and Singer S. J. Iron–dextran antibody conjugates: General method for simultaneous staining of two components in high-resolution immunoelectron microscopy. *Proc. Natl Acad. Sci. USA* 1979, **76**, 3392–3396.

16. Tokuyasu K. T. Immuno-cryoultramicrotomy. In: Polak J. M. and Varndell I. M., eds., *Immunolabelling for Electron Microscopy*. Amsterdam, Elsevier Science Publishers, 1984: 71–82.

17. Cuello A. C., Priestley J. V. and Milstein C. Immunocytochemistry with internally labelled monoclonal antibodies. *Proc. Natl Acad. Sci. USA* 1982, **79**, 665–669.

18. Descarries L. and Beaudet A. The use of autoradiography for investigating transmitter-specific neurons. In: Bjorklund A. and Hökfelt T., eds., *Handbook of Chemical Neuroanatomy. Vol. 1: Methods in Chemical Neuroanatomy*. Amsterdam, Elsevier, 1983: 286.

19. Ross M. E., Park D. H., Teitelman G., Pickel V. M., Reis D. J. and Joh T. H. Immunohistochemical localization of choline acetyltransferase using a monoclonal antibody: An autoradiographic method. *Neuroscience* 1983, **10**, 907–922.

20. Cuello A. C., Milstein C. and Galfré G. Immunocytochemistry with monoclonal antibodies. In: Cuello A. C., ed., *Immunohistochemistry*. Chichester, John Wiley & Sons, 1983: 215–255.

21. Pelletier G. Identification of endings containing dopamine and vasopressin in the rat posterior pituitary by a combination of radioautography and immunocytochemistry at the ultrastructural level. *J. Histochem. Cytochem*. 1983, **31**, 562–564.

22. MacMillan F. M., Sofroniew M. V., Sidebottom E. and Cuello A. C. Immunocytochemistry with monoclonal anti-immunoglobulin as a developing agent: Application to immunoperoxidase staining and radioimmunocytochemistry. *J. Histochem. Cytochem*. 1984, **32**, 76–82.

23. Pickel V. M. and Beaudet A. Combined use of autoradiography and immunocytochemical methods to show synaptic interactions between chemically defined neurons. In: Polak J. M. and Varndell I. M., eds., *Immunolabelling for Electron Microscopy*. Amsterdam, Elsevier Science Publishers, 1984: 259–266.

24. Sternberger L. A., Donati E. J., Cuculis J. J. and Petrali J. P. Indirect immunouranium technique for staining of embedded antigen in electron microscopy. *Exp. Mol. Pathol*. 1965, **4**, 112–125.

25. Hanker J. S., Deb C., Wasserkrug H. L. and Seligman A. M. Staining tissue for light and electron microscopy by bridging metals with multidentate ligands. *Science* 1966, **152**, 1631–1634.
26. Seligman A. M., Wasserkrug H. L. and Hanker J. S. A new staining method (OTO) for enhancing contrast of lipid-containing membranes and droplets in osmium tetroxide-fixed tissue with osmiophilic thiocarbohydrazide (TCH). *J. Cell Biol.* 1966, **30**, 424–432.
27. Sternberger L. A., Donati E. J., Hanker J. S. and Seligman A. M. Immuno-diazothioether-osmium tetroxide (immuno-DTO) technique for staining embedded antigen in electron microscopy. *Exp. Mol. Pathol. Suppl.* 1966, **3**, 36–43.
28. Yamamoto N. Application of metal labeled antibody method and a consideration of the fixatives suitable for immunohistochemical study. *Acta Histochem. Cytochem.* 1977, **10**, 246–252.
29. Kendall P. A. Labelling of thiolated antibody with mercury for electron microscopy. *Biochim. Biophys. Acta* 1965, **97**, 174–176.
30. Kendall P. A. Antibody labelling for electron microscopy: Immunochemical properties of immunoglobulin G heavily labelled with methylmercury after thiolation by homocysteine thiolactone. *Biochim. Biophys. Acta* 1972, **257**, 101–110.
31. Schurer J. W., Hoedemaeker P. J. and Molenaar I. Polyethyleneimine as tracer particle for (immuno) electron microscopy. *J. Histochem. Cytochem.* 1977, **25**, 384–387.
32. Antakly T. W., Tanaka S., Ohkawa K.-I. and Bernhard W. Cytochemical localization of peroxidase and peroxidase-labelled antibodies in ultrathin frozen sections. *Cell. Mol. Biol.* 1979, **24**, 205–212.
33. Geuze, H. J., Slot J. W., Van der Ley P. A. and Scheffer R. C. T. Use of colloidal gold particles in double labelling immunoelectron microscopy of ultrathin frozen tissue sections. *J. Cell Biol.* 1981, **89**, 653–665.
34. Slot J. W. and Geuze H. J. The use of protein A–colloidal gold (PAG) complexes as immunolabels in ultrathin frozen sections. In: Cuello A. C., ed., *Immunohistochemistry*. Chichester, John Wiley & Sons, 1983: 323–346.
35. Slot J. W. and Geuze H. J. Gold markers for single and double immunolabelling of ultrathin cryosections. In: Polak J. M. and Varndell I. M., eds., *Immunolabelling for Electron Microscopy*. Amsterdam, Elsevier Science Publishers, 1984: 129–142.
36. Tokuyasu K. T. and Singer S. J. Improved procedures for immunoferritin labeling of ultrathin frozen sections. *J. Cell Biol.* 1976, **71**, 894–906.
37. Tokuyasu K. T. Immunochemistry on ultrathin frozen sections. *Histochem. J.* 1980, **12**, 381–403.
38. Geuze H. J., Slot J. W., Strous G. J. A. M., Codish H. F. and Schwartz A. L. Immunocytochemical localization of the receptor for asialoglycoprotein in rat liver cells. *J. Cell Biol.* 1982, **92**, 865–870.
39. De Waele M., De Mey J., Moeremans M., De Brabander M. and Van Camp B. Immunogold staining method for the detection of cell surface antigens with monoclonal antibodies. In: Bullock G. R. and Petrusz P., eds., *Techniques in Immunocytochemistry, Vol. 2*. New York, Academic Press, 1983: 1–23.
40. De Waele M. Haematological electron immunocytochemistry. Detection of cell surface antigens with monoclonal antibodies. In: Polak J. M. and Varndell I. M., eds., *Immunolabelling for Electron Microscopy*. Amsterdam, Elsevier Science Publishers, 1984: 267–288.
41. Matutes E. and Catovsky D. The fine structure of normal lymphocyte subpopulations — a study with monoclonal antibodies and the immunogold technique. *Clin. Exp. Immunol.* 1982, **50**, 416–425.
42. Robinson D. S. F., Tavares de Castro J., Polli N., O'Brien M. and Catovsky D. Simultaneous demonstration of membrane antigens and cytochemistry at ultrastructural level: A study with the immunogold method, acid phosphatase and myeloperoxidase. *Br. J. Haematol.* 1984, **56**, 617–631.
43. Langley O. K., Ghandour M. S., Vincendon G. and Gombos G. An ultrastructural immunocytochemical study of nerve-specific protein in rat cerebellum. *J. Neurocytol.* 1980, **9**, 783–798.
44. De Mey J., Moeremans M., Geuens G., Nuydens R. and De Brabander M. High resolution light and electron microscopic localization of tubulin with the IGS (ImmunoGold Staining) method. *Cell Biol. Int. Rep.* 1981, **5**, 889–899.
45. Pinto da Silva P. Freeze-fracture cytochemistry In: Polak J. M. and Varndell I. M., eds., *Immunolabelling for Electron Microscopy*. Amsterdam, Elsevier Science Publishers, 1984: 179–188.
46. Wolosewick J., De Mey J. and Meininger V. Ultrastructural localization of tubulin and actin in polyethylene glycol-embedded rat seminiferous epithelium by immunogold staining. *Biol. Cell* 1983, **49**, 219–226.

47. Small J. V. Polyvinylalcohol, a water soluble resin suitable for electron microscope immunocytochemistry. In: Csanady A., Röhlich P. and Szabo D., eds., *Proceedings of the Eighth European Congress Electron Microscopy. Vol. 3*. Budapest, 1984: 1799.
48. Varndell I. M., Tapia F. J., Probert L., Buchan A. M. J., Gu J., De Mey J., Bloom S. R. and Polak J. M. Immunogold staining procedure for the localisation of regulatory peptides. *Peptides* 1982, **3**, 259–272.
49. Larsson L.-I. Peptide immunocytochemistry. *Prog. Histochem. Cytochem.* 1981, **13(4)**, 1–85.
50. Larsson L.-I. Labelled antigen detection methods. In: Polak J. M. and Varndell I. M., eds., *Immunolabelling for Electron Microscopy*. Amsterdam, Elsevier Science Publishers, 1984: 123–128.
51. Roth J. The protein A–gold technique for antigen localisation in tissue sections by light and electron microscopy. In: Polak J. M. and Varndell I. M., eds., *Immunolabelling for Electron Microscopy*. Amsterdam, Elsevier Science Publishers, 1984: 113–121.
52. Polak J. M. and Varndell I. M. (eds) *Immunolabelling for Electron Microscopy*. Amsterdam, Elsevier Science Publishers, 1984.
53. Bullock G. R. and Petrusz P. (eds) *Technqiues in Immunocytochemistry, Vol. 1*. New York, Academic Press, 1982.
54. Bullock G. R. and Petrusz P. (eds) *Techniques in Immunocytochemistry, Vol. 2*. New York, Academic Press, 1983.
55. King J. C., Lechan R. M., Kugel G. and Anthony E. L. P. Acrolein: A fixative for immunocytochemical localization of peptides in the central nervous system. *J. Histochem. Cytochem.* 1983, **31**, 62–68.
56. McLean I. W. and Nakane P. K. Periodate–lysine–paraformaldehyde fixative. A new fixative for immuno-electron microscopy. *J. Histochem. Cytochem.* 1974, **22**, 1077–1083.
57. Somogyi P. and Takagi H. A note on the use of picric acid–paraformaldehyde–glutaraldehyde fixative for correlated light and electron microscopic immunocytochemistry. *Neuroscience* 1982, **7**, 1779–1783.
58. Doerr-Schott J. and Garaud J.-C. Ultrastructural identification of gastrin-like immunoreactive ιerve fibers in the brain of *Xenopus laevis* by means of colloidal gold or ferritin immunocytochemical methods. *Cell Tissue Res.* 1981, **216**, 581–589.
59. Craig S. and Goodchild D. J. Postembedding immunolabelling. Some effects of tissue preparation on the antigenicity of plant proteins. *Eur. J. Cell Biol.* 1982, **28**, 251–256.
60. Bendayan M. and Zollinger M. Ultrastructural localization of antigenic sites on osmium-fixed tissues applying the protein A–gold technique. *J. Histochem. Cytochem.* 1983, **31**, 101–109.
61. McNeill T. H. and Sladek C. D. The effect of tissue processing on the retention of vasopressin in neurons of the neurohypophyseal system. *J. Histochem. Cytochem.* 1980, **28**, 604–605.
62. Dudek R. W., Childs G. V. and Boyne A. F. Quick-freezing and freeze-drying in preparation for high quality morphology and immunocytochemistry at the ultrastructural level: Application to pancreatic beta cell. *J. Histochem. Cytochem.* 1982, **30**, 129–138.
63. Dudek R. W., Varndell I. M. and Polak J. M. Combined quick-freeze and freeze-drying techniques for improved electron immunocytochemistry. In: Polak J. M. and Varndell I. M., eds., *Immunolabelling for Electron Microscopy*. Amsterdam, Elsevier Science Publishers, 1984: 235–248.
64. Causton B. E. The choice of resins for electron immunocytochemistry. In: Polak J. M. and Varndell I. M., eds., *Immunolabelling for Electron Microscopy*. Amsterdam, Elsevier Science Publishers, 1984: 29–36.
65. Roth J., Bendayan M., Carlemalm E., Villiger W. and Garavito M. Enhancement of structural preservation and immunocytochemical staining in low temperature embedded pancreatic tissue. *J. Histochem. Cytochem.* 1981, **29**, 663–671.
66. Bendayan M. and Shore G. C. Immunocytochemical localization of mitochondrial proteins in the rat hepatocyte. *J. Histochem. Cytochem.* 1982, **30**, 139–147.
67. Roth J. and Berger E. G. Immunocytochemical localization of galactosyltransferase in Hela cells: codistribution with thiamine pyrophosphatase in trans Golgi cisternae. *J. Cell Biol.* 1983, **93**, 223–229.
68. Thorens B., Roth J.,Perrelet A., Norman A. W. and Orci L. Immunocytochemical localization of the vitamin D-dependent calcium binding protein in chick duodenum. *J. Cell Biol.* 1982, **94**, 115–122.
69. Varndell I. M. and Polak J. M. Double immunostaining procedures: Techniques and applications. In: Polak J. M. and Varndell I. M., eds., *Immunolabelling for Electron Microscopy*. Amsterdam, Elsevier Science Publishers, 1984: 155–177.

70. Morgan C. The use of ferritin-conjugated antibodies in electron microscopy. *Int. Rev. Cytol.* 1972, **32**, 291–326.
71. Hunt S. P., Kelly J. S. and Emson P. C. The electron microscopic localisation of methionine enkephalin within the superficial layers (I and II) of the spinal cord. *Neuroscience* 1980, **5**, 1871–1890.
72. Leranth C. S. and Frotscher M. Commissural afferents to the rat hippocampus terminate on VIP-like immunoreactive non-pyramidal neurons. An EM immunocytochemical degenerative study. *Brain Res.* 1983, **276**, 357–361.
73. Sumal K. K., Blessing W. W., Joh T. H., Reis D. J. and Pickel V. M. Synaptic interaction of vagal afferents and catecholaminergic neurons in the rat nucleus tractus solitarius. *Brain Res.* 1983, **277**, 31–40.
74. Somogyi P., Freund T. F., Wu J.-Y. and Smith A. D. The section-Golgi impregnation procedure. 2. Immunocytochemical demonstration of glutamate decarboxylase in Golgi-impregnated neurons and in their afferent synaptic boutons in the visual cortex of the cat. *Neuroscience* 1983, **9**, 475–490.
75. Ruda M. A. Opiates and pain pathways: demonstration of enkephalin synapses on dorsal horn projection neurons. *Science* 1982, **215**, 1523–1525.
76. Larsson L.-I. Simultaneous ultrastructural demonstration of multiple peptides in endocrine cells by a novel immunocytochemical method. *Nature* 1979, **282**, 743–746.
77. Roth J. The preparation of protein A–gold complexes with 3 nm and 15 nm gold particles and their use in labelling multiple antigens on ultrathin sections. *Histochem. J.* 1982, **14**, 791–801.
78. Bendayan M. Double immunocytochemical labelling applying the protein A–gold technique. *J. Histochem. Cytochem.* 1982, **30**, 81–85.
79. Tapia F. J., Varndell I. M., Probert L., De Mey J. and Polak J. M. Double immunogold staining method for the simultaneous ultrastructural localisation of regulatory peptides. *J. Histochem. Cytochem.* 1983, **31**, 977–981.
80. Varndell I. M. and Polak J. M. The use of immunogold staining procedures in the demonstration of neurochemical coexistence at the ultrastructural level. In: Osborne N. N., ed., *Dale's Principle and Communication between Neurones.* Oxford, Pergamon Press, 1983: 179–200.
81. Varndell I. M. and Polak J. M. The use of double immunogold-staining procedures at the ultrastructural level for demonstrating neurochemical coexistence. In: Chan-Palay V. and Palay S. L., eds., *Coexistence of Neuroactive Substances in Neurons.* New York, John Wiley and Sons, 1984: 279–303.
82. Sternberger L. A. *Immunocytochemistry*, 2nd ed. New York, John Wiley and Sons, 1979.
83. Griffiths G., Brands R., Burke B., Louvard D. and Warren G. Viral membrane proteins acquire galactose in trans Golgi cisternae during intracellular transport. *J. Cell Biol.* 1982, **95**, 781–792.
84. Keller G. A., Tokuyasu K. T., Dutton A. H. and Singer S. J. An improved procedure for immunoelectron microscopy: Ultrathin plastic embedding of immunolabeled ultrathin frozen sections. *Proc. Natl Acad. Sci. USA* 1984, **81**, 5744–5747.
85. Sikri K. L., Varndell I. M., Goodman R. H., Benoit R., Diani A. R. and Polak J. M. Mammalian somatostatin-containing D cells exhibit rat somatostatin cryptic peptide (RSCP) immunoreactivity. *Cell Tissue Res.* Submitted for publication.

10   *Lectin Histochemistry*

A. Leathem

Carbohydrates are widely distributed throughout tissues: predominantly at cell surfaces, to a variable exent within the cytoplasm (depending on the functional activity of a cell) and in the extracellular matrix. The majority of cell sugars are glycoconjugates, bound to proteins (as proteoglycans and glycoproteins) and to lipids (glycolipids). The recent upsurge in carbohydrate interest results both from our appreciation of the role of carbohydrates as important recognition markers on the surface of cells, and from the application of lectins as very simple and selective tools for the detection of carbohydrate groups.

Lectins were probably first described by Stillmark[1] in his thesis of 1888 (two years before antibodies were described) and have been defined recently[2] as sugar-binding proteins or glycoproteins of non-immune origin which agglutinate cells and/or precipitate glycoproteins. They must therefore have multivalent binding sites. Most lectins that have been reported originate from plants (phytohaemagglutinins) but, increasingly, lectins from animals are being described (*see*[3] for review).

The cell surface is covered by a forest of sugar chains or oligosaccharides which are important in cell–cell recognition and the particular value of lectins lies in their ability to identify minor differences among the trees and to distinguish cell populations and behaviour related to sugar expression.

1. STRUCTURE OF CARBOHYDRATES AND GLYCOCONJUGATES

The basic monosaccharide carbohydrate structure is a six-carbon ring or chain with L and D configurations. Of hundreds of such sugar groups which exist, only about 7 are commonly found in mammals; they are: mannosyl- (Man), glucosyl- (Glc), galactosyl- (Gal), fucosyl- (Fuc), acetylgalactosyl- (GalNAc), acetylglucosaminyl- (Glc$N$Ac), and sialic acid groups, but the specificity conferred by sugars results from their arrangement in chains. Polymers of two or more (up to 18) monosaccharides form complex chains with branches, called 'oligosaccharides'. Very long chains (polysaccharides or glycans) can exist and may contain a single repeated unit (homoglycans), e.g. glucose in cellulose, or up to 7 different units (heteroglycans). The possible permutations conferred by such a system are enormous, with additional variations permitted by L or D forms and alpha or beta linkages to different carbons of sugar units, e.g. 1 to 3, 1 to 4. Carbohydrate chains start life attached to a protein and even the long extracellular polymers usually retain some peptides. Complex carbohydrates or polysaccharides with a minor protein core are known as proteoglycans, and are frequently sulphated; they form the connective tissue and extracellular matrix. Proteins with minor sugar are known as glycoproteins (or glycopeptides), but the distinction is not always clear. Oligosaccharides linked to lipids (glycolipids) are major constituents of the cell membrane.

A common linkage of protein to oligosaccharide chains is by the amide group of asparagine; these are known as N-glycans and commonly link through acetylglu-cosamine (*see Fig.* 10.1). Linkage through the hydroxyl group to amino acids serine and threonine produces O-glycans, where acetylgalactosamine is commonly involved. Such linkages appear to influence lectin binding.

Asparagine-N-Acetylglucosamine = Galactose-Sialic acid

Fig. 10.1. Example of unbranched oligosaccharide chain of 3 monosaccharide units; probably the type of chain recognized by a lectin.

1.1. Neoglycoproteins

These are synthetic glycoconjugates composed of a known protein or peptide conjugated to a mono- or polysaccharide chain, an example being mannose groups attached to bovine serum albumin. They are a relatively new group of substances, generally limited and expensive, which appear to be of great potential interest in determination of lectin-binding sites, as the oligosaccharide can be of pure specified composition.

1.2. Immobilized Sugars

A wide selection of monosaccharide groups attached in different spatial arrangements to beads of agarose, sepharose and polyacrylamide gels are available. These provide a simple affinity chromatography system for the isolation of lectins from crude extracts and also a means of controlling lectin binding to tissues; if the lectin solution is incubated with beads of immobilized sugar, then centrifuged and tissue sections are incubated with the supernatant, no binding should occur. The lectin can be recycled by elution from the beads. A simple immobilized sugar is probably available in many laboratories in the form of Sephadex beads (glucose polymers) and Sepharose (galactose polymers). Labelled lectins can be assessed for both lectin binding and activity of the label by incubation with immobilized sugar, washing the beads, and examining them either by fluorescence microscopy or by addition of enzyme substrate.

2. OCCURRENCE OF LECTINS

Lectins are widely distributed in animals and plants, in particularly large amounts in the seeds of peas and beans[3] and extracts have been used to cross-link oligosaccharides on erythrocyte surfaces and identify blood groups. The functions of most lectins are unknown but the large concentration in the seeds of legumes gives rise to speculation that they may have a controlling role in processes such as germination, or perhaps in protection against predators, as some lectins are very poisonous. From experiments such as the inhibition of sponge cell aggregation by carbohydrates and the development of sporing bodies by slime moulds, lectins appear to play a role in cell–cell recognition and, although few lectins have been described from higher organisms, it seems probable that lectin systems are important not only in cell and glycoconjugate recognition but also in the regulation of cell growth.

3. STRUCTURE AND NOMENCLATURE OF LECTINS

From their ability to cross-link cells and glycoconjugates, all lectins (except specially prepared subunits) have two or more sugar-binding sites (probably up to 18 on each molecule). Concanavalin A has been the most intensively investigated and serves as a good model. It is a tetramer at neutral/slightly acid pH, at more alkaline pH larger polymers form and, at acid pH, dimers. The binding activity, as with other lectins, is partly pH dependent, but is also determined by the presence of heavy metal ions, in particular Ca^{2+} (but Mn, Mg, Co, Ni and Zn have also been described as prerequisites). Removal of heavy metals by sequestrating agents abolishes the binding capacity and even the absence of heavy metal from diluting solutions weakens the binding. We try to overcome this problem, inherent in many lectins, by diluting and washing in a 'lectin buffer' (*see below*) containing a dilute cocktail of heavy metals. It may also account for the weak binding encountered using phosphate as a buffering system (as in phosphate-buffered saline), the phosphate reacting with and sequestering heavy metals.

The majority of lectins take their name from the plant or animal from which they are derived, such as peanut lectin or peanut agglutinin (abbreviated to PNA) where the popular name is used, or *Bandeiraea simplicifolia* agglutinin (abbreviated to BSA lectin). Such abbreviations give rise to confusion, for

Table 10.1. **Lectins and their inhibiting simple sugars**

Lectin source	Inhibiting simple sugar	Blood group identified
Abrus precatorius (abrin)	Gal-	B,O > A
Anguilla anguilla (eel)	L-Fucose-	O
Arachis hypogaea (peanut)	Gal-β-L-3GalNAc-	'T'
Bandeiraea (*Griffonia*)	I Gal-	B
simplicifolia	II β-GlcNAc-	Tk
Bauhinia purpurea	GalNAc-	All ABO
Concanavalia ensiformis (Con A)	α-Man-, α-Glc-	
[Succinyl Con A	α-Glc-]	
Datura stramonium	Acetyl-lactosamine-	
Dolichos biflorus	α-GalNAc-	A(A$_1$)
Glycine max (soyabean)	GalNAc-	
Helix aspersa (snail)	GalNAc-	A
Helix pomatia (snail)	GalNAc-	A
Lens culinaris (lentil) A,B	α-Man-, α-Glc-	
Limax flavus (slug)	Sialic acid	All
Limulus polyphemus	Sialic acid	All
(horseshoe crab)		
Lotus tetragonolobus	α-L-Fucose	O
Maclura pomifera	Gal-	
Phaseolus limensis	GalNAc-	A
Phaseolus vulgaris E.	GalNAc?	?All
Phaseolus vulgaris L.		
Phytolacca (pokeweed)	β-L- 4-GlcNAc-	
Pisum sativum	GlcNAc-, Man-	
Ricinus communis I	Gal-	
Ricinus communis II	Gal-, GalNAc-	
Solanum tuberosum (potato)	GlcNAc-	
Triticum (wheatgerm)	GlcNAc-polymers	
Ulex europaeus I	α-L-Fucose	O
Ulex europaeus II	GlcNAc-polymers	
Vicia fava	Glc-, Man-	
Vicia graminea		MM/NN
Viscum album (mistletoe)	Gal-	

Bandeiraea is also known as *Griffonia*, and some abbreviations are made on historical grounds, such as PHA (standing for phytohaemagglutinin). *Table* 10.1 shows commonly available lectins omitting abbreviations.

4. WHY USE LECTINS?

Tissue carbohydrates and mucosubstances have been comprehensively reviewed[4,5] and conventional histochemical identification methods described.[6–8] Such methods as the periodic-acid–Schiff reaction are invaluable; they are easily performed, if rather crude probes for detecting carbohydrate-related substances. Lectins offer a more sensitive and more selective means of distinguishing minor sugar components and slight differences between structures, but much work is needed to know the distribution of lectin-binding sites in normal human tissues, what they signify, and how they change in disease.

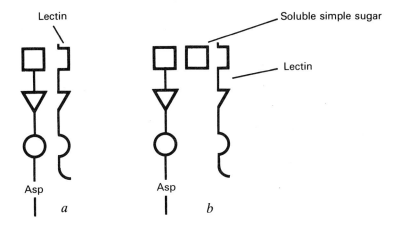

Fig. 10.2. Oligosaccharide chain fitting into binding site on a lectin (*a*) and (*b*) blocked from binding by a soluble simple sugar.

Lectins are naturally occurring receptors for carbohydrates and probably provide exquisite sensitivity for the recognition of very minor oligosaccharide differences between cells — at least as good as the specificity and sensitivity achieved with monoclonal antibodies, and perhaps more practical. The lock and key interaction between lectins and oligosaccharides is illustrated in *Fig.* 10.2 and, as with a lock and key, a missing key pin causes failure to engage the lock, so the fine selectivity achieved by lectins is scarcely appreciated if the binding site is thought of as a simple sugar, merely because binding may be inhibited by that sugar. Binding inhibition of different lectins by the same simple sugar (e.g. of concanavalin A, lentil and *Pisum sativum* by mannosyl-) gives no prediction of the likely binding sites or tissue distribution (lentil lectin binds to α-mannose-fucosyl groups at dilutions where concanavalin A shows no binding), as quite different oligosaccharides are preferentially bound by each lectin. Lectins prefer to bind to complex rather than to simple sugars, e.g. wheatgerm agglutinin prefers polymers of *N*-acetylglucosamine to simple *N*-acetylglucosamine.

Undoubtedly only a small fraction of lectins which exist have been described, and these are of a limited type, for we expect a substance extracted from a plant and which agglutinates red blood cells (for this is the usual assay) to be useful in distinguishing animal cell populations. It is equally naive to assume that one species produces only one lectin ('peanut lectin', 'soyabean lectin') and a vast range of lectins must remain undiscovered. Meanwhile, there is a large variety of lectins available, their particular attractions to the histologist being: ready availability commercially; reasonable cost; good stability; easily used, and most important, almost certain reaction with some mammalian cells or products.

5. ISOLATION OF LECTINS

Aqueous extracts of many plants agglutinate red blood cells and such crude extracts are available commercially for blood grouping. If such a crude haemagglutinating extract were to be labelled e.g. by fluorescein isothiocyanate

(FITC), it might be of value, certainly as an initial screening for binding to tissues. But many (? all) plants contain isolectins, with different sugar binding specificity, e.g. *Bandeiraea (Griffonia) simplicifolia* has at least 2 major isolectins, I binding to galactose and II binding to *N*-acetyl-D-glucosamine. Therefore some further purification step is needed and some understanding of the processes is valuable, even for those purchasing lectins, since commercial preparations vary enormously in their purity. Using affinity chromatography on beads coated with insolubilized sugars, lectins and isolectins can be purified with relative ease: e.g.

Step 1. Saline extract of fresh beans.
Step 2. Ammonium sulphate precipitation to concentrate.
Step 3. Molecular weight separation by gel exclusion chromatography.
Step 4. Affinity chromatography on immobilized sugar.

Each step can be monitored using haemagglutination (washed fresh or neuraminidase-treated cells). Lectin is eluted from immobilized sugar using the soluble simple sugar, e.g. at 0·1 M concentration. Homogeneity can be determined by polyacrylamide gel electrophoresis. The purity required is determined by the application but for most purposes the purest and most active preparation is preferable or interpretation of results is confused.

6. METHODS FOR DETECTION OF LECTIN BINDING TO TISSUES

Although a huge variety of methods can be applied to demonstrate lectin binding, four main approaches are described here, each having different advantages or disadvantages; the method selected probably depends on the purpose intended and individual experience in histochemistry:

   i. *Direct method*: Using lectin conjugated to a label such as fluorescein isothiocyanate (FITC), peroxidase or alkaline phosphatase.
  ii. *Antibody method*: Using antibodies to the native lectin, either as an indirect antibody, peroxidase–anti-peroxidase (PAP) or similar amplification system.
 iii. *Avidin–biotin method*: Using biotinylated lectin and avidin-conjugated label.
  iv. *Carbohydrate-conjugated label*: Where native lectin acts as a bridge to link glycosylated label to tissues.

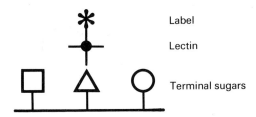

Fig. 10.3. Direct lectin conjugate method.

6.1. Direct Lectin Binding

PRINCIPLE. Lectin conjugated to label is incubated on the tissue, the fluorescent or enzyme label is then revealed (*see Fig.* 10.3).

This is a very simple and fast technique, particularly applicable to frozen sections, tissue cultures, cytological smears/suspensions and other unprocessed tissue. It also frequently works on paraffin sections.

   i. Tissue sections air dried (with or without acetone or methanol 'fixation').
  ii. Lectin conjugate applied at optimal dilution/time.
 iii. Slide washed, enzyme substrate added or examined with fluorescence microscopy.

COMMENTS. This is an excellent approach for fresh and unprocessed tissue and my first choice for starting work on a new lectin or tissue system. There is a high chance of success, not only with frozen sections but also with paraffin sections. On receiving a new batch of lectin conjugate it is necessary to determine the optimum dilution and time of incubation; the dilutions used in many publications seem extravagant and the times very short. These variables can best be determined by doubling dilutions (with a range of, for example, 200, 100, 50, 20, 10, 5 μg lectin/ml solution) for such times as 30, 60 and 120 min at room temperature. In my experience, FITC-conjugated lectins are more reliable, can be used at higher dilutions and are cheaper to buy than are peroxidase-conjugated lectins. (I have found many commercial peroxidase conjugates exhibit reduced binding properties and shopping around is expensive in time and funding.) The optimum conditions determined for frozen sections are not those for paraffin sections, this probably results from the two-fold attenuation by:

  a. Sequestration of reactive sites during fixation.
  b. Removal of glycolipids during processing.

Paraffin sections require higher concentrations of lectin or amplification to show detectable binding, particularly when 'good' fixatives (such as buffered formol--saline) are used.

Direct lectin conjugate staining is excellent for biological research but since most human surgical histopathology is performed on fixed and processed tissue we have to accept the suboptimal material and adapt to it. This means we probably fail to detect a large section of the glycolipid material which is lost in the processing fluids. Glycolipids are important components of the cell membrane, e.g. many blood group substances and xylene, chloroform etc. may dissolve them out. A comparison between lectin binding to air-dried and to acetone-'fixed' cryostat sections can be dramatic! To detect remaining glycans/glycoproteins, the following methods are more sensitive and may be necessary for paraffin sections, since the use of native lectin allows tighter binding.

6.2. Antibody Methods

PRINCIPLE. Native lectin is incubated and bound to tissue sections; after washing it

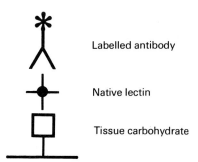

Labelled antibody

Native lectin

Tissue carbohydrate

Fig. 10.4. Labelled antibody method.

is detected by an antibody (raised to that lectin) either conjugated to a label or a further step, such as the PAP method etc. is used to amplify the immune reaction (*see Fig.* 10.4).

   i. Lectin incubated on slide at optimal dilution/time, e.g. 10 µg/ml for 1 h. Washed.
  ii. Rabbit antibody to lectin incubated, at optimal dilution/time, e.g. 10 µg/ml for 1 h. Washed.
 iii. Peroxidase-conjugated antibody to rabbit IgG incubated at optimal dilution/time, e.g. 1:50 for 30 min. Washed.
 iv. Substrate added, e.g. DAB 500 µg/ml, H_2O_2 30 vol. 10 µl/ml, for 10 min.

COMMENTS. Antibodies to lectins have only recently appeared commercially. It is possible to raise one's own antibodies (if in possession of an animal licence) with the caveat that some lectins are extremely toxic. Nick Atkins and I have raised antibodies to many lectins using this protocol:

   i. Lectin dissolved in Tris-buffered saline to 1 mg/ml.
  ii. Soluble simple sugar added (1 M) to occupy binding sites.
 iii. Mixture incubated for 1 h at 60 °C to denature/reduce toxicity.
 iv. Mixture homogenized with Freund's adjuvant (ratio 1 part lectin : 2 parts adjuvant).
  v. Rabbits immunized subcutaneously at 2–3 week intervals.

Antibody in the form of purified IgG fraction is preferable to whole antiserum unless the whole serum is used at dilutions of 1 : 100 or more; this is because glycoproteins in the serum compete with the lectins for binding sites on the tissues and may elute off the lectin. Similarly, sections should *not* be treated with normal swine serum or similar in order to reduce background staining. No pre-treatment may be necessary or, if there is non-specific background (as from hydrophobic bonding), sections may be pre-treated with, and lectins diluted in, bovine serum albumin (a protein fairly free from glycosylated groups).

The antibody method is both simple and sensitive.

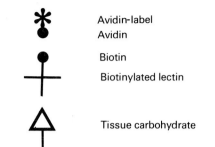

Avidin-label
Avidin

Biotin

Biotinylated lectin

Tissue carbohydrate

Fig. 10.5. Lectin–avidin–biotin method.

6.3. Biotinylated Lectin Method

PRINCIPLE. Lectin conjugated to biotin is incubated on the section, avidin conjugated to a label (FITC or peroxidase) binds to the lectin (*see Fig.* 10.5).

i. Biotinylated-lectin incubated on section, e.g. 10 µg/ml for 1 h. Washed.
ii. Avidin-conjugated to label (FITC or peroxidase) incubated on section, e.g. 1–10 µg/ml for 10 min. Washed.
iii. DAB/H_2O_2 substrate used for peroxidase, or FITC visualized under fluorescence microscope.

COMMENTS. This appears a potentially good system. Several biotinylated lectins are comercially available and the conjugation procedure is straightforward to anyone with experience in protein techniques. Avidin is available conjugated to FITC, peroxidase and alkaline phosphatase. A practical problem is the binding of avidin to endogenous tissue biotin, particularly in, for example, kidney, but this may be overcome using unlabelled avidin.[9]

6.4. Carbohydrate-conjugated Label

PRINCIPLE. Native lectin is incubated on tissue; most lectins have multiple binding sites, not all of which are occupied on the tissue section. Label conjugated to a sugar (which will bind to unoccupied sites on the lectin) is added. The label is then revealed (*see Fig.* 10.6).

i. Native lectin incubated on tissue, e.g. 10 µg/ml for 1 h. Washed.
ii. Glycosylated label incubated on tissue, e.g. 20 µg/ml for 1 h. Washed.
iii. Enzyme or FITC label visualized.

COMMENTS. This system, introduced by Keida et al.[10] is similar to Wachsmuth's MAGIC enzyme technique.[11] Potentially a very versatile and sensitive approach, it is somewhat limited by the poor range of glycosylated markers available at a reasonable price. The most widely available glycosylated enzyme is horseradish

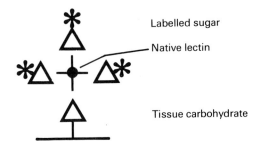

Fig. 10.6. Carbohydrate-conjugated label.

peroxidase, which has a large number of mannosyl- residues. This immediately lends itself to exploring lectins with mannose binding properties, e.g. concanavalin A, lentil lectin. Thus:

   i. Concanavalin A incubated on section, e.g. 20 μg/ml for 1 h. (or an excess to ensure spare binding sites). Washed.
  ii. Horseradish peroxidase incubated on section, e.g. 20 μg/ml for 10 min. Washed.
 iii. DAB or similar substrate incubated on section.

COMMENT. Concanavalin A binds very tightly to peroxidase; this can be demonstrated by mixing concanavalin A with peroxidase, incubating for varying times then adding DAB/H_2O_2 substrate. The concanavalin A progressively reduces the activity of peroxidase, presumably through tighter and tighter binding to mannosyl-, stereologically reducing the active enzyme sites on the peroxidase. Thus timing can be fairly important with this technique. Also, as peroxidase is a glycoprotein, it may be capable of displacing lectin from tissue sections at high concentrations.

7. CONTROLS

7.1. Positive

The majority of lectins, whether available as purified or crude extracts, react with erythrocyte glycoconjugates (generally non-specifically) and steps in purification are monitored by haemagglutination. The reaction varies from being non-species specific to requiring neuraminidase-treated erythrocytes. A wide spectrum of erythrocytes (fresh and preserved) are available commercially and are ideal for testing the activity of any new batch of lectin. Using drops of a 2 per cent suspension of washed erythrocytes (washed to remove competing plasma glycoconjugates) mixed on a white tile with doubling dilutions of lectin, the activity (and partial specificity) of each lectin can be determined. This is a very useful exercise, for I have found that a concentration of lectin which agglutinates red cells under these conditions is most likely to be suitable for tissue sections. The sensitivity of the system can be increased by addition of antibody (e.g. 1 : 100 to

1 : 1000) to the lectin/erythrocyte mixture to further cross-link the cells; this aids in determination of activity/specificity of antibody batches (again using either highly diluted antiserum or IgG fraction to avoid competitive glycoproteins in plasma).

If labelled lectins are used, the label, e.g. FITC, may be detected after washing the cells; this tests both the lectin binding and its label activity. Peroxidase of course produces a problem owing to endogenous erythrocyte peroxidases.

7.1.1. Tissue Controls

The object of this is to demonstrate lectin binding to tissue components, and there are frequently built-in positive controls in the form of erythrocytes and endothelium, to which many lectins bind. I have previously thought the kidney to be an excellent positive control tissue for a wide range of lectins and this is still so with two caveats:

i. The kidney has such accessible terminal oligosaccharides that it is extremely easy to stain. If a checkerboard of dilutions against times is used to determine optimum conditions on kidney sections, then such optimum dilutions and times may not be optimal on another tissue. Other more difficult tissues to stain, such as prostate and salivary glands, should be tried as well.
ii. Polymorphism in carbohydrate expression may occur within one tissue from different population groups. This has been shown superbly by the work of Ponder[12] using mouse chimeras. There may be polymorphic differences in human tissues, detected using novel lectins.

Overall, for any one lectin, it is probably worth having two positive control tissues, one which works very well, e.g. kidney, to show the system is working; one which works less well to test the limits of the system.

7.2. Negative Controls

The precise oligosaccharide binding configurations for most lectins are not known and are probably complex, made more so by:

a. Binding forces that exist (other than lock and key) between lectin and sugar, e.g. hydrophobic bonding.
b. The longer a lectin is incubated with an oligosaccharide, the tighter becomes the binding, particularly to complex carbohydrates, this means the more difficult to displace by a soluble simple sugar.

Attempts to displace lectins from tissues to which they have bound meet with variable degrees of success using soluble simple sugars, particularly if trying to displace a lectin from a complex oligosaccharide, even using molar concentrations of sugar. Displacement is easier using a more complex soluble carbohydrate or neoglycoprotein. However, it is certainly worth trying to elute off lectin with a soluble simple sugar as this provides a potentially good negative control. It may seem surprising that, whereas soluble simple sugars can be used to displace lectins from chromatography columns of immobilized sugars, this approach is less successful on tissue sections until one considers that the immobilized sugar in affinity chromatography is generally a simple mono- or disaccharide; the oligosaccharide chains in tissues are longer and branching.

8. USE OF ENZYMES

These are conveniently divided into two groups:

a. Non-specific — needed to aid in detection/demonstration.
b. Specific — to cleave off specific carbohydrate groups which, by their removal, help confirm the identity of that group, or else allow the identification of an adjacent or subterminal group.

8.1. Non-specific (Trypsin, Pronase etc.)

The value of such enzymes depends very much on the treatment the tissue has received, fixed tissue requiring enzyme treatment proportional to the efficiency of the fixative. As with antigen detection, the effect of brief proteolysis is probably to loosen the tissue structure and allow penetration of macromolecules. Short proteolytic treatment may dramatically enhance the demonstration of carbohydrates in paraffin sections and it is our usual practice to use lectins with and without proteolysis.

The attachment of oligosaccharides to a protein core, although to a limited range of amino acids (mostly to asparagine, serine, threonine), leaves them vulnerable to cleaving off during proteolysis. However, since many oligosaccharides have a protein core embedded in or through a membrane, cleaving off is perhaps not significant under mild conditions. Trypsin hydrolyses carboxyl groups of lysine and arginine and contaminants (such as chymotrypsin which hydrolyses tyrosine and phenylalanine) may cleave off peptides closer to the oligosaccharide chain.

Often only a whiff of trypsin is needed to dramatically enhance lectin binding to paraffin sections. Each new tissue, lectin or fixative requires testing at different trypsin times such as 0, 1, 2, 5, 10, 20, 40 min.

8.2. Specific Enzyme Cleavage

There is enormous potential for this procedure both at a biochemical and histological level. By digesting off carbohydrate groups separately, it is possible to dissect the structure of a glycoconjugate by determination of lectin binding before and after cleavage. Glycosidases for this fall into two groups: endoglycosidases (which break into the middle of oligosaccharides) and exoglycosidases (which remove specific terminal carbohydrates) (see Fig. 10.7).

By seeking binding to lectins before and after enzyme digestion, sugars which are chopped off can be identified and the next terminal sugar can then be identified.

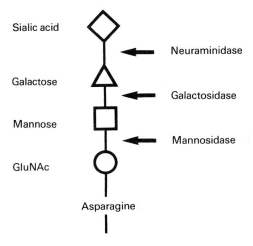

Fig. 10.7. Sequential hydrolysis by exoglycosidases.

There is a wide variety of exoglycosidases and endoglycosidases commercially available, exhibiting considerable selectivity; unfortunately many are prohibitively expensive. Perhaps greater availability will come with cloning of genes responsible for their synthesis.

8.2.1. Neuraminidase

The specificity of glycosidases varies very much according to source, this is particularly so for neuraminidase, e.g. from such sources as *Clostridium perfringens*, *Vibrio cholera*, influenza virus, *Arthrobacter ureafaciens*, *Streptococcus pneumoniae*. The sialic acids of different tissues are variably sensitive to these. There is enormous potential here for investigation of sialation in normal and neoplastic tissues.

8.2.2. Glycosidases

The exoglycosidases are better enzymes to start with than endoglycosidases, partly on grounds of cost and partly availability. The purest reagent should be used and the manufacturers approached to determine what impurities (particularly other enzymes) might be present. *Diplococcus pneumoniae* provides a good source for several exoglycosidases, particularly for galactosidase and acetylglucosaminidase. The advantage of this source is that the enzymes will work in the same buffer system. For example, digestion works well on sections using milliunits of each glycosidase in 0·05 M cacodylate buffer at pH 6·0 for 1–2 h at 37 °C.

Caution is needed in the interpretation of results using enzymes. The susceptibility of a sugar group to hydrolysis or cleavage depends upon the adjacent linkages and glycosidases from different sources, but with apparently identical substrates, will produce different results. Hence the problem in comparing results from different laboratories.

The time taken for different glycosidases to hydrolyse tissue oligosaccharides varies greatly and considerable experimentation is needed; the complete hydrolysis sought by biochemists can be very aggressive, expensive and unnecessary for tissue sections. Much shorter incubations (10–60 min) and more dilute solutions are suitable on tissue sections.

9. INTERPRETATION OF BINDING

Despite the distribution of carbohydrates in extracellular material, the binding patterns with lectins are generally clean and highly defined in normal tissues; background staining is more a function of hydrophobic bonding. The frequent finding of erythrocyte/endothelium binding can be a useful indication that the method has worked and, except for such lectins as *Ulex europaeus* I (which binds strongly to endothelium), is not dominant.

A rewarding start to any lectin work is a comparison on adjacent sections of conventional histochemical methods and lectin binding. But a frustrating stage is reached when, for example, a periodic acid–Schiff stain is strongly positive but lectin binding is negative. A further panel of lectins might be screened (as are commercially available) or attempts made to reveal sequestered or subterminal oligosaccharides by using enzymes.

A major feature of lectin binding is the enormous heterogeneity in carbohydrate expression which may be seen in morphologically identical and adjacent cells (*see Figs.* 10.8–10.14). The reproducibility of such heterogeneity suggests that we are demonstrating not only variations in cell activity and function, but probably constant cell subpopulations which have differentiated along separate pathways.

There is a vast potential for the exploration of oligosaccharide changes in health and disease. If the prospect is too daunting, much valuable information can be obtained screening one tissue with a panel of lectins, or a panel of tissues with one lectin. The different expressions of carbohydrate by apparently homogeneous cell populations, e.g. liver, epithelia, can be explored and related to other known markers. The cytological expression of lectins (? membrane, ? cytoplasmic, ? where in the cytoplasm) requires investigation. The use of enzymes, although fascinating, introduces such a wealth of complex data that it may not be a tool for beginners.

10. APPLICATIONS OF LECTIN HISTOCHEMISTRY

The expression of oligosaccharides during development, differentiation, growth, dysplasia and neoplasia, and the detection of these changes is a new approach to cell biology and pathology; lectins provide an excellent tool for their exploration. There is a temptation to take a selection or kit of commercial lectins to test on a wide range of tissues, but this is less likely to be useful than trying a selection of lectins against a single tissue or a single lectin against a range of tissues. Results achieved using living cell–lectin interactions do not readily translate to tissue

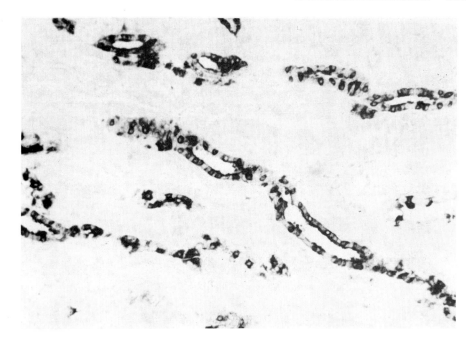

Fig. 10.8. *Dolichos biflorus* lectin on normal human kidney, binding to distal and collecting tubule epithelium only. Formalin-fixed paraffin section, indirect immunoperoxidase.

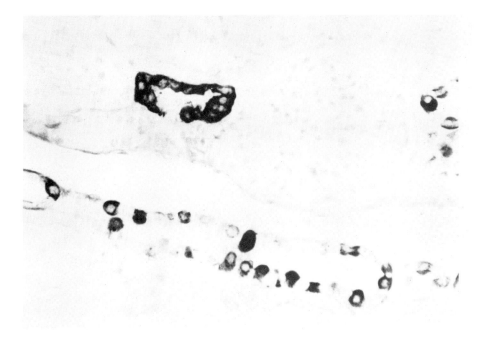

Fig. 10.9. As *Fig.* 10.8. High power shows heterogeneity in binding to morphologically similar cells.

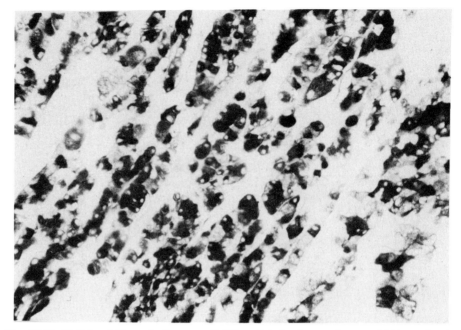

Fig. 10.10. *Dolichos biflorus* lectin on well-differentiated adenocarcinoma of kidney showing heterogeneity in binding to tumour cells. Formalin-fixed paraffin section, indirect immunoperoxidase.

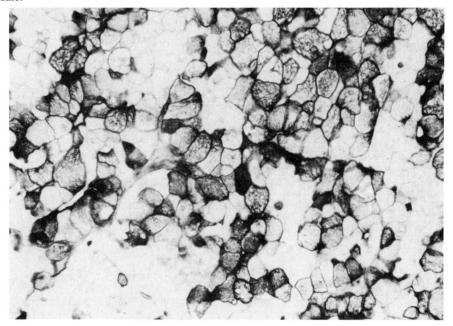

Fig. 10.11. *Dolichos biflorus* lectin on poorly differentiated carcinoma of kidney showing membrane and cytoplasmic binding to majority of cells. Formalin-fixed, paraffin section, indirect immunoperoxidase.

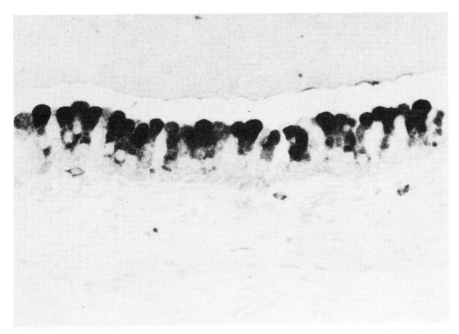

Fig. 10.12. Peanut agglutinin on normal human prostate epithelium after neuraminidase digestion showing heterogeneity in binding. Formalin-fixed paraffin section. avidin–biotin method.

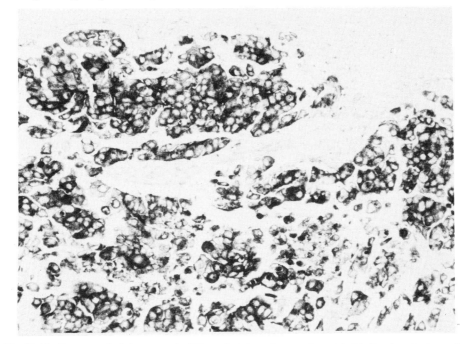

Fig. 10.13. Peanut agglutinin on poorly differentiated carcinoma of prostate binding to cytoplasm of most tumour cells. Formalin-fixed paraffin section, avidin–biotin method.

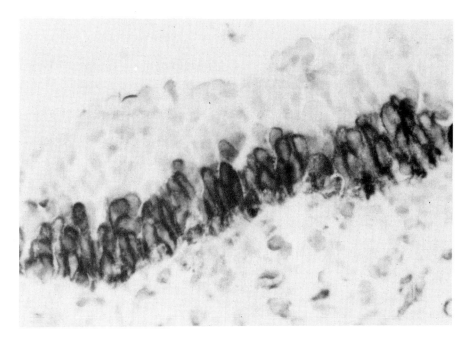

Fig. 10.14. Peanut agglutinin on human nasal epithelium showing intercellular and cytoplasmic binding to the basal layer of epithelial cells. Formalin-fixed paraffin section, avidin–biotin method.

sections and comparatively few results have been published on the histological binding patterns of lectins on different tissues; a useful review has been written by Cooper[13] on the tissue distribution of peanut lectin binding sites.

As lectins have been applied by immunologists to distinguish cell populations by differences in surface sugars on lymphoid cells, malignancies of the lymph node have been studied with lectins more than other tissues. It has been suggested that certain binding patterns of concanavalin A on the tissues of patients with Hodgkin's disease can be related to an unfavourable prognosis.[14] In addition, lectins may distinguish neoplastic from normal and reactive cell populations[15,16] and use of a panel of lectins showed a heterogeneity of binding to different populations of Reed–Sternberg cells.[17]

A difference in carbohydrate expression between normal and malignant epithelium has been reported in the breast[18] and in the urinary bladder.[19] Changes in lectin-binding patterns may be seen within a tissue type which reflect differences in morphological and functional differentiation; this has been described particularly in the urinary bladder.[20] In the gut there are lectin binding differences between pre-neoplastic, benign and malignant epithelium.[21]

The binding of *Ulex europaeus* I lectin to vascular endothelium is a useful marker and may be valuable for detecting vascular invasion by tumours. The lectin of *Ulex* I has been used to identify tumours of endothelial origin;[22] this lectin is possibly a more reliable means of identifying endothelial cells than is factor VIII.[23]

The interpretation of changes occurring in pathological tissue requires preliminary work on the expression of sugars in normal tissues. This is notably absent from the literature, an exception being the useful paper by Bell and Skerrow on factors affecting binding of lectins to normal human skin.[24]

11. TOXICITY OF LECTINS

I have no direct experience of human toxic reactions during 6 years use of many lectins but some are undoubtedly extremely poisonous, in particular abrin, ricin and perhaps mistletoe lectins. Working through catalytic inactivation of ribosomal factors, one molecule of ricin is likely to be sufficient to kill one cell, thus 10 µg of ricin would be sufficient to kill every cell in a human (Georgi Markov, a Bulgarian journalist, was assassinated in 1978 with a minute ball-bearing containing ricin fired through the tip of an umbrella). Toxins such as ricin and abrin consist of two chains, one for binding (β) and the other being the active toxin (α); when dissociated, the α chain cannot bind and is therefore no longer toxic. There is considerable current interest in attaching such a chain to tumour-directed antibodies to produce a 'magic bullet' in the treatment of cancers.

For safety, lectins are best stored in solution to avoid dust inhalation.

Acknowledgements

I am particularly grateful to Nick Atkins for his advice, Andrew Aylott for his technical help and Susan Brooks for assistance in the preparation of this chapter.

Appendix

I. SOME PRACTICAL POINTS

A. Lectin Buffer

Tris-buffered saline (TBS) is a very useful buffer for biochemistry (Tris 60·57 g, NaCl 87 g, H_2O to 10 l, pH 7·4 corrected with 0·1 M HCl). Traces of several heavy metals needed to maintain active binding sites on many lectins are added to batches of TBS as a cocktail (1 mM calcium, manganese and magnesium chlorides from a concentrated stock) for diluting and washing solutions.

B. pH of Solutions

The activity and tightness of binding of lectins varies with pH and although we use a pH of 7·4 most often, it is likely that this is not the optimum pH for many lectins and further work is needed.

C. Trypsin Digestion

Trypsin (400 mg) (Sigma grade II), 400 mg $CaCl_2$, warm Tris-buffered saline (TBS) 400 ml. Filter, use at once at 37 °C.

D. Neuraminidase Digestion

One unit neuraminidase (purest affordable) dissolved in 10 ml warm sodium acetate buffer pH 5·5, add 100 μl of 1 per cent $CaCl_2$. Incubate at 37 °C.

E. Simple Sugar Solutions

Where possible use the sugar as a glycoside, e.g. methyl mannoside instead of mannose, dissolved in TBS; 0·1 M is a useful concentration for most purposes.

REFERENCES

1. Stillmark H. Über Ricin, ein giftiges Ferment aus den Samen von *Ricinus communis* L. und einigen anderen Euphorbiaceen. Thesis. Dorpat, Estonia, 1888. Quoted by Oppenheimer C. *Toxins and Antitoxins*. London, Griffin, 1906. (English translation of *Toxine und Antitoxine*. Jena, Fischer, 1904.)
2. Goldstein I., Hughes R. C., Monsigny M., Osawa T. and Sharon N. What should be called a lectin? *Nature* 1980, **285**, 66.
3. Gold E. and Balding P. *Receptor-specific Proteins*. Amsterdam, Excerpta Medica, 1975.
4. Pearse A. G. E. *Histochemistry*, 3rd ed. Edinburgh, Churchill Livingstone, 1968.
5. Pearse A. G. E. *Histochemistry*, 4th ed., Vol 2. Edinburgh, Churchill Livingstone, 1985.
6. Drury R. A. B. and Wallington E. A. *Carleton's Histological Technique*, 5th ed. Oxford, Oxford Medical Publications, 1980.
7. Bancroft J. and Stevens A. *Theory and Practice of Histological Techniques*, 2nd ed. Edinburgh, Churchill Livingstone, 1982.
8. Bancroft J. and Cook H. *Manual of Histological Techniques*, 2nd ed. Edinburgh, Churchill Livingstone, 1984.
9. Wood G. S. and Warnke R. Suppression of endogenous avidin-binding activity in tissues and its relevance to biotin–avidin detection systems. *J. Histochem. Cytochem.* 1981, **29**, 1196–1204.
10. Kieda C., Delmotte F. and Monsigny M. Preparation and properties of glycosylated cytochemical markers. *FEBS Lett.* 1977, **76**(2), 257–261.
11. Wachsmuth E. D. The localisation of enzymes in tissue sections by immunohistochemistry: conventional antibody and mixed aggregation techniques. *Histochem. J.* 1976, **8**, 253–270.
12. Ponder B. A. J. Lectin histochemistry. In: Polak J. M. and Van Noorden S., eds., *Immunocytochemistry, Practical Applications in Pathology and Biology*. Bristol, Wright. PSG, 1983: 129–142.
13. Cooper H. Lectins as probes in histochemistry and immunohistochemistry: the peanut (*Arachis hypogaea*) lectin. *Human Pathol.* 1984, **15**, 904–906.
14. Ree H. J. and Crowley J. P. Concanavalin A binding histiocytes in Hodgkin's disease and their relation to clinicopathologic features of the disease. *Cancer* 1983, **52**, 252–257.
15. Collins R. D., Jacobson W. and Stoddart R. W. Lectin staining of carbohydrates of haemic cells. III, The cells of Hodgkin's disease and other lymphomas. *Histopathology* 1982, **6**, 601–616.
16. Ree H. J. and Hsu S.-M. Lectin histochemistry of malignant tissues: I. PNA receptors in follicular lymphoma and follicular hyperplasia: an immunohistochemical study. *Cancer* 1983, **51**, 1631–1638.

17. Morris C. S. and Stuart A. E. The lectin binding affinity of Reed–Sternberg cells. *J. Pathol.* 1983, **139**, 151–158.
18. Leathem A., Dokal I. and Atkins N. Lectin binding to normal and malignant breast tissue. *Diagn. Histopathol.* 1983, **6**, 171–180.
19. Daly J., Prout G., Hagen K., Lin J., Kamali H., Plotkin C. and Wolf G. Identification of malignant transitional cells by the binding of fluorescein-labelled lectin. *Surg. Forum* 1979, **30**, 565–567.
20. Lehman T., Cooper H. and Grant Mullholland S. Peanut lectin binding sites in transitional cell carcinoma of the urinary bladder. *Cancer* 1984, **53**, 272–277.
21. Klein P. J., Osmers R., Vierbuchen M., Ortmann M., Kania J. and Uhlenbruck G. The importance of lectin binding sites and carcinoembryonic antigen with regard to normal, hyperplastic, adenomatous and carcinomatous colonic mucosa. *Recent Results in Cancer Research*, 1981, **79**, 1–9.
22. Miettinen M., Holthofer H., Lehto V., Miettinen A. and Virtanen I. *Ulex europaeus* I lectin as a marker of tumours derived from endothelial cells. *Am. J. Clin. Pathol.* 1983, **79**, 32–36.
23. Ordonez N. and Batsakis J. Comparison of *Ulex europaeus* I lectin and factor VIII-related antigen in vascular lesions. *Arch. Pathol. Lab. Med.* 1984, **108**, 129–132.
24. Bell C. and Skerrow C. Factors affecting the binding of lectins to normal human skin. *Br. J. Dermatol.* 1984, **111**, 517–526.

11

The Localization of Receptors and Binding Sites with Reference to Steroids

R. A. Walker

Evidence for the endocrine regulation of human breast cancer extends from the nineteenth century. However, an understanding of the mechanisms involved and the development of methods which could identify endocrine-responsive breast carcinomas did not come until the 1960s when tritiated oestrogens of high specific activity became available. By using these it was shown[1] that oestrogen, on entering a target cell, binds to a high affinity receptor present within the cytoplasm; this oestrogen–receptor complex then translocates to the nucleus and activates the transcription of factors which are important for the growth and metabolism of the cell. Initial studies of human breast carcinomas utilized the uptake of tritiated oestradiol by tumour slices *in vitro* with subsequent autoradiography[2] to identify the receptors, but this approach was generally replaced by biochemical assays which rely upon the homogenization of the tumour tissue.[3] The commonest method employed is the dextran-coated charcoal assay with Scatchard plot; sedimentation analytical methods are also used.

The clinical significance of receptor assays is well recognized. Approximately half of the women whose tumours have detectable oestrogen receptor will obtain objective remission from some form of endocrine therapy, and this number increases to three-quarters when progesterone receptor, an oestrogen-induced protein, is included.[4]

There are, however, restrictions with the biochemical assays. They are expensive to perform, requiring specialized equipment, such as ultracentrifuges, and well-qualified staff, which probably accounts for the small number of centres in the United Kingdom that can undertake such assays. Selection of the tissue sample is important since homogenization of normal tissue as well as tumour could well affect results. Differences have been observed between centres where the surgeon selects the tissue and ones where the pathologist is involved; the pathologist is a better judge.[5] Other major problems relating to homogenization are that the cellularity of carcinomas may affect the results, and this factor cannot be determined. Similarly, heterogeneity cannot be predicted. If carcinomas are composed of varying mixtures of hormone-sensitive and hormone-independent cells this may well affect the response to endocrine-based treatments. Unless the nuclei are analysed separately there is no evidence as to whether the receptor complex has translocated to the nucleus; failure to do so may account for the failure to respond to endocrine therapy in some cases. Many of these problems could be overcome with the use of histological methods, and this has stimulated interest in the development of such approaches, which are summarized in *Table 11.1*.

Table 11.1. Histological methods for the localization of oestrogen binding sites and receptors

Binding sites	
Immunocytochemical	17-β-Oestradiol Anti-oestradiol antiserum Indirect immunofluorescence[6,7] or PAP technique [8,9,12]
Cytochemical	17-β-Oestradiol–fluorescein[17,18] 17-β-Oestradiol–BSA–fluorescein[19,20] 17-β-Oestradiol–BSA–peroxidase[9,23,25]
Receptors	
Immunocytochemical	Polyclonal oestrogen receptor antibody[39] Monoclonal oestrogen receptor antibody[43–45] + Indirect immunofluorescence or PAP technique

1. IMMUNOCYTOCHEMICAL LOCALIZATION

The initial immunocytochemical studies were undertaken using cell suspensions derived from tumours which were incubated with 2×10^{-7} M 17-β-oestradiol.[6] This was then localized within the cells using an antiserum to oestradiol and an indirect immunofluorescence method. By adjusting the times and temperatures, the site of fluorescence could be seen to change within the cell, suggesting translocation. The immunofluorescence approach was then applied to frozen sections of breast carcinomas but polyoestradiol phosphate was used.[7] Correlation with biochemical assays was reported to be 90 per cent. Mercer et al.[8] and Walker et al.[9] also examined frozen sections, but using monomeric oestradiol at concentrations of 1×10^{-7} and 5×10^{-8} M, respectively, followed by a two-stage[8] or three-stage[9] immunoperoxidase method. Mercer et al. noted a reaction in 75 per cent of cases, but considered it unlikely that specific high affinity receptors had been detected. We found that only a small proportion of cells within a tumour stained which seemed unlikely to represent the hormone receptor status, and concluded that the method seems unsuitable for the detection of steroid binding in tissue sections.

Other workers have utilized fixed, paraffin-embedded tissue. Prebound endogenous oestrogen was detected in routine sections of 10 out of 25 breast carcinomas by Ghosh et al.[10] using an antiserum to oestradiol and the PAP method. Kurzon and Sternberger[11] developed a method for studying oestrogen binding in rat uterus which has not been applied to human tissues. Slices of tissue were incubated with 17-β-oestradiol prior to fixation in picric acid–formaldehyde and processing. They considered that the oestradiol protected the specific binding sites during fixation, but that it was lost during embedding. Sections were then re-exposed to oestradiol after dewaxing, prior to the application of antiserum to oestradiol and the PAP method. Taylor et al.[12] examined routinely formalin-fixed, paraffin-embedded breast carcinomas using polyoestradiol phosphate at a concentration of 1 mg/ml, followed by the immunoperoxidase method, and observed a correlation with biochemical assays in 60 per cent of cases.

The major concern with several of these approaches is the use of fixed, embedded tissue and the concentration of oestradiol, which is outside the range for high affinity binding, and these considerations will be further discussed below. The other critical factor for these immunocytochemical methods is that studies concerned with the abilities of antibodies to 17-β-oestradiol to recognize the hormone when receptor bound have concluded that they cannot.[13–15] Morrow et al.[15] studied polyoestradiol phosphate in particular and considered that it interacted with non-specific rather than high-affinity receptor protein; anti-oestradiol antibody could not react with the hormone when complexed to these proteins. Underwood et al.[16] have suggested that inaccessability of receptor-bound oestradiol to oestradiol antibodies is because the receptor–hormone interaction is of a clathrate type.

2. CYTOCHEMICAL STUDIES

As an alternative to the immunocytochemical methods several groups have tried simpler approaches using oestradiol linked to fluorescein directly[17,18] or through bovine serum albumin (BSA)[19,20] with coupling at either the 6 or 17 position of oestradiol. The relative binding affinities (RBAs) of different conjugates for the receptor have been assessed; coupling at the 17 position leads to a higher RBA than at the 6 position,[21] and is preferable, whilst the effect of using BSA is to give a much lower RBA.[22] The use of simple, direct-labelled oestradiol conjugates would seem preferable with regard to better tissue penetration due to low molecular weight and affinity for the receptor. If peroxidase is used as a label in place of fluorescein,[9,23–25] BSA has to be employed as a link. Besides this leading to a larger molecular weight conjugate, the method employed for coupling the peroxidase can also affect the final molecular size. Although the periodate conjugation method[26] results in a higher percentage of binding of both

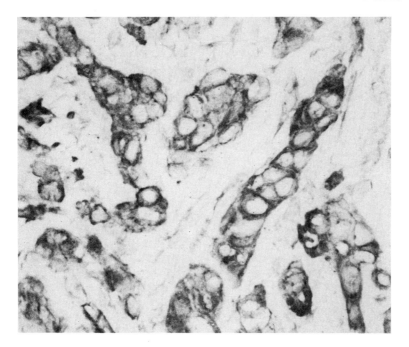

Fig. 11.1. Breast carcinoma: cryostat section post-fixed in acetone for 3 min, then treated with oestradiol-6-carboxymethyloxime–BSA–peroxidase. The peroxidase was developed using the diaminobenzidine–hydrogen peroxide method. Many of the cells have cytoplasmic staining.

oestradiol–BSA and peroxidase than the glutaraldehyde technique,[27] it leads to the formation of polymerized complexes, so further increasing the size of the molecule.[24,25] These problems encountered with peroxidase conjugates have to be balanced against the advantages obtained by having a permanent preparation, from which an accurate assessment of the percentage of reactive cells can be made. There is also some evidence that oestradiol–BSA conjugates are unstable in aqueous solutions, and that free oestrogen is present which will saturate receptor sites.[28]

In spite of these problems/criticisms there have been several reports of a good agreement between the qualitative results obtained by cytochemical and biochemical methods[9,20] indicating that the conjugates do have a specificity for target tissues. Certainly, using peroxidase-labelled oestradiol,[23–25] evidence of binding can be observed in 80 per cent of breast carcinomas examined, with the percentage of cells staining within individual tumours ranging from 20 to 85 per cent. The site of reaction within tumour cells can be defined as cytoplasmic or nuclear. Some carcinomas will be composed of cells showing either cytoplasmic or nuclear staining, others will be entirely cells with a cytoplasmic reaction (*Fig.* 11.1) and in a small percentage only nuclear staining (*Fig.* 11.2) will be seen. Comparisons with biochemical assays have shown that it is those tumours with nuclear staining only and specimens in which sampling could be a problem which have been the source of differences between the cytochemical and biochemical methods. However, there is only a poor correlation between the concentration of

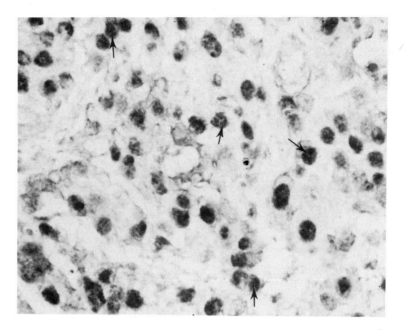

Fig. 11.2. Breast carcinoma reacted as above but in this only nuclear staining is seen (arrowed).

oestrogen receptor within a tumour, as assayed, and the percentage of cells reacting. This is surprising since the amount of oestrogen receptor in a carcinoma is considered to be a useful indicator of the likely response to endocrine therapy, and suggests that there are differences in what is being detected by the two methods.

The concentrations of oestradiol–BSA–peroxidase that are required to achieve staining have ranged from 1×10^{-7} M for 17-β-oestradiol-17-hemisuccinate–BSA–peroxidase, 2×10^{-7} M for 17-β-oestradiol-6-carboxymethyloxime–BSA–peroxidase prepared by the glutaraldehyde method, and 5×10^{-7} M to 1×10^{-6} M when the periodate method of conjugation has to be used. These levels are outside the range defined for the high affinity binding characteristics of oestrogen receptor, and will be discussed further in Section 3.

Evidence for the specificity of binding has been achieved by competing with agents that will also interact with receptor/binding sites, such as tamoxifen or diethylstilbestrol at 100 times excess, and this has successfully been achieved by many groups, although some have experienced difficulty.[22]

Fluorescein–oestrogen conjugates have been applied to the analysis of breast cancer cells derived by fine needle aspiration,[29] although again the concentrations have been in the 10^{-7} M range. However, by utilizing a fluorescence-activated cell sorter, binding of N-fluoresceino-N^1-(17-β-oestradiol-hemisuccinamide) thiourea to breast tumours can be detected in the nanomolar concentration range,[30] indicative of high affinity binding. Similar criteria apply when inherently fluorescent ligands (coumestrol, 12-oxo-oestradiol) are applied to cells providing that an image intensification system is linked to the fluorescence microscope.[31]

The majority of the studies concerned with 'receptor' localization have concentrated on oestrogen-specific sites in breast cancer. Further information regarding possible response to hormone therapy can be gained by knowledge about progesterone receptor status. There have been reports of studies using 11-hydroxyprogesterone–BSA–fluorescein,[32] but care has to be taken to prevent non-specific binding of the agent to glucocorticoid or androgen receptor sites. Good correlation with biochemical assays was found. Pertschuk et al.[32] have also applied the cytochemical methods to prostatic carcinoma, and Bergqvist et al.[33] to human endometrium, apparently both with favourable results.

3. PROBLEMS OF THE HISTOLOGICAL METHODS

As indicated in several places there have been criticisms of the histological methods. Those relating to the immunological detection of oestrogen when bound to its receptor have been discussed. The other two main areas of dispute relate to tissue preparation, and the specificity of the methods.

3.1. Tissue Preparation

Oestrogen receptor, being a cytosolic protein, readily diffuses in aqueous solutions;[34] it is also labile, and rapid cooling of tissue samples to below 0 °C and storage at very low temperatures is important.[35] It would seem unlikely that receptors could retain their binding capacities at the temperatures tissues are exposed to during paraffin embedding, let alone the effects induced by fixation. Extensive studies[23] of the effects on frozen sections of air-drying procedures and exposure to acetone, alcohols and formol–saline has shown that only brief air-drying or acetone fixation is suitable for histological methods, and that other treatments result in diffusion and loss of reactivity. Penney and Hawkins[22] assessed the effect on oestrogen receptor activity of tissue sectioning and exposure to aqueous media with and without fixation. They found that receptor activity is very readily lost from unfixed tissues into the media, that acetone fixation appears to destroy at least 50 per cent of receptor activity and that fixation with ethanol after incubation as employed by some groups[7,20] abolishes steroid binding. From these studies it would appear that brief fixation of frozen sections by acetone might allow some 25 per cent of oestrogen receptor within the tissue to remain active; a compromise rather than an ideal situation.

3.2. Specificity

As indicated, the oestrogen receptor exhibits high affinity binding, i.e. all receptor sites will be saturated by concentrations of oestradiol of 10^{-9} M. This receptor has been designated the true or type I binding site. It is because of the strict criteria that define the nature of oestrogen receptor that histological methods have been criticized,[35] since the range of oestradiol concentrations required are outside the limits for high affinity binding. Other types of binding sites have been identified. Type II binding sites have been described in rat uterus[35] and human breast cancers.[37] These have a lower affinity but a higher capacity for oestradiol. Below 10^{-9} M oestradiol, few of these sites will be occupied but most will be saturated at around 10^{-7} M. Some of the histological studies could therefore be detecting type II binding sites. Oestradiol will also bind to soluble molecules such as albumin,[35] and above 10^{-7} M binding to these sites may become substantial. These substances have been labelled type III binding sites, and they may be significant for several of the histological methods.[15]

The critical evaluation of these methods is whether they can predict response to endocrine therapy; Pertschuk et al.[32] found a correlation between response and tumours showing oestrogen binding in 60 per cent of cases, which increased to 72 per cent when progesterone binding was included, figures which are similar to those observed with biochemical assays.

4. ANTIBODIES TO RECEPTORS

Due to the problems relating to the specificity of the histological methods, an obvious development is the use of antibodies directed against the receptors. Raam et al.[38] raised polyclonal antibodies against receptors isolated from human breast carcinomas and, using frozen sections that were progressively dehydrated then rehydrated, were able to detect cytoplasmic receptors in 12 out of 16 carcinomas with a positive correlation with the dextran-coated charcoal assay in 9 cases.[39] Nuclear staining was seen in only one tumour. Greene et al.[40–42] have developed monoclonal antibodies to oestrogen receptor. The antibodies against oestrogen receptor protein from the MCF-7 breast cancer cell line have been shown to react with the cytostolic oestradiol–receptor complex and the nuclear oestradiol–receptor complex, and they can recognize receptor whose steroid-combining site is either occupied or unoccupied. Each monoclonal antibody appears to recognize non-overlapping antigenic determinants on the receptor molecule. Immunohistological methods have been undertaken to localize receptor in human uterus[43] and breast tumours.[44,45] The studies on human uterus used frozen sections but it was found that fixation was necessary since incubation of sections in buffer resulted in a progressive loss of specific staining. However, prolonged fixation also adversely affected reactions. The best results were achieved with 4 per cent paraformaldehyde, periodate–lysine–paraformaldehyde, picric acid–paraformaldehyde and acetone, but fixation beyond an hour was unsuitable for all. The interesting finding in this study, which used two monoclonal antibodies, was that all staining was nuclear. The same observation was made in studies of frozen sections of human breast tumours and MCF-7 cells,[44] using five monoclonal antibodies and similar fixation techniques. Heterogeneity of staining was a common finding.

Of even more relevance to routine diagnostic histopathology are the reports[45] describing the detection of oestrogen receptor in Bouin's-fixed, paraffin-embedded tissue using a mixture of two monoclonal antibodies. As with the other studies using the same antibodies, the staining has been nuclear; by using a higher concentration of the antibodies one out of 35 cases was shown to have cytoplasmic staining. The staining was correlated with the dextran-coated charcoal analysis of the tumours; 4 out of 25 had a nuclear reaction but no evidence of detectable receptor by the biochemical assay and 20 out of 43 cases were negative by immunohistology but had detectable receptor by the biochemical method. The number of negative cases was reduced to 6 out of 16 cases when the antibody concentration was increased. It could be that the determinants for the antibodies in these tumours may have been obscured/modified by fixation and processing. Details as to the clinical response of these patients are not available.

The findings with the monoclonal antibodies raise many questions as to the orthodox teaching about receptor mechanisms. Similar reservations have been raised from studies of progesterone receptors in chick oviduct using a combination of polyclonal and monoclonal antibodies.[46] However, whether the nuclear staining indicates an 'intact' or functional receptor mechanism has to be resolved. Similarly, it is difficult from present studies to evaluate the merits of applying monoclonal antibodies to fixed, embedded tissue. It could well be that in some tumours, receptor may be affected by fixation, the affinity of the receptor for the antibody may vary from tumour to tumour, and the antigenic nature of the receptor may vary between carcinomas necessitating the use of several antibodies. This is an area that obviously requires further investigation. Unless the antibodies to oestrogen receptor do give information about the functional nature of the receptor, as with biochemical assays, the detection of progesterone receptor or oestrogen-regulated proteins[47] will also have to be undertaken for the methods to be of clinical value.

5. CONCLUSIONS

Histological methods for the detection of steroid receptors are obviously of value in that they can be applied to a wider group of patients than is presently possible with the biochemical assays, and that they can provide information about heterogeneity of tumours which could be clinically important. The problem with several of the approaches described relates to the specificity of the methods, in that type II and III oestrogen-binding sites are probably detected rather than the type I, high affinity receptor, but the introduction of antibodies to oestrogen receptor can overcome this. However, the data obtained so far from several studies utilizing the same group of antibodies are contrary to the theories related to receptor mechanisms; the development of other monoclonal antibodies to the receptor, e.g. from other cell lines, is therefore awaited with interest. The stabilization of receptor proteins within tissue sections is at the present time one of the major problems of all histological methods and is an area for further consideration.

Acknowledgements

I am grateful to Mrs G. Holmes for secretarial help.

REFERENCES

1. Toft D. and Gorski J. A receptor molecule for estrogens: isolation from the rat uterus and preliminary characterisation. *Proc. Natl Acad. Sci. USA* 1966, **55**, 1574–1581.
2. Jensen E. V., Block G. E., Smith S., Kyser K. A. and DeSombre E. R. Estrogen receptors and breast cancer response to adrenalectomy. *Natl Cancer Inst. Monograph* 1971, **34**, 55–70.

3. McGuire W. L. Estrogen receptors in human breast cancer. *J. Clin. Invest.* 1973, **52**, 73–77.

4. DeSombre E. R. Steroid receptors in breast cancer In: McDivitt R. W., Oberman H. A., Ozello L. and Kaufman N., eds., *The Breast*, Baltimore, Williams and Wilkins, 1984: 149–174.

5. Nicolo G., Carbone A., Esposito M. and Santi L. Sampling and storage of breast cancer tissue for steroid receptor assays. In: Leclercq G., Toma S., Paridaens R. and Heuson J. C., eds., *Clinical Interest in Steroid Hormone Receptors in Breast Cancer*. Berlin, Springer-Verlag, 1984: 3–11.

6. Nenci I., Beccati M. D., Piffanelli A. and Lanza G. Detection and dynamic localization of oestradiol–receptor complexes in intact target cells by immunofluorescent technique. *J. Steroid Biochem.* 1976, **7**, 505–510.

7. Pertschuk L. P., Tobin E. H., Brigati D. J., Kim D. S., Bloom N. D., Gaetjens E., Berman P. J., Carter A. C. and Degenshein G. A. Immunofluorescent detection of estrogen receptors in breast cancer: comparison with dextran-coated charcoal and sucrose gradient assays. *Cancer* 1978, **41**, 907–911.

8. Mercer W. D., Lippman M. E., Wahl T. M., Carlson C. A., Wahl D. A., Lezotte D. and Teague P. O. The use of immunocytochemical techniques for the detection of steroid hormones in breast cancer cells. *Cancer* 1980, **46**, 2859–2868.

9. Walker R. A., Cove D. H. and Howell A. Histological detection of oestrogen receptor in human breast carcinomas. *Lancet* 1980, **i**, 171–173.

10. Ghosh L., Ghosh B. C. and Das Gupta T. K. Immunocytological localisation of estrogen in human mammary carcinoma cells by horseradish–anti-horseradish peroxidase complex. *J. Surg. Oncol.* 1978, **10**, 221–224.

11. Kurzon R. M. and Sternberger L. A. Estrogen receptor immunocytochemistry. *J. Histochem. Cytochem.* 1978, **26**, 803–808.

12. Taylor C. R., Cooper C. L., Kurman R. J., Goebelsmann U. and Markland F. S. Detection of estrogen receptor in breast and endometrial carcinoma by the immunoperoxidase technique. *Cancer* 1981, **47**, 2634–2640.

13. Fishman J. and Fishman J. H. Competitive binding assay for estradiol receptor using immobilized antibody. *J. Clin. Endocrinol. Metab.* 1974, **39**, 603–606.

14. Casteneda E. and Liao S. The use of anti-steroid antibodies in the characterisation of steroid receptors. *J. Biol. Chem.* 1975, **250**, 883–888.

15. Morrow B., Leav I., DeLellis R. A. and Raam S. Use of polyestradiol phosphate and anti-17-β-estradiol antibodies for the localisation of estrogen receptor in target tissues: a critique. *Cancer* 1980, **46**, 2872–2879.

16. Underwood J. C. E., Sher E., Reed M., Eisman J. A. and Martin, T. J. Biochemical assessment of histochemical methods for oestrogen receptor localisation. *J. Clin. Pathol.* 1982, **35**, 401–406.

17. Barrows G. H., Stroupe S. B. and Riehm J. D. Nuclear uptake of ethyl-succinamide bridged 17-β-estradiol–fluorescein as a marker of estrogen receptor activity. *Am. J. Clin. Pathol.* 1980, **73**, 330–339.

18. Nenci I., Dandliker W. B., Meyers C. Y., Marchetti E., Mazola A. and Fabris G. Estrogen receptor cytochemistry by fluorescent oestrogen. *J. Histochem. Cytochem.* 1980, **28**, 1081–1088.

19. Lee S. H. Cancer cell estrogen receptor of human mammary carcinoma. *Cancer*, 1979, **44**, 1–12.

20. Pertschuk L. P., Gaetjens E., Carter A. C., Brigati D. J., Kim D. S. and Fealey T. E. An improved histochemical method for the detection of estrogen receptors in mammary carcinoma. *Am. J. Clin. Pathol.* 1979, **71**, 504–508.

21. Dandliker W. B., Brawn R. J., Hsu M.-L., Brawn P. N., Levin J., Meyers C. Y. and Kolb V. M. Investigation of hormone-receptor interactions by means of fluorescent labelling. *Cancer Res.* 1978, **38**, 4212–4224.

22. Penney G. C. and Hawkins R. A. Histochemical detection of oestrogen receptor: a progress report. *Br. J. Cancer*, 1982, **45**, 237–246.

23. Walker R. A. The cytochemical demonstration of oestrogen receptors in human breast carcinoma. *J. Pathol.* 1981, **135**, 237–247.

24. Walker R. A. The use of peroxidase-labelled hormones in the study of steroid binding in breast carcinomas. In: Kaiser E., Gabl F., Muller M. H. and Bayer M., eds., *Eleventh International Congress of Clinical Chemistry*. Berlin, De Gruyter, 1982: 507–511.

25. Walker R. A. Peroxidase labelling of estrogen binding sites in breast cancer. In: Lee S. H. and

Pertschuk L. P., eds., *Localization of Putative Steroid Receptors*, Vol. II. Florida, CRC Press, 1985: in press.

26. Nakane P. K. and Kawaoi A. Peroxidase-labelled antibody a new method of conjugation. *J. Histochem. Cytochem.* 1974, **22**, 1084–1091.

27. Avrameas S. and Ternynck T. Peroxidase labelled antibody and Fab conjugates with enhanced intracellular penetration. *Immunochemistry* 1971, **8**, 1175–1179.

28. Binder M. Oestradiol–BSA conjugates for receptor histochemistry: problems of stability and interactions with cytosol. *Histochem. J.* 1984, **16**, 1003–1023.

29. Gunduz N., Zheng S. and Fisher B. Fluoresceinated estrone binding by cells from human breast cancers obtained by needle aspiration. *Cancer* 1983, **52**, 1251–1256.

30. Van N. T., Raber M., Barrows G. H. and Barlogie B. Estrogen receptor analysis by flow cytometry. *Science* 1984, **224**, 876–879.

31. Martin P. M., Magdelenat H. P., Benyahia B., Rigand O. and Katzenellenbogen J. A. New approach for visualizing estrogen receptors in target cells using inherently fluorescent ligands and image intensifications. *Cancer Res.* 1983, **43**, 4956–4965.

32. Pertschuk L. P., Tobin E. H., Tanapat P., Gaetjens E., Carter A. C., Bloom N. D., Macchia R. J. and Eisenberg K. B. Histochemical analyses of steroid hormone receptors in breast and prostatic carcinoma. *J. Histochem. Cytochem.* 1980, **28**, 799–810.

33. Bergqvist A., Carlström K. and Ljungberg O. Histochemical localization of estrogen and progesterone receptors: evaluation of a method. *J. Histochem. Cytochem.* 1984, **32**, 493–500.

34. McCarty K. S. Jr, Woodward B. H., Nichols D. E., Wilkinson W. and McCarty K. S. Sr Comparison of biochemical and histochemical techniques of estrogen receptor analyses in mammary carcinoma. *Cancer* 1980, **46**, 2842–2845.

35. Chamness G. C., Mercer W. D. and McGuire W. L. Are histochemical methods for estrogen receptor valid? *J. Histochem. Cytochem.* 1980, **28**, 792–798.

36. Clark J. H. and Markaverich B. M. Relationship between type I and type II estradiol binding sites and estrogen induced responses. *J. Steroid Biochem.* 1981, **15**, 49–54.

37. Panko W. B., Watson C. S. and Clark J. H. The presence of a second, specific estrogen binding site in human breast cancer. *J. Steroid Biochem.* 1981, **14**, 1311–1316.

38. Raam S., Peters L., Rafkind I., Putmun E., Longcope C. and Cohen J. L. Simple methods for production and characterization of rabbit antibodies to human breast tumor estrogen receptors. *Mol. Immunol.* 1981, **18**, 143–156.

39. Raam S., Nemoth E., Tamura H., O'Briain D. S. and Cohen J. L. Immunohistochemical localization of estrogen receptors in human mammary carcinoma using antibodies to the receptor protein. *Eur. J. Cancer Clin. Oncol.* 1982, **18**, 1–12.

40. Greene G. L., Fitch F. W. and Jensen E. V. Monoclonal antibodies to estrophilin: probes for the study of estrogen receptors. *Proc. Natl Acad. Sci. USA* 1980, **77**, 157–161.

41. Greene G. L. and Jensen E. V. Monoclonal antibodies as probes for the detection and study of estrogen receptor. *J. Steroid Biochem.* 1982, **16**, 353–359.

42. Greene G. L., Nolan C., Engler J. P. and Jensen E. V. Monoclonal antibodies to human estrogen receptor. *Proc. Natl Acad. Sci USA* 1980, **77**, 5115–5119.

43. Press M. F. and Greene G. L. An immunocytochemical method for demonstrating estrogen receptor in human uterus using monoclonal antibodies to human estrophilin *Lab. Invest.* 1984, **50**, 480–486.

44. King W. J. and Greene G. L. Monoclonal antibodies localise oestrogen receptor in the nuclei of target cells. *Nature* 1984, **307**, 745–747.

45. Skovgaard Poulsen H., Ozello L., King W. J. and Greene G. L. The use of monoclonal antibodies to estrogen receptors (ER) for immunoperoxidase detection of ER in paraffin sections of human breast cancer tissue. *J. Histochem. Cytochem.* 1985, **33**, 87–92.

46. Gasc J-M., Renmoir J-M., Radanyi C., Joab I., Tuohimaa P. and Baulieu E. E. Progesterone receptor in the chick oviduct: an immunohistochemical study with antibodies to distinct receptor components. *J. Cell Biol.* 1984, **99**, 1193–1201.

47. Adams D. J., Hajj H., Edwards D. P., Bjercke R. J. and McGuire W. L. Detection of a Mr 24,000 estrogen-regulated protein in human breast cancer by monoclonal antibodies. *Cancer Res.* 1983, **43**, 4297–4301.

12

Hybridization Histochemistry: Analysis of Specific mRNAs in Individual Cells

J. L. Roberts and J. N. Wilcox

Events surrounding the biosynthesis of proteins localized to specific tissues are of very great interest to scientists today. Hundreds of specific proteins have been identified in various tissues and cell types. However, due to the tremendous movement of proteins throughout the body, either by the systemic circulation, axonal transport, or other means, it is not always clear whether, when a protein is identified in a specific tissue, it was actually synthesized there. Thus, it is possible that the presence of a protein in a specific tissue is a result of its uptake or transport there and not due to local synthesis. Therefore, it is important, particularly when wishing to study the regulation of protein biosynthesis, to establish that a protein is actually synthesized in the tissue in which it has been identified.

The classical way of showing the synthesis of a protein in a particular tissue has been to use either *in vitro* or *in vivo* radioactive labelling techiques. However, in complex tissues there may be more than one cell type expressing a particular gene. Standard techniques of labelling, homogenizing the tissue, and subsequent analysis by protein fractionation and/or immunoprecipitation characterize the proteins but do not tell you which cell type they came from. Currently, there is no method for directly identifying synthesis of a specific protein in an individual cell. However, an indirect method is available based on the localization of that protein's mRNA in the same cell.

For a cell to be able to synthesize a protein it must have in its cytoplasm the mRNA encoding that protein. Thus, by identifying the mRNA and its coded protein in a specific cell, one can make a good argument for the fact that the protein is indeed synthesized in that cell. In a series of biosynthetic studies done in model systems where the cell type expressing a specific protein product is known to be homogeneous, it has been shown that the level of synthesis of the protein reflects fairly well the level of mRNA present in that cell type. Thus, a technique that detects more or less of a specific mRNA in an individual cell, would also identify possible changes in synthesis of the protein encoded by the mRNA.

The technique that identifies specific mRNAs in individual cells is in situ cDNA–mRNA hybridization or hybridization histochemistry. Below, we would like to discuss some general aspects of the technique directing the reader to several very good methodology papers on the subject, as well as identifying some of the interesting findings in recent applications of this technique in the past few years. Although our interest has been primarily in the production of peptide hormones, the discussion presented here is applicable to any specific protein product, be it an enzyme, structural protein, or membrane protein. (*See also* Chapter 17).

1. TECHNIQUE

Basically the procedure of in situ cDNA–mRNA hybridization is similar to that of immunohistochemistry with the exception that radioactive or chemically tagged nucleic acids are used to identify specific sequences of nucleotides in mRNAs fixed in the tissue section. Many different procedures are available for this technique and we have reviewed them recently in detail.[1] Our basic procedure is outlined in the Appendix and several of the methodological considerations are discussed below. An example of the procedure is shown in *Fig.* 12.1.

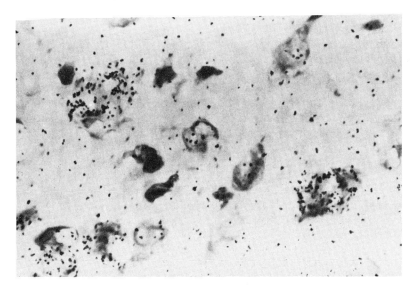

Fig. 12.1. This is an example of a typical in situ hybridization using the procedure described in the Appendix with a c[³H]DNA POMC probe on a 10-μm thick section in the medial region of the rat arcuate nucleus. Note the larger number of positive grains over several of the cells. × 300.

The tissues for study can be fixed using a variety of different methods, either perfusion fixation with paraformaldehyde, which we have found best for our studies on neuropeptides, or immersion fixation in a whole variety of other fixatives.[1-4] Specific mRNAs have been localized to individual cells in either whole tissue sections or monolayer cultures. After fixation, most researchers use some type of deproteinizing permeabilization technique.[1-3] In addition to exposing mRNA nucleotide sequences by removing protein bound to the mRNA in the tissue, the permeabilization protocol appears to enhance the ability of the probe to enter into the tissue section.

The size of the radioactive nucleic acid probe has a dramatic effect on the resulting autoradiographic signal.[1-3] The basic observation is that small probes, even down to the size of 20–30 bases, give much stronger signals than the larger probes (200–500).[1,3] The best interpretation of this is that the smaller probe has a greater ability to diffuse into the tissue matrix. The effects of salt concentration, temperature of hybridization wash and melting temperature of the in situ cDNA–mRNA hybrids have also been analysed.[2,4] These studies suggest that the nucleic acid hybrids that are formed have similar properties to nucleic acid hybrids resulting from solution or filter hybridizations. Thus, the specificity of the cDNA–mRNA hybrids can be inferred from their biophysical characteristics in situ in the same way as after filter or solution hybridization.

After hybridization of the cDNA probe to the mRNA, unhybridized material is removed by extensive washing in salt solutions. Prolonged washing of the sections is necessary to allow the unhybridized labelled cDNA probe to diffuse out of the tissue section. The resulting stable hybrids are then visualized either by autoradiography, when radioactive probes are used, or by immunohistochemistry when biotin-tagged probes are used. Thus the presence of autoradiographic grains or immunohistochemical stains over individual cells identifies the presence of the mRNA complementary to the probe.[5]

Cell architecture can also play an important role with regard to the efficiency of detection of the hybridization signal resulting from in situ hybridization. An example of this is the pituitary pro-opiomelanocortin (POMC) system. The copy number of POMC mRNA is essentially the same in the intermediate and anterior pituitary POMC cells, as we know from solution hybridization studies and knowledge of the number of POMC-expressing cells.[6] However, the in situ hybridization signal for POMC mRNA in the anterior lobe is about five times weaker than is the signal from the intermediate lobe cells.[3] Interestingly, when animals are injected with colchicine, the anterior lobe cells become as easy to detect as cells in the intermediate lobe. One interpretation of this observation is that the cell architecture changes associated with the colchicine treatment have made the POMC mRNA easier to detect. While there are no changes in POMC mRNA levels after colchicine treatment,[3] the corticotroph cells become more rounded, losing the stellate processes and thus condensing the POMC mRNA in a more confined area.

2. APPLICATIONS AND RECENT DISCOVERIES

As mentioned above, many people have used this technique to identify the sites of expression of specific gene products. Using the high energy isotope X-ray film method of detecting cDNA–mRNA hybrids, researchers have shown gross tissue localization of a whole variety of mRNAs.[2,7] Using lower energy isotopes or biotin-labelled probes, the exact cells of expression have been identified for a variety of peptide hormone and structural protein gene products.[8–10] These studies have not only confirmed the concept of local synthesis suggested by earlier biosynthetic studies, identifying the exact cellular sites of synthesis, but have also shown surprising previously unknown sites of synthesis of some proteins, such as pro-epidermal growth factor in rodent kidney.[11]

Studies in situ have also been used for developmental analysis of the ontogeny of expression of specific gene products.[4,12] Indeed, expression of a specific mRNA must precede the expression of the protein it codes for. Thus, it should be possible to detect the mRNA several days before the protein is detected by immunohistochemistry. Such studies have been used elegantly in the case of identifying peptide hormones in the mollusc *Aplysia*, showing that egg-laying hormone mRNA can be detected early during development before the precursor bag cells have even migrated from the body wall.[4]

cDNA–mRNA hybridization in situ has also been combined with other anatomical techniques to enhance interpretation of its results. By combining in situ hybridization with immunohistochemistry specific for a different protein, mRNA levels in subpopulations of cells within a cluster of expressing cells can be analysed independently.[13,14] This represents a tremendous advance in that antibodies exist for a variety of substances, and this will enhance our ability to identify specific functionally related subpopulations of cells showing positive in situ hybridization. The labelling of nerve cell bodies by retrograde axonal transport of fluorescent dyes is also compatible with in situ hybridization. *Plate* 9 shows the co-localization of the fluorescent dye Fast Blue retrogradely transported from the pre-optic area to the medial basal hypothalamus with POMC mRNA as detected by in situ hybridization. By counting silver grains in only those cells containing Fast Blue and POMC mRNA, it is therefore possible to estimate POMC gene expression in only that subpopulation of POMC neurones that project to the pre-optic area. Another recent development allows for the detection of two different mRNAs on the same tissue section. This technique uses cDNA probes labelled with different energy isotopes; the silver grains resulting from each isotope are then distinguished by differential colour processing.[15] This technique is very important in that it will allow the positive identification of cells co-expressing two different peptides, a question we are not able to resolve by current methods.

The usefulness of the in situ hybridization technique is rapidly expanding as the resolution of hybridization is refined and the technique is coupled with other classical anatomical techniques to further characterize the cells expressing the gene of interest. The most exciting potential of the procedure is the possibility of using in situ hybridization as a semiquantitative assay for specific mRNAs at the level of the individual cell. Early studies indicated that the number of viral genome copies per cell showed a linear relationship to the density of silver grains after in situ hybridization.[2] The technique has also been used to estimate relative amounts of POMC[3] or globin[16] mRNAs in mammalian tissue. We have observed statistically significant differences in the number of silver grains over individual cells in culture where there are larger than two-fold changes in specific mRNA levels as determined by parallel solution hybridization experiments. However, these studies have used in situ hybridization to measure regulation of mRNAs in homogeneous systems. The real value of the technique is its ability to resolve cDNA–mRNA hybrids at the level of the individual cell and, therefore, its potential for answering questions of regulation where the mRNA of interest is heterogeneously regulated in multiple cell types. If it is not possible to dissect out a homogeneously responding subpopulation of cells, then analysis by in situ hybridization is virtually the only way to analyse regulation of gene expression in such a population of cells. This is generally true in the brain and this technology will surely contribute significantly to knowledge of molecular neuroendocrinology.

Appendix

PROTOCOL FOR IN SITU HYBRIDIZATION IN RAT BRAIN WITH cDNA PROBES

This procedure has been optimized for the POMC system in the hypothalamus, but does work for other probes in other regions of the brain and other tissues, such as pituitary or liver.

1. Anaesthetize animal and rinse the systemic circulation via cardiac perfusion with 20 ml saline followed by 20 min ($\approx$200 ml) perfusion with ice-cold 4 per cent paraformaldehyde in 0·1 M sodium phosphate (pH 7·4) at 100–120 mmHg pressure.
2. Tissues of interest are removed and placed in sterile 15 per cent sucrose in PBS (2·7 mM KCl, 1·5 mM KH_2PO_4, 137 mM NaCl, 8·1 mM $Na_2HPO_4 \cdot 7H_2O$ pH 7·0) containing 0·1 per cent sodium azide, for no longer than 60 min at 4 °C.
3. Tissues are embedded in OCT compound (American Scientific Products) frozen in liquid nitrogen, and stored at −70 °C until sectioned.
4. 10-μm sections are taken on a cryostat. Sections are quickly thaw-mounted on gelatin-coated slides at room temperature and immediately refrozen in slide boxes kept in the cryostat. Sections may be stored at −70 °C with dessicant for up to 6 weeks.
5. Sections are removed from freezer and immediately covered with 5 μg/ml proteinase-K (E.M. Reagents, American Scientific Products) in sterile water and incubated for 10 min at 30 °C. Rapid treatment with proteinase K destroys any endogenous RNAses released by the freeze–thaw process and increases permeation of the probe into the tissue.
6. Slides washed for 10 min at room temperature in 0·2 × SSC (*see below*).
7. Dry around section with fibre-free tissue.
8. Section covered with 50 μl hybridization buffer (*see below*) without probe for 1–2 h at 30 °C in air-tight hybridization boxes containing filter paper saturated with 4 × SSC + 50 per cent formamide.
9. Dry around section with tissue, removing as much of the hybridization buffer as possible.

10. Section covered with 25 µl hybridization buffer containing 20 000–40 000 counts/min c[^{3}H]DNA probe (approximately 0·5–1·0 ng cDNA insert).
11. Incubate 15–20 h in hybridization box at 30 °C.
12. Wash 2 × 10 min in 2·0 × SSC, room temperature.
13. Wash 3 h to overnight in 0·5 × SSC at room temperature. (Longer washes tend to give lower background.)
14. Dehydrate 2 min each in graded alcohols (50, 70, 90 per cent) containing 0·3 M ammonium acetate.
15. Dry 30 min in vacuum dessicator and store slides at room temperature with dessicant until autoradiography.
16. Dip slides in 1 : 1 solution Kodak NTB2 nuclear emulsion and water; dry 2 h in dark.
17. Expose slides at 4 °C in dark with dessicant for 2–8 weeks.
18. Develop slides as follows: Kodak D19 developer diluted 1 : 1 with water for 3 min, water for 20 s, fix for 3 min in Kodak Rapid Fixer.
19. Wash slides 3 × 5 min in water and counterstain with haematoxylin and eosin.

Hybridization buffer
   5 × SSC
   0·02 per cent bovine serum albumin (Sigma, Fraction 5)
   0·02 per cent polyvinyl-pyrrolidone (Sigma, Type 360)
   0·02 per cent Ficoll (Sigma, Type 400)
   50 per cent deionized formamide

1 × SSC (sodium chloride, sodium citrate)
   150 mM NaCl
   15 mM sodium citrate pH 7·0

REFERENCES

1. Wilcox J. N., Gee C. E. and Roberts J. L. *In situ* cDNA: mRNA hybridization: development of a technique to measure mRNA levels in individual cells. *Methods Enzymol.* 1985, in press.
2. Brahic M. and Haase A. T. Detection of viral sequences of low reiteration frequency by in situ hybridization. *Proc. Natl Acad. Sci. USA* 1978, **75**, 6125–6129.
3. Gee C. E. and Roberts J. L. In situ hybridization histochemistry: a technique for the study of gene expression in single cells. *DNA* 1983, **2**, 157–163.
4. McAllister L. B., Scheller R. H., Kandel E. R. and Axel R. In situ hybridization to study the origin and fate of identified neurons. *Science* 1983, **222**, 800–808.
5. Kelsey J. E., Watson S. J., Burke S., Akil H. and Roberts J. L. Characterization of a target of *in situ* hybridization. *J. Neurosci.* 1985, in press.
6. Herbert E., Birnberg N., Lissitsky J. C., Civelli O. and Uhler M. Proopiomelanocortin: a model for the regulation of neuropeptides in pituitary and brain tissue. *Neurosci. Newslett.* 1981, **12**, 16–21.
7. Hudson P., Penschow J., Shine J., Ryan G., Niall H. and Coughlan J. Hybridization histochemistry: use of recombinant DNA as a 'homing probe' for tissue localization of specific mRNA populations. *Endocrinology* 1981, **108**, 353–356.
8. Gee C. E., Chen C. L. C., Roberts J. L., Thompson R. and Watson S. J. Identification of proopiomelanocortin neurons in rat hypothalamus by in situ cDNA–mRNA hybridization. *Nature* 1983, **306**, 374–376.

9. Singer R. H. and Ward D. C. Actin gene expression visualized in chicken muscle tissue culture by using in situ hybridization with a biotinated nucleotide analog. *Proc. Natl Acad. Sci. USA* 1982, **79**, 7331–7335.

10. Hafen E., Levine M., Garber R. L. and Gehring W. J. An improved in situ hybridization method for the detection of cellular RNAs in drosophila tissue sections and its application for localizing transcripts of the homeotic antennapedia gene complex. *EMBO J.* 1983, **2**, 617–623.

11. Rall L. B., Scott J. and Bell G. I. Mouse prepro-epidermal growth factor synthesis by the kidney and other tissues. *Nature* 1985, **313**, 228–231.

12. Angerer L. M. and Angerer R. Detection of poly A+ RNA in sea urchin eggs and embryos by quantitative in situ hybridization. *Nucleic Acids Res.* 1981, **9**, 2819–2840.

13. Griffin W. S. T., Alegos M., Nilaver G. and Morrison M. R. Brain protein and messenger RNA identification in the same cell. *Brain Res. Bull.* 1983, **10**, 597–601.

14. Brahic M., Haase A. T. and Cash E. Simultaneous in situ detection of viral RNA and antigens. *Proc. Natl Acad. Sci. USA* 1984, **81**, 5445–5448.

15. Haase A. T., Walker D., Stowring L., Ventura P., Geballe A., Blum H., Brahic M., Goldberg R. and O'Brien K. Detection of two viral genomes in single cells by double-label hybridization in situ and color microradioautography. *Science* 1985, **227**, 189–192.

16. Dormer P., Korge E. and Hartenstein R. Quantitation of globin mRNA in individual human erythroblasts by in situ hybridization. *Blut* 1981, **43**, 79–83.

13

Quantitative Analysis: Computer-assisted Morphometry and Microdensitometry applied to Immunostained Neurones

L. F. Agnati, K. Fuxe, A. M. Janson,
M. Zoli and A. Härfstrand

Semi-automatic and automatic image analysis has recently made it possible to carry out a quantitative morphofunctional characterization of transmitter-identified neurones.[1-10] Using these new procedures, single nerve cells and entire nerve cell groups can be described in quantitative terms, as can objective criteria for their existence. Methods have also been developed for the characterization of terminals and dendrites of transmitter-identified neurones,[5] for the three-dimensional reconstruction of transmitter-identified nerve cell groups and for the analysis of coexistence of neuroactive substances in nerve cell bodies, using morphometrical programs and, in nerve terminal systems, microdensitometrical programs.[2,5,8,9] These methods have been used, amongst other things, in the characterization of degenerative and regenerative features of the ascending mesostriatal dopamine neurones[11] and of ageing processes in transmitter-identified neurones.[12-14]

In this chapter these various methods will be reviewed, together with new modifications which, by means of automatic image analysis, allow the rapid identification of nerve cell groups. Finally, we will also summarize the method developed for semiquantitative evaluation of the antigen content in immunocytochemical preparations using saturation analysis in combination with microdensitometry.[8,15]

1. MORPHOMETRICAL CHARACTERIZATION OF TRANSMITTER-IDENTIFIED NERVE CELL GROUPS

Frontal sections are used and marked with rectangular (Cartesian) coordinates. The y-axis is often represented by the midline, and the x-axis is often a tangent to the ventral border of the brain. Using this system the position of each transmitter-identified nerve cell body can be identified by means of its x- and y-coordinates. By plotting these positions, a 'scatter' diagram can be obtained to illustrate the location of the nerve cell population (*Fig.* 13.1*a*). Furthermore, a quantitative description of the nerve cell group is obtained by the use of a grid of unitary squares superimposed on the Cartesian plane followed by plotting the number of nerve cell bodies falling in each of the squares measured ('density map', *see Fig.* 13.1*b*). The size of the square (grid step) is usually five times the mean of the maximal diameter of the profiles present in the frontal section. The area of the analysis is determined by displacement of lines parallel to the x- and y-axis towards the area of the transmitter-identified nerve cell population, until the first positive square is reached.[5] On the density map it can be tested whether the number of transmitter-identified nerve cell bodies in each square follows a Poissonian distribution and whether the frequency polygons obtained differ from a random distribution. Evidence for the existence of a cluster (cell group) of transmitter-identified neurones is obtained, when it is shown that a random distribution does not exist (chi-squared test) and that a clumped distribution is present (coefficient of variation is larger than 1). The possible co-distribution of two cell groups of transmitter-identified neurones can be well illustrated in a quantitative way by plotting the density maps from two adjacent sections in the same figure (*Fig.* 13.2). Each square has an identical position in the two maps and is divided into white and black triangles. The numbers in the respective triangles show how many nerve cell bodies exist in that square. In the figure the co-distribution of tyrosine hydroxylase (TH)- and cholecystokinin (CCK)-immunoreactive nerve cell bodies in the substantia nigra is shown in the 3-month-old adult male rat. The figure gives an immediate evaluation of the putative co-distribution of TH- and CCK-immunoreactive cell bodies in discrete regions of the substantia nigra. In *Fig.* 13.3 the same type of double density plot has been used to show the distribution and coexistence of TH and glucocorticoid receptor (GR) immunoreactivity in the mediobasal hypothalamus based on two-colour immunocytochemistry on the same section. The GR immunoreactivity is mainly present in the nuclei of the nerve cell bodies and shows a much higher density than the TH immunoreactive profiles, which represent perikarya.

In analysis of the density maps, the squares containing two or more nerve cell bodies are always considered, while empty squares are only considered when completely surrounded by non-empty squares. Squares with only one nerve cell body are taken into account only when they are in continuity with a square having two or more nerve cell bodies.[5] In this way it becomes possible to define the

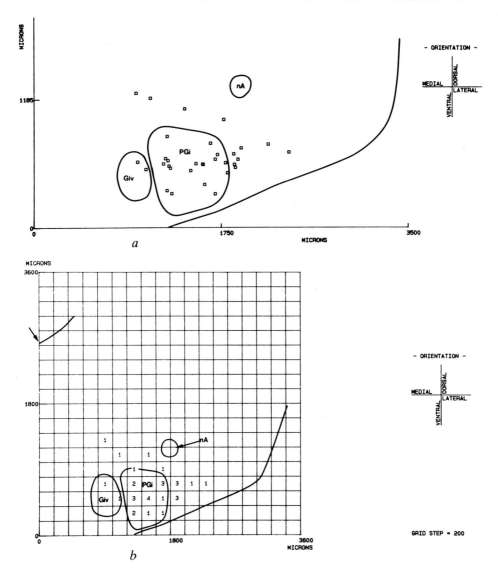

Fig. 13.1 Scatter diagram (*a*) and density map (*b*) of PNMT-immunoreactive (IR) nerve cell bodies (group C1) in the ventrolateral reticular formation of the medulla oblongata of the male rat (1·7 mm rostral to obex). PAP procedure. Giv = nuc. gigantocellularis reticularis, pars ventralis; PGi = nuc. paragigantocellularis reticularis; nA = nuc. ambiguus. Origin = intersection of the midline with the ventral surface. Arrow shows the intersection of the midline with the dorsal surface.

boundaries of the nerve cell group in an objective way (*see above*). Subgroups have been objectively defined by means of differences in nerve cell density and on the presence of a bimodal distribution of the nerve cell bodies as shown in a frequency distribution analysis.[3] Using the cell density approach, a nerve cell cluster can sometimes be split into two clusters by considering only those squares which contain 30 per cent or more of the peak density value per square observed in

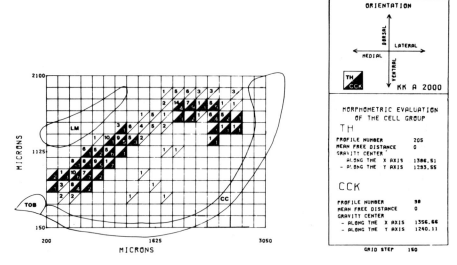

Fig. 13.2. TH- and CCK-immunoreactive neurones of the substantia nigra of the 3-month-old adult male rat are plotted as density maps at the intermediate level of the substantia nigra. PAP procedure. *See* right upper panel for symbols.

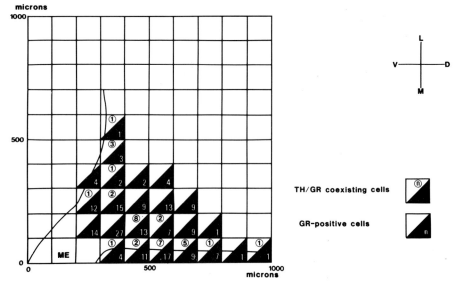

Fig. 13.3. Density map of glucocorticoid receptor (GR)- and TH-immunoreactive (IR) profiles in the mediobasal hypothalamus obtained from the same coronal section of male rat after a double immunostaining procedure showing 'blue' staining (α-chloronaphthol) for TH immunoreactivity in the perikaryon and 'brown' staining (DAB) for GR immunoreactivity in the nucleus. PAP procedure. Coexistence is indicated by the ring around the figure for TH in the white triangle (*see*[24]). Origin = ventral midline surface. Semi-automatic image analysis. Antisera diluted 1 : 1000.

the nerve cell population studied. Thus, computer-assisted morphometry allows the detection of topological heterogeneities in a transmitter-identified nerve cell group.

In describing the nerve cell groups of transmitter-identified cells the following cell group parameters have been used:

a. *Gravity centre.* The gravity centre location shows where the weight of the group is concentrated and is obtained by considering the means of all the x- and y-coordinates of all the transmitter-identified immunoreactive profiles within the nerve cell group. The standard deviation and especially the coefficient of variation of the M_x and M_y values of the gravity centre give a measure of the dispersion of the transmitter-identified nerve cell group along the mediolateral and dorsoventral axes, respectively.

b. *Mean free distance.* This parameter is the mean of all the distances of each nerve cell from all the other nerve cells present in the transmitter-identified group. The parameter measures the compactness of the transmitter-identified nerve cell group.

c. *The homogeneity index.* The index represents the ratio between the mean free distance within the subgroup and the mean free distance observed within the entire nerve cell body population. In this way it is possible to obtain a value for the degree of compactness in the subgroup versus that in the entire transmitter-identified nerve cell group.

d. *The slope of the major axis of the transmitter-identified cell group.* This parameter gives the orientation of the cells with respect to the x and y axes. It is obtained by means of interpolation of the best fit line through the Cartesian coordinates of the entire cell group.

e. *The volume fraction.* This value is the fraction of the cell group volume occupied by the cell bodies of the transmitter-identified cell group. By means of the Delesse formula this value is obtained by considering at each rostrocaudal level analysed, the total area of the nerve cell bodies in relation to the area defined by the accepted positive squares. In this way it is possible to evaluate the dilution of the nerve cell bodies containing a known transmitter with other types of neurones as well as with glia cells, fibres and neuropil.

1.1. Automatic Image Analysis in the Demonstration of Nerve Cell Clusters of Transmitter-identified Neurones

Our previous work on the description of groups and subgroups of transmitter-identified nerve cell populations was performed by semi-automatic image analysis, which is a time-consuming procedure.[3] By means of the IBAS image analyser we have therefore recently introduced automatic image analysis in order to obtain a complementary, but more rapid, method for the objective demonstration of groups (clusters) of transmitter-identified nerve cell bodies. In this analysis, usually one-half of the brain is analysed and the image is improved by means of averaging and normalization programs. Averaging entails repeated acquisition of the image to improve the signal-to-noise ratio; noise is distributed according to a Gaussian distribution and should, in the averaging procedure, reach the mean grey value of 255 (white). Normalization is a way of extending the grey tones present in the image over the entire range of grey tones, i.e. this function increases the contrast of the image by mapping the occupied grey values linearly in the range 0–255.

The neuro-anatomical landmarks are marked on the screen. The image is then converted into a binary image by selecting a grey tone level, which demonstrates only the immunoreactive nerve cell bodies. Grey tone levels below this critical value will be considered as white, while grey tone levels above it will be considered as black. In the 'editor' function artefacts can be removed and the neuro-anatomical landmarks can be added to the binary image which is then plotted on the printer. The neuro-anatomical landmarks are then erased in the editor function and by means of the identification and measuring programs, the mean area and counts of, for example, tyrosine hydroxylase-immunoreactive nerve cell bodies, are obtained. By means of the 'close' function it now becomes possible to evaluate objectively whether the immunochemically identified nerve cell bodies are scattered or form a cluster. By means of this function, groups of closely associated objects are united to form larger individual objects. In this way immunoreactive nerve cell bodies which are closely packed will unite to form a big cluster, while cells far apart will remain scattered. In the editor function, we subsequently determine the area and shape factor of the five largest clusters formed and also determine their angle (angle x) with the x-axis. After these measurements, performed in the editor function, the remaining scattered cells are removed. Subsequently with the 'Boolop' function the newly formed image with the clusters is compared with the original image, so that the overlap area is demonstrated. This overlap area demonstrates the original field area of nerve cell bodies, from which the big clusters are formed with the help of the close function. When considering the ratio of the field area of each cell group and the area formed by the close function, the degree of compactness of the group can be evaluated. The higher the ratio, the higher the degree of compactness of a cell group. Furthermore, the field area of each respective group and the mean area of each individual nerve cell body are also considered in order to evaluate the number of immunoreactive cell bodies in each cluster. Finally, the images of the united nerve cell groups are plotted, so that their gravity centre locations can be determined by semi-automatic image analysis, using a magnetic tablet coupled to an Apple II computer, equipped with suitable morphometrical programs. Examples of these types of analysis are shown in *Figs*. 13.4–13.6 and in *Table* 13.1. Phenylethanol-amine-*N*-methyltransferase (PNMT)-immunoreactive nerve cell bodies have been analysed in coronal sections of the medualla oblongata. The main cluster formed is found in the lateral part of the adrenaline C1 group of the ventrolateral rostral reticular formation of the medualla oblongata (*see Table* 13.1).

1.2. Three-dimensional Description of Transmitter-identified Nerve Cell Groups

The cell group exists in a rostrocaudal frame as long as the two-dimensional analysis of the coronal sections reveals a clumped distribution of positive squares significantly deviating from a random distribution. As an arbitrary criterion for the existence of a cell group we have suggested a volume fraction of 2 per cent or above.[16]

Another parameter of a cell group is the total number of nerve cell bodies within it. The total number of cells can be obtained by considering the rostrocaudal length of the cell group, the thickness of the sections and the mean

Fig. 13.4–13.6. Automatic image analysis to demonstrate cell clusters.

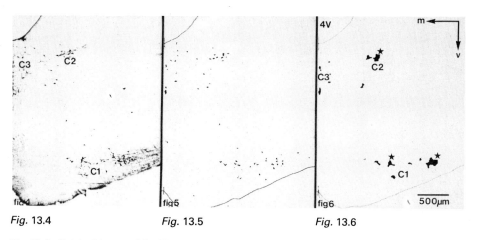

Fig. 13.4   Fig. 13.5   Fig. 13.6

Fig. 13.4. Original image of the PNMT-immunoreactive cell body profiles in the medulla oblongata 1·7 mm rostral to obex of the male rat (groups C1–C3).

Fig. 13.5. Discriminated image with the ventral and dorsal surface outlined (binary image), showing only the PNMT-immunoreactive cell body profiles.

Fig. 13.6. Formation of cell clusters by the 'close' function in the IBAS image analyser. Cell bodies 30–40 μm apart are united by the 'close' function. Accepted clusters are marked with a star (*see Table 13.1*).

number of profiles in the section analysed. In our analysis, the Königsmark formula has been used[17] taking into account only profiles with a diameter equal to or above 30 per cent of the maximal diameter in the cell group population. The nerve cell bodies are, however, ellipsoids and not spheres, and therefore only approximate values are obtained. Another parameter which must be considered, is the estimation of the volume fraction of the entire cell group. This value is obtained by determining the volume fraction at each rostrocaudal level. The value of the volume fraction is obtained, according to the Delesse formula, by comparing the area of the nerve cell bodies with that of the entire cell group at that level as defined by the accepted square in the density map (*see above*).

The transmitter-identified nerve cell group can be defined three-dimensionally in several ways. One way is to plot the gravity centre location along the rostrocaudal axis, both with regard to its medial and lateral location (M_x values) and with regard to its ventral and dorsal location (M_y values). The three-dimensional reconstruction of the transmitter-identified cell group can be made by showing the gravity centre location as well as the boundaries of the cell group in a three-dimensional frame (*see Fig.* 13.7). The dispersion around the gravity centres can be plotted by showing the coefficients of variation of the *x*-and *y*-coordinates of the gravity centre at the various rostrocaudal levels analysed.

Table 13.1. Demonstration of PNMT-immunoreactive cell groups of the male rat in coronal sections by image analysis

Section	Cell body parameters		Cell group parameters							
Rostrocaudal level (mm rostral (+) or caudal (−) to obex)	Mean area (Ā), μm²	Total counts (N)	Area (A), μm²	Shape factor (SF) 0 to 1	X-angle, degrees	Field area (FA₂), μm²	Gravity centres GCₓ, μm	GCᵧ, μm	FA/A	FA/Ā*
1. +1·7 mm	537	61								
Cluster I accepted (lateral C1)			18 200	0·339	131	5631	1953	661	0·309	11
Cluster II accepted (medial C1)			6404	0·191	159	3045	1242	644	0·475	6
Cluster III accepted (lateral C2)			9546	0·278	148	3045	1036	2436	0·319	6
Cluster IV not accepted (medial C2)			4302	0·179	111	1861			0·433	4
Cluster V not accepted (medial C1 area)			3673	0·157	63	1668			0·454	3
2. +1·5 mm	421	42								
Cluster I accepted (lateral C1)			14 770	0·348	139	6211	1880	652	0·421	15
Cluster III not accepted (lateral C2)			2489	0·159	27	1329			0·534	3
3. +1·0 mm	507	35								
Cluster I accepted (lateral C1)			42 990	0·490	76	11 020	1799	744	0·256	22
Cluster II not accepted (medial C1)			2465	0·142	122	1450			0·588	3
4. +0·9 mm	428	11								
Cluster I not accepted (lateral C1)			2779	0·164	111	991			0·357	2
Cluster II not accepted (medial C1)			1088	0·105	76	580			0·533	1
5. −0·2 mm	185	21								
Cluster VI accepted (part of nucleus solitarius)			18 650	0·371	58	3504	549	2550	0·188	19

* Estimated number of cells in each cluster.

By means of the IBAS image analyser (see Figs. 13.4–13.6, and text), using the 'close' function, PNMT-immunoreactive cell bodies lying within a distance of 65 μm from each other have been united to form clusters. The five largest clusters found at each rostrocaudal level analysed have been selected and analysed. Clusters have been accepted when

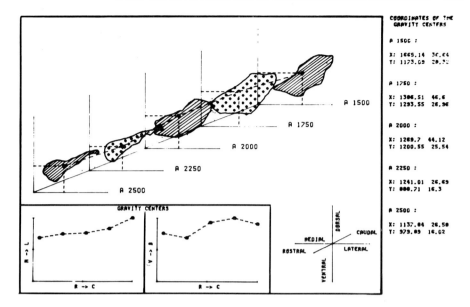

Fig. 13.7. Three-dimensional reconstructions of TH-immunoreactive profiles visualized by means of the PAP technique at various rostrocaudal levels of the substantia nigra. On the *x*-axis of the five plots (five different König and Klippel levels), the mediolateral coordinate has been recorded, while on the *y*-axis the ventrodorsal coordinate has been recorded. The boundary of the TH cell population has been plotted with the respective gravity centre at each rostrocaudal level. The values on the right give the values of the *x* and *y* coordinates (with respective standard deviation) for the gravity centre at each König and Klippel level considered.

2. MAIN MORPHOMETRICAL FEATURES OF TRANSMITTER-IDENTIFIED NEURONES AT THE LEVEL OF THE SINGLE NEURONE

2.1. Morphometrical Analysis of Nerve Cell Bodies

In our analysis[1,3,5] we have mainly considered: area perimeter, maximal diameter, Ferret diameter and shape factor. The shape factor is equal to 1 for a perfect circle and 0 for a segment of a straight line. The Ferret diameters are the projections of the diameters of the profile along the *x*- and *y*-axis.[18]

2.2. Morphometrical Analysis of the Neuropil

2.2.1. Dendrites

As in the morphometrical analysis of nerve cell bodies, the density distribution of dendrites has been demonstrated by the use of a grid superimposed on a Cartesian plane. This grid is made up of unitary squares 100 μm per side. The length of the dendrites and the number of branchings can then be determined in each square.[1] A density map is thus obtained. Standard statistical techniques, such as Lorenz curves and Gini indices, then allow the quantitation of the density and uniformity of distribution, of the number of branchings and the density and uniformity of the dendrite length. The Gini index gives an evaluation of the degree of unevenness in quantitative terms.

2.2.2. Nerve Terminals

In this analysis also, density maps can be made as described for cell bodies and dendrites by superimposing 100-μm squares on the Cartesian plane.[1,4] Again the uniformity of the density distribution of the terminals can be analysed by Lorenz curves and by the use of the Gini index, which allows a quantitative value for the degree of concentration of nerve terminals in the area analysed. When a very high density of transmitter-identified nerve terminals exists the method must be modified. In this case the systematic sampling of each unitary square involves the use of a mask consisting of a grid of circles of diameter 1 μm and intercentre distance, 4 μm. In each square only those circles are considered that are completely filled by the transmitter-identified nerve terminals. An image analyser registers the completely filled circles. This type of mask has been shown to have the best power of discrimination. In the image analyser, it is also possible to remove all grey tones except the high tones in the nerve terminal networks analysed (computer-assisted microdensitometry), making it possible to quantitate by this procedure not only densities but also the antigen contents of densely packed transmitter-identified nerve terminals.[1,4]

3. QUANTITATIVE METHODS FOR THE EVALUATION OF THE ENTITY OF COEXISTENCE OF NEUROACTIVE SUBSTANCES IN TRANSMITTER-IDENTIFIED NERVE CELL BODIES AND NERVE TERMINALS USING COMPUTER-BASED MORPHOMETRY AND MICRODENSITOMETRY

3.1. Entity of Coexistence of Neuroactive Substances in Nerve Cell Bodies

3.1.1. The Occlusion Method

This method allows an overall quantitative evaluation of the entity of coexistence of neuroactive substances in nerve cell bodies.[2] It is based on the analysis of three adjacent sections stained in a random way with antisera against A and B, and with both antisera together. This method has the advantage that an x-and y-axis can be omitted. In addition, no elution of immunoreactivity is necessary. (Such elution usually impairs the degree of immunoreactivity of the coexisting antigen.) Another advantage of this method is that, in combination with computer-assisted microdensitometry, it can be used to give an overall evaluation of coexistence in nerve terminal systems also (*see below*).[8,9,19] The entity of coexistence is obtained by subtracting from the sum of the number of immunoreactive nerve cell bodies demonstrated with antiserum A alone and antiserum B alone, the number of immunoreactive nerve cell bodies demonstrated with the combined use of antisera A and B (*Fig.* 13.8). This procedure is based on the fact that coexisting cells are demonstrated both with antiserum A alone and antiserum B alone as well as with the combined use of the two antisera. This procedure has been used, amongst other things, to characterize coexistence of neurotensin and TH immunoreactivity in the ventral midbrain (*Fig.* 13.9).

3.1.2. The Overlap Method

In this analysis[5] the image analyser records the position and the perimeter of each nerve cell body demonstrated by an antiserum A in a Cartesian plane. In an

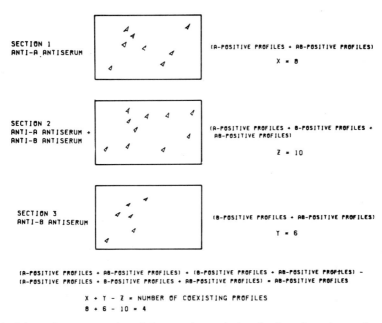

Fig. 13.8. Schematic representation of the morphometrical evaluation of coexistence by means of the occlusion method. For further details, *see text.*

adjacent section, or in the same section following mild elution of the A immunoreactivity, staining for immunoreactivity B is performed. The image analyser records the position and the perimeter of the B immunoreactive nerve cell bodies in the same Cartesian plane (*Fig.* 13.10). The computer considers coexistence to have taken place when the A- and B-immunoreactive areas overlap by at least 30 per cent. In the case of adjacent sections a correction factor must be calculated, since a number of A- and B-immunoreactive areas overlap by at least 30 per cent. In the case of adjacent sections a correction factor must be calculated, since a number of A- and B-immunoreactive nerve cells disappear as you move from one section to the other. This method can also be used with binary coding of the images to reach the quantitative evaluation of three sets: the A-positive population, the B-positive population and the A + B-positive population.[20] This method has the advantage that coexistence of neuroactive substances can be assessed in each nerve cell body of a nerve cell group. On the other hand, this method cannot be used to analyse coexistence in a nerve terminal population. To obtain absolute values on coexistence using the overlap method, the same section should be used with mild electrophoretic elution, avoiding impairment of the immunoreactivity of the coexisting antigen.

3.2. Quantitative Methods for the Evaluation of the Entity of Coexistence of Neuroactive Substances in Nerve Terminal Populations using the Occlusion Principle

3.2.1. The Synaptosomal Method

The procedure[9] involves the preparation of a brain homogenate using 0·32 M

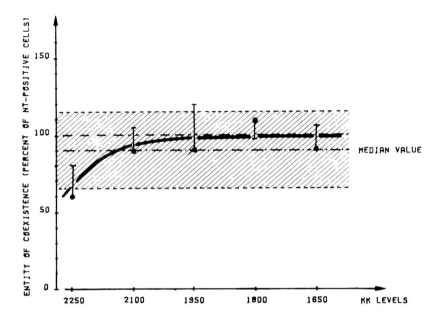

Fig. 13.9. The entity of coexistence between TH and neurotensin immunoreactivity has been evaluated in the lateral ventral tegmental area (VTA) of the rat by means of the occlusion method, using triplets of sections stained for TH/(neurotensin + TH)/neurotensin. The values of coexistence (means ± s.e.m. of 4 animals) are given, as well as the confidence interval for the median (Nair's test, confidence intervals for the median in samples from any continuous population).[23] All the neurotensin-immunoreactive cells present at several levels are TH-immunoreactive. However at the most rostral level (KK 2250), only about 60 per cent coexistence is observed.

sucrose containing 2mM ethylenediaminetetraacetic acid (EDTA) and 0·5 mM MgCl$_2$. The fixation procedure involves the resuspension of the partially purified synaptosomal fraction for 90 min in 2 per cent formaldehyde dissolved in phosphate-buffered saline. With the indirect immunofluorescence method, the best results were obtained with fluorescein isothiocyanate (FITC)-labelled immunoglobulins.

In this analysis it is important to use saturating concentrations of the respective antisera to obtain the correct value for the entity of coexistence. For the evaluation of coexistence, a smear of the synaptosomal preparation is made followed by a schematic sampling of the immunoreactive dots using a sampling grid. We have also been able to study the number of immunoreactive dots directly in the fluorescence microscope using a TV camera highly sensitive to fluorescent light. The TV camera is coupled to the IBAS image analyser, so that dots within the sampled image can be counted directly.

3.2.2. The Section Method

The analysis[9] is carried out by means of the IBAS image analyser, using computer-assisted microdensitometry. The procedure is summarized in *Fig.* 13.11.

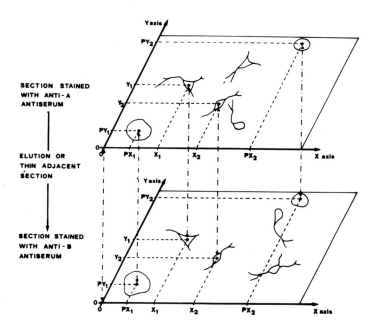

Fig. 13.10. Schematic representation of the morphometrical evaluation of coexistence by means of the overlap method. The neuro-anatomical landmarks are indicated by the coordinates $P(x_1, y_1)$. The coordinates are given only for the cells in which there is coexistence (i.e. they are positive in both sections). For further details, *see text.*

The analysis is based on defining the background by initiating the program for the grey tone histogram. All sections can then be discriminated in the same way by always using the mean grey tone level of the background minus two standard deviations. The same area is then analysed in three adjacent sections and the specific immunoreactive areas are determined for the adrenaline-synthesizing enzyme, PNMT alone, neuropeptide Y (NPY) alone, and for PNMT plus NPY. The area of the single terminal is obtained by analysing the histogram of the areas of the various single objects, making up the total specific immunoreactive area in the region analysed in the respective sections. The minimum value in the histogram is presumed to represent the diameter of the single terminal and was found to be 1·5 μm. The number of terminals with coexistence is finally obtained by dividing the immunoreactive area for NPY and PNMT immunoreactivity by the area of the single terminal (SAT), based on the value of the minimum diameter (*see above*). The percentage coexistence is obtained by expressing the number of terminals in which coexistence is present as a percentage of the number of, for example, PNMT-immunoreactive terminals. As an example it may be mentioned that the entity of coexistence of TH and CCK immunoreactivity within the dorsal part of the caudal area of the nucleus accumbens is reduced in the ageing brain from 45 to 25 per cent together with a preferential reduction of

The occlusion for the quantitative determination of coexistence in terminals

1. Background definition : Graytone histogram   ($\bar{X}$: SD)
2. Discrimination by using the graytone level   ($\bar{X}$-2. SD)
3. Assessment of the specific area for NPY(SA_{NPY})
4. Assessment of the specific area for PNMT(SA_{PNMT})
5. Assessment of the specific area for NPY $\bullet$ PNMT($SA_{NPY \bullet PNMT}$)
6. Evaluation of the specific   area of distribution
   of the single objects :
   The minimum value is the area of the single terminal   (SA_T)

$$SA_{NPY} \bullet SA_{PNMT} - SA_{NPY \bullet PNMT} = SA_{NPY/PNMT}$$

$(SA_{NPY/PNMT})/(SA_T) = N_{NPY/PNMT}$ = Number of terminals with coexistence

$(SA_{PNMT})/(SA_T) = N_{PNMT}$ = Number of PNMT positive terminals

$((N_{NPY/PNM})/(N_{PNMT})) \cdot 100$ = Percent of coexistence

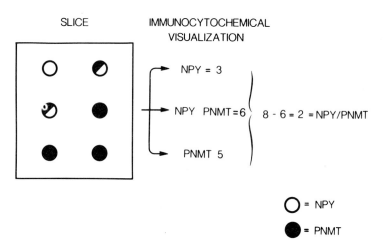

Fig. 13.11. Schematic illustration of the procedures used to determine coexistence in nerve terminal networks present in sections by means of the occlusion method. SA = the size of the specific area evaluated by means of the IBAS image analyser. The specific area has been obtained for PNMT, NPY and NPY + PNMT-like immunoreactivity. A_T = area of the single terminal obtained by means of the histogram of the single objects present in the image. The principle of the occlusion method is also indicated in the lower part of the figure.

CCK-immunoreactive profiles. These results are of great interest since they imply that in the aged brain, peptide co-modulators may be preferentially diminished leading to a loss of modulation and thus of plasticity within central synapses.

4. SEMIQUANTITATIVE ESTIMATIONS OF ANTIGEN CONTENT USING IMMUNOCYTOCHEMISTRY IN COMBINATION WITH COMPUTER-ASSISTED MICRODENSITOMETRY

Recently, a procedure has been introduced which will give a reliable value of the relative contents of the antigen in the structures analysed.[8,15] In an immunocytochemical procedure, especially when employing a monoclonal antibody, the antigen–antibody reaction is assumed to be controlled by the law of mass action. Saturation analysis is therefore performed in this procedure using different dilutions of the monoclonal antibody. In the subsequent Scatchard analysis a straight line will be obtained, if there is no cooperativity between sites and no heterogeneity of sites. Provided a straight line is obtained, the calculated B_{max} value of the antigen content will give a reliable value of the relative amounts of antigen present in the analysed structures, which can also be obtained by non-linear fitting analysis (*Fig.* 13.12). The measurements of the immunoreactivity at the various dilutions of the antibody can be performed either by quantitative microfluorimetry, when the indirect immunofluorescence method is used, or by microdensitometry in the image analyser, when the unlabelled immunoperoxidase procedure is used. Studies on central 5-hydroxytryptamine (5-HT) neurones and on substance P (SP)-immunoreactive nerve terminal

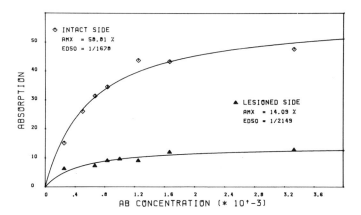

Fig. 13.12. Semiquantitive evaluation of substance P-like immunoreactivity within the substantia nigra of the male rat following partial hemitransection in front of the substantia nigra. For further details, *see text*. A saturation analysis using different concentrations of the substance P antiserum was performed on the intact and lesioned sides. A non-linear fitting procedure was used. It is shown that the B_{max} values on the lesioned side are markedly reduced compared with the corresponding values on the intact side. However, the ED_{50} values are of a similar magnitude indicating no change in the affinity of the antibody for the antigen substance P.

networks, especially in the substantia nigra, have demonstrated that this new procedure represents a reliable semiquantitative way of determining the immunoreactivity in transmitter-identified nerve cell body and nerve terminal populations (*Plate* 10), providing new ways of performing functional studies on transmitter-indentified neurones.[8] When analysing the 5-HT content with two different antibodies to 5-HT, the same B_{max} value was obtained but with different K_D values. Furthermore, Scatchard analysis has always given a straight line as the best fit line. In *Plate* 10 similar results are shown using the present immunocytochemical procedures and radioimmunoassay determinations of SP-like immunoreactivity in the substantia nigra. Thus, the ratio of the SP-like immunoreactivity in the lateral to that in the medial part of the substantia nigra was similar with the two procedures. Following lesions of the descending strionigral SP pathway it can be demonstrated that there is a marked reduction in the B_{max} value of SP-like immunoreactivity in the substantia nigra without any modulation of the K_D value, i.e. of the affinity of the antibody for its antigen. It is envisaged that this new method will provide a powerful tool in the analysis of discrete transmitter-identified nerve cell body and nerve terminal systems.

5. ON THE USE OF COMPUTER-ASSISTED MORPHOMETRY AND MICRODENSITOMETRY IN STUDIES ON AGEING PROCESSES IN TRANSMITTER-IDENTIFIED NEURONES

A summary of the age-induced changes in monoaminergic and peptidergic nerve cell bodies and nerve terminal systems obtained by the morphometrical methods described above[12–14] is given in *Table* 13.2. In the substantia nigra there is a preferential disappearance of CCK-immunoreactive nerve cell bodies compared with the TH-immunoreactive nerve cell bodies and, in the nucleus accumbens, there is a 50 per cent disappearance of CCK-immunoreactive nerve terminals and of terminals with CCK/TH coexistence but only a 20 per cent disappearance of TH-immunoreactive terminals. In the raphe nuclei there is a preferential disappearance of SP-like immunoreactivity compared with 5-HT-like immuno-reactivity and the coexistence of 5-HT/SP-like immunoreactivity is reduced by 40 per cent. The noradrenaline cell bodies of the locus coeruleus and the adrenaline cell bodies of the paragigantocellular reticular nucleus are relatively unaffected by ageing as seen from the small reduction in the number of TH-immunoreactive cell bodies in these areas. However, NPY immunoreactivity in the locus coeruleus was markedly reduced. Taken together, these results support our hypothesis that the peptide co-modulators in the monoamine systems are preferentially affected by ageing, leading to a loss of plastic responses in the monoamine neurones (*see also*[21]).

However, large reductions in the number or peptidergic neurones are also observed in other areas unrelated to the monoamine neurones (*see also below*). Thus, in the posterior part of the paraventricular hypothalamic nucleus, a marked reduction in the number of oxytocin and vasopressin neurones is observed, while the lateral magnocellular part is fairly unaltered, showing local differences also in the vulnerability of the peptidergic neurones to the ageing process. Widespread reductions in the number of peptide-immunoreactive cell bodies and terminals are also observed in the substantia gelatinosa of the spinal trigeminal nucleus (*see*

*Table 13.2. **Changes in transmitter-identified neurones of the 24-month-old male rat as discovered by image analysis***

Area	Cell bodies, mean percentage change in number	Terminals, mean percentage change in number	Absorbance in nerve terminals, mean percentage change
1. Nuc. accumbens (dorsomedial part)			
CCK IR		−50†	
Coexisting CCK/TH terminals		−50†	
TH IR		−20*	
2. Nuc. paraventricularis hypothalami			
A. Posterior part			
Oxytocin IR	−40†		
Vasopressin IR	−65†		
B. Lateral magnocellular part			
Oxytocin IR	−10		−10
Vasopressin IR	−10		+100†
3. Substantia nigra			
TH IR	−30*		
CCK IR	−60†		
4. Peri-aqueductal grey			
CCK IR	+2		
5. Locus coeruleus			
TH IR	−7		
NPY IR	−60†		
6. Nuc. paragigantocellularis reticularis medullae oblongata			
TH IR (adrenaline cells)	−15		
Enkephalin IR	−50†		
7. Raphe nuclei of the medulla oblongata (B1–B3 groups)			
SP IR	−45†		
5-HT IR	−15		
Coexistence 5-HT/SP IR	−40†		
Enkephalin IR	−73†		
8. Substantia gelatinosa of the caudal spinal trigeminal nuc.			
NPY	−30*		−29*
Enkephalin	−34†		−7
Somatostatin	−32*		−57†
Neurotensin	−70*		−50†
Substance P			−32†
Cholecystokinin			−51†
5-HT			−5

The results are shown as mean changes taken as percentage of the number of profiles found in the adult 3-month-old male rat ($n = 4$).

*$P < 0.05$.

†$P < 0.01$.

Student's t-test.

IR = immunoreactivity.

In the absorbance measurements, no standard curve was used. Therefore, the relationship between the change in antigen content and the change in absorbance is unknown. Consequently, the percentage change in relative antigen content may be different from that for absorbance.

From the literature.[12–14,22]

below) (*Table* 13.2). Thus, many but not all peptide neurones of different types appear to be especially vulnerable to the ageing process, indicating marked losses of peptidergic transmission in selected brain regions. Reduced peptide synthesis may be the major cause of the changes observed. The marked disappearance of enkephalin-immunoreactive nerve cell bodies within the nucleus raphe magnus during the ageing process, further underlines the vulnerability of the peptidergic neurones to ageing, with regard to the formation and metabolism of biologically active neuropeptides. In contrast, monoamine synthesis appears to be more easily maintained in the aged brain, probably because it is enzymatic and not dependent on transcription and translation processes like peptide synthesis. The reduced peptide synthesis may also contribute to degeneration processes in the brain in view of the trophic and metabolic role of peptides.

In the analysis of a large number of peptidergic nerve cell bodies and terminals within the substantia gelatinosa of the caudal part of the spinal trigeminal nucleus, widespread changes are demonstrated using the morphometrical and microdensitometrical programs of the image analyser (*Table* 13.2). In this analysis the image analyser TESAK VDC501 equipped with the digital computer PDP 11 was used with the hardware and software specifications as described by Agnati et al.[5] The counts of the nerve cell bodies were carried out automatically after proper selection of the diameter of the neuronal structures to be measured. The semiquantitative evaluation of the absorbance of the nerve terminal systems of the substantia gelatinosa was, in this case, performed at a single antibody concentration, based on previous antibody dilution curves. Immunoreactivity was converted into the grey tone scale of the image analyser (0 = black; 256 = white).

NPY-, neurotensin-, enkephalin- and somatostatin-immunoreactive nerve cell body systems of the substantia gelatinosa of the caudal part of the spinal trigeminal nerve are all significantly reduced during the ageing process. The most marked reduction is found in the number of neurotensin-immunoreactive nerve cell bodies, which now only amounted to 20 per cent of the adult neurotensin-immunoreactive nerve cell population. It is important to emphasize that following colchicine treatment, which causes an accumulation of neuropeptide immunoreactivity in the nerve cell bodies, there is no longer a significant reduction in the number of NPY-and enkephalin-immunoreactive nerve cells of the substantia gelatinosa. In the analysis of the immunoreactive nerve cell bodies of the substantia gelatinosa, all objects with a diameter smaller than 4 μm were excluded. These results therefore indicate that the reduction in the number of neuropeptide nerve cell bodies in the substantia gelatinosa of the caudal spinal trigeminal nucleus mainly reflects a deficiency in the synthesis of the respective neuropeptide or an enhancement of peptide metabolism.

The results obtained on the amount of immunoreactivity in the various peptidergic and monoaminergic systems of the substantia gelatinosa are also summarized in *Table* 13.2. A significant reduction is noted in the SP-, CCK-, somatostatin- and neurotensin-immunoreactive nerve terminal networks, while the immunoreactivity in the enkephalin and 5-HT systems is unaffected by ageing. In this semiquantitative evaluation, the specific immunoreactivity was obtained by subtracting the grey tone value of the background from the grey tone value of the measured immunoreactivity. The measurements were performed at four different rostrocaudal levels.

These results illustrate that age-induced changes within transmitter-identified neurones can be quantitatively defined in a simple way by means of automatic image analysis.

Acknowledgements

This work was supported by grants from the National Institute of Health, USA (MH25504), the Swedish Medical Research Council (04X-715), L. Osterman's Stiftelse, Stockholm, Sweden, and by a CNR International Grant. We are grateful for the excellent secretarial assistance of Mrs Anne Edgren.

REFERENCES

1. Agnati L. F., Fuxe K., Calza L., Hökfelt T., Johansson O., Benfenati F. and Goldstein M. A morphometric analysis of transmitter identified dendrites and nerve terminals. *Brain Res. Bull.* 1982, **9**, 53–60.
2. Agnati L. F., Fuxe K., Locatelli V., Benfenati F., Zini I., Panerai A. E., El Etreby M. F. and Hökfelt T. Neuroanatomical methods for the quantitative evaluation of coexistence of transmitter in nerve cells. Analysis of the ACTH- and beta-endorphin immunoreactive nerve cell bodies of the mediobasal hypothalamus of the rat. *J. Neurosci. Methods*, 1982, **5**, 203–214.
3. Agnati L. F., Fuxe K., Zini I., Calza L., Benfenati F., Zoli M., Hökfelt T. and Goldstein M. Principles for the morphological characterization of transmitter identified nerve cell groups. *J. Neurosci. Methods*, 1982, **6**, 157–167.
4. Agnati L. F., Fuxe K., Zini I., Calza L., Benefenati F., Zoli M., Hökfelt T. and Goldstein M. A new approach to quantitate the density and antigen contents of high densities of transmitter-identified terminals, immunocytochemical studies on different types of tyrosine hydroxylase immunoreactive nerve terminals in nucleus caudatus putamen of the rat. *Neurosci. Lett.* 1982, **32**, 253–258.
5. Agnati L. F., Fuxe K., Benfenati F., Zini I., Zoli M., Fabbri L. and Härfstrand A. Computer assisted morphometry and microdensitometry of transmitter identified neurons with special reference to the mesostriatal dopamine pathway. I. Methodological aspects. *Acta Physiol. Scand., Suppl.* 1984, **532**, 5–36.
6. Unnerstall J. R., Niehoff D. L., Kuhar M. J. and Palacios J. M. Quantitative receptor autoradiography using ^{3}H-ultrofilm: application to multiple benzodiazepine receptors. *J. Neurosci. Methods*, 1982, **6**, 59–73.
7. Fuxe, K., Calza L., Benfenati F., Zini I. and Agnati L. F. Quantitative autoradiographic localization of ^{3}H-imipramine binding sites in the brain of the rat: Relationship to ascending 5-hydroxytryptamine neuron systems. *Proc. Natl Acad. Sci. USA*, 1983, **80**, 3836–3840.
8. Fuxe K., Agnati L. F.,Andersson K., Zoli M., Benfenati F., Eneroth P. and Cuello C. Quantitative microfluorimetry and semiquantitative immunocytochemistry as tools in the analysis of transmitter identified neurons. In: Agnati L. F. and Fuxe K., eds., *Wenner–Gren Symposium on Quantitative Neuroanatomy in Transmitter Research*, Stockholm May 3–4, 1984. London, Macmillan, 1985.
9. Fuxe K., Agnati L. F., Zoli M., Härfstrand A., Grimaldi R., Bernardi P., Camurri M., Tucci F. and Goldstein M. Development of quantitative methods for the evaluation of the entity of coexistence of neuroactive substances in nerve terminal populations in discrete areas of the central nervous system: Evidence for hormonal regulation of cotransmission. In: Agnati L. F. and Fuxe K., eds., *Wenner–Gren Symposium on Quantitative Neuroanatomy in Transmitter Research*, Stockholm May 3–4, 1984. London, Macmillan, 1985.
10. Agnati L. F. and Fuxe K. (Eds.) *Quantitative Neuroanatomy in Transmitter Research*, Wenner–Gren Symposium Series, Vol. 42. London, Macmillan, 1985.
11. Agnati L. F., Fuxe K., Calza L., Goldstein M., Toffano G., Giardino L. and Zoli M. Computer assisted morphometry and microdensitometry of transmitter identified neurons with special reference to the mesostriatal dopamine pathway. II. Further studies on the effects of the GM1 ganglioside on the degenerative and regenerative features of mesostriatal dopamine neurons. *Acta Physiol. Scand., Suppl.* 1984, **532**, 37–44.

12. Agnati L. F., Fuxe K., Benfenati F., Toffano G., Cimino M., Battistini N., Calza L. and Merlo Pich E. Computer assisted morphometry and microdensitometry of transmitter identified neurons with special reference to the mesostriatal dopamine pathway. III. Studies on aging processes. *Acta Physiol. Scand., Suppl.* 1984, **532**, 45–66.

13. Agnati L. F., Fuxe K., Giardino L., Calza L., Zoli M., Battistini N., Benfenati F., Vanderhaeghen J. J., Guidolin D., Ruggeri M. and Goldstein M. Evidence for cholecystokinin–dopamine receptor interactions in the central nervous system of the adult and old rat. Studies on their functional meaning. *Ann. N.Y. Acad. Sci.* 1985, **448**, 231–254.

14. Agnati L. F., Fuxe K., Calza L., Giardino L., Zini I., Toffano G., Goldstein M., Marraa P., Gustafsson J.-Å., Yu Z.-Y., Cuello A. C., Terenius L., Lang R. and Ganten D. Morphometrical and microdensitometrical studies on monoaminergic and peptidergic neurons in the aging brain. In: Agnati L. F. and Fuxe K., eds., *Wenner–Gren Symposium on Quantitative Neuroanatomy in Transmitter Research*, Stockholm, May 3–4, 1984. London, Macmillan, 1985.

15. Fuxe K., Agnati L. F., Hökfelt T., Calzá L., Benfenati F., Mascagni F. and Goldstein M. Immunocytochemistry of central neurons. In: Bignami A., Bloom F. E., Bolis C. L. and Adeloye A. eds., *Central Nervous System Plasticity and Repair*. New York, Raven Press, 1985: 47–56.

16. Agnati L. F., Fuxe K., Calza L., Zini I., Hökfelt T., Steinbusch A. and Verhofstad A. A method for rostrocaudal integration of morphometric information from transmitter-identified cell groups. A morphometrical identification and description of 5-HT cell groups in the medulla oblongata of the rat. *J. Neurosci. Methods*, 1984, **10**, 83–101.

17. Königsmark B. W. Methods for the counting of neurons. In: Nauta W. J. H. and Ebbesson S. O. E. eds., *Contemporary Research Methods in Neuroanatomy*. New York, Springer Verlag, 1970.

18. Weibel E. R. *Stereological Methods*. London, Academic Press, 1979.

19. Zoli M., Fuxe K., Agnati L. F., Härfstrand A., Terenius L., Toni R. and Goldstein M. Computer assisted morphometry of transmitter-identified neurons. New openings for the understanding of peptide–monoamine interactions in the mediobasal hypothalamus. In: Panula P., Päivärinta H. and Soinila S. eds., *Neurohistochemistry Today. Neurology and Neurobiology*, New York, Alan R. Liss Inc; 1985: in press.

20. Jonsson G., Hallman H. and Luthman J. Quantitation of nerve terminal networks of transmitter-identified neurons after selective neurotoxic lesions. In: Agnati L. F. and Fuxe K., eds., *Wenner–Gren Symposium on Quantitative Neuroanatomy in Transmitter Research*. London, Macmillan, 1985.

21. Black I. B., Adler J. E., Dreyfus C. F., Jonakait G. M., Katz D. M., LaGamma E. F. and Markey K. M. Neurotransmitter plasticity at the molecular level. *Science*, 1984, **225**, 1266–1270.

22. Calza L., Agnati L. F. and Fuxe K. Aging induced changes in the peptide neurons of the substantia gelatinosa caudal part of the spinal trigeminal nucleus. *Acta Physiol. Scand.* 1985, submitted for publication.

23. Colquhoun D. *Lectures on Biostatistics*. Oxford, Clarendon Press, 1971.

24. Fuxe K., Wikström A-C., Okret S., Agnati L. F., Härfstand A., Yu Z-Y., Granholm L., Zoli M., Vale W. and Gustafsson J-Å. Mapping out glucocorticoid receptor immunoreactive neurons in the rat tel- and diencephalon using a monoclonal antibody against rat liver glucocorticoid receptor. *Endocrinology* 1985, in press.

14 *Photomicrography*

S. Bradbury

In any field of microscopy, whether in materials or biological science, a great deal of effort and skill goes into the preparation of the material and into the actual study of the specimen. All this may well be useless, however, unless some record of the results can be made for publication and discussion. In the past, this usually took the form of a written description, occasionally supplemented by a freehand drawing which would be reproduced in a journal or book by means of a copper plate engraving or by lithography. Today, reports of microscopical studies are invariably illustrated by means of photomicrographs which may be either monochrome (reproduced as half-tone plates) or full-colour illustrations. The use of photography to illustrate the microscopical image is now relatively easy (in comparison with freehand drawing) and it has the merit of being absolutely objective. Photomicrographs do, on the other hand, often suffer from the fact that the film reproduces *all* the features in a complex image, with the result that on occasion out-of-focus detail may detract from the relevant parts of the image and confuse the picture. Again, the structures of interest (for example, the processes of a neurone) may pursue a devious course throughout the thickness of a section and so pass out of the restricted focal plane of high power objectives. Problems such as these are common and require some thought on the part of the microscopist in order to produce the best possible illustrative material in any given situation.

At the same time, it is desirable to consider carefully whether the required photomicrographs should be taken in colour or in black and white. The latter has the merit of cheapness and most laboratories have the necessary facilities for developing and printing such micrographs so that the results are quickly available. Again, the operator has a close control on the quality at all stages. Monochrome photographs are easy to reproduce and most journals cater for this medium alone, purely on the grounds of the high printing and blockmaking costs incurred in the production of full-colour illustrations. Colour, especially in the form of 35-mm transparency material, is much in demand for the illustration of lectures; in fluorescence microscopy, in histochemistry and immunocytochemistry, colour is often essential in order to convey adequately the required scientific information. The introduction of automatic photomicrographic cameras and the ready availability of fast, fine-grain colour emulsions has brought the production of both monochrome and colour micrographs within the capabilities of all microscopists. Despite these advantages, it is very common to see good articles illustrated by poor photomicrographs, with faults only too readily apparent in both the microscopy and the photography. In order to produce a good micrograph, it is necessary to:

 a. Produce a good, well-stained slide
 b. Produce from it a good microscopical image
 c. Record that image faithfully on photographic film and paper.

The first of these is self evident and will not be considered further in this chapter. The second requirement is for correct microscopy and the importance of this stage cannot be overemphasized. Unless the instrument is correctly adjusted it will not give a high quality image and no photographic skill can remedy this fundamental defect. It is not possible here to provide a full text of microscopy; the books listed in the bibliography may be consulted for details,[1-3] but one or two major points are worth repeating here.

1. CHOICE AND USE OF MICROSCOPE OBJECTIVES

The microscope objective is perhaps the most important single component of the system, as it not only provides most of the primary magnification of the image, but, much more importantly, is responsible for the resolution or ability to see fine detail in the image. If the resolution of a system is high, then points in the object which lie very close together will be seen in the image as separate points. If the resolution is low, then these same points will not be seen as separate. Magnification of an image in which there is resolution of fine detail will enable that detail to be seen more easily and will thus help in the interpretation of the image; magnification of an image which lacks resolution, however, only results in the production of a larger blurred image, which does not provide any further information.

   The resolution of a microscope objective depends on the numerical aperture of the lens and on the wavelength of the light used to form the image. The numerical aperture (NA), a term introduced by Abbe in the late nineteenth century, has been used for many years as an indicator of the performance of an objective. It is defined by the formula

$$NA = n \sin \alpha$$

where n is the refractive index of the medium between the front lens of the objective and the specimen and α is half the angle of acceptance of the lens (*Fig.* 14.1*a*). The importance of NA in resolution is apparent from the expression

$$\rho = \frac{0.61\lambda}{NA}$$

In this equation, 0·61 is a constant, λ is the wavelength of the light and NA the numerical aperture of the objective. It is clear that in order to increase the resolution, i.e. make the value of ρ *small*, we must either use an objective of a high NA and/or use light of a short wavelength. There are obvious limitations on decreasing the wavelength of the illuminating light, as the eye loses much of its sensitivity at short wavelengths and, if photography is used, exposure times at short wavelengths become prohibitively long. It is, therefore, usually better if increased resolution is needed to try and use an objective of a higher numerical aperture. With 'dry' lenses, i.e. those in which air forms the medium between the front element of the objective and the cover glass, the maximum value of the NA is about 0·95. At such high apertures the objectives become very sensitive to the influence of aberrations (especially spherical aberration) and are difficult to use. This problem is largely overcome by the use of immersion objectives in which a fluid such as water, glycerol or an oil is placed between the front lens of the objective and the coverslip (*Fig. 14.1b*). With these lenses the NA is increased as the factor *n* in the definition of NA now becomes greater than 1. The maximum NA for an oil immersion lens is about 1·4 and, with such a lens, operated in green light at a wavelength of 5500 nm, the maximum resolution under favourable conditions would be approximately 0·25 μm. Traditionally, oil immersion objectives would have an initial magnification of about 100 ×; many oil or glycerol immersion objectives are now available with much lower initial magnifications. Such lenses are often very valuable for photomicrography for three main reasons. As they are of the immersion type the influence of spherical aberration introduced by, for example, a slight variation from the correct cover glass thickness is minimized. Secondly, low power immersion objectives have a much larger field of view than high power lenses; this is related to their lower initial magnifying power, so that more of the specimen may be photographed at a time. Finally, the use of low magnification immersion lenses is valuable because they have higher numerical apertures than comparable dry lenses. High NA not only improves the resolving power, but it also means that the brightness of the image is greatly increased. (In a transmitted light objective image brightness is related to the square of NA, whilst with an incident light system in which the objective serves as its own condenser, the brightness is related to the fourth power of the NA of the objective.) A bright image will, of course, reduce the photographic exposure times, so helping to avoid problems with mechanical vibrations in the system blurring the image and reducing the **risk** of reciprocity failure becoming significant.

Objectives are commonly available in several differing degrees of correction, with a wide range of numerical apertures and initial magnifications. The cheapest, called achromats, are corrected for primary axial chromatic aberration, i.e. the red and blue rays are brought to the same focal point, whilst their spherical aberration is corrected for an intermediate wavelength, usually in the green. Images produced by achromats may show a residual colour fringe around contrasty objects; if this is to be removed then more highly corrected (and therefore much more expensive) objectives must be used. Such increased

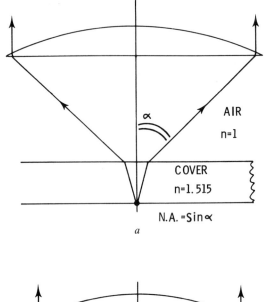

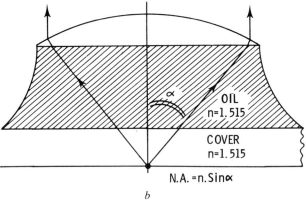

Fig. 14.1. (*a*) The path of rays entering a 'dry' objective from an object mounted beneath a coverslip of refractive index *n* = 1·515. The half-angle of acceptance of the lens is denoted by the symbol α. Numerical aperture (NA) for this lens is equal to sin α, as the refractive index (RI) of the medium between the coverslip and the front lens is air. (*b*) The path of rays entering an immersion objective. Here the value of the half-angle of acceptance (α) is increased; the NA of this lens is equal to *n* sin α where *n* is the RI of the immersion oil, usually taken as 1·515.

corrections necessitate the use of lens elements made of either fluorite (fluorspar) or the newer, specially formulated, rare-earth optical glasses. Fluorite or semi-apochromatic objectives, as they are often called, usually possess a larger numerical aperture than achromats of corresponding focal length and therefore the image which they give is not only of higher contrast due to the increased optical corrections, but is also brighter and has a higher resolution of detail. The ultimate in microscope objectives is found in lenses which are termed apochromatic. Here the chromatic aberration has been eliminated almost entirely, so that there is no perceptible residual colour; the spherical aberration is corrected for two colours rather than for one and the numerical aperture has been increased still further for any given magnification. Such lenses, correctly used on suitable specimens, will give images of the very highest quality. Apochromats are, however, of very complex construction and are, therefore, very expensive; they often have a very short working distance and may possess considerable curvature of field. This latter defect is inherent in all the three types of objective mentioned and obviously is very undesirable in a lens intended for photomicrography as only the centre of the picture will be in sharp focus. Most makers now provide objectives, in all three categories, especially computed to minimize field curvature and these should be used for photomicrography wherever possible. Although plan-apo objectives are the best for photo work, nevertheless, extremely good results may be obtained from plan-fluorite objectives and these are much more tolerant of operator error. This, therefore, makes them the lenses of choice for most work.

2. THE IMPORTANCE OF THE SUBSTAGE CONDENSER IN TRANSMITTED LIGHT MICROSCOPY

Although many microscopists devote much expense to obtaining the very best in the way of objectives, very few bother to consider what is equally important in obtaining a good image — the substage condenser. If one is working in incident light, then the objective serves as its own condenser and, provided that the aperture and field diaphragms are correctly set, a good image will usually be obtained. With transmitted light microscopy, however, a separate substage condenser is necessary and often this is of inferior quality and is incapable of allowing the objective to operate at its maximal efficiency. The substage condenser has three major functions:

a. It serves to concentrate light onto the object with uniform intensity over the whole illuminated field
b. It controls the aperture of the illuminating cone of light, matching this to the numerical aperture of the objective
c. It provides special types of illumination such as hollow cones for dark ground and phase contrast.

The condenser often provided with microscopes is of the so-called Abbe type. This is not chromatically corrected and suffers from excessive spherical aberration. In consequence it cannot provide a sharp image of the field diaphragm which thus appears blurred and fringed with colours. Much better results are obtained from the use of a highly corrected aplanatic/achromatic condenser which will give a sharp, colourless image of the field diaphragm in front of the lamp collector lens.

The condenser should be adjusted to give correct Köhler illumination. With this an evenly illuminated object field may be obtained from a non-homogeneous lamp filament and maximal control of aperture and contrast is ensured. When setting up Köhler illumination, the condenser must be correctly centred to the optical axis of the microscope and focused so that an image of the field diaphragm in front of the lamp collector lens is in sharp focus in the plane of the specimen. At the same time, the aperture iris diaphragm in the front focal plane of the condenser must be correctly adjusted. The best setting is such as to illuminate about 80–90 per cent of the diameter of the back focal plane of the objective with a solid cone of light. This may be easily judged by removing the microscope eyepiece and inspecting (with a focusing telescope inserted into the microscope tube) the image of the back focal plane of the objective. An image of the aperture diaphragm will be seen in this plane and it is then a simple matter to adjust the opening of the diaphragm to the correct value. If the objective aperture is not filled with light then the working aperture is too small and valuable resolution is lost and detail obscured by diffraction (*Fig.* 14.2*a,b*). Note, however, that closing the aperture diaphragm (and hence reducing the working aperture) *below* the optimum is permissible on occasion in order either to increase the contrast of the micrograph or to increase the depth of field, provided that the consequent loss of detail and resolution is acceptable. If the aperture diaphragm is opened too much, detail again tends to be lost, this time by excess glare in the image.

It is worth stressing at this point the importance of using coverslips of correct thickness. All dry objective lenses of large aperture are sensitive to the effects of cover thickness. If the cover is either too thick, or too thin, then spherical aberration will be introduced into the image, which degrades its sharpness. This exactly parallels the effect noticed when there is grease or other contamination on the front lens of the objective. Some high aperture dry lenses have what is termed a 'correction collar' which may be rotated to compensate for any variation in coverglass thickness from the correct value; it requires considerable skill to use such a correction collar and it is better to ensure that the correct thickness of cover is used from the outset. At the same time, it must be emphasized that the mimimum thickness of mountant should be used. Most objectives are corrected for minimal spherical aberration at a cover thickness of 0·17 mm (corresponding to a No. 1 1/2 coverslip), but there is much variation in thickness within any box of slips and if really critical work is contemplated then the thickness of each slip should be measured.[4]

It must be emphasized that good photomicrographs cannot be produced unless the microscope image is itself first class. Time taken to improve the microscopy is well spent, no matter how modern and expensive the photomicrographic camera in use. Study of some of the reference works on microscopy listed in the bibliography will help improve one's use of the microscope.

3. TRANSFERRING THE IMAGE ONTO FILM

Over the years, several ways have been devised for converting the virtual image of the microscope, nominally located at infinity, into a real image located at some finite distance from the microscope eyepiece where a photographic film is placed. Such devices range from the very simple, using a normal photographic camera with its own integral lens, through the use of such a camera minus its own lens but fitted with a microscope adapter, to the very sophisticated, fully automated photomicroscope. Most laboratories which take photomicrographs only on an intermittent basis, do this with an eyepiece camera attachment. This consists of a film transport device and shutter mounted on a beam-splitter unit. This latter also carries a small telescope to allow visualization of the image and to allow this to be framed and

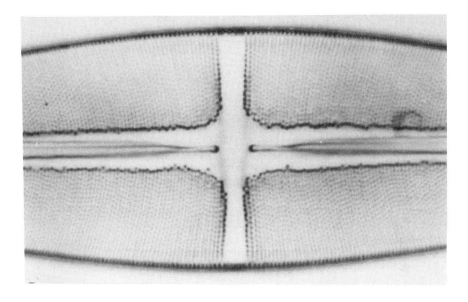

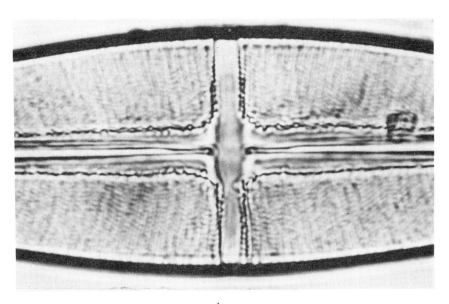

b

Fig. 14.2. (*a*) A photomicrograph of a diatom taken with the × 40 microscope objective working at its optimum. The aperture diaphragm was opened to give a solid cone of illumination equivalent to 90 per cent of the NA of the objective. Note that the openings in the frustule of the diatom are clearly resolved into individual dots. (*b*) The same diatom, photographed with the same objective lens but with the aperture diaphragm closed to restrict the illuminating cone to <10 per cent of the objective NA. Note that (i) the resolution of detail has vanished, (ii) there are diffraction fringes and details and (iii) the image contrast is increased.

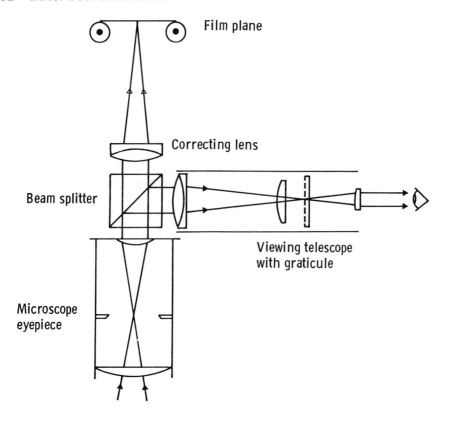

Fig. 14.3. A diagram of the basic components in a modern photomicrographic camera.

focused (*Fig.* 14.3). Such cameras are now often incorporated as an integral part of large research microscopes (*Fig.* 14.4) and provide fully automatic control over such matters as exposure, shutter operation and film wind-on; they are available from most manufacturers and provide equipment which is convenient to use, is constantly available and enables photomicrographs to be obtained with the minimum of technical knowledge.

An older type of system uses a microscope which is fitted with a long extension bellows and shutter unit mounted on a strong support pillar. There is a ground glass focusing screen together with a cut-film or roll-film holder fitted at the upper end of the bellows. Such a system allows the use of large format film material (e.g. 4 × 5″ cut film), so giving a larger image which requires less enlargement in order to produce the final print. At the same time the exposures are made on individual sheets of film, so that a small number of exposures may be processed and given separate, possibly different, treatments.

With all microscope camera systems, there is the problem of introducing extra spherical aberration (and thereby destroying the image sharpness) due to the refocusing of the microscope which must take place in order to throw a real image into a plane at a finite distance from the eyepiece. This is especially a problem with high aperture 'dry' objectives. The traditional solution is either to use a special

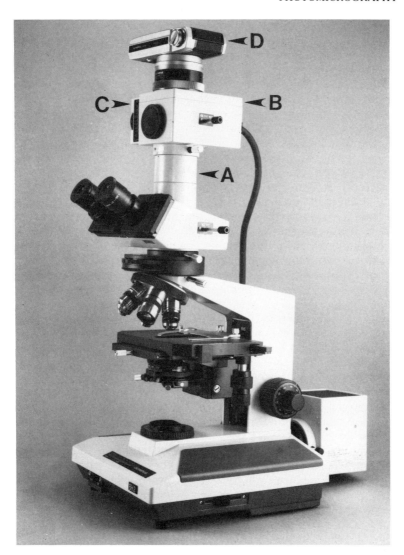

Fig. 14.4. A typical research microscope fitted with a photomicrographic attachment. Above the basic stand with its binocular eyepieces may be seen the photo tube (A) containing an eyepiece, the beam splitter housing (B), a colour-temperature meter (C) which enables the light to be adjusted for correct colour renderings, and camera film transport mechanism (D).

projection eyepiece which allows the eye lens to be separated from the field lens by a sufficient distance to compensate for the increased tube length or to raise the eyepiece in the microscope tube by means of a special collar. A modern solution is to fit a special negative lens into the system above the microscope eyepiece (*Fig.* 14.3) and so provide the necessary correction for any spherical aberration introduced by the formation of the image in the film plane.

4. THE CHOICE OF FILM SIZE AND MATERIAL

In many cases the choice of the size of film to be used will be dictated by the equipment available in the laboratory; in most cases this will be designed to accept 35-mm film which gives a picture area of 24 × 36 mm. If a choice is available, however, then there are advantages in considering the use of a larger format such as 120 roll film, producing negative images 6 × 6, 6 × 7 or 6 × 9 cm (according to the film back in use) or of cut film in the 4 × 5″ size; 35 mm has the advantage that the material is cheap, and a wide range of photographic emulsions is available covering all the needs of the microscopist. Most of the commonly used 35-mm emulsion types are available ready loaded into cassettes in either 20 or 36 exposure reels. It is unfortunate that this tends to encourage indiscriminate exposure and leads to wastage of film. It is possible to cut off and process short lengths from a longer cassette length, but this is cumbersome and is not often attempted. With most emulsions, however, it is possible to purchase the film in 30-metre units or in even longer lengths which are then stored in a light-tight device known as a 'bulk loader'; from this it is possible to load any required number of exposures into cassettes for use in the photomicroscope. Such a method of operation is very economical and allows film wastage to be reduced to a minimum.

With care, enlargements of high quality may be produced from negatives made on 35-mm film. The small image size does place a considerable emphasis on the darkroom skill of the photographer and it is essential that clean working is adopted as the slightest speck of dust on a 35-mm negative will cause a marked blemish on a print which has been enlarged ten times or more.

Thirty-five mm is certainly the cheapest format if colour reversal film (which allows the production of transparencies) is to be used, and if the illustration of lectures is the prime reason for taking micrographs, then 35 mm is probably the format of choice.

With large-format films, the greater area of the image makes the degree of enlargement needed much less and this makes it much easier to obtain sharp, clean prints. A microscopist who is only concerned with the production of prints (in either black and white or colour) and wishes to obtain the very highest photographic quality, would certainly be advised to use the largest film size available to him. In addition, the ability to expose and process individual sheets is valuable. Not only is this conducive to a careful consideration of exactly how many micrographs are to be taken, it also ensures that each shot is carefully composed and the microscopical conditions are exactly optimal for each shot. It is also possible to vary the photographic processing for each exposure in order to allow the manipulation of, for example, contrast and photographic density.

Most formats are now also available in the so-called 'instant picture films'. These, available in both monochrome and colour, eliminate the delay between exposing the micrograph and viewing the result. Many different types of emulsion are now available, of differing speeds and contrasts, and recently these materials have been introduced in the 35-mm format. At present it seems that their main disadvantage is their cost!

Before a monochrome emulsion can be chosen for any particular task we need to know about several of its characteristics. Among these are:

a. Contrast
b. Resolving power
c. Sensitivity to light — the 'speed'
d. Sensitivity to colour.

The first three of these are directly related to the size and size distribution of silver halide grains in the gelatin base of the emulsion, whilst the fourth is a consequence of treatment of the emulsion ('sensitizing') with certain dyes during its manufacture.

The contrast is a measure of the range of grey tones which an emulsion can reproduce, from absolute black through to pure white. A contrasty emulsion would have no capability of reproducing an extensive range of grey tones and would have a silver halide grain population in which the grains were small and of an even size distribution. If the grains were of mixed sizes, then the emulsion would be able to reproduce a wide range of grey tones. It should be remembered that the contrast of an emulsion is also affected by its development. High resolving power (the ability to reproduce fine detail) is more important in an emulsion intended for photomicrography than in one intended for general pictorial use. Resolving power increases as the grain size diminishes and as the number of superimposed layers of silver halide grains in the emulsion is reduced, i.e. by manufacturing thinner emulsion coatings. Resolving power is *reduced* if there is any significant reflection of light which has passed through the emulsion back from the base material into the halide layer. This 'halation' is minimized by adding a backing of light-absorbing dye which absorbs most of the superfluous light. For general photomicrography, we should choose a fine-grained, thinly coated emulsion which has an efficient anti-halation backing. Such emulsions are often relatively insensitive to light and are said to be 'slow'. The more responsive an emulsion is to light, then the less light is required to obtain a negative of satisfactory density and the shorter would be the exposure time. An emulsion of this type would be said to be 'fast', as the speed of an emulsion is inversely proportional to the exposure required for a given density. Different emulsions may be compared by means of the various rating systems which are in use; these apply numbers to films so that the higher the number, the greater the speed.

Two major systems are in use today. In the arithmetical scheme favoured by the American and British manufacturers, a doubling of the speed numbe (the ASA number) denotes a doubling of the speed. The German and Continental makers on the other hand use a logarithmic scale in which a doubling of the speed is indicated by an increase of three in the so-called DIN number. An emulsion such as Kodak Technical Pan would be rated (according to the development which it is to receive) at about 125 ASA which is comparable to a DIN rating of 22. An emulsion which was twice as fast would therefore be ASA 250 or DIN 25. As mentioned above, the speed rating may be profoundly affected by the type of development which is given to the film and many developers are available which are claimed to enhance the speed of a film. Similarly, increasing the development time will also effect some increase in the speed as well as increasing the contrast. It is also worth noting that the colour of the light will also affect the effective speed of the film. This is more marked with blue sensitive emulsions which appear to be faster to daylight than to the more reddish tungsten artificial light.

If a film is given very short (say less than 1/1000 s) or very long (say over 1 s) exposures, then its effective speed is also reduced as the emulsion does not respond to exposure to light to the predicted extent. This failure of the film to obey the reciprocal relationship between intensity of light and the time of exposure is called 'reciprocity failure' and may well be very significant with the longer exposures often needed in photomicrography, as even if the correct exposure is given as indicated by the meter, the film may well be markedly underexposed. Some of the more expensive photomicrographic cameras have a control which can be set to compensate automatically for the reciprocity failure of any given film.

The spectral responses of photographic emulsions may vary dramatically and seldom do they correspond with the sensitivity of the human eye. The basic silver halide emulsions are sensitive to ultraviolet, violet and blue wavelengths up to about 5000 nm and are called 'blue-sensitive' or 'ordinary' emulsions. They are not affected by the longer wavelengths, so that green, orange and red objects do not affect the emulsion and so reproduce dark in the print (*Fig.* 14.5*b*). It was discovered late in the nineteenth century that by treating the emulsion with certain dyes during its manufacture, the colour response of the emulsion could be extended into the green and sometimes into the yellow. Such emulsions are called 'orthochromatic' and whilst now sensitive to green light, are totally insensitive to orange and red wavelengths beyond about 5900 nm. This means that they may be handled and processed by a deep red safelight. It also means that they will reproduce reds and oranges very dark on the print and, because tungsten light is lacking in these longer wavelengths, they tend to have a much slower speed rating in artificial light than in daylight. Modern emulsions have had their colour sensitization still further extended into the red or even into the infrared; these 'panchromatic' emulsions form the majority of the film stock in use today. They will reproduce colours reasonably faithfully (*Fig.* 14.5*a*), although red and violet tend to appear lighter and greens tend to appear darker, than they do to the eye. For photomicrography it is best therefore to choose a fairly slow, fine-grain panchromatic film for general monochrome work and most makers have a suitable film in this category. If high speed is more important, e.g. in fluorescence microscopy or with high power phase or Nomarski contrast, then suitable films are available, but at the expense of losing fineness of the grain and the higher degrees of contrast. It is always wise to investigate carefully the technical data of all the films made by the manufacturer of your choice, as often a specialist professional film will give far better results for photomicrography than one of the standard emulsions sold for amateur pictorial work. Whichever emulsion is finally chosen, it is good advice to stick to it and to one carefully chosen developer so that their full potentialities may be learned and exploited. In this way your chance of producing good photomicrographs is greatly increased.

5. CONTRAST CONTROL IN BLACK AND WHITE PHOTOMICROGRAPHY

It should be our aim to produce negatives which will give prints with a full tonal range on a standard grade of printing paper. Sometimes, because of the nature of the microscopical specimen or for other reasons, the contrast of the microscopical image may be low and a straight photomicrograph would give a very poor

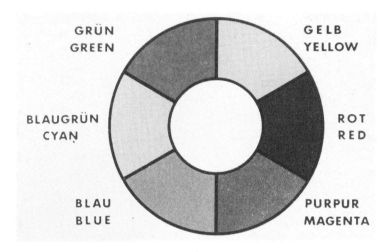

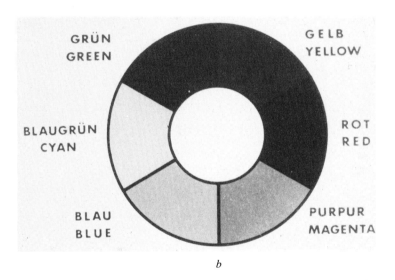

b

Fig. 14.5. (*a*) A colour wheel, composed of segments of the stated colours, photographed on panchromatic film. Note that most colours are reproduced correctly. (*b*) The same colour wheel photographed with a blue sensitive emulsion. Note that green, orange and red wavelengths do not affect the film and hence reproduce very dark in the print. (Reproduced from *Photomicrography* by Reichert with permission.)

negative. In such cases it is desirable to increase the contrast at the stage of making the exposure or in the film processing. If the specimen is strongly coloured then contrast may be manipulated photographically either by the careful choice of emulsion or by the use or panchromatic film and colour filters. If there are red-stained areas, for example, which must be contrasted against blue-stained parts, then this is easily done by the use of an orthochromatic film stock.

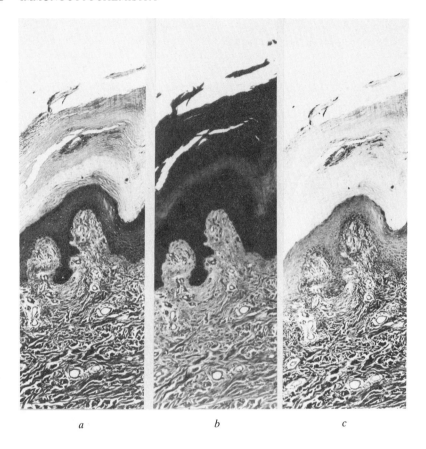

a   *b*   *c*

Fig. 14.6. A trichrome-stained section of human skin. The keratin layers at the top of the micrograph are stained red, whilst the collagen fibres in the dermis at the bottom are stained green.

(*a*) Photographed on panchromatic film with no filter. (*b*) Photographed on panchromatic film with a blue-green filter. The keratin is now very dark and the dermal fibres are lighter in tone. (*c*) Photographed on panchromatic film through a deep red filter. Now the red keratin is light in tone and the green dermal fibres strongly contrasted.

Alternatively, a panchromatic film may be used and the exposure made through a colour filter. These are available in sets for photomicrography and their use has been surveyed.[5] If we wish to *increase* the contrast of a given colour then the exposure should be made through a filter of the complementary colour, i.e. to increase the contrast of a red-stained object use a green filter and to increase the contrast of a blue object use an orange or yellow filter (*Fig.* 14.6). Conversely, if a given colour is to be rendered light in tone, in order to reproduce detail, then the exposure should be through a filter of the *same* colour as the object of interest. If the object is not stained and its contrast has to be increased on a photograph, then we must either give the shortest possible exposure and increase the development time, use a more active, 'contrasty' developer, or change to a slower, finer-grained film which would naturally possess more contrast. It may even be necessary in extreme circumstances to adopt all of these suggestions.

Occasionally the converse situation presents itself and it is necessary to *decrease* the contrast. In such cases it is advisable to take the micrograph through a colour filter of the same colour as the majority of the specimen. It is also advisable to give a very generous exposure (say 50 per cent increase) and reduce the development time by about one-third. It is also possible to use one of the specially formulated soft-working developers, e.g. Kodak's Technidol LC, intended for use with their Technical Pan film; if still less contrast is needed, use a faster film, as these are generally less contrasty then the slower emulsions.

6. COLOUR FILMS

Most microscopists will need to take micrographs in colour at some time; the immediate question is usually whether the pictures should be taken on colour negative film (which normally is used to obtain colour prints) or on reversal film which provides colour transparencies. If the pictures are for the illustration of talks only then reversal film is the obvious choice, but if there is also the question of using them for the illustration of displays or printed matter then the answer is not so clearcut. There are many variables to take into consideration, but on balance most microscopists prefer to use reversal film. It is easier to obtain good colour prints from positive transparencies than to obtain good transparencies from a colour negative and it is this consideration which usually prevails. There is now little to choose in other respects, as most film speeds and both daylight and artificial light balanced film are now easily available in both colour negative and reversal form. As with monochrome films, it is good advice to choose one film type and persevere in its use, so that its performance under varying conditions may be fully appreciated and the best possible results thereby obtained. If other considerations do not rule out such a film, a medium speed colour film (say of about 100 ASA) will generally be found the most satisfactory for photomicrography. Very slow films may result in the exposures becoming so long that reciprocity failure ensues and causes colour shifts in the transparency. On the other hand, the very fast colour films which are now available (of about 600 ASA or even faster) may be required if photographs of sensitive material, e.g. weakly fluorescent compounds, are being taken; such very fast colour films do, however, show more graininess in the structure of their image.

It is often not realized that light varies very much in the proportion of the differing wavelengths which it contains. Daylight is rich in the shorter wavelengths, e.g. in blue light, whilst tungsten light is exactly the opposite, i.e. richer in red light. The eye is very adaptable to such differences and unless they are extreme, both will appear to us as 'white' light. Colour film, however, is not adaptable, and this difference must be taken into account by the manufacturers. The nature of light in this respect is expressed as a property known as 'colour temperature' which represents the temperature in kelvins to which a black body must be heated in order to cause it to emit light of the same characteristic colour. Two types of colour film are made. Daylight film is corrected to give a correct colour rendition when exposed by light with a colour temperature of 5500 K; artificial light film gives correct rendering of colours when exposed to light of 3200 K (*Plate* 11a). If daylight film is used under tungsten light the pictures will have a marked orange cast whilst an artificial light film exposed to daylight (or electronic flash) will show a strong bluish cast. In microscopy such departures from the true colour may not be as apparent as in general photography, but it is desirable to balance the colour temperature of the illumination as closely as possible to that for which the film is corrected. Whether daylight or artificial light film is used in the microscope is very much a matter of personal preference as it is usual to correct for discrepancies by the use of a filter. If daylight film is used with a tungsten light source (much the most common combination), then a blue filter such as the 80A or 80B would be used. If less correction is needed, then paler blue filters such as the 80C, 82C or 82A could be tried. In many cases the choice of filter remains one of trial and error, but at least one maker produces a photomicroscope attachment which has a colour temperature meter incorporated into it. This works well and allows very exact matching of the colour temperature of the light source to that for which the film is corrected. It should be mentioned that if no colour temperature meter is used any light source such as Q.I. bulb or a low wattage tungsten bulb should be operated at the maximal voltage. Any reduction in light *intensity* which is needed should be carried out by inserting neutral density filters into the light path. These attenuate the light without changing its colour temperature; if the intensity is regulated by reducing the voltage passing through the bulb then the colour temperature of the light falls and the micrographs will take on an orange or a yellow cast (*Plate* 11*b*).

If reciprocity failure occurs in colour film, it is usually shown in the form of a marked colour cast in the transparency. This is because of differing response from the three layers (sensitive to red, green and blue light, respectively) which comprise the emulsion of a modern colour film. Usually empirical trials will be needed to see if such reciprocity failure is taking place under the conditions obtaining in any given system. If so, it is often possible to obtain good results by adding a colour-correcting filter in the illuminating system. Colour correction filters come in six different colours (yellow, magenta, cyan, red, green and blue) and in a series of varying strengths. In order to remove a colour cast add a filter of the complementary colour, i.e.

To remove: Yellow   *add*   Blue
            Red   *add*   Cyan
            Magenta   *add*   Green
            Blue   *add*   Yellow
            Cyan   *add*   Red
            Green   *add*   Magenta

If the conditions are standardized as far as possible, then it should be relatively easy, following a trial series of exposures under controlled conditions, to decide whether a colour-correcting filter is needed and, if so, what strength it should have.

Colour film is much less tolerant than monochrome to deviations from the correct exposure. It is essential to run a series of tests with any equipment in order to determine the correct setting of the meter on the apparatus, as there are often quite significant variations between equipment provided by different makers. In many cases it will be found that the most pleasing results, with strongly saturated colours, are obtained with the ASA setting somewhat higher than that printed in the maker's literature. Colour film will not tolerate much overexposure, as the colours appear pale and 'washed-out'; some makes will allow a little underexposure, but as far as possible one should aim for correct exposure. If it is essential to obtain colour pictures and there clearly is insufficient light for an acceptably short exposure, then it is worth exposing the film at double its normal ASA rating and asking the processing laboratory to compensate by increasing the first development time. This may be done with most types of colour film without significant loss of picture quality. Indeed, with some makes of colour film it is possible to *quadruple* the indicated film speed and still obtain pictures, provided that the inevitable resulting loss of quality is acceptable.

7. EXPOSURE DETERMINATION

It is clearly very important that the exposure given to the film should be correct. With most automatic photomicrographic cameras this is no problem once the correct setting of the ASA speed rating for any given film has been checked by a sequence of test exposures. Some care is needed, however, if integrated or whole field metering is in use rather than spot metering in which only a small portion of the illuminated field is used for measurement of the light intensity. Integrated meter readings can give erroneous results if the subject only occupies a very small area of the field of view or if the subject is brightly luminous on a dark background. This may well be so in the case of either fluorescence or darkground microscopy. In such cases the tendency is for the automatic camera to overexpose the film. Many of the modern photomicroscopes now incorporate an extra control to compensate for such extreme cases, but it is often possible to make an empirical adjustment manually. In any case film is relatively cheap compared to the cost of the time spent by the skilled microscopist in finding the relevant part of the specimen and taking the pictures, so that it is sensible practice to take a series of exposures and, if there is any doubt, to 'bracket' the exposure with one or two which are given less than the indicated time and one or two which are given more.

With some of the older eyepiece camera attachments, no automatic metering facilities are provided. In such cases it is necessary to standardize the lighting conditions as far as possible and to carry out a series of test exposures. If a large format system is in use, these may be made by withdrawing the dark slide of the film holder in a series of steps and making multiple exposures on the same sheet of film. If roll-film or 35-mm cameras are in use then a series of separate exposures should be made, increasing the exposure time for each subsequent frame. A suitable series of increments for either method is 2, 4, 8 and 16 units. In the case of a large format system, the dark slide is removed and an exposure of 2 s (say) given. The slide is then inserted to cover one-quarter of the film and a second exposure of 2 s given. The slide is further inserted to cover one-half of the film and an exposure of 4 s given, followed by insertion of the slide to cover three-quarters of the film when a final 8-s exposure is given. On development it will be found that the individual segments of the film have been exposed for the required 2, 4, 8 and 16 units of time. Inspection will then allow the correct exposure to be determined for the conditions of the test. With modern films such a series will almost certainly require much shorter time intervals; one possible sequence on a modern shutter might be 1/15 + 1/15 + 1/8 + 1/4 s, giving a series of 1/15, 1/8, 1/4, 1/2 s for the four incremental steps.

If at all possible, the use of a photoelectric exposure meter is advised. Many types are commercially available which have accessories (or which can be adapted in a departmental workshop) to measure the intensity of the light on either the ground glass of a large-format camera or in the eyepiece of the microscope when a 35-mm attachment camera is in use. Most meters will, of course, be calibrated in f numbers for use with normal cameras, but it is easy to obtain a reading, carry out a series of test exposures as indicated above and so obtain a 'nominal f aperture' of the system which can be used on the meter as an index for the determination of the correct exposure in future. For large-format cameras an integrating exposure meter or a spot meter fitted with a movable probe are commercially available for insertion into the film plane. These are excellent pieces of equipment which give a very accurate measurement of the required exposure, although possibly their high cost may prevent their purchase unless a large amount of large-format work is regularly carried out.

REFERENCES

1. Bradbury S. *An Introduction to the Optical Microscope. R.M.S. Handbook No. 01*. London, Oxford University Press/Royal Microscopical Society, 1984.
2. Hartley W. G. *Hartley's Microscopy*. Oxford, Senecio Publishing, 1981.
3. Spencer M. *Fundamentals of Light Microscopy*. Cambridge, Cambridge University Press, 1982.
4. Bradbury S. A commercial gauge suitable for the measurement of cover glass thickness. *Microscopy*, 1969, **31**, 214–216.
5. Bradbury S. Filters in microscopy. *Proceedings of the Royal Microscopical Society* 1985, **20**, 83–91.

GENERAL BIBLIOGRAPHY ON PHOTOGRAPHIC TECHNIQUE AND PHOTOMICROGRAPHY

Arnold C. R., Rolls P. J. and Stewart J. C. J. *Applied Photography*. London, Focal Press, 1971.
Engel C. E. *Photography for the Scientist*. London, Academic Press, 1968.
Kodak manual *Photography through the Microscope*. Rochester, Eastman Kodak, 1980.
Langford M. J. *Basic Photography*. London, Focal Press, 1972.
Langford M. J. *Advanced Photography*. London, Focal Press, 1972.
Loveland R. P. *Photomicrography — A Comprehensive Treatise*, Vols 1 and 2. New York, John Wiley & Sons, 1970.

Applications

15

Immunocytochemistry of Cell and Tissue Cultures

K. R. Jessen

This chapter will deal exclusively with some technical and biological aspects of the combined use of cultured cells and immunohistochemical techniques. The discussion is based chiefly on experience with nervous tissue in culture and in situ using indirect immunofluorescence methods. The essential points made, however, should be generally applicable to other cell systems and to peroxidase as well as fluorescence methods.

1. CULTURE TYPES AND GENERAL PRINCIPLES

Cells can be maintained alive in a variety of ways after removal from the body. In the context of immunohistochemical studies, it is essential to distinguish between two situations. Firstly, parts of organs, constituting relatively large pieces of tissue, can be maintained as explants (organ culture or organotypic tissue culture). Immunohistochemical studies on such systems are usually carried out in the same manner as on freshly dissected tissue, i.e. by fixation and/or freezing of the cultured organ fragment followed by sectioning and application of antibodies to free floating or mounted sections. For example, antibodies have been used in this way to visualize intrinsic peptide-containing neurones in the gut, using hemisections of the gut wall maintained in culture for 2–3 weeks to achieve extrinsic denervation.[1] Since such studies are, as far as the immunohistochemical method is concerned, identical to those on freshly dissected tissues, they will not be dealt with further in this chapter. Secondly, very small tissue fragments or single cells may be cultured (tissue or cell culture, respectively). For immunohistochemistry these preparations differ in one important respect from those mentioned above, since in this case the antibodies are applied to intact cells, living, fixed or frozen, sitting on a glass or plastic surface; no sectioning is therefore involved in the procedure. The localization of antigens in this type of preparation is in general straightforward and does not pose any major new technical problems. They are particularly suitable for the visualization of cell surface molecules, in which case the antibodies are applied directly to the living cells, while excellent results have also been obtained on the localization of intracellular antigens. For such studies, the cell membrane must be made permeable to large molecules by fixation, lipid extraction, or freezing, prior to the application of antibodies. The next two sections of this chapter deal with methodological aspects of antigen localization in this type of culture system. In the last section the advantages of the combined use of immunohistochemical and tissue and cell culture techniques for solving biological problems will be illustrated by two examples from neurobiology.

Prior to a more detailed discussion, however, it is useful to raise two general cautionary points. Firstly, although cells often retain their differentiation and specific properties to a remarkable degree in culture, as the voluminous literature on the subject demonstrates (see Federoff and Hertz[2] and Fischbach and Nelson[3]), it is important to remember that there are many examples to the contrary. An example from our own studies is provided by a surface antigen found on glial cells from the peripheral nervous system, termed rat neural antigen-1 (RAN-1).[4] This antigen is defined by a mouse antiserum raised against a rat neural tumour. It has proved useful for cell type identification in culture, since, under culture conditions it is expressed by all types of peripheral glial cells, while being largely absent from fibroblastic cells.[5,6] The expression of RAN-1 in culture, however, does not always reflect its expression in situ. The clearest example is provided by the glial cells found within the myenteric plexus of the gut, the enteric glial cells. Although these cells, like other peripheral glia, show bright, RAN-1 immunostaining in culture (Fig. 15.1), no RAN-1 is in fact detectable on these cells prior to culturing.[6] To a lesser extent, Schwann cell expression of RAN-1 is also enhanced under culture conditions. Judging from the intensity of immunostaining, Schwann cells express only relatively low levels of RAN-1 in situ while, again, in culture the RAN-1 labelling is bright and unambiguous. Our observations on another glial surface antigen, rat neural antigen-2 (RAN-2)[7] suggest that in contrast to RAN-1, those peripheral glia that express this antigen in situ lose it during maintenance in culture.[6] These two examples show that in some instances, care is needed in extrapolating from immunohistochemical findings obtained in culture, to mature cells in vivo.

The second point is technical. Although variation exists, both between different antigens and cell types, it holds true in general that the quality and reproducibility of immunostaining increases as the cell density in the culture decreases. To obtain immunostaining of high quality in cultures of dissociated cells, the cells should be relatively sparsely seeded and preferably not allowed to grow to confluence before being used for antigen localization. Similarly, in tissue cultures, clearest localization of antigens is obtained at the borders of the outgrowth zone around the explant, while the immunostaining near, and especially inside, the explant is generally unsatisfactory. The importance of cell density on the quality of immunostaining is illustrated in Figs. 15.1 and 15.2.

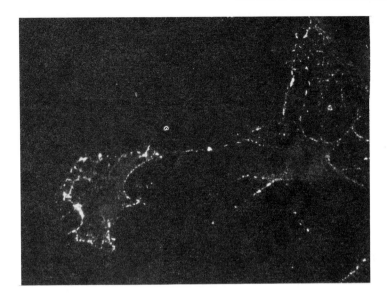

Fig. 15.1. Unambiguous speckled RAN-1 immunostaining on the surface of three cultured enteric glial cells lying at the border of the outgrowth zone. Rat myenteric plexus, 7 days in culture.

Fig. 15.2. Unsatisfactory RAN-1 immunostaining in the same culture as illustrated in *Fig*. 15.1, showing the staining pattern observed in culture areas where cell density is high.

2. LOCALIZATION OF INTRACELLULAR ANTIGENS

As mentioned above, antibody will not penetrate into cells if applied directly to living cultures. For the localization of intracellular antigens, therefore, the first consideration is to find a suitable method for rendering cell membranes permeable to the antibodies, thus allowing access to the antigen. This can be achieved either by chemical agents or by freezing.

2.1. Chemical Permeabilization

In principle, any method can be used which (a) effectively permeabilizes the cell membrane; (b) does not interfere with the configuration of the antigen in question to the extent of reducing antibody binding to unworkable levels; (c) does not lead to high non-specific background fluorescence, e.g. by reacting with substances such as polylysine or collagen, that in some culture systems are used to coat culture surfaces, or by being autofluorescent itself and (d) preserves cell morphology adequately. Chemical agents in common use for immunohistochemical studies on cultured cells are discussed below, while *Table* 15.1 shows various possible combinations of these treatments. It is worth stressing that the procedure. that gives optimal results depends on the antigen, cell type and culture system in question, and is best determined empirically for each new situation. The information provided below and in *Table* 15.1 should provide a useful starting point for efforts in that direction (*see* Osborn[8] for useful discussion and Pearse[9] concerning the chemistry of fixatives and precipitants).

2.1.1. Cross-linking Reagents

FORMALDEHYDE. This frequently used fixative reacts in a complex manner with tissue proteins resulting in the formation of links between adjacent protein chains. Weak formaldehyde fixation in an aqueous solution effectively permeabilizes cell membranes to antibody molecules, while a more thorough fixation, e.g. in higher concentrations of formaldehyde or in glutaraldehyde, may be considerably less successful in achieving high permeability. Most frequently 1–3 per cent formaldehyde is made up in 0·1 M phosphate-buffered saline (PBS) at neutral pH, and fixation is carried out for 5–20 min at room temperature. Formaldehyde fixation is commonly used, in combination with a subsequent acetone or alcohol treatment, in the localization of actin, myosin, and intermediate filament proteins (Treatment 3, *Table* 15.1).[10] It can also be employed alone for the localization of neuropeptides (Treatment 2, *Table* 15.1), although in our hands benzoquinone gives better results in this case, and it gives good preservation of cell morphology.

BENZOQUINONE. As with formaldehyde, this compound exhibits a variety of reactions with proteins, and appears to act as a weak bifunctional cross-linking reagent. In concentrations of 0·4–0·5 per cent in PBS it has proved superior to a number of other fixatives for immunostaining of peptide antigens in endocrine tissue.[11] Fixation in this solution for 30 min at room temperature (Treatment 2,

Table 15.1. **Chemical permeabilization**

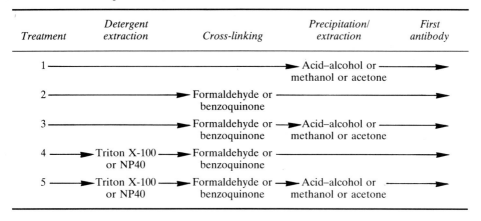

Treatment	Detergent extraction	Cross-linking	Precipitation/ extraction	First antibody
1			➤ Acid–alcohol or methanol or acetone	➤
2		➤ Formaldehyde or benzoquinone		➤
3		➤ Formaldehyde or benzoquinone	➤ Acid–alcohol or methanol or acetone	➤
4	➤ Triton X-100 or NP40	➤ Formaldehyde or benzoquinone		➤
5	➤ Triton X-100 or NP40	➤ Formaldehyde or benzoquinone	➤ Acid–alcohol or methanol or acetone	➤

Table 15.1) has been successfully used in the visualization of neuropeptides[12] and other small molecules such as serotonin[13] in cultured neurones.

2.1.2. *Protein Precipitants/Lipid Extractants*

METHANOL, ETHANOL AND ACETONE. These agents remove lipid from tissues and severely distort the tertiary structure of proteins, causing protein precipitation. They are, therefore sometimes referred to, together with other compounds having similar effects, as 'precipitant fixatives'. They are most commonly used as 100 per cent methanol, 5 per cent acetic acid in ethanol (acid–alcohol) and 100 per cent acetone respectively, at −10 to −20 °C for 10 to 20 min. It is important to wash the cultures thoroughly between this treatment and application of the first antibody. This can be achieved by dipping the coverslip carrying the cultures repeatedly in several beakers of washing solution, e.g. tissue culture medium such as Eagle's minimum essential medium buffered with HEPES buffer (MEM-HEPES). When using acid–alcohol, the reaction between the acid and the colour of pH indicator in the washing solution provides a useful way of judging the effectiveness of the washes. These agents are often used in combination with cross-linking reagents (Treatment 3, *Table* 15.1). For the localization of intracellular proteins such as cytoskeletal intermediate filaments and myelin proteins, both methanol and acid–alcohol also yield excellent results when used on their own (Treatment 1, *Table* 15.1).[5] In this case, the preservation of morphology tends to vary considerably from area to area in the culture, although it is often very good and background fluorescence is low.

2.1.3. *Detergents*

TRITON X-100 AND NP40. The use of these gentle non-ionic detergents in conjunction with immunohistochemistry falls into two categories. Firstly, they can be used prior to fixation, for extraction of cell membranes, an approach primarily adopted for studies on cytoskeletal antigens. This can result in the total removal of intracellular membranes as well as the plasma membrane, although the degree to which this takes place can to some extent be controlled by varying the detergent concentration and incubation time. Since cytoskeletal proteins are often sensitive to their ionic environment and also because of the presence of proteolytic enzymes in unfixed cells, the type of buffer employed during the extraction is an important consideration, and may have to be determined empirically for each new type of experiment. A fairly safe general procedure is to use 0·1–0·5 per cent detergent made up in PIPES buffer containing 1 mM EGTA for the removal of Ca^{2+}, and 2 mM Mg^{2+}; incubation time most often varies from 1 to 3 min. This is generally followed by cross-linkage (Treatment 4, *Table* 15.1) or cross-linkage and precipitation fixation (Treatment 5, *Table* 15.1).[13] Secondly, detergents are often used subsequent to fixation in order to remove remaining lipids (not shown in *Table* 15.1).[12] Since the proteins are now fixed, the choice of buffer is not critical. A common procedure is an overnight wash in 0·3 per cent detergent made up in PBS.

2.2. Permeabilization by Freezing

Sometimes it is important to avoid any loss in the level of antibody–antigen binding, which often occurs during chemical permeabilization. In these instances, freezing may be the method of choice. This is achieved by placing the coverslip carrying the culture, with the cell side facing upwards, on a metal plate cooled with solid CO_2. The cells freeze in a few seconds, judged by the surface of the coverslip turning white, due to the freezing of the thin film of culture medium covering the cells. The coverslip is now removed from the plate and allowed to thaw completely during a few minutes. This procedure is repeated two or three times before application of the first antibody. While leading to optimal preservation of some antigens, a modification of this method may be necessary if the antigen is freely soluble. In the absence of fixation for anchoring or precipitating such molecules, they may diffuse out of the cells during the freeze–thaw cycles. This can be effectively prevented by covering the culture with the first layer of antibodies before the start of the freeze–thaw procedure. At the end of the last thaw, some more antibody solution may be added; the coverslip soon reaches room temperature and incubation with first antibody is now continued in the same way as following chemical permeabilization.

2.3. Comments

At the end of permeabilization by chemical agents or freezing, the immunostaining and subsequent mounting does not differ in principle from the procedures used for other material, e.g. freeze-mounted tissue sections. It will often be found, however, that washing time after first and second antibody layers can be shortened considerably. Provided that 5 or 6 changes in volume of about 25 ml each or larger, are employed, 30–45 s are often adequate. An example of intracellular immunostaining is provided in *Fig.* 15.3, showing the localization of substance P to nerve fibres in explant cultures of the myenteric plexus from the guinea-pig.

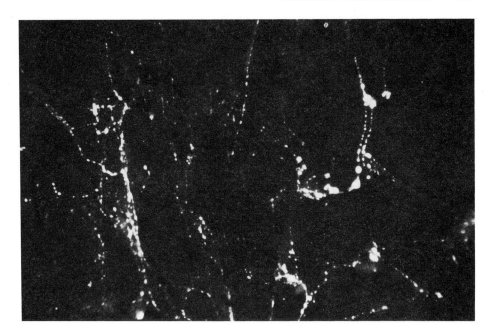

Fig. 15.3. Substance P-like immunoreactivity inside varicose nerve fibres in an explant culture from the enteric nervous system. Guinea-pig myenteric plexus, 10 days in culture.

Localization of intracellular antigens in culture offers, in some respects, technical advantages over immunostaining of in situ material, e.g. tissue sections, and can be a very useful adjunct to in situ studies. For instance, cultures can be used to answer unambiguously questions about which tissue elements are responsible for immunostaining observed in sections, otherwise not answerable except by electron microscopical immunohistochemistry. Two examples are provided in the localization of glial fibrillary acidic protein (GFAP) and glutamine synthetase in the gastrointestinal tract.[6,15,16] When frozen sections of the gut wall are treated with antibodies against GFAP or glutamine synthetase, and processed according to the indirect immunofluorescence technique, high quantities of both antigens are revealed inside the small ganglia of the myenteric plexus, as judged by the intense fluorescence (*Figs.* 15.4a and 15.5a). However, the question as to which cellular elements inside the ganglion carry these antigens cannot be answered with any certainty from this type of observation. In the absence of other information, they could be glial, neuronal or both, and it is not clear from the staining pattern, whether the antigens are intracellular or reside on cell surface membranes. To answer these questions, immunostaining was carried out on explant cultures of the rat myenteric plexus (*Figs.* 15.4b and 15.5b). In these cultures, individual neurones, glial cells and their processes can be clearly observed, both with phase contrast and fluorescence optics. It was found that both of these antigens were localized in the glial cells, where they were present both in the cell bodies and in the processes. Further, neither antigen could be visualized

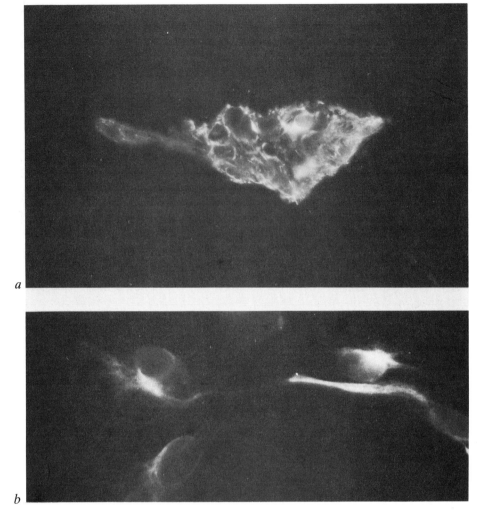

Fig. 15.4. GFAP-like immunoreactivity. (*a*) Frozen section from the rat colon, showing one ganglion of the myenteric plexus; the localization within the ganglion of the intense immunoreactivity is difficult to determine. (*b*) Four immunoreactive enteric glial cells in culture; rat myenteric plexus, 15 days in culture. (Reproduced by permission from *Nature*, **286**, no. 5744, 736–737. Copyright © 1980, Macmillan Journals Limited.)

first unless the cells had previously been permeabilized. This shows that both of these antigens are exclusively intracellular in this cell type (*see below*).

3. LOCALIZATION OF CELL SURFACE ANTIGENS

In the case of cell surface antigens, permeabilization of the cell membrane prior to the application of antibodies is unnecessary. It is best avoided, unless there are

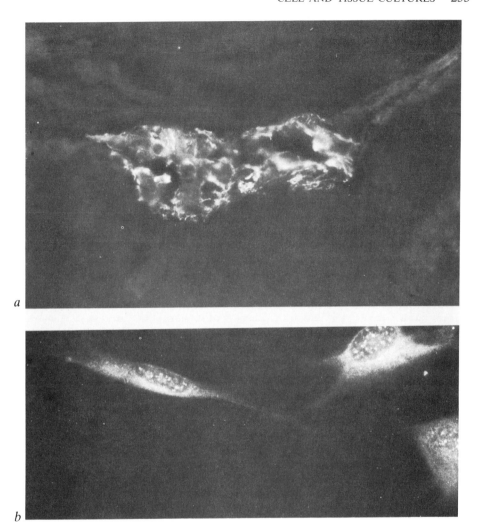

Fig. 15.5. Glutamine synthetase-like immunoreactivity. (*a*) Frozen section from the rat colon showing one myenteric ganglion embedded between two muscle layers. Bright immunoreactivity can be seen inside the ganglion, although its cellular localization can not be clearly defined. (*b*) Three immunopositive enteric glial cells in culture. Rat myenteric plexus 10 days in culture.

specific reasons to the contrary, since it will in most cases lead to weaker immunostaining and may cause higher background fluorescence, through non-specific antibody binding to intracellular components. The first layer of antibody can be applied to the cells immediately after removal from the tissue culture medium or following a brief wash in medium without serum or in PBS. Since the cells are unfixed, incubation time in the antibody is usually short, 20–60 min at room temperature, and overnight incubation, commonly used for intracellular antigens, is not recommended. Washes following first and second antibody layers proceed in a similar manner to that described for intracellular localization. Unless the cells are to be observed and photographed immediately following the immunostaining, treatment with formaldehyde or acid–alcohol is recommended to preserve cell morphology. This is most commonly done after the wash which follows the second layer of antibody. Alternatively, it can be carried out between the first and second layers, immediately after the wash following the first layer of antibody. An example of surface immunostaining is shown in *Plate* 12. This figure, as well as *Fig.* 15.1, illustrates clearly the characteristically uneven or speckled appearance of the fluorochrome. In most cases this will be caused by patching of membrane molecules, which can take place in unfixed membranes during the staining procedure.[17] This phenomenon may serve to enhance the sensitivity of the immunostaining, since aggregates of a fluorochrome are more easily visualized and distinguished from background fluorescence, than the same quantity of the fluorochrome evenly distributed over the cell membrane.

As with intracellular antigens, the immunohistochemical localization of cell surface antigens in culture often provides clearcut answers to questions about antigen localization. This applies both to the determination of which cell type carries the antigen and to extracellular versus intracellular localization: an antigen which can be visualized without permeabilization must reside on the cell surface, while intracellular localization is indicated if permeabilization is a prerequisite for antigen detection. These features are illustrated in studies on an antigen defined by a monoclonal antibody designated 38D7.[18] This antibody, raised by immunizing with dorsal root sensory ganglia, produced a staining pattern in peripheral ganglia which could not have been interpreted with any degree of certainty without electron microscopical immunohistochemistry or tissue culture studies. The antigen distribution in a frozen section of the rat dorsal root ganglion is shown in *Fig.* 15.6. The precipitate, indicating the presence of the antigen, is found surrounding, or associated with the periphery of, neuronal cell bodies. On the basis of this observation, the antigen could reside in at least four different locations: inner or outer aspect of neuronal membranes, or inner or outer aspect of the membrane of the satellite cells that surround the neuronal perikarya. This problem was easily solved by immunostaining of dissociated cell cultures from the dorsal root ganglia. The antigen was restricted to the outer aspect of neuronal cell membranes (*Fig.* 15.7). The results of similar experiments on a variety of cell and explant cultures from the central and peripheral nervous system suggest that the antigen 38D7 is restricted to the surface membranes of peripheral neurones while being absent from central neurones and nearly all non-neuronal cells. It would thus represent the first chemical marker to comprehensively distinguish between peripheral and central neurones.

4. APPLICATION OF IMMUNOHISTOCHEMISTRY AND CULTURE TECHNIQUES TO PROBLEMS IN NEUROBIOLOGY

The combination of tissue culture methods and immunological technology represents a novel approach of considerable promise in studies on the nervous system. Below, two recent examples are described briefly. The first deals with

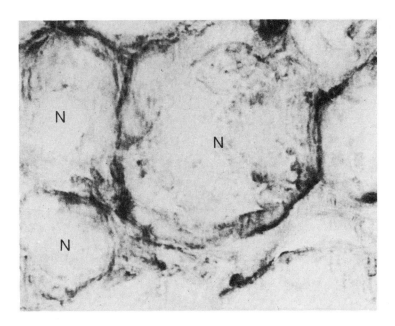

Fig. 15.6. 38D7 immunoperoxidase labelling associated with the peripheral areas ot sensory neurones (N). Frozen section from a rat dorsal root sensory ganglion. (Photograph kindly supplied by Dr Rhona Mirsky, University College, London, and reproduced by permission from *Nature*, **291**, no. 5814, 418–420. Copyright © 1981 Macmillan Journals Limited.)

neuroglia interactions in the peripheral nervous system, while the second addresses questions concerning brain development.

4.1. Neuronal Control of Myelin Synthesis

Schwann cells and oligodendrocytes elaborate multilayered myelin sheaths around axons in the peripheral and central nervous systems, respectively. Since some serious diseases involve the failure of these cells to form or maintain the myelin sheath, it is of great importance to understand the factors that control myelin synthesis and maintenance. One of the central questions in this context concerns the control of synthesis, within the Schwann cell or oligodendrocyte cytoplasm, of the major biochemical constituents of myelin, such as galactocerebroside (GC), sulphatide (S), and myelin basic protein (BP). This problem was studied by Mirsky and her colleagues, using antibodies against GC, S and BP, and dissociated cell cultures from parts of the peripheral and central nervous systems as a source of Schwann cells and oligodendrocytes, respectively.[19] It was found that when myelin-forming Schwann cells were dissociated from the axons they normally surround and put into culture, GC, S, and BP were all immunohistochemically demonstrable in the Schwann cells for periods of up to 16–20 h after they had been put into culture (*Fig.* 15.8). Interestingly, however, the number of positive cells fell rapidly after longer culture periods, so that after 5–6 days none of the Schwann cells contained a detectable quantity of these substances. Mere

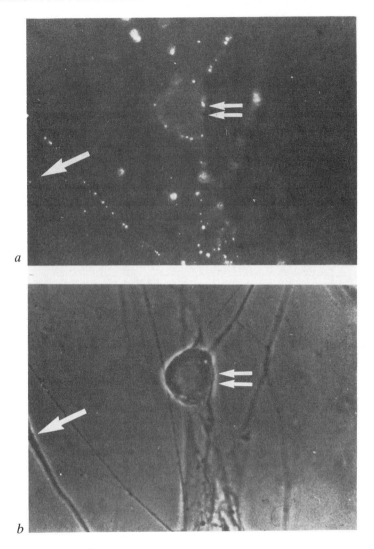

Fig. 15.7. 38D7 immunofluorescence staining of a culture containing sensory neurones and glia. Note immunopositive neuronal cell body (double arrow) and fibres, while a glial cell (single arrow) is negative. (*a*) and (*b*) represent fluorescence and phase contrast view respectively, of the same field. (Photographs kindly supplied by Dr Rhona Mirsky, University College London.)

contact with axons in Schwann cell cultures of this type was not enough to prevent this disappearance of myelin molecules, as shown by co-culturing the Schwann cells with peripheral neurones. Oligodendrocytes on the other hand behaved differently. GC, S and BP, all of which were readily detectable by immunohistochemistry from the onset of culturing, were fully maintained in the oligodendrocytes when grown in dissociated cultures without neurones for periods of up to several weks. These results suggest that, surprisingly, the organization of myelin synthesis differs between these two cell types, in that rat Schwann cells require a

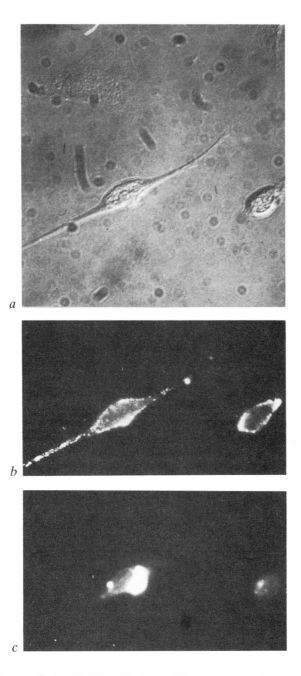

a

b

c

Fig. 15.8. Two Schwann cells in a 16–20 h cell culture of the newborn rat sciatic nerve. The culture was incubated with mouse anti-RAN-1 antibodies followed by goat anti-mouse Ig conjugated to rhodamine and, after fixation, with rabbit-anti-BP followed by goat anti-rabbit Ig conjugated to fluorescein. The cells were viewed by Nomarski optics (*a*), rhodamine fluorescence (*b*) and fluorescein fluorescence (*c*). Both Schwann cells are RAN-1 positive and one of them contains readily demonstrable basic protein. (Photographs kindly supplied by Dr Rhona Mirsky, University College, London, and reproduced from *The Journal of Cell Biology* 1980, **84**, 483–494, by copyright permission of the Rockefeller University Press.)

continuing signal from appropriate axons to make detectable amounts of the major chemical components of myelin, while oligodendrocytes do not.[19]

4.2. Biological Clocks Versus Positional Information in Brain Development

In a study of brain development, Abney et al.[20] used immunohistochemical markers to define and identify the major cell types present in dissociated cell cultures from the developing rat brain: neurones, defined by their expression of tetanus toxin receptors; astrocytes, defined by the filament protein GFAP; oligodendrocytes defined by their content of galactocerebroside (GC). By immunostaining freshly prepared cell suspensions from embryonic and neonatal rat brain, a regular sequence of appearance of these molecules in normal development was observed: neurones that bound tetanus toxin were present already in 10-day-old embryos (the earliest time tested), GFAP-positive astrocytes appeared at 15–16 days *in utero*, while galactocerebroside-positive oligodendrocytes were first detected 2–3 days after birth. These authors now addressed the question as to whether this regular timing had been programmed into the respective precursor cells already by the tenth day of embryonic life, or whether extrinsic signals, determined by the position of the precursor cells within the developing tissue, were required as triggers (positional information, *see* Wolpert[21]). Dissociated cell cultures were prepared from brains of 10-day-old embryos and maintained *in vitro* for up to several weeks. The appearance of the molecules identifying each cell type was monitored by immunostaining the cells after varying periods *in vitro*. This revealed a pattern which exactly matched the normal developmental timing. Thus, neurones binding tetanus toxin were found from the onset of culturing; GFAP-immunofluorescent astrocytes first appeared after 5–6 days in culture (10-day-old embryo +5–6 days in culture equalling 15–16 days *in utero*); and galactocerebroside-containing oligodendrocytes were first seen after 13–14 days of culturing (10-day-old embryo +13–14 days in culture equalling second and third day after birth, since the gestation period is 21 days). It was concluded 'that biological clocks are more important than positional information in gliogenesis between 10 days of gestation and 1 week after birth in the rat'.[20]

Appendix

SOURCES OF ANTIBODIES USED FOR THE ILLUSTRATED PREPARATIONS

RAN-1	*see*[4]
RAN-2	*see*[7]
38D7	*see*[18]
GFAP	Dr B. Pruss National Institutes of Health Bethesda, Md, USA

Glutamine synthetase	Dr O. S. Jorgensen Copenhagen University Denmark
Galactocerebroside	see[22]
α-BP	Dr McFarlin National Institutes of Health Bethesda, Md, USA
Conjugated second- layer antibodies	Cappel Laboratories and Nordic Laboratories

REFERENCES

1. Schultzberg M., Dreyfus C. F., Gershon M. D., Hökfelt T., Elde R., Nilsson F., Said S. and Goldstein M. VIP, enkephalin, substance P and somatostatin-like immunoreactivity in neurons intrinsic to the intestine: immunohistochemical evidence from organotypic tissue cultures. *Brain Res.* 1978, **155**, 239–248.
2. Federoff S. and Hertz L. (eds) *Cell Tissue and Organ Cultures in Neurobiology.* New York, Academic Press, 1977.
3. Fischbach G. D. and Nelson P. G. Cell culture in neurobiology. In: Brookhart J. M. and Mountcastle V. B., eds., *Handbook of Physiology: The Nervous System.* Bethesda, American Physiological Society, 1977: 719–774.
4. Fields K. L., Gosling C., Megson M. and Stern P. L. New cell surface antigens in rat defined by tumours of the nervous system. *Proc. Natl Acad. Sci. USA* 1975, **72**, 1286–1300.
5. Raff M. C., Fields K. L., Hakomori S., Mirsky R., Pruss R. M. and Winter J. Cell-type-specific markers for distinguishing and studying neurones and the major classes of glial cells in culture. *Brain Res.* 1979, **174**, 283–308.
6. Jessen K. R. and Mirsky R. Astrocyte-like glia in the peripheral nervous system: an immunohistochemical study of enteric glia. *J. Neurosci.* 1983, **3**, 2206–2218.
7. Bartlett P. F., Noble M. D., Pruss R. M., Raff M. C., Rattray S. and Williams C. A. Rat neural antigen-2 (RAN-2): a cell surface antigen on astrocytes, ependymal cells, Müller cells and leptomeninges defined by a monoclonal antibody. *Brain Res.* 1981, **204**, 339–352.
8. Osborn M. Localization of proteins by immunofluorescent techniques. *Techniques in Cellular Physiology P107.* Amsterdam, Elsevier/North Holland Scientific Publishers Ltd, 1981: 1–28.
9. Pearse A. G. E. *Histochemistry: Theoretical and Applied,* Vol. 1. Edinburgh, Churchill Livingstone, 1980.
10. Gröschel-Stewart U., Chamley J. H., McConnell J. and Burnstock G. Comparison of the reaction of cultured smooth and cardiac muscle cells and fibroblasts to specific antibodies to myosin. *Histochemistry* 1975, **43**, 215–224.
11. Polak J. M. and Pearse A. G. E. Bifunctional reagents as vapour and liquid-phase fixatives. *Histochem. J.* 1975, **7**, 179–186.
12. Jessen K. R., Saffrey M. J., Van Noorden S., Bloom S. R., Polak J. M. and Burnstock G. Immunohistochemical studies of the enteric nervous system in tissue culture and in situ: localization of vasoactive intestinal polypeptide (VIP), substance-P and enkephalin immuno-reactive nerves in the guinea-pig gut. *Neuroscience* 1980, **5**, 1717–1735.
13. Saffrey M. J. Personal communication, 1982.
14. Osborn M. and Weber K. The detergent-resistant cytoskeleton of tissue culture cells includes the nucleus and the microfilament bundles. *Exp. Cell Res.* 1977, **106**, 339–349.
15. Jessen K. R. and Mirsky R. Glial cells in the enteric nervous system contain glial fibrillary acidic protein. *Nature* 1980, **286**, 736–737.
16. Jessen K. R. and Mirsky R. Immunohistochemical methods define a new type of glial cell. *Neurosci. Lett.* 1981, Suppl. 7, 396.
17. de Petris S. and Raff M. C. Normal distribution, patching and capping of lymphocyte surface immunoglobulin studied by electron microscopy. *Nat. New Biol.* 1973, **241**, 257.

18. Vulliamy T., Rattray S. and Mirsky R. Cell surface antigen distinguishes sensory and autonomic peripheral neurons from central neurons. *Nature* 1981, **291**, 418–420.
19. Mirsky R., Winter J., Abney E. R., Pruss R. C., Gavrilovic J. and Raff M. C. Myelin-specific proteins and glycolipids in rat Schwann cells and oligodendrocytes in culture. *J. Cell Biol.* 1980, **84**, 483–494.
20. Abney E. R., Bartlett P. P. and Raff M. C. Astrocytes ependymal cells and oligodendrocytes develop on schedule in dissociated cell cultures of embryonic rat brain. *Dev. Biol.* 1981, **83**, 301–310.
21. Wolpert U. Positional information and the spatial pattern of differentiation. *J. Theoret. Biol.* 1969, **25**, 1–47.
22. Ranscht B., Clapshaw P. A., Price J., Noble M. and Siefert W. Development of oligodendrocytes and Schwann cells studied with a monoclonal antibody against galactocerebroside. *Proc. Natl Acad. Sci. USA* 1982, **72**, 2709–2713.

16

Immunocytochemistry of Cell Lines

B. A. Gusterson and P. Monaghan

In this chapter the use of tumour cell lines is considered in relation to the study of aspects of differentiation and phenotypic changes associated with malignancy. The examples given relate predominantly to stratified squamous epithelia and the human breast, but the techniques and approaches outlined are equally applicable to other systems.

Immunocytochemical examination has been used to study the expression of surface and cytoplasmic antigens in cell lines and the relationship between these findings and those seen *in vivo*.

1. GENERAL COMMENTS

Although there have been many definitions proposed for a continuous 'cell line', there is an implicit assumption that the cells have undergone a change which gives them an infinite life span. This will be the case both with cells established from tumours and with normal cells which have been 'transformed' in culture. Because such cells can be grown for long periods of time and for many generations, there are obviously heavy selective pressures with a tendency toward the emergence of more phenotypically homogeneous populations. Unfortunately, in many instances selection is for the faster growing cells which are not necessarily characteristic of the tissue as a whole, and not capable of exhibiting differentiated features. It is thus often difficult to characterize cultured cells and to relate them to functionally defined cell types in their tissue of origin. With the introduction of new tissue culture techniques and the recognition that different cell types have specific nutritional requirements it has, however, been possible to establish normal human epidermal cells in culture,[1] where they will stratify and produce a skin-like structure. Squamous carcinoma cell lines have also been established[2–5] which when grown under the same conditions also differentiate. The introduction of three-dimensional matrix supports in both epidermal[6] and breast systems[7,8] has further advanced our ability to obtain continuous cultured cells that retain differentiated features.

To dissect tissue culture systems, stable markers are required to identify specific cell types and, in differentiating systems, to define cells at various points of the differentiation pathway. With the expansion of hybridoma technology,[9] many reagents are now available which are assisting in the characterization of cultured cells and defining the functional parameters of these cells. A major advantage of cell lines which differentiate is the ability to produce unlimited numbers of cells both for biochemical analysis and to provide reproducible systems to examine extrinsic and intrinsic factors controlling the differentiation process. In this chapter, the approaches used to study some of these problems are considered.

2. DIFFERENTIATED FUNCTIONS OF CELLS IN CULTURE AND IDENTIFICATION OF CELL TYPES

The advances produced by the ability to serially culture keratinocytes from normal human skin,[1] and to re-implant these cells in immune deprived animals,[10] has laid the foundation for detailed studies into the mechanisms controlling differentiation. The parallel establishment of cell lines from squamous carcinomas has also provided the material for comparative studies between the normal cells and their malignant counterparts.

During the process of keratinization, there are a number of identifiable cytoplasmic and membrane changes that occur morphologically. These have been shown to have their biochemical correlates in expression of different keratin genes,[11] the synthesis of the histidine-rich protein filaggrin which is located in the keratohyalin granules[12] and the production of certain specific membrane proteins.[13] Antibodies against these different components have thus provided the tools both to characterize *in vitro* models and to examine the changes in expression of these determinants that may occur in pathological states.

The first objective with any tissue culture system is to establish the cell types that are being cultivated. The presence of cytokeratin, for example, is specific for the epithelial phenotype. Similarly, the presence of neurofilaments can define neuronal cells, glial fibrillary acidic protein is found in glial cells, and desmin in muscle cells.[14] Vimentin in culture, however, is not a dependable marker of mesenchymal cells as many epithelial cells can express vimentin when grown *in vitro*. In our laboratory, in order to characterize a cell as epithelial, we use a broadly cross-reacting rabbit polyclonal antiserum which produces a characteristic cytoskeletal pattern (*Fig.* 16.1).

Although keratinocytes and squamous carcinoma cell lines will stratify when grown on a plastic substrate, improved results are obtained when these cells are grown on floating collagen gels. Normal cells grown on these gels demonstrate a very ordered stratification with mitotically active basal cells abutting the collagen and flattening of the cells towards the surface where they form a typical cornified envelope and have a keratohyalin granular layer.[5] Similar results can be obtained

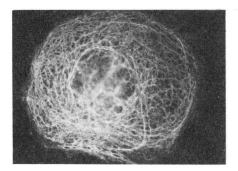

Fig. 16.1.   *Fig.* 16.2.

Fig. 16.1. Human breast carcinoma cell line PMC42 stained with a polyclonal anti-keratin antiserum using the indirect technique as seen by immunofluorescence microscopy. The antibody demonstrates a dense pattern of filaments which exhibit perinuclear condensation. Cultures were fixed for 2 h in methanol at 4°C prior to staining.

Fig. 16.2. Squamous carcinoma cell line LICR-LON-HN5 growing on collagen gel. Note the localization of a keratohyaline granule associated antigen using the indirect peroxidase technique (arrow). *See* Appendix, Section I for method.

with squamous carcinoma cell lines, although the degree of disorganization of the cells and the degree of pleomorphism characterizes the cultured cells as being malignant.[5] These cultured cells retain their ability to produce keratohyalin granule-associated proteins (*Fig.* 16.2), thus providing a system in which to examine the factors affecting the synthesis of this protein. When grown as xenografts in immunologically incompetent mice these cells reproduce the morphology of their cells of origin, with the normal keratinocytes producing an epidermal cyst and the malignant cells a squamous cell carcinoma.[4,5,10]

Collagen gels have also found their utility in the study of morphogenetic aspects of the normal human breast where isolated breast ducts, when embedded in collagen gels, will migrate through the gel and produce new tubular structures, which in places are double layered with a luminal lining cell and an outer cell layer analogous to the situation *in vivo*.[8] Cloned rodent mammary tumour cell lines will produce similar 'duct-like' structures.[15] Thus, techniques have been recently developed which enable cell biologists to study the morphology of both normal and malignant human epithelial cells in culture. In order to dissect the mechanisms controlling differentiation, however, it is necessary to use a combined approach whereby a detailed analysis is made of well-characterized cellular markers *in vivo* and their expression in a parallel *in vitro* system where the effects of external modulating culture conditions can be examined. In this context, we will consider the use of keratins in the epidermal system and a combination of membrane glycoproteins and basement membrane products in the breast.

There are now known to be at least 19 human keratin proteins,[16] distinguishable on the basis of their molecular size, isoelectric point and cellular distribution *in vivo*. The biochemical extraction analyses and immunocytochemical results with monoclonal antibodies that recognize specific keratins indicate that the basal cells of the epidermis can be delineated by their expression of certain keratins.[17-19] Thus monoclonal antibodies, such as LICR-LON-16a, which recognize a

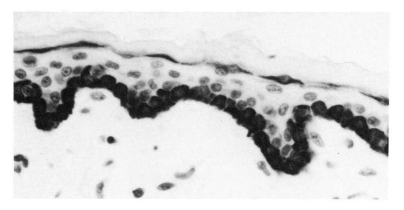

Fig. 16.3. Normal skin from the chest wall: methacarn-fixed, paraffin section stained with monoclonal LICR-LON-16a. There is intense staining of the epidermal basal cells using the indirect alkaline phosphatase technique. (Reproduced by permission from *Journal of Pathology*, John Wiley and Sons Ltd, Chichester, Sussex.)

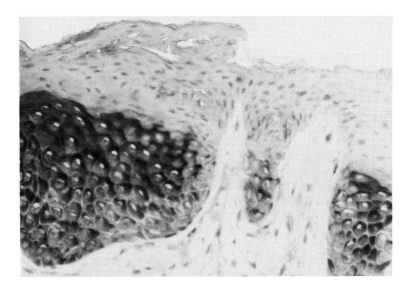

Fig. 16.4. Eczematous skin: methacarn-fixed, paraffin section stained with monoclonal antibody LICR-LON-16a using the indirect alkaline phosphatase technique. There is strong staining of both basal and spinous cells. (Reproduced by permission from *Journal of Pathology*, John Wiley and Sons Ltd, Chichester, Sussex.)

cytokeratin doublet of 45–46 kdal by immunoblotting against epidermal keratins, specifically stain the basal layer of normal skin[19] (*Fig.* 16.3). This specificity is, however, lost in reactive conditions such as eczema (*Fig.* 16.4) and psoriasis with staining now moving to the superficial layers. The change in expression of this particular marker finds its parallel in cultured epidermal cells where basal cell specificity is lost and there is staining of all layers of the cultured epithelium (*Fig.* 16.5). In squamous cell carcinomas, the undifferentiated 'basaloid' cells are not

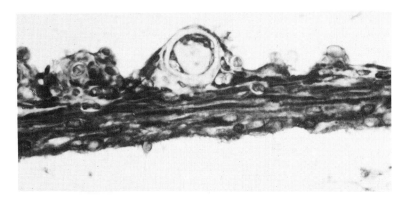

Fig. 16.5. Multilayered culture of human epidermal keratinocytes stained with LICR-LON-16a using the indirect peroxidase technique. There is intense staining of all cell layers. *See* Appendix, Section I for method.

Fig. 16.6. Electron micrograph of a vertical section of a culture of R401, a cell line derived from normal rat breast stained by immunoperoxidase for the presence of extracellular collagen IV (arrow). *See* Appendix, Section I for method.

stained and a similar staining pattern is seen in cultured carcinoma cells. Thus the normal cultured keratinocytes are more closely related to hyperproliferative skin conditions and probably reflect a wound healing situation, while the *in vitro* retention of the growth pattern seen in squamous cell carcinomas *in vivo* indicates that this is an appropriate model in which to examine altered keratin gene expression.

In the case of the breast, primary organ cultures in collagen gels have aided studies of morphogenesis, but human cell lines have been rather unrewarding in terms of their ability to differentiate. In the rat, however, it has been possible to isolate putative breast epithelial stem cells (Rama 25) which, under certain culture conditions, give rise to cells with luminal and myoepithelial characteristics (Rama 29).[20] Myoepithelium-like lines have also been established from normal rat breast cultures (Rama 401),[21] where the cells synthesize basement membrane components such as laminin and type IV collagen (*Fig.* 16.6). Such systems are therefore of value in studies of the synthesis of basement membrane components by myoepithelial cells and the abnormalities in basement membrane production that are associated with malignancy.[22]

Human breast carcinoma cell lines have been used to analyse antigenic heterogeneity in the normal human breast and in breast carcinomas using antibodies raised to the human milk fat globule membrane.[23] These antibodies, for example LICR-LON-M8, recognize carbohydrate determinants on membrane glycoproteins and produce a heterogeneous staining pattern in the normal breast duct both *in vivo* and *in vitro*.[8] Using cell lines which express these determinants, it is possible to demonstrate that antigenic expression is not correlated with the cell cycle, a study that would not be possible *in vivo*. It is also apparent that a population expressing one antigen, after continuous cultivation, will regenerate a mixed (positive and negative) population.[23] A similar situation appears to be true for both normal and malignant cultured breast cells. This apparent reversibility in the nature of the heterogeneity is an important phenomenon as it suggests that any use of such antibodies for therapeutic drug targeting is unlikely to be successful. The question of whether these antibodies are in fact recognizing cells in different states of differentiation or merely variations in carbohydrate determinants in otherwise similar cells is not known. Using two-colour fluorescence with two antibodies directed against different sugar determinants, it is now possible to demonstrate that populations of cultured breast cells can be stained with a single antibody while other cell populations bear a different carbohydrate structure (*Plate* 13).[23] These data, while demonstrating conclusively the presence of carbohydrate-directed cell surface heterogeneity in both normal and tumour cell populations, do not negate the potential utility of these reagents for diagnostic purposes, such as the detection of micrometastases.[24]

From this brief summary of some of the investigations that are being carried out on epithelial systems, it can be seen that cell lines may be used both to dissect cell populations and to look at certain aspects of cell differentiation.

3. DEMONSTRATION OF RECEPTORS IN CELL LINES

In this section we will consider studies demonstrating receptors on tumour cell lines, with particular reference to the receptors for oestrogen and the receptor for epidermal growth factor (EGF). The staining protocols used are presented in the Appendix (*see also* Chapter 11.)

The mechanism whereby oestrogens regulate growth in reproductive tissues is unknown, but the responses in target cells are thought to be mediated by high-affinity, specific binding proteins that are generally referred to as receptors. In the breast, initial observations suggested that the oestrogen receptor was predominantly extranuclear.[25] Recent studies, however, with monoclonal antibodies to the oestrogen receptor have demonstrated purely *nuclear* staining.[26] These antibodies have all been raised to receptor extracts derived from the human breast cell line MCF-7.[26] The nuclear localization of the receptor both in the cell line (*Fig.* 16.7) and in the surgical biopsies of breast carcinomas[26] was a surprising result, but supports the view that many previous claims to demonstrate cytoplasmic receptor immunocytochemically were erroneous and

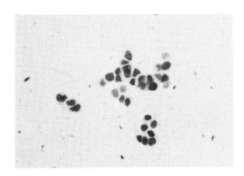
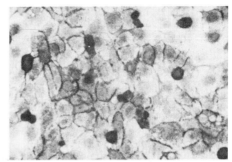

Fig. 16.7.   *Fig.* 16.8.

Fig. 16.7. Immunocytochemical localization of oestrogen receptor in MCF-7 human breast cancer cells. Note the nuclear localization. (Stained using a diagnostic kit supplied by Abbot, USA-Delkenheim.) *See* Appendix, Section III for method.

Fig. 16.8. Squamous carcinoma cell line LICR-LON-HN10 growing as a monolayer stained with monoclonal EGFR1 using the indirect peroxidase technique. Note the heterogeneous staining with strong positivity of the cells in mitosis. *See* Appendix, Section I for method.

the investigators were probably identifying an oestrogen receptor-related protein. Thus, although the human breast cell line, MCF-7, is considered to be unrepresentative of breast carcinomas in general, its oestrogen receptor content makes it a useful source of material for biochemical, molecular biological and immunocytochemical analyses.

Epidemal growth factor receptors, in contrast, are located predominantly in the cell membrane and, using immunohistochemical techniques and a monoclonal antibody (EGFR1)[27] directed against the epidermal growth factor receptor, it has been possible to demonstrate the receptor on a wide range of normal human epithelial tissues and tumours.[28] Both the tissue distribution and a number of *in vitro* studies suggest that epidermal growth factor may be involved in the control of proliferation and possibly differentiation of surface epithelia.[29] The strong staining on some tumour cell lines, and in particular lines derived from squamous cell carcinomas of the oropharynx and larynx, indicates that there is an increased expression of receptors in these tumours as a function of malignancy. Using the monoclonal antibody EGFR1, it has been possible to demonstrate that the number of receptors, as demonstrated by binding studies on squamous carcinoma cell lines, is paralleled by the intensity of staining.[29] The binding studies, however, cannot be related to the distribution of receptor, but, using immunocytochemistry, it is possible to demonstrate the apparent heterogeneity of receptor levels on different cell populations (*Fig.* 16.8).

Using such cell lines with high levels of receptor, whether for epidermal growth factor or for other ligands such as transferrin, it is possible to examine the mechanisms of internalization of receptor proteins and their ligands through the mechanism of coated pits (*Fig.* 16.9).[30] Such techniques, usually using immuno-gold, give very good localization and make it possible to carry out time-course experiments whereby receptors can be followed. At the present time, the role of epidermal growth factor *in vivo* is still ambiguous, but cell lines provide the tools

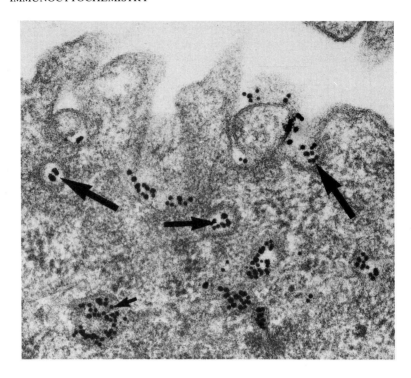

Fig. 16.9. Squamous carcinoma cells A431 were incubated for 30 min at 5 °C and 22 °C with an antibody to the transferrin receptor labelled with 12-nm gold colloid. They were then fixed in dilute Karnovsky's fluid, postfixed in 2 per cent osmium tetroxide, dehydrated and embedded in Epon. Sections were contrasted with aqueous uranyl acetate and lead citrate. Large arrows indicate coated regions where the antibody has bound to the plasma membrane and entered peripheral cisternae (small arrow). (Electron micrograph kindly supplied by Professor Colin R. Hopkins, University of Liverpool.)

with which it should be possible to assess the distribution of the receptor in relation to differentiation and to examine the induced synthesis of molecules by EGF.

4. FUTURE DEVELOPMENTS

Culture systems that permit the expression of differentiated functions, make it possible to evaluate the mechanisms controlling differentiation and to obtain large preparations of cells on which to carry out both molecular and biochemical analyses. Antibodies can also be generated to recognize surface determinants on individual cell types, such as basal cells in the skin,[31] thus permitting cell separation studies. It is therefore envisaged that within the next few years there will be rapid advances in our knowledge about the processes of differentiation in the epidermis. Immunocytochemical localization of desmoplaquins in cultured keratinocytes has already demonstrated that there is probably a requirement for new desmosomal protein synthesis during this process.[32]

The separation of basal cells obviously permits the future investigation of the growth capacities of individual cells within the basal layer. It will also be possible, by following basal cells as they differentiate, to look at new surface and cytoplasmic components that are associated with the keratinization process. It would appear likely that further developments of culture systems will not only add rapidly to our knowledge of the changes that take place in malignancy, but also to those processes underlying benign skin conditions.

Having established those moieties which have importance as markers of differentiation, it will then be possible by molecular biological techniques to use these purified cell populations to clone the genes for these proteins. Such cloned genes then provide the probes for in situ hybridization, a technique that on cultured cells will further expand our knowledge of the control of gene expression. Similarly, using cDNA probes for and antibodies directed against different parts of receptors, it will be possible to examine both their cellular localization and how they are controlled.

Thus well-characterized cell lines can provide the systems for studying many aspects of differentiation and malignancy. They also enable us to develop antibodies and molecular probes that will have potential utility in both cell biology and pathology.

Acknowledgements

We would like to thank Mr John Ellis for his excellent photographic assistance.

Appendix

I. INDIRECT IMMUNOPEROXIDASE ON LIVE TISSUE CULTURE CELLS

Localization of membrane or extracellular antigens. For electron microscopy, cells are grown on Thermanox coverslips. All washes and incubations are at $0\,°C$ on ice to prevent internalization of antibodies.

Wash cells in washing buffer (WB):
    RPMI 1640 plus HEPES
    1 per cent BSA,
    1 per cent sodium azide and
    0·02 per cent fetal calf serum.
First antibody, diluted in WB:
    EGFR1, 1 : 1000 *or*   1 h
    Anti-collagen IV, 1 : 50   1 h
Wash: 3 × 5 min in WB   15 min
Second antibody, diluted in WB:
    Peroxidase conjugate, 1 : 25   1 h

Develop:

50 µg 3,3-diaminobenzidine tetrahydrochloride (DAB) in 100 ml PBS plus 100 µl 30 per cent H_2O_2	5 min

For light microscopy, mount in polyvinyl alcohol. For electron microscopy fix in 2 per cent glutaraldehyde, and postfix in 1 per cent osmium tetroxide. Both fixatives are phosphate buffered (pH 7·2, 350 mosmol). Dehydrate in ethanol, embed in Epon/Araldite and polymerize at 60 °C for 16 h. Peel away the coverslips and replace with fresh resin.

II. DOUBLE IMMUNOFLUORESCENCE ON TISSUE CULTURE CELLS

Monoclonal antibodies LICR-LON-M8 (IgG1) and LICR-LON-M18 (IgM) were localized simultaneously on the surface of the breast tumour cell line MCF-7. MCF-7 grown on petriperm and stained at ~ 80 per cent confluence during the log phase of growth. All reagents, washing buffer (WB) and incubations are carried out on ice.

Wash in WB: 3 × 5 min	15 min
First antibodies, applied together:	
LICR-LON-M8 and	
LICR-LON-M18 at 1 : 50 in WB	1 h
Wash in WB: 3 × 5 min	15 min
Second antibodies, fluorescent conjugates, applied together:	
FITC goat anti-mouse IgG1 and	
RITC goat anti-mouse IgM at 1 : 5 in WB	1h
Wash in WB: 3 × 5 min	15 min
Mount in polyvinyl alcohol.	

FITC = fluorescein isothiocyanate
RITC = tetramethylrhodamine isothiocyanate

III. LOCALIZATION OF OESTROGEN RECEPTOR (ER) ON MCF-7 CELLS IN TISSUE CULTURE

Fix MCF-7 cells on *glass coverslips* with 3·6 per cent formol/PBS	10 min
Wash in PBS	10 min
Cold methanol (−20 °C)	4 min
Cold acetone (−20 °C)	1 min
Wash in PBS	2 × 5 min
2 per cent normal goat serum in PBS	15 min
First antibody: anti-ER H222 antibody at 20 µg/ml in PBS	30 min
Wash in PBS	2 × 5 min
Bridge antibody: goat anti-rat IgG diluted 1 : 100 in PBS	30 min
Wash in PBS	2 × 5 min
PAP complex: diluted 1 : 100 in PBS	30 min

Wash in PBS	2×5 min
Develop : DAB (0·625 mg/ml) plus	6 min
hydrogen peroxide (0·06 per cent) in PBS	
Wash in distilled water	
Counterstain in 1 : 100 Harris haematoxylin	5 min
Wash in tap water	5 min
Dehydrate and mount in xylene-soluble mountant.	

IV. REAGENTS

First antibodies :   Polyclonal anti-keratin and type IV collagen antisera were kindly provided by Dr M. Warburton at the Ludwig Institute for Cancer Research. The anti-EGF receptor monoclonal EGFR1 was provided by Dr B. Ozanne, UTHSCD, Dallas, USA.
The monoclonal antibodies LICR-LON-16a and LICR-LON-32a are available on request from Dr B. Gusterson at the Ludwig Institute for Cancer Research.
Antibody ER H222 and immunocytochemistry kit were supplied by Abbot USA-Delkenheim.

Second antibodies : Peroxidase and alkaline phosphatase conjugates were obtained from Sigma. FITC and RITC conjugates were obtained from Dako.

REFERENCES

1. Rheinwald J. G. and Green H. Serial cultivation of strains of human epidermal keratinocytes: the formation of keratinizing colonies from single cells. *Cell* 1975, **6**, 331–344.
2. Rheinwald J. G. and Beckett M. A. Tumorigenic keratinocyte lines requiring anchorage and fibroblast support cultured from human squamous cell carcinomas. *Cancer Res.* 1981, **41**, 1657–1663.
3. Easty D. M., Easty G. C., Carter R. L., Monaghan P. and Butler L. J. Ten human carcinoma cell lines derived from squamous carcinomas of the head and neck. *Br. J. Cancer* 1981, **43**, 772–785.
4. Monaghan P., Knight J., Cowley G. and Gusterson B. Differentiation of a squamous carcinoma cell line in culture and tumorigenicity in immunologically incompetent mice. *Virchows Arch. Abt. B Zellpathol.* (*Cell Pathol*). 1983, **400**, 87–95.
5. Knight J., Gusterson B. A., Cowley G. and Monaghan P. Differentiation of normal and malignant squamous epithelium *in vivo* and *in vitro* — morphological study. Ultrastruct. Pathol. 1985. In press.
6. Fusenig N. E., Breitkreutz D., Dzarlieva R. T., Boukamp P., Bohnert A. and Tilgen W. Growth and differentiation characteristics of transformed keratinocytes from mouse and human skin *in vitro* and *in vivo*. *J. Invest. Dermatol.* 1983, **81**, 168s–175s.
7. Emerman J. T., Enam I. J., Pitelka D. R. and Nandi S. Hormonal effects on intracellular and secreted casein in cultures of mouse mammary epithelial cells on floating collagen membranes. *Proc. Natl Cancer Inst.* 1977, **74**, 4466–4470.
8. Foster C. S., Smith C. A., Dinsdale E. A., Monaghan P. and Neville A. M. Human mammary gland morphogenesis *in vitro*: the growth and differentiation of normal breast epithelium in collagen gel cultures defined by electron microscopy, monoclonal antibodies, and autoradiography. *Dev. Biol.* 1983, **96**, 197–216.
9. Köhler G. and Milstein C. Continuous cultures of fused cells secreting antibody of predefined specificity. *Nature* 1975, **256**, 495–497.

10. Cowley G., Gusterson B. A. and Knight J. Growth in agar and tumor formation in immunologically incompetent mice as criteria for keratinocyte transformation. *Cancer Lett.* 1983, **21**, 95–104.

11. Fuchs E. and Green H. Changes in keratin gene expression during terminal differentiation of the keratinocyte. *Cell* 1980, **19**, 1033–1042.

12. Harding C. R. and Scott I. R. Histidine-rich proteins (Filaggrins): Structural and functional heterogeneity during epidermal differentiation. *J. Mol. Biol.* 1983, **170**, 651–673.

13. Rice R. H. and Green H. Presence in human epidermal cells of a soluble protein precursor of the cross-linked envelope: Activation of the cross-linking by calcium ions. *Cell* 1979, **18**, 681–694.

14. Lazarides E. Intermediate filaments as mechanical integrators of cellular space. *Nature* 1980, **283**, 249–256.

15. Ormerod E. J. and Rudland P. S. Mammary gland morphogenesis *in vitro*: Formation of branched tubules in collagen gels by a cloned rat mammary cell line. *Dev. Biol.* 1982, **91**, 360–375.

16. Moll R., Franke W. W., Schiller D. L., Geiger B. and Krepler R. The catalog of human cytokeratins: patterns of expression in normal epithelia, tumors and cultured cells. *Cell* 1982, **31**, 11–24.

17. Banks-Schlegel S. P., Schlegel R. and Pinkus G. S. Keratin protein domains within the human epidermis. *Exp. Cell Res.* 1981, **136**, 465–469.

18. Sun T-T., Eichner R., Nelson W. G., Scheffer B. A., Tseng C. G., Weiss R. A., Jarvinen M. and Woodcock-Mitchell J. Keratin classes: molecular markers for different types of epithelial differentiation. *J. Invest. Dermatol.* 1983, **81**, 109s–115s.

19. Knight J., Gusterson B. A., Russell-Jones R., Landells W. and Wilson P. Monoclonal antibodies specific for subsets of epidermal keratins: biochemical and immunocytochemical characterization — application in pathology and cell culture. *J. Pathol.* 1985, In press.

20. Bennett D. C., Peachey L. A., Durbin H. and Rudland P. S. A possible mammary stem cell line. *Cell* 1978, **15**, 283–298.

21. Warburton M. J., Ormerod E. J., Monaghan P., Ferns S. and Rudland P. S. Characterization of a myoepithelial cell line derived from a neonatal rat mammary gland. *J. Cell Biol.* 1981, **91**, 827–836.

22. Gusterson B. A., Warburton M. J., Mitchell D., Ellison M., Neville A. M. and Rudland P. S. Distribution of myoepithelial cells and basement membrane proteins in the normal breast and in benign and malignant breast diseases. *Cancer Res.* 1982, **42**, 4763–4770.

23. Edwards P. A. W. Heterogenous expression of cell-surface antigens in normal epithelia and their tumours, revealed by monoclonal antibodies. *Br. J. Cancer* 1985, **51**, 149–160.

24. Dearnaley D. P., Sloane J. P., Ormerod M. G., Steele K., Coombes R. C., Clink H. McD., Powles T. J., Ford H. T., Gazet J.-C. and Neville A. M. Increased detection of mammary carcinoma cells in marrow smears using antisera to epithelial membrane antigen. *Br. J. Cancer* 1981, **44**, 85–90.

25. Jensen E. V., Block G. E., Smith S., Kyser K. and DeSombre E. R. Estrogen receptors and breast cancer response to adrenalectomy. *Natl Cancer Inst. Monogr.* 1971, **34**, 55–70.

26. Greene G. L., Sobel N. B., King W. J. and Jensen E. V. Immunochemical studies of estrogen receptors. *J. Steroid Biochem.* 1984, **20**, 51–56.

27. Waterfield M. D., Mayes E. L. V., Stroobant P., Bennett P. L. P., Young S., Goodfellow P. N., Banting G. S. and Ozanne B. A monoclonal antibody to the human epidermal growth factor receptor. *J. Cell Biochem.* 1982, **20**, 149–161.

28. Gusterson B. A., Cowley G., Smith J. A. and Ozanne B. Cellular localisation of human epidermal growth factor receptor. *Cell Biol. Int. Rep.* **8**, 649–658.

29. Cowley, G., Smith J. A., Gusterson B., Hendler F. and Ozanne B. The amount of EGF receptor is elevated on squamous cell carcinomas. In: Levine A. J., Vande Woude G. F., Topp W. C. and Watson J. D., eds., *Cancer Cells 1. The Transformed Phenotype*. Cold Spring Harbor Symposium, 1984: 5–10.

30. Hopkins C. R. and Trowbridge I. S. Internalization and processing of transferrin and the transferrin receptor in human carcinoma A431 cells. *J. Cell Biol.* 1983, **97**, 508–521.

31. Morhenn V. B., Wood G. S., Engleman E. G. and Oseroff A. R. Selective enrichment of human epidermal cell subpopulations using monoclonal antibodies. *J. Invest. Dermatol.* 1983, **81**, 127s–131s.

32. Watt F. M., Mattey D. L. and Garrod D. R. Calcium-induced reorganization of desmosomal components in cultured human keratinocytes. *J. Cell Biol.* 1984, **99**, 2211–2215.

17   *Immunocytochemistry of Oncogenes*

G. Gastl, J. M. Ward and U. R. Rapp

Oncogenes are the transforming versions of certain cellular genes, the proto-on-cogenes, some of which have been identified as components in the growth regulatory network of normal cells.[1–6] The alteration of proto-oncogenes to transforming genes is the consequence of somatic mutations which occur during the development of natural tumours in experimental animal systems and in man.[3,4,7] Because of the essential role of proto-oncogenes in normal cellular physiology, as well as their subversion to act as stimulators of unregulated growth in tumour cells, it has become important to develop techniques for detection of their products in individual cells of various tissues.

Immunocytochemical and in situ hybridization approaches have recently become applicable to the study of oncogenes due to improvements in methods of tissue preparation and the availability of molecularly cloned oncogene probes as well as oncogene-specific antisera. Using these reagents, oncogene proteins and mRNA were readily detected in cell lines and tissue extracts by immunoprecipitation, immunofluorescence, Northern blotting and quantitative RNA dot blotting and these findings have provided us with the first clues as to their subcellular distribution (*Fig.* 17.1) and relative abundance in different tissues. While most of the data that were obtained with these techniques were average values for mixtures of different cell types, they did suggest, at least for some oncogenes, that their normal precursors were associated with the development and growth state of specific cell lineages.[1,2] Moreover, some tumour types were found to consistently express altered (transforming) forms of specific proto-oncogenes.[3] Because of the cellular complexity of tissues such as bone marrow, where expression of a variety of proto-oncogenes has been observed, as well as cell heterogeneity in natural tumours, a re-examination of oncogene expression at the level of individual cells is

Oncogenes

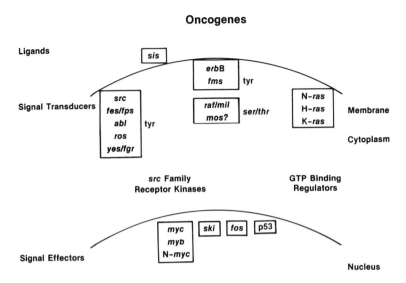

Fig. 17.1. Oncogenes. Genes are grouped according to their amino acid sequence relatedness and location in the cell.

essential. The availability of in situ methods for oncogene detection is also of clinical importance as several blood and epithelial cancers are associated with expression of specific oncogenes that could be monitored during treatment regimens.

1. IN SITU HYBRIDIZATION

In situ hybridization has been applied by several investigators for detecting a variety of mRNAs encoding viral and cellular proteins.[8-33] The ultimate usefulness
of this technique depends on the sensitivity of the method and the accuracy with which signals reflect local concentrations of target RNAs. A major difficulty in the in situ hybridization methods arises from the requirement that the essentially labile target molecules (i.e. RNA) must be retained in a suitable state for hybridization while avoiding artefacts and destruction of cellular morphology. Another factor limiting hybrid formation is the low concentration of mRNA in the cell and steric hindrance from the cell membrane and intracellular structures. Since the technical problems of in situ hybridization are independent of the target mRNA, we present here an overview on methodological data obtained with a variety of probes. (*See also* Chapter 12.)

1.1. Probes

Most recent in situ hybridization studies have used nick-translated recombinant DNAs as probes,[34] whereas earlier studies utilized cDNA probes synthesized from purified mRNA templates.[35] Cox demonstrated a dramatic increase in hybridization efficiency by using asymmetric RNA probes in which only the mRNA coding strand is represented.[30] To prepare large quantities of probes, most investigators used the SP6 bacterial *in vitro* transcription system.[30,36] Single-stranded RNA or DNA probes offer the advantage that there is no re-annealing of probe in solution to compete with hybridization to cellular mRNA. However, the stability of DNA and the general availability of plasmid recombinants make double-stranded DNA probes sufficient for most applications. Using probes labelled with two different isotopes, ^{3}H and ^{35}S, double labelling and colour microradioautography have been successfully employed to identify visna virus RNA in individual cells.[40] In regard to isotopes, there are two advantages of ^{35}S-labelled probes over ^{3}H probes: first the specific activity increases about ten-fold[28] and second the efficiency of grain development is higher for ^{35}S (0·5 grain per disintegration) than for ^{3}H (0·1 grain per disintegration).[39] Fluorochrome-labelled and biotinylated probes are not commonly employed for in situ hybridizations because of increases in background of three to twenty-fold and the fact that sensitivity decreases relative to radioactively labelled probes.[13,32,38]

1.2. Hybridization Conditions

Prehybridizing probes in solution with rRNA, tRNA and addition of single-stranded heterologous DNA to the hybridization mixture reduces background of in situ hybridization.[3,28,32] Hayashi et al. have demonstrated that acetylation of the slide also reduces non-specific background hybridization.[37] Non-specific binding of probes to cell components and the glass slide can also be diminished by pre-treating slides with serum albumin, Ficoll and polyvinylpyrrolidone and by including these components in the hybridization medium.[9,28]

Most in situ hybridizations have been performed in 50 per cent formamide with high salt concentrations at temperatures between 40 and 50 °C. However, Brahic et al.[9] have reported in situ hybridizations using nick-translated cDNA probes in 50 per cent formamide/high salt at temperatures ranging between 20 and 30 °C, yielding a twenty-fold higher hybridization efficiency than at 45 °C without loss of specificity. Using asymmetric RNA probes, Cox has shown a higher thermal stability of RNA–RNA hybrids than of heterologous hybrids with an optimal temperature of 50 °C (4–8 h optimum hybridization time) for in situ hybridization.[30]

The kinetics of DNA–RNA hybrid formation in situ are pseudo-first order with respect to the concentration of probe. The results obtained with asymmetric RNA probes differ somewhat as RNA–RNA hybridization reactions terminate far below saturation of available targets, perhaps due to removal of probe by self-reassociation.[30] For both asymmetric and symmetric probes, kinetic studies have revealed that hybridization is complete within 3–4 h when the reactions are carried out at 37 °C in probe excess (0·6 µg/ml).[30,32]

1.3. Pre-treatment of Cells

To preserve cellular RNA and cell morphology and to maximize the diffusion of probe throughout the cytoplasm, various fixatives such as ethanol/acetic acid, glutaraldehyde, paraformaldehyde, and Carnoy's fixative have been used.[9,13,32,33]

A standard part of almost all published in situ hybridization procedures has been postfixation treatment of cells with proteinase allowing greater penetration of the probe throughout the cellular matrix and better accessibility to mRNA for hybridization.[9] Lawrence and Singer found no advantage to proteolytic digestion after fixation with paraformaldehyde[32] and in fact observed that incubations as short as 5 min in proteinase K (5 µg/ml) could cause loss of more than half of the total cellular RNA. Proteolytic digestion does become necessary, however, if glutaraldehyde is used as a fixative which, while protecting RNA, apparently cross-links the cytoplasm such that RNA is less accessible for hybridization.[32] Following this treatment, samples should be washed in 2 × sodium chloride, sodium citrate (SSC) (150 mM NaCl + 15 mM sodium citrate pH 7·0) in 50 per cent formamide (at 37 °C) for at least 90 min to reduce non-specific background.[32]

1.4. Detection of Hybrid

For autoradiography, slides are dipped in photographic emulsion at 45 °C. Although in principle RNA–DNA as well as RNA–RNA hybrids should be stable under these conditions, hybrids may melt in water at room temperature.[9] Therefore, most investigators dehydrate slides in 300 mM ammonium acetate in ethanol prior to exposure to the photographic emulsion.

1.5. Sensitivity

The number of grains developed after in situ hybridization has been shown to be proportional and in a linear relationship to the amount of hybrid formed, which at saturation is equivalent to the amount of RNA sequences detected.[1,24] By dot-blot determinations of c-myc RNA concentrations in total cellular RNA, Pfeifer-Ohlson et al. demonstrated that c-myc transcripts constituted 0·04–0·05 per cent of the total poly(A) RNA in a 30-day-old human placenta, a concentration that could be readily detected by in situ hybridization using either $d[^{32}P]CTP$ or $d[^{125}I]CTP$ probes with specific activities of $(2–5) \times 10^8$ dis./min per µg DNA.[29]

Southern et al. calculated that the sensitivity of in situ hybridization to tissue sections is reduced two to ten-fold relative to the dot-blot method using ^{32}P probes with specific activities of $(1–5) \times 10^8$ counts/min per µg.[31] Lawrence and Singer[32] found that a $c[^{32}P]DNA$ probe used for in situ detection of actin mRNAs (fragment size ~ 450 base-pairs) with specific activities in the range of $(0·8–2·2) \times 10^8$ counts/min per µg (Cerenkov counts) gave optimal signals at 110 pg probe per 10^5 cells. Thus, these authors estimate, for a sample consisting of 10^5 cells, a detection limit of 20 copies of a 2-kilobase message per cell. With high specific radioactivities of probes labelled with ^3H, ^{125}I, ^{32}P or ^{35}S and with improvements in the hybridization technology itself, sensitivities in the range of 10^{-18}g of specific nucleotide sequence per cell have been accomplished.[32]

In conclusion, the sensitivity of the in situ hybridization allows localization of mRNA transcripts originating from oncogene sequences even though these are present at low abundance. Moreover, this technique offers the possibility to investigate oncogene expression in cells and tissues that are difficult to isolate in pure form or in large quantities. As few as 1000 cells per sample are required, the effort of isolating RNA from each sample is avoided, numerous samples can be assayed simultaneously, results are obtained in quantitative form, and the morphology of single cells can be evaluated in the same samples.

1.6. Applications

Few data about oncogene mRNA expression in single cells are available at present. Applying in situ hybridization techniques, *myc* transcription has been investigated in two different cell systems. Schwab et al. have reported enhanced expression of the cellular oncogene N-*myc* in human neuroblastomas where this gene is amplified.[41] In this study a 2·0-kilobase-pair EcoRI fragment containing the N-*myc* oncogene was labelled with ^{125}I by using random primers and reverse transcriptase. Frozen neuroblastoma tumours were cut into sections, fixed with ethanol/acetic acid, treated with proteinase K and hybridized essentially as described by Brahic and Haase.[9] Tumours exhibited marked heterogeneity in cell morphology and N-*myc* expression, whereas small neuroblastic cells showed amplified N-*myc* expression. No N-*myc* expression was detected in control sections of normal kidney from the area surrounding the neuroblastoma. In contrast, in a neuroblastoma line (Kelly) N-*myc* expression was found uniformly enhanced.[41]

Cellular *myc* oncogene expression has also been analysed in developing human placenta, where *myc* genes were presumably not amplified, using a combination of in situ, dot-blot and Northern blot hybridization.[29] For in situ hybridizations, Pfeifer-Ohlsson used a ^{125}I-labelled DNA probe to demonstrate that cytotrophoblasts from only the first trimester, have active c-*myc* genes. Furthermore, c-*myc* expression correlated with the distribution of replicating cytotrophoblasts identified by [³H]thymidine labelling of explants.

2. IN SITU IDENTIFICATION AND LOCALIZATION OF ONCOGENE-ENCODED PROTEINS BY IMMUNOHISTOCHEMISTRY

2.1. Raising Antibodies to Oncogene-encoded Proteins

The isolation of oncogenes by molecular cloning and characterization of their nucleotide sequences has opened up new strategies for raising antibodies directed against oncogene proteins, which provide exquisite tools for their detection by immunohistochemical and cytochemical techniques. These strategies include the use of defined oligopeptides representing one of a few epitopes of oncogene proteins,[42,43] oncogene protein produced in *Escherichia coli*,[44–47] and the inoculation of cells transformed by a particular oncogene via virus infection or DNA transfection.[48,49] While these reagents are routinely used for oncogene protein detection in established cell lines by immunoprecipitation and indirect immunofluorescence techniques,[1,2,48] immunohistochemical detection in tissue sections and primary cells has only recently been reported for some oncogenes[49,52]

2.2. Antibody Specificity

The most common obstacles encountered using in situ procedures to identify oncogene proteins include (a) cross-reaction of monoclonal and polyclonal antisera with related and unrelated antigens, (b) antisera directed against 'cryptic' determinants, (c) post-translational processing of oncogene products by partial degradation, glycosylation, phosphorylation, acetylation, ribosylation or ubiquinization, and (d) alteration of antigenic determinants by fixation which may affect antibody binding in two ways, either by enhancing specific reactivity because of the unfolding of cryptic epitopes or by abolishing reactivity due to denaturation.[42,51] It is therefore desirable to use a panel of well-characterized antisera for any one of the oncogenes under study in order to assume high sensitivity and specificity. Moreover, several controls are required in order to differentiate non-specific from specific staining with the avidin–biotin complex (ABC) and peroxidase–anti-peroxidase (PAP) methods as is illustrated below.

Focal non-specific staining may be a characteristic of certain antibody preparations and these focal patterns may also occur using specific isotypes of non-immune sera. For example, specific focal staining for an oncogene protein was apparently found in human colon carcinomas (Plate 14), and was not seen with a single control serum, an isotypic IgG, used at the same concentration, yet when a normal serum from another species was utilized, an identical focal pattern was seen. However, normal colon epithelium on the same section never stained with either of the three preparations. It appeared that the isotypic antibody did not produce non-specific staining to the same degree as the 'specific' antibody did at the same IgG concentrations. Unfortunately, absorbed antiserum was not available as a control. In addition, the antibody specific for tumour cells was directed against a cell surface antigen, but did not clearly stain tumour cell membranes, even though we demonstrated that other antibodies against another oncogene protein in these cell membranes were immunoreactive with tumour cell membranes.

Another non-specific staining reaction occurred in one case of human colon carcinoma. Abnormal cells (degenerate, hyperplastic, or neoplastic) frequently appear to bind serum proteins or immunoglobulins non-specifically more than do normal cells. In this case, antibodies to a membrane oncogene protein stained tumour cell membranes (Plate 15), while the isotypic non-specific antibody at the same IgG concentration did not; however, a non-immune serum from another species was reactive in an identical pattern.

We have also noted focal non-specific staining using a variety of antisera to many types of antigen. Appropriate and multiple controls should always be used, especially when characterizing a new antibody. These controls should include the use of (a) affinity-purified antibody or absorbed sera where the former is not available; (b) isotypic immunoglobulin at the same Ig concentration; (c) other antibodies specific for other membrane or non-membrane antigens in the same tissue or slide.

When using the ABC technique and the Vectastain kit (Vector Laboratories, Inc., Burlingame, CA) with some antisera, the membranes of neutrophils or other haematopoietic cells of rodents were sometimes stained brown, even without primary antibody (*Plate* 17). This reaction cannot be blocked by increasing the pH to 9·4, and was sometimes more prominent in Bouin's fixed tissues. Non-specific neutrophil staining was found to be associated with specific batches of the Vectastain kit and perhaps also the use of the secondary antibody.

2.3. Applications

2.3.1. Localization of pp60[c-src] Protein

Recently, Levy et al. demonstrated elevated levels of pp60[c-src] in heart, brain and other neural tissues of chick during embryogenesis.[49] They reacted fixed paraffin tissue sections of chick embryo with an antibody that was raised against pp60[v-src] purified from *E. coli* expressing a cloned v-src gene. A modification of the ABC technique was used to localize pp60[c-src] immunoreactivity.[50,51] Furthermore, immunohistochemical analysis showed that normal pp60[src] appears to be expressed in developing retinal neurones at the onset of differentiation and also in postmitotic mature retinal neurones.[52]

2.3.2. In Situ Demonstration of ras Oncogene Products

Hand et al. have performed ABC immunoperoxidase assays on formalin-fixed paraffin-embedded tissues of human mammary and colon carcinomas using monoclonal antibodies with specificities for Hu-ras[T24] and Hu-ras[Ha] peptides (RAP monoclonal antibodies).[53] They found markedly increased levels of cytoplasmic *ras* protein in single cells and within tissues obtained from tumours compared to the vast majority of benign lesions. Formalin (10 per cent) or glutaraldehyde (2 per cent) fixation was essential to alter the configuration of p21 and thus expose the determinants detected by these monoclonal antibodies. Only weak binding of several anti-*ras* antibodies to undenatured tumour cell extracts in solid phase radioimmunoassays (RIAs) was seen, suggesting poor binding to epitopes in native p21.[53] Furthermore, RAP monoclonal antibodies could not discriminate among Hu-ras products (i.e. Ha-, Ki-, and N-ras) and were therefore classified as 'group-specific' in contrast to 'type-specific' reagents. Using the RAP monoclonal antibodies and the ABC immunoperoxidase method differential *ras* expression was also shown in malignant and benign colonic diseases.[54]

2.3.3. Demonstration of Ras[Ha] p21 in Murine Virus-induced Tumours

Using a polyclonal sheep antiserum against a 20 amino acid peptide of *Ras*[Ha] p21, we were successful in demonstrating cell surface and cytoplasmic *Ras*[Ha] p21 in sarcoma cells induced by Harvey sarcoma/Moloney leukaemia virus in

BALB/c mice (*Plate* 16).[55] This antibody was affinity purified and used successfully at dilutions of up to 1 : 100 (0·7 μg IgG/ml) on tissues fixed in formaldehyde or Bouin's fixatives or on frozen sections. Antigen was not found in mouse fetuses, any normal adult tissues, or in chemically induced tumours, even those containing a transfected *Ras*^Ha oncogene. In contrast, p21 was found in up to 50 per cent of the sarcoma cells in virus-infected mice on the cell membrane and in the cytoplasm (*Plate* 18). In tumours induced in mice by injection of *Ras*^Ha oncogene-transfected 3T3 cells, less than 1 per cent of the cells were immunoreactive (*Plate* 16). In virus-infected mice, a few erythroblasts in the spleen contained membrane p21 (*Plate* 19) and cell membranes of reticular cells in lymph nodes draining the tumours were also occasionally immunoreactive (*Plate* 20). Using a double-staining technique,[56] we also demonstrated Rauscher leukaemia virus p30 in the cytoplasm of splenic megakaryocytes and *Ras*^Ha p21 in erythroblasts of virus-infected mice (*Plate* 21).

2.3.4. Demonstration of v-raf and v-myc Proteins

Using rabbit and goat polyclonal antisera to a C-terminal synthetic peptide (SP-63)[4,57] or the C-terminal two-thirds of the v-*raf*-transforming protein produced in *E. coli* (α PV),[47] the *raf* antigen was localized to tumours only in Bouin's-fixed tissues. Formalin-fixed tumours were not immunoreactive. The protein was localized in the cytoplasm of *raf/myc* recombinant virus-induced lymphoma lymphoblasts[58] (*Plate* 22) and sarcoma cells induced by the v-*raf* oncogene carrying 3611 MSV[58] but was also found on the cell membrane of a small proportion of tumour cells in hepatocellular adenomas induced in mice by diethylnitrosamine (*Plate* 23). Using the same antisera, high levels of *raf* protein expression were also detected in 50 per cent of human lung adeno- and small-cell carcinoma cell lines and biopsies (Linnoila J. D., Minna J. D. and Rapp U. R., 1985, unpublished observations). The *myc*-oncogene protein was very difficult to demonstrate and was found in the nucleus only in a few lymphoma cells induced by the *raf/myc* recombinant J2 virus[58] (*Plate* 17). These proteins were never found in tissues of uninfected control mice.

3. CONCLUSION

The in situ detection of mRNAs and proteins encoded by transforming genes has already accomplished practical applicability for histopathological investigations. At present, RNA and DNA probes provide the most specific and sensitive tools for in situ identification and localization of activated oncogenes. By in situ hybridization techniques even minute amounts of oncogene transcripts can be detected. For application in clinical and diagnostic medicine it might be expected that the sensitivity of hybridization techniques which do not employ isotopes can be considerably increased in the near future.

Poly- and monoclonal antibodies directed against oncogene proteins can be more easily applied for in situ detection of oncogenes. But specificity might be reduced by cross-reactivity and non-specific binding and should therefore be proved by a panel of specific and non-specific antibodies. Thus we might expect that the in situ investigation of oncogenes will further elucidate the role and interaction of potential transforming genes in proliferation, differentiation and transformation of cells from a variety of tissues.

Acknowledgements

We thank J. L. Cleveland for discussions and help with the manuscript.

REFERENCES

1. Bishop J. M. Cellular oncogenes and retroviruses. *Annu Rev. Biochem.* 1983, **52**, 301–354.
2. Bishop J. M. and Varmus H. E. Functions and origins of retroviral transforming genes. In: Weiss R., Teich N., Varmus H. and Coffin J., eds., *Molecular Biology of Tumor Viruses RNA Tumor Viruses.* New York, Cold Spring Harbor Press, 1984: 999–1108.
3. Varmus H. E. The molecular genetics of cellular oncogenes. *Annu. Rev. Genet.* 1984, **18**, 553–612.
4. Rapp U. R., Bonner T. I., Moelling K., Jomsen H. W., Bister K. and Ihle J. Genes and gene products involved in growth regulation of tumor cells. In: *Recent Results in Cancer Research.* Heidelberg, Springer Verlag, 1985: in press.
5. Ihle J. N., Keller J., Rein A., Cleveland J. and Rapp U. Interleukin 3 regulation of the growth of normal and transformed hematopoietic cells. In: Feramisco J., Ozanne B. and Stiles C., eds., *Cancer Cells, Growth Factors and Transformation.* Cold Spring Harbor Conference, 1984: 211–219.
6. Aaronson S. A., Robbins K. C. and Tronick S. R. Human proto-oncogenes, growth factors, and cancer. In: Forol R. J. and Maizel A. L. eds., *Mediators in Cell Growth and Differentiation.* New York, Raven Press, 1985: 241–255.
7. Duesberg P. H. Retroviral transforming genes in normal cells (?). *Nature* 1983, **304**, 219–226.
8. Harrison M. F., Conkie D., Affara N. and Paul J. In situ localization of globin messenger RNA formation. I. During mouse fetal liver development. *J. Cell Biol.* 1974, **63**, 401–413.
9. Brahic M. and Haase A. T. Detection of viral sequences of low reiteration frequency by in situ hybridization. *Proc. Natl Acad. Sci. USA* 1978, **75**, 6125–6129.
10. Saber M. A., Zern M. A. and Shafritz D. Use of in situ hybridization to identify collagen and albumin mRNAs in isolated mouse hepatocytes. *Proc. Natl Acad. Sci. USA* 1983, **80**, 4017–4020.
11. Pachmann K., Pech M., Pachmann U. and Dörmer P. Identification of the messenger (m)RNA coding for the constant fragment (cμ) of the heavy chain with cloned DNA in single cells by in situ hybridization. *Blut* 1983, **46**, 107–110.
12. Hudson P., Penschow J., Shine J., Ryan G., Niall H. and Coghlan J. Hybridization histochemistry: Use of recombinant DNA as 'a homing probe' for tissue localization of specific mRNA populations. *Endocrinology* 1981, **108**, 353–356.
13. Brigati D. J., Myerson D., Leary J. J., Spalholz B., Travis S. Z., Fong C. K. Y., Hsiung G. D. and Ward D. C. Detection of viral genomes in cultured cells and paraffin-embedded tissue sections using biotin-labeled hybridization probes. *Virology* 1983, **126**, 32–50.
14. DeLeon D. V., Cox K. H., Angerer L. M. and Angerer R. C. Most early-variant histone mRNA is contained in the pronucleus of sea urchin eggs. *Dev. Biol.* 1983, **100**, 197–206.
15. Lynn D. V., Angerer L. M., Bruskin A. M., Klein W. H. and Angerer R. C. Localization of a family of mRNAs in a single cell type and its precursors in sea urchin embryos. *Proc. Natl Acad. Sci. USA* 1983, **80**, 2656–2660.
16. Jeffery W. R., Tomlinson C. R. and Brodeur R. D. Localization of actin messenger RNA during early ascidian development. *Dev. Biol.* 1983, **99**, 408–417.
17. Gee C. E., Chen C. C., Roberts J. L., Thompson R. and Watson S. J. Identification of propiomelanocortin neurones in rat hypothalamus by in situ cDNA–RNA hybridization. *Nature* 1983, **306**, 314–316.
18. Edwards M. K. and Wood W. B. Location of specific messenger RNAs in *Caenorhabditis elegans* by cytological hybridization. *Dev. Biol.* 1983, **97**, 375–390.

19. Dzaidek M. A. and Andrews G. K. Tissue specificity of alpha-fetoprotein messenger RNA expression during mouse embryogenesis. *EMBO J.* 1983, **2**, 544–549.

20. Yasuda K., Okuyama K. and Okada T. S. The accumulation of δ-crystallin mRNA in transdetermination and transdifferentiation of neural retina cells into lens. *Cell Differ.* 1983, **12**, 177–183.

21. Berge- Lefranc J. L., Carlouzou G., Bignon C. and Lissitzki S. Quantitative in situ hybridization of ³H-labeled complementary deoxyribonuceic acid (cDNA) to the messenger ribonucleic acid. *J. Clin. Endocrinol. Metab.* 1983, **57**, 470–476.

22. Saber M. A., Shaftriz D. A. and Zern M. A. Changes in collagen and albumin mRNA in liver tissue of mice infected with *Schistosoma mansoni* as determined by in situ hybridization. *J. Cell Biol.* 1983, **97**, 986–992.

23. Benditt E. P., Barrett T. and McKougall J. K. Viruses in the etiology of atherosclerosis. *Proc. Natl Acad. Sci. USA* 1983, **80**, 6386–6389.

24. Jeffery W. R. and Wilson L. J. Localization of messenger RNA in the cortex of *Chaetopterus* eggs and early embryos. *J. Embryol. Exp. Morphol.* 1983, **75**, 225–239.

25. Angerer L. M., DeLeon D. V., Angerer R. C., Showman R. M., Wells D. E. and Raff R. A. Delayed accumulation of maternal histone mRNA during sea urchin oogenesis. *Dev. Biol.* 1984, **101**, 477–484.

26. Gubits R. M., Lynch K. R., Kulkarni A. B., Dolan K. P., Gresik E. W., Hollander P. and Feigelson P. Differential regulation of α 2u globulin gene expression in liver, lachrymal gland, and salivary gland. *J. Biol. Chem.* 1984, **259**, 12803–12809.

27. Jeffery W. R. Spatial distribution of messenger RNA in the cytoskeletal framework of ascidian eggs. *Dev. Biol.* 1982, **103**, 482–492.

28. Zawatzky R., De Maeyer E. and De Maeyer-Guignard J. Identification of individual interferon-producing cells by in situ hybridization. *Proc. Natl Acad. Sci. USA* 1985, **82**, 1136–1140.

29. Pfeifer-Ohlsson S., Goustin A. S., Rydnert J., Waldeström T., Bjersing L., Stehelin D. and Ohlsson R. Spatial and temporal pattern of cellular *myc* oncogene expression in developing placenta: Implications for embryonic cell proliferation. *Cell* 1984, **38**, 585–596.

30. Cox K. H., DeLeon D. V., Angerer L. M. and Angerer R. C. Detection of mRNAs in sea urchin embryos by in situ hybridization using asymmetric probes. *Dev. Biol.* 1984, **101**, 485–502.

31. Southern P. J., Blount P. and Oldstone M. B. A. Analysis of persistent virus infections by in situ hybridization to whole-mouse sections. *Nature* 1984, **312**, 555–558.

32. Lawrence J. B. and Singer R. H. Quantitative analysis of in situ hybridization methods for the detection of actin gene expression. *Nucleic Acids Res.* 1985. **13**, 1777–1799.

33. Breborowicz J. and Tamaoki T. Detection of messenger RNAs of α-fetoprotein and albumin in a human hepatoma cell line by *in situ* hybridization. *Cancer Res.* 1985, **45**, 1730–1736.

34. Rigby P. W. J., Dieckmann M., Rhodes C and Berg B. Labeling deoxyribonucleic acid to high specific activity in vitro by nick-translation with DNA polymerase I. *J. Mol. Biol.* 1977, **113**, 237–251.

35. Harding J. D., McDonald R. J., Przybyla A. E., Chirgwin J. M., Pictet R. L. and Rutter W. J. Changes in the frequency of specific transcripts during development of the pancreas. *J. Biol. Chem.* 1977, **252**, 7391–7397.

36. Butler E. and Chamberlin M. J. Bacteriophage SP6-RNA polymerase. Isolation and characterization of the enzyme. *J. Biol. Chem.* 1982, **257**, 5772–5778.

37. Hayashi S., Gilliam I. C., Delaney A. D. and Tener G. M. Acetylation of chromosome squashes of *Drosophila melanogaster* decreases the background of autoradiographs from hybridization with ¹²¹I-labeled RNA. *J. Histochem. Cytochem.* 1978, **26**, 677–679.

38. Baumann J. G. J., Wiegant J. and Van Duijn P. Cytochemical hybridization with fluorochrome-labeled RNA. II. Applications. *J. Histochem. Cytochem.* 1981, **29**, 238–246.

39. Rogers A. W. *Techniques of Autoradiography*, 3rd edn. New York, Elsevier/North Holland, 1979.

40. Haase A. T., Walker D., Stowring L., Venluru P., Gabella A., Blum H., Brahic M., Goldberg R. and O'Brien K. Detection of two viral genomes in single cells by double-label hybridization in situ and color microradioautography. *Science* 1985, **227**, 189–192.

41. Schwab M., Ellison J., Busch M., Rosenau W., Varmus H. E. and Bishop J. M. Enhanced expression of the human gene N-*myc* consequent to amplification of DNA may contribute to malignant progression of neuroblastoma. *Proc. Natl Acad. Sci. USA* 1984, **81**, 4940–4944.

42. Lerner R. A. Tapping the immunological repertoire to produce antibodies of predetermined specificity. *Nature* 1982, **299**, 592–596.

43. Walter G., Scheidtmann K. H., Carbone A., Laudano A. P. and Doolittle R. F. Antibodies specific for the carboxy- and amino-terminal regions of simian virus 40 large tumor antigen. *Proc. Natl Acad. Sci. USA* 1980, **77**, 5197–5200.

44. Privalsky M. L., Sealy L., Bishop J. M., McGrath J. P. and Levinson A. D. The product of the avian erythroblastosis virus *erbB* locus is a glycoprotein. *Cell* 1983, **32**, 1257–1267.

45. Klempnauer K. H., Ramsay G., Bishop J. M., Giovanella M., Moscovici C., McGrath J. P. and Levinson A. The product of the retroviral transforming gene *v-myb* is a truncated version of the protein encoded by the cellular oncogene *c-myb*. *Cell* 1983, **33**, 345–355.

46. Gilmer T. M. and Erikson R. L. Development of anti-pp60^{src} serum antigen produced in *E. coli*. *J. Virol.* 1983, **45**, 462–465.

47. Rapp U. R., Schultz A. and Oppermann H. Preparation of *raf* oncogene precipitating antisera with *raf* protein produced in *E. coli*. Submitted for publication.

48. Hunter, T. The proteins of oncogenes. *Sci. Am.* 1984, **251**, 70–79.

49. Levy B. T., Sorge L. K., Meymandi A. and Maness P. F. pp^{c-src} kinase is in chick and human embryonic tissues. *Dev. Biol.* 1984, **104**, 9–17.

50. Hsu S., Raine L. and Fanger H. Use of avidin–biotin–peroxidase complex (ABC) in immunoperoxidase techniques: A comparison between ABC and unlabeled antibody (PAP) procedures. *J. Histochem. Cytochem.* 1981, **29**, 577–580.

51. Towle A. C., Lauder J. M. and Joh T. H. Optimization of tyrosine hydroxylase immunocytochemistry in paraffin sections using pretreatment with proteolytic enzymes. *J. Histochem. Cytochem.* 1985, in press.

52. Sorge L. K., Levy B. T. and Maness P. F. pp60^{c-src} is developmentally regulated in the neural retina. *Cell* 1984, **36**, 249–257.

53. Hand P. H., Thor A., Wunderlich D., Muraro R., Caruso A. and Schlom J. Monoclonal antibodies of predefined specificity detect activated *ras* gene expression in human mammary and colon carcinomas. *Proc. Natl Acad. Sci. USA* 1984, **81** 5227–5231.

54. Thor A., Hand P. H., Wunderlich D., Caruso A., Muraro R. and Schlom J. Monoclonal antibodies define differential *ras* gene expression in malignant and benign colonic disease. *Nature* 1984, **311**, 562–565.

55. Ward J. M., Pardue R., Takahashi K., Shih T. and Weislow O. Immunocytochemical localization of *Ras*^{Ha} p21 in fixed tissue sections of experimentally induced tumors of mice and rats. *Proc. Am. Assoc. Res. Cancer* 1985, **26**, 62.

56. Hsu S. M. and Soban E. Color modification of diaminobenzidine (DAB) precipitation by metallic ions and its application for double immunohistochemistry. *J. Histochem. Cytochem.* 1982, **30**, 1079–1082.

57. Schultz A. M., Copeland T. D., Mark G. E., Rapp U. R. and Oroszlan S. Detection of the myristilated *gag-raf* transforming protein with *raf*-specific antipeptide sera. *J. Virol.* 1985, in press.

58. Rapp U. R., Cleveland J. L., Fredrickson T. N., Holmes K. L., Morse H. C. III, Jansen H. W., Patschinsky T. and Bister K. Rapid induction of hemopoietic neoplasms in newborn mice by a *raf (mil)/myc* recombinant murine retrovirus. Submitted for publication.

18

Immunocytochemical Localization of Intrinsic Amines

A. A. J. Verhofstad and
H. W. M. Steinbusch

The intrinsic amines dopamine, noradrenaline, adrenaline, serotonin and histamine are found in many mammalian as well as non-mammalian species. In order to assess their biological significance a brief introductory review will be given of the distribution of cells containing these compounds in mammals.

1. DISTRIBUTION IN MAMMALS

Dopamine and Noradrenaline

Both dopamine and noradrenaline have been found in the central and peripheral nervous systems. In the central nervous system they are localized in neurones.[1-3]

From the cell bodies which are concentrated in a limited number of areas of the brain stem, nerve processes are distributed over the brain and spinal cord. In the peripheral nervous system noradrenaline is present in the sympathetic postganglionic neurones[4] with cell bodies in the ganglia of the sympathetic trunk and the prevertebral autonomic nerve plexuses distributing axons to organs controlled by the sympathetic nervous system. Small ganglionic cells (small intensely fluorescent cells) containing dopamine or noradrenaline have been described in the autonomic ganglia.[5-6] Outside the nervous system, high concentrations of noradrenaline have been found in the adrenal medulla of many species (in some species, e.g. rabbit and guinea-pig, concentrations are extremely low) as well as in extra-adrenal medullary tissues (chromaffin tissues or chromaffin paraganglia).[7]

The carotid body and related structures (non-chromaffin paraganglia) seem to contain substantial amounts of dopamine and noradrenaline.[8,9]

Adrenaline

Adrenaline is mainly detected in the adrenal medulla. Thus the adrenal medulla is able to store adrenaline as well as noradrenaline. Remarkable species differences with regard to the adrenaline:noradrenaline ratio have been reported,[7,10] correlating well with the separate adrenaline and noradrenaline medullary cells that can be distinguished by several methods.[10,11]

Recently, adrenaline-containing neurones have been demonstrated in the central nervous system, their cell bodies being located in the lower brain stem.[12]

Serotonin

Serotonin is also found in several organs. In the central nervous system serotoninergic cell bodies have been described in the brain stem,[1,2,13,14] their processes widely spread over the brain and spinal cord. A limited number of serotonin-containing neurones are supposed to be involved in the innervation of the gastrointestinal tract.[15–17] However, most of the gut serotonin is stored in the enterochromaffin cells.[18–20]

There is also evidence of serotonin in the so-called 'neuroepithelial bodies' of the lung[21,22] and in endocrine organs, e.g. in the thyroid gland.[23] Finally, in certain species (rat, mouse), mast cells contain serotonin.[24]

Histamine

Histamine has been demonstrated in the central nervous system, especially in the hypothalamus. Neurochemical studies on brain tissue indicate that it is probably localized in neuronal cells.[25] Biochemical as well as histochemical procedures have shown the presence of histamine in mast cells and in the gastrointestinal mucosa, particularly in the acid-producing part of the stomach (enterochromaffin-like cells).[26]

Thus dopamine, noradrenaline, adrenaline, serotonin and histamine are distributed throughout the organism, stored in different cell types. Consequently, one might expect these compounds to be involved in numerous regulatory functions, either as neurotransmitters or as hormones or hormone-like substances.

In addition, tumours have been described that contain either serotonin[27] (carcinoids) or noradrenaline and/or adrenaline[28] (phaeochromocytomas). These tumours are supposedly derived from the enterochromaffin cells or from the adrenal medulla and extra-adrenal medullary tissues, respectively.

It is now generally accepted that *dopamine, noradrenaline* and *adrenaline* have common precursors and are derived from the same amino acid, L-tyrosine[29,30] (*Fig.* 18.1). Consequently, cells synthesizing noradrenaline, for example, contain tyrosine, dopa (3,4-dihydroxyphenylalanine) and dopamine (3,4-dihydroxy-

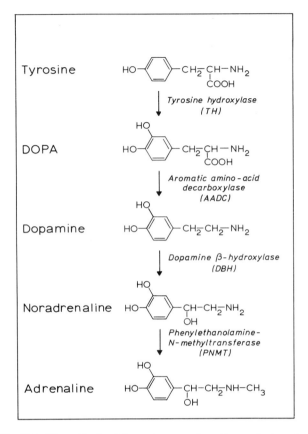

Fig. 18.1. Biosynthesis of dopamine, noradrenaline and adrenaline.

phenylethylamine), as well as the biosynthetic enzymes tyrosine hydroxylase, aromatic L-amino acid decarboxylase (dopa decarboxylase) and dopamine β-hydroxylase, while adrenaline and the corresponding enzyme, phenylethanol-amine *N*-methyltransferase, are absent. On the other hand, cells synthesizing adrenaline contain all components of the biosynthetic pathway. Thus, noradrena-line is present in the adrenaline-synthesizing as well as in the noradrenaline-syn-thesizing cells. In the first case it forms an intermediate step to adrenaline, in the second the end-product. Therefore, one might expect the noradrenaline concentration in the noradrenaline-synthesizing cells to be much higher than in the adrenaline-synthesizing cells. Comparable remarks can be made concerning the dopamine concentration in cells synthesizing dopamine, noradrenaline or adrenaline.

The biosynthetic pathway of *serotonin* is indicated in *Fig.* 18.2. Serotonin is the end-product of a pathway starting with the conversion of the amino acid L-tryptophan into 5-hydroxytryptophan.[31] Finally, 5-hydroxytryptophan is converted to serotonin (5-hydroxytryptamine).

It has been claimed that both pathways (*Figs.* 18.1 and 18.2) have one enzyme in common,[32] i.e. aromatic L-amino acid decarboxylase (dopa decarboxylase).

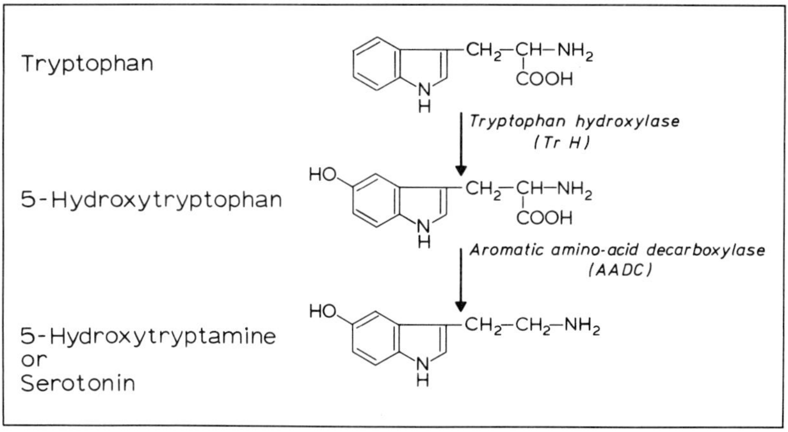

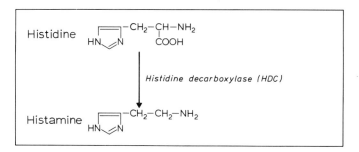

Fig. 18.2. Biosynthesis of serotonin.

Fig. 18.3. Biosynthesis of histamine.

However, there are other reports indicating that two different types of decarboxylase exist.[33,34]

Histamine is derived from the amino acid L-histidine (*Fig.* 18.3).[25] This conversion is catalysed by a specific enzyme, L-histidine decarboxylase, which is different from L-amino acid decarboxylase.

2. IDENTIFICATION OF CELLS SYNTHESIZING INTRINSIC AMINES

Cells synthesizing dopamine, noradrenaline, adrenaline, serotonin, or histamine can now be identified by applying antibodies either to the biosynthesizing enzymes or to the end-products.

2.1. The First Approach

The first approach (antibodies to the synthesizing enzymes) was introduced in the late sixties and early seventies[35-40] and has remained in use.[41-43] Keeping in mind the biosynthetic pathway, it will be evident that at least three types of antibodies are required to determine whether a particular cell or cell group contains *dopamine, noradrenaline or adrenaline*. In practice, three adjacent serial sections are stained with an antibody to tyrosine hydroxylase, dopamine β-hydroxylase and phenylethanolamine *N*-methyltransferase, respectively.[39] If all reactions are positive, one is dealing with an adrenaline-synthesizing cell. Alternatively, cells only immunoreactive to tyrosine hydroxylase and dopamine β-hydroxylase supposedly synthesize noradrenaline. Cells only immunoreactive to tyrosine hydroxylase probably produce dopamine. For the *serotonin-synthesizing cells* antibodies to both tryptophan hydroxylase[43–45] and aromatic L-amino acid decarboxylase[38] have been used. The latter antibody also recognizes the decarboxylating enzyme of the dopamine, noradrenaline and adrenaline pathways and can therefore only be used as a marker for serotonin synthesis if other indicators are also available. Recently, antibodies to histidine decarboxylase have allowed the detection of *histamine-synthesizing cells*.[46,47] Application of antibodies to the synthesizing enzymes has proved to be a great achievement. However, there are several practical and fundamental limitations. Firstly, in order to raise antibodies, enzymes have to be isolated and purified as far as possible without disturbing their immunogenic properties. Secondly, antibodies to the biosynthesizing enzymes tend to be species specific.[48] Thirdly, immunoreactivity solely indicates that a particular compound can be synthesized. Whether it is really produced cannot be determined.

2.2. The Second Approach

The second approach, i.e. the use of antibodies to the end-products, was first reported in 1978 (antibodies to serotonin).[49] Since then, immunocytochemical applications of antibodies to other amines have been described. Although not all methodological problems have been solved, so far the results obtained clearly show the major advantage of this approach. Immunoreactions have been observed in a variety of species, demonstrating that application of these types of antibodies is not hampered by species specificity. Thus, one of the drawbacks of the first approach has been overcome. However, it will be obvious that antibodies to dopamine, noradrenaline, adrenaline, serotonin and histamine can only be used to reveal the presence or absence of these compounds. Whether these compounds are really synthesized at the site or merely stored cannot be determined, unless, in addition, antibodies to the biosynthesizing enzymes are used. Thus, the first and second approaches are complementary to each other. Ideally, they should be used simultaneously.

3. IMMUNOCYTOCHEMISTRY USING ANTIBODIES TO DOPAMINE, NORADRENALINE, ADRENALINE, SEROTONIN AND HISTAMINE

The detection of compounds of low molecular weight by antibodies shares problems generally encountered in immunocytochemistry. However, there are two problems to be mastered in particular, namely the production of specific antibodies, and the proper preservation of the substances to be demonstrated. This section is aimed at summarizing the results obtained so far.

3.1. Preparation of Antisera

Since dopamine, noradrenaline, adrenaline, serotonin and histamine are present in many species, they are not regarded as foreign materials by the immune apparatus. In addition, the molecules are too small to be immunogenic, but might present immunogenic determinants after linkage to a foreign carrier protein[50] as for many non- or weakly immunogenic substances. The specificities of antibodies raised to noradrenaline, adrenaline, serotonin and histamine have been tested by bioassay, complement fixation tests, immunoprecipitation, immunoelectrophoresis and/or radioimmunoassay.[51–61] The first reports on the applicability of these antibodies in immunocytochemistry appeared in 1978 (serotonin[49]) and 1980 (noradrenaline and adrenaline[62]). Since then, other reports dealing with the development of immunocytochemical techniques for serotonin, dopamine and histamine have been published.[63–67] Obviously, conjugates prepared by linkage of haptens to a carrier protein carry different types of antigenic determinants, related partly to the hapten and partly to the original carrier protein. In some papers the proposed structure of the immunogen synthesized was indicated.[52,53–57,59–61,68,69] However, a detailed characterization of the immunogens currently used for the raising of antibodies is lacking.

So far, the best immunocytochemical results have been obtained when immunogens were prepared using formaldehyde, glutaraldehyde or mixtures of formaldehyde and glutaraldehyde as coupling reagents. Different carrier proteins have been employed, e.g. bovine serum albumin, thyroglobulin and haemocyanin. Usually, rabbits were used for immunization. If the antibodies are raised by bovine serum albumin conjugates, antibodies to the carrier protein should be removed by liquid phase absorption to eliminate unwanted staining of collagen-rich tissues, e.g. blood vessel walls.

3.2. Preparation of the Tissues

Successful immunostaining requires antibodies of high titre and avidity and of well-defined specificity. The antigen must not suffer loss of concentration or antigenicity and the tissue structure must be preserved with as little unwanted attachment of immunoglobulins as possible.

So far, two preparatory steps seem to be crucial, namely *fixation* and (except for isolated cells) *embedding*. Acceptable results have been obtained with both indirect immunofluorescence[70] and immunoperoxidase.[71,72]

3.2.1. *Fixation*

Unfixed tissues (cryostat sections) did not reveal any immune reactions. Staining results were only obtained with fixatives containing the same aldehyde(s) utilized as a coupling reagent during the preparation of the immunogen, although different tissues may show different fixation requirements. Thus, epithelial cells containing serotonin, noradrenaline or adrenaline showed intense cytoplasmic staining if fixed with 4 per cent paraformaldehyde in 0·1 M sodium phosphate buffer pH 7·3. Neuronal cells containing dopamine, noradrenaline or histamine were best demonstrated following fixation with mixtures of paraformaldehyde, glutaraldehyde and picric acid in the same buffer.[73] Neuronal cells containing serotonin stained equally well when fixed in this mixture or in buffered 4 per cent paraformaldehyde. Although perfusion fixation is preferable, immersion fixation of small tissue samples can be used as well. None of the above fixatives seemed to retain a sufficient quantity of adrenaline in neuronal cells indicating that the matter of fixation is not yet fully understood.

Following fixation and rinsing (in the present study a buffer containing 5 per cent sucrose was used) a number of methods can be used to produce *tissue sections*, namely frozen, paraffin or plastic sections and sections cut with a vibrating knife microtome. Examples are illustrated in *Figs*. 18.4–18.11.

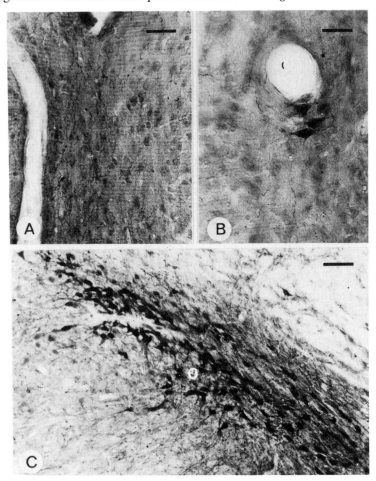

Fig. 18.4. Photomicrographs of rat brain fixed by perfusion with 4 per cent paraformaldehyde–0·05 per cent glutaraldehyde–0·2 per cent picric acid. Transverse Vibratome sections (thickness 50 μm) stained by the peroxidase–anti-peroxidase technique using an antiserum to dopamine. Dopamine-immunoreactive fibres in the nucleus raphe dorsalis (*A*), dopaminergic cell bodies and varicose fibres in the periventricular region of the thalamus (*B*) and the substantia nigra (*C*). Note in (*A*) the lack of staining of cell bodies known to contain serotonin. Bars: (*A*) 35 μm, (*B*) 85 μm and (*C*) 15 μm.

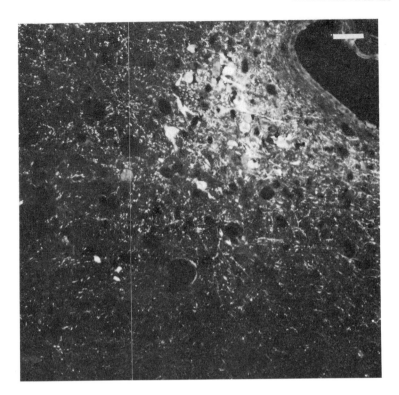

Fig. 18.5. Photomicrograph of rat brain fixed by perfusion with 4 per cent paraformaldehyde–0·05 per cent glutaraldehyde–0·2 per cent picric acid. Transverse cryostat section (thickness 10 μm) stained by indirect immunofluorescence with an antiserum to noradrenaline. Noradrenaline-immunoreactive cell bodies and varicose fibres in the area of the locus coeruleus. Bar: 85 μm.

3.2.2. Frozen Sections

Most of the present observations are based upon study of cryostat sections (*Figs.* 18.5, 18.6, 18.7, 18.8*A*, 18.8*B*, 18.9, 18.11*A*, 18.11*B*). Tissues can be frozen on cryostat chucks either with powdered carbon dioxide or with isopentane cooled by liquid nitrogen (used at a temperature of −150 °C). Pending use they can be stored at −70 °C or lower.

Serial sections of 7–10 μm in thickness, with reasonably good structural preservation, can be obtained easily. During freezing and sectioning cellular membranes are disrupted. This improves penetration by antibodies during staining. However, one must also consider the possible loss of antigenic substances during the fixation, sectioning and staining procedure. Using model experiments, Schipper and Tilders[74] estimated that about 50 per cent of serotonin was retained during fixation and rinsing whereas other substances were completely washed out. Frozen sections cut in a cryostat or on a freezing microtome should be as thin as possible, firstly to aid penetration of antibodies and, secondly, because under high power thick sections appear less brilliant and

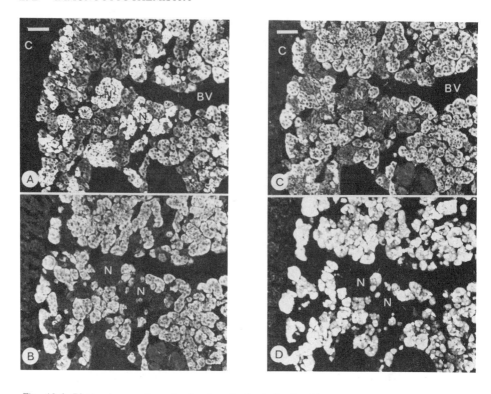

Fig. 18.6. Photomicrographs of rat adrenal gland fixed with 4 per cent paraformaldehyde. Consecutive cryostat sections (thickness 7 μm) stained by indirect immunofluorescence using antisera to noradrenaline (*A*), adrenaline (*B*), bovine dopamine β-hydroxylase (*C*) and bovine phenylethanolamine *N*-methyltransferase (*D*). The adrenal medulla shows noradrenaline-storing cells [positive in (*A*) and (*C*), negative in (*B*) and (*D*); two groups are indicated by N], and adrenaline-storing cells [positive in (*A*)–(*D*)]. C, cortex; BV, blood vessels. Bars: 35 μm.

distinct than thin sections, due to the relatively low depth of the field, precluding detailed high power observations.

3.2.3. *Paraffin Sections*

It would be most rewarding if tissues could be embedded in paraffin without loss of immunostaining properties. Serial cryostat sections of small and fragile tissues are hard to achieve and, in general, cellular details are badly preserved. Hence, if different cell types occur in a particular area, it might be impossible to estimate which particular cells contain the immunoreactive material. Finally, the ability to carry out retrospective pathology would be helpful. Preliminary results show that

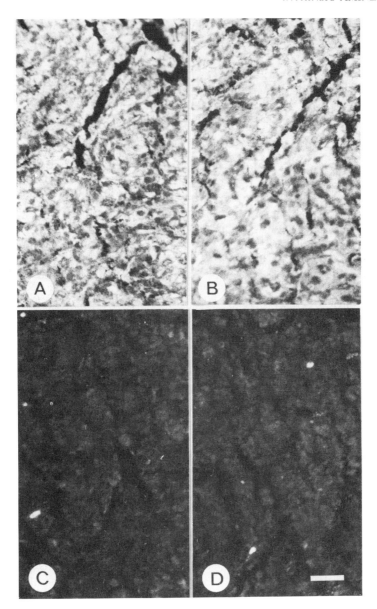

Fig. 18.7. Photomicrographs of an extra-adrenal phaeochromocytoma (presumably a tumour of the organs of Zuckerkandl) fixed by immersion with 4 per cent paraformaldehyde. Consecutive cryostat sections (thickness 7 μm) stained by indirect immunofluorescence with antisera to bovine dopamine β-hydroxylase (*A*), noradrenaline (*B*), bovine phenylethanolamine *N*-methyltransferase (*C*) and adrenaline (*D*). Tumour cells seem to store only noradrenaline [only (*A*) and (*B*) are positive]. Bar: 43 μm.

deparaffinized sections of epithelial tissues containing noradrenaline, adrenaline or serotonin give acceptable results if properly fixed, rinsed etc. (*Fig.* 18.10).

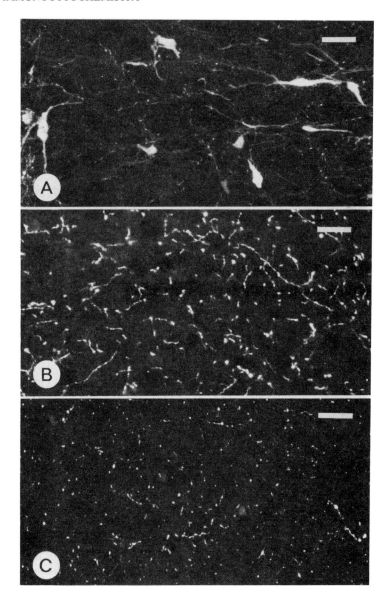

Fig. 18.8. Photomicrographs of rat brain fixed by perfusion with 4 per cent paraformaldehyde. Cryostat sections (thickness 7 μm; *A* and *B*) or Epon section (thickness 1 μm; *C*) stained by indirect immunofluorescence using an antiserum to serotonin. Immunoreactive cell bodies in the brain stem (*A*) and nerve fibres (*B,C*). Bars: (*A*) 35 μm and (*B,C*) 14 μm.

However, much less immunoreactivity can be observed in noradrenergic or serotoninergic neurones in paraffin-embedded tissues, perhaps because antigenic material is lost and/or its antigenicity is impaired during processing. Further exploration of the possible use of paraffin sections is needed.

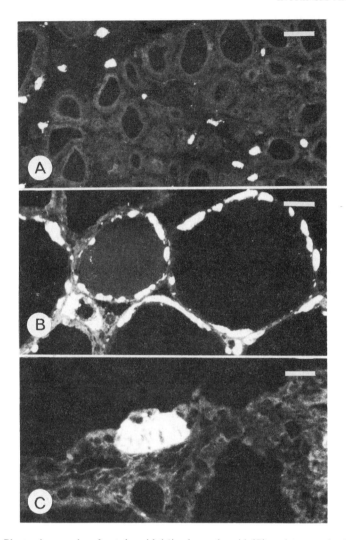

Fig. 18.9. Photomicrographs of rat thyroid (*A*), sheep thyroid (*B*) and neonatal rabbit lung (*C*). Tissues fixed by perfusion (*A,C*) or immersion (*B*) with 4 per cent paraformaldehyde. Cryostat sections (thickness 7 μm) stained by indirect immunofluorescence with an antiserum to serotonin. Note that in the rat thyroid (*A*) only mast cells are stained, while in the thyroid gland of the sheep (*B*) parafollicular cells are serotonin-immunoreactive. A neuroepithelial body is shown in the bronchiolar epithelium of the lung (*C*). Bars: (*A,B*) 43 μm and (*C*) 28 μm.

3.2.4. Plastic Sections

If the antigen survives the preparatory steps, thin sections (1–2 μm) of plastic-embedded tissues immunostained after removal of the plastic show structural preservation and resolution superior to that of paraffin sections and may allow a combined light and electron microscopical study on the same tissue block. In the present study, paraformaldehyde-fixed, Epon-embedded tissues

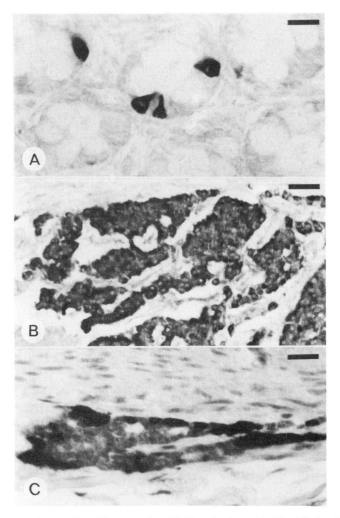

Fig. 18.10. Photomicrographs of human colon (*A*,*B*) and urinary bladder (*C*) fixed by immersion with 4 per cent paraformaldehyde. Paraffin sections (thickness 10 μm) stained according to the peroxidase–anti-peroxidase technique using an antiserum to serotonin. Section (*C*) counterstained with haematoxylin. Typical serotonin-immunoreactive (enterochromaffin) cells are present in the mucosa (*A*). Immunoreactivity in carcinoid tissue invading the external muscular layer. (*B*). Tumour cells in the muscular layer of the urinary bladder (*C*). These tumour cells are only partly serotonin-immunoreactive. Bars: (*A*,*C*) 17 μm and (*B*) 44 μm.

were stained following the removal of Epon by sodium methoxide.[75] Initial experiments revealed serotonin immunoreactivity in neuronal (*Fig.* 18.8*C*) as well as in epithelial cells known to contain serotonin.

3.2.5. *Sections cut with a Vibrating Knife Microtome*

A vibrating knife microtome is a combination of a tissue slicer and a microtome.

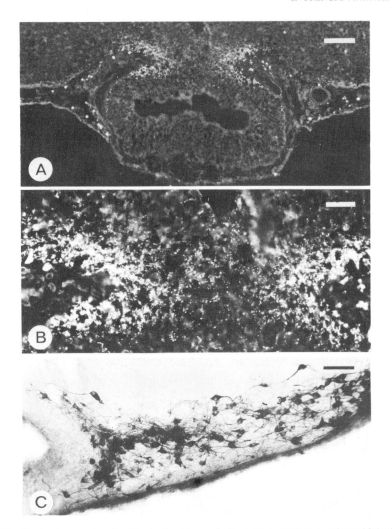

Fig. 18.11. Photomicrographs of rat brain fixed by perfusion with 4 per cent paraformaldehyde–0·05 per cent glutaraldehyde–0·2 per cent picric acid. Transverse cryostat (*A*,*B*; thickness 10 μm) or Vibratome (*C*; thickness 50 μm) sections stained by indirect immunofluorescence (*A*,*B*) or the peroxidase–anti-peroxidase technique (*C*) with an antiserum to histamine. Histamine-immunoreactive varicose fibres in the dorsal labium of the infundibulum (*A*,*B*) and histamine-immunoreactive cell bodies in the nucleus caudalis magnocellularis (*C*). Bars: (*A*) 15 μm, (*B*) 35 μm and (*C*) 85 μm.

By means of a vibrating knife (usually a disposable safety razor blade) serial sections of unfrozen and unembedded tissues can be cut along a desired plane of sectioning. This procedure avoids disruption of cellular membranes by freezing and thawing, and the loss of antigenic material during dehydration etc. Unfortunately, producing sections is time consuming. Moreover, not all tissues are easily cut and thin sections are hard to produce. As a consequence, in immunocytochemistry, vibrating knife microtome sections are mainly used for studying the spatial distribution of immunoreactive material (*Figs.* 18.4, 18.11*C*).

Immunoperoxidase-stained sections (50–100 μm) appeared extremely useful for examining the dendritic pattern of serotoninergic neurones of the brain stem.[76] Sections cut with a vibrating knife microtome also proved to be suitable for the demonstration of serotoninergic nerve processes by electron microscopial immunocytochemistry (pre-embedding technique).[77]

3.3. Staining Methods

In the present study both indirect immunofluorescence[70] and immunoperoxidase (peroxidase–anti-peroxidase)[71,72] techniques were employed.

3.4. Specificity

An antiserum may only be used as a specific marker substance, if cross-reactivity to tissue components other than the ones to be studied has been ruled out. Consequently, besides immunological controls, tissues either containing or devoid of a particular substance should be examined. Since fixation, embedding etc. might change the antigenicity of the tissue, the specificity of an antiserum is only relevant under the conditions employed during tissue preparation.

Inevitably, antisera raised by complexes obtained by linkage of haptens to a carrier protein are heterogeneous. Only those antibodies recognizing the haptens or hapten-derived compounds are relevant. As was indicated earlier, routinely, all sera raised with bovine serum albumin conjugates were treated with the carrier protein (5 mg bovine serum albumin per ml antiserum) to absorb antibodies immunoreactive to the carrier substance. Several *immunological control* tests were performed. *Firstly*, pre-immune (non-immune) instead of immune serum was used. *Secondly*, serum samples were absorbed with different concentrations of conjugates of dopamine, noradrenaline, adrenaline, serotonin or histamine. *Thirdly*, in the case of serotonin antisera, the binding capacity of different haptens was examined by blocking (inhibition) experiments. Both absorption and blocking experiments were done by incubating serum samples with different concentrations of the conjugates or haptens for 1 h at 37 °C, followed by incubation for 18 h at 4 °C. Before use, all sera were centrifuged for 20 min at 20 000 × g and 4 °C to remove any precipitate. All three controls were tested by indirect immunofluorescence on appropriately fixed rat tissues known to contain either dopamine, noradrenaline, adrenaline, serotonin or histamine. The results clearly indicate that the presented staining results are due to antibody binding [method specificity, i.e. no staining with pre-immune (non-immune) sera]. From the absorption experiments (*Table* 18.1), it seems that the antibodies raised are highly specific to the hapten portion of the immunogen (antibody specificity). However, as is shown by an example of blocking experiments (*Table* 18.2), serotonin-immunocytochemical staining can be blocked by several substances including dopamine. It is questionable whether compounds like 5-methoxytrypt-amine are real constituents of the brain.[78] Since no staining was observed in areas containing dopaminergic neurones (*see below*), cross-reactivity to dopamine does not seem to be of importance in immunocytochemistry, at least with the present methods of fixation etc. Interestingly, 6-hydroxy-1,2,3,4-tetrahydro-β-carbo-

Table 18.1. *Absorption experiments*

Dopamine-antisera, tested on				☐ Rat medulla oblongata	
				☐ Rat hypothalamus	
Absorbent*	BSA–DA	BSA–NA	BSA–A	BSA–5-HT	BSA–Hist
Fluorescence	negative	normal	normal	normal	normal

Noradrenaline-antisera, tested on				☐ Rat medulla oblongata	
				☐ Rat ductus deferens	
				☐ Rat adrenal medulla	
Absorbent*	BSA–NA	BSA–DA	BSA–A	BSA–5-HT	BSA–Hist
Fluorescence	negative	normal	normal	normal	normal

Adrenaline-antisera, tested on				☐ Rat adrenal medulla	
Absorbent*	BSA–A	BSA–DA	BSA–NA	BSA–5-HT	BSA–Hist
Fluorescence	negative	normal	normal	normal	normal

Serotonin-antisera, tested on				☐ Rat medulla oblongata	
				☐ Rat duodenum	
Absorbent*	BSA–5-HT	BSA–DA	BSA–NA	BSA–A	BSA–Hist
Fluorescence	negative	normal	normal	normal	normal

Histamine-antisera, tested on				☐ Rat hypothalamus	
Absorbent*	BSA–Hist	BSA–DA	BSA–NA	BSA–A	BSA–5-HT
Fluorescence	negative	normal	normal	normal	normal

*2 mg/ml optimally diluted antiserum.
BSA, bovine serum albumin.
DA, dopamine.
NA, noradrenaline.
A, adrenaline.
5-HT, 5-hydroxytryptamine (serotonin).
Hist, histamine.

line, the formaldehyde derivative of serotonin, appeared to be a much more effective blocking agent. Similar observations were obtained by other investigators using model experiments.[74,79,80] All evidence available now indicates that the immunocytochemical specificity of the currently used hapten antibodies is due to identical alterations to the hapten molecule during linkage to the carrier protein and during fixation.

Tissue controls involved staining for dopamine, noradrenaline, adrenaline, serotonin and histamine in sections through the rat brain and, for noradrenaline, sections of the duodenum and ductus deferens of the rat and of the adrenal medulla of rat, hamster and guinea-pig. Rat and hamster adrenals possess separate noradrenaline-storing cells, while in the guinea-pig only adrenaline-storing cells can be found.[10] Moreover, in the rat the noradrenaline-storing cells are without a preferential position, whilst in the hamster they are located at the cortical–medullary boundary.[81] Adrenaline antisera were tested on adrenals of the same species. Serotonin antisera were tested on sections of rat brain stem, rat duodenum (enterochromaffin cells) and rat and sheep thyroid glands, since in the rat thyroid, serotonin occurs in mast cells, while in the sheep serotonin is stored in the parafollicular cells.[82] Rat parafollicular cells do not seem to contain

Table 18.2. **Blocking (inhibition) experiments**

Antibody	Concn of antigen (μ M)												
	2	4	8	16	32	64	128	256	512	1024	2048	4096	8192
6-OHTHBC	+	−	−	−	−	−	−	−	−	−	−	−	−
5-HT	++	+	+	+	−	−	−	−	−	−	−	−	−
5-HTP							+++	+++	+++	+++	+++	+++	+++
L-Trp							+++	+++	+++	+++	+++	+++	+++
DA							++	++	++	+	+	+	+
NA							+++	+++	+++	+++	+++	+++	+++
A							+++	+++	+++	+++	+++	+++	+++
5-MT				++	++		+	+	+	−	−	−	−
Oct							+++	+++	+++	+++	+++	+++	+++
Syn							+++	+++	+++	+++	+++	+++	+++
Hist							+++	+++	+++	++	++	++	++

Fluorescence intensity in comparable sections of *rat medulla oblongata* after adding different concentrations of several substances to samples of serotonin antiserum, previously absorbed with bovine serum albumin (dilution 1:500).
+++ normal fluorescence.
++ some reduction.
+ strong reduction.
− no fluorescence.
6-OHTHBC, 6-hydroxy-1,2,3,4-tetrahydro-β-carboline.
5-HT, 5-hydroxytryptamine (serotonin).
5-HTP, 5-hydroxytryptophan.
L-Trp, L-tryptophan.
DA, dopamine.
NA, noradrenaline.
A, adrenaline.
5-MT, 5-methoxytryptamine.
Oct, octapamine.
Syn, synephrine.
Hist, histamine.

serotonin. The immunocytochemical findings agreed well with previous histochemical and biochemical data. Thus, cells of the substantia nigra, known to contain dopamine, were stained by the antiserum to dopamine (*Fig.* 18.4C), whereas cells of the locus coeruleus, known to contain noradrenaline, were stained only by the antiserum to noradrenaline (*Fig.* 18.5). These cell bodies are innervated by serotoninergic nerve terminals. This could easily be demonstrated by application of antiserum to serotonin. Raphe nuclei containing serotoninergic perikarya (*Figs.* 18.4A. 18.8A) were only stained by serotonin antisera. No staining was observed in the substantia nigra known to contain dopaminergic cell bodies. Duodenum and ductus deferens showed noradrenaline-immunoreactive nerve fibres with a distribution that conforms to that of the sympathetic postganglionic axons. In the adrenal medulla of rat and hamster, the majority of the parenchymal cells were immunoreactive to both noradrenaline and adrenaline antisera (adrenaline-storing cells), while a few groups of limited numbers of cells were only immunoreactive to noradrenaline antisera (noradrenaline-storing cells). In the rat (*Fig.* 18.6A,B) these groups were not limited to certain sites whilst in the hamster they were located at the boundary between cortex and medulla. In the guinea-pig adrenal medulla, all parenchymal cells were immunoreactive to both noradrenaline and adrenaline antisera. Enterochromaffin cells of the rat duodenum were only stained by antisera to serotonin (*Fig.* 18.10*A*). Thyroids of rats and sheep showed patterns of immunoreactivity for serotonin as expected, namely mast cells in the rat (*Fig.* 18.9*A*) and parafollicular cells in the sheep (*Fig.* 18.9*B*). Histamine-immunoreactive neurones were found in the rat brain, in particular the posterior part of the hypothalamus (*Fig.* 18.11). Mast cells in the central nervous system as well as the gastrointestinal mucosa were also stained.

In a second series of tissue experiments both antibodies to the synthesizing enzymes and antibodies to the end-products were applied. Cells immunoreactive for dopamine β-hydroxylase but negative for phenylethanolamine *N*-methyltransferase, i.e. synthesizing noradrenaline, were only immunoreactive to noradrenaline antisera (*Fig.* 18.6*A–D*). In the adrenal medulla of rat and hamster these cells were distributed as described above. In addition, cells immunoreactive for both enzymes, i.e. synthesizing adrenaline, were immunoreactive for noradrenaline as well as for adrenaline (*Fig.* 18.6*A–D*). In the guinea-pig adrenal medulla, all parenchymal cells were stained by all four antisera. Likewise, in the rat brain stem, serotoninergic perikarya were immunoreactive for serotonin as well as for the synthesizing enzyme dopa decarboxylase.[84] In general, in the rat brain, staining results obtained by antibodies to histamine and the synthesizing enzyme histidine decarboxylase are in agreement.[87]

Thirdly, serotonin immunoreactivity in the rat brain and spinal cord was studied following treatment with neurotoxins 5,6- and 5,7-dihydroxytryptamine, known to cause a depletion of serotonin. Under these conditions serotonin immunoreactivity was strongly reduced.[84–86] A similar approach might give additional information about the specificity of antisera to noradrenaline and adrenaline. However, such experiments have not yet been performed.

Both 'immunological' and 'tissue' controls clearly indicate that antisera raised by immunogens prepared by linkage of dopamine, noradrenaline, adrenaline, serotonin and histamine to carrier proteins can be used as specific markers for cells storing these compounds.

3.5. Sensitivity

In order to assay the sensitivity of the staining methods, the immunofluorescent staining results on cryostat sections were compared with known histochemical and biochemical data. For dopamine and noradrenaline in neuronal cells there is a good correlation between the immunocytochemical results and findings achieved by, for example, induced fluorescence. However, adrenaline-containing neurones cannot be stained immunocytochemically, although the antibody techniques seem to be highly sensitive for both noradrenaline and adrenaline stored in epithelial cells. Thus, studying the development of the noradrenaline and adrenaline-storing cells of the rat adrenal medulla, their earliest appearance as estimated immunocytochemically corresponds quite well with biochemical data.[88] Serotonin immunoreactivity was observed in neuronal and epithelial cells positively stained by existing histochemical techniques for serotonin. However, additionally, highly immunoreactive nerve terminals were revealed in previously negative-seeming brain areas, such as the cerebral cortex, known to contain substantial amounts of serotonin.[14,89] These findings indicate a considerable increase in sensitivity for the visualization of serotonin compared with the older techniques. Antibodies to histamine have been introduced quite recently. Nevertheless, immunocytochemical data obtained so far, show a good agreement with previously reported biochemical findings on the distribution of histamine.[87]

3.6. Applications in Biology and Pathology to Date

New histochemical techniques render it possible to solve questions left by previous ones or will open new fields of research. Thus, detailed studies of the distribution of the serotoninergic and histaminergic neurone systems in the rat brain and spinal cord have been published.[14,87,89] The presence of a serotoninergic innervation of the gastrointestinal tract could be demonstrated convincingly.[16,17,90] Likewise, in the lung it could be proved that the so-called 'neuroepithelial bodies' do indeed store serotonin.[91] The superior cervical ganglion of the rat appeared to contain serotonin-, noradrenaline- and histamine-immunoreactive small ganglion cells,[6,92] In a combined immunocytochemical and biochemical study on the rat medulla, it was shown that the adrenaline-storing cells store serotonin as well as adrenaline.[93] Likewise, the co-storage of opioid peptides and noradrenaline or adrenaline in the adrenal medulla[94] as well as the presence of intrinsic amines and other neuroactive principles in the same neurone have been studied.[84–86] In addition, several reports show that the presently described immunocytochemical techniques, if combined with the retrograde axonal tracing procedures, can be used to examine the projections of neurones containing intrinsic amines.[76,95] Finally, antibodies to noradrenaline, adrenaline or serotonin appear to be useful tools in the diagnosis of tumours supposedly derived from adrenal or extra-adrenal medullary tissues or from epithelial cells storing serotonin.[96,97]

4. CONCLUDING REMARKS

a. Two immunocytochemical procedures can be used to examine cells containing the intrinsic amines dopamine, noradrenaline, adrenaline, serotonin or histamine, i.e. antibodies to the synthesizing enzymes and those to the amines themselves. Ideally, both approaches should be used simultaneously.

b. Antibodies to intrinsic amines are important immunocytochemical tools to elucidate the localization of these compounds in neural and neuroendocrine systems both at light and electron microscopical levels. Application of these antibodies will lead to a better understanding of these systems in the normal as well as the diseased state, especially if combined with other techniques such as biochemistry.

Appendix

SOURCES OF IMMUNE REAGENTS

All primary antisera were raised by the authors.

Fluorescein-conjugated sheep anti-rabbit Ig	Statens Bakteriologiska Lab. Stockholm, Sweden
Fluorescein-conjugated rabbit anti-sheep Ig	Nordic, Tilburg, The Netherlands
Unconjugated goat anti-rabbit Ig (Fc-specific)	Nordic, Tilburg, The Netherlands
Unconjugated swine anti-sheep Ig	Cappel Labs, Cochrane, USA
Sheep PAP	Cappel Labs, Cochrane, USA
Rabbit PAP	Dakopatts, Copenhagen, Denmark

REFERENCES

1. Dahlström A. and Fuxe K. Evidence for the existence of monoamine-containing neurons in the central nervous system. I. Demonstration of monoamines in the cell bodies of brain stem neurons. *Acta Physiol. Scand. Suppl.* 1964, **62** (232).
2. Fuxe K. Evidence for the existence of monoamine neurons in the central nervous system. IV. Distribution of monoamine nerve terminals in the central nervous system. *Acta Physiol. Scand. Suppl.* 1965, **64**, 39–85.
3. Lindvall O. and Björklund A. Dopamine- and norepinephrine-containing neuron systems: Their anatomy in the rat brain. In: Emson P. C., ed., *Chemical Neuroanatomy*. New York, Raven Press, 1983: 229–255.
4. Norberg K. A. Transmitter histochemistry of the sympathetic adrenergic nervous system. *Brain Res.* 1967, **5**, 125–170.
5. Williams T. and Jew J. Monoamine connections in sympathetic ganglia. In: Elfvin L-G., ed., *Autonomic Ganglia*. Chichester, Wiley & Sons, 1983: 235–264.
6. Verhofstad A. A. J., Steinbusch H. W. M., Penke B., Varga J. and Joosten H. W. J. Serotonin-immunoreactive cells in the superior cervical ganglion of the rat: evidence for the existence of separate serotonin- and catecholamine-containing small ganglionic cells. *Brain Res.* 1981, **212**, 39–49.
7. Holzbauer M. and Sharman D. F. The distribution of catecholamines in vertebrates. In: Blaschko H. and Muscholl E., eds., *Catecholamines: Handbook of Experimental Pharmacology*, Vol. 33. Berlin, Springer, 1972: 110–185.
8. Hansen J. T. and Christie D. S. Rat carotid body catecholamines determined by high performance liquid chromatography with electrochemical detection. *Life Sci.* 1981, **29**, 1791–1795.
9. Böck P. The paraganglia. In: Oksche A. and Vollrath L., eds., *Handbuch der mikroskopischen Anatomie des Menschen*, Band VI, Teil 8. Berlin, Springer, 1982.
10. Coupland R. E. The adrenal medulla. In: Beck F. and Lloyd J. B., eds., *The Cell in Medical Science. Vol. 3: Cellular Specialization*. London, Academic Press, 1975: 193–242.
11. Coupland R. E. Observations on the form and size distribution of chromaffin granules and on the identity of adrenaline and noradrenaline-storing chromaffin cells in vertebrates and man. *Mem. Soc. Endocrinol.* 1971, **19**, 611–635.
12. Hökfelt T., Fuxe K., Goldstein M. and Johansson O. Immunohistochemical evidence for the existence of adrenaline neurons in the rat brain. *Brain Res.* 1974, **66**, 235–251.
13. Fuxe J. and Jonsson G. Further mapping of central 5-hydroxytryptamine neurons: studies with the neurotoxic dihydroxytryptamines In: Costa E., Gessa G. L. and Sandler M., eds., *Serotonin:*

new Vistas. Histochemistry and Pharmacology: Advances in Biochemical Psychopharmacology, Vol. 10. New York, Raven Press, 1974: 1–12.

14. Steinbusch H. W. M. Distribution of serotonin-immunoreactivity in the central nervous system of the rat—Cell bodies and terminals. *Neuroscience* 1981, **6**, 557–618.

15. Gershon M. D. Properties and development of peripheral serotonergic neurons. *J. Physiol. (Paris)* 1981, **77**, 257–265.

16. Furness J. B. and Costa M. Neurons with 5-hydroxytryptamine-like immunoreactivity in the enteric nervous system: their projections in the guinea pig small intestine. *Neuroscience* 1982, **7**, 341–349.

17. Costa M., Furness J. B., Cuello A. C., Verhofstad A. A. J., Steinbusch H. W. M. and Elde R. P. Neurons with 5-hydroxytryptamine-like immunoreactivity in the enteric nervous system: their visualization and reactions to drug treatment. *Neuroscience* 1982, **7**, 351–363.

18. Barter R. and Pearse A. G. E. Mammalian enterochromaffin cells as the source of serotonin (5-hydroxytryptamine). *J. Pathol. Bacteriol.* 1955, **69**, 25–31.

19. Vialli M. Histology of the enterochromaffin cells. In: Erspamer V., ed., *5-Hydroxytryptamine and Related Indolealkylamines: Handbook of Experimental Pharmacology*, Vol. 19. Berlin, Springer, 1966: 1–65.

20. Penttilä A. Histochemical reactions of the enterochromaffin cells and the 5-hydroxytryptamine content of the mammalian duodenum. *Acta Physiol. Scand. Suppl.*, 1966, **69**(281).

21. Lauweryns J. M. and Peuskens J. C. Neuroepithelial bodies (neuroreceptor or secretory organs?) in human infant bronchial and bronchiolar epithelium. *Anat. Rec.* 1972, **172**, 471–482.

22. Lauweryns J. M., Cokelaere M. and Theunynck P. Serotonin producing neuroepithelial bodies in the rabbit respiratory mucosa. *Science* 1973, **180**, 410–413.

23. Paasonen M. K. 5-Hydroxytryptamine in mammalian thyroid gland. *Experientia* 1958, **14**, 95–96.

24. Parrat J. R. and West G. B. 5-Hydroxytryptamine and tissue mast cells. *J. Physiol.* 1957, **137**, 169–178.

25. Schwartz J-C., Pollard H. and Quach T. T. Histamine as a neurotransmitter in the mammalian brain: neurochemical evidence. *J. Neurochem.* 1980, **35**, 26–33.

26. Håkanson R., Larsson L-I., Liedberg G. and Sundler F. The histamine-storing enterochromaffin-like cells of the rat stomach. In: Coupland R. E. and Fujita T., eds., *Chromaffin, Enterochromaffin and Related Cells*. Amsterdam, Elsevier, 1976: 243–263.

27. Kaplan E. L. The carcinoid syndromes. In: Friesen S. R., ed., *Surgical Endocrinology: Clinical Syndromes*. Philadelphia, Lippincott, 1978: 120–147.

28. Manger W. M. and Gifford R. W. *Pheochromocytomas*. New York, Springer, 1977.

29. Von Euler U. S. Synthesis, uptake and storage of catecholamines in adrenergic nerves. In: Blaschko K. and Muscholl E., eds., *Catecholamines: Handbook of Experimental Pharmacology*, Vol. 33. Berlin, Springer, 1972: 186–230.

30. Stjärne L. The synthesis, uptake and storage of catecholamines in the adrenal medulla. The effects of drugs. In: Blaschko H. and Muscholl E., eds., *Catecholamines: Handbook of Experimental Pharmacology*, Vol. 33. Berlin, Springer, 1972: 231–269.

31. Hagen P. B. and Cohen L. H. Biosynthesis of indolealkylamines. Physiological release and transport of 5-hydroxytryptamine. In: Erspamer V., ed., *5-Hydroxytryptamine and Related Indolealkylamines: Handbook of Experimental Pharmacology*, Vol. 19. Berlin, Springer, 1966: 182–211.

32. Christenson J. G., Dairman W. and Udenfriend S. On the identity of dopa decarboxylase and 5-hydroxytryptophan decarboxylase. *Proc. Natl Acad. Sci. USA* 1972, **69**, 343–347.

33. Sims K. L., Davis G. A. and Bloom F. E. Activities of 3,4 dihydroxy-L-phenylalanine and 5-hydroxy-L-tryptophan decarboxylases in rat brain: assay characteristics and distribution. *J. Neurochem.* 1973, **20**, 449–464.

34. Rahman M. K., Nagatsu T. and Kato T. Aromatic L-amino acid decarboxylase activity in central and peripheral tissues and serum of rats with L-DOPA and L-5-hydroxytryptophan as substrates. *Biochem. Pharmacol.* 1981, **30**, 645–649.

35. Geffen L. B., Livett D. G. and Rush R. A. Immunohistochemical localization of protein components of catecholamine storage vesicles. *J. Physiol.* 1969, **204**, 593–605.

36. Fuxe K., Goldstein M., Hökfelt T. and Joh T. H. Immunohistochemical localization of dopamine β-hydroxylase in the peripheral and central nervous system. *Res. Commun. Chem. Pathol. Pharmacol.* 1970, **1**, 627–636.

37. Goldstein M., Fuxe K., Hökfelt T. and Joh T. H. Immunohistochemical studies on phenylethanolamine N-methyltransferase, dopadecarboxylase and dopamine β-hydroxylase. *Experientia* 1971, **27**, 951–952.

38. Hökfelt T., Fuxe K. and Goldstein M. Immunohistochemical localization of aromatic L-amino acid decarboxylase (DOPA decarboxylase) in central dopamine and 5-hydroxytryptamine nerve cell bodies of the rat. *Brain Res.* 1973, **53**, 175–180.
39. Hökfelt T., Fuxe K., Goldstein M. and Joh T. H. Immunohistochemical studies of three catecholamine synthesizing enzymes: aspects on methodology. *Histochemie* 1973, **33**, 231–254.
40. Hartman B. K. Immunofluorescence of dopamine β-hydroxylase. Application of improved methodology to the localization of the peripheral and central noradrenergic nervous system. *J. Histochem. Cytochem.* 1973, **21**, 312–332.
41. Swanson L. W. and Hartman B. K. The central adrenergic system. An immunofluorescence study of the localization of cell bodies and their efferent connections in the rat utilizing dopamine β-hydroxylase as a marker. *J. Comp. Neurol.* 1975, **163**, 467–506.
42. Fuxe K., Hökfelt T., Agnati L. F., Johansson O., Goldstein M., Pérez de la Mora M., Possani L., Tapia R., Teran L. and Palacios R. Mapping out central catecholamine neurons: immunohistochemical studies on catecholamine-synthesizing enzymes. In: Lipton M. A., DiMasco A. and Killman K. F., eds., *Psychopharmacology. A Generation of Progress.* New York, Raven Press, 1978: 69–94.
43. Pickel V. M., Joh T. H. and Reis D. J. Monoamine-synthesizing enzymes in central dopaminergic, noradrenergic and serotoninergic neurons. Immunocytochemical localization by light and electron microscopy. *J. Histochem. Cytochem.* 1976, **24**, 792–806.
44. Joh T. H., Shikimi T., Pickel V. M. and Reis D. J. Brain tryptophan hydroxylase: purification, production of antibodies to, and cellular and ultrastructural localization in serotonergic neurons of the rat midbrain. *Proc. Natl Acad. Sci. USA* 1975, **72**, 3575–3579.
45. Gershon M. D., Dreyfus C. F., Pickel V. M., Joh T. H. and Reis D. J. Serotonergic neurons in the peripheral nervous system: identification in gut by immunohistochemical localization of tryptophan hydroxylase. *Proc. Natl Acad. Sci. USA* 1977, **74**, 3086–3089.
46. Watanabe T., Taguchi Y., Shiosaka S., Tanaka J., Kubota H., Terano Y., Tohyama M. and Wada H. Distribution of the histaminergic neuron system in the central nervous system of rats. A fluorescent immunohistochemical analysis with histidine decarboxylase as a marker. *Brain Res.* 1984, **295**, 13–25.
47. Pollard H., Pachot I. and Schwartz J-C. Monoclonal antibody against L-histidine decarboxylase for localization of histaminergic cells. *Neurosci. Lett.* 1985, **54**, 53–58.
48. Grzanna R. and Coyle J. T. Rat adrenal dopamine β-hydroxylase: purification and immunological characteristics. *J. Neurochem.* 1976, **27**, 1091–1096.
49. Steinbusch H. W. M., Verhofstad A. A. J. and Joosten H. W. J. Localization of serotonin in the central nervous system by immunohistochemistry: description of a specific and sensitive technique and some applications. *Neuroscience* 1978, **3**, 811–819.
50. Landsteiner K. *The Specificity of Serological Reactions.* Cambridge, Harvard University Press, 1947.
51. Went S., Kesztyüs L. and Szilágyi T. Weitere Untersuchungen über die physiologische Wirkung des Adrenalylazoprotein-Antikörper. *Naunyn-Schmiedeberg's Arch. Pharmakol. Exp. Pathol.* 1943, **201**, 143–149.
52. Fillipp G. and Schneider H. Die Synthese des Serotoninazoproteins. Experimente zur Frage der Serotoninimmunität. *Acta Allergol.* 1964, **19**, 216–228.
53. Ranadive N. S. and Sehon A. H. Antibodies to serotonin. *Can. J. Biochem.* 1967, **45**, 1701–1710.
54. Peskar B. and Spector S. Serotonin: radioimmunoassay. *Science* 1973, **179**, 1340–1341.
55. Spector S., Berkowitz B., Flynn E. J. and Peskar B. Antibodies to morphine, barbiturates and serotonin. *Pharmacol. Rev.* 1973, **25**, 281–291.
56. Grota L. J. and Brown G. M. Antibodies to indolealkylamines: serotonin and melatonin. *Can. J. Biochem.* 1974, **52**, 196–202.
57. Grota L. J. and Brown G. M. Antibodies to catecholamines. *Endocrinology* 1976, **98**, 615–622.
58. Kellum J. M. and Jaffe B. M. Validation and application of radioimmunoassay for serotonin. *Gastroenterology* 1976, **70**, 516–522.
59. Miwa A., Yoshioka M., Shirahata A. and Tamura Z. Preparation of specific antibodies to catecholamines and L-3,4 dihydroxyphenylalanine. I. Preparation of the conjugates. *Chem. Pharm. Bull. (Tokyo)* 1977, **25**, 1904–1910.
60. Miwa A., Yoshioka M. and Tamura Z. Preparation of specific antibodies to catecholamines and L-3,4 dihydroxyphenylalanine. III. Preparation of antibody to epinephrine for radioimmunoassay. *Chem. Pharm. Bull. (Tokyo)* 1978, **26**, 3347–3352.
61. Went S. and Kesztyüs L. Histaminazoproteinversuche. I. Tierexperimentelle Untersuchungen

mit Histaminazoprotein. *Acad. Med. Sci. Hung.* 1951, **2**, 89–102.
62. Verhofstad A. A. J., Steinbusch H. W. M., Penke B., Varga J. and Joosten H. W. J. Use of antibodies to norepinephrine and epinephrine. In: Eränkö O., Soinila S. and Päivärinta H., eds., *Histochemistry and Cell Biology of Autonomic Neurons, SIF cells and Paraneurons: Advances in Biochemical Psychopharmacology*, Vol. 25. New York, Raven Press, 1980: 185–193.
63. Facer P., Polak J. M., Jaffe B. M. and Pearse A. G. E. Immunocytochemical detection of serotonin with monoclonal antibodies. *Histochem. J.* 1979, **11**, 117–121.
64. Consolazione A., Milstein C., Wright B. and Cuello A. C. Immunocytochemical detection of serotonin with monoclonal antibodies. *J. Histochem. Cytochem.* 1981, **29**, 1425–1430.
65. Takeuchi J., Kimura H. and Sano Y. Immunohistochemical demonstration of the distribution of serotonin neurons in the brain stem of the rat and cat. *Cell Tissue Res.* 1982, **224**, 247–267.
66. Geffard M., Buijs R., Seguela P., Pool C. W. and Le Moal M. First demonstration of highly specific and sensitive antibodies against dopamine. *Brain Res.* 1984, **294**, 161–165.
67. Geffard M., Seguela P. and Heinrich-Rock A-M. Antisera against catecholamines: Specificity studies and physicochemical data for antidopamine and anti-*p*-tyramine antibodies. *Mol. Immunol.* 1984, **21**, 515–522.
68. Miwa A., Yoshioka M. and Tamura Z. Preparation of specific antibodies to catecholamines and L-3,4-dihydroxyphenylalanine. II. The site of attachment on catechol moiety in the conjugates. *Chem. Pharm. Bull. (Tokyo)* 1978, **26**, 2903–2905.
69. Steinbusch H. W. M., Verhofstad A. A. J. and Joosten H. W. J. Antibodies to serotonin for neuroimmunocytochemical studies on the central nervous system. Methodological aspects and applications. In: Cuello A. C., ed., *IBRO Handbook Series: Methods in the Neurosciences. Vol. 2: Neuroimmunocytochemistry*. Chichester, Wiley & Sons, 1983: 193–214.
70. Coons A. H. Fluorescent antibody methods. In: Danielli J. F., ed., *General Cytochemical Methods*. New York, Academic Press, 1958: 399–422.
71. Sternberger L. A., Hardy P. H., Cuculis J. J. and Meyer H. G. The unlabeled antibody enzyme method of immunohistochemistry. Preparation and properties of soluble antigen–antibody complex (horseradish peroxidase–antihorseradish peroxidase) and its use in identification of spirochetes. *J. Histochem. Cytochem.* 1970, **18**, 315–333.
72. Sternberger L. A. *Immunocytochemistry*. New York, Wiley & Sons, 1979.
73. Somogyi P. and Takagi H. A note on the use of picric acid–paraformaldehyde–glutaraldehyde fixative for correlated light and electron microscopic immunocytochemistry. *Neuroscience* 1982, **7**, 1779–1783.
74. Schipper J. and Tilders F. J. H. A new technique for studying specificity of immunocytochemical procedures: Specificity of serotonin-immunostaining. *J. Histochem. Cytochem.* 1983, **31**, 12–18.
75. Mayor H. D., Hampton J. C. and Rosario B. A simple method for removing the resin from epoxy-embedded tissue. *J. Biophys. Biochem. Cytol.* 1961, **9**, 909–910.
76. Steinbusch H. W. M., Nieuwenhuys R., Verhofstad A. A. J. and Van der Kooy D. The nucleus raphe dorsalis of the rat and its projection upon the caudatoputamen. A combined cytoarchitectonic, immunohistochemical and retrograde transport study. *J. Physiol. (Paris)* 1981, **77**, 157–174.
77. Maxwell D. J., Leranth C. and Verhofstad A. A. J. Fine structure of serotonin-containing axons in the marginal zone of the rat spinal cord. *Brain Res.* 1983, **266**, 253–259.
78. Bosin T. R., Jonsson G. and Beck O. On the occurrence of 5-methoxytryptamine in the brain. *Brain Res.* 1979, **173**, 79–88.
79. Milstein C., Wright B. and Cuello A. C. The discrepancy between the cross-reactivity of a monoclonal antibody to serotonin and its immunohistochemical specificity. *Mol. Immunol.* 1983, **20**, 113–123.
80. Peressini S., Brusco A. and Pecci Saavedra J. Basis for the specificity of anti-5-HT antisera in immunocytochemistry applied to the central nervous system. *Histochemistry* 1984, **80**, 597–601.
81. Eränkö O. Histochemical demonstration of noradrenaline in the adrenal medulla of the hamster. *J. Histochem. Cytochem.* 1956, **4**, 11–13.
82. Falck B., Larson B. V., Mecklenburg C., Rosengren E. and Svenaeus K. On the presence of a second specific cell system in mammalian thryoid gland. *Acta Physiol. Scand.* 1964, **62**, 491–492.
83. Pickel V. M., Joh T. H. and Reis D. J. A serotoninergic innervation of noradrenergic neurons in locus coeruleus: demonstration by immunocytochemical localization of the transmitter specific enzymes tyrosine and tryptophan hydroxylase. *Brain Res.* 1977, **131**, 197–214.

84. Hökfelt T., Ljungdahl A., Steinbusch H. W. M., Verhofstad A. A. J., Nilsson G., Brodin E., Pernow B. and Goldstein M. Immunohistochemical evidence of substance P-like immunoreactivity in some 5-hydroxytryptamine-containing neurons in the rat central nervous system. *Neuroscience* 1978, **3**, 517–538.
85. Johansson O., Hökfelt T., Pernow B., Jeffcoate S. L., White N., Steinbusch H. W. M., Verhofstad A. A. J., Emson P. C. and Spindel E. Immunohistochemical support for three putative transmitters in one neuron: coexistence of 5-hydroxytryptamine, substance P and thryotropin releasing hormone-like immunoreactivity in medullary neurons projecting to the spinal cord. *Neuroscience* 1981, **6**, 1857–1881.
86. Gilbert R. F. T., Emson P. C., Hunt S. P., Bennet G. W., Marsden C. A., Sandberg B. E. B., Steinbusch H. W. M. and Verhofstad A. A. J. The effects of monoamine neurotoxins on peptides in the rat spinal cord. *Neuroscience* 1982, **7**, 69–87.
87. Steinbusch H. W. M. and Mulder A. H. Immunohistochemical localization of histamine in neurons and mast cells in the rat brain. In: Björklund A., Hökfelt T. and Kuhar M. J., eds., *Classical Transmitters and Transmitter Receptors in the CNS*. Part II, *Handbook of Chemical Neuroanatomy*, Vol. 3. Amsterdam, Elsevier, 1984: 126–140.
88. Verhofstad A. A. J., Coupland R. E., Parker T. R. and Goldstein M. Immunohistochemical and biochemical study on the development of the noradrenaline- and adrenaline-storing cells of the rat adrenal medulla. *Cell Tissue Res.* 1985, **242**, 233–243.
89. Lidov H. G. W., Grzanna R. and Molliver M. E. The serotonin innervation of the cerebral cortex in the rat — an immunohistochemical analysis. *Neuroscience* 1980, **5**, 207–227.
90. Kurian S. S., Ferri G-L., De Mey J. and Polak J. M. Immunocytochemistry of serotonin-containing nerves in the human gut. *Histochemistry* 1983, **78**, 523–529.
91. Lauweryns J. M., de Bock V., Verhofstad A. A. J. and Steinbusch H. W. M. Immunohistochemical localization of serotonin in intrapulmonary neuroepithelial bodies. *Cell Tissue Res.* 1982, **226**, 215–223.
92. Häppölä O., Soinila S., Päivärinta H., Panula P. and Eränkö O. Histamine-immunoreactive cells in the superior cervical ganglion and in the coeliac-superior mesenteric ganglion complex of the rat. *Histochemistry* 1985, **82**, 1–3.
93. Verhofstad A. A. J. and Jonsson G. Immunohistochemical and biochemical evidence for the presence of serotonin in the adrenal medulla of the rat. *Neuroscience* 1983, **10**, 1443–1453.
94. Kobayashi S., Ohashi T., Uchida T., Nakao K., Imura H., Yanaihara N. and Verhofstad A. A. J. Co-storage of adrenaline and noradrenaline with met-enkephalin-Arg[6]-Gly[7]-Leu[8] and met-enkephalin-Arg[6]-Phe[7] in chromaffin cells of hamster adrenal medulla. *Arch. Histol. Jap.* 1984, **47**, 319–336.
95. Skirboll L., Hökfelt T., Norell G., Phillipson O., Kuypers H. G. J. M., Bentivoglio M., Catsman-Berrevoets C. E., Visser T. J., Steinbusch H. W. M., Verhofstad A. A. J., Cuello A. C., Goldstein M. and Brown M. A method for specific transmitter identification of retrogradely labeled neurons: immunofluorescence combined with fluorescence tracing. *Brain Res. Rev.* 1984, **8**, 99–127.
96. Bosman, F. T., Brutel de la Riviere A., Giard R. W. M., Verhofstad A. A. J. and Cramer-Knijnenburg G. Amine and peptide hormone production by lung carcinoid: A clinicopathological and immunocytochemical study. *J. Clin. Pathol.* 1984, **37**, 931–936.
97. Wells C. A., Taylor S. M. and Cuello A. C. Argentaffin and argyrophil reactions and serotonin content of endocrine tumours. *J. Clin. Pathol.* 1985, **38**, 49–53.

19

Immunocytochemistry and Evolutionary Studies with particular reference to Peptides

M. C. Thorndyke

Immunocytochemistry has probably contributed more than any other technique to the promotion of studies in peptide evolution. Invertebrate investigations have gained considerable benefit from immunocytochemistry, particularly since many invertebrates are small and not well suited to the large scale extraction and purification techniques used with larger vertebrate species. There are exceptions, notably the work on fly extracts discussed in a later section. Generally speaking, however, invertebrates lend themselves more easily to histological study. There are, of course, certain limitations inherent in this approach. Firstly, one is only looking at the surviving members of ancient groups and many of these will have become very specialized and perhaps not ideal representatives of their forebears. This point is, in fact, valid whatever techniques are used.

It is also important to remember that immunocytochemistry tells us only of the presence of an immunologically similar molecule. Certainly there may be an implication of function, for example that of a transmitter if located in the CNS. Frequently, however, there is insufficient evidence from location studies to determine anything of function. Here, studies on receptors and sites of action should ideally accompany immunocytochemical investigations. In a similar way, radioimmunoassay (RIA) of tissue extracts is a useful adjunct to immunocyto-chemistry and can produce additional pertinent data, particularly with regard to chemical characterization and molecular heterogeneity. Indeed RIA is frequently used alongside immunocytochemistry and although this review is concerned primarily with the latter, relevant RIA data will be referred to whenever necessary. The methods used do not differ from those applied in other studies and will not be discussed here.

One of the difficulties shared by immunocytochemistry and radioimmunoassay is that of specificity. A full assessment of this problem may be found elsewhere in this volume, suffice it to say here that one must be aware of the possibility of antibody binding to a totally unrelated molecule which, by chance, shares some antigenic determinants with the peptide antigen. The crab gastrin/cholecystokinin (CCK)-like peptide referred to later is a pertinent case in point. Notwithstanding these problems, with the judicious use of a range of antisera, as well as the careful use of controls, immunocytochemistry represents an important component in an effective armoury for investigations into the mysteries of phylogenesis.

1. COELENTERATES

These are the survivors of the most primitive of multicellular organisms with a well-developed internal signalling system and, since they are diploblastic (ectodermal and endodermal layers only), without a circulatory system, it is not surprising that the signalling system takes the form of a primitive and largely ectodermal nerve net. The cells comprising the nerve net may be either electrically or chemically coupled.[1,2] The relative proportion of each depends both on species and stage in the life cycle of that species since the coelenterate cycle includes sessile (polyp) and swimming (medusoid) stages.

The great bulk of the immunocytochemical work on coelenterates has been concentrated on the hydrozoan *Hydra* sp., which is unusual in having only a polypoid form. *Hydra* sp. follows the basic pattern established for coelenterate nervous systems having an ectoderm-associated nerve net with regional concentrations of cell bodies and processes in the hypostome (mouth) and peduncle (foot or basal disc). Following the ultrastructural identification of dense-cored, peptide-like, synaptic vesicles, a series of immunocytochemical studies by Grimmelikhuijzen,[3–6] has established the presence of a number of neuropeptides in this primitive nervous system.

Interestingly, these investigations show a characteristic pattern of distribution with very clearcut regional concentrations (*Fig.* 19.1). The immunocytochemical studies were carried out alongside RIA of hydra extracts and produced an interesting pattern of results. Thus, for each of the peptides, cholecystokinin (CCK), substance P, neurotensin and bombesin, a number of antisera used in RIA were able to detect the presence of peptide-like immunoreactive material in hydra extracts. In striking contrast, for each of these peptides, only a single antiserum was effective in immunocytochemical studies. This raises the interesting possibility that the native peptide is quite dissimilar from its mammalian counterpart, at least while it is within the cells in which it is produced, i.e. the processing/packaging is different. Analyses for oxytocin-like and vasopressin-like activity produced the reverse situation. Here, several antisera were effective in immunocytochemistry while neither peptide was detectable in RIA. Only with FMRF (phenylalanine-methionine-arginine-phenylalanine) amide did immunocytochemical results parallel those of RIA in that antisera were equally potent in either test. Double labelling with a mixture of oxytocin and vasopressin antisera, raised in different species (rabbit or guinea-pig) suggested to the

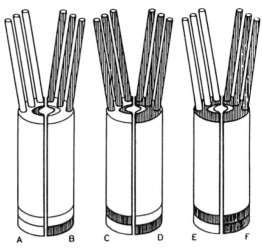

Fig. 19.1. Diagrammatic representation of the distribution of peptide immunoreactivity in *Hydra* sp. (A) CCK; (B) substance P; (C) FMRF amide; (D) bombesin; (E) neurotensin; (F) oxytocin.

investigator the coexistence of these peptides. However, an alternative possibility exists, that of antiserum cross-reactivity in immunocytochemistry, particularly in view of the sequence similarities between oxytocin and vasopressin. A further possibility would be the presence of a single native peptide with sequences common to both oxytocin and vasopressin, e.g. arginine vasotocin or other lower vertebrate neurohypophyseal peptide. In addition to these 'vertebrate' peptides, native hydra peptides have been detected. These are head activator and foot activator and the suggestions are that they are distinct from any of the peptides described above. Head activator is of particular interest since it has been found (by RIA) in mammalian brain extracts.[7]

Hydra sp. is to some extent an atypical, although convenient, example of the coelenterate phylum and it is unfortunate that other examples of this important group have not yet been investigated in such detail. Only in the case of FMRF amide-like peptides has a thorough screening of other species been carried out. Here abundant peptidergic neurones are described in all coelenterate groups including Hydrozoa, Anthozoa and Scyphozoa.[3]

Notwithstanding the possibility that some, or even all, of these described peptide immunoreactivities could well represent cross-reactivity with a smaller number of as yet uncharacterized native peptides which have antigenic determinants in common with the known 'vertebrate' peptides, there can be no doubt that these studies clearly indicate that a neuronal peptide signalling system is already established in the most primitive of diploblastic organisms. As such it provides a firm foundation for phylogenesis.

2. ANNELIDS AND PLATYHELMINTHES

Along with a number of other lower invertebrate phyla, including the nematodes, these two large groups have received comparatively little attentiom from the point of view of peptide regulators and it was not until the immunochemical demonstration of pancreatic polypeptide (PP)-like and vasoactive intestinal peptide (VIP)-like material in the nervous system of *Lumbricus terrestris*[8] that any morphological correlates were described.

The general endocrinology and neurendocrinology of these groups have been well covered in one recent[9] and one not so recent[10] review as well as being discussed as part of a general appraisal of peptide neuroregulation in invertebrates by Haynes.[11] In spite of the sparsity of information, there can be little doubt that peptides are important regulatory agents in these groups of animals. Immunocytochemical studies have revealed the presence of members of the β-lipotrophic hormone (LPH) family in the nervous system of representatives from both annelids and platyhelminthes. Alumets and colleagues[12] showed the presence of both enkephalin and β-endorphin in *Lumbricus* sp., while endorphin was indicated in *Dendrobaena* sp.[13] Enkephalin-like immunoreactivity has also been found in the nervous system of two leeches, *Hirudo* sp.[14] and *Haemopsis* sp.,[15] while ACTH-like material has been detected in the nervous system of the platyhelminth *Dugesia* sp.,[16] a species where somatostatin-like immunoreactivity was also described.

A number of other peptide antisera have been tested and here a wide range have been utilized in the leech, *Hirudo* sp.[14] to show the presence of bombesin-like, gastrin/CCK-like, substance P-like and VIP-like molecules. This is an important demonstration since the leech is used widely in neurobiological studies and may well represent a useful tool for the investigation of basic peptidergic mechanisms. Study of parasitic helminthes is limited to the gull tapeworm *Diphyllobothrium dendritium*. Here FMRF amide-, vasotocin-, enkephalin-, neurotensin-, PP-, peptide histidine isoleucine (PHI)-, growth hormone-releasing factor (GRF)-, gastrin-releasing peptide (GRP)- and gastrin-like activity have been located in various parts of the nervous system.[17] There are also the single observations of CRF-immunoreactive neurones in *Dendrobaena* sp.[18] and of CCK-like material in the nervous system of the polychete *Nereis diversicolor*.[19] In addition to immunocytochemical evidence, there are indications from a number of studies using RIA, bioassay and receptor assay, that peptide regulators are present throughout these more primitive invertebrates.[20]

3. MOLLUSCS

The Mollusca is one of two invertebrate phyla to have received considerable and detailed attention with regard to the immunocytochemical localization of peptides. The other phylum is the Arthropoda. Molluscan studies are of particular interest for two main reasons. First, molluscan organization and physiological sophistication in many ways approaches that seen in higher vertebrates and is especially marked amongst the highly developed cephalopods. Second, the molluscs were one of the first groups of invertebrates to reverse the trend of 'vertebrate' peptides being found in invertebrates with identification and sequencing of FMRF amide[21] and the subsequent description of immunologically similar material in vertebrates.[22,23] It is, however, pertinent to point out that some of these findings may represent cross-reactivities with related enkephalin-like products[24] or even totally unrelated peptides.[25] A further example of a molluscan peptide with a mammalian counterpart is provided by eledoisin, from the salivary gland of the cephalopod *Eledone* sp., a peptide with similarities to vertebrate tachykinins.[26]

There is now a good deal of information available with respect to peptides native to molluscs. This includes hormones concerned with growth, egg laying and associated behaviours, with these factors being detected in several neuronal as well as non-neuronal sites.[24] In spite of this rather healthy state of affairs within the molluscs, including the complete identification, sequencing and cDNA studies on aplysia egg-laying hormones,[27] little effort has gone into the investigation of possible extra-molluscan production of these peptides, or even possible variations amongst other molluscan groups. There is, therefore, little information available on the evolution or phylogenesis of such factors. On the other hand there is a wealth of compelling evidence for the presence of both neuronal and endocrine peptides with the immunological characteristics of vertebrate neurohormonal peptides. In contrast to many other invertebrate studies, the early work on molluscan vertebrate-like peptides centred on insulin-like material identified on the basis of bioactivity.[28] Only later did immunocytochemical studies confirm its presence in the intestine.[29,30] Further studies have established the presence of a wide range of vertebrate-like peptides in molluscs (*Table* 19.1.)[31–38] Although these investigations have been limited to a relatively small number of species, it would now seem highly likely that the widespread use of peptides as neurohormonal regulators was already well established in their primitive molluscan antecedants.

4. ARTHROPODS

Although this section will consider the part played by immunocytochemistry in the identification and characterization of peptide-containing cells in the Arthropoda, it should be recognized at once that this work has been almost entirely restricted to the insects. Studies of crustaceans are as yet of little direct relevance to the study of evolution and phylogeny. Here, investigations have concentrated on the localization of native hormones concerned with the regulation of carbohydrate metabolism, moulting and changes in pigmentation.[39] A few of these studies, notably those examining the control of pigment granule movement, have established a relationship between crustacean red pigment concentrating hormone (RPCH) and insect adipokinetic hormone (AKH), the latter involved in fat mobilization during flight.[40] Other native arthropod hormones are well known,[24] but relatively little attention has been paid to their significance in evolutionary terms. The underlying reason for this is the lack of suitable antisera for the study of their wider distribution.

With respect to the occurrence of vertebrate-like peptides in crustaceans, evidence is so far limited to the identification of gastrin/CCK-like material in a crab, *Cancer magister*,[41,42] to be considered later, as well as the description of substance P-like and enkephalin-like material in lobster eyestalks[43] and a somatostatin-like factor in isopod neurones.[44] Insects are now almost certainly the most thoroughly investigated invertebrate group with respect to the occurrence and distribution of 'vertebrate'-type peptides, although even here detailed work has been limited to a handful of species in particular the blow fly,

Calliphora vomitoria; silk worm, *Bombyx mori*; locust, *Locusta migratoria*; tobacco hornworm, *Manduca sexta*; hoverfly, *Eristalis aeneus* and cockroach, *Periplaneta americana*. The pattern which has emerged is one of a widespread distribution of peptides showing close immunological similarity to established vertebrate peptides, with the nervous system representing the major site of involvement. In contrast, gut peptides are comparatively few and far between or, at least, thorough investigations of the gastrointestinal tract in insects are rare.[45–48] Despite the apparent wealth of information available, comparative appraisals of group relationships are limited to rather few peptide families. Thus pancreatic polypeptide-like (PP-like) immunoreactivity has been detected throughout the nervous system in *Periplaneta* sp.,[46] in the brain of *Calliphora* sp.[49] and *Bombyx* sp.,[50] as well as the fused ventral ganglia of *Eristalis* sp.[51] Since it is also found associated with the gut in both *Periplaneta* sp. and *Calliphora* sp.,[49] it would seem highly likely that this is a common neurohormonal regulator in insects. Those other peptides which have received most attention in insects, similarly belong to the gut–pancreatic axis and include glucagon which is found in a range of species.[46,51] Recent studies on *Manduca* sp.[52] suggest there might be a different distribution of glucagon-like material, with some neuronal perikarya elaborating a peptide with N-terminal and others with C-terminal immunoreactivity. It is, however, the presence of insulin-like material which has attracted most attention. Here its presence was first suggested by two physiological studies on carbohydrate regulation in flies[53,54] together with a single RIA investigation.[55] This work was followed rapidly by a number of immunocytochemical studies which defined the source of this insulin-like material as the median neurosecretory cells in *Calliphora* sp.[56] and in the brain of *Bombyx* sp.[50] and eristalis larvae.[51] Later work on brain extracts from *Calliphora* sp.[57] confirmed the presence of insulin-like immunoreactivity and bioactivity. More recent studies on *Manduca* sp.[52] suggest that immunoreactivities for A-chain and B-chain reside in separate cell populations. The real significance of these findings has only recently been put into perspective, with the complete identification of prothoracotrophic hormone (PTTH) in *Bombyx* sp.[58] PTTH has been known for some time as a native insect peptide synthesized in brain neurosecretory cells and concerned with stimulation of ecdysone secretion from the prothoracic glands. The purified hormone has been shown to have in the region of 50 per cent sequence homology with mammalian insulin and it would seem highly likely that many, if not all, of the reports on the presence of insulin-like material referred to above, in fact reflect the identification of PTTH sources and activity. This finding of sequence similarity between PTTH and insulin has clear evolutionary significance. Could it be that PTTH represents a phylogenetic ancestor of insulin, or are both perhaps derived from an even more ancestral molecule? Equally, this could be an example of parallel or convergent evolution. Clearly, cDNA methods and gene characterization will be of great importance in the determination of the precise relationship between PTTH and insulin. It should be quite evident here that immunocytochemistry has played a central part in the emergence of this insulin story in insects.

One or two other peptides have also been looked at in some detail. Gastrin/CCK-like immunoreactivity has been described in several insect species by immunocytochemistry and RIA[50,59,60] and will be considered in a later section of this chapter.

Table 19.1. Occurrence of vertebrate peptides in some molluscan species

Peptide	Species					
	Lymnaea[31,33]	Cerastoderma[32]	Achatina[20]	Helix[38]	Octopus[34,35]	Aplysia[36,37]
ACTH	+					
Calcitonin	+					
Leu/Met enkephalin	+		+	+	+	
Gastrin/CCK (C-term)	+					+[36]
Glucagon	+	+				
GIP	+					
Insulin	+					
αMSH	+				+	
Neurotensin			+			
Oxytocin	+					
Pancreatic polypeptide	+					
PHI			+			
Somatostatin	+			+		
Secretin	+	+	+			
Substance P	+[33]		+			
TRH	+					
Vasopressin	+				+	
Vasotocin	+					+[37]
VIP	+					

Table 19.2. *Distribution of immunoreactive peptide-like factors in Ascidians*

Peptide	Ciona intestinalis[68,75*]		Styela clava/ Ascidiella aspersa[71,76*]		Styela plicata[77–79]	
	Brain	Gut	Brain	Gut	Brain	Gut
ACTH	+[75]				+	+
Bombesin	+	+				
Calcitonin	+	+	+*	+		
Gastrin/CCK (C-term)	+*	+	+[71]	+		
Calcitonin gene-related peptide	+*		+*			
Endorphin	+[75]					
Enkephalin	+[75]					
LHRH	+[75]					
αMSH	+				+	
Motilin	+					
Neurotensin	+	+				
Pancreatic polypeptide	+	+				
Prolactin	+					
Secretin	+	+	+*	+*	+	+
Somatostatin	+	+	+*		+	+
Substance P	+	+	+*			
VIP	+		+*	+*		

*Thorndyke, 1982 unpublished observations.

Finally, it is worth noting that materials with opiate-like immunoreactivity have also been shown in *Calliphora* sp.,[61] *Bombyx* sp.[62] and *Leucophaea* sp.[63] while antibodies raised against vertebrate neurohypophyseal peptides, including vasopressin, oxytocin and vasotocin, have also been used to demonstrate the distribution of factors showing similarity to these peptides.[62,64–66]

5. PROTOCHORDATES

Of the remaining invertebrate groups only the protochordates have been investigated in any detail while the echinoderms have received only scant attention with the suggestion of insulin-like and substance P-like activity by bioassay. Immunocytochemical tests in this group have so far proved negative.[20] One of the reasons underlying the detailed study of protochordates is, of course, their important and widely recognized position as the living representatives of basal chordate stock. In particular, attention has been focused on selected tunicates,[67–71] and the cephalochordate *Branchiostoma* sp.[72–74] The tunicates are of particular significance in that they are the earliest chordates in which immunocytochemistry has established the presence of a clearly defined brain–gut axis. Thus a wide range of peptides has been found in neurones in the brain as well as in gut endocrine cells and often with a dual distribution (*Table* 19.2).[75–79] Notwithstanding the problems associated with specificity and cross-reactivity when using mammalian antisera in heterologous species, it is quite clear that these immunocytochemical studies indicate that, even at the relatively simple level of organization found in the tunicate CNS, there exists an underlying complexity of peptide regulators. Furthermore, in addition to their presence in the CNS at this pre-vertebrate stage (*Plate* 24), similar peptide-like materials are also produced in significant quantities by the gut endocrine cells typical of these species.

For many years the neural ganglion/neural gland complex of ascidians was considered a homologue of the vertebrate pituitary. Although a few still cling to this view, immunocytochemical studies by Goosens[80] have gone some way to displace this idea. Experiments using antisera raised against vasotocin, mesotocin and neurophysins I and II all failed to produce a positive localization.

In the other protochordate group, the cephalochordata, however, the concept of a pituitary homologue has rather more convincing support. Here, an immunocytochemical study of *Branchiostoma* sp. has shown that Hatschek's pit, a structure first alluded to as a pituitary precursor by morphologists almost a century ago, includes a cell population with luteinizing hormone (LH)-like activity. Further, steroid hormone-like activity is claimed to be stimulated following exposure to LH.[81] In addition to this specific illustration, there are a number of recent investigations which suggest that the brain–gut axis established in tunicates is maintained in cephalochordates. Thus cells with insulin-, glucagon-and gastrin-like activity were first described in branchiostoma gut by Van Noorden and Pearse.[72] Following this, Reinecke[73] and Van Noorden[20] indicated the occurrence of calcitonin-, neurotensin-, pentagastrin-, PP-, secretin-, somatostatin- and VIP-like activity in gut endocrine cells. In a similar way other studies on the nervous system have suggested neuronal CCK-like[74] (*Fig.* 19.2), calcitonin-, FMRF amide- and PP-like[20] immunoreactivity.

6. CHORDATES

Those primitive vertebrates considered most closely related to protochordates, the agnathans, have also received close attention. A series of immunochemical investigations have established a level of organization which is clearly more sophisticated than that seen in protochordates, more nearly approaching that seen in higher vertebrates.[82-87] Furthermore, more than 10 years has passed since the sequencing of hagfish insulin.[88]

Peptide regulators are patently important components of the central and peripheral nervous system in agnathans and many of the elements of the full vertebrate endocrine system are already present. Thus, the pattern upon which the peptide regulatory system of higher vertebrates is based is quite clearly evident in these lowly, albeit specialized, animals.

7. EVOLUTION OF PEPTIDE FAMILIES

It would take several chapters to give full credit to a detailed consideration of the evolution and phylogeny of vertebrate neurohormonal peptide regulators. For this reason, the remaining section will consider one system in some detail in an attempt to throw some light on potential evolutionary relationships that might exist between invertebrates and vertebrates, at the same time highlighting the problems and pitfalls which beset this area.

7.1. Gastrin/Cholecystokinin

In spite of the shortcomings of immunocytochemical identification alluded to in the introduction, the meticulous use of carefully characterized panels of antisera can yield a wealth of information useful in the assessment of phylogenetic relationships. The particular problems of precise identity are of course associated with the similar chemistries of gastrin and CCK as well as distantly similar peptides such as FMRF amide. A discussion of the chemistry of this group is outside the scope of this chapter and has been admirably covered in a recent review.[89] Specific identity is, perhaps, not initially of absolute and overwhelming importance in evolutionary studies, since one might in any case reasonably expect some differences in primary structure, although this need not always be so. However, given the necessity and use of suitable controls and collections of region-specific antisera, then presence, location and with this, hints as to possible functions, are not totally without significance. Indeed they are often a useful first step in any study of peptide evolution.

At the same time, it is imperative to remember that target cells/receptors are equally open to the pressures of evolution and this must be borne in mind in any speculative exercise.

The evidence from the most primitive group of animals, the coelenterates, points to an early neuronal origin for the gastrin/CCK family.[3] Interestingly, FMRF amide is also present in significant quantities in coelenterates, perhaps indicating that this has always been a unique product in spite of limited sequence similarities. Indeed, in view of current knowledge of the gastrin genome,[90] convergence or the simple convenience of suitable amino acid groupings is more likely to be the answer to this superficial question. The CCK-like peptide present in *Hydra* sp. is concentrated in the region around the mouth and it could be significant that even at this early stage in evolutionary history CCK is implicated in feeding strategies. CCK/gastrin-like peptides remain as components of the nervous system in annelids, being found in the CNS of both polychaetes[19] and oligochaetes.[76] In addition it is here for the first time found associated with the gut and this has even led some authors to postulate a role in the control of digestive activity.[91]

Fig. 19.2. Gastrin/CCK-immunoreactive cells in the brain of the Pink Bollworm, *Pectinophora gossypiella* (Insecta). Bouin's-fixed, paraffin section, indirect immunofluorescence method. (From the unpublished work of K. Kamel.)

Many insects also show gastrin/CCK-like immunoreactivity, particularly in neurones (*Fig.* 19.2). A recent and detailed mapping study of neurohormonal gastrin/CCK-like peptides in the blowfly *Calliphora vomitoria*[92] has shed some light on the universal distribution of this material and given some basis for the determination of its possible functions. Interestingly, as has been found in the majority of other invertebrates studied in this way, the material localized reacts most favourably with antisera specific for the common C-terminus, rather than the N-terminus.[60,76,89,92] In *Calliphora* sp. the gastrin/CCK-like material is found specifically in the fused thoracic and abdominal ganglia[92] as well as in certain median neurosecretory cells[59] from where it may be transported to the corpus cardiacum.[93] In view of its location within the neuropil of brain and thoracic ganglia, these authors speculate that here the gastrin/CCK-like material may subserve a neuromodulatory/neurotransmitter role, while its presence in the abdomen and associated nerves is thought to indicate a participation in gut functions such as secretory or motor control. Clearly these ideas require experimental confirmation although the investigations considered above patently illustrate the value of a careful immunocytochemical study. An additional and most evolutionarily relevant observation in these studies was the coexistence of gastrin/CCK-like immunoreactivity with secretin-like and, on occasion, PP-like, factors. A major discussion of primitive genomes and post-translational processing is obviously beyond the scope of this chapter but this immunocytochemical finding quite clearly points to an area in need of close attention. The rewards from such scrutiny are likely to be considerable.

Crustaceans, the remaining major arthropod group have brought forth a most interesting development. In a fascinating series of papers, Vigna and his colleagues[41,42,76,94] describe the distribution of a peptide detected by antibodies raised against gastrin/CCK. This particular peptide has been localized in the pro-cuticle as well as in non-endocrine gastric epithelial cells in the fore gut of the Dungeness crab, *Cancer magister*. No material was detectable in the nervous system. Present indications are that although this peptide is sufficiently similar to vertebrate gastrin/CCK to react with C-terminally directed antisera and vertebrate gastrin/CCK receptors,[41,42] it is otherwise unrelated chemically.[94,95] This work spectacularly highlights the need for care and attention when describing results from immunochemical studies. Gastrin/CCK-like material is also widely distributed in the molluscan nervous system,[28,36] and here too it has been localized to specific, often giant, neurones.[38] As in the arthropods, this may represent an opportunity for the careful investigation of functions.

As one approaches the invertebrate near relatives of chordates, the dual distribution of gastrin/CCK-like material in brain and gut clearly begins to manifest itself. Thus gastrin/CCK-immunoreactive fibres have been demonstrated in all parts of the ascidian neural ganglion,[71] while other workers have shown the presence of gastrin-like material in gut endocrine cells[67] and in gut extracts.[76] The presence of gastrin/CCK-like factors in the gut of the cephalochordate *Branchiostoma* sp. and the intestine of the agnathans *Lampetra fluviatilis* and *Myxine glutinosa* was established some years ago in a series of investigations which also suggested a possible coexistence with glucagon-like material[72,82,83,96,97] (*Fig. 19.3*). This coexistence is supported in a more recent mapping study of branchiostoma gut.[7] CCK-like material is also found in the CNS of *Branchiostoma* sp.[74] (*Fig. 19.4*) and in extracts from lamprey brain.[84] Once again, the implications are of a factor with C-terminal rather than N-terminal immunoreactivity.

In the remaining vertebrates, through to mammals, current discussion centres around the origin, in both chemical and cellular terms, of gastrin and CCK as separate products. A detailed appraisal of this problem has been covered recently by Dockray and Dimaline.[89] Endocrine cells with gastrin/CCK-like immunoreactivity have been found in the gut of all lower vertebrates studied so far. Thus it has been described in fish intestine,[85,98,99] while, uniquely, Langer and colleagues[100] suggest the presence of gastrin-like material in the fish midgut following the use of antisera directed against the N-terminal region of gastrin. It is also clear that gastrin/CCK-like factors are present in fish brain.[101] For many years in the amphibians, the situation has been complicated by the presence of caerulein, a peptide found in the skin of certain frogs which shares its C-terminal sequence with gastrin and CCK. In the past it has been suggested that this, or a very similar peptide, represents the midvertebrate 'gastrin/CCK-equivalent' and the forerunner of higher vertebrate gastrin and CCK molecules.[98] The situation has been clarified recently,[89,102] with the evidence that caerulein is restricted to the skin while the peptide in amphibian brain and gut is more like CCK-8. Gastrin-like material is found in higher vertebrates (reptiles and birds) although it is suggested that it may be quite different from its mammalian

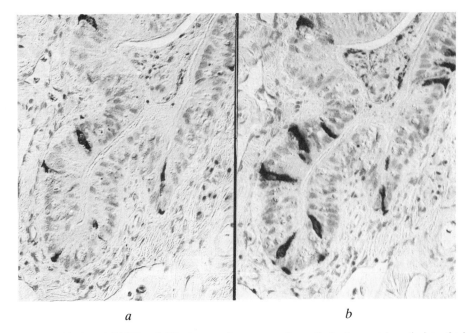

a   b

Fig. 19.3. (*a*) Gastrin/CCK and (*b*) glucagon-immunoreactive cells in the post-hepatic intestinal epithelium of larval (9 cm) lamprey, *Lampetra planeri* (Cyclostomata). Bouin's-fixed, serial 2-μm paraffin sections. PAP method. All the gastrin/CCK-like cells correspond to glucagon-like cells. (Photomicrograph by courtesy of S. Van Noorden.)

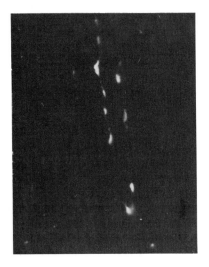

Fig. 19.4. Gastrin/CCK-immunoreactive fibres in the nerve cord of *Branchistoma lanceolatum* (Cephalochordata). Bouin's-fixed, paraffin section, indirect immunofluorescence.

counterpart.[89] What is certain, however, is that whenever gastrin-like material has been localized it is consistently found in a region just anterior to the small intestine. With regard to brain gastrin/CCK-like material, in higher vertebrates, as in lower vertebrates, it appears to resemble CCK-8, with a location which varies according to species studied, although a hypothalamic location is quite common.[89] Thus, as far as gastrin/CCK-like peptides are concerned, immunocytochemistry together with RIA studies has, in spite of some of the difficulties which arise with the use of antibodies, made a significant contribution towards our understanding of petide evolution. There is, therefore, good reason to believe that gastrin/CCK-like molecules have been important signalling agents for multicellular organisms since the earliest times of animal history.

There are, of course, several other avenues in evolutionary studies where immunocytochemistry has played an important part and, of these, the phylogeny of the pancreatic islets stands out as an area where a major advance has resulted from its use. A detailed appraisal of this subject may be found in two recent reviews.[87,103] Finally, it is worth noting that immunocytochemistry has not only had an impact in the study of peptide hormone evolution. In an earlier study on the origin of thyroid follicle cells in protochordates, this author[104] utilized immunocytochemistry to localize material immunologically similar to mammalian thyroglobulin in the iodine-binding, pre-thyroidal cells of the ascidian endostyle.

8. CONCLUDING THOUGHTS

From the evidence presently available, there is now good reason to believe that peptide regulators were first utilized in multicellular animals as messengers in the nervous system. Thus, the coelenterate nerve net apparently makes full use of peptide molecules as signalling agents. This does not rule out, of course, the use of peptides by ancient, but non-surviving organisms, i.e. those with no living representatives, as simple cell-to-adjacent-cell communicators in a primitive paracrine fashion.

The real significance of the presence of peptide regulators in coelenterates is perhaps not *what* is present, but the simple fact that they *are* present, even in these simple, primitive, organisms. In other words, the crux of the matter is that it seems highly likely that in the region of 600 million years ago, multicellular animals had already developed the 'concept' of peptides as messenger molecules. Their precise identity is not crucial to this thesis, although it is quite clear that some peptide sequences make remarkably good messengers and have been conserved in the genome of probably all animals for some time. We should not be surprised by this; if a particular chemical is effective and useful, it tends to be conserved. What one must not lose sight of is that neurohormonal peptides do not evolve in isolation; receptor evolution also takes place and therefore functions deserve equal consideration. It is clear that in some instances location alone can be a guide in general terms to function. For example, if a peptide is found in a central neuronal site, then clearly it is likely to have a role as a transmitter or modulator. Other locations make such assessments rather more difficult. In some instances function may perhaps be conserved. In this way there are indications that insulin or at least a molecule very like it, may play a part in the control of carbohydrate metabolism both in invertebrates and vertebrates,[87,103] although the findings on

PTTH[58] could bring this into doubt. In a similar way there are some data, amongst that discussed earlier, to suggest that CCK-like molecules have long been associated with feeding mechanisms in the broadest sense. However, it would be misleading to imply that peptide function may always be conserved. In fact there is abundant evidence which leads one to quite the opposite conclusion. It should be expected that the peptides have a variety of roles, determined by their anatomical and phylogenetic distribution. If they are efficient regulators, and they patently are, then one might predict that they will be retained with the genome and drawn upon when selective pressures demand.

Appendix

SOURCES OF REAGENTS USED FOR PREPARATIONS ILLUSTRATED

Rabbit anti-gastrin/CCK no. L48	G. J. Dockray, Liverpool
Fluorescein-conjugated goat anti-rabbit Ig (*Figs*. 19.2, 19.4, *Plate* 24)	Nordic
Rabbit anti-gastrin/CCK no. 899	Hammersmith Hospital
Rabbit anti-glucagon no. 499	Hammersmith Hospital
Unconjugated swine anti-rabbit Ig	Dako
Rabbit PAP (*Fig*. 19.3)	Miles

REFERENCES

1. Spencer A. N. The parameters and properties of a group of electrically coupled neurons in the central nervous system of a hydrozoan jellyfish. *J. Exp. Biol.* 1981, **93**, 33–50.
2. Martin S. M. and Spencer A. N. Neurotransmitters in coelenterates. *Comp. Biochem. Physiol.* 1983, **74c**, 1–14.
3. Grimmelikhuijzen C. J. P. Peptides in the nervous system of coelenterates. In: Falkmer, S., Håkanson R. and Sundler F., eds., *Evolution and Tumour Pathology of the Neuroendocrine System*. Amsterdam, Elsevier, 1984: 7–38.
4. Grimmelikhuijzen C. J. P., Sundler F. and Rehfeld J. F. Gastrin/CCK-like immuno-reactivity in the nervous system of coelenterates. *Histochemistry*, 1980, **69**, 61–68.
5. Grimmelikjuijzen C. J. P., Balfe A., Emson P. C., Powell D. and Sundler F. Substance P-like immunoreactivity in the nervous system of hydra. *Histochemistry*, 1981, **71**, 325–333.
6. Grimmelikhuijzen C. J. P., Carraway R. E., Rokaeus A. and Sundler F. Neurotensin-like immunoreactivity in the nervous system of hydra. *Histochemistry*, 1981, **72**, 199–209.
7. Bodenmüller H. and Schaller H. C. Conserved amino acid sequence of a neuropeptide, the head activator, from coelenterates to humans. *Nature* 1981, **293**, 579–580.
8. Sundler F., Håkanson R., Alumets J. and Walles B. Neuronal localization of pancreatic polypeptide (PP) and vasoactive intestinal peptide (VIP) in the earthworm (*Lumbricus terrestris*). *Brain Res. Bull.* 1977, **2**, 61–65.
9. Olive P. J. W. Endocrine adaptations in Annelida. In: Barrington, E. J. W., ed., *Hormones and Evolution*. London, Academic Press, 1979: 73–118.

10. Golding D. W. Neuroendocrine phenomena in non-arthropod invertebrates. *Biol. Rev.*, 1974, **49**, 161–224.
11. Haynes L. W. Peptide neuroregulators in invertebrates. *Prog. Neurobiol.* 1980, **15**, 205–245.
12. Alumets J., Håkanson R., Sundler F. and Thorell J. Neuronal localisation of immunoreactive enkephalin and β-endorphin in the earthworm. *Nature* 1979, **279**, 805–806.
13. Rémy C. and Dubois M. P. Localisation par immunofluorescence de peptides analogues a l'-endorphine dans les ganglions infra-oesophagiens du Lombricide *Dendrobaena subrubicunda* Eisen. *Experienta*, 1979, **35**, 137–138.
14. Osborne N. N., Patel S. and Dockray G. J. Immunohistochemical demonstration of peptides, serotonin and dopamine-β-hydroxylase-like material in the nervous system of the leech *Hirudo medicinalis*. *Histochemistry* 1982, **75**, 573–583.
15. Zipser B. Identification of specific leech neurones immunoreactive to enkephalin. *Nature* 1980, **283**, 857–858.
16. Schilt J., Richoux J. P. and Dubois M. P. Demonstration of peptides immunologically related to vertebrate hormones in *Dugesia lugubris* (Turbellaria, Tricladida). *Gen. Comp. Endocrinol.* 1981, **43**, 331–335.
17. Falkmer S., Gustafsson M. K. S. and Sundler F. Personal communications.
18. Rémy C., Tramu G. and Dubois M. P. Immuno-histological demonstration of a CRF-like material in the central nervous system of the annelid *Dendrobaena*. *Cell Tissue Res.* 1982, **227**, 569–575.
19. Engelhardt R. P., Dhainaut-Courtois N. and Tramu G. Immunohistochemical demonstration of a CCK-like peptide in the nervous system of a marine annelid worm, *Nereis diversicolor*. *Cell Tissue Res.* 1982, **227**, 401–411.
20. Van Noorden S. The neuroendocrine system in protostomian and deuterostomian invertebrates and lower vertebrates. In: Falkmer S., Håkanson R. and Sundler F., eds., *Evolution and Tumour Pathology of the Neuroendocrine System*. Amsterdam, Elsevier, 1984: 7–38.
21. Price D. A. and Greenberg M. J. The structure of a molluscan cardioexcitatory neuropeptide. *Science* 1977, **197**, 670–671.
22. Boer H. H., Schot L. P. C., Veenstra J. A. and Reichelt D. Immunocytochemical identification of neural elements in the central nervous systems of a snail, some insects, a fish and a mammal with an antiserum to the molluscan cardioexcitatory tetrapeptide FMRF-amide. *Cell Tissue Res.* 1980, **213**, 21–27.
23. Dockray G. J., Vaillant C., Williams R. G., Gayton R. J. and Osborne N. N. Vertebrate brain-gut peptides related to FMRF amide and met-enkepehalin Arg[6] Phe[7]. *Peptides* 1981, **2**, Suppl. 2, 25–30.
24. Greenberg M. J. and Price D. A. Invertebrate neuropeptides: native and naturalized. *Annu. Rev. Physiol.* 1983, **45**, 271–288.
25. Dockray G. J., Reeve J. R., Shively J., Gayton R. J. and Barnard C. S. A novel active pentapeptide from chicken brain identified by antibodies to FMRF amide. *Nature* 1983, **305**, 328–330.
26. Erspamer V. and Anastasi A. Structure and pharmacological actions of eledoisin the active endecapeptide of the posterior salivary gland of *Eledone*. *Experientia* 1962, **181**, 58–59.
27. Scheller R. H., Jackson J. F., McAllister L. B., Schwartz J. M., Kandel E. R. and Axel R. A family of genes that codes for ELH, a neuropeptide eliciting a stereotyped pattern of behaviour in *Aplysia*. *Cell* 1982, **28**, 707–719
28. Davidson J. K., Falkmer S., Mehrotra B. K. and Wilson S. Insulin assays and light microscopical studies of digestive organs in protostomian and deuterostomian species and in coelenterates. *Gen. Comp. Endocrinol.* 1971, **17**, 388–401.
29. Fritsch H. A. R., Van Noorden S. and Pearse A. G. E. Cytological and immunofluorescence investigations on insulin-like producing cells in the intestine of *Mytilus edulis* L. (Bivalvia). *Cell Tissue Res.* 1976, **165**, 365–369.
30. Plisetskaya E., Kazakov V. K., Solbitskaya L. and Leibson L. G. Insulin-producing cells in the gut of freshwater bivalve molluscs *Anodonta cygnea* and *Unio pictorum* and the role of insulin in the regulation of their carbohydrate metabolism. *Gen. Comp. Endocrinol.* 1978, **35**, 133–145.
31. Schot L. P. C., Boer H. H., Swaab D. F. and Van Noorden S. Immunocytochemical demonstration of peptidergic neurons in the central nervous system of the pond snail *Lymnaea stagnalis* with antisera raised to biologically active peptides of vertebraes. *Cell Tissue Res.* 1981, **216**, 273–291.

32. Banks I., Sloan J. M. and Buchanan K. D. Glucagon-like (GLI) and secretin-like immunoreactivity (SLI) within the digestive system of the bivalve mollusc *Cerastoderma edule. Reg. Peptides* 1980, **1**, (Suppl. 1), 7.

33. Grim-Jørgensen Y. Immunoreactive thyrotropin-releasing factor in a gastropod: distribution in the central nervous system and hemolymph of *Lymnaea stagnalis. Gen. Comp. Endocrinol.* 1978, **35**, 387–390.

34. Martin R., Frosch D., Weber E. and Voigt K. H. Mer-enkephalin-like immunoreactivity in a cephalopod neurohemal organ. *Neurosci. Lett.* 1979, **15**, 253–257.

35. Martin R., Frosch D. and Voigt K. H. Immunocytochemical evidence for melanotropin- and vasopressin-like material in a cephalopod neurohemal organ. *Gen. Comp. Endocrinol.* 1980, **42**, 235–243.

36. Vigna S. R., Morgan J. L. M. and Thomas T. M. Localization and characterization of gastrin/cholecystokinin-like immunoreactivity in the central nervous system of *Aplysia californica. J. Neurosci.* 1984, **4** 1370–1377.

37. Moore G. J., Thornhill J. A., Gill V., Lederis K. and Lukowiak K. An arginine vasotocin-like neuropeptide is present in the nervous system of the marine mollusc *Aplysia californica. Brain Res.* 1981, **206**, 213–218.

38. Osborne N. N., Cuello A. C. and Dockray G. J. Substance P and cholecystokinin-like peptides in Helix neurons and cholecystokinin and serotinin in a giant neuron. *Science*, 1982, **216**, 409–411.

39. Kleinholz L. H. and Keller R. Endocrine regulation in crustacea. In: Barrington E. J. W., ed., *Hormones and Evolution*. London, Academic Press, 1979: 159–213.

40. Mordue W. Adipokinetic hormone and related peptides. In: Farner D. S. and Lederis K., eds., *Neurosecretion: Molecules, Cells, Systems*. New York, Plenum Press, 1981: 391–401.

41. Larson B. A. and Vigna S. R. Identification of specific leech neurones immunoreactive to enkephalin. *Nature* 1983, **283**, 857–858.

42. Scalise F. W., Larson B. A. and Vigna S. R. Localization of a peptide identified by antibodies to gastrin/CCK in the gut of *Cancer magister. Cell Tissue Res.* 1984, **283**, 113–119.

43. Mancillas J. R., McGinty J. F., Selverston A. I., Karten H. and Bloom F. E. Immunocytochemical localization of enkephalin and substance P in retina and eyestalk neurones of lobster. *Nature* 1981, **293**, 576–578.

44. Martin G. and Dubois M. P. A somatostatin-like antigen in the nervous system of an isopod *Porcellio dilatatus* Brandt. *Gen. Comp. Endocrinol.* 1981, **45**, 125–130.

45. Endo Y. and Nishiisutsuji-Uwo J. Gut endocrine cells in insects: the ultrastructure of the gut endocrine cells of the lepidopterous species, *Biomed. Res.* 1981, **2**, 270–280.

46. Fujita T., Yui R., Iwanaga T., Nishiitsutsuji-Uwo J., Endo Y. and Yanaihara N. Evolutionary aspects of brain-gut peptides: an immunohistochemical study. *Peptides* 1981, **2**, Suppl. 2, 123–131.

47. Iwanaga T., Fujita T., Nishiitsutsuji-Uwo J. and Endo Y. Immunohistochemical demonstration of PP-, somatostatin-, enteroglucagon- and VIP-like immunoreactivities in the cockroach midgut. *Biomed. Res.* 1981, **2**, 202–207.

48. Endo Y., Iwanaga T., Fujita T. and Nishiitsutsuji-Uwo J. Localization of pancreatic polypeptide (PP)-like immunoreactivity in the central and visceral nervous systems of the cockroach *Periplaneta. Cell Tissue Res.* 1982, **227**, 1–9.

49. Duve H. and Thorpe A. The distribution of pancreatic polypeptide in the nervous system and gut of the blowfly, *Calliphora vomitoria* (Diptera). *Cell. Tissue Res.* 1982, **227**, 67–77.

50. Yui R., Fujita T. and Ito S. Insulin-, gastrin-, pancreatic polypeptide-like immuno-reactive neurons in the brain of the silkworm, *Bombyx mori. Biomed. Res.* 1980, **1**, 42–46.

51. El Salhy M., Abou-El-Ela R., Falkmer S., Grimelius L. and Wilander E. Immunohistochemical evidence of gastro-entero-pancreatic neurohormonal peptides of vertebrate type in the nervous system of the larva of a dipteran insect, the hoverfly, *Eristalis aeneus. Reg. Peptides* 1980, **1**, 187–204.

52. El Salhy M., Falkmer S., Kramer K. J. and Speirs R. D. Immunohistochemical investigations of neuropeptides in the brain, corpora cardiaca, and corpora allata of an adult lepidopteran insect, *Manduca sexta* (L). *Cell Tissue Res.* 1983, **232**, 295–317.

53. Chen A. C. and Friedman S. Hormonal regulation of trehalose metabolism in the blowfly *Phormia regina*: Interaction between hypertrehalosemic and hypotrehalosemic hormones. *J. Insect Physiol.* 1977, **23**, 1223–1232.

54. Duve H. The presence of a hypoglycemic and hypotrehalocemic hormone in the neuro-secretory system of the blowfly *Calliphora erythrocephala. Gen. Comp. Endocrinol.* 1978, **36**, 102–110.

55. Ishay J., Glitter S., Galun R., Doron M. and Laron Z. The presence of insulin in and some effects of exogenous insulin on hymenoptera tissues and body fluids. *Comp. Biochem. Physiol.* 1976, **54a**, 203–206.

56. Duve H. and Thorpe A. Immunofluorescent localization of insulin-like material in the median neurosecretory cells of the blowfly, *Calliphora vomitoria* (Diptera). *Cell Tissue Res.* 1979, **200**, 187–191.

57. Duve H., Thorpe A. and Lazarus N. R. Isolation of material displaying insulin-like immunological and biological activity from the brain of the blowfly *Calliphora vomitoria*. *Biochem. J.* 1979, **184**, 221–227.

58. Nagasawa H., Kamito T., Fugo H., Suzuki A. and Ishizaki H. Amino-terminal amino-acid sequence of the silkworm prothoracicotropic hormone: Homology with insulin. *Science* 1984, **226**, 1344–1345.

59. Duve H. and Thorpe A. Gastrin/cholecystokinin (CCK)-like immunoreactive neurones in the brain of the blowfly, *Calliphora erythrocephala* (Diptera). *Gen. Comp. Endocrinol.* 1981, **43**, 381–391.

60. Dockray G. J., Duve, H. and Thorpe A. Immunochemical characterization of gastrin/cholecystokinin-like peptides in the brain of the blowfly, *Calliphora vomitoria*. *Gen. Comp. Endocrinol.* 1981, **45**, 491–496.

61. Duve H. and Thorpe A. Immunocytochemical identification of α endorphin-like material in neurones of the brain and corpus cardiacum of the blowfly, *Calliphora vomitoria* (Diptera). *Cell Tissue Res.* 1983, **233**, 415–426.

62. Rémy C., Girardie J. and Dubois M. P. Vertebrate neuropeptide-like substances in the suboesophageal ganglion of two insects: *Locusta migratoria* R. et F. (Orthoptera) and *Bombyx mori* L. (Lepidoptera). Immunocytochemical investigation. *Gen. Comp. Endocrinol.* 1979, **37**, 93–100.

63. Hansen B. L., Hansen G. N. and Scharrer B. Immunoreactive material resembling vertebrate neuropeptides in the corpus cardiacum and corpus allatum of the insect *Leucophaea maderae*. *Cell Tissue Res.* 1982, **225**, 319–329.

64. Strambi C., Strambi A., Cupo A., Rougon-Rapuzzi G. and Martin N. Étude des taux d'une substance apparentée à la vasopressine dans le système nerveux de grillons soumis à différentes conditions hygrométriques, *C.R. Acad. Sci. France* 1978, **287D**, 1227–1230.

65. Rémy C. and Girardie J. Anatomical organization of two vasopressin–neurophysin-like neurosecretory cells throughout the central nervous system of the migratory locust. *Gen. Comp. Endocrinol.* 1980, **40**, 27–35.

66. Verhaert P., Geysen J., De Loof A. and Vandesande F. Immunoreactive material resembling vertebrate neuropeptides and neurophysins in the brain, suboesophageal ganglion, corpus cadiacum and corpus allatum of the dictyopteran *Periplaneta americana* L. *Cell Tissue Res.* 1984, **238**, 55–59.

67. Fritsch H. A. R., Van Noorden S. and Pearse A. G. E. Localisation of somatostatin- and gastrin-like immunoreactivity in the gastrointestinal tract of *Ciona intestinalis* L. *Cell Tissue Res.* 1978, **186**, 181–185.

68. Fritsch H. A. R., Van Noorden S. and Pearse A. G. E. Gastro-intestinal and neurohormonal peptides in the alimentary tract and cerebral complex of *Ciona intestinalis* (Ascidiaceae). *Cell Tissue Res.* 1982, **223**, 369–402.

69. Bevis P. J. R. and Thorndyke M. C. Endocrine cells in the oesophagus of the ascidian *Styela clava*, a cytochemical and immunofluorescence study. *Cell Tissue Res.* 1978, **187**, 153–158.

70. Thorndyke M. C. and Probert L. Calcitonin-like cells in the pharynx of the ascidian *Styela clava*. *Cell Tissue Res.* 1979, **203**, 301–309.

71. Thorndyke M. C. Cholecystokinin (CCK)/gastrin-like immunoreactive neurones in the cerebral ganglion of the protochordate ascidians *Styela clava* and *Ascidiella aspersa*. *Reg. Peptides* 1982, **3**, 281–288.

72. Van Noorden S. and Pearse A. G. E. The localization of immunoreactivity to insulin, glucagon and gastrin in the gut of *Amphioxus (Branchiostoma) lanceolatus*. In: Grillo T. A. I., Liebson L. and Epple A., eds., *The Evolution of Pancreatic Islets*. New York, Pergamon, 1976: 163–178.

73. Reinecke M. Immunohistochemical localization of polypeptide hormones in endocrine cells of the digestive tract of *Branchistoma lanceolatum*. *Cell Tissue Res.* 1981, **219**, 445–456.

74. Thorndyke M. C. Peptide systems in protochordates. I. The brain-gut axis. In: Lofts B. and Holmes W. N., eds., *Current Trends in Comparative Endocrinology*. Hong Kong, Hong Kong University Press, 1985: 1041–1045.

75. Georges D. Presence of vertebrate-like hormones in the nervous system of a tunicate, *Ciona intestinalis*. In: Lofts B. and Holmes W. N., eds., *Current Trends in Comparative Endocrinology*. Hong Kong, Hong Kong University Press, 1985: 55–57.

76. Larson B. A. and Vigna S. R. Species and tissue distribution of cholecystokinin/gastrin-like substances in some invertebrates. *Gen. Comp. Endocrinol.* 1983, **50**, 469–475.

77. Pestarino M. Cytochemical investigations on endocrine-like cells of the digestive tract of the ascidian *Styela plicata*. *Bas. Appl. Histochem.* 1982, **26**, 107–116.

78. Pestarino M. Nervous and gastric localization of polypeptide hormone-producing cells in the ascidian *Styela plicata*. *Bas. Appl. Histochem.* 1982, **26**, Suppl., LVIII.

79. Pestarino M. Occurrence of different secretin-like cells in the digestive tract of the ascidian *Styela plicata* (Urochordata, Ascidiacea). *Cell Tissue Res.* 1982, **226**, 231–235.

80. Goosens N. Immunohistochemistry of the neural complex of Tunicates. *Biol. Jb. Dodonaea* 1977, **45**, 138–140.

81. Chang C., Liu Y. and Zhu H. The sex steroid hormones and their functional regulation in Amphioxus (*Branchiostoma belchrigray*). In: Lofts B. and Holmes W. N., eds., *Current Trends in Comparative Endocrinology*. Hong Kong, Hong Kong University Press, 1985: 205–207.

82. Östberg Y., Van Noorden S. and Pearse A. G. E. Cytochemical, immunofluorescence and ultrastructural investigations on polypeptide hormone localisation in the islet parenchyma and bile duct mucosa of a cyclostome, *Myxine glutinosa*. *Gen Comp. Endocrinol.* 1975, **25**, 274–291.

83. Östberg Y., Van Noorden S., Pearse A. G. E. and Thomas N. Cytochemical, immunofluorescence and ultrastructural investigations on polypeptide hormone localisation in the intestinal mucosa of a cyclostome, *Myxine glutinosa*. *Gen. Comp. Endocrinol.* 1976, **28**, 213–227.

84. Holmquist A. L., Dockray G. J., Rosenquist G. L. and Walsh J. H. Immunochemical characterization of cholecystokinin-like peptides in lamprey gut and brain. *Gen. Comp. Endocrinol.* 1979, **37**, 474–481.

85. Vigna S. R. Distinction between cholecystokinin-like and gastrin-like biological activities extracted from gastrointestinal tissues of some lower vertebrates. *Gen. Comp. Endocrinol.* 1979, **39**, 512–520.

86. Vigna S. R. and Gorbman A. Stimulation of intestinal lipase secretion by porcine cholecystokinin in the hagfish, *Eptatretus stouti*. *Gen. Comp. Endocrinol.* 1979, **38**, 356–359.

87. Van Noorden S. and Falkmer S. Gut–islet endocrinology — some evolutionary aspects. *Invest. Cell Pathol.* 1980, **3**, 21–35.

88. Peterson J. D., Steiner D. F., Emdin S., Östberg Y. and Falkmer S. Isolation, composition and amino acid sequence of the insulin from a primitive vertebrate (hagfish, *Myxine glutinosa*). *Fed. Proc.* 1973, **32**, 577.

89. Dockray G. J. and Dimaline R. Evolution of the gastrin/CCK family. In: Falkmer S., Håkanson R. and Sundler S. *Evolution and Tumour Pathology of the Neuroendocrine System*. Amsterdam, Elsevier, 1984: 59–82.

90. Yoo J. O., Powell C. T. and Agarwal K. L. Molecular cloning and nucleotide sequence of full-length cDNA coding for porcine gastrin. *Proc. Natl Acad. Sci. USA*, 1982, **79**, 1049–1053.

91. Rzasa P., Kaloustrian K. V. and Prokop E. K. Immunochemical evidence for a gastrin-like peptide in the intestinal tissues of the earthworm *Lumbricus terrestris*. *Comp. Biochem. Physiol.* 1982, **71A**. 631–634.

92. Duve H. and Thorpe A. Immunocytochemical mapping of gastrin/CCK-like peptides in the neuroendocrine system of the blowfly *Calliphora vomitoria* (Diptera). *Cell Tissue Res.* 1984, **237**, 309–320.

93. Duve H., Thorpe A. and Strausfeld J. J. Cobalt-immunocytochemical identification of peptidergic neurons in *Calliphora* innervating central and peripheral targets. *J. Neurocytol.* 1983, **12**, 847–861.

94. Larson B. A., Scalise F. W., Reeve J. R. Jr and Vigna S. R. Immunochemical characterization and localization of gastrin/cholecystokinin-like immunoreactants in the Dungeness crab, *Cancer magister*. *Am. Zool.* 1982, **22**, 949.

95. Vigna S. R. 1984, Personal communication.

96. Van Noorden S., Greenberg J. and Pearse A. G. E. Cytochemical and immunofluorescence investigations on polypeptide hormone localization in the pancreas and gut of larval lamprey. *Gen. Comp. Endocrinol.* 1971, **19**, 192–199.

97. Van Noorden S. and Pearse A. G. E. Immunoreactive polypeptide hormones in the pancreas and gut of the lamprey. *Gen Comp. Endocrinol.* 1974, **23**, 311–324.

98. Larsson L.-I. and Rehfeld J. F. Evidence for a common evolutionary origin of gastrin and cholecystokinin. *Nature* 1977, **269**, 335–338.
99. Rombout J. H. W. M. and Taverne-Thiele J. J. An immunocytochemical and electron-microscopical study of endocrine cells in the gut and pancreas of a stomachless teleost fish, *Barbus conchonius* (Cyprinidae). *Cell Tissue Res.* 1982, **227**, 577–593.
100. Langer M., Van Noorden S., Polak J. M. and Pearse A. G. E. Peptide hormone-like immunoreactivity in the gastrointestinal tract and endocrine pancreas of eleven teleost species. *Cell Tissue Res.* 1979, **199**, 493–508.
101. Notenboom C. D., Garaud J. C., Doerr-Schott J. and Terlou M. Localization by immunofluorescence of a gastrin-like substance in the brain of the Rainbow Trout, *Salmo gairdneri*. *Cell Tissue Res.* 1981, **214**, 247–255.
102. Dimaline R. Different distributions of CCK- and caerulein-like peptides in brain and gut of two amphibians. *Reg. Peptides* 1982, **4**, 360.
103. Falkmer S., El Salhy M. and Titlbach M. Evolution of the neuroendocrine system in vertebrates. A review with particular reference to the phylogeny and postnatal maturation of the islet parenchyma. In: Falkmer S., Håkanson R. and Sundler F., eds., *Evolution and Tumour Pathology of the Neuroendocrine System*. Amsterdam, Elsevier, 1984: 59–87.
104. Thorndyke M. C. Evidence for a 'mammalian' thyroglobulin in endostyle of the ascidian *Styela clava*. *Nature* 1978, **271**, 61–62.

20

Immunocytochemistry of the Diffuse Neuroendocrine System

J. M. Polak and S. R. Bloom

It was first thought that endocrinology dealt exclusively with hormones produced in ductless glands, later including scattered endocrine cells, and acting via the circulation. However, peptide immunocytochemistry has been instrumental in revolutionizing these concepts.[1] Thus, using specific antibodies to substance P, the simultaneous localization of substance P in some endocrine argentaffin cells and in enteric nerves of the rodent gut mucosa was demonstrated.[2,3] A year or two later, peripheral autonomic nerves were found to contain some other peptides, vasoactive intestinal polypeptide (VIP),[4,5,] somatostatin[6] and the enkephalins,[7] the latter two also being localized by immunocytochemistry to typical endocrine cells of the gut,[8-10] pancreas,[8] thyroid[11,12] and adrenal.[13,14] It is now widely accepted that many active peptides, all accommodated under the broad term 'regulatory peptides', are present both in the typical endocrine cells of Feyrter's diffuse endocrine[15] or Pearse's APUD (*a*mine or amine *p*recursor *u*ptake and *d*ecarboxylation) system[16] and in central[17] and peripheral nerves.[18]

A new aspect of endocrinology has thus emerged. Unlike classical glandular endocrinology, dealing with circulating hormones and their excess or deficiency following malfunction of particular endocrine glands, this newly recognized endocrine system is characterized by the production and release of a number of regulatory peptides, only some of which act as circulating hormones. Others act only in the area of their release from endocrine cells or nerves, as potent modulators or as neurotransmitters. In addition, it is possible that several regulatory peptides are released simultaneously and act in an orchestrated (antagonistic or synergistic) manner.

1. THE DIFFUSE NEUROENDOCRINE SYSTEM

Feyrter first introduced the idea of a 'diffuse endocrine system' when he described the presence of a large number of 'clear' cells (weakly stained by conventional histological reagents) to which he attributed a local endocrine (paracrine) role.[15] These cells were described in most organ systems, including the gut, pancreas, skin, adrenals, lung, pituitary and genitourinary system. The modern concept of neuroendocrinology was introduced more than 30 years ago.[19] The recent finding of identical active peptides in neurones and endocrine cells has unified these two systems into a single and versatile diffuse neuroendocrine system.[20] Support for this concept has been mostly derived from morphological (principally immunocytochemical) observations. Physiological proof of its existence as a uniform and coordinated sytem must await the further development of modern technology.

This massive system of regulatory peptides was first found in the mammalian gut and brain, similar peptides being present in the skin of amphibia, and it is now known to be present in almost every peripheral system. These peptides are of very ancient origin, being found throughout the animal kingdom (*see* Chapter 19) and even in plants. Our knowledge of their chemistry, actions and other properties, at least in vertebrates, has grown considerably in the last few years (*see* several review articles and books dealing with the subject,[1,21–25]).

2. IDENTIFICATION OF REGULATORY PEPTIDES

We have recently studied the tissue distribution and precise quantities of a number of regulatory peptides both in the central nervous system and in most peripheral tissues, using a combination of immunocytochemistry and radioimmunoassay.[22] Technical advances have greatly improved the reliability and consistency of these two methods. The recognition that peptide hormones are water soluble led to the development of a wide variety of fixatives suitable for peptide immunocytochemistry.[26] They are principally based on cross-linking agents (e.g. diethylpyrocarbonate or *p*-benzoquinone) which make the peptide insoluble and yet preserve its antigenic sites for subsequent reaction with antibodies. In addition, improved immunization procedures have resulted in the production of many antibodies (poly- or monoclonal) of high affinity, avidity and titre for immunocytochemistry (*see* Chapters 1, 2). Lastly, the availability of pure (synthetic) peptides or their fragments has not only improved the specificity of the antibodies, it has also permitted the accurate identification, in terms of the specific antibody, of a particular segment of the peptide antigen's amino acid sequence. The combination of correct fixation, e.g. cross-linking agents used in solution or as vapour fixatives, a variety of methods of tissue preparation, e.g. cryostat sections, freeze-drying followed by vapour fixation and paraffin embedding, and the use of a panel of region-specific antibodies has greatly aided the accurate localization of regulatory peptides (*see* Chapters 3, 22). Although these are optimal procedures, the quality of antibodies now available usually permits the immunostaining of endocrine cells in parafffin sections from routinely fixed tissues. Immunostaining of nerves can also be performed in formalin-fixed tissues, preferably on cryostat sections.

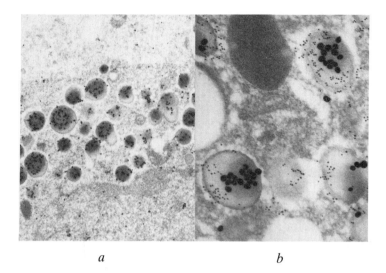

a   *b*

Fig. 20.1.(*a*) GLP$_1$ immunoreactivity localized to the core of secretory granules from a human pancreatic A cell. Immunogold staining procedure with 20-nm gold particles. Uranyl acetate and lead citrate counterstain. × 20 000. (*b*) GLP$_2$ immunoreactivity (40-nm gold particles) localized to the core, and glicentin immunoreactivity (10-nm gold particles) restricted to the halo of secretory granules from human pancreatic A cell. GLP$_1$, GLP$_2$ and glicentin are different peptide fragments of the proglucagon molecule. Double immunogold staining procedure. Uranyl acetate and lead citrate counterstain. × 48 000.

3. MORPHOLOGY

Regulatory peptide-containing cells are either grouped together in glandular structures, e.g. pancreatic islets, adrenal medulla, or diffusely spread, intermingled with non-endocrine elements, e.g. gut, thyroid. Endocrine cells stain weakly by conventional methods but are frequently argyrophil or argentaffin. By electron microscopy, endocrine cells are characterized by intracytoplasmic dense-cored granules of variable size, electron density and position of limiting membrane. These features permit the distinction of different endocrine cell types.[27] The application of electron microscopical immuno-cytochemistry (*see* Chapter 9) to the study of endocrine cells (*Fig.* 20.1) has been instrumental in attaching a functional slant to the purely descriptive electron microscopical classification of these cells.

Peptide-containing nerves are characterized in general by their beaded appearance. The nerve cell origin of these fibres, their nature (sensory/autonomic) and the possible coexistence of two or more neurotransmitters/neuromodulators in the same nerve, is best investigated by various neurosurgical and neurochemical procedures including denervation and the use of neurotoxins (capsaicin, colchicine, 6–hydroxydopamine, reserpine etc.), as well as retro- and anterograde labelling procedures in combination with immunocytochemistry (*see* Chapter 22). All these novel procedures have redefined neuroanatomy in a novel *neurochemical* way. Nerves containing classical neurotransmitters can now also be visualized by immunocytochemistry. The use of antibodies to converting enzymes, including tyrosine hydroxylase and dopamine-β-hydroxylase (*see* Chapter 18) and choline acetyltransferase, is gradually replacing other histochemical procedures, e.g. cholinesterase staining or formaldehyde-induced fluorescence. Like endocrine cells, nerve terminals and cell bodies contain electron microscopically defined neurosecretory granules which can be further defined by the use of electron microscopical immunocytochemistry (*see* Chapter 9). This technique made possible, for instance, the recognition of the heterogeneity of the p-type granules (*see* Section 5.1.2.) and will doubtless be used to answer the still-open question of coexistence of neurotransmitters in the same or separate secretory granules within the same nerve terminal.

4. IMMUNOCYTOCHEMICAL MEANS OF VISUALIZING THE DIFFUSE NEUROENDOCRINE SYSTEM IN ITS ENTIRETY

Several histological stains have been used for the separate demonstration of endocrine (APUD) cells and nerves. Some of these (silver impregnation,[28] lead haematoxylin[29]) became quite popular because of their simplicity and satisfying results. However, not all endocrine or neural cell types can be readily stained and immunocytochemistry of regulatory peptides has played a key role in the demonstration of individual components of the diffuse neuroendocrine system. The simultaneous staining of both components of this system was not achieved until recently.

Several general markers have recently been proposed for the demonstration of endocrine cells and/or the innervation.

4.1 Neuron-specific Enolase

Neuron-specific enolase (NSE) was one of the first neuroendocrine markers to be immunocytochemically identified. NSE is an isomer of the glycolytic enzyme, enolase, thought first to be present exclusively in neurones — hence its name. However, it was later found to be present in endocrine (APUD) cells and central and peripheral nerves.[30] Ultrastructural immunocytochemical studies localize NSE to the non-granular (cytosolic) portion of the cytoplasm;[31] thus this enzyme is the best available marker, not only for the simultaneous visualization of both nerves and endocrine cells, but also for neuroendocrine tissue that is low in peptide/amine storage and thus poorly granulated.

Comparative immunocytochemical studies, using antibodies to NSE and to a variety of peptides/amines/enzymes, show that some endocrine cells and nerves reactive to NSE remain uncharacterized in respect of their peptide or other secretory products. This suggests that the diffuse neuroendocrine system is larger than the sum of the components that can presently be shown by specific antibodies other than to NSE.

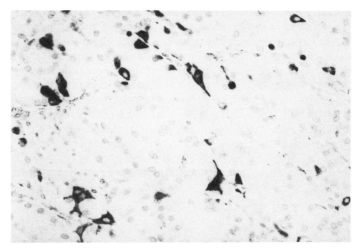

Fig. 20.2. Chromogranin-immunoreactive endocrine cells in human fundus. Tissue freeze-dried and fixed in benzoquinone vapour; 5-µm paraffin section, immunostained by the peroxidase–anti-peroxidase method.

4.2. Intermediate Filaments and Other Cytoskeletal Proteins

Considerable attention has recently been paid to the demonstration of the various components of the cytoskeleton by the use of a variety of monoclonal and polyclonal antibodies (*see* Chapters 23–25). All classes of nerve can be demonstrated by antibodies to neurofilaments.[32,33] Neurofilaments are insoluble proteins and thus resistant to poor or non-optimal fixation which makes them reliable markers in histopathology. Immunocytochemistry of these cytoskeletal components of nerves, combined with that for the cytosolic metabolic enzyme, NSE, has great potential for demonstrating the anatomical and functional state of nerves. The visualization of supporting (glial/Schwann) cells can also be achieved by immunocytochemistry using monoclonal or polyclonal antibodies to glial fibrillary acidic protein (GFAP) or S-100, another brain protein.[32,33]

4.3. Chromogranin

Chromogranin is a large-molecular-weight protein originally extracted from the granules of the adrenal medulla.[34,35] Antibodies and antisera to chromogranin are now available and work well on conventionally fixed tissue at both light and electron microscopical level. Monoclonal antibodies to chromogranin stain almost all identifiable endocrine cells of the gut, pancreas, thyroid and adrenal[36,37] (*Fig.* 20.2). Ultrastructural studies indicate that the protein is present in secretory granules[38] and, therefore, poorly granulated endocrine cells stain weakly or not at all. Some polyclonal antibodies have been shown to stain neural tissue also.[39]

Thus, it is now possible to demonstrate the so-called 'diffuse neuroendocrine system' as an anatomical entity, using newly available markers. It is important, however, to be aware of the different staining properties of these markers in order to use them correctly.

5. DISTRIBUTION OF REGULATORY PEPTIDES IN SIX ORGAN SYSTEMS

In this section, the regulatory peptide system in six peripheral organs will be described. It must, however, be pointed out that only a very general description will be given. Details of species variations, for instance, which are considerable, cannot be undertaken in this short review. The central nervous system and thyroid are now covered under separate chapters (*see* Chapters 21, 22, 32). Regulatory peptides of the pituitary were described in the first edition of this book.[40]

5.1. Gastrointestinal Tract

The gastrointestinal tract contains the largest number of regulatory peptides; in fact, a significant proportion of them were first discovered in the gut. Their localization in mucosal endocrine cells or enteric nerves has been fully elucidated by immunocytochemistry. Those regulatory peptides localized in mucosal endocrine cells have a limited and well-defined anatomical distribution in the bowel, whereas those localized in enteric nerves are found in almost every area and layer of the gut wall.[41]

5.1.1. Endocrine Cells

The distribution of the different endocrine cell types in the various anatomical regions of the gut matches well with the information obtained by radioimmunoassay of tissue extracts from similar areas.[42]

Mucosal endocrine cells intermingled with non-endocrine epithelial cells (*Fig. 20.3*) show a characteristic pyramidal shape with the apex in connection with the intestinal lumen and crowned with microvilli and the base resting on the basal lamina of the epithelium. A characteristic dendritic process or tail elongation along the basement membrane has been described in several types of gut endocrine cell.[43]

5.1.2. Nerves

Peptide-containing enteric nerves are very well demonstrated by immunocytochemistry.[44] Nerves containing substance P and vasoactive intestinal polypeptide (VIP)/peptide histidine isoleucine (methionine) [PHI(M)] are the most abundant, while nerves containing somatostatin, the enkephalins, neuropeptide with tyrosine (NPY), calcitonin gene-related peptide(CGRP) and

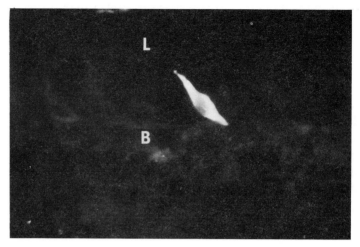

Fig. 20.3 Glucagon-immunoreactive endocrine cell in human colon. Tissue fixed in benzoquinone solution, 10-μm cryostat section immunostained by the indirect immunofluorescence technique. L = lumen, B = basement membrane.

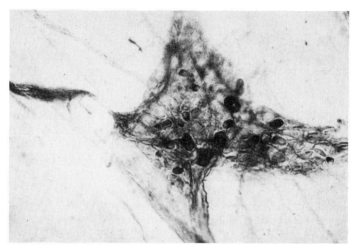

Fig. 20.4. VIP-immunoreactive ganglion cells and nerves in the submucous plexus of human gut. Tissue fixed in benzoquinone solution; whole mount preparation immunostained by the peroxidase–anti-peroxidase method.

galanin are found in lesser numbers. Peptidergic nerves form an intricate and complex network (*Fig.* 20.4) in every layer of the bowel wall, each nerve beaded in the classical manner. Numerous and extensive studies have been carried out on this rich peptide-containing network.[45] Delicate surgical procedures, e.g. myotomy, extrinsic denervation and separate cultures of enteric plexus,[46] have permitted the accurate mapping of individual peptide pathways, their predominant intrinsic origin from neurones located in the myenteric or submucous plexus and their putative extrinsic pathways.[47]

It has been known for some time that regulatory peptides in enteric nerves are in general stored in large p-type electron-dense granules of a non-adrenergic non-cholinergic type.[48,49] The application of immunocytochemistry at ultrastructural level has confirmed the reputed heterogeneity of these p-type neurosecretory granules[50] and allowed the subclassification of this large peptidergic component of the enteric nervous system.[50] Using the immunogold staining procedure (see Chapter 9) and specific antibodies to substance P, VIP, enkephalin and somatostatin, we have recently demonstrated the localization of these peptides in separate subclasses of large dense p-type neurosecretory granule recognizable by their differences in size and electron density. Many other subclasses of peptidergic granule remain unlabelled by the above procedure, suggesting that a number of other regulatory peptides, e.g. NPY, CGRP, remain to be ultrastructurally localized.

5.1.3. Non-tumour Pathology of Endocrine Cells

5.1.3.1. HYPERPLASIA OF GASTRIN (G) CELLS G-cell hyperplasia was first recognized by immunocytochemistry.[51] Subsequently, functional studies were also carried out and this entity is now recognized as G-cell hyperplasia/hyperfunction.[52] G-cell hyperplasia can be found in severe cases of duodenal ulcer with symptoms closely resembling those of the Zollinger–Ellison syndrome (no tumour is found) or in the opposite condition, lack of acid secretion, as in atrophic gastritis with or without pernicious anaemia.[53]

5.1.3.2. HYPERPLASIA OF EC-LIKE CELLS This cell, which is present in the fundic mucosa, has a characteristic elongated shape, and does not reach the lumen. It was named enterochromaffin-like (EC-like or ECL) cell.[54] The peptide product of these cells, contained in characteristic secretory granules, is as yet unknown, but EC-like cells in rodents have been shown to produce histamine, which can now be demonstrated by specific antibodies, or by antibodies to the converting enzyme histidine decarboxylase. Before the advent of immunocytochemistry, and even today, EC-like cells are specifically stained by the Sevier–Munger silver impregnation method.[28] These cells become hyperplastic in conditions of elevated gastrin levels, due either to hyperfunction of G-cells or to a gastrinoma. Furthermore, hypergastrinaemia due to lack of gastric acid inhibition of gastrin (pernicious anaemia), can lead to EC-like cell hyperplasia, formation of micronodules and, subsequently, carcinoids.[55] These findings, which used to be somewhat rare, are becoming more common as drugs, such as ATPase inhibitors, that produce maximum gastric acid suppression are the current treatment for duodenal ulcer and other hypersecretory conditions.

5.1.3.3. HYPERPLASIA OF INTESTINAL ENDOCRINE CELLS The presence of numerous immunoreactive endocrine cells in coeliac disease has frequently been reported.[56,57] The findings of so-called secretin- and cholecystokinin (CCK)-cell hyperplasia are particularly relevant in coeliac disease as it has been recognized for some time that anatomical abnormalities of the duodenal mucosa lead to an impairment of hormone release and subsequently to defective pancreatic bicarbonate and enzyme secretion. Thus an apparent hyperplasia (excess intracytoplasmic peptide storage with brighter and more visible immunostained cells) is observed. Whether or not this is a true hyperplasia is yet to be established.

5.1.4. Non-tumour Pathology of Enteric Nerves

5.1.4.1. HYPERPLASIA Crohn's disease is characterized by strikingly increased and highly abnormal VIPergic nerves in the bowel. Interestingly, these changes are particularly marked in the mucosa and submucosa of both the granulomatous and non-granulomatous areas, thus making the immunocyto-chemical examination of endoscopic biopsies a potentially useful diagnostic tool.[43]

5.1.4.2. HYPOPLASIA Peptidergic nerves, in particular those containing VIP and substance P, are significantly decreased in diseases of the bowel,[44] e.g. Chagas' and Hirschsprung's diseases in man, and grass sickness disease in the horse, all three conditions being associated with intractable chronic constipation and absence or degeneration of intrinsic neuronal cell bodies.

5.2. Pancreas

The immunolocalization of four regulatory peptides (insulin, glucagon, pancreatic polypeptide and somatostatin) in four different endocrine cell types of the pancreatic islets is beyond dispute (*Fig. 20.5*).[58] A fifth endocrine cell type, recognized at the ultrastructural level by the presence of small (mean diameter 100–110 nm) round, dense (D_1 type) secretory granules, still awaits the elucidation of its peptide product. At least 8 regulatory peptides (VIP[59]/PHI,[60] substance P,[61] enkephalin,[61] gastrin/CCK,[61] NPY,[62] CGRP[63] and bombesin[64]) have been localized by immunocytochemistry to nerves of the pancreas. Of these, one of the most convincingly demonstrated is VIP. VIP nerves are present in the exocrine pancreas and in and around the pancreatic islets. In this latter localization, they are particularly concentrated in dense networks. The immunolocalization of VIP in the pancreas[59] is of interest in view of the reported actions of VIP in the exocrine pancreas, as a stimulant of pancreatic bicarbonate secretion and as a potent insulinotrophic peptide. NPY-immunoreactive nerves are mostly found in the exocrine tissue, and originate from extrinsic sympathetic neurones. NPY coexists with noradrenaline in the same neurones. CGRP-immunoreactive nerves are found around blood vessels of the exocrine and endocrine pancreas and in some species coexist with substance P.

5.2.1. Non-tumour Pathology

5.2.1.1. HYPERPLASIA A condition previously known as 'nesidioblastosis' or

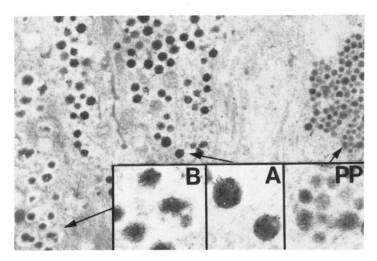

Fig. 20.5. Human infant pancreatic endocrine cells immunostained with an antiserum directed against glucagon. The secretory granules in only one cell type (inset: *A*) are stained. Insulin (inset: B) and pancreatic polypeptide (inset: PP)-containing secretory granules remain unlabelled. Immunogold staining procedure using 10-nm gold particles. Uranyl acetate and lead citrate counterstain. × 14 000; insets × 40 000.

intractable neonatal hypoglycaemia is associated with β-cell hyperplasia. A broad spectrum of lesions ranging from mild hyperplasia and β-cell hyperfunction to the presence of a clearcut adenoma has been described in pancreatic tissue associated with neonatal hypoglycaemia.[65]

5.2.1.2. HYPOPLASIA Somatostatin cells in certain cases of intractable hypoglycaemia are markedly hypoplastic.[66] Somatostatin is known to suppress the release of insulin, and thus a lack of somatostatin could lead to β-cell overactivity and hyperplasia.

5.3. Respiratory Tract

Nine regulatory peptides, VIP, PHI, substance P, bombesin, CGRP, galanin, NPY and possibly somatostatin and CCK have been shown to be present in the respiratory tract of man and other mammals.[67–72] VIP and PHI are found in autonomic nerves, most of which originate locally from neuronal cell bodies found in the tracheal wall. VIP- and PHI-containing nerves are found principally in the upper respiratory tract, including the nasal mucosa, in close contact with seromucous glands and blood vessels. This anatomical distribution fits well with the proposed vasodilatory and secretomotor actions of VIP and PHI. Substance P is found in nerves throughout the bronchial tree, in particularly close contact with the bronchial epithelium and smooth muscle as well as around blood vessels in the same area. Substance P nerve fibres originate outside the lung, from sensory neurones in the nodose ganglion. Substance- P is a putative sensory neurotransmitter, as well as displaying potent broncho- and vasoconstrictor actions. Bombesin or gastrin-releasing peptide is predominantly localized to a subclass of mucosal endocrine cells, most abundant in the human fetal or neonatal lung. The distribution of NPY-immunoreactive nerves, as in most other tissue, parallels that of noradrenergic neurones. Galanin-containing nerves are most abundant in the upper respiratory tract. These nerves originate from local cell bodies, found in particular in the tracheal wall, which also contain VIP and PHI. CGRP is a peptide present in sensory nerves with a distribution identical to that of substance P. CGRP also originates mostly from primary sensory neurones of the nodose ganglion, some of which also contain substance P. In the rat, CGRP is also contained in mucosal endocrine cells (*Fig.* 20.6).

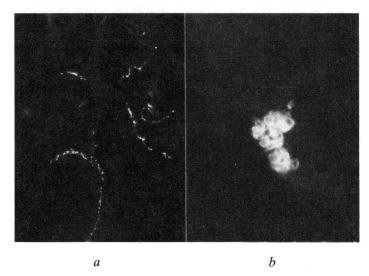

a   b

Fig. 20.6. CGRP-immunoreactive fibres (*a*) and cells (*b*) in rat lung. Tissue fixed in benzoquinone solution; 15-μm cryostat section, immunostained by the indirect immunofluorescence method.

5.3.1. Non-tumour Pathology

5.3.1.1. HYPERPLASIA Nerves containing the vasodilatory peptide VIP are markedly hyperplastic in the nasal mucosa of a subgroup of patients with vasomotor rhinitis.[73]

5.3.1.2. HYPOPLASIA The bombesinergic system of the neonatal lung is markedly depleted in babies with hypoplastic lungs associated with the respiratory distress syndrome.[69]

5.4. Genital Tract

The genital tract contains endocrine cells and nerves. Endocrine cells are found in the prostate, urethra and uterine cervix and endometrium.[74] The existence of these endocrine cells has been shown by the use of silver impregnations, but as yet no peptide has been confidently demonstrated in them.

In addition, both the female and male genital tracts are richly innervated by peptidergic nerves, as demonstrated by immunocytochemistry and radioimmunoassay.[75] VIP, PHI, NPY, substance P and CGRP are the most abundant peptides, although lesser concentrations of somatostatin and galanin can also be detected. The distribution of peptidergic nerves, as in other tissues, fits well with the proposed sets of actions of the peptides. Thus, the sensory neurotransmitters substance P and CGRP are found in particularly high concentrations beneath the vaginal epithelium, and around the sensory corpuscles of the glans penis,[76] whereas VIP and PHI, known for their powerful vasodilatory and secretory properties, are found in the erectile tissue of the penis and in the cervicovaginal region, in close proximity to mucous glands and blood vessels. NPY-immunoreactive nerves are abundant around blood vessels and beneath the epithelium of the fallopian tube, prostate and seminal vesicles. Galanin-immunoreactive nerves are abundant in and around the smooth muscle. VIP, galanin and PHI nerves originate principally from local neurones located, for instance, in the paracervical ganglion[77] and the prostate.

5.4.1. Non-tumour Pathology

The abundant VIPergic innervation of the erectile tissue of the penis encouraged speculation that VIP nerves may be involved in erection. This hypothesis has been confirmed by a number of neurophysiological experiments[78] and also by data obtained from human pathology. VIPergic nerves of the corpus cavernosum are markedly depleted in impotent patients when compared with age-matched controls.[79]

5.5. Urinary System

Peptidergic nerves of the bladder and ureter include those containing substance P, bombesin, i.e. in the rat bladder, VIP/PHI, NPY, galanin and CGRP.[80] As in the genital system, substance P and CGRP occur in sensory areas, e.g. beneath the epithelium and in the smooth muscle. Both peptides are found in peripheral branches of sensory neurones, running along the pelvic and hypogastric nerves and originating in primary sensory neurones located in the dorsal root ganglion. VIP/PHI, and possibly galanin, originate in local neurones of the pelvic ganglion innervating blood vessels of the submucosa and the smooth muscle. NPY is also found around blood vessels and in smooth muscle and appears to have a dual origin from noradrenergic neurones of the sympathetic chain, via the hypogastric nerve, and from local cell bodies of the pelvic ganglion. Peptidergic nerves are also seen beneath the bladder epithelium.

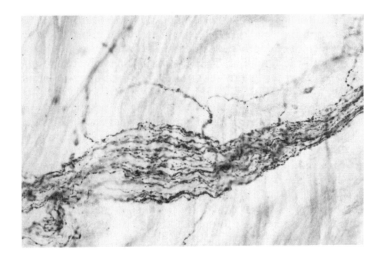

Fig. 20.7. CGRP-immunoreactive nerves in the myocardium of the ventricle. Tissue fixed in benzoquinone solution; 30-μm cryostat section, immunostained by the peroxidase–anti-peroxidase method.

5.5.1. Non-tumour Pathology

The potent muscle relaxant properties of VIP and the marked VIPergic innervation of the bladder smooth muscle prompted an investigation into the state of VIP-containing nerves of the bladder in patients with over-reactive 'unstable' bladder. VIPergic nerves were found to be markedly depleted, particularly in the smooth muscle of the bladder wall, in patients with 'unstable' bladder, in keeping with the potent modulatory effects of VIP on bladder smooth muscle.[81]

5.6. Cardiovascular System

Until recently, few peptides were demonstrated convincingly and in high concentrations in the heart. However, increasing evidence indicates that the whole cardiovascular system of man and other mammals contains large numbers of powerfully active peptides. The most abundantly present are substance P[82], CGRP,[83] and NPY.[84,85] Somatostatin and VIP are present in lesser concentrations. Primary afferent fibres, containing substance P and CGRP, are found throughout the mammalian cardiovascular system. In the heart these sensory nerve fibres occur around coronary blood vessels, in the endocardium, myocardium (*Fig.* 20.7) and epicardium and are associated with the conduction system. Blood vessels also receive a rich supply of afferent fibres containing these peptides. NPY is found in the heart and around blood vessels and is the most abundant neuropeptide found in the human heart. NPY is partly co-stored with noradrenaline in the same neurones, although an independent NPY-containing cardiac plexus has recently been described. NPY nerves are found around coronary and other blood vessels, including those of the circle of Willis. The afferent and efferent arterioles of the juxtaglomerular apparatus of the kidney, as well as the interlobular and renal arteries, are heavily innervated by NPY-containing fibres present in the catecholaminergic system. The cardiac atrium also contains large quantities of the peptide family called atrial natriuretic peptide (ANP).[86] The release from a distended atrium of a factor to regulate water and sodium load from the kidney was long postulated. This fitted well with the finding of numerous electron-dense secretory granules in atrial cardiocytes. Application of molecular biology has now permitted analysis of the nucleotide sequence and deduction of the amino acid sequence of the ANP family of peptides. ANP is extractable in considerable concentrations from the right atrium, and immunocytochemistry localizes it to specialized cardioendocrine cells and to the electron-dense secretory granules.[87]

5.6.1. Non-tumour Pathology

Nerves, in particular of the juxtaglomerular apparatus and renal artery, containing the potent vasoconstrictor NPY have recently been found to be markedly depleted in experimentally induced hypertension (ligation of the abdominal aorta).[92]

6. NEUROENDOCRINE NEOPLASMS

Although endocrine tumours (APUDomas) can be regarded as rare, our understanding of them has advanced greatly since the advent of immunocytochemistry and electron microscopy.[88,89] Among these tumours, the most frequently found are those of pancreatic origin. At least four characteristic clinical syndromes are nowadays well recognized as due to tumours supplying markedly increased circulating levels of the corresponding hormone. They are the insulinoma syndrome, the glucagonoma syndrome, the VIPoma [Verner–Morrison or watery diarrhoea hypokalaemia and achlorhydria (WDHA)] syndrome, the gastrinoma or Zollinger–Ellison syndrome and the syndrome associated with a tumour producing growth hormone-releasing factor (GRF). Although many of these pancreatic tumours produce more than one hormone (and are nowadays called 'islet cell tumours'), one regulatory peptide is produced predominantly as shown by immunocytochemistry (*Fig.* 20.8). These features are further validated by electron microscopy. Although several endocrine cell types can be distinguished in a mixed tumour, one of them, usually corresponding to the hormone responsible for the clinical features, is in the majority. In addition, immunocytochemistry for neuron-specific enolase (NSE) (*see* Section 4.1) can be of excellent diagnostic value. Neuroendocrine tumours of the lung are quite commonly found, in particular the small cell carcinoma of the lung.[90] Small cell carcinomas have been shown to produce and release several of the regulatory peptides, in particular the trophic peptide bombesin or gastrin-releasing peptide (GRP). They are poorly granulated tumours and thus the general neuroendocrine marker neuron-specific enolase is of excellent diagnostic value (*Fig.* 20.9). Chromogranin is also a good marker for neuroendocrine differentiation, but is especially useful in highly granulated tumours. Neuroendocrine tumours of the

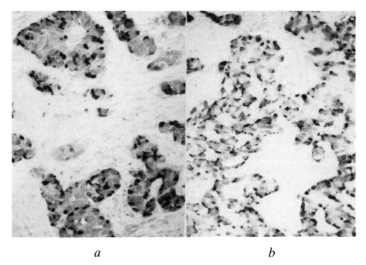

a   *b*

Fig. 20.8. Glucagon-immunoreactive cells (*a*) and pro-glucagon immunoreactive cells (*b*) in human pancreatic endocrine tumour. Tissue freeze-dried and fixed in benzoquinone vapour; 5-µm paraffin section, immunostained by the peroxidase–anti-peroxidase method.

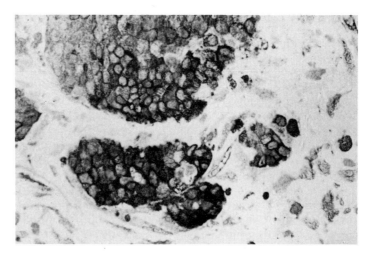

Fig. 20.9. Neuron-specific enolase-immunoreactive cells in a small cell carcinoma of the lung. Formalin fixation; 5-µm paraffin section, immunostained by the peroxidase–anti-peroxidase method.

thryoid (medullary carcinoma) are discussed in Chapter 32. Phaeo-chromocytomas, like normal adrenal medullary tissue, frequently produce regulatory peptides, including NPY, the enkephalins and ACTH. The presence of neurotensin-like material has lately been shown in a phaeochromocytoma cell line maintained in culture. The endocrine nature of the primary neuroendocrine tumour of the skin (the Merkel cell tumour), like the small cell carcinoma of the lung, is best demonstrated by the use of antibodies to neuron-specific enolase. These tumours are also frequently found to be poorly granulated; argyrophilic silver impregnation in consequence is of little or unpredictable value. Neuroendocrine tumours of other regions, including the oesophagus, larynx and uterine cervix, have frequently been described.[91] Peptides have been demonstrated by immunocytochemistry to be produced in extrapulmonary small cell carcinomas but again neuron-specific enolase has been shown to be the best marker for diagnosis of these tumours.

7. CONCLUSION

The finding of identical active peptides in both the endocrine and the neural parts of the diffuse neuroendocrine system has added a new dimension to endocrinology. The leading role played by immunocytochemistry in most of these new and exciting discoveries is beyond dispute. The increasing availability of monoclonal antibodies and pure peptides or their derivative fragments for ensuring specificity, and the possibility of correlating immunocytochemical findings with precise quantitative analysis of tissue peptides makes the investigation of regulatory peptide distribution both rewarding and enjoyable. The demonstration of abnormal peptide distribution in a considerable number of common human disorders adds respectability to our efforts and there is no doubt that immunocytochemistry of regulatory peptides will not only be with us for many years to come, but will even exceed our expectations as the essential tool for these investigations.

Appendix

SOURCES OF IMMUNE REAGENTS FOR THE ILLUSTRATED PREPARATIONS

Mouse anti-chromogranin* (monoclonal)	Dr R. V. Lloyd, University of Michigan Medical School, Michigan, USA
Guinea-pig anti-bovine insulin	Miles
Rabbit anti-rat CGRP	Hammersmith Hospital
Rabbit anti-VIP	Hammersmith Hospital
Rabbit anti-glucagon	Hammersmith Hospital
Rabbit anti-GLP-1	Hammersmith Hospital
Rabbit anti-GLP-2	Hammersmith Hospital
Rabbit anti-human NSE†	Dr P. J. Marangos, NIMH Bethesda, MD, USA
Rabbit anti-bovine PP	Eli Lilly Co., USA

*A polyclonal antibody to chromogranin is available from RIA (UK)/Immunonuclear Corp.
†An antibody to bovine NSE is available from Dako and to sheep NSE from Pharmacia.

Peroxidase-conjugatd rabbit anti-   Miles
   guinea-pig Ig
Fluorescein-conjugated goat anti-   Miles
   rabbit Ig
Unconjugated goat anti-rabbit Ig and   Miles
   goat anti-mouse Ig
Rabbit PAP and mouse PAP   Miles
Gold-adsorbed goat anti-rabbit and   Janssen Pharmaceutica
   goat anti-guinea-pig Ig

REFERENCES

1. Van Noorden S. and Polak J. M. Immunocytochemistry of regulatory peptides. In: Bullock G. R. and Petrusz P., eds. *Techniques in Immunocytochemistry*, Vol. III. New York, Academic Press, 1985: 116–154.
2. Pearse A. G. E. and Polak J. M. Immunocytochemical localisation of substance P in mammalian intestine. *Histochemistry* 1975, **41**, 373–375.
3. Nilsson G., Larsson L.-I., Brodin E., Pernow P. and Sundler F. Localisation of substance P-like immunoreactivity in mouse gut. *Histochemistry* 1975, **43**, 97–99.
4. Bryant M. G., Bloom S. R., Polak J. M., Albuquerque R. H., Modlin I. and Pearse A. G. E. Possible dual role for VIP as gastrointestinal hormone and neurotransmitter substance. *Lancet* 1976, **i**, 991–993.
5. Larsson L.-I., Fahrenkrug J., Schaffalitzky de Muckadell O., Sundler F., Håkanson R. and Rehfeld J. F. Localisation of vasoactive intestinal polypeptide (VIP) to central and peripheral neurons. *Proc. Natl Acad. Sci. USA* 1976, **73**, 3197–3200.
6. Hökfelt T., Johansson O., Efendic S., Luft R. and Arimura A. Are there somatostatin-containing nerves in the rat gut? Immunohistochemical evidence for a new type of peripheral nerves. *Experientia* 1975, **31**, 852–854.
7. Elde R., Hökfelt T., Johansson O. and Terenius L. Immunohistochemical studies using antibodies to leucine-enkephalin. Initial observations on the nervous system of the rat. *Neuroscience* 1976, **1**, 349–351.
8. Polak J. M., Pearse A. G. E., Grimelius L., Bloom S. R. and Arimura A. Growth-hormone release-inhibiting hormone in gastrointestinal and pancreatic D cells. *Lancet* 1975, **i**, 1220–1225.
9. Polak J. M., Sullivan S. N., Bloom S. R., Facer P. and Pearse A. G. E. Enkephalin-like immunoreactivity in the human gastrointestinal tract. *Lancet* 1977, **i**, 972–974.
10. Alumets J., Håkanson R., Sundler F. and Chang K. J. Leu-enkephalin-like material in nerves and enterochromaffin cells in the gut. *Histochemistry* 1978, **56**, 187–196.
11. Hökfelt T., Efendic S., Hellerström C., Johansson O., Luft R. and Arimura A. Cellular localisation of somatostatin in endocrine cells and neurons of the rat with special reference to the A₁ cells of the pancreatic islet and to the hypothalamus. *Acta Endocrinol Suppl.* 200, 1975, **80**, 5–41.
12. Van Noorden S., Polak J. M. and Pearse A. G. E. Single cellular origin of somatostatin and calcitonin in the rat thyroid gland. *Histochemistry* 1977, **53**, 243–247.
13. Schultzberg M., Hökfelt T., Lundberg J. M., Terenius L., Elfvin L. G. and Elde R. Enkephalin-like immunoreactivity in nerve terminals in sympathetic ganglia and adrenal medulla and in adrenal medullary gland cells. *Acta Physiol. Scand.* 1978, **103**, 475–478.
14. Varndell I. M., Tapia F. J., De Mey J., Rush R. A., Bloom S. R. and Polak J. M. Electronimmunocytochemical localisation of enkephalin-like material in catecholamine-containing cells of the carotid body, the adrenal medulla, and in phaeochromocytomas of man and other mammals. *J. Histochem. Cytochem.* 1982, **30**, 682–690.
15. Feyrter F. *Über diffuse endokrine epitheliale Organe.* Liepzig, J. A. Barth, 1938: 6–17.
16. Pearse A. G. E. The cytochemistry and ultrastructure of polypeptide hormone-producing cells of the APUD series, and the embryonic, physiologic and pathologic implications of the concept. *J. Histochem. Cytochem.* 1969, **17**, 303–313.
17. Roberts G. W., Crow T. J. and Polak J. M. Neuropeptides in the brain. In: Bloom S. R. and Polak J. M. eds., *Gut Hormones*, 2nd ed. Edinburgh, Churchill Livingstone, 1981: 457–463.

18. Polak J. M. and Bloom S. R. Peripheral localization of regulatory peptides as a clue to their function. *J. Histochem. Cytochem.* 1980, **28**, 918–924.
19. Scharrer E. Principles of neuroendocrine integration. *Res. Publ. Assoc. Nerv. Ment. Dis.* 1966, **43**, 1–35.
20. Polak J. M. and Bloom S. R. The diffuse neuroendocrine system. *J. Histochem. Cytochem.* 1979, **27**, 1398–1400.
21. Solcia E., Capella C., Buffa R., Usellini L., Fiocca R. and Sessa F. Endocrine cells of the digestive system. In: Johnson L. R., ed.,*Physiology of the Gastrointestinal Tract*. New York, Raven Press, 1981: 39–58.
22. Polak J. M. and Bloom S. R. Regulatory peptides: key factors in the control of bodily functions. *Br. Med. J.* 1983, **286**, 1461–1466.
23. Walsh J. H. Gastrointestinal hormones and peptides. In: Johnson L. R., ed., *Physiology of the Gastrointestinal Tract*. New York, Raven Press, 1981: 59–144.
24. Bloom S. R. and Polak J. M. Introduction. In: Bloom S. R. and Polak J. M., eds., *Gut Hormones*. Edinburgh, Churchill Livingstone, 1981: 3–9.
25. Ballesta J., Bloom S. R. and Polak J. M. Distribution and localization of regulatory peptides. *CRC Crit. Rev. Clin. Lab. Sci.* 1985, **22**, 185–217.
26. Pearse A. G. E. and Polak J. M. Bifunctional reagents as vapour and liquid fixatives for immunocytochemistry. *Histochem. J.* 1975, **7**, 179–186.
27. Solcia E., Polak J. M., Larsson L.-I., Buchan A. M. J. and Capella C. Update on Lausanne classification of endocrine cells. In: Bloom S. R. and Polak J. M., eds., *Gut Hormones*, 2nd ed. Edinburgh, Churchill Livingstone, 1981: 96–100.
28. Grimelius L. and Wilander E. Silver impregnation and other non-immunocytochemical staining methods. In: Polak J. M. and Bloom S. R., eds., *Endocrine Tumours*. Edinburgh, Churchill Livingstone, 1985: 95–115.
29. Solcia E., Vasallo G. and Capella C. Lead-haematoxylin as a stain for endocrine cells. *Histochemie*, 1969, **20**, 116–126.
30. Polak J. M. and Marangos P. J. Neuron-specific enolase, a marker for neuroendocrine cells. In: Falkmer S., Håkanson R. and Sundler F., eds., *Evolution and Tumour Pathology of the Neuroendocrine System*. Amsterdam, Elsevier Science Publishers, 1984: 433–480.
31. Vinores S. A., Herman M. M., Rubinstein L. J. and Marangos P. J. Electron microscopic localization of neuron-specific enolase in rat and mouse brain. *J. Histochem. Cytochem.* 1984, **32**, 1295–1302.
32. Bishop A. E., Carlei F., Lee C., Trojanowski J., Marangos P. J., Dahl D and Polak J. M. Combined immunostaining of neurofilaments, neuron specific enolase, GFAP and S-100. *Histochemistry* 1985, **82**, 93–97.
33. Hacker G. W., Polak J. M., Springall D. R., Ballesta J., Cadieux A., Gu J., Trojanowski J. Q., Dahl D. and Marangos P. J. Antibodies to neurofilament protein and other brain proteins reveal the innervation of peripheral organs. *Histochemistry*, 1985, **82**, 581–593.
34. Blaschko, H., Comline, R. S., Schneider, F. H., Silver, M. and Smith, A D. Secretion of a chromaffin granule protein, chromogranin, from the adrenal gland after splanchnic stimulation. *Nature* 1967, **215**, 58–59.
35. O'Connor, D. T., Frigon, F. P. and Sokoloff, R. L. Human chromogranin A: purification and characterization from catecholamine storage vesicles of pheochromocytoma. *Hypertension* 1983, **6**, 2–12.
36. Wilson, B. S. and Lloyd, R. V. Detection of chromogranin in neuroendocrine cells with a monoclonal antibody. *Am. J. Pathol.* 1984, **115**, 458–468.
37. Facer, P., Bishop, A. E., Lloyd, R. V., Wilson, B. S., Hennessy, R. and Polak, J. M. Chromogranin A: a newly recognised marker for endocrine cells of the human gastrointestinal tract. *Gastroenterology*, in press.
38. Varndell I. M., Lloyd R. V., Wilson B. S. and Polak J. M. Ultrastructural localisation of chromogranin: a potential marker for the electron microscopic recognition of endocrine cell secretory granules. *Histochem. J.* 1985, **17**, 981–992.
39. Somogyi P., Hodgson A. J., DePotter R. W., Fischer-Colbrie R., Schober M., Winkler H. and Chubb I. W. Chromogranin immunoreactivity in the central nervous system. Immunochemical characterisation, distribution and relationship to catecholamine and enkephalin pathways. *Brain Res. Rev.* 1984, **8**, 193–203.
40. Petrusz P. and Ordronneau P. Immunocytochemistry of pituitary hormones. In: Polak J. M. and Van Noorden S., eds., *Immunocytochemistry, Practical Applications in Pathology and Biology*. Bristol. Wright. PSG, 1983: 212–232.

41. Polak J. M. and Bloom S. R. Localisation of regulatory peptides in the gut. *Bri. Med. Bull.* 1983, **38**, 303–308.
42. Bloom S. R. and Polak J. M. Gut hormone overview. In: Bloom S. R. and Polak J. M., eds., *Gut Hormones*. Edinburgh, Churchill Livingstone, 1978: 3–19.
43. Larsson L.-I., Somatostatin cells. In: Bloom S. R. and Polak J. M., eds., *Gut Hormones*, 2nd ed. Edinburgh, Churchill Livingstone, 1981: 350–353.
44. Bishop A. E., Ferri G. L., Probert L., Bloom S. R. and Polak J. M. Peptidergic nerves. In: Polak J. M., Bloom S. R., Wright N. A.and Daley M. J., eds., *Structure of the Gut. Basic Science in Gastroenterology*. England, Glaxo Group Research Limited, 1982: 221–237.
45. Costa M. and Furness J. B. Immunohistochemistry on whole mount preparations. In: Cuello A. C., ed., *Immunohistochemistry*. Chichester, John Wiley & Son, 1983: 373–398.
46. Jessen K. R., Saffrey M. J., Van Noorden S., Bloom S. R., Polak J. M. and Burnstock G. Immunohistochemical studies of the enteric nervous system in tissue culture and in situ: localization of vasoactive intestinal polypeptide (VIP), substance P and enkephalin-immunoreactive nerves in the guinea pig gut. *Neuroscience* 1980, **5**, 1717–1735.
47. Schultzberg M., Hökfelt T., Nilsson G., Terenius L., Rehfeld J. F., Brown M. Elde R., Goldstein M. and Said S. Distribution of peptide and catecholamine-containing neurons in the gastrointestinal tract of rat and guinea pig. Immunohistochemical studies with antisera to substance P, vasoactive intestinal polypeptide, enkephalins, somatostatin, gastrin/cholecystokinin, neurotensin and dopamine β-hydroxylase. *Neuroscience* 1980, **3**, 689–744.
48. Baumgarten H. G., Holstein A. F. and Owman C. H. Auerbach's plexus of mammals and man: electron microscopical identification of three different types of neuronal processes in myenteric ganglia of the large intestine from rhesus monkeys, guinea pigs and man. *Z. Zellforsch Mikrosk. Anat.* 1970, **106**, 376–397.
49. Cook R. D. and Burnstock G. The ultrastructure of Auerbach's plexus in the guinea-pig. 1. Neuronal elements. *J. Neurocytol.* 1976, **5**, 171–194.
50. Probert L., De Mey J. and Polak J. M. Ultrastructural localization of four different neuropeptides within separate populations of p-type nerves in the guinea pig colon. *Gastroenterology* 1983, **85**, 1094–1104.
51. Polak J. M., Stagg B. and Pearse A. G. E. The two types of Zollinger–Ellison syndrome. Immunofluorescent, cytochemical and ultrastructural studies of the antral and pancreatic gastrin cells in different clinical states. *Gut* 1972, **13**, 510–512.
52. Lewin K. J., Yang K., Ulich T., Elashoff J. D. and Walsh J. Primary gastrin cell hyperplasia. *Am. J.Surg. Pathol.* 1984, **8**, 821–832.
53. Polak J. M., Hoffbrand V., Reed P. I. and Pearse A. G. E.Qualitative and quantitative studies of antral and fundic G cells in pernicious anaemia. *Scand. J. Gastroenterol.* 1973, **8**, 361–367.
54. Håkanson R., Larsson L.-I., Liedberg G. and Sundler F. The histamine-storing enterochromaffin-like cells of the rat stomach. In: Coupland R. E. and Fujita T., eds., *Chromaffin, Enterochromaffin and Related Cells*. Amsterdam, Elsevier Scientific Publishing Co., 1976: 243–263.
55. Larsson L.-I., Rehfeld, J. F. and Stockbrügger R. Mixed endocrine gastric tumors associated with hypergastrinemia of antral origin. *Am. J. Pathol.* 1978, **93**, 53–68.
56. Polak J. M., Pearse A. G. E., Van Noorden S., Bloom S. R. and Rossiter M. A. Secretin cells in coeliac disease. *Gut* 1973, **14**, 870–874.
57. Sjolund K., Alumets J., Berg N.-O., Håkanson R. and Sundler F. Enteropathy of coeliac disease in adults. *Gut* 1982. **23**, 42–48.
58. Klöppel G. and Lenzen S. Anatomy and physiology of the endocrine pancreas. In: Klöppel G. and Heitz P. U., eds., *Pancreatic Pathology*. Edinburgh, Churchill Livingstone, 1984: 133-153.
59. Bishop A. E., Polak J. M., Green I. C., Bryant M. G. and Bloom S. R. The location of VIP in the pancreas of man and rat. *Diabetologia* 1980, **18**, 73–78.
60. Yiangou Y., Christofides N. D., Evans J. E. and Bloom S. R. The presence of peptide histidine isoleucine in human, dog, guinea-pig and rat pancreas. *Diabetologia* 1983, **25**, 125–127.
61. Larsson L.-I., Innervation of the pancreas by substance P, enkephalin, vasoactive intestinal polypeptide and gastrin/CCK immunoreactive nerves. *J. Histochem. Cytochem.* 1979, **27**, 1283–1284.
62. Carlei F., Allen J. M., Bishop A. E., Bloom S. R. and Polak J. M. Occurrence, distribution and nature of NPY in the rat pancreas. *Experientia* 1985, in press.
63. Mulderry P. K., Ghatei M. A., Bishop A. E., Allen Y. S., Polak J. M. and Bloom S. R. Distribution and chromatographic characterisation of CGRP-like immunoreactivity in the brain and gut of the rat. *Regulat. Pept.* 1985, **12**, 133–143.

64. Ghatei M. A., George S. K., Major J. H., Carlei F., Polak J. M. and Bloom S. R. Bombesin-like immunoreactivity in the pancreas of man and other mammalian species. *Experientia* 1984, **40**, 884–886.

65. Heitz P., Klöppel G., Hacki W. H., Polak J. M. and Pearse A. G. E. Nesidioblastosis: the pathological basis of persistent hyperinsulinaemic hypoglycaemia in infants. Morphological and quantitative analysis of seven cases based on specific immunostaining and electron microscopy. *Diabetes*, 1977, **26**, 632–642.

66. Bishop A. E., Polak J. M., Garin Chesa P., Timson C. M., Bryant M. G. and Bloom S. R. Decrease of pancreatic somatostatin in neonatal nesidioblastosis. *Diabetes* 1981, **30**, 122–126.

67. Polak J. M. and Bloom S. R. Regulatory peptides and neuron specific enolase in the respiratory tract of man and other mammals. *Exp. Lung Res.* 1982, **3**, 313–328.

68. Polak J. M. and Bloom S. R. Regulatory peptides of the respiratory tract: a newly discovered control system. In: Martini L. and Ganong W. E., eds., *Frontiers in Neuroendocrinology*, Vol. 8. New York, Raven Press, 1984: 199–221.

69. Polak J. M. and Bloom S. R. Distribution of regulatory peptides in the respiratory tract of man and mammals. In: Bloom S. R., Polak J. M. and Lindenlaub E., eds., *Systemic Role of Regulatory Peptides*, Symposium Oosterbeek, Netherlands, May 2nd–6th, 1982. Stuttgart, F. K. Schattauer Verlag, 1982: 241–269.

70. Sheppard M. N., Polak J. M., Allen J. M. and Bloom S. R. Neuropeptide tyrosine (NPY): a newly discovered peptide is present in the mammalian respiratory tract. *Thorax* 1984, **39**, 326–330.

71. Cadieux A., Springall D. S., Mulderry P. K., Rodrigo J., Ghatei M. A., Terenghi G., Bloom S. R. and Polak J. M. Occurrence, distribution and ontogeny of CGRP immunoreactivity in the rat lung: effect of capsaicin treatment and surgical denervations. *Neuroscience*, Submitted for publication.

72. Cheung A., Polak J. M., Bauer F. E., Christofides N. D., Cadieux A., Springall D. R. and Bloom S. R. The distribution of galanin immunoreactivity in the respiratory tract of pig, guinea pig, rat and dog. *Thorax*, in press.

73. Kurian S. S., Blank M. A., Sheppard M. N., Stanley P. J., Mackay I. S., Cole P. J., Bloom S. R. and Polak J. M. Vasoactive intestinal polypeptide (VIP) in vasomotor rhinitis. *IRCS Medical Science II* 1983, 425–426.

74. Fetissof F., Berger G., Dubois M. P., Arbeille-Brassart B., Lansac J., Sam-Giao M. and Jobard P. Endocrine cells in the female genital tract. *Histopathology* 1985, **9**, 133–145.

75. Huang W. M., Gu J., Blank M. A., Allen J. M., Bloom S. R. and Polak J. M. Peptide-immunoreactive nerves in the mammalian female genital tract. *Histochem. J.* 1984, **16**, 1297–1310.

76. Polak J. M. Gu J., Mina S. and Bloom S. R. VIP ergic nerves in the penis. *Lancet* 1981, **ii**, 217–219.

77. Inyama C. O., Hacker G. W., Gu J., Dahl D., Bloom S. R., and Polak J. M. Cytochemical relationships in the paracervical ganglion (Frankenhäuser) of rat studied by immunocytochemistry. *Neurosci. Lett.* 1985, **55**, 311–316.

78. Larsen J. J., Ottensen B. and Fahrenkrug L. Vasoactive intestinal polypeptide (VIP) in the male genitourinary tract, concentration and motor effect. *Invest. Urol.* 1981, **19**, 211–213.

79. Gu J., Lazarides M., Pryor J. P., Blank M. A., Polak J. M., Morgan R., Marangos P. J. and Bloom S. R. Decrease of vasoactive intestinal polypeptide (VIP) in the penises from impotent men. *Lancet* 1984, **ii**, 315–318.

80. Polak J. M. and Bloom S. R. Localisation and measurement of VIP in the genitourinary system of man and animals. *Peptides* 1984, **5**, 225–230.

81. Gu J., Restorick J. M., Blank M. A., Huang W. M., Polak J. M., Bloom S. R. and Mundy A. R. Vasoactive intestinal polypeptide in the normal and unstable bladder. *Br. J. Urol.* 1983, **55**, 645–647.

82. Wharton J., Polak J. M., McGregor G. P., Bishop A. E. and Bloom S. R. The distribution of sustance P-like immunoreactive nerves in the guinea pig heart. *Neuroscience* 1981, **6**, 2193–2204.

83. Mulderry P. K., Ghatei M. A., Rodrigo J., Allen J. M., Rosenfeld M. G., Polak J. M. and Bloom S. R. Calcitonin gene-related peptide in cardiovascular tissues of the rat. *Neuroscience* 1985, **14**, 947–954.

84. Gu J., Polak J. M., Allen J. M., Huang W. M., Sheppard M. N., Tatemoto K. and Bloom S. R. High concentrations of a novel peptide, neuropeptide Y, in the innervation of mouse and rat heart. *J. Histochem. Cytochem.* 1984, **32**, 467–472.

85. Gu J., Polak J. M., Adrian T. E., Allen J. M., Tatemoto K. and Bloom S. R. Neuropeptide tyrosine (NPY) — a major cardiac neuropeptide. *Lancet* 1983, **i**, 1008–1010.
86. Palluk R., Gaida W. and Hoefke W. Atrial natriuretic factor. *Life Sci.* 1985, **36**, 1415–1427.
87. Cantin M., Gutkowska J., Thibault G., Milne R. W., Ledoux S., Min Li S., Chapeau C., Garcia R., Hamet P. and Genest J. Immunocytochemical localisation of atrial natriuretic factor in the heart and salivary glands. *Histochemistry* 1984, **80**, 113–127.
88. Polak J. M. and Bloom S. R. Pathology of peptide-producing neuroendocrine tumours. *Br. J. Hosp. Med.* 1985, 78–88.
89. Polak J. M. and Bloom S. R. (eds.) *Endocrine Tumours*. Edinburgh, Churchill Livingstone, 1985.
90. Sheppard M. N., Corrin B., Bloom S. R. and Polak J. M. Lung endocrine tumours. In: Polak J. M. and Bloom S. R., eds., *Pathology of Peptide-producing Neuroendocrine Tumours*. London, *British Journal of Hospital Medicine* 1985: 209–228.
91. Ibrahim N. B. N., Briggs J. C. and Corbishley C. M., Extrapulmonary oat cell carcinoma. *Cancer* 1984, **54**, 1645–1661.
92. Ballesta J., Polak J. M., Allen J. M., Pals D. T., Lawson J. A., Luden J. H., Slizgi G. R. and Bloom S. R. NPY immunoreactivity in experimental hypertension (aorta-ligated and DOCA treated rats). *Regulat. Pept.* 1984, **9**, 323.

21

Immunocytochemistry of Brain Neuropeptides

G. W. Roberts and Y. S Allen

The discovery of small peptide molecules (the enkephalins) as the endogenous ligands for opiate receptors in the central nervous system was the initial impetus for the intensive neuropeptide research of the last decade. During this time more than 40 pharmacologically active peptides have been localized to mammalian brain and spinal cord and studies of their distribution, receptors, functions and behavioural effects have radically altered our perception of the neurochemistry of brain function.[1]

A brief summary of the distribution of some of these peptides in the central nervous system is given in *Table* 21.1. A detailed review of peptide distribution patterns is beyond the scope of this chapter but many recent reviews on the topic are available.[2,3]

The distribution patterns of peptides in the nervous system and the mounting evidence that they may act as neurotransmitters or neuromodulators in the brain encourages speculation that they may be involved in the pathogenesis of nervous diseases. However, the inaccessibility of the human central nervous system (CNS) severely limits the direct analysis of peptide content in living patients –there are few indications for cerebral biopsy and available tissue in such cases is usually limited to the superficial cortex. Fortunately, the finding that many enzymes and chemical markers of neurotransmitter systems are stable after death, has permitted postmortem studies of the peptide and classical transmitter systems in tissue extracts and on sections. To date most studies of the neuropeptide content of the human CNS have been made on postmortem tissue despite its inevitably less-than-perfect structural preservation.

Studies using radioimmunoassay techniques have shown that peptide concentrations are affected in some neurological diseases such as Huntington's

Table 21.1. *Summary of distribution of neuropeptides in the mammalian CNS*

Neuropeptide	Cortex	Striatum	Hypo-thalamus	Brain stem	Spinal cord
Pituitary peptides					
Corticotrophin	−	−	++	++	+
Growth hormone	−	−	++	++	−
Lipotrophin	−	−	++	++	−
αMSH	−	−	++	++	−
Oxytocin	−	−	++	++	+
Prolactin	−	−	++	++	−
Vasopressin	−	−	++	+	+
Gastrointestinal peptides					
Cholecystokinin	++	++	++	++	++
Gastrin	−	−	−	++	+
Glucagon	+(?)	−	++	+	−
Insulin	+(?)	−	++	+	−
Motilin	++(?)	−	−	−	−
Neurotensin	+	++	++	++	++
Secretin	−	−	++	+	−
Somatostatin	++	++	++	++	++
Substance P	+	++	++	++	++
Vasoactive intestinal polypeptide	++	++	++	++	+
Opioid peptides					
Dynorphin	++	++	++	++	++
β-Endorphin	−	−	++	++	+
Met-enkephalin	+	++	++	++	++
Leu-enkephalin	+	++	++	++	++

Neuropeptide	Cortex	Striatum	Hypo-thalamus	Brain stem	Spinal cord
Hypothalamic releasing peptides					
Corticotrophin-releasing factor	++	++	++	++	−
Growth hormone-releasing factor	−	−	++	−	−
Luteinizing hormone-releasing hormone	−	−	++	−	−
Thyrotrophin-releasing hormone	−	−	++	++	+
Miscellaneous peptides					
Angiotensin	−	+	++	++	−
Bombesin	−	−	++	++	+
Bradykinin	−	+	++	++	−
Calcitonin	−	−	++	+	−
Calcitonin gene-related peptide	−	−	++	−	−
Galanin	+	++	++	++	++
Neuropeptide Y	++	++	++	++	++
Neurokinin A (substance K)	+	+	++	+	+

++ Cell bodies and fibres.
+ Fibres.
? Unconfirmed.
− Not reported.

chorea (substance P, cholecystokinin, met-enkephalin, somatostatin neuropeptide Y), Parkinson's disease (met-enkephalin, cholecystokinin, substance P), Alzheimer's disease (somatostatin) and Riley–Day syndrome (substance P).[4]

In this chapter we will show how immunocytochemistry using formalin-fixed, paraffin-embedded or Vibratome sections and the PAP technique can be used to supplement and sometimes gain advantage over the standard histological and neurochemical techniques and the particular contribution that only immunocytochemistry can make in dissecting the morphology and transmitter specifity of the structures involved in pathological processes. We will confine ourselves to discussion of two neurodegenerative conditions–Alzheimer's dementia (AD) and Huntington's chorea (HC) and examine the role of two peptides, somatostatin and neuropeptide Y (NPY) in these conditions.

1. SOMATOSTATIN AND NEUROPEPTIDE Y

Somatostatin was isolated and sequenced from brain extracts in 1972; NPY was discovered in a similar manner 10 years later. The two peptides share no similarity in amino acid sequence. Both peptides are present in many CNS areas in relatively large amounts,[5,6] have similar distribution patterns (as shown by radioimmuno-assay) and their concentrations in subcortical areas are positively correlated (not correlated in cortical areas).[7] Whilst a considerable volume of data has accumulated on the physiological actions of somatostatin[8] in the CNS, the study of NPY physiology is still in its infancy.[9]

Immunocytochemical studies have demonstrated both NPY and somatostatin-like immunoreactivity to be present in large numbers of neurones, fibres and terminals throughout the human brain. Their distributions are very similar (by RIA studies) and this prompted investigations to see if these peptides are co-localized or closely connected in some way. Using immunocytochemistry, co-localization of somatostatin and NPY has been demonstrated in rat, monkey and human brain to varying degrees in different brain regions. There may be almost total co-localization of the two peptides in the rat striatum and cortex,[10] about 25 per cent co-localization in monkey cortex[11] and an indeterminate but significant amount of co-localization in human cortex.[12,13] On this latter point the lack of positive correlation between somatostatin and NPY content in cortical areas may indicate that the degree of co-localization in human cortex is not great. Both peptides are present in extensive local fibre networks and in the long pathways connecting distant brain areas (stria terminalis, fornix, corpus callosum etc.).[14]

How are these two peptides (which vary in their degree of co-localization according to brain area) affected in neurodegenerative disease? Two conditions have been chosen to illustrate this:

a. Alzheimer's disease — in which the areas affected are mainly the cortex and hippocampus (where the degree of co-localization may be small)
b. Huntington's chorea — in which the area affected is mainly the striatum (where somatostatin/NPY co-localization is almost total).

2. ALZHEIMER'S DISEASE

Senile dementia of the Alzheimer type (AD) is a chronic, progressive, neurodegenerative condition affecting about 15 per cent of the population over 65 years old in a moderate or severe form. It is characterized clinically by memory and cognitive impairment which leads to global intellectual impairment.[15] Histologically, AD manifests characteristic pathological changes, the most prevalent of which are the presence of senile or neuritic plaques and neurofibrillary tangles[16] which can be demonstrated by a variety of histological, and more recently immunocytochemical, techniques (*Fig.* 21.1). Neurofibrillary tangles are found within neurones and consist of paired helical filaments that take on the form of tangled masses, loops or coils. Senile plaques have an amyloid (protein)- and aluminium-containing core[17] surrounded by degenerating axons and other processes. Many of the constituents of the surrounding neurites are derived from both local and distant cells and contain substances found in presynaptic boutons. It has been hypothesized that plaques result from degenerated nerve endings.

These morphological changes (plaques and tangles) are also seen to varying degrees in normal aged and non-demented patients, those with Pick's disease, Creutzfeldt–Jakob disease and parkinsonism and in such apparently unrelated neurological conditions as Down's syndrome and punch-drunk syndrome (dementia pugilistica).[16] They are, however, more prevalent in AD and the severity of the dementia has been related directly to the number of neurofibrillary tangles and senile plaques.[18] The relationship between tangles and plaques is unknown, though it is tempting to speculate that the axons and dendrites of abnormal (tangle-containing) neurones terminate in abnormal bulbous sproutings which constitute the bulk of the plaque. There is no direct evidence for this, however, and it is possible that the nerve terminals forming the plaque only arise in neurones unaffected by tangle formation. Furthermore, the physical distribution of plaques and tangles (they do not always occur together) suggests that they may not be closely associated structurally and may arise independently.

Biochemical analysis of the various transmitter parameters in AD brain tissue has established that there is a loss of cholinergic and monoaminergic projections to the cerebral cortex and a lowered cholinergic content of many brain areas.[19] Peptide content of AD brains has also been investigated and shows substantial loss of cortical somatostatin (with few changes in subcortical regions) with preservation of cortical NPY, CCK and VIP (all present in intrinsic cortical neurones).[5,7] The decrease in somatostatin content is positively correlated with decreases in the cholinergic system[5] and serotonin[55] receptors in the temporal cortex[20] and inversely correlated with plaque count and degree of dementia.[21]

Immunocytochemical techniques have been used to examine the distribution of somatostatin and NPY-like immunoreactivity in AD brain and to try to relate the staining pattern seen to the pathological markers and neurochemical changes of the disease.[22,23] In control tissue (formalin-fixed, paraffin-embedded), somatostatin and NPY immunoreactivity were abundant in the frontal and temporal cortex. The neurones were localized in cortical layers II, III, V and VI and were predominantly of the non-pyramidal type (stellate, multipolar and bitufted). Numerous somatostatin- and NPY-containing immunoreactive structures were also found in other regions as previously reported. On examination, sections (formalin-fixed, paraffin-embedded) from cases of AD showed marked decreases in somatostatin immunoreactivity. In superficial layers (II and III) of the temporal cortex, parahippocampal gyrus and uncus, both neuronal and fibre staining were substantially depleted, some regions showing no observable immunoreactive neurones and a concomitant decrease in local fibre innervation. In addition to the decreases in immunoreactive material, many somatostatin-positive cells, particularly in deeper cortical layers, showed morphological changes consistent with neuronal degeneration and analogous to those obtained in silver-impregnated sections (*Fig.* 21.1). NPY-immunoreactive neurones and fibres did not appear to be as severely affected, although degenerate neuronal profiles could be seen in the deeper cortical layers of severe cases.

Cell counts show that there is a marked decrease (45 per cent) in the number of large somatostatin-immunoreactive neurones (those with a cross-sectional area greater than 100 μm^2) in AD compared with aged controls.[24] No cell counts have yet been done on comparable NPY-stained material.

Somatostatin- and NPY-positive structures were also associated with the plaques and tangles which are the hallmarks of AD. In silver-impregnated sections which were subsequently immunostained, some neurones with argyrophilic tangles were observed to be somatostatin-positive (*Plate* 25). This association was not so marked with NPY-positive material in general, although a few tangle-containing neurones positive for NPY were found in the deeper cortical layers of the most severe cases. Somatostatin-immunoreactive fibres were also found to be associated with plaques (*Fig.* 21.1) with somatostatin-positive axons and dendrites forming large dystrophic clusters and participating with other elements in the plaque formation. NPY-, VIP-, CCK- and substance P-immunoreactive fibres are found to be similarly related to plaques.

The immunocytochemical work described above demonstrates the relationship between the pathological markers of dementia (plaques and tangles) and peptide immunoreactivity. The plaques are obseved to contain heterogeneous peptide-immunoreactive elements. Most of the peptides associated with plaques (NPY, CCK, VIP, substance P) are not decreased in AD or correlated with severity of dementia. Somatostatin content is decreased in AD and is found to be associated with tangles in addition to plaques. This implies that tangles and plaques may arise independently and that plaques are neurochemically non-specific pathological structures, whereas tangles may be highly selective.

On this basis it is a reasonable step to the hypothesis that somatostatin-containing neurones (and probably to a lesser degree somatostatin/NPY neurones) are a primary focus for the disease process and that the loss of somatostatin content or formation of tangles in somatostatin neurones is an early event in the overall pathology of dementia.

The relationship of somatostatin loss to cell loss in cortical areas and atrophy of cholinergic cells in the basal nucleus of Meynert[19] is unknown. It has been reported that cortical lesions will lead to atrophy in the basal nucleus of Meynert,[25] a finding that leads to the conclusion that the cholinergic deficits in AD may be a secondary consequence of the loss of cortical (somatostatin-containing?) neurones. With this knowledge of the degeneration in

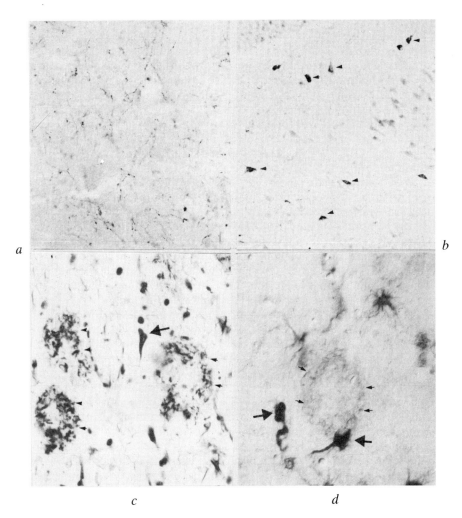

Fig. 21.1. Formalin-fixed, paraffin sections of human brain, taken postmortem. The rabbit antibody to somatostatin was from RIA(UK) and the second antibody and PAP from Miles. (*a*) Somatostatin-positive fibres in normal temporal cortex — PAP. (*b*) Somatostatin-immunoreactive cells (arrowed) in the hilus of the dentate gyrus — PAP. (*c*) Silver-stained section of temporal cortex from a case of Alzheimer's dementia showing tangles (large arrows) and plaques (small arrow) — Bielchowsky. (*d*) Somatostatin-positive neurones in the temporal cortex of a case of Alzheimer's dementia. The degenerate profiles with thickened and twisted dendrites (large arrows) and the somatostatin-immunoreactive plaque (small arrows) can be seen. Note the astrocyte which also shows some somatostatin immunoreactivity — PAP.

transmitter-specific subsets of neurones in Alzheimer's disease, it should now be possible to develop animal models to investigate in detail the neurochemical and behavioural consequences of such lesions and to explore the possibilities of pharmacological amelioration of the condition.

3. HUNTINGTON'S CHOREA

Huntington's chorea is an autosomal dominant inherited neurodegenerative disorder characterized clinically by severe dyskinesia and eventual dementia. Neuropathologically, severe atrophy of the caudate nucleus along with moderate cell loss in many brain regions is evident. Neurochemically, significant depletions of the neurotransmitter-synthesizing enzymes glutamic acid decarboxylase and choline acetyltransferase are found in the striatum (see[26] for review). Recently, the concentrations of a number of neuropeptides (substance P,[27] methionine enkephalin and CCK[28]) have been shown to be decreased. This suggests an indiscriminate cell loss in the caudate, as substance P, met-enkephalin and gamma-aminobutyric acid (GABA) are known to be contained in projection neurones whereas subpopulations of GABA-containing cells and cholinergic neurones are clearly intrinsic.[29]

Early RIA reports indicated that somatostatin and NPY are present in elevated concentrations in samples of HC caudate nucleus,[30,31] suggesting that neuronal elements containing these two peptides persist during the disease process. Since both NPY and somatostatin are present in intrinsic cells within the caudate, this would suggest that cell loss in HC may be more selective than originally thought. Why such cells survive has become an intriguing question.

Immunocytochemical studies on Vibratome sections designed to produce an almost 'Golgi-like' image[32] (with the obvious advantage of transmitter specificity) have shown that in the rat both somatostatin and NPY antisera recognize a subpopulation of neurones united by two common attributes: their size (always falling between 15 and 20 μm in diameter) and their spine-free dendrites. Accordingly they have been classified as medium-sized aspiny type III and IV neurones. [33,34] They are believed to be local circuit neurones and there is substantial evidence that, in the vast majority of these neurones, somatostatin and NPY are co-localized. However, there is also some evidence for independent subpopulations of NPY cells.[34]

Immunocytochemical investigations have shown that, contrary to expectations from RIA data, neurones containing somatostatin immunoreactivity are not significantly affected in HC striatum but there is a marked increase in the concentration of neuronal processes containing somatostatin immunoreactivity in the striatum.[35] This may account for reports of increased somatostatin levels.

It is not known whether the discernibly increased somatostatin-immunoreactive processes are derived from intrinsic striatal neurones or extrinsic afferents. Intrinsic neurones are not likely to account for the change and it is possible that disproportionate maintenance of the afferent axonal arborizations, despite loss of target cells or the accumulation of peptide in afferent axons (possibly from nucleus accumbens or ventral striatum), is responsible. A similar situation may account for the reported increase in NPY immunoreactivity. However, investigations have revealed that NPY-containing neurones are significantly increased in density in choreic caudate and show no signs of morphological degeneration when compared to caudate tissue from neurologically normal controls[36] (Plate 26 and Fig. 21.2).

This suggests a selective preservation of NPY-containing cells in the disease condition and may reflect the survival potential of a particular cell type (the medium-sized aspiny neurone) over the many other striatal cell types present.

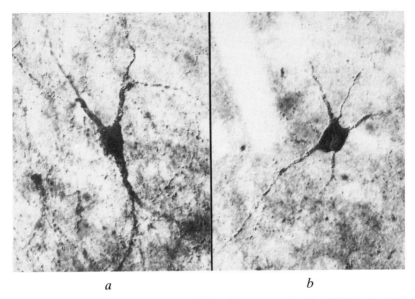

Fig. 21.2. Formalin-fixed Vibratome section (100 μm) immunostained for NPY by the 'Golgi-like' immunoperoxidase technique.[32] The NPY antibody was raised at the Hammersmith Hospital. NPY-immunoreactive cells in the caudate nucleus from (*a*) control tissue and (*b*) Huntington's chorea show essentially the same neuronal characteristics with no obvious signs of degeneration.

Indeed, it has been suggested that it is investiture of spines that confers a vulnerability to destruction in this disease.[30]

Another approach to the question of cell survival in HC has been to examine the reaction of somatostatin and NPY cells to chemical lesions of the striatum in animals. Kainic acid and other excitotoxic glutamate analogues are chemicals capable of mimicking the neurochemical and morphological changes of the disease,[37] thereby encouraging speculation by some that neuronal degeneration in HC may be caused by a similar glutamate-related endogenous toxin.[38,39] However, both somatostatin-[40] and NPY-containing[36] cells in the rat striatum have been shown to be destroyed by kainate lesions with no observable increase in somatostatin or NPY content, at variance with the observations in human disease shown by immunocytochemistry and RIA.

A number of conclusions can be drawn from this: firstly, that kainate neurotoxicity is unlikely to be related to the disease process, casting some doubt on the endogenous toxin hypothesis of HC mentioned earlier; secondly, that duplication of neuropathology is not the only criterion in animal models of neurological disease, the relevant neurochemical parameters must also be altered.

4. CONCLUSION

Neuropeptide research in psychiatric disease is still in its infancy. However, great strides have already been made by immunocytochemistry in the transmitter characterization of defined neuropathological lesions. As knowledge steadily accrues, new conceptual frameworks can be formulated to relate transmitters, morphology, circuitry and pathology with the disturbances of function to reach beyond a mere neurochemical formulation of disease and begin to approach a real understanding of the 'blooming buzzing confusion that is the brain'.

Acknowledgements

This work was supported by the Medical Research Council, UK and the Research Reagent Programme of RIA UK Ltd.

REFERENCES

1. Hökfelt T., Johansson O. and Goldstein M. Chemical anatomy of the brain. *Science* 1984, **225**, 1326–1334.
2. Emson P. C. (ed.) *Chemical Neuroanatomy*. New York, Plenum Press, London, 1984.
3. Iverson L. L., Iverson S. D. and Snyder S. H. *Handbook of Psychopharmacology*, Vol 15. New York, Plenum Press, 1982.
4. Rosser M. N. and Emson P. C. Neuropeptides in degenerative diseases of the central nervous system. *Trends Neurosci.* 1982, **4**, 399–401.
5. Ferrier I. N., Cross A. J., Johnson J. A., Roberts G. W., Crow T. J., Corsellis J. A. N., Lee Y. C., O'Shaughnessy D., Adrian T. E., McGregor G. P., Bacarese-Hamilton A. J. and Bloom S. R. Neuropeptides in Alzheimer Type Dementia. *J. Neurol Sci.* 1983, **62**, 159–170.
6. Adrian T. E., Allen J. M., Bloom S. R., Ghatei M. A., Roberts G. W., Crow T. J., Tatemoto K. and Polak J. M. Neuropeptide Y in human brain—high concentrations in basal ganglia. *Nature* 1983, **306**, 584–586.
7. Allen J. M., Ferrier I. N., Roberts G. W., Cross A. J., Adrian T. E., Crow T. J. and Bloom S. R. Elevation of neuropeptide Y (NPY) in substantia innominata in Alzheimer's Type dementia. *J. Neurol Sci.* 1981, **64**, 325–331.
8. Arimura A. Recent progress in somatostatin research. *Biomed. Res.* 1981, **2**, 233–257.
9. Allen J. M., Birchain P. M. M., Edwards A. V., Tatemoto K. and Bloom S. R. Neuropeptide Y (NPY) reduces myocardial perfusion and inhibits the force of contraction of the isolated perfused rabbit heart. *Regul Pep.* 1983, **6**, 247–254.
10. Vincent S. R., Skirboll L., Hökfelt T., Johansson O., Lundberg J. M., Elde R. P., Terenius L. and Kimmel J. Co-existence of somatostatin and avian pancreatic polypeptide (APP)-like immunoreactivity in some forebrain neurons. *Neuroscience* 1982, **7**, 439–446.
11. Hendry S. H. C., Jones E. G., DeFelipe J., Schmechel D., Brandon C. and Emson P. C. Neuropeptide-containing neurons of the cerebral cortex are also GABAergic. *Proc. Natl Acad. Sci. USA* 1984, **81**, 6526–6530.
12. Vincent S. R., Johansson O., Hökfelt T., Meyerson B., Sachs C., Elde R. P., Terenius L. and Kimmel J. Neuropeptide co-existence in human cortical neurons. *Nature* 1982, **298**, 65–67.
13. Chronwall B., Chase T. N. and O'Donohue T. Coexistence of neuropeptide Y and somatostatin in rat and human cortical and rat hypothalamic neurons. *Neurosci Lett.* 1984, **52**, 213–217.
14. Roberts G. W., Polak J. M. and Crow T. J. Peptide circuitry of the limbic system. In: Trimble M. R. and Zarifian N. E., eds., *Psychopharmacology of the Limbic System*. Oxford, Oxford University Press, 1984: 226–243.
15. Katzman R. (ed.) *Biological Aspects of Alzheimer's Disease*. Banbury Report Vol. 15. New York, Cold Spring Harbour Laboratory, 1983.
16. Tomlinson B. E. and Corsellis J. A. N. Ageing and the dementias. In: Hume-Adams J., Corsellis J. A. N. and Duchen L. W., eds., *Greenfield's Neuropathology*. London, Edward Arnold, 1984: 951–1025.
17. Candy J. M., Oakley A. E. Atack J., Perry R. H., Perry E. K. and Edwardson J. A. New observations on the nature of senile plaque cores. In: Vizi E. S. and Magyar K, eds., *Regulation of Transmitter Function*. Proceedings of the 5th Meeting of the European society of Neurochemistry. 1984, 301–304.

18. Wilcock G. K. The temporal lobe in dementia of the Alzheimer type. *Gerontology* 1983, **29**, 320–324.
19. McGeer P. L. Aging, Alzheimer's disease and the cholinergic system. *Can. J. Physiol. Pharmacol* 1984, **62**, 741–754.
20. Cross A.J., Crow T. J., Ferrier I. N., Johnson J. A., Bloom S. R. and Corsellis J. A. N. Serotonin receptor changes in dementia of the Alzheimer type. *J. Neurochem.* 1984, **43**, 1574–1581.
21. Perry R. H., Candy J. M. and Perry E. K. Some observations and speculations concerning the cholinergic system and neuropeptides in Alzheimer's disease. In: Katzman R., ed., Banbury Report Vol. 15 *Biological Aspects of Alzheimer's disease.* New York, Cold Spring Harbour Laboratory, 1983: 351–362.
22. Roberts G. W., Crow T. J. and Polak J. M. Location of neuronal tangles in somatostatin neurons in Alzheimer's disease. *Nature* 1985, **314**, 92–94.
23. Chan-Palay V., Long W., Allen Y. S., Hasler U. and Polak J. M. (1985) Cortical neurons immunoreactive with antisera against neuropeptide Y and its precursor peptide are altered in Alzheimer's dementia. *J. Comp. Neurol.* 1985, in press.
24. Joynt R. J. and McNeil T. H. Neuropeptides in aging and dementia. *Peptides* 1984, **5**, Supplement 1, 269–272.
25. Pearson R. C. A., Gatter K. C. and Powell T. P. S. Retrograde cell degeneration in the basal nucleus in monkey and man. *Brain Res.* 1983, **261**, 321–326.
26. Bird T. D. and Spokes E. G. S. (1982) Huntington's Chorea. In: T. J. Crow ed., *Disorders of Neurohumoural Transmission.* London, Academic Press, 1982: 145–182.
27. Gale J. S., Bird E. D., Spokes E. G., Iverson L. L. and Jessell T. M. Human brain substance P: distribution in controls and in Huntington's Chorea. *J. Neurochem.* 1978, **30**, 633–634.
28. Emson P. C., Rehfeld J. F., Langevin M. and Rossor M. (1980) Reduction in CCK-like immunoreactivity in the basal ganglia in Huntington's disease. *Brain Res.* 1980, **198**, 497–500.
29. Graybiel A. M. and Ragsdale C. W., Biochemical anatomy of the striatum. In: Emson P. C., ed., *Chemical Neuroanatomy*, New York, Raven Press, 1983: 427–504.
30. Aronin N., Cooper P. E., Lorenz L. J., Bird E. D., Sagar S. M., Leeman S. E. and Martin J. B. Somatostatin is increased in the basal ganglia in Huntington's disease. *Ann. Neurol.* 1983, **13**, 519–527.
31. Dawbarn D. and Emson P. C. Increased brain concentration of neuropeptide Y in Huntington's chorea. In: Peptides and Neurological Disease, Cambridge, August 1983.
32. Sofroniew M. V. and Glasman W. Golgi-like immunoperoxidase staining of hypothalamic magnocellular neurons that contain vasopressin, oxytoxin or neurophysin in the rat. *Neuroscience* 1981, **6**, 619–643.
33. Takagi H., Somogyi P., Somogyi J. and Smith A. D. Fine structural studies on a type of somatostatin-immunoreactive neuron and its synaptic connections in the rat neostriatum: a correlated light and electron microscopical study. *J. Comp. Neurol.* 1983, **214**, 1–16.
34. Allen Y. S., Crow T. J. and Polak J. M. Neuropeptide Y-immunoreactive neurons in the rat forebrain: characterisation by Golgi-like immunoperoxidase staining. 1985, in preparation.
35. Marshall P. C. and Landis D. M. D. Huntington's disease is accompanied by changes in the distribution of somatostatin-containing neuronal processes. *Brain Res.* 1985, **329**, 71–82.
36. Allen Y. S., Cole G.; Crow T. J., Brownell B. and Polak J. M. Neuropeptide Y-containing cells in Huntington's Chorea and in the kainic acid lesioned striatum as an animal model of the disease. In: *Program of the 14th Annual Meeting, Society for Neuroscience* 1984, 242.
37. Coyle J. T. Excitatory amino acid neurotoxins. In: Iversen L. L., Iversen S. D., Snyder S. M., eds., *Handbook of Psychopharmacology*, Vol. 15. New York, Plenum Press, 1983: 237–269.
38. Olney J. W. and DeGubareff T. Glutamate neurotoxicity and Huntington's Chorea. *Nature* 1978, **271**, 557–559.
39. McBean G. J. and Roberts P. J. Chronic infusion of L-glutamate causes neurotoxicity in rat striatum. *Brain Res.* 1984, **290**, 372–375.
40. Beal M. F. and Martin J. B. Effects of lesions on somatostatin-like immunoreactivity in the rat striatum. *Brain Res.* 1983, **266**, 67–73.

22

Neurochemistry of the Spinal Cord

S. J. Gibson and J. M. Polak

1. NEUROTRANSMITTER CANDIDATES IN THE SPINAL CORD

The spinal cord is a vital link between the internal and external environment and the central nervous system. The primary role of the spinal cord is to receive afferent stimuli, in the form of action potentials, and to relay the information to and from the brain. In particular, the mechanism of pain transmission has attracted much active research. The classical hypothesis that the effect of nerves on target organs is mediated by chemicals released from those nerves was first studied in the peripheral nervous system, but proved to be a valid concept in the central nervous system also.[1,2] One of the most important advances in the field of neurochemistry has been the discovery of chemically defined pathways in the central nervous system. Twenty years ago seven chemically identified neurotransmitters had been documented. The first of these classical neurotransmitters to be identified in the spinal cord was acetylcholine which is excitatory and is released by motoneurones at their central synapses with Renshaw cells.[3] Others include the catecholamines, dopamine and noradrenaline, the monoamine, serotonin, the inhibitory amino acids, γ-aminobutyric acid (GABA) and glycine, and glutamic acid which is a powerful excitatory neurotransmitter.[4,5] Many other transmitters have now been identified, some of which are designated 'putative' because their involvement in synaptic transmission in the central nervous system is still unresolved, for example histamine, taurine and aspartic acid. Furthermore, the discovery of a new family of chemicals, neuroactive peptides, in the brain and spinal cord has enormously expanded the list of candidate neurotransmitters or neuromodulators.[6-8]

2. EARLY METHODS FOR IDENTIFICATION OF NEUROTRANSMITTER/MODULATOR SUBSTANCES

Prior to the advent of immunocytochemistry, several direct and indirect histochemical methods were used to visualize and map transmitters or enzymes specifically related to transmitter metabolism in the central nervous system. The well-established acetylcholinesterase stain indicated the presence of cholinergic systems.[9-12] Extralysosomal, fluoride-resistant acid phosphatase is produced by primary sensory neurones and transported to their central terminals in the dorsal horn and is visualized with remarkable efficiency in rodent spinal cord using the Gomori method.[13] The best known method for the demonstration of monoamines is the Falck–Hillarp induced fluorescence method.[14-16] This approach is based on the reaction between monoamine and the formaldehyde-based fixative which forms strongly fluorescent compounds. In addition to the induced fluorescence method for demonstration of monoamines, certain oxidizing agents (e.g. potassium permanganate) form electron-dense precipitates and allow visualization at the electron microscopical level.[17] These two methods can be used alone or in combination with pharmacological manipulation to localize specific classes of amine, i.e. noradrenaline, dopamine or serotonin. Finally, the transmitter re-uptake mechanism at the neuronal cell membrane can be utilized to localize transmitter substances autoradiographically, using radioactive ligands such as γ–[^{3}H]aminobutyric acid, [^{3}H]catecholamines or receptor ligands, for example, the putative cholinergic nicotinic receptor ligand, α–[^{125}I]bungarotoxin.[18,19]

3. IMMUNOCYTOCHEMISTRY

Advances in technology have increased the precision of biochemical purification or synthesis (subsequent to isolation and chemical characterization) of many substances including classical and putative neurotransmitters. This refinement has encouraged production of a wide range of antibodies. Improvements in antiserum quality and immunocytochemical techniques have allowed a more accurate identification of specific neurochemical pathways, providing insight into possible functional roles of their constituents. Antibodies are now available not only to peptides and their precursor molecules, but also to enzymes involved in catecholamine synthesis, (e.g. tyrosine hydroxylase, dopamine β-hydroxylase, dopa decarboxylase, phenylethanolamine N-methyltransferase), acetylcholine synthesis (choline acetyltransferase) and the γ-aminobutyric acid-synthesizing enzyme, glutamic acid decarboxylase, as well as to serotonin and dopamine. Even more recently, antisera to general cytoskeletal elements, e.g. glial fibrillary acidic protein and neurofilament protein, (see Chapters 24 and 25), have gained universal application[20] (Figs. 22.1 and 22.2).

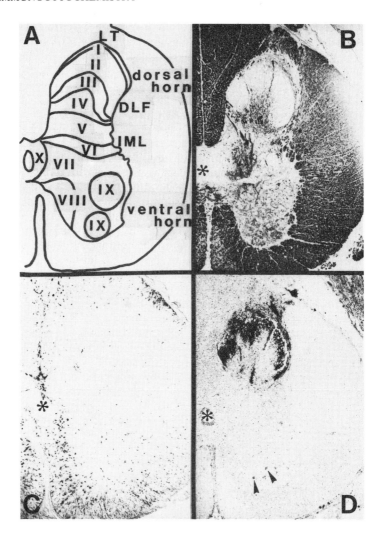

Fig. 22.1 (*A*) Diagrammatic representation of half a transverse section through the lumbar spinal cord of man. The grey matter is classically divided into regions or laminae (I–X) according to cell size. Central terminations of primary sensory neurones enter the dorsal horn via Lissauer's tract (LT) and end mainly in the substantia gelatinosa (laminae II). DLF, = dorsolateral funiculus, IML = intermediolateral cell column region. (*B*) Antiserum to neurofilament protein marks the majority of neurones in the spinal cord, in particular the myelinated axons and motoneurones. * = central canal. Bouin's fixation, wax section (10 μm) peroxidase–anti-peroxidase technique. Magnification × 16. (*C*) Serial section to (*B*), immunostained for glial fibrillary acidic protein. This marks the non-neuronal supporting cells, astrocytes in the spinal cord. * = central canal. Peroxidase–anti-peroxidase technique. Magnification × 16. (*D*) Serial section to (*C*), immunostained for galanin. Galanin-immunoreactivity is most concentrated in fibres and terminals in laminae I–III of the dorsal horn and Lissauer's tract. Some weakly immunostained motoneurones are also visible (arrows). * = central canal. Immunogold–silver intensification method. Magnification × 16.

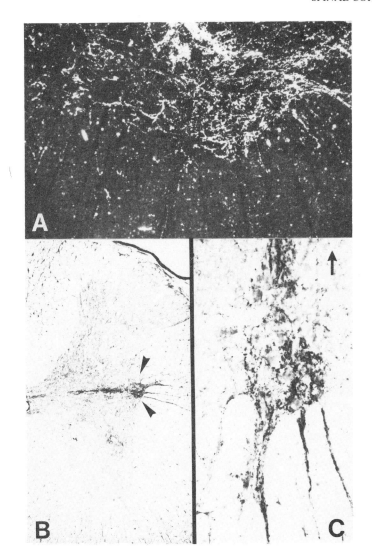

Fig. 22.2. (*A*) Serotonin-immunoreactive fibres in the lumbar ventral horn of the rat spinal cord. Paraformaldehyde fixation, cryostat section (20 μm) peroxidase–anti-peroxidase method. Dark field illumination. Magnification × 135. (*B*) Tyrosine hydroxylase-immunoreactive fibres are mainly concentrated in the intermediolateral cell columns (arrows) of thoracic levels of the rat spinal cord. Paraformaldehyde fixation, cryostat section (20 μm) peroxidase–anti-peroxidase method. Magnification × 40. (*C*) High power photomicrograph of (*B*) tyrosine hydroxylase immunoreactivity in the intermediolateral cell column. Arrow points towards the central canal. Magnification × 250.

4. SPINAL CORD TECHNOLOGY

4.1. Dissection

The spinal cord is an extremely friable tissue and is well protected by the vertebral column and the meningeal coverings in the living animal. The exposed spinal cord is thus very susceptible to damage and requires gentle handling during dissection. A laminectomy (excision of the posterior vertebral arch) is the most commonly used method of dissection. Identification of at least some segments is best made while the cord is in situ, e.g. the sciatic nerve in the rat terminates in segments L4, L5, and L6, and the identified dorsal roots should be tied with cotton prior to removal of the tissue and placement in liquid fixative or storage solution.

In small animals, the spinal cord can be ejected from the vertebral column using applied hydraulic pressure; this offers an alternative rapid method of removal.[21] However, the end-result is not always perfect as there is a tendency for spinal roots to tear from adjoining regions of the spinal cord during the forced exit.

4.2. Fixation

To preserve antigens during immunocytochemical procedures it is necessary to expose the spinal cord to fixative solutions such as alcohols, formalin and glutaraldehyde. All of these have a detrimental effect on antigenicity so a compromise between retention of antigenicity and morphological preservation has to be found. Ordinary formalin-based fixatives, e.g. paraformaldehyde, Zamboni's or Bouin's solutions, are the most frequently employed, preferably by perfusion of the experimental animal, although p-benzoquinone[22] is an acceptable alternative and in many instances offers a clearer visualization of peptide-containing nerve fibres.

In man and the larger mammals, where fixation by perfusion is not possible, spinal cord can be satisfactorily fixed by immersion providing small slices (< 1 cm width) are used. In all cases, to allow optimal penetration of fixative to the spinal cord, the dura should be split. After sufficient time has been allowed for completion of fixation, spinal cord is stored in a buffer containing sucrose and fungicide.

4.3. Immunostaining

Spinal cord lends itself well to cryostat, paraffin, resin or Vibratome sections. For general light microscopical studies, sections are usually cut at 10–30 µm thickness and mounted on slides coated with chrome–alum, formol–gelatin, albumen or poly(L-lysine).[23] We have found poly(L-lysine) to be the most reliable of these for secure section adhesion throughout the long washing steps necessary during immunostaining procedures. Thicker samples are most frequently processed as free-floating sections and mounted after completion of immunostaining. All immunocytochemical methods (immunofluorescence, peroxidase, peroxidase–anti-peroxidase, avidin–biotin and immunogold–silver — see Chapter 5) used on other tissues can be equally successfully applied to spinal cord. Modifications of the methods to improve penetration of antisera and hence visualization of antigenic sites, include dehydration and rehydration of sections prior to application of primary antibody and addition of Triton X–100 to the antisera and/or washing buffer. Finally, spinal cord is a highly vascular tissue, thus blocking of endogenous peroxidase, especially in non-perfused tissue, is a prerequisite for all techniques involving peroxidase reactions.

5. CLASSICAL NEUROTRANSMITTERS

Considerable progress has been made in recent years in characterizing known neurotransmitter substances and mapping their distribution and origin in the spinal cord.[18,24,25] The availability of antibodies to amines, e.g. serotonin, dopamine, and enzymes involved in their biosynthesis, has led immunocytochemistry to rival aldehyde-induced fluorescence techniques as a method of localization. However, autoradiographic demonstration of labelled amines and amino acids, e.g. glutamic acid, glycine, is frequently employed but, in general, resolution of nerves and their processes is low compared with the immunocytochemical procedure. Much elegant work on the localization of serotonin, and the enzymes choline acetyltransferase, glutamic acid decarboxylase and tyrosine hydroxylase is documented in the literature. The distribution of these substances in the spinal cord is only briefly summarized in *Table* 22.1 since the bias of this chapter is towards the new class of candidate transmitters, peptides (*Fig.* 22.2).

Table 22.1. Immunocytochemistry of classical neurotransmitters in the spinal cord

System	Antisera to	Localization in spinal cord	
		Dorsal	Ventral
Acetylcholine	Choline acetyltransferase	Fibres	Fibres, motoneurones preganglionic neurones[26]
Serotonin	Serotonin	Fibres	Fibres, dense in ventral horn and intermediolateral cell columns of thoracic regions[27]
γ-Aminobutyric acid	Glutamic acid decarboxylase	Fibres and cell bodies	Cell bodies and some motoneurones[18,28]
Dopamine/ noradrenaline/ adrenaline	Tyrosine hydroxylase	Fibres	Fibres, dense in ventral horn and intermediolateral cell columns of thoracic regions[29]

5.1. Peptides as Neurotransmitter Candidates

Neuropeptides in the spinal cord form a vast, rapidly expanding group of chemical families many of which may qualify as candidates for neurotransmitter or neuromodulator functions.[6-8] However, several biochemical, physiological and pharmacological criteria must be fulfilled before any substance can seriously be considered a neurotransmitter, namely (*a*) synthesis and storage in a neurone and transport to nerve terminals, (*b*) release, and (*c*) mimicry of the effects supposedly mediated by the naturally occurring substance at the postsynaptic receptors.[30]

In practice it has not proved an easy task to demonstrate all the above points for any one peptide. To date, the strongest putative neurotransmitter is substance P first hypothesized in 1953 (but *see* [31]). Much excitement was shown when this peptide was localized to primary sensory neurones and a Ca^{2+}-dependent release could be evoked from the central terminals of primary sensory afferent fibres in the spinal cord. Fervent study was devoted to this peptide and subsequently many more peptides were isolated and, using immunocytochemistry, localized to specific regions of the central nervous system.

More than 20 peptides have now been localized to the spinal cord. Those of major importance are listed in *Table* 22.2. It is of interest that their distribution is widespread, not only in the central and peripheral nervous systems, but also in some endocrine cells. The peptides are grouped in families according to similarities in their amino-acid sequence, e.g. vasoactive intestinal polypeptide and peptide histidine isoleucine. In general, each family has a similar distribution in the spinal cord. It is possible to expand the groups of chemically related peptides by adding the miscellaneous peptides which also have similiar distribution patterns, e.g. calcitonin gene-related peptide and cholecystokinin/gastrin-like peptide could join the substance P family. However, such strict classifications are not advisable as most of the peptides (except those which co-locate) have some unique distribution feature (*see* Section 7).

The rapid discovery of new peptides and peptide families in the spinal cord (the most recent being galanin) challenges physiologists to define their modes of action as those of neurotransmitters, neuromodulators or trophic agents.

6. MEANS OF INVESTIGATING THE ORIGIN OF PEPTIDE–CONTAINING NERVES

There are three possible sources of peptide-containing nerves in the spinal cord (*Fig.* 22.3). One is extrinsic, from afferent fibres derived from neurones within the dorsal root ganglia; two are intrinsic, from local cell bodies and interneurones or axons and collaterals from descending systems (*Fig.* 22.3). Immunocytochemistry, alone or combined with pharmacological, neurochemical and neurosurgical procedures, has given significant insight into the various transmitter and peptide-containing pathways in the spinal cord. The methods which are most frequently employed to determine the origin of peptide-containing nerves in spinal cord of experimental animals are described below.

6.1. Extrinsic Systems

6.1.1. *Capsaicin*

In recent years, capsaicin (8-methyl-*N*-vanillyl-6-nonenamide), extracted from red peppers, has become a powerful tool for the neurochemist.[46,47] In man, application of capsaicin to the skin induces an initial burning sensation followed by a prolonged period of desensitization which is thought to represent the release

*Table 22.2. **Peptides in the mammalian spinal cord***

Peptide family	Isolated	Brain	Spinal cord	Main action in spinal cord (where known)
Substance P	Horse hypothalamus	*	*	Excitatory[8]
Neurokinin A[35]	Pig spinal cord	*	*	?
Neurokinin B[35]	Pig spinal cord	†	†	?
Vasoactive intestinal polypeptide	Pig intestine	*	*	Excitatory[8]
Peptide histidine isoleucine[36]	Pig intestine	*	*	?
Peptide histidine methionine[36]	Human neuroblast-oma cells	*	*	?
Bombesin	Frog skin	*	*	Excitatory[6]
Gastrin releasing pep-tide	Pig antrum	*	*	?
Neuromedin B[37,38]	Pig spinal cord	*	*	?
Adrenocorticotrophic hormone	Pituitary	*	*	?
Dynorphin	Pituitary	*	*	Inhibitory[8]
Endorphin	Pituitary	*	*	Inhibitory[8]
Methionine-enkephalin	Pig brain	*	*	Inhibitory[8]
Leucine-enkephalin	Pig brain	*	*	Inhibitory[8]
FMRF amide-like peptide[39]	Mollusc CNS	*	*	?
Vasopressin	Cow/ox hypotha-lamus	*	*	Excitatory[32]
Oxytocin and related neurophysins	Cow/ox hypotha-lamus	*	*	Excitatory[32]
Neurotensin	Cow/ox hypotha-lamus	*	*	Excitatory[33]
Neuromedin N[40]	Pig spinal cord	†	†	?
Miscellaneous peptides				
Cholecystokinin/gas-trin-like peptide	Pig intestine	*	*	Excitatory[8]
Somatostatin	Cow/ox hypotha-lamus	*	*	Inhibitory[6]
Neuropeptide Y[41]	Pig brain	*	*	?
Calcitonin gene-related peptide[42]	Medullary carcino-ma; rat thyroid	*	*	?
Corticotrophin releas-ing factor[43]	Sheep hypothalamus	*	*	?
Angiotensin	Hypothalamus	*	*	Excitatory[6]
Galanin[44,45]	Pig intestine	*	*	?
Thyrotrophin-releasing hormone	Hypothalamus	*	*	Excitatory[34]

* = present.
† = not yet localized by immunocytochemistry.
? = unknown.
Relatively newly discovered peptides only are referenced. For others *see* the literature.[6,7,8,18]

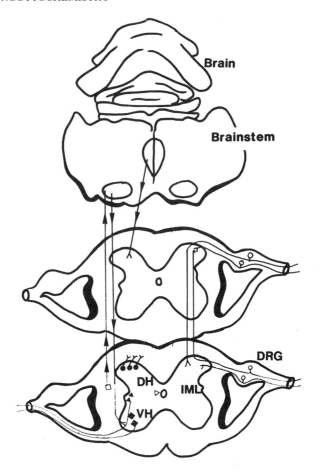

Fig. 22.3. The terminals of primary sensory neurones end in the upper dorsal horn and represent an extrinsic source of peptide. Descending fibres derived from peptide-containing cell bodies in the brain-stem nuclei or brain is one intrinsic source. Colchicine pre-treatment of spinal cord reveals numerous local cell bodies and interneurones (●□▲); some give rise to ascending fibres (□), others (▲) are localized to parasympathetic and sympathetic intermediolateral columns (IML) and to motoneurone soma (■).

and subsequent depletion of 'nociceptive substance' from sensory nerve terminals. This neurotoxin selectively affects the small-sized dorsal root ganglion cells, some of which synthesize peptides. The action of capsaicin depends on the route of administration chosen. Systemic administration to neonatal rats induces a permanent degeneration of small dorsal root ganglion cells, while in the adult the effect is reversible. A more localized action can be produced by applying capsaicin perineurally directly to a peripheral nerve, e.g. the sciatic. This does not produce any apparent degeneration, but blocks transport of peptide and induces long-term changes in the peptide-containing terminals in the spinal cord as well as in the periphery (*Fig.* 22.4).

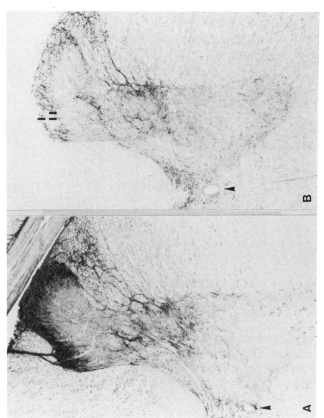

Fig. 22.4. Neonatal administration of capsaicin to rats induces a marked decrease of substance P immunoreactivity from primary sensory neurones and their terminals in the spinal cord. There is a dramatic loss of substance P from laminae I and II and from a small region ventral to the central canal (arrow) in the treated rat (*B*) compared to the control (*A*). In other regions, including the ventral horn, substance P immunoreactivity remains unaffected. Cryostat section (20 μm) thoracic spinal cord. *p*-Benzoquinone fixation, peroxidase–anti-peroxidase method. Magnification × 98.

6.1.2. Ganglionectomy and Dorsal Rhizotomy

Surgical deafferentation can be achieved by removing the dorsal root ganglion (ganglionectomy) or sectioning the dorsal and ventral roots (rhizotomy) (*Fig.* 22.5). Ventral rhizotomy results in loss of motor control from respective dermatomes and is thus rarely performed for ethical reasons. However, presence of substantial numbers of afferent fibres has been reported in the ventral roots. Removal of the ganglion eliminates afferents derived from cells in the dorsal root ganglion from both dorsal and ventral roots (cf. dorsal rhizotomy which only interrupts afferents travelling via the dorsal roots to the spinal cord) (*Fig.* 22.3). Thus, ganglionectomy allows an approximation of the number of peptide-containing afferent fibres in the ventral root. For effective removal of afferent input to any given segment of the spinal cord, several ganglia or spinal roots should be treated since not all afferent fibres end in their segment of entry but may ascend or descend over several segments before terminating.

6.1.3. Peripheral Nerve Lesions

While manipulation of peripheral nerves does not strictly qualify as a deafferentation procedure, sectioning or ligation of a peripheral nerve may result in a degenerative atrophy of the central terminals of primary sensory neurones[13] (*Fig.* 22.5). Thus, in theory, the topography and relative numbers of peptide-containing afferent fibres projecting to the spinal cord can be calculated for any given nerve from the region of decreased immunostaining.

6.2. Intrinsic Systems

6.2.1. Colchicine and Vinblastine

In the spinal cord, peptide-containing cell bodies are less frequently observed than fibres and terminals. Cell bodies often contain low levels of peptide/transmitter in comparison to the nerve endings. Techniques which halt transport of peptide/transmitter from the cell body and induce a build-up of levels are necessary to enhance visualization. Inhibitors of mitosis such as colchicine and vinblastine and other alkaloids from *Vinca* species are commonly used in experimental animals to arrest axonal transport.[48] For demonstration of cell bodies in the spinal cord, colchicine (or other like substances) may be administered systemically, intraventricularly or injected directly into the cord (*Fig.* 22.6), the last possibility being the most widely used and successful method. Survival times of 2–3 days are normally long enough to allow sufficient build-up of peptide/transmitter within the perikaryon.

6.2.2. Hemisection and Transection

The presence of peptide-containing ascending and descending pathways in the spinal cord can be determined by examining the effect of spinal cord lesions, namely hemisections or transections, on peptide levels (build-up or depletion)

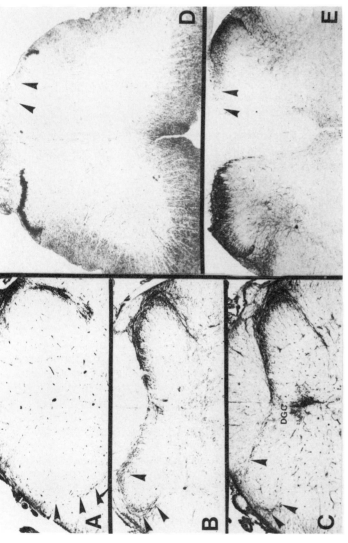

Fig. 22.5. (*A*) There is a dramatic loss of vasoactive intestinal polypeptide immunoreactivity from the ipsilateral sacral (S₂) dorsal horn of the cat (arrows) following unilateral sacral ganglionectomy. *p*-Benzoquinone fixation, cryostat section (20 μm), peroxidase–anti-peroxidase method. Magnification × 22. (*B, C*) Serial sections of substance P (*B*) and calcitonin gene-related peptide (*C*) immunoreactivity in the sacral (S₃) spinal cord of the cat following unilateral rhizotomy. There is a marked depletion of both peptides from the ipsilateral dorsal horn (arrows) and dorsal grey commissure (DGC) in (*C*). *p*-Benzoquinone fixation, cryostat section (20 μm), peroxidase–anti-peroxidase method. Magnification × 22. (*D, E*) Fluoride-resistant acid phosphatase (FRAP) activity and substance P immunoreactivity are decreased from the ipsilateral medial two thirds of the lumbar (L4) dorsal horn (arrows) of the rat after sectioning of the sciatic nerve. The lateral third corresponds to afferent terminations from other nerves. (*D*) Paraformaldehyde fixation, frozen section (50 μm), Gomori method; (*E*) *p*-benzoquinone fixation, cryostat section (20 μm), peroxidase–anti-peroxidase method. Magnification (*D*) × 32. (*E*) × 95.

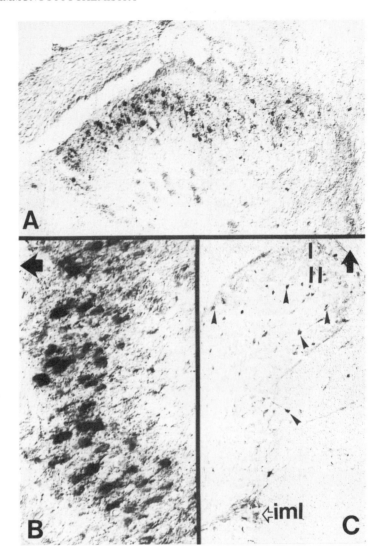

Fig. 22.6 Colchicine pre-treatment allows visualization of numerous peptide-containing cell bodies: (*A*) neurotensin-immunoreactive cell bodies in the dorsal horn of rat. (*B*) High power photomicrograph of neurotensin in the upper dorsal horn, immunoreactive cells are numerous and densely packed. Arrow points dorsally. (*C*) Vasoactive intestinal polypeptide-immunoreactive neurones (some arrowed) in the thoracic spinal cord of rat are scattered throughout the dorsal horn and in the intermediolateral cell columns (iml). Large arrow points dorsally. *p*-Benzoquinone fixation, cryostat section (20 μm), peroxidase–anti-peroxidase method. Magnification: (*A*) × 150, (*B*) × 310, (*C*) × 52.

rostral and caudal to the lesion. Complete transection of the spinal cord, although drastic, ensures that possible contributions from fibre tracts which cross from one side of the cord to the other are not overlooked. In the ventral spinal cord, where peptides originate mainly from intrinsic sources, the effect of hemisection/transection is quite clear. However, in the dorsal horn where, in some cases, there is a substantial amount of peptide in afferent fibres and terminals, small changes may be masked. In this case, removal of the afferent input (pre-treatment with capsaicin, rhizotomy or ganglionectomy) may be advantageous prior to lesioning the spinal cord.

6.2.3. Brain and Brain Stem Lesions

Peptides and transmitter-containing cell bodies are numerous in the cortex, hypothalamus and medulla and some project to the spinal cord. Study of changes in staining patterns in the spinal cord following knife-cut or electro-coagulation lesions to discrete brain areas determines which nuclei have descending projections to the spinal cord. In addition, chemical lesioning of brain stem nuclei has proved a useful tool. Serotoninergic neurones in the medullary raphe complex are known to project to the spinal cord. After lesioning these descending fibres with the neurotoxin 5,6-or 5,7-dihyroxytrypta-mine, not only serotonin-containing, but also substance P-containing and thyrotrophin-releasing hormone-containing fibres disappear from the ventral spinal cord. This was proof not only of a descending substance P and thyrotrophin-releasing hormone system but also of coexistence between the three substances.[18]

6.2.4 Tracing and Immunocytochemistry

The advent of anterograde and retrograde tracing methods has demonstrated that nerves from many brain and brain stem regions project to the spinal cord and has also helped to elucidate many afferent and efferent connections between the spinal cord and peripheral nervous system.[49] In addition, intracellular filling of spinal cord neurones with tracers following recording has allowed a classification of morphological features with electrophysiological characteristics.

Combination of immunocytochemistry with tracing techniques has proved an extremely powerful and precise tool for demonstrating the origin of peptide/transmitter in the spinal cord. At present many tracers are available. Horserad-ish peroxidase (HRP) and conjugates including wheat-germ agglutinin (WGA) and the fluorescent dyes are the most frequently used. Choice of tracer predetermines the immunocytochemical procedure. Originally HRP was used to map pathways; this necessitated demonstration of monoamine or peptide as a fluorescent product, followed by photography and histochemical processing for HRP. Detection of a monoamine or peptide by an immunoperoxidase method using one chromogen and of tracer HRP with another permitted simultaneous visualization of both reaction products. WGA can now be used as a tracer and demonstrated immunocytochemically. All methods involving peroxidase chemistry for both tracer and cell product require rigorous control procedures to avoid false positives. The use of fluorescent dyes, e.g. True Blue, combined with aldehyde-induced fluorescence, acetylcholinesterase staining and, in particular, the indirect immunofluorescence method for detection of monoamine and peptide antigenic sites has revolutionized neurochemical tracing.[50] The technique is comparatively simple, compared to the HRP method, and allows visualization of two fluorescent markers (emitting light of different wavelengths) in the same section.[49,51]

6.2.5. Cultures

The complexity and heterogeneity of the central nervous system has hindered study of the action of neurones in the mammalian spinal cord. Cell culture (*see* Chapter 15) has given much insight into a possible role of peptides as candidate neurotransmitters in the spinal cord. The decreased complexity and increased accessibility of dissociated cultured neurones offers an alternative opportunity for examining events in detail. Although, as monolayers, cell cultures depart from the situation *in vivo*, they are useful models for assay of pharmacological actions and the physiology of neurotransmitters, peptides and related substances at synapses between neurones of the dorsal horn of the spinal cord.[52,53]

7. DISTRIBUTION

7.1. Dorsal Root Ganglia

The cells in the dorsal root ganglia were originally divided into two classes according to their diameters.[54] The large cells give rise to large myelinated fibres and the small cells (primary sensory neurones) to the unmyelinated (C) and finely myelinated (Aδ) fibres. The central terminations of primary afferent fibres derived from the small cells end mainly in the substantia gelatinosa (lamina II, *see Fig.* 22.1). The enzyme, fluoride-resistant acid phosphatase (FRAP) and several peptides including calcitonin gene-related peptide, cholecystokinin/gastrin-like peptide, galanin, gastrin-releasing peptide/bombesin-like peptide, somatostatin, substance P and vasoactive intestinal polypeptide have been localized to a subpopulation of small dorsal root ganglion cells.

At the ultrastructural level, several attempts at further classification of the cells in the dorsal root ganglia have been made on the basis of the Nissl substance and the distribution of the Golgi apparatus. However, according to the peptide staining, there appears to be a case for a third subtype of cell, an intermediate-sized neurone (20–40 μm in diameter).[42,55] In these cells, substance P, calcitonin gene-related peptide, cholecystokinin/gastrin-like peptide and vasoactive intestinal polypeptide have been identified; in general the immunoreaction appears less intense and more particulate in the cytoplasm than in the smaller-sized cells (15–20 μm). To date, calcitonin gene-related peptide is the most abundant peptide in the dorsal root ganglion and, like FRAP, is easily detected (*Figs.* 22.7 and 22.8). Colchicine injections to the ganglia or ligation of the peripheral process of the spinal nerve, markedly improve visualization of peptide-containing cell bodies, in particular those immunoreactive for vasoactive intestinal polypeptide.

With so many peptides present in the primary sensory neurones of the dorsal root ganglia, it is hardly suprising that some co-locate (*Fig.* 22.8) and indeed it is possible that more than two peptides will be found within one cell in the future. It is not known whether there is preferential or simultaneous release from the

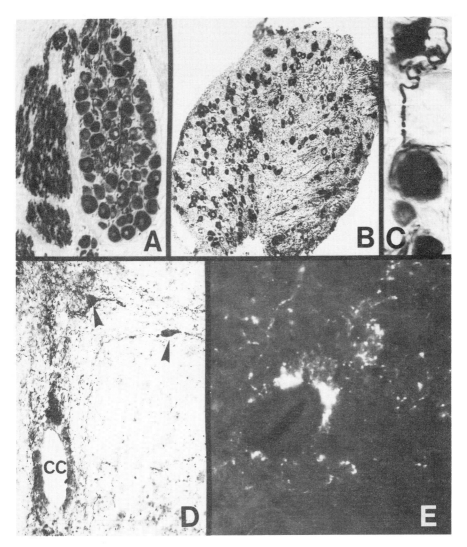

Fig. 22.7. (*A*) Antiserum directed to neurofilament protein marks all cells in the dorsal root ganglion. Human lumbar dorsal root ganglion. Formalin fixation, wax section (5 μm) peroxidase–anti-peroxidase method. Magnification × 90. (*B*) In the dorsal root ganglion calcitonin gene-related peptide-immunoreactive cells are the most abundant of all the peptides. The majority of small sized and some intermediate sized cells are positively stained. No large cells are immunoreactive. Rat dorsal root ganglion, Bouin's fixation, wax section (5 μm). Immunogold–silver intensification method. Magnification × 150. (*C*) High power photomicrograph of calcitonin gene-related peptide-immunoreactive primary sensory neurones in the rat dorsal root ganglion. *p*-Benzoquinone fixation, cryostat section (20 μm), peroxidase–anti-peroxidase method. Magnification × 415. (*D*) Neuropeptide Y-immunoreactive fibres and cell bodies (arrows) in the dorsal grey commissure directly dorsal to the central canal (cc) in the sacral spinal cord of the rat. Paraformaldehyde fixation, cryostat section (20 μm) peroxidase–anti-peroxidase method. Magnification × 110. (*E*) Neurophysin-immunoreactive fibres in lamina X, the area around the central canal. Rat thoracic spinal cord. *p*-Benzoquinone fixation, cryostat section (20 μm), indirect immunofluorescence method. Magnification × 125.

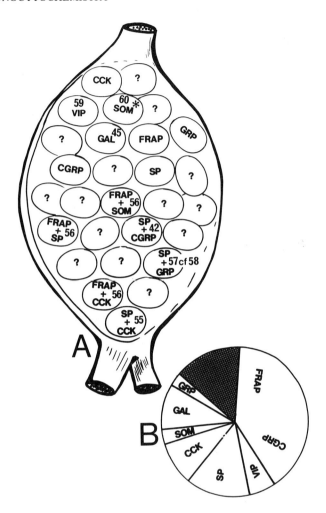

Fig. 22.8 (*A*) Neuronal populations in the rat dorsal root ganglia. Some degree of coexistence has been demonstrated in certain subpopulations (*see* the literature). *It should be noted that somatostatin is present in a separate population of cells from those which contain substance P.[60] (*B*) Relative proportion of peptide and FRAP-containing cells (15–45 μm diameter) so far identified in the dorsal root ganglion. Gastrin-releasing peptide/bombesin-like peptide and somatostatin represent small subset, 5–15 per cent, cholecystokinin/gastrin-like peptide, galanin, substance P and vasoactive intestinal polypeptide, 20–40 per cent and calcitonin gene-related peptide and FRAP, 40–50 per cent of the total. The dark hatched area is reserved for cells which may contain as yet unidentified peptide(s).

CGRP, calcitonin gene-related peptide; CCK, cholecystokinin/gastrin-like peptide; GAL, galanin; GRP, gastrin-releasing peptide/bombesin-like peptide; SOM; somatostatin; SP, substance P; VIP, vasoactive intestinal polypeptide.

terminals of the peptides which are co-stored, although there is some evidence for a behavioural effect in the lumbar spinal cord induced by substance P and calcitonin gene-related peptide together, which is different from that produced by either peptide alone.[61]

7.2. Dorsal Horn

The dorsal horn is particularly rich in peptides and, indeed, every peptide so far discovered in the spinal cord, with the exception of thyrotrophin-releasing hormone, is concentrated in the upper dorsal horn laminae I–III[7,18,42,62] (*Fig.* 22.1). This region of the spinal cord contains thousands of closely packed neurones and its complexity derives from its role in processing and sorting incoming information. The majority of terminals of afferent fibres end in laminae I–II and consequently many peptide-containing fibres and terminals are concentrated here; substance P, cholecystokinin/gastrin-like peptide and calcitonin gene-related peptide are some of the most abundant. Rhizotomy and capsaicin experiments have aided in determining the extrinsic (afferent) origin of peptide-containing nerves in the dorsal horn. Following these manipulations the residual peptide is derived from local cell bodies, ascending axons from interneurones or supraspinal tracts. Other peptides localized to the upper dorsal horn which are not found in afferents and are thought to be intrinsic to the central nervous system, include the enkephalins, neurotensin and neuropeptide Y (*see Table* 22.3).

Table 22.3. *Peptide distribution in dorsal horn of rat*

Peptide	Laminae			Cell bodies	Neonatal capsaicin or rhizotomy
	I	*II*	*III*		
Adrenocorticotrophic hormone[62]	+	+	+	n.r.	n.r.
Angiotensin[7]	+	++	+	n.r.	↓
Calcitonin gene-related peptide[42]	++++	++++	+	n.r.	↓
Cholecystokinin/gastrin-like peptide[62,63]	+++	+++	+	oo	↓
Corticotrophin releasing hormone[43,64]	+	++	+	oo	↓
Dynorphin[39]	+	++	+	oo	n.r.
Enkephalin[18,62,39]	++	+++	+	oo	↔
FMRF amide-like peptide[39]	+	++	+	oo	↔
Galanin[44,45]	++	++	+	ooo	↓
Growth hormone releasing peptide/ bombesin[58]	++	+++	+	n.r.	↓
Neuromedin B[38]	+	++	n.r.	n.r.	n.r.
Neurotensin[18,39,62]	++	++	+	oooo	↔
Neuropeptide Y[41]	+	++	+	ooo	↔
Neurophysin-oxytocin[62]	+	n.r.	n.r.	n.r.	↔
Peptide histidine isoleucine[36,63]	+	n.r.	n.r.	oo	↓
Somatostatin[7,65,66]	+	++	+	oooo	↓
Substance P[18,62]	++++	++++	++	ooo	↓
Thyrotrophin-releasing hormone[18]	n.r.	n.r.	n.r.	n.r.	n.r.
Vasopressin[62]	n.r.	n.r.	n.r.	n.r.	n.r.
Vasoactive intestinal polypeptide[65,67]	+	n.r.	n.r.	oo	↓
Fluoride-resistant acid phosphatase*[13]	n.r.	+++	n.r.	oo	↓

Fibres: ++++ very dense, +++ dense, ++ moderate, + sparse.
Cell bodies: oo moderate (5–10/section), ooo numerous (10–20/section), oooo abundant (20/section).
n.r. = not reported, ↓ decrease, ↔ no change.
*The enzyme, fluoride-resistant acid phosphatases is also included here.

While it may appear that there is little organization in the distribution of peptide-containing fibres and terminals in the dorsal horn, there are certain notable differences; for example, vasoactive intestinal polypeptide and peptide histidine isoleucine nerves are distributed mainly to lamina I; neuromedin B and somatostatin to lamina II; substance P, cholecystokinin/gastrin-like peptide and enkephalin to laminae I–III etc. In addition, since some peptides are known to be co-located in dorsal root ganglion cells, a certain degree of overlap in the distribution of central terminals in the spinal cord may be expected.[36,42]

In general, the distribution of any given peptide in the dorsal horn is similar throughout the entire length of the spinal cord, but there is a trend for some peptides (substance P, cholecystokinin/gastrin-like peptide, galanin, somatostatin) to increase in density in sacral segments. This is particularly exaggerated in the case of vasoactive intestinal polypeptide and peptide histidine isoleucine which are thought to mark preferentially a population of pelvic visceral afferent fibres.[68,69]

Most of the peptide immunoreactivity in the dorsal horn is localized to fibres and terminals although neurotensin-, galanin-, neuropeptide Y- and corticotrophin-releasing factor-immunoreactive cell bodies can be identified (*Fig.* 22.6). However, after colchicine injection, many more peptide-containing cell bodies are revealed in lamina I through to lamina III. For a more detailed study of the morphology and laminar distribution of peptide-containing cell bodies in the dorsal horn *see* Hunt et al.[70]

Interpretation of the presence of peptides in descending and ascending pathways in the dorsal horn is complicated by possible contributions from afferent fibres and local cell bodies. Peptides not present in the latter locations include oxytocin, vasopressin and their related neurophysins; their origin from the paraventricular nucleus can be clearly shown, whilst, with regard to substance P, it can only be estimated that 20 per cent of the immunoreactivity in the dorsal horn derives from descending fibres. Ascending vasoactive intestinal polypeptide and corticotrophin-releasing factor projections have recently been described originating from cell bodies located near the dorsal horn in the dorsolateral funiculus.[43,71]

7.3. Central Canal

The region around the central canal (lamina X) and the dorsally adjacent area (dorsal grey commissure) is also thought to receive information concerning noxious events in the periphery.[72] Indeed, most peptides found in primary afferent fibres can be localized to lamina X. In particular, substance P, cholecystokinin/gastrin-like peptide, calcitonin gene-related peptide, vasoactive intestinal polypeptide and gastrin-releasing peptide/bombesin-like peptide are located to a small longitudinal bundle of fibres ventral to the central canal and all are depleted from this bundle following neonatal capsaicin pre-treatment (*Fig.* 22.4). Other fibres found in the central canal region include those immunoreactive for enkephalin, neurotensin, neuropeptide Y and corticotrophin-releasing factor which are possibly derived from intrinsic sources. In common with the dorsal horn, numerous small-sized, peptide-containing cell bodies have been identified in this region of the spinal cord, particularly in sacral levels, as well as clusters of large perikarya immunoreactive for cholecystokinin/gastrin-like peptide and galanin in lumbar segments.

7.4. Ventral Spinal Cord

In the ventral spinal cord (laminae VI, VII, VIII, and IX), the relative abundance of peptide is small compared to the dorsal horn or central canal region.

Fibres immunoreactive for substance P, cholecystokinin/gastrin-like peptide, somatostatin, neuropeptide Y and enkephalin are the most numerous. The remaining peptides localized to the dorsal horn (*Table* 22.3) are represented only as occasional fibres in the ventral cord. In addition to fibres, some somatic motoneurones have been identified as calcitonin gene-related peptide-,[42] galanin-[45] or somatostatin-immunoreactive[66] (*Fig.* 22.9).

In certain regions of the ventral spinal cord, peptide-immunoreactive fibres tend to aggregate around specific groups of motoneurones. In lumbar segments of the rat, substance P-containing fibres appear to be concentrated around a mediolateral motor nucleus and to form a peptidergic link between the ventral and dorsal spinal cord[73] (*Fig.* 22.9). Somatostatin-, enkephalin- and neuropeptide Y-immunoreactive fibres are particularly notable in sacral segments in close association with the pudendal motor nucleus (Onuf's nucleus) which controls striated muscles involved in micturition, defaecation and ejaculation (*Fig.* 22.9).

The origin of many peptides (from descending or intrinsic systems) in the ventral cord remains to be established. Hemitransection or transection markedly reduces levels of substance P (*Fig.* 22.9) and thyrotrophin-releasing hormone caudal to the lesion, while no change in enkephalin, neurotensin, or neuropeptide Y immunoreactivity has yet been reported.

7.5. Autonomic Centres

The autonomic centres of the spinal cord are located in the intermediolateral cell columns of thoracic and high lumbar segments and the sacral segments of the ventral spinal cord. The sympathetic system is localized to thoracolumbar regions and the parasympathetic to the sacral spinal cord. Together these constitute the autonomic efferent outflow. Many peptide-containing fibres and cell bodies are prevalent in the intermediolateral cell columns of the spinal cord[39,74] (*Table* 22.4; *Fig.* 22.10). Using the technique of retrograde tracing from sympathetic nerves, e.g. hypogastric, and parasympathetic nerves, e.g. pelvic, combined with immunocytochemistry, it has been possible to demonstrate the presence of some peptides in preganglionic, visceromotoneurones (cholecystokinin/gastrin-like peptide,[75] somatostatin[66] and enkephalin[69]). To date only cholecystokinin/gastrin-like peptide has been localized to both sympathetic and parasympathetic neurones. In general, the intermediolateral cell columns of sacral segments (parasympathetic sacral nucleus) are more densely innervated than those of thoracic regions (*Fig.* 22.10).

Apart from a direct source of peptide from the preganglionic neurones, other sources of peptide-containing fibres to the intermediolateral cell columns include the following: visceral afferent fibres, some of which contain substance P, cholecystokinin/gastrin-like peptide and vasoactive intestinal polypeptide, which terminate near to the preganglionic cells; supraspinal systems, substance P, enkephalin, oxytocin and thyrotrophin-releasing hormone, which are partly

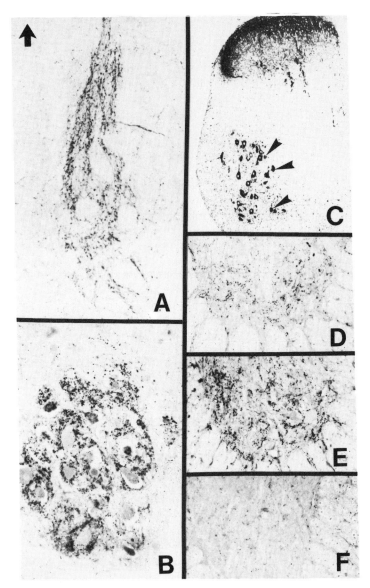

Fig. 22.9. (*A*) Substance P-immunoreactive fibres aggregate around a motor nucleus in the upper lumbar ventral spinal cord of the rat. Fibres emanate from or extend to the dorsal horn. Arrow points dorsally. *p*-Benzoquinone fixation, cryostat section (20 µm) peroxidase–anti-peroxidase method. Magnification × 120. (Reproduced from Gibson et al.[73] with permission.)

(*B*) Somatostatin-immunoreactive fibres in and around the pudendal motor nucleus (Onuf's nucleus) in the sacral (S$_2$) segment of human spinal cord. Formalin fixation, wax section (5 µm), peroxidase–anti-peroxidase method. Magnification × 275. (Courtesy of Dr T Katagiri.)

(*C*) Calcitonin gene-related peptide-immunoreactive motoneurones in the lumbar spinal cord of the rat (some arrowed). Bouin's fixation, wax section (10 µm), immunogold–silver intensification method. Magnification × 50.

(*D, E, F*) Substance P-immunoreactive fibres in the ventral horn of the rat spinal cord rostral (*D*), just rostral (*E*) and (*F*) caudal to transection of the mid-thoracic level. Note build-up of immunoreactivity in (*E*) and comparative loss below the lesion in (*F*). *p*-Benzoquinone fixation, cryostat section (20 µm), peroxidase–anti-peroxidase method. Magnification × 120.

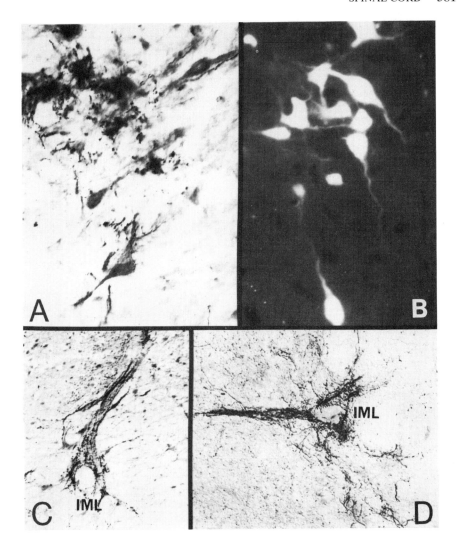

Fig. 22.10. (*A*) Calcitonin gene-related peptide-immunoreactive neurones in the sacral interme-
diolateral cell column of the rat. Magnification × 315.
(*B*) Preganglionic neurones in the rat parasympathetic nucleus labelled retrogradely with the tracer
True Blue following injection in the pelvic nerve. Magnification × 320.
(*C*) Peptide histidine isoleucine-immunoreactive fibres in the lateral dorsal horn of the guinea-pig.
The fibres terminate in close proximity to cells of the intermediolateral cell column (IML).
Magnification × 50.
(*D*) Cholecystokinin/gastrin-like immunoreactive fibres in the thoracic intermediolateral cell
column (IML) of the rat. Magnification × 135.

derived from descending fibres and a final possible source is from cell bodies
located in the intermediomedial region, dorsal grey commissure or dorsal horn.

Table 22.4. **Peptide distribution in autonomic centres**

Peptide	Fibres in IML	Cells in IML	Preganglionic neurones
Calcitonin gene-related peptide	*	*	n.r.
Cholecystokinin/gastrin-like peptide	*	*	*
Corticotrophin releasing hormone	*	*	n.r.
Dynorphin	*	*	n.r.
Enkephalin	*	*	*
FMRF-amide-like peptide	*	*	n.r.
Galanin	*	n.r.	n.r.
Neurotensin	*	*	n.r.
Neuropeptide Y	*	*	n.r.
Neurophysin, oxytocin vasopressin	*	n.r.	n.r.
Peptide histidine isoleucine	*	*	n.r.
Somatostatin	*	*	*
Substance P	*	*	n.r.
Thyrotrophin-releasing hormone	*	n.r.	n.r.
Vasoactive intestinal polypeptide	*	*	n.r.
Fluoride-resistant acid phosphatase*	*	*	n.r.

*Present.
n.r. = not reported.
IML = intermediolateral cell columns.
*The enzyme, fluoride-resistant acid phosphatase is also included here.

8. CONCLUSION

The use of immunocytochemistry has permitted the demonstration of a highly organized, complex system of peptide-containing pathways in the spinal cord. Peptides are found in all areas, in particular those which control sensory, motor and integrative functions. Specific functional roles have not yet been attributed to all neuropeptides. However, the enkephalins and substance P appear to be intimately involved in pain processing and nociception,[8,76–78] and vasoactive intestinal polypeptide is very likely to be a neurotransmitter/modulator of visceral afferents in the genito-urinary tract.[69] While study of peptides has enlarged our concept of neural transmission in the central nervous system, the complexity of peptide-containing systems and their interactions with the established neurotransmitters is still far from being understood.

Acknowledgements

We thank Professor S. R. Bloom, Professor P. D. Wall, Professor J. S. Kelly, Dr J. F. B. Morrison and Mr G. van Aswegen for their valuable and continuing collaboration.

Note

Since the preparation of this chapter, oxytocin-, vasopressin-, and some enkaphalin-containing cells have been demonstrated in the dorsal root ganglion.[79,80] Furthermore, other examples of peptide coexistence have appeared in the literature[81,82] and cells containing up to four different peptides have recently been reported in cholchicine-treated cats.[83]

Appendix

1. CHOICE OF FIXATIVE

Paraformaldehyde, formol saline and Bouin's solutions are all suitable fixatives for visualization of, not only peptide-containing nerves, but also those immuno-reactive for glial fibrillary acidic and neurofilament proteins, tyrosine hydroxylase and serotonin. In our experience, however, the use of *p*-benzoquinone as a fixative allows equally good detection of all the above excepting serotonin. In addition, localization of some peptides (substance P, neurotensin, neurophysin and vasoactive intestinal polypeptide) is markedly enhanced in *p*-benzoquinone-fixed samples of spinal cord.

II.   *p*-BENZOQUINONE FIXATION BY PERFUSION (SMALL MAMMALS)

1. Anaesthetize animal and lay out on a dissection board ventral side uppermost. Securely fix limbs to the board.
2. Expose heart. Insert a needle or cannula (preferably with a rounded end) which is attached to perfusion pump tubing, through the left ventricle into the aorta (you can feel with your hand). Fix the position of the needle by clamping the ventricle with a pair of haemostats. Cut the right atrium.
3. Start perfusion pump. Flush out the blood at a slow rate (10–20 ml/min) with 0·01 M phosphate-buffered saline (PBS) pH 7·1 – 7·4, approximately 50 ml/100 g body weight. Continue until the liver and kidneys become pale in colour.
4. Change solutions. Pass a freshly prepared solution of 0·4 per cent *p*-benzoquinone in PBS (200 ml/100 g body weight). The tissues should become a dark yellow/brown colour.
5. Dissect out spinal cord and place in a fresh solution of *p*-benzoquinone for 0·5–2 h or until tissue is the colour of milk chocolate.
6. Wash well in PBS and store in PBS containing 15 per cent sucrose and 0·01 per cent sodium azide at 4 °C.
7. Prepare cryostat sections 10–30 μm thick. Mount on poly(L-lysine)-coated slides and allow to dry for 2 h prior to use.

III. INHIBITION OF ENDOGENOUS PEROXIDASE

When using peroxidase immunocytochemistry it is necessary to inhibit endoge-nous peroxidase prior to immunostaining. This is particularly important when processing tissues which have been fixed by immersion not perfusion, since perfusion flushes out the blood which is a source of confusing enzyme activity.

1. Prior to application of primary antiserum place slides in a solution of PBS containing 0·3 per cent hydrogen peroxide for 30 min. If a large amount of endogenous peroxidase is present, a vigorous reaction occurs and can cause bubbles to form under the sections, eventually lifting them from the slides. A gentler alternative is to use a solution of methanol containing 0·3 per cent hydrogen peroxide.
2. Was in tap water, rinse in distilled water.

IV. AIDS TO ANTIBODY PENETRATION

Spinal cord contains a large amount of myelin which can act as a barrier to the antiserum. Several methods, either alone or together, can be used to aid the penetration of the antiserum into the tissue section.

1. Dehydration of the sections through graded alcohols and into trichloroethane or xylene followed by rehydration prior to immunostaining.
2. Addition of 0·2 per cent Triton X-100 to the antiserum diluent. In practice, however, it is advisable to omit Triton X-100 from the primary antiserum (if applied to sections mounted on slides). Triton X-100 is a surfactant and this property coupled with long incubation (> 10 h) causes the antiserum to 'run off' the section which in turn leads to uneven staining.
3. Addition of 0·2 per cent Triton X-100 to the washing buffers.

V. SOURCE OF ANTISERA AND CHEMICALS

Antiserum	Source
Calcitonin gene-related peptide	H/H
Cholecystokinin/gastrin-like peptide	H/H
Galanin	H/H
Glial fibrillary acidic protein	Dr P. Woodhams, MRC Neurobiology Unit, London, UK
Neurofilament protein	Dr D. Dahl, West Roxbury Veterans Administration Hospital, Boston, USA
Neuropeptide Y	H/H
Neurophysin	Dr W. B. Watkins, University of Auckland, New Zealand
Neurotensin	H/H
Peptide histidine isoleucine	H/H
Serotonin	Dr J. De Mey, Janssen Life Sciences Products, Beerse, Belgium

Somatostatin	RIA (UK) Inc., USA
Substance P	H/H
Tyrosine hydroxylase	Professor J. Thibault, Collège de France, Biochemie Cellulaire, Paris, France
Vasoactive intestinal polypeptide	H/H
Goat anti-rabbit Ig	Miles Laboratories, Slough, UK
Goat anti-rabbit Ig– fluorescein isothiocyanate	Miles Laboratories, Slough, UK
Goat anti-rabbit Ig–immunogold (G5)	Janssen Life Sciences Products, Beerse, Belgium
Peroxidase–anti-peroxidase complex	Miles Laboratories, Slough, UK
Chemicals	
Capsaicin	Fluka AG, Switzerland (Fluorochem Ltd, Derbyshire, UK)
Colchicine	Sigma, Dorset, UK
True Blue	Sigma, Dorset, UK
Triton X-100	Sigma, Dorset, UK
Poly(L-lysine)	Sigma, Dorset, UK
p-Benzoquinone	Köch Light Labs Ltd, Berks, UK

H/H = Departments of Histochemistry and Medicine, Hammersmith Hospital, Du Cane Road, London, UK.

REFERENCES

1. Loewi O. Über humorale Übertragborrkeit der Hernervenwirkung — II. Mitteilung Pflügers. *Arch. Ges. Physiol.* 1921, **293**, 201–203.
2. Dale H. H. Pharmacology and nerve-endings. Walter Ernest Dixon Memorial Lecture. *Proc. R. Soc. Med. Therap.* 1934, **Sect. 28**, 319–332.
3. Eccles J. C. (ed.) *The Physiology of Nerve Cells.* Baltimore, The Johns Hopkins Press, 1957.
4. Cooper J. R., Bloom F. E. and Roth R. H. (eds.) *The Biochemical Basis of Neuropharmacology,* 3rd ed. New York, Oxford University Press, 1975.
5. Burgen A. S. V. and Mitchell J. F. (eds). *Gaddum's Pharmacology,* 8th ed. London, Oxford University Press, 1978.
6. Salt T. E. and Hill R. G. Neurotransmitter candidates of somatosensory primary afferent fibres. *Neuroscience* 1983, **10**, 1083–1103.
7. Hökfelt T., Johansson O., Ljungdahl A., Lundberg J. M. and Schultzberg M. Peptidergic neurons. *Nature* 1980, **284**, 515–518.
8. Emson P. C. Peptides as neurotransmitter candidates in the mammalian CNS. *Prog. Neurobiol.* 1979, **13**, 61–116.
9. El-Badawi A. and Schenk E. A. Histochemical methods for separate consecutive and simultaneous demonstration of acetylcholinesterase and norepinephrine in cryostat sections. *J. Histochem. Cytochem.* 1967, **15**, 580–588.
10. Marchand R. and Barbeau H. Vertically orientated alternating acetylcholinesterase rich and poor terminals in laminae VI, VII, VIII of the lumbosacral cord of the rat. *Neuroscience.* 1982, **7**, 1197–1202.
11. Gerebtzoff M. A. In: Alexandra P. and Bacq Z. M., eds., *International Series of Monographs on Pure and Applied Biology.* Vol.3: *Cholinesterases,* London Pergamon Press, 1959.
12. Butcher L. L. Acetylcholinesterase histochemistry. In: Björklund A. and Hökfelt T., eds, *Handbook of Chemical Neuroanatomy.* Vol I, *Methods in Chemical Neuroanatomy.* New York, Elsevier Science Publishers, 1983: 1–49
13. Knyihar E. and Csillik B. FRAP. Histochemistry of the primary nociceptive neurons. *Prog. Histochem. Cytochem.* 1981, **14**, 1–137.

14. Falck B., Hillarp N.-A., Thieme G. and Torp A. Fluorescence of catecholamines ad related compounds condensed with formaldehyde. *J. Histochem. Cytochem.* 1962, **10**, 348–354.
15. Hökfelt T. and Ljungdahl A. Modification of the Falck–Hillarp formaldehyde fluorescence method using the vibratome: Simple rapid and sensitive localisation in sections of unfixed or formalin fixed brain tissue. *Histochemie* 1972, **29**, 325–339.
16. Björklund A. Fluorescence histochemistry of biogenic amines. In: Björklund A. and Hökfelt T., eds., *Handbook of Chemical Neuroanatomy*. Vol. 1, *Methods in Chemical Neuroanatomy*. New York, Elsevier, 1983: 50–121.
17. Richards G. Ultrastructural visualisation of biogenic monamines. In: Björklund A. and Hökfelt T., eds., *Handbook of Chemical Neuroanatomy*. Vol 1, *Methods in Chemical Neuroanatomy*. New York, Elsevier, 1983: 122–146.
18. Hunt S. P. Cytochemistry of the spinal cord. In: Emson P. C., ed., *Chemical Neuroanatomy*. New York, Raven Press, 1983: 53–84.
19. Iversen L. L. and Bloom F. E. Studies of the uptake of [³H]GABA and [³H]glycine in slices and homogenates of rat brain and spinal cord by microscopic autoradiography. *Brain Res.* 1972, **41**, 131–143.
20. Eng L. F. and DeArmond S. J. Immunocytochemical studies of astrocytes in normal development and disease. In: Fedoroff S. and Hertz L., eds., *Advances in Cellular Neurobiology*, Vol. 3. London, Academic Press, 1982: 145–165.
21. de Sousa B. N and Horrocks L. A. Development of rat spinal cord 1. Weight and length with a method for rapid removal. *Dev. Neurosci.* 1979, 2, 115–121.
22. Pearse A. G. E. and Polak J. M. Bifunctional reagents as vapour and liquid-phase fixatives for immunhistochemistry. *Histochem. J.* 1975, 7, 179–186.
23. Huang W. M., Gibson S. J., Facer P., Gu J., and Polak J. M. Improved section adhesion for immunocytochemistry using high molecular weight polymers of lysine as a slide coating. *Histochemistry* 1983, **77**, 275–279.
24. Hökfelt T., Fuxe K., and Goldstein M. Applications of immunocytochemistry to studies on monoamine cell systems with special reference to nervous tissues. In: Hijmans W. and Schaeffer M., eds., *Fifth International Conference on Immunofluorescence and Related Staining Techniques. Part XII. Hormones and enzymes.* New York, New York Academy of Science, 1975: 407–432.
25. Verhofstad A. A.-J., Steinbusch H. W. M., Joosten H. W. J., Penke B., Varga J. and Goldstein M. Immunocytochemical localisation of noradrenaline, adrenaline and serotonin. In: Polak J. M. and Van Noorden S., eds., *Immunocytochemistry: Practical Applications in Pathology and Biology*. Bristol, Wright-PSG, 1983: 143–165.
26. Barber R. P., Phelps P. E., Houser C. R., Crawford G. D., Salvaterra P. M. and Vaughn J. E. Morphology and distribution of neurons containing choline acetyltransferase in the adult rat spinal cord: An immunohistochemical method. *J. Comp. Neurol.* 1984. **229**, 329–346.
27. Steinbusch H. W. M. Distribution of serotonin-immunoreactivity in the central nervous system of the rat—cell bodies and terminals. *Neuroscience*, 1981, **6**, 557–618.
28. Barber R. P., Vaughn J. E. and Roberts E. The cytoarchitecture of gabaergic neurons in rat spinal cord. *Brain Res.* 1982, **238**, 305–328.
29. Ljungdahl A., Hökfelt T., Nilsson G. and Goldstein M. Distribution of substance P-like immunoreactivity in the central nervous system of the rat—II. Light microscopic localisation in relation to catecholamine-containing neurons. *Neuroscience*, 1978, **3**, 945–976.
30. Werman R. Criteria for identification of a central nervous system transmitter. *Comp. Biochem. Physiol.* 1966, **18**, 745–766.
31. Wall P. D. and Fitzgerald M. If substance P fails to fulfil the criteria as a neurotransmitter in somatosensory afferents, what might be its function? In: Porter R. and O'Conner. M., eds., *Substance P in the Nervous System*. London, Pitman, 1982: 249–266.
32. Backman S. B. and Henry J. L. Effects of oxytocin and vasopressin on thoracic sympathetic preganglionic neurones in the cat. *Brain Res. Bull.* 1984, **13**, 679–684.
33. Miletic V. and Randic M. Neurotensin excites cat spinal neurones located in laminae I-III. *Brain Res.* 1979, **169**, 600–604.
34. Ono H. and Fukuda H. Ventral root depolarization and spinal reflex augmentation by a TRH analog in rat spinal cord. *Neuropharmacology* 1982, **21**, 39–44.
35. Kimura S., Okada M., Sugita Y., Kanazawa H. and Mimekata E. Novel neuropeptides, neurokinin α and ß isolated from porcine spinal cord. *Proc. Jap. Acad.* 1983, **59**, Ser. B, 101–104.

36. Anand P., Gibson S. J., Yiangou Y., Christofides N. D., Polak J. M. and Bloom S. R. PHI-like immunoreactivity co-locates with the VIP-containing system in the human lumbosacral spinal cord. *Neurosci. Lett.* 1984, **46**, 191–196.
37. Minamino N., Kangawa K. and Matsuo H. Neuromedin B: A novel bombesin-like peptide identified in porcine spinal cord. *Biochem. Biophys. Res. Commun.* 1983, **114**, 541–548.
38. Namba M., Ghatei M. A., Gibson S. J., Polak J. M. and Bloom S. R. Distribution and localisation of neuromedin B-like immunoreactivity in pig, cat and rat spinal cord. *Neuroscience.* 1985, **15**, 1217–1226.
39. Sasek C. A., Seybold V. S. and Elde R. P. The immunohistochemical localisation of nine peptides in the sacral parasympathetic nucleus and the dorsal grey commissure in the rat spinal cord. *Neuroscience* 1984, **12**, 855–873.
40. Minamino N., Kangawa K. and Matsuo H. Neuromedin N: a novel neurotensin-like peptide identified in porcine spinal cord. *Biochem. Biophys. Res. Commun.* 1984, **122**, 542–549.
41. Gibson S. J., Polak J. M., Adrian T. E., Allen J. M., Kelly J. S. and Bloom S. R. The distribution and origin of a novel brain peptide neuropeptide Y in the spinal cord of several mammals. *J. Comp. Neurol.* 1984, **227**, 78–92.
42. Gibson S. J., Polak J. M., Bloom S. R., Sabate I. M., Mulderry M. A., McGregor G. P., Morrison J. F. B., Kelly J. S., Evans R. M. and Rosenfeld M. G. Calcitonin gene-related peptide (CGRP) immunoreactivity in spinal cord of man and eight other species. *J. Neurosci.* 1984, **4**, 3101–3111.
43. Merchenthaler I., Hynes M. A., Vigh S., Shally A. V. and Petrusz P. Immunocytochemical localisation of corticotrophin releasing factor (CRF) in the rat spinal cord. *Brain Res.* 1983, **275**, 373–377.
44. Rokaeus A., Melander T., Hökfelt T., Lundberg J. M., Tatemoto K., Carlquist M. and Mutt J. A galanin-like peptide in the central nervous system and intestine of the rat. *Neurosci. Lett.* 1984, **47**, 161–166.
45. Ch'ng J. L. C., Christofides N. D., Anand P., Gibson S. J., Allen Y. S., Su H. C., Tatemoto K., Morrison J. F. B., Polak J. M. and Bloom S. R. Distribution of galanin-immunoreactivity in the central nervous system and the responses of galanin-containing neural pathways to injury. *Neuroscience.* 1985, in press.
46. Fitzgerald M. Capsaicin and sensory neurons—a review. *Pain.* 1983, **15**, 109–130.
47. Nagy J. I., Hunt S. P., Iversen L. L. and Emson P. C. Biochemical and anatomical observations on the degeneration of peptide-containing primary afferent neurons after neonatal capsaicin. *Neuroscience.* 1981, **6**, 1923–1934.
48. Dahlström A. Effects of vinblastine and colchicine on monoamine-containing neurons of the rat with special regard to the axoplasmic transport of amine granules. *Acta Neuropathol.* 1971, **5**, 226–237.
49. Heimer L. and Robards M. J. (eds.) *Neuroanatomical Tract-tracing Methods.* New York, Plenum Press, 1983.
50. Skirboll L., Hökfelt T., Norell G., Phillipson O., Kuypers H. G. J. M., Bentivoglio M., Catsman-Bevrevoets C. E., Visser T. J., Steinbusch H., Verhofstad A., Cuello A. C., Goldstein M. and Brownstein M. A method for specific transmitter identification of retrogradely labelled neurons: Immunofluorescence combined with fluorescent tracing. *Brain Rev.* 1984, **8**, 99–127.
51. Björklund A. and Hökfelt T. (eds.) *Handbook of Chemical Neuroanatomy.* Vol 1, *Methods in Chemical Neuroanatomy.* New York, Elsevier, 1983.
52. Jessell T. M. and Yamamoto M. Identification and interactions of substance P, enkephalin and serotonin neurons *in vivo* and in dissociated cell culture. In: Fink G. and Whalley L. J., eds., *Neuropeptides: Basic and Clinical Aspects*, Proceedings of the XIth Pfizer International Symposium, 1981. London, Churchill Livingstone, 1981: 73–87.
53. Haynes L. W., Smyth D. G. and Zakarian S. Immunocytochemical localisation of α-endorphin (lipotropin C-fragment) in the developing rat spinal cord and hypothalamus. *Brain Res.* 1982, **232**, 115–128.
54. Lieberman A. R. Sensory ganglia. In: Landon D. N., ed., *The Peripheral Nerve.* London, Chapman and Hall, 1976: 182–278.
55. Tuchscherer M. M. and Seybold V. S. Immunohistochemical studies of substance P, cholecystokinin-octapeptide and somatostatin in dorsal root ganglia of rat. *Neuroscience.* 1985, **14**, 593–603.
56. Dalsgaard C-J., Ygge J., Vincent S. R., Ohrling M., Dockray G. J. and Elde R. Peripheral projections and neuropeptide coexistence in a subpopulation of fluoride-resistant acid phosphatase reactive spinal primary sensory neurons. *Neurosci. Lett.* 1984, **51**, 139–144.

57. Fuxe K., Agnati L. F., McDonald T., Locatelli V., Hökfelt T., Dalsgaard C-J., Battistini N., Yanaihara N., Mutt V. and Cuello A. C. Immunohistochemical indications of gastrin releasing peptide-bombesin-like immunoreactivity in the nervous system of the rat. Codistribution with substance P-like immunoreactive terminal systems and coexistence with substance P-like immunoreactivity in dorsal root ganglion cell bodies. *Neurosci. Lett.* 1983, **37**, 17–22.

58. Panula P., Hadjiconstantinou M., Yang -Y. T. and Costa E. Immunohistochemical localisation of bombesin/gastrin releasing peptide and substance P in primary sensory neurons. *J. Neurosci.* 1983, **3**, 2021–2029.

59. Lundberg J. M., Hökfelt T., Nilsson G., Terenius L., Rehfeld J., Elde R., and Said S. Peptide neurons in the vagus, splanchnic and sciatic nerves. *Acta Physiol. Scand.* 1979, **104**, 499–501.

60. Hökfelt T., Elde R., Johansson O., Luft P., Nilsson G. and Arimura A. Immunohistochemical evidence for separate populations of somatostatin-containing and substance P-containing primary afferent neurons in the rat. *Neuroscience.* 1979, **1**, 131–136.

61. Wiesenfeld-Hallin Z., Hökfelt T., Lundberg J. M., Forssman W. G., Reinecke M., Tschoop F. A. and Fischer J. Immunoreactive calcitonin gene-related peptide and substance P coexist in sensory neurons and interact in spinal behavioral responses. *Neurosci. Lett.* 1984, **52**, 199–204.

62. Gibson S. J., Polak J. M., Bloom S. R. and Wall P. D. The distribution of nine peptides in rat spinal cord with special emphasis on the substantia gelatinosa and on the area around the central canal (lamina X). *J. Comp. Neurol.* 1981, **201**, 65–79.

63. McGregor G. P., Gibson S. J., Sabate I. M., Blank M. A., Christofides N. D., Wall P. D., Polak J. M. and Bloom S. R. Effect of peripheral nerve section and nerve crush on spinal cord neuropeptides in the rat; increased VIP and PHI in the dorsal horn. *Neuroscience.* 1984, **13**, 207–216.

64. Skofitch G., Hamill G. S. and Jacobowitz D. M. Capsaicin depletes corticotrophin-releasing factor like immunoreactive neurons in the rat spinal cord and medulla oblongata. *Neuroendocrinology* 1984, **38**, 514–517.

65. Jancsó G., Hökfelt T., Lundberg J. M., Kiraly E., Halasz N., Nilsson G., Terenius L., Rehfeld J., Steinbusch H., Verhofstad A., Elde R., Said S. and Brain M. Immunohistochemical studies on effect of capsaicin on spinal and medullary peptide and monoamine neurons using antisera to substance P gastrin/CCK, somatostatin, VIP, enkephalin, neurotensin and 5– hydroxytryptamine. *J. Neurocytol.* 1981, **10**, 963–980.

66. Schrøder H. D. Somatostatin in the caudal spinal cord: an immunohistochemical study of the spinal centres involved in the innervation of pelvic organs. *J. Comp. Neurol.* 1984, **223**, 400–413.

67. Gibson S. J., Polak J. M., Anand P., Blank M. A., Morrison J. F. B., Kelly J. S. and Bloom S. R. The distribution and origin of VIP in the spinal cord of six mammalian species. *Peptides*, 1984, **5**, 201–207.

68. Anand P., Gibson S. J., McGregor G. P., Blank M. A., Ghatei M. A., Bacarese-Hamilton A. J., Polak J. M. and Bloom S. R. A VIP-containing system concentrated in the lumbosacral region of the human spinal cord. *Nature*, 1983, **305**, 143–145.

69. de Groat W. C., Kawantani M., Hisamitsu T., Lowe I., Morgan C., Roppolo J., Booth A. M., Nadelhaft I., Kuo D. and Thor K. The role of neuropeptides in the sacral autonomic reflex pathways of the cat. *J. Auton. Nerv. Syst.* 1983, **7**, 339–350.

70. Hunt S. P., Kelly J. S., Emson P. C., Kimmel J. R., Miller R. and Wu J-Y. An immunohistochemical study of neuronal populations containing neuropeptides or GABA within the superficial layers of the rat dorsal horn. *Neuroscience.* 1981, **6**, 1883–1898.

71. Fuji K., Senba E., Ueda Y. and Yokyama M. Vasoactive intestinal polypeptide (VIP)-containing neurons in the spinal cord of the rat and their projections. *Neurosci. Lett.* 1983, **37**, 51–55.

72. Nahin R. L., Madsen A. M. and Giesler G. J. Jr. Anatomical and physiological studies of the grey matter surrounding the spinal cord central canal. *J. Comp. Neurol.* 1983, **220**, 321–335.

73. Gibson S. J., Bloom S. R. and Polak J. M. A novel substance P pathway linking the dorsal and ventral horn in the upper lumbar segments of the rat spinal cord. *Brain Res.* 1984, **301**, 243–251.

74. Anand P. and Bloom S. R. Neuropeptides are selective markers of spinal cord autonomic pathways. *Trends Neurosci.* 1984, **7**, 267–268.

75. Schrøder D. A. Localisation of cholecystokinin-like immunoreactivity in the rat spinal cord with particular reference to the autonomic innervation of the pelvic organs. *J. Comp. Neurol.* 1983, **217**, 176–186.

76. Jessell T. and Iversen L. L. Opiate analgesics inhibit substance P release from rat trigeminal nucleus. *Nature* 1977, **268**, 549–551.

77. Hunt S. P., Ninkovic M., Gleave J. R. W., Iversen S. D. and Iversen L. L. Interelationship between enkephalin and opiate receptors in the spinal cord. In: Fink G. and Whalley L. J., eds., *Neuropeptides: Basic and Clinical Aspects.* Proceedings of the XIIth Pfizer International Symposium. London, Churchill Livingstone, 1983: 13–23.

78. Porter R. and O'Connor M. (eds.) Substance P in the nervous system. *Ciba Foundation Symposium.* London, Pitman, 1982.

79. Kai-Kai M. A., Swann R. W. and Keen P. Localization of chromatographically characterised oxytocin and arginine-vasopressin to sensory neurones in the rat. *Neurosci. Lett.* 1985, **55**, 83–88.

80. Roppollo J. R., Lowe I. P. and de Groat W. C. Immunohistochemical identification of leucine enkephalin in dorsal root ganglion cells of the rhesus monkey. *Soc. Neurosci. Abstracts* 1984, **10**, 993.

81. Gibson S. J., McCrossan M. V. and Polak J. M. A sub-population of calcitonin gene-related peptide (CGRP)-immunoreactive neurones in the dorsal root ganglia also display substance P, somatostatin or galanin immunoreactivity. *XIIth International Anatomical Congress*, 1985, p. 232.

82. de Groat W. C., Kawantani M., Houston M. B. and Erdman S. L. Colocalisation of VIP, substance P, CCK, somatostatin and enkephalin immunoreactivity in lumbosacral dorsal root ganglion cells of the cat. *Soc. Neurosci. Abstracts* 1984, **10**, 48.

83. Leah J. D., Cameron A. A. and Snow P. J. Neuropeptides in physiologically identified mammalian sensory neurons. *Neurosci. Lett.* 1985, **56**, 257–263.

23

Immunocytochemistry of Intermediate Filament Proteins

F. Ramaekers, J. Broers and G. P. Vooijs

Eukaryotic cells contain a fibrillar intracytoplasmic matrix (the cytoskeleton) consisting of different types of filamentous structure. Within this matrix, three different types of filament can be distinguished on the basis of their ultrastructural appearance and biochemical composition. Besides microfilaments (5–7 nm) and microtubules (22 nm), filaments measuring 8–11 nm in diameter are commonly seen in mammalian cells and often constitute a considerable part of the cellular cytoskeleton. These so-called 'intermediate-sized filaments' (*Fig.* 23.1) are extremely insoluble and show a protein composition which is completely different from that of microfilaments and microtubules.

1. INTERMEDIATE FILAMENTS SHOW A TISSUE-SPECIFIC PROTEIN COMPOSITION

Biochemical and immuno(cyto)chemical studies have demonstrated that five different types of intermediate filament proteins (IFPs) can be distinguished in mammalian cells.[1] These include the cytokeratins, vimentin, desmin, glial fibrillary acidic protein (GFAP) and the neurofilament triplet. Most important for the principles described in this chapter is the fact that these different types of IFP are distributed in a tissue-specific manner. From *Table* 23.1 it may be obvious that the subdivisions made between tissues on the basis of their IFP content correspond strikingly well with the major tissue classification based on histological principles.

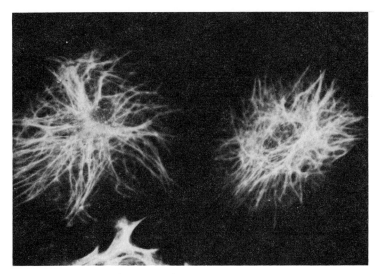

Fig. 23.1. Intermediate filament (cytokeratin) pattern in cultured hepatocellular carcinoma cells from guinea-pig.

Table 23.1. **Tissue and tumour specificity of intermediate filament proteins**

Type of IFP	Tissue type	Tumour type
Cytokeratins	Epithelial tissues	Carcinomas
Vimentin	Mesenchymal tissues	Lymphomas, sarcomas, melanomas
Desmin	Muscle tissues	Myosarcomas
GFAP	Astroglial cells	Astrocytomas
Neurofilament proteins	Nerve tissue	Some neural tumours

Antibodies (either monoclonal or polyclonal) to the different types of IFP have been prepared by several groups working in the field.[1-5] Most of these antibodies could be shown to react in a tissue-specific manner, meaning that antibodies to cytokeratin react with epithelia, antibodies to desmin only with muscle tissues, GFAP antibodies react only with glial cells and neurofilament antibodies react with neural cells. Antibodies to vimentin normally react only with cells and tissues of mesenchymal origin, but for this type of IFP some exceptions are known (*see below*). In this way the different types of tissue denoted in *Table* 23.1 can be distinguished by immunocytochemical methods, using specific antibodies to IFP.

Exceptions to the rule as outlined in *Table* 23.1 have also been described. These include some epithelioid tissues, such as the seminiferous epithelium, lens epithelium, pigment epithelium in the retina, ciliary epithelium and epithelial cells covering the iris, which do not contain cytokeratins but possess vimentin IFP. Some types of neurone have been shown to be completely devoid of intermediate filaments, while some types of vascular smooth muscle contain either vimentin alone or vimentin as well as desmin.

Furthermore, it should be kept in mind that certain tissue types in developing embryos may contain IFP different from those present in their adult counterparts. For example, myoblasts contain vimentin instead of desmin and neonatal glial cells may contain vimentin in addition to GFAP. Some types of kidney epithelial cell transiently express vimentin during embryogenesis.[6] Finally, cells brought into tissue culture commonly express vimentin as well as their cell-specific IFP.[1]

2. USE OF POLYCLONAL AND MONOCLONAL ANTIBODIES TO INTERMEDIATE FILAMENT PROTEINS IN TUMOUR DIAGNOSIS

Tumour diagnosis in surgical pathology is still largely based on the morphological aspects of the tissues and cells to be examined. Depending on the experience of the pathologist, many tumours can be typed by routine histological staining techniques applied to formalin-fixed and paraffin-embedded sections. The remaining tumours, which can be typed only partly or not at all, are given prefixes such as anaplastic, pleomorphous, small cell, spindle cell etc. It is often impossible to relate such (malignant) tumours to a particular type of tissue (or to an organ, in the case of metastases) on the basis of routine histology and experience, or even on the basis of more specific histological staining techniques. Yet an accurate diagnosis of a malignant tumour is obviously of paramount importance in planning treatment and estimating prognostic factors. Antibodies to IFP have been tested elaborately for use in tissue characterization in the last few years and from the data obtained in these studies it is obvious that IFP can be used as powerful markers in tumour typing (*see* literature for reviews[4,5] and references therein). Most importantly, it has been shown that generally tumour cells retain their original IFP, with the exception of a few types of neoplasm. Therefore, as depicted in *Fig.* 23.2 and *Plate* 27, and summarized in *Table* 23.1, carcinomas can be stained exclusively with antibodies to cytokeratins while mesenchymal tumours, such as lymphomas, soft tissue sarcomas, malignant fibrous histiocytomas, melanomas, seminomas and schwannomas are negative for cytokeratin, but positive for vimentin.

Tumours derived from striated or smooth muscle tissues, i.e. rhabdomyosarcomas and leiomyosarcomas, can be differentiated from other soft tissue tumours because of their reactivity with desmin antibodies. Moreover, astrocytomas are strongly positive for GFAP, while ganglioneuroblastomas, and some of the neuroblastomas are positive with neurofilament antisera. However, some neuroendocrine tumours (pulmonary carcinoids, oat cell carcinomas, phaeochromocytomas and Merkel cell tumours) have also been described as containing neurofilament proteins.[7]

Antibodies directed against cytokeratins and vimentin can be particularly helpful in differentiating between an anaplastic carcinoma and a lymphoma in the case of a potential metastasis. When routine staining procedures do not permit diagnosis, a positive reaction for cytokeratin clearly points to a carcinoma, whereas a positive reaction for vimentin, without a positive reaction for cytokeratin, is suggestive of a lymphoma.

Routine histological diagnosis of small-round-cell tumours of childhood (including lymphomas, neuroblastomas, Ewing's sarcomas, nephroblastomas and

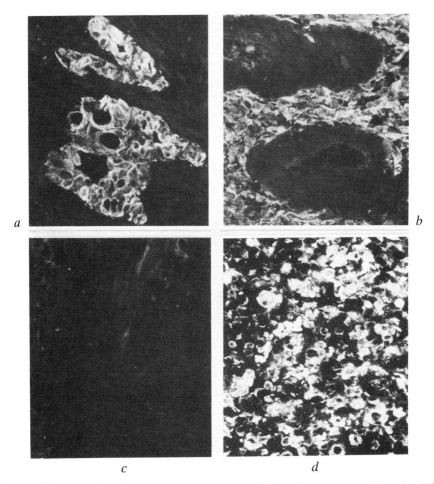

Fig. 23.2. Frozen sections of a cervical adenocarcinoma (*a*), a colon carcinoma (*b*) and a diffuse centrocytic lymphoma (*c, d*) stained by immunofluorescence with polyclonal rabbit antisera to human skin keratins (*a, c*) or to bovine lens vimentin (*b, d*). Note the mutual exclusiveness of the reaction patterns of the antisera, keratin antibodies reacting only with the epithelial tumour cells (*a*) and not with the lymphoma or stroma tissue (*c*) and vimentin antibodies reacting with the lymphoma (*d*) and stromal cells but not with the carcinoma cells (*b*).

embryonal rhabdomyosarcomas) is often difficult or impossible due to the lack of differentiation of the tumour cells. With the aid of antibodies directed against IFP, however, considerable progress can be made. *Table* 23.2 shows the reaction patterns of the various types of childhood tumour with IFP antibodies. Since treatment protocols for these tumours differ, an accurate diagnosis is imperative for optimal therapy. In the case of spindle cell tumours too, antibodies to cytokeratin, vimentin and desmin can be helpful in diagnosis. *Plate* 27 illustrates three cases of spindle cell tumour, all three initially diagnosed as

Table 23.2. *Intermediate filament protein typing in childhood tumours*

Tumour type	Vimentin	Desmin	Cytokeratin	Neurofilament
Lymphoma	+	−	−	−
Ewing's sarcoma	+	−	−	−
Embryonal rhabdomyosarcoma	+	+	−	−
Nephroblastoma	+*	+/−†	+/−*	−
Neuroblastoma	−	−	−	+/−‡

*Blastoma cells in nephroblastoma (Wilms' tumour) contain either cytokeratin and vimentin or only vimentin.
†The myoid component sometimes occurring in nephroblastomas can be stained with antibodies to desmin.
‡Neuroblastomas may contain neurofilaments or be completely devoid of IFP.

malignant fibrous histiocytoma (MFH). Incubation of these tumours with IFP antibodies, however, revealed in one case strong cytokeratin positivity as well as a reaction with a vimentin antibody (*Plate 27a, b*) and in another case reactivity with a desmin antibody (*Plate 27e, f*). These tumours were therefore reclassified as a mesothelioma and a rhabdomyosarcoma, respectively. Real MFH is only positive for vimentin and does not react with antibodies to cytokeratin or desmin (*Plate 27c, d*).

3. COEXPRESSION OF DIFFERENT TYPES OF INTERMEDIATE FILAMENT PROTEIN IN CERTAIN TUMOURS

In contrast to normal adult human tissues, certain neoplastic tissues may coexpress different types of IFP. Although most malignancies do not acquire an additional IF system, vimentin IFs have been found next to cytokeratin IFs in pleomorphic adenomas of the parotid gland, in adenoid cystic carcinomas of the salivary gland and in the lung, in mesotheliomas of the spindle cell type, in renal cell carcinomas (Grawitz tumour) and nephroblastomas (Wilms' tumour), and in some endometrial carcinomas. Furthermore, metastatic carcinoma cells growing in body cavity fluids (ascites or pleural fluid) may obtain an additional vimentin cytoskeleton.[8]

Coexpression of vimentin and desmin is observed in most if not all rhabdomyosarcomas and in leiomyosarcomas. We have observed that astrocytomas contain both GFAP and vimentin. From the foregoing it may be obvious that when tumour cells contain more than one IFP type, in most cases vimentin is one of them. However, coexpression of cytokeratins and neurofilament proteins has also been observed. This is the case in some neuroendocrine tumours, such as Merkel cell tumours[7] and some lung tumours as well as in cell lines derived from such neoplasms.

The fact that certain types of tumour cell express vimentin as well as their tissue-specific IFP can be of help in immunocytochemical identification of these malignancies. For example, coexpression of vimentin and cytokeratins in renal cell tumours (especially Grawitz tumour) allows the distinction of clear cell-type renal tumours from other epithelial tumours with the same morphology.

4. CYTOKERATINS ARE DISTRIBUTED IN SPECIFIC COMBINATIONS THROUGHOUT EPITHELIAL TISSUES

As described above, conventional polyclonal antisera to cytokeratins can be useful in the identification of epithelial tumours and in distinguishing these carcinomas from other non-epithelial tumours, especially in those cases where routine histological techniques fail to do so. Antisera to cytokeratins are commonly produced against keratin proteins derived from human stratum corneum, but they have also been directed against hepatocyte cytokeratins or urinary bladder cytokeratins. These conventional rabbit antisera generally show a rather broad cross-reactivity staining virtually all epithelial tissues, although some antibodies raised in rabbits against isolated cytokeratin polypeptides show some specificity in their reaction and distinguish, for example, keratinizing from non-keratinizing epithelia. These reaction patterns find their basis in the fact that cytokeratins are a family of IFP. Thus, in contrast to most other types of intermediate filament, cytokeratin filaments are characterized by a remarkable biochemical diversity, represented in human tissues by at least 19 different cytokeratin polypeptides. These polypeptides are not expressed randomly throughout epithelia but occur in cell-type-specific combinations. The diverse types of carcinoma also differ in their cytokeratin polypeptide content. The polypeptide patterns of epithelial tumours are either identical with the cytokeratin pattern present in the cell of origin or at least closely related to it.[9] The cytokeratin polypeptide patterns (as obtained by two-dimensional gel electrophoresis) can be used for the identification or at least a subclassification of epithelial tissues and carcinomas. Moll et al.[9] have published a description of all the human cytokeratins and designated them 1 to 19, cytokeratin 1 being the polypeptide with the highest molecular weight and highest isoelectric pH and cytokeratin 19 being the polypeptide with the lowest molecular weight and a low isoelectric pH. 'Simple' epithelial tissues, such as liver, exhibit a simple cytokeratin pattern, only containing polypeptides 8 and 18.

The transitional epithelium of the urinary bladder contains cytokeratins 7, 8, 13, 18 and 19. Squamous epithelia are characterized by the absence of cytokeratins 7, 8 and 18 but contain the more basic and high-molecular-weight polypeptides 1 to 6 in addition to polypeptides 9 to 14.

Accordingly, the presence or absence of specific cytokeratin polypeptides in certain tumours can be used to sustain differential diagnosis. For example, in the female genital tract, the presence of cytokeratin 17 in cervical adenocarcinomas may help in distinguishing these malignancies from endometrial carcinomas. When applied to lung cancer, large cell undifferentiated bronchus carcinoma can be distinguished from other lung tumours by the presence of cytokeratins 6 and 7. Furthermore, in many cases anaplastic or undifferentiated squamous cell carcinomas can be distinguished from anaplastic or undifferentiated adenocarcinomas. In summary, we can state that studying of the cytokeratin patterns of individual epithelial tumours is helpful in improving tumour diagnosis and classification, especially in cases of undifferentiated primary tumours and carcinoma metastases.

5. USE OF MONOCLONAL ANTIBODIES TO CYTOKERATINS IN THE SUBCLASSIFICATION OF CARCINOMAS

The observation of tissue-specific expression of 19 different cytokeratin polypeptides in epithelium-derived tumours, together with the fact that specific cytokeratin patterns are retained during malignant growth and metastasis, serves as a basis for a novel approach to tumour characterization, with cytokeratins as differentiation markers. Although gel electrophoresis of cytoskeletal proteins from small biopsies is possible in a specially devised laboratory, these techniques offer several disadvantages and problems for routine diagnostic use. First, two-dimensional gel electrophoresis is rather complicated and time consuming for routine purposes and is certainly not a standard technique in pathology laboratories at this moment. Second, a certain amount of tissue is needed for isolation of the cytoskeletal proteins of a tumour. Gel electrophoresis of cytoskeletal proteins from small biopsies is feasible, but involves especially sensitive protein staining procedures. Moreover, a certain tumour preparation may consist, for a large part, of stromal and inflammatory cells and may contain only a few tumour cells. Third, tumours are often heterogeneous in histology. Gel electrophoresis will often not detect this fact and, therefore, diagnosis will be doubtful.

For these reasons, the use of monoclonal antibodies specific to certain cytokeratin polypeptides is a more practical strategy for histodiagnosis of carcinoma cells based on their cytokeratin content.

Several monoclonal antibodies to different cytokeratins have been prepared and different reactivities of such antibodies with different subtypes of epithelia and epithelial tumours have been noted by several authors.[5,6,10–15]

On the basis of their immunocytochemical reaction patterns with frozen tissue sections, four main types of monoclonal antibody to cytokeratin can be distinguished so far:[5,6,10–15]

a. Broadly cross-reacting monoclonal cytokeratin antibodies such as $K_2G8.13$, AE_1 and AE_3, RCK 102 (*Fig.* 23.3), and PKK1. These antibodies have a reactivity pattern similar to that of a polyclonal cytokeratin antiserum.

b. Monoclonal cytokeratin antibodies, such as LE 61, LE 65, CK_1–CK_4, RGE 53 (the latter being directed exclusively to cytokeratin 18) which will react only with simple, glandular epithelia and mesothelial cells and their tumours (adenocarcinomas and mesotheliomas), but not with squamous epithelia or squamous cell carcinomas.

c. Monoclonal cytokeratin antibodies reacting only with epidermal (suprabasal) cells, such as AE_2 and RKSE 60. The latter antibody also reacts with keratinizing cells in squamous cell carcinomas.

d. Monoclonal cytokeratin antibodies which have a more restricted reaction pattern, such as LE 41 which is directed against cultured kidney cells.

Some of these antibodies have been used in tumour characterization[5,6,14,15] and have shown that they can be used to subdivide groups of carcinomas. For example

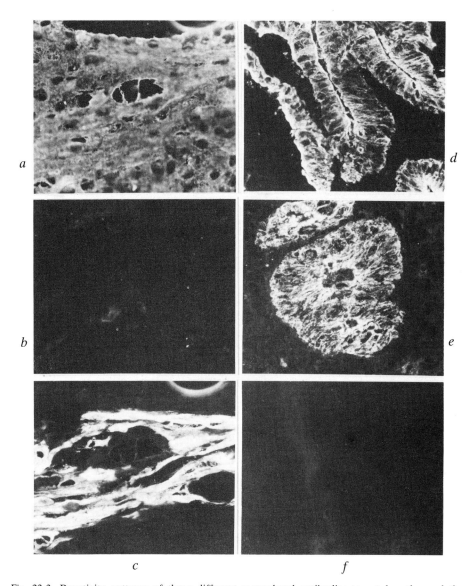

Fig. 23.3. Reactivity patterns of three different monoclonal antibodies to cytokeratins and the polyclonal keratin antiserum with frozen sections of a well-differentiated keratinizing squamous cell carcinoma of the vulva (a, b, c) and an adenocarcinoma of the colon $(d, e f)$. (a) Polyclonal rabbit antiserum to skin keratins; (b, e) RGE 53; (c, f) RKSE 60; (d) RCK 102. RCK 102 is a broadly cross-reacting antibody, staining virtually all epithelial tissues, while RGE 53 specifically reacts with glandular and columnar epithelial tissues and RKSE 60 only stains keratinizing squamous epithelial cells.

RGE 53, CK_1–CK_4 and LE 61/65 can distinguish adenocarcinomas from squamous cell carcinomas. In cases of anaplastic carcinoma metastases, in particular, these antibodies can be of great help for tumour diagnosis. They can not only refine carcinoma typing in frozen sections of surgically obtained material (*Fig.* 23.3), but can also be applied successfully to cytopathological preparations of thin needle aspirates obtained from palpable lymph nodes.[16]

Thus, in the future the use of an appropriate collection of monoclonal cytokeratin antibodies will not only allow tumours of epithelial origin to be distinguished from tumours of non-epithelial origin, but, in addition, will allow a further subdivision of carcinomas in relation to their histological origin. Furthermore, such antibodies will be useful probes for the identification of subpopulations of tumour cells in heterogeneous neoplasms and for histogenetic studies.

6. PITFALLS IN THE APPLICATION OF INTERMEDIATE FILAMENT PROTEIN ANTIBODIES FOR TISSUE RECOGNITION

It is a well-known fact that formalin fixation of tissues and subsequent paraffin embedding may drastically decrease the immunocytochemical detectability of IFP, especially the cytokeratins, vimentin and desmin. In general, this means that these IFPs can be detected only when present in relatively high concentrations or after proteolytic predigestion of paraffin sections. False negative results may, however, be obtained not only as a result of fixation, but also because of limited cross-reactivity of certain IFP antibodies. This may be especially true for cytokeratin antisera. For example, antisera to human skin keratins often do not react or react only weakly with certain types of adenocarcinoma. Therefore it is recommended that frozen sections are used whenever possible and that antisera are elaborately tested for cross-reactivity with several different types of tissue. Furthermore, the use of a combination of IFP antisera (for example cytokeratin and vimentin antibodies) may help to avoid false negative (or false positive) results.

The use of monoclonal IFP antibodies may present special problems. First, as for the polyclonal antisera, only some of the monoclonal IFP antibodies described so far seem to react in routine formalin-fixed paraffin sections. Second, selective masking of a certain epitope at specific stages of the cell cycle or during neoplastic progression may cause false negative results. And finally, epitopes recognized by monoclonal antibodies may be sensitive not only to formalin fixation, but also to acetone or methanol fixation.

However, when used with care and evaluated critically, immunocytochemical staining of IFP provides a powerful tool for the characterization of normal and malignant tissues.

Appendix

I. PREPARATION OF FROZEN SECTIONS FOR IMMUNOSTAINING

Cryostat sections are air dried, then fixed for 5–10 min in methanol at −20 °C.

They are then rapidly transferred to acetone for 5 s and allowed to dry before application of the primary antibody; the site of reaction is detected by the indirect method.

II. SOURCES OF ANTIBODIES USED FOR THE ILLUSTRATED PREPARATIONS

Primary antibodies	*See* the literature cited in the text
Fluorescein- and peroxidase- conjugated secondary antibodies	Nordic
Texas Red-conjugated sheep anti- mouse IgG (F(ab)$_2$ fraction)	New England Nuclear

REFERENCES

1. Franke W. W., Schmid E., Schiller D. L., Winter S., Jarasch E. D., Moll R., Denk H., Jackson B. W. and Illmensee K. Differentiation-related patterns of expression of proteins of interme-diate-sized filaments in tissues and cultured cells. *Cold Spring Harbor Symp. Quant. Biol.* 1981, **46**, 431–453.
2. Lazarides E. Intermediate filaments as mechanical integrators of cellular space. *Nature* 1980, **283**, 249–256.
3. Anderton B. Intermediate filaments: a family of homologous structures. *J. Muscle Res. Cell Motil.* 1981, **2**, 141–166.
4. Osborn M. and Weber K. Tumor diagnosis by intermediate filament typing: A novel tool for surgical pathology. *Lab. Invest.* 1983, **48**, 372–394.
5. Ramaekers F. C. S., Puts J. J. G., Moesker O., Kant A., Huysmans A., Haag D., Jap P. H. K., Herman C. J. and Vooijs G. P. Antibodies to intermediate filament proteins in the immunocytochemical identification of human tumours: an overview. *Histochem. J.* 1983, **15**, 691–713.
6. Holthöfer H., Miettinen A., Paasivuo R., Lehto V.-P., Lidner E., Alfthan O. and Virtanen I. Cellular origin and differentiation of renal carcinomas. *Lab. Invest.* 1983, **49**, 317–326.
7. Muijen v. G. N. P., Ruiter D. J. and Warnaar S. O. Intermediate filaments in Merkel cell tumours. *Hum. Pathol.* 1985, **16**, 590–595.
8. Ramaekers F. C. S., Haag D., Kant A., Moesker O., Jap P. H. K. and Vooijs G. P. Coexpression of keratin- and vimentin-type intermediate filaments in human metastatic carcinoma cells. *Proc. Natl Acad. Sci. USA* 1983, **80**, 2618–2622.
9. Moll R., Franke W. W., Schiller D. L., Geiger B. and Krepler R. The catalog of human cytokeratins: Patterns of expression in normal epithelia, tumors and cultured cells. *Cell* 1982, **31**, 11–24.
10. Lane E. B. Monoclonal antibodies provide specific intramolecular markers for the study of epithelial tonofilament organization. *J. Cell Biol.* 1982, **92**, 665–673.
11. Tseng S. C. G., Jarvinen M. J., Nelson W. G., Huang J-W., Woodcock-Mitchell J. and Sun T.-T. Correlation of specific keratins with different types of epithelial differentiation: monoclonal antibody studies. *Cell* 1982, **30**, 361–372.
12. Gigi O., Geiger B., Eshhar Z., Moll R., Schmid E., Winter E., Schiller D. L. and Franke W. W. Detection of a cytokeratin determinant common to diverse epithelial cells by a broadly cross-reacting monoclonal antibody. *EMBO J.* 1982, **1**, 1429–1437.
13. Debus E., Weber K. and Osborn M. Monoclonal cytokeratin antibodies that distinguish simple from stratified squamous epithelia: characterization on human tissues. *EMBO J.* 1982, **1**, 1641–1647.

14. Ramaekers F. C. S., Huysmans A., Moesker O., Kant A., Jap P. H. K., Herman C. and Vooijs G. P. Monoclonal antibodies to keratin filaments, specific for glandular epithelia and their tumors. Use in surgical pathology. *Lab. Invest.* 1983, **49**, 353–361.
15. Debus E., Moll R., Franke W. W., Weber K. and Osborn M. Immunohistochemical distinction of human carcinomas by cytokeratin typing with monoclonal antibodies. *Am. J. Pathol.* 1984, **114**, 121–130.
16. Ramaekers F., Haag D., Jap P. and Vooijs P. Immunochemical demonstration of keratin and vimentin in cytological aspirates. *Acta Cytologica* 1984, **28**, 385–392.

24

Intermediate Filaments and Differentiation in the Central Nervous System

D. Dahl and A. Bignami

The recent surge of interest in cytoskeletal 10-nm filaments is partly due to the cell specificity of their protein subunits. Although intermediate filaments have similar, if not identical, appearance regardless of their location, intermediate filament proteins are probably the only proteins fitting conventional histological classifications and may be thus considered as taxonomic characters for the principal tissues of the body, i.e. epithelia (keratins), muscle (desmin), mesenchyma (vimentin), neurones (neurofilament protein 'triplet') and astroglia (glial fibrillary acidic [GFA] protein) (for recent reviews *see*[1,2]). This finding has practical applications in development since it allows the cell type to be identified independently of morphological criteria. As a specific example, some of the traditional staining methods for astrocytes and neurones have been superseded by immunohistological methods with GFA protein and neurofilament antibodies in studies of CNS development.

1. GLIAL FILAMENTS

Glial filaments are heteropolymers of GFA protein and vimentin, the mesenchymal-type intermediate filament protein.[3] As a general rule, the cell-specific intermediate filament protein (GFA protein) is the major intermediate filament subunit in mature astrocytes, the same being true for vimentin in immature glia.[4-7] There are, however, some exceptions, mainly involving neuroglia maintaining in adult brain the radial orientation characteristic of the embryo.[8] In Müller glia of

401

rat retina and Bergmann glia of chicken cerebellum, filaments are of the vimentin type and do not stain with anti-GFA protein.[9,10] In fish, both Bergmann glia and Müller glia are GFA protein positive.[11,12] The species variations in this respect should be noted since they suggest that intermediate filament proteins are interchangeable to some extent.

Studies of GFA protein immunoreactivity in tissue sections as well as the electrophoretic analysis of brain filament preparations during CNS development, indicate that most astrocytes are formed at the time of myelination.[13–15] Myelination is heralded by a period of intense glial proliferation, a phenomenon first reported by Roback and Scherer in 1935[16] and called 'myelination gliosis'. Before myelination gliosis, glial cells are relatively few in white matter and do not stain with anti-GFA protein. Following the burst of glial proliferation GFA protein positive fibres first appear and become progressively more numerous until the framework supporting myelinated axons is completed.

Astrocytes are present in brain and spinal cord well before myelination. In the rat, astrocyte fibres appear on the surface on the cerebral cortex at birth and on day 9 a continuous glia limitans is formed.[17] Bergmann glia in rat cerebellum become GFA protein positive on day 4.[13] In rat cerebrum, the period of rapid accumulation of myelin-specific basic proteins does not start before day 12.[18] In human and rhesus monkey, radial glia, the main type of glia in the embryo, are already GFA protein positive by the tenth and sixth week of gestation, respectively.[19,20] In neonatal murine brain, radial glia express GFA protein as a response to injury[21] and it remains to be seen whether GFA protein expression under these conditions is regulated at the transcriptional or translational level. A clone encoding mouse GFA protein[22] may allow studies of hybridization in situ to determine whether GFA protein-negative murine radial glia contain GFA protein mRNA.

Raff et al.[23] have shown that early- and late-appearing astrocytes express different antigens and thus constitute two phenotypically distinct cell populations, type I and type II astrocytes. It was also shown that, depending on the tissue culture conditions and without undergoing cell division, a vimentin-positive precursor originates either an oligodendrocyte or a type II astrocyte.[24] This suggests that mitotic cells in myelinating white matter are glioblasts undergoing terminal differentiation into type II astrocytes or oligodendrocytes. The study of Schmechel and Rakic[25] showing transitional forms between embryonal radial glia and astrocytes, suggests that the former are the precursors of type I astrocytes.

2. NEUROFILAMENTS

Several studies have been reported in the literature on the first appearance of the neurone-specific intermediate filament proteins, the neurofilament (NF) protein triplet, in brain development.[26–31] These studies have shown that with one possible exception (to be discussed later) NF expression is an early event in ontogeny. It was also noted[27,29] that immunodecoration with NF antibodies was similar in distribution to the staining of neurofibrils, as reported in the classical papers on the development of the nervous system conducted with reduced silver methods.

The co-localization of NF and vimentin in early differentiating neurones has also been reported.[28,29,31,32] In the rat, vimentin and NF coexist in neurones for a limited period of embryonal development (days 12–14) both *in vivo* and in primary culture.[28] Since in murine brain neurones are generated throughout embryonal development and for at least one week after birth, the finding suggests that neuronal precursors later in development are committed to the neuronal lineage and as such do not express the mesenchymal intermediate filament protein. The finding of NF immunoreactivity in mitotic neuroblasts derived from 18-day rat embryo brain supports this interpretation.[33]

2.1. The High-molercular-weight Neurofilament Protein (NF 200 K)

The three NF polypeptides are named according to their molecular weight as determined by sodium dodecylsulphate–polyacrylamide gel electrophoresis (SDS–PAGE) (NF 70 K, NF 150 K, NF 200 K). Although the method overestimates the molecular weight of the NF triplet,[34] to avoid confusion we maintain the conventional terminology.

Studies conducted with monoclonal and polyclonal antibodies indicate that the appearance of the NF 200 K is markedly delayed compared to the other two proteins of the triplet.[35–37] In rat and rabbit it only became detectable postnatally. According to another study, NF 200 K was coexpressed with NF 150 K and NF 70 K, i.e. at a much earlier stage of development.[29]

Recent observations in this laboratory[38] suggest that the late appearance of NF immunoreactivity is related to post-translational modification rather than to late expression, since it was only observed with axon-specific antibodies (*Figs 24.1 and 24.2*). As an example, with axon-specific antibodies NF immunoreactivity was first observed on day 10 in the chicken embryo compared to day 3 with antibodies decorating NF in both axons and perikarya.[27]

Perhaps more interesting, the time of appearance varied considerably according to the system, thus suggesting that post-translational modification of NF proteins[39] plays a role in axonal maturation. Sensory nerves were still negative on day 14 in the chicken embryo (*Fig. 24.3*). In the optic nerve, immunoreactivity first appeared on day 17 and only the temporal bundle of the retina stained with the antibodies in newborn chicken (*Fig. 24.4*). With conventional NF antibodies, i.e. antibodies decorating NF in both axons and perikarya, immunoreactivity first appears on day 4 in sensory ganglia and on day 5 in the optic fibres of the retina.[27]

It is interesting to note that axonal specificity so far has been observed only with antibodies reacting with NF 150 K and NF 200 K.[40–43] The studies of Weber and his collaborators[44,45] suggest a possible explanation for these findings. Although both proteins are filament proteins rather than filament-associated proteins (NF 150 K forms filaments *in vitro* and NF 200 K is also capable of self-assembly[46]), their large molecular weight compared to other intermediate filament proteins is due to a carboxy terminal extension of unique amino acid composition sticking out from the filament and probably mediating cytoplasmic interactions. This, we believe, is the part of the molecule more likely to undergo post-translational modification in development. It remains to be seen whether the modified epitopes recognized by axon-specific antibodies are phosphorylated[47] and whether incubation with phosphatase[48] abolishes the immunoreactivity. This, in fact,

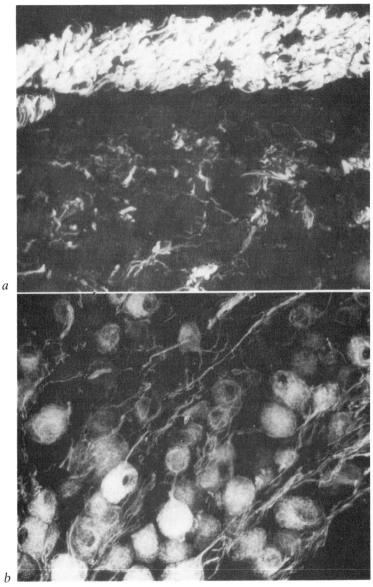

Fig. 24.1. Reactivity of axon-specific and conventional NF antibodies with sensory neurones in the chicken. The same results were obtained in rat. (*a*) Axon-specific antibodies (NF 200 K monoclonal, *see Fig.* 24.2) strongly decorate axons in the posterior root ganglion and adjacent nerve bundle (on the top of the figure). Neuronal perikarya are not stained. (*b*) Both axons and neuronal perikarya are stained by conventional NF monoclonal antibodies[51] in the posterior root ganglion. Indirect immunofluorescence staining on cryostat sections. *See*[51] for details of antibodies and methods. (From Bignami et al.[51] with permission.)

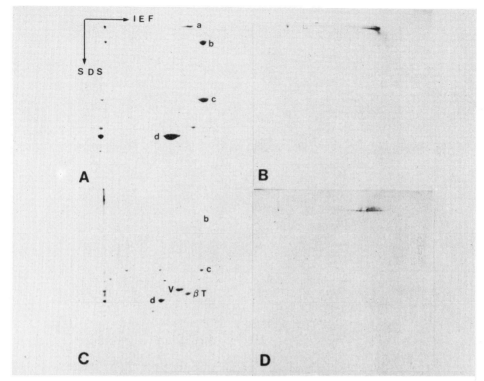

Fig. 24.2. Reactivity of axon-specific NF monoclonal antibodies with 2 M urea bovine spinal cord extract (*A* and *B*) and rat optic nerve SDS extract (*C* and *D*) analysed by two-dimensional gel electrophoresis. (*A*) and (*C*) are Coomassie blue-stained gels. Only NF 200 K reacts with the antibodies on the immunoblots (*B* and *D*). Identified spots in (*A*) and (*C*): (*a*) NF 200 K; (*b*) NF 150 K; (*c*) NF 70 K; (*d*) GFA protein; V, vimentin; βt, β-tubulin. Note the absence of the vimentin and β-tubulin spots in the bovine spinal cord extract (*A*). In the optic nerve extract NF 200 K is only identified by immunoblotting (*C*). (From Bignami et al.[51] with permission.)

appeared to be the case for the NF 200 K monoclonal illustrated in this review (*Figs* 24.1*a*, 24.2, 24.3*a*, 24.4*a*).Dilution of the antibodies with sodium phosphate[48] and incubation of the sections with phosphatase[48] completely abolished the axonal staining.[49] Conversely, NF antibodies decorating axons and perikarya were not affected by this treatment.

2.2. Transient NF Expression in Purkinje Cell Development

With some exceptions, e.g. axon-specific NF antibodies, there is a good correlation between NF immunoreactivity, silver impregnation and electron microscopy. In the cerebellar cortex, basket axons which are packed with filaments at the electron microscopical level stain intensely with both silver and NF antisera. Conversely, Purkinje cell perikarya and dendrites are NF negative by all these criteria, i.e. electron microscopy, silver impregnation and immuno-staining.

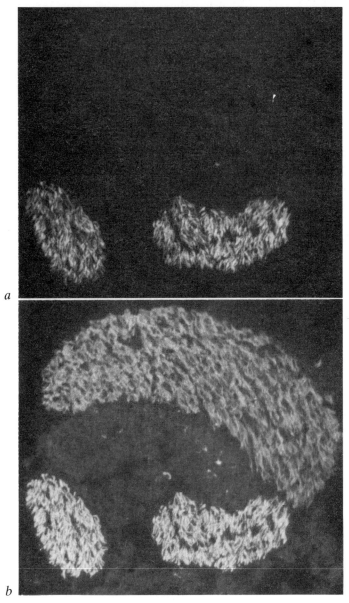

Fig. 24.3. Three peripheral nerves in a 14-day-old chick embryo are double stained with the axon-specific NF antibodies shown in *Fig.* 24.2 (*a*, fluorescein optics) and with conventional NF antisera[41] raised in rabbit (*b*, rhodamine optics). The nerve that is unstained in (*a*) is probably a sensory nerve, because axons emerging from sensory ganglia were NF negative with the same antibodies at this stage of development. Indirect immunofluorescence on cryostat section. (From Dahl and Bignami[49] with permission.)

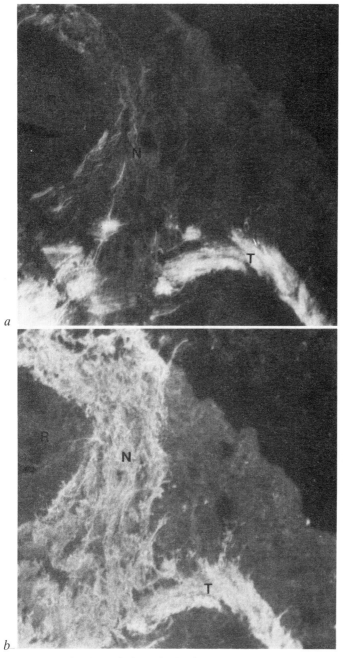

Fig. 24.4. Horizontal section through the optic disc (origin of optic nerve in the eye) of a newborn chicken double stained with the axon-specific NF antibodies shown in *Fig.* 24.2 (*a*, fluorescein optics) and with conventional NF antisera[41] (*b*, rhodamine optics). The bundle on the temporal side of the disc, T, is well stained with both antibodies (*a* and *b*). Conversely, the bundle on the nasal side of the disc, N, contains only few immunoreactive fibres in (*a*). R, retina. Indirect immunofluorescence on cryostat section. (From Dahl and Bignami[49] with permission.)

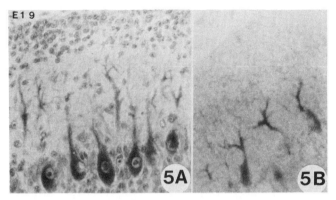

Fig. 24.5. Appearance of chicken Purkinje cells on embryonal day 19 (E19). (*A*) Haematoxylin counterstain. (*B*) No counterstain. Immunoperoxidase staining of paraffin sections with conventional NF antisera.[41] (From Bignami et al.[51] with permission.)

The observation that immature Purkinje cells in the chicken embryo are intensely argyrophil,[50] led to a detailed immunohistological study of the phenomenon[51] (*Figs* 24.5, 24.6). The selectivity of reduced silver staining for NF is not dependent on the presence of filaments but is based on a property of NF proteins. The three polypeptides forming the filament are intensely argyrophil in spinal cord and peripheral nerve extracts analysed by SDS–PAGE.[52]

An interesting finding in this study was the close correlation between the appearance of NF-positive baskets and the disappearance of NF immunoreactivity in Purkinje cells, thus suggesting a contact-dependent phenomenon such as the expression of glycerol-3-phosphate dehydrogenase by Bergmann glia in the mouse. A positive correlation between the expression of the glial enzyme and Purkinje cell presence has been reported in mutant mice cerebellum.[53]

The absence of neurofilaments in immature Purkinje cells[51] was a surprising finding since it apparently contradicted the general rule as to the close correlation between electron microscopy and immunohistology. The absence of 10-nm filaments in immature Purkinje cells containing intermediate filament proteins is not a unique phenomenon, however. Similar observations have been reported in C6 glioma[54] *in vitro* and in optic nerve axons before myelination.[35] A study on the synthesis of intermediate filament proteins in cultured astrocytes suggests that the subunits under these conditions may form shorter filaments not identifiable by electron microscopy, rather than reside in the cytoplasm as soluble monomers.[55] In short-term incorporation experiments, astrocyte intermediate filament proteins in the Triton-insoluble fraction were labelled within 5 minutes after exposure to radioactive leucine, thus suggesting that incorporation of proteins into the cytoskeletal fraction occurred concurrently with the completion of the polypeptide chain.

An intriguing possibility suggested by the work of Sternberger and Sternberger[48] is that NF protein may persist in mature Purkinje cells but in a form not recognized by our antibodies. It was reported that monoclonal antibodies reacting with the NF 200 K and NF 150 K polypeptide, decorated Purkinje cell perikarya and dendrites rather than baskets in rat cerebellum. However, the

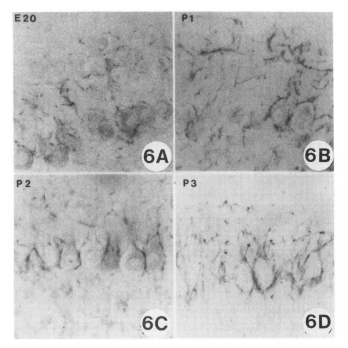

Fig. 24.6. Disappearance of NF immunoreactivity in chicken Purkinje cells and formation of NF-positive baskets at hatching. (*A*) Embryonal day 20 (E20). (*B*) Postnatal day 1 (P1). (*C*) Postnatal day 2 (P2). (*D*) Postnatal day 3 (P3). Immunoperoxidase staining with conventional NF antisera.[41] No counterstain. (From Bignami et al.[51] with permission.)

staining changed to the traditional pattern (selective decoration of the baskets) following incubation with trypsin and phosphatase.

REFERENCES

1. Dahl D. and Bignami A. Intermediate filaments in nervous tissue. In: Shay, J. W., ed., *Cell and Muscle Motility*, Vol. VI. New York, Plenum, 1985: 75–96.
2. Osborn M., Altmannsberger M., Debus E. and Weber K. Conventional and monoclonal antibodies to intermediate filament proteins in human tumor diagnosis. In: Levine, A. J., Vanderwoude, G. F., Topp, W. C. and Watson, J. D., eds., *Cancer Cells 1/The Transformed Phenotype*. Cold Spring Harbor Laboratory, 1984: 191–200.
3. Quinlan R. A. and Franke W. W.. Molecular interactions in intermediate-sized filaments revealed by chemical crosslinking. Heteropolymers of vimentin and glial filament protein in cultured human glioma cells. *Eur. J. Biochem.* 1983, **132**, 477–484.
4. Dahl D., Rueger D. C., Bignami A., Weber K. and Osborn M. Vimentin, the 57 000 dalton protein of fibroblast filaments, is the major cytoskeletal component in immature glia. *Eur. J. Cell Biol.* 1981, **24**, 191–196.
5. Schnitzer J., Franke W. W. and Schachner M. Immunocytochemical demonstration of vimentin in astrocytes and ependymal cells of developing and adult mouse nervous system. *J. Cell Biol.* 1981, **90**, 435–447.
6. Tapscott S. J., Bennett G. S., Toyama Y., Kleinbart F. and Holtzer H. Intermediate filament proteins in the developing chick spinal cord. *Dev. Biol.* 1981, **86**, 40–54.

7. Yen S.-H. and Fields K. L. Antibodies to neurofilament, glial filament and fibroblast intermediate filament proteins bind to different cell types of the nervous system. *J. Cell Biol.* 1981, **88**, 115–126.

8. Dahl D., Bignami A., Weber K. and Osborn M. Filament proteins in rat optic nerves undergoing Wallerian degeneration. Localization of vimentin, the fibroblastic 100-Å filament protein, in normal and reactive astrocytes. *Exp. Neurol.* 1981, **73**, 496–506.

9. Shaw G. and Weber K. The intermediate filament complement of the retina—a comparison between different mammalian species. *Eur. J. Cell Biol.* 1984, **33**, 95–104.

10. Debus E., Weber K. and Osborn M. Monoclonal antibodies specific for glial fibrillary acidic (GFA) protein and for each of the neurofilament triplet polypeptides. *Differentiation*, 1983, **25**, 193–203.

11. Dahl D., Crosby C. J., Sethi J. S. and Bignami A. Glial fibrillary acidic (GFA) protein in vertebrates. Immunofluorescence and immunoblotting study with monoclonal and polyclonal antibodies. *J. Comp. Neurol.*, 1985, **239**, 75–88.

12. Bignami A. Glial fibrillary acidic (GFA) protein in Müller glia. Immunofluorescence study of the goldfish retina. *Brain Res.* 1984, **300**, 175–178.

13. Bignami A. and Dahl D. Differentiation of astrocytes in the cerebellar cortex and the pyramidal tracts of the newborn rat. An immunofluorescence study with antibodies to a protein specific to astrocytes. *Brain Res.* 1973, **49**, 393–402.

14. Dahl D. The vimentin–GFA protein transition in rat neuroglia cytoskeleton occurs at the time of myelination. *J. Neurosci. Res.* 1981, **6**, 741–748.

15. Dahl D., Strocchi P. and Bignami A. Vimentin in the central nervous system. A study of the mesenchymal-type intermediate filament-protein in Wallerian degeneration and in postnatal rat development by two-dimensional gel electrophoresis. *Differentiation*, 1982, **22**, 185–190.

16. Roback H. N. and Scherer H. J. Über die feinere Morphologie de frühkindlichen Hirnes unter besonderer Berücksichtigung der Gliaentwicklung. *Virchows Arch. Pathol. Anat.* 1935, **294**, 365–413.

17. Bignami A. and Dahl D. Astrocyte-specific protein and neuroglial differentiation. An immunofluorescence study with antibodies to the glial fibrillary acidic protein. *J. Comp. Neurol.* 1974, **153**, 27–38.

18. Cohen S. R. and Guarnieri M. Immunochemical measurement of myelin basic protein in developing rat brain: an index of myelin synthesis. *Dev. Biol.* 1976, **49**, 294–299.

19. Antanitus D. S., Choi B. H. and Lapham L. W. The demonstration of glial fibrillary acidic protein in the cerebrum of the human fetus by indirect immunofluorescence. *Brain Res.* 1976, **103**, 613–616.

20. Levitt P. and Rakic P. Immunoperoxidase localization of glial fibrillary acidic protein in radial glial cells and astrocytes of the developing rhesus monkey brain. *J. Comp. Neurol.* 1980, **193**(3), 417–448.

21. Bignami A. and Dahl D. Astrocyte-specific protein and radial glia in the cerebral cortex of newborn rat. *Nature*, 1974, **252**, 55–56.

22. Lewis S. A., Balcarek J. M., Krek V., Shelanski M. and Cowan N. J. Sequence of a cDNA clone encoding mouse glial fibrillary acidic protein: structural conservation of intermediate filaments. *Proc. Natl Acad. Sci. USA*, 1984, **81**, 2743–2746.

23. Raff M. C., Williams B. P. and Miller R. H. The *in vitro* differentiation of a bipotential glial progenitor cell. *EMBO J.* 1984, **3**, 1857–1864.

24. Temple S. and Raff M. Differentiation of a bipotential glial progenitor cell in single cell microculture. *Nature*, 1985, **313**, 223–225.

25. Schmechel D. E. and Rakic P. A Golgi study of radial glial cells in developing monkey telencephalon: Morphogenesis and transformation into astrocytes. *Anat. Embryol.* 1979, **156**, 115–152.

26. Bennett G. S., Tapscott S. J., DiLullo C. and Holtzer H. Differential binding of antibodies against the neurofilament triplet proteins in different avian neurons. *Brain Res.* 1984, **304**, 291–302.

27. Bignami A., Dahl D. and Seiler M. W. Neurofilaments in the chick embryo during early development. I. Immunofluorescent study with antisera to neurofilament protein. *Develop. Neurosci.* 1980, **3**, 151–161.

28. Bignami A., Raju T. R. and Dahl D. Localization of vimentin, the non-specific intermediate filament protein, in embryonal glia and in early differentiating neurons. *Dev. Biol.* 1982, **91**, 286–295.

29. Cochard P. and Paulin D. Initial expression of neurofilaments and vimentin in the central and peripheral system of the mouse embryo *in vivo. J. Neurosci.* 1984, **4**, 2080–2094.
30. Raju T. R., Bignami A. and Dahl D. *In vivo* and *in vitro* differentiation of neurons and astrocytes in the rat embryo. Immunofluorescence study with neurofilament and glial filament antisera. *Dev. Biol.* 1981, **85**, 344–351.
31. Tapscott S. J., Bennett G. S. and Holtzer H. Neuronal precursor cells in the chick neural tube express neurofilament proteins. *Nature*, 1981, **292**, 836–838.
32. Jacobs M., Choo Q. L. and Thomas C. Vimentin and 70 K neurofilament protein co-exist in embryonic neurons from spinal ganglia. *J. Neurochem.* 1982, **38**, 969–977.
33. Asou H., Iwasaki N., Hirano S. and Dahl D. Mitotic neuroblasts in dissociated cell cultures from embryonic rat cerebral hemispheres express neurofilament protein. *Brain Res.* 1985, **332**, 355–357.
34. Kaufmann E., Geisler N. and Weber K. SDS–PAGE strongly overestimates the molecular masses of the neurofilament proteins. *FEBS Lett.* 1984, **170**, 81–84.
35. Pachter J. S. and Liem R. K. H. The differential appearance of neurofilament triplet polypeptides in the developing rat optic nerve. *Dev. Biol.* 1984, **103**, 200–210.
36. Shaw G. and Weber K. Differential expression of neurofilament triplet proteins in brain development. *Nature*, 1982, **298**, 277–279.
37. Willard W. and Simon C. Modulations of neurofilament axonal transport during the development of rabbit retinal ganglion cells. *Cell*, 1983, **35**, 551–559.
38. Dahl D. and Bignami A. Axon specific neurofilament epitopes in development. An index of nerve maturation. *J. Neuropathol. Exp. Neurol.* 1985, (Abstr.) **44**, 328.
39. Nixon R. A., Brown B. A. and Marotta C. A. Post-translational modification of a neurofilament protein during axoplasmic transport: implications for regional specialization of CNS axons. *J. Cell Biol.* 1982, **94**, 150–158.
40. Dahl D., Bignami A., Bich N. T. and Chi N. H. Immunohistochemical localization of the 150 K neurofilament protein in the rat and the rabbit. *J. Comp. Neurol.* 1981, **195**, 659–666.
41. Dahl D. Immunohistochemical differences between neurofilaments in perikarya, dendrites and axons. Immunofluorescence study with antisera raised to neurofilament polypeptides (200 K, 150 K, 70 K) isolated by anion exchange chromatography. *Exp. Cell Res.* 1983, **149**, 397–408.
42. Debus E., Flügge G., Weber K. and Osborn M. A monoclonal antibody specific for the 200 K polypeptide of the neurofilament triplet. *EMBO J.* 1982, **1**, 41–45.
43. Shaw G., Osborn M. and Weber K. An immunofluorescence microscopical study of the neurofilament triplet proteins, vimentin and glial fibrillary acidic protein within the adult brain. *Eur. J. Cell Biol.* 1981, **26**, 68–82.
44. Geisler N., Fischer S. Vandekerckhove J., Plessman U. and Weber K. Hybrid character of a large neurofilament protein (NF-M): intermediate filament type sequence followed by a long acidic carboxyterminal extension. *EMBO J.* 1984, **3**, 2701–2706.
45. Geisler N., Kaufmann E., Fischer S., Plessman U. and Weber K. Neurofilament architecture combines structural principles of intermediate filaments with carboxyterminal extensions in size between triplet proteins. *EMBO J.* 1983, **8**, 1295–1302.
46. Gardner E. E., Dahl D. and Bignami A. Formation of 10-nanometer filaments from the 150 K dalton neurofilament protein *in vitro. J. Neurosci. Res.* 1984, **11**, 145–155.
47. Wong J., Hutchison S. B. and Liem R. K. H. An isoelectric variant of the 150,000-dalton neurofilament polypeptide. Evidence that phosphorylation state affects its association with the filament. *J. Biol. Chem.* 1984, **259**, 10867–10874.
48. Sternberger L. A. and Sternberger N. H. Monoclonal antibodies distinguish phosphorylated and nonphosphorylated forms of neurofilaments *in situ. Proc. Natl Acad. Sci. USA*, 1983, **80**, 6126–6130.
49. Dahl D. and Bignami A. Neurofilament phosphorylation in development. A sign of axonal maturation? *Exp. Cell Res.* 1985, in press.
50. Levi Montalcini R. Growth and differentiation in the nervous system. In: Allen J., ed., *The Nature of Biological Diversity*. New York, McGraw-Hill, 1963: 261–295.
51. Bignami A., Grossi M. and Dahl D. Transient expression of neurofilament protein without filament formation in Purkinje cell development. Immunohistological and electron microscopic study of chicken cerebellum. *Int. J. Dev. Neurosci.*, 1985, **3**, 365–377.
52. Gambetti, P., Autilio-Gambetti L. and Papasozomenos S. Ch. Bodian's silver method stains neurofilament polypeptides. *Science*, 1981, **213**, 1521–1522.
53. Fisher M. Neuronal influence on glial enzyme expression: evidence from mutant mouse cerebella. *Proc. Natl Acad. Sci. USA*, 1984, **81**, 4414–4418.

54. Bissell M. G., Rubinstein L. J., Bignami A. and Herman M. M. Characteristics of the rat C-6 glioma maintained in organ culture systems. Production of glial fibrillary acidic protein in the absence of gliofibrillogenesis. *Brain Res.* 1974, **82**, 77–89.
55. Chiu F.-C. and Goldman J. E. Synthesis and turnover of cytoskeletal proteins in cultured astrocytes. *J. Neurochem.* 1984, **42**, 166–174.

25

Neurofilaments and Glial Filaments in Neuropathology

J. Q. Trojanowski

The development of molecular probes, such as monoclonal antibodies and cDNA probes, has ushered in a new era in the study of human diseases. Now it is possible to search routinely for molecular abnormalities in human tissue. Several molecules restricted to different cell types in the central (CNS) and peripheral (PNS) nervous systems have been discovered.[1-4] One of the first cell type-specific proteins identified was glial filament (GF) protein.[5,6] Antibodies to GF protein were shown to be useful diagnostic reagents nearly 10 years ago,[7,8] and it is now recognized that antibodies to GF and other intermediate filament (IF) proteins can be used in immunohistochemical studies of diseases.[1-4,6,9,10]

The critical importance of rigorously determining the specificity of antibody probes cannot be overemphasized. Immunohistochemistry provides indirect evidence (reaction product in tissue) for the identity of an antigen in situ. Without accurate data on the specificity of an antibody, and well-executed controls, reaction product in a tissue section is meaningless. Because of homologies among IF proteins,[4,11] this is of fundamental importance for immunohistochemical studies of these proteins in disease.

This chapter will focus on the two major IF proteins found in the CNS and PNS, i.e. GF protein and neurofilament (NF) subunits. The use of NF- and GF-specific antibodies to study CNS and PNS tumours and neurodegenerative disorders is then considered.

413

1. INTERMEDIATE FILAMENT POLYPEPTIDES: A FAMILY OF PROTEINS

Intermediate filaments, filamentous structures 10 nm in diameter, form part of the cytoskeleton of nearly all mammalian cells.[2-4] Their function is unknown. Five different classes of IFs (NFs, GFs, desmin, keratin and vimentin filaments) are distinguished biochemically and immunochemically; they cannot be distinguished by conventional electron microscopy. NFs are present in neurones and GFs are found in glial cells. Desmin, keratin and vimentin filaments are restricted to muscle, epithelial and mesenchymal cells, respectively. However, vimentin can be coexpressed with any of the other four classes of IF, and it appears in glial and neuronal precursor cells prior to GF or NF proteins.[12]

Early investigative work on IF proteins was limited by the extensive homologies among IF polypeptides. Monoclonal antibodies can recognize a single IF protein, or two or more classes of IFs.[4,11,13,14] Antiserum raised to one IF protein may thus contain antibodies which recognize other classes of IF.

Studies of the role of IFs in pathological processes are limited because information concerning important aspects of the cell biology of IF proteins is incomplete. For example, the structure of IFs is not known, and it is unclear to what extent IF proteins differ across species. Further, some IFs are modified after synthesis by phosphorylation or other mechanisms under normal and pathological circumstances and these phenomena are ill defined.[9,10,15-20]

2. NEUROFILAMENTS

NFs are heteropolymers composed of three different subunits with apparent molecular weights of 68, 150 and 200 kdal,[22-25] referred to here as NF68, NF150 and NF200, respectively. Although NF proteins may be transiently expressed in non-neuronal avian cells,[26,27] among mammals only mature neurones express these polypeptides. NF triplet proteins are immunochemically distinct,[14] and are the products of separate genes.[28] It is controversial whether or not neurones express all three NF subunits.[29,30] Recent studies[30] (Trojanowski et al.[54]) indicate that CNS and PNS axons and PNS neuronal perikarya rarely, if ever, lack any of the three NF subunits. Most CNS neuronal perikarya express all of the NF triplet proteins, but many do not. Because of microheterogeneity[15] among NF proteins (due to phosphorylation, proteolysis, conformational rearrangements of the filament structure etc.), the failure to detect NF subunits in a population of neurones may reflect methodological limitations rather than a real absence of these antigens.

Tissue preparative variables and the immunohistochemical procedures significantly affect the detectability of NF antigens.[30] Furthermore, false negative and positive results are likely to continue to be a problem in immunohistochemical studies of normal and diseased tissues until more is known about such things as the modifications NF proteins undergo following translation, and the significance of species differences among NF proteins.

3. GLIAL FILAMENTS

The first polypeptide to be identified as a major component of an IF class was GF protein, a 50–52-kdal protein isolated by Eng and coworkers.[5,6] In the mature normal CNS, GF protein is expressed by astrocytes, rare ependymal cells, and radial glia in the molecular layer of the cerebellum.[6] Mature oligodendrocytes lack IFs and do not express GF protein except during development.[31] Controversy exists regarding the expression of GF in some cells outside the CNS, such as in Schwann cells.[32] Recently, a monoclonal antibody to GF produced by Lee et al.[11] was shown to recognize GF antigens in astrocytes, but not the IFs of Schwann cells under normal or pathological circumstances[20,33] (*see also Fig. 25.1*).

Immunohistochemical and immunochemical studies using well-characterized species-specific monoclonal antibodies to GF are needed to resolve these controversies. We have begun such studies in bovine tissues, since bovine immunogens were used by Lee et al.[11] to produce and characterize NF- and GF-specific monoclonal antibodies. Species differences may be critical in the demonstration of an IF antigen in tissue.[34] For example, the detection of GF protein in radial glia of cerebellum is species as well as technique dependent (*Fig. 25.1*).

4. NEUROFILAMENT AND GLIAL FILAMENT PROTEINS IN TUMOURS OF THE CENTRAL AND PERIPHERAL NERVOUS SYSTEM

The initial studies of GF protein expression in glial neoplasms[6–8] led to the use of antibodies to other IF proteins to study a variety of neoplasms[2–4,33] (*see also* Chapter 23). We have based our studies of human neoplasms with monoclonal antibodies to NF and GF on the following hypotheses which we continue to test:

a. IF proteins are immunochemically distinct
b. IF proteins, except vimentin, are cell type specific
c. IF proteins are developmentally regulated
d. Neoplastic cells express the same IF proteins as their progenitor cells.

These hypotheses require further validation. For example, data on the expression of IF antigens in the developing organism are essential for the interpretation of immunohistochemical studies of neoplastic and non-neoplastic disease of the nervous system using antibodies to these proteins. A number of developmental studies on the expression of IF proteins has been published[12,27,35–37] (*see also* Chapter 24), but information on this subject is still incomplete.

Furthermore, immunohistochemical studies alone do not establish the identity of an antigen stained by an antibody; isolation of the antigen is required. Antibodies may have surprising cross-reactions. For example, antiserum raised to synthetic α-melanocyte-stimulating hormone (α-MSH) cross-reacts with NF150 as shown recently with the immunoblot method.[38,39] However, the immunoblot method also suffers from limitations, such as the inability to

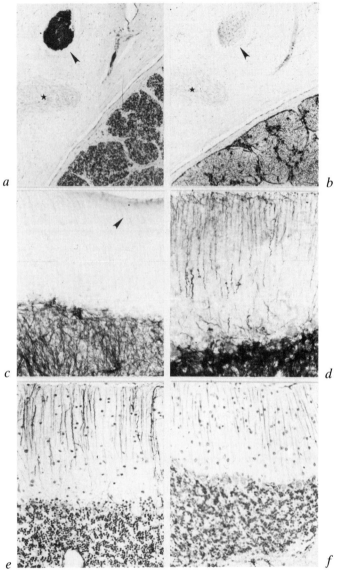

Fig. 25.1. Identically prepared paraffin sections of Bouin's-fixed bovine optic nerve immunostained with monoclonal antibodies to bovine NF (*a*) or bovine GF (*b*) proteins. The optic nerve and PNS nerves (arrow heads) are immunostained in (*a*); only astrocytes and not Schwann cells are stained in (*b*). The same blood vessel (asterisk) is identified in (*a*) and (*b*). In (*c*) and (*d*), identically fixed (Bouin's fixative) Vibratome sections of bovine cerebellum are immunostained with the monoclonal antibody to GF. Radial glia are faintly stained in (*c*) (arrow head); after treatment with trypsin (0·4 mg/ml for 10 min) they are more prominent (*d*). Paraffin sections of Bouin's-fixed human (*e*) and rat (*f*) cerebellum were not treated with trypsin, yet radial glia are intensely GF positive with the same monoclonal antibody to GF as in (*c*) and (*d*). The monoclonal antibody to NF (4.3F9) used in *Figs.* 25.1–25.3 recognizes human and rat NF200; in bovine tissues it recognizes both NF200 and NF150. The same monoclonal antibody to GF (2.2B10) was used in *Figs.* 25.1–25.3. All sections are counterstained with haematoxylin.

Table 25.1.   CNS and non-CNS tumours with NF and/or GF antigens

Tumour	NF positive	GF positive
Aesthesioneuroblastoma	+	−
Astroblastoma	−	+
Astrocytoma (grades I–IV)	−	+
Carcinoid	+	−
Cerebral neuroblastoma	+	+
Choroid plexus papilloma	−	+
Ependymoma	−	+
Ganglioglioma	+	+
Ganglioneuroblastoma	+	−
Ganglioneuroma	+	−
Gemistocytic astrocytoma	−	+
Haemangioblastoma	−	+
Medulloblastoma	+	+
'Merkel's cell' tumour of skin	+	−
Mixed glioma	−	+
Neuroblastoma	+	−
Oat cell carcinoma	+	−
Oligodendroglioma	−	+
Paraganglioma	+	−
Phaeochromocytoma	+	−
Pineoblastoma	+	+
Pineocytoma	−	+
Pituitary adenoma	−	+
Pleomorphic adenoma	−	+
Retinoblastoma	−	+
Subependymal giant cell astrocytoma	−	+
Subependymoma	−	+
Teratoma	+	+

ascribe immunobands to single cells in situ. Also, the presence of NF breakdown products in tumours may complicate analysis. Furthermore, since NF antigens are ubiquitous in nerves, it may be impossible to examine the IFs of some tumours independent of normal NF proteins. The molecular identity of the antigens stained by antibodies to NF and GF in neoplastic tissues has rarely been proven.[3]

Numerous studies have described the expression of GF protein in human neoplasms; few have employed antibodies to NF proteins.[1–4,6–8,33,40–42] The CNS and extra-CNS tumours known to contain cells which express NF and/or GF proteins are listed in *Table* 25.1. However, it has become clear in recent years that the interpretation of immunohistochemical studies of neoplasms, based on the use of antibodies to IF, requires a more detailed understanding of the cell biology of IF proteins.

NF and GF are not tumour-specific antigens; thus antibodies to these antigens do not establish that the immunoreactive cells are neoplastic. The abnormal appearance of the IF of neoplastic cells may aid in the differentiation of reactive from neoplastic cells. For example, aggregates of immunoreactive NF proteins are seen in diseased but not in normal neurones.[9,18,33,42,43]

Antibodies to NF and GF proteins help to identify cell types in a tumour, but this may not resolve the question of its histogenesis. For example, we have noted the presence of both NF and GF antigens, the absence of either, or the presence of one or the other of these antigens in medulloblastomas[33] (Tremblay et al.[55]). However, the majority of cells in medulloblastomas do not express NF or GF antigens. Our studies probably reveal the capacity of medulloblastomas to undergo neuronal or glial differentiation and do not define the histogenesis of these tumours. Additional studies must determine whether the absence of NF and GF antigens in medulloblastomas is a tissue processing artefact, a result of modifications of NF and GF antigens in neoplastic cells etc.

Future studies must address the issue of whether or not the detection of cell type-specific markers in other kinds of tumour reflects the histogenesis of a tumour or the ability of neoplastic cells to differentiate. This problem is further illustrated by two extra-CNS tumours shown to express NF antigens, i.e. oat cell carcinomas of the lung and so-called 'Merkel cell tumours' of the skin.[3,40-42] Thus far, the NF-containing precursor cells of these tumours have not been identified, and normal Merkel cells appear to lack NF proteins.[44] Despite uncertainties on these issues, the identification of NF and GF antigens in tumours may be of diagnostic and prognostic significance.[2-4,33]

Since the introduction of IF class-specific antibodies for the diagnostic evaluation of human tumours, it has become evident that they allow a more objective basis for determining the cell types within a given tumour. However, the interpretation of this data will be limited until the additional information discussed here becomes available. Studies which address these issues will permit validation of the hypotheses mentioned earlier.

5. NEUROFILAMENT AND GLIAL FILAMENT PROTEINS IN NEURODEGENERATIVE DISORDERS

Neurodegenerative disorders are nervous system diseases of unknown aetiology. Senile dementia of the Alzheimer type (SDAT) is prototypical of such diseases. Although neurones are primarily affected in these diseases, astrocytes are also involved. Gliosis is a common response to a CNS lesion, and it is nearly always associated with an increase in GF immunoreactivity.[6] Schwann cells proliferate in response to peripheral nerve injury, but they do not express a polypeptide biochemically identical to GF protein. The response to injury of other extra-CNS glial cells, such as enteric glia, is not well defined. The expression of GF protein by astrocytes following injury has recently been reviewed by Eng and DeArmand.[6]

Alterations in human NF proteins due to disease have not been extensively examined. Accumulations of filaments 10 nm in diameter in neurones have been observed in experimental animals exposed to a number of chemicals, such as acrylamide, aluminum, maytansine etc.[45,46] In only a few cases have NF antigens been demonstrated in these abnormal filaments.[45]

NF immunoreactivity is altered in a complex manner in SDAT. NFs have been implicated in the two pathological hallmarks of this disease, i.e. neurofibrillary tangles and senile plaques. Since other antigens are found in these lesions too, it is uncertain if neurofibrillary tangles and senile plaques are composed primarily of altered NFs, or if they are formed from other components to which NF proteins bind.[9,10,43,45,47-50] As summarized in *Table* 25.2, abnormal patterns of

Table 25.2. *Neurodegenerative disorders with abnormal patterns of NF immunoreactivity*

Disease	Abnormal NF pattern
Amytotrophic lateral sclerosis	Spheroids
Down's syndrome	Neurofibrillary tangles, senile plaques
Hallervorden–Spatz disease	Spheroids, bizarre neurones
Neuroaxonal dystrophy	Spheroids
Neuronal intranuclear inclusion disease	Intranuclear NF inclusions
Olivo-ponto-cerebellar atrophy	Torpedos
Parkinson's disease	Lewy bodies
Pick's disease	Pick bodies, neurofibrillary tangles
Progressive supranuclear palsy	Neurofibrillary tangles
Senile dementia of the Alzheimer type	Neurofibrillary tangles, senile plaques

immunoreactivity are seen in a number of other neurodegenerative conditions.[43,51–53]

Abnormal patterns of NF immunoreactivity in a disease do not necessarily implicate NF proteins in the aetiology of the disorder; such patterns can occur when NFs are secondarily affected by a disease process. For example, alterations in NF immunoreactivity are also observed in peripheral nerves subjected to nerve transection or nerve crush.[16,19,20] NF immunoreactivity appears granular (*Fig. 25.2*) in the distal portion of transected nerves,[20] while in the parent perikaryon NF staining increases.[16] We have seen NF-positive neurones of bizarre appearance in the CNS associated with mass lesions, infarcts or heterotopias. In Hallervorden–Spatz disease, the same monoclonal antibody used to detect NF antigens in the transected nerves also stains spheroids (*Fig. 25.2c*). Longitudinally sectioned axons in this case contain granular NF immunoreactivity (*Fig. 25.2d*) similar to that seen in nerves undergoing wallerian degeneration (*Fig. 25.2b*).

Sections of cortex from a case of SDAT immunostained with the same monoclonal antibody to NF 200 used in *Fig. 25.2* detects this antigen in senile plaques and in several abnormal profiles consistent with neurofibrillary tangles (*Fig. 25.3*).

However, the distinction between normal NF-positive neurones and neurones with NF-positive neurofibrillary tangles may be difficult since NF antigens, are present in normal neuronal perikarya.

The significance of finding NF antigens in the lesions associated with neurodegenerative disorders is unclear. Do NF antigens accumulate in neurofibrillary tangles and senile plaques because they are bystanders in the pathological process or does the aetiology of SDAT have to do with some fundamental abnormality in the synthesis, assembly and turnover of the neuronal cytoskeleton?

The use of molecular probes such as the monoclonal antibodies discussed here will permit these questions to be addressed. Immunohistochemical studies of neurodegenerative disorders with monoclonal antibodies to NF and GF antigens, as well as to other cytoskeletal proteins, will allow hypotheses concerning the role IFs play in these diseases to be tested. Such probes will also lead to improvements in the diagnosis of these diseases.

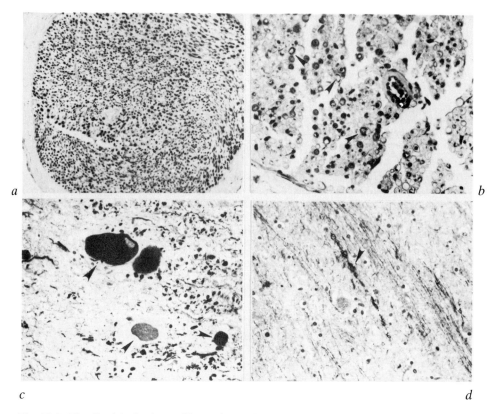

Fig. 25.2. The Bouin's-fixed paraffin sections of rat nerve seen here were stained with the monoclonal antibody to NF. (*a*) and (*b*) are normal and transected sciatic nerve, respectively. The arrow heads in (*b*) identify granular reaction product in degenerating nerves. NF-positive spheroids (small arrow heads in *c*) and degenerating NF-positive axons (arrow head in *d*) are seen in formalin-fixed tissue from a patient with Hallervorden–Spatz disease. An abnormal NF-positive neurone is seen in (*c*) (large arrow head).

Acknowledgements

Appreciation is expressed to Drs V. M.-Y. Lee and W. W. Schlaepfer for helpful discussions and collaboration on many aspects of the work reviewed here. Ms T. Schuck provided expert technical help. Supported in part by NIH grants CA36245, NS18616 and Teacher Investigator Development Award NS00762.

Appendix

I. METHODS

Tissue from experimental animals was fixed in Bouin's fluid for 4–6 h. It was then

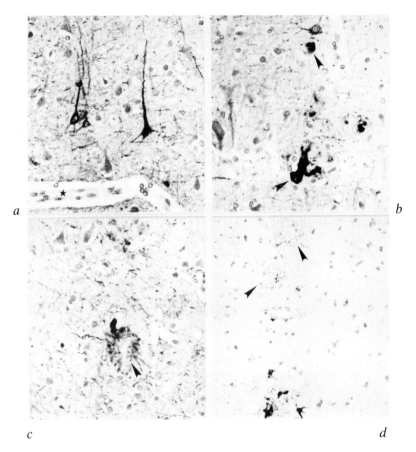

Fig. 25.3. Paraffin sections of formalin-fixed cortex from a case of SDAT. The anti-NF monoclonal antibody stains normal neurones in (*a*) (the asterisk identifies a blood vessel), neurofibrillary tangles (arrow heads) in (*b*), and the periphery of a senile plaque (arrow head) in (*c*). The monoclonal antibody to GF stains reactive astrocytes in (*d*). The processes of GF-positive astrocytes rim the perimeter of senile plaques (arrow heads in *d*).

transferred to phosphate-buffered saline (PBS) prior to paraffin embedding and sectioning. Unembedded tissue was cut on a Vibratome and sections were collected in PBS.

Treatment of sections with enzymes (trypsin, alkaline phosphatase) or detergents (Triton X-100, sodium dodecylsulphate) prior to immunostaining has a variable effect on the immunoreactivity of NF and GF epitopes, depending on the monoclonal antibody used and the phosphorylation state of the epitope recognized.[17,20,21,30].

Standard PAP or ABC-peroxidase immunostaining procedures were used for paraffin sections mounted on chrome–alum-treated slides or Vibratome free-floating sections.

Endogenous peroxidase was blocked with 1 part 30 per cent H_2O_2 to 5 parts methanol for 30 min.

Non-specific binding was blocked with 2 per cent fetal calf serum in buffer (this solution could be re-used for up to one week if kept at 4 °C).

Incubation with the primary antibody in spent hybridoma supernatant was overnight at 4 °C or for 30–120 min at room temperature.

Peroxidase was developed in 0·05 per cent DAB in 0·1 M Tris buffer pH 7·6, with 0·03 per cent H_2O_2 and 0·01 M imidazole.

II. REAGENTS

Rat monoclonal antibodies to bovine NF and GF proteins used in these studies were raised by V. M-Y. Lee and coworkers.[11,14,21,30]

ABC reagents	Vectastain, Vector Laboratories
Rabbit anti-rat immunoglobulins	Cappel Worthington, Cooperbiomedical Inc.
Rat PAP	Sternberger-Meyer Immunochemicals

REFERENCES

1. Bonnin J. M. and Rubinstein L. J. Immunohistochemistry of central nervous system tumors. Its contributions to neurosurgical diagnosis. *J. Neurosurg.* 1984, **60**, 1121–1133.
2. Ramaekers F. C. S., Puts J. J. G., Moesker O., Kant A., Huysmans A., Haag D., Jap P. H. K., Herman C. J. and Vooijs G. P. Antibodies to intermediate filament proteins in the immunohistochemical identification of human tumors: An overview. *Histochem. J.* 1983, **15**, 691–713.
3. Virtanen I., Miettinen M., Lehto V.-P., Kariniemi A.-L. and Paasivuo R. Diagnostic application of monoclonal antibodies to intermediate filaments. *Ann. NY Acad. Sci.* 1985, in press.
4. Vogel A. M. and Gown A. M. Monoclonal antibodies to intermediate filament proteins: Use in diagnostic surgical pathology. In: Shay J. W., ed., *Cell and Muscle Motility*, Vol. 5. New York, Plenum Publishing Corp., 1984: 379–402.
5. Eng L. J., Vanderhaeghen J. J., Bignami A. and Gerstel B. An acidic protein isolated from fibrous astrocytes. *Brain Res.* 1971, **28**, 351–354.
6. Eng L. F. and DeArmond S. J. Immunochemistry of the glial fibrillary acidic protein. *Prog. Neuropathol.* 1983, **5**, 19–40.
7. Duffy P. E., Graf L. and Rapport M. M. Identification of glial fibrillary acidic protein by the immunoperoxidase method in human brain tumors. *J. Neuropathol. Exp. Neurol.* 1977, **36**, 645–652.
8. Deck J. H. N., Eng L. F., Bigbee J. and Woodcock S. M. The role of glial fibrillary acidic protein in the diagnosis of central nervous system tumors. *Acta Neuropathol.* 1978, **42**, 183–190.
9. Autilio-Gambetti L., Gambetti P. and Crane R. Paired helical filaments: Relatedness to neurofilaments shown by silver staining and reactivity with monoclonal antibodies. In: Katzman, R., ed., *Biological Aspects of Alzheimer's Disease*. Banbury Report Series, Cold Spring Harbor Laboratory, 1984: 117–124.
10. Schlaepfer, W. W. Neurofilaments and the abnormal filaments of Alzheimer's disease. In: Katzman, R., ed., *Biological Aspects of Alzheimer's Disease*. Banbury Report Series, Cold Spring Harbor Laboratory, 1984: 107–115.
11. Lee V. M.-Y., Page C., Wu H.-L. and Schlaepfer W. W. Monoclonal antibodies against gel excised glial filament proteins and their reactivity with other intermediate filament proteins. *J. Neurochem.* 1984, **42**, 25–32.
12. Tapscott S. J., Bennett G. S., Toyama Y., Kleinbart F. and Holzer H. Intermediate filament proteins in the developing chick spinal cord. *Dev. Biol.* 1981, **86**, 40–54.
13. Pruss R. M., Mirsky R. and Raff M. C. All classes of intermediate filaments share a common antigenic determinant defined by a monoclonal antibody. *Cell* 1981, **27**, 419–428.

14. Lee V. M.-Y., Wu H.-L. and Schlaepfer W. W. Monoclonal antibodies recognize individual neurofilament triplet proteins. *Proc. Natl Acad. Sci. USA* 1982, **79**, 6089–6092.

15. Goldstein M. E., Sternberger L. A. and Sternberger N. H. Microheterogeneity ("neurotypy") of neurofilament proteins. *Proc. Natl Acad. Sci. USA* 1983, **80**, 3101–3105.

16. Moss T. J. and Lewkowicz S. J. The axon reaction in motor and sensory neurons of mice studied by a monoclonal antibody marker of neurofilament protein. *J. Neurol. Sci.* 1983, **60**, 267–280.

17. Sternberger L. A. and Sternberger N. H. Monoclonal antibodies distinguish phosphorylated and nonphosphorylated forms of neurofilaments in situ. *Proc. Natl Acad. Sci. USA* 1983, **80**, 6126–6130.

18. Lee V. M.-Y. and Page C. Dynamics of nerve growth factor induced neurofilament and vimentin filament expression and organization in PC12 cells. *J. Neurosci.* 1984, **4**, 1705–1714.

19. Schlaepfer W. W., Lee C., Trojanowski J. Q. and Lee V. M.-Y. Persistence of immunoreactive neurofilament protein breakdown products in transected rat sciatic nerve. *J. Neurochem.* 1984, **43**, 857–864.

20. Trojanowski J. Q., Lee V. M.-Y. and Schlaepfer W. W. Neurofilament breakdown products in degenerating rat and human peripheral nerves. *Ann. Neurol.* 1984, **16**, 349–355.

21. Carden M. J., Schlaepfer W. W. and Lee V. M.-Y. The structure, biochemical properties and immunogenicity of neurofilament peripheral regions are determined by phosphorylation. *J. Biol. Chem.* 1985, **260**, 9805–9817.

22. Liem R. K. H., Yen S.-H., Salomon G. B. and Shelanski M. L. Intermediate filaments in nervous tissue. *J. Cell Biol.* 1979, **79**, 637–645.

23. Schlaepfer W. W., Lee V. M.-Y. and Wu H.-L. Assessment of immunological properties of neurofilament triplet proteins. *Brain Res.* 1981, **226**, 259–272.

24. Willard M. and Simon C. Antibody decoration of neurofilaments. *J. Cell Biol.* 1981, **89**, 198–205.

25. Schlaepfer W. W. and Freeman L. A. Neurofilament proteins of rat peripheral nerve and spinal cord. *J. Cell Biol.* 1978, **78**, 653–662.

26. Granger B. L. and Lazarides E. Expression of the major neurofilament subunit in chicken erythrocytes. *Science* 1983, **221**, 553–556.

27. Bennett G. S. and DiLullo, C. Transient expression of a neurofilament protein by replicating cells of the embryonic chick brain. *Dev. Biol.* 1985, **107**, 107–127.

28. Czosnek H., Soifer D. and Wisniewski H. M. Studies on the biosynthesis of neurofilament proteins. *J. Cell Biol.* 1980, **85**, 726–734.

29. Dahl D. Immunohistochemical differences between neurofilaments in perikarya, dendrites and axons. *Exp. Cell Res.* 1983, **149**, 397–408.

30. Trojanowski J. Q., Obrocka M. A. and Lee V. M.-Y. Distribution of neurofilament subunits in neurons and neuronal processes: Immunohistochemical studies of bovine cerebellum with subunit-specific monoclonal antibodies. *J. Histochem. Cytochem.* 1985, **33**. 557–563.

31. Choi, B. H. and Kim R. C. Expression of glial fibrillary acidic protein in immature oligodendroglia. *Science* 1984, **223**, 407–409.

32. Dahl D., Chi N. H., Miles L. E., Nguyen B. T. and Bignami A. Glial fibrillary acidic (GFA) protein in Schwann cells: Fact or artifact. *J. Histochem. Cytochem.* 1982, **30**, 912–918.

33. Trojanowski J. Q., Lee V. M.-Y. and Schlaepfer W. W. An immunohistochemical study of human central and peripheral nervous system tumors using monoclonal antibodies against neurofilaments and glial filaments. *Human Pathol.* 1984, **15**, 248–257.

34. Shaw G. and Weber K. The intermediate filament complement of the brain: A comparison between different mammalian species. *Eur. J. Cell Biol.* 1984, **33**, 95–104.

35. Shaw G. and Weber K. Differential expression of neurofilament triplet proteins in brain development. *Nature* 1982, **292**, 836–838.

36. Trojanowski J. Q., Gordon D., Obrocka M. A. and Lee V. M.-Y. Developmental expression of neurofilament and glial filament proteins in the developing human pituitary gland. *Dev. Brain Res.* 1984, **13**, 229–239.

37. Lee, V. M.-Y. Neurofilament protein abnormalities in PC12 cells: comparison with neurofilament proteins of normal cultured rat sympathetic neurons. *J. Neurosci.*, 1985, **33**, 900–904.

38. Drager U. C., Edwards D. L. and Kleinschmidt J. Neurofilaments contain α-melanocyte-stimulating hormone (α-MSH)-like immunoreactivity. *Proc. Natl Acad. Sci. USA* 1983, **80**, 6408–6412.

39. Trojanowski J. Q., Stone R. A. and Lee V. M.-Y. Presence of an alpha-MSH-like epitope in the 150,000 Dalton neurofilament subunit from diverse regions of the CNS: An immunohistochemical and immunoblot study in guinea pig. *J. Histochem. Cytochem.* 1985, **33**, 900–904.

40. Lehto V.-P., Stenman S., Miettinen M., Dahl D. and Virtanen I. Expression of a neural type of intermediate filament as a distinguishing feature between oat cell carcinoma and other lung cancers. *Am. J. Pathol.* 1983, **110**, 113–118.

41. Blobel G. A., Gould V. E., Moll R., Lee I., Huszar M., Geiger B. and Franke W. Coexpression of neuroendocrine markers and epithelial cytoskeletal proteins in bronchopulmonary neuroendocrine neoplasms. *Lab. Invest.* 1985, **52**, 39–51.

42. Leff E., Brooks J. J. and Trojanowski J. Q. Expression of neurofilaments and neuron specific enolase in small cell tumors of skin using immunohistochemistry. *Cancer* 1985, **56**, 625–631.

43. Nakazato Y., Sasaki A., Hirato J. and Ishida Y. Immunohistochemical localization of neurofilament protein in neuronal degeneration. *Acta Neuropathol.* 1984, **64**, 30–36.

44. Saurat J.-H., Merot Y., Didierjean L. and Dahl D. Normal rabbit Merkel cells do not express neurofilament proteins. *J. Invest. Dermatol.* 1984, **82**, 641–642.

45. Ghetti B. and Gambetti P. Comparative immunocytochemical characterization of neurofibrillary tangles in experimental maytansine and aluminum encephalopathies. *Brain Res.* 1983, **277**, 388–393.

46. Griffin J. W., Price D. L. and Hoffman P. N. Neurotoxic probes of the axonal cytoskeleton. *Trends Neurosci.* 1983, **6**, 490–495.

47. Powers J. M., Schlaepfer W. W., Willingham M. C. and Hall B. J. An immunoperoxidase study of senile cerebral amyloidosis with pathogenetic considerations. *J. Neuropathol. Exp. Neurol.* 1981, **40**, 592–612.

48. Anderton B. H., Breinburg D., Downes M., Green P. J., Tomlinson B. E., Ulrich J., Wood J. N. and Kahn J. Monoclonal antibodies show that neurofibrillary tangles and neurofilaments share antigenic determinants. *Nature* 1982, **298**, 84–86.

49. Dahl D., Selkoe D. J., Pero R. T. and Bignami A. Immunostaining of neurofibrillary tangles in Alzheimer's senile dementia with a neurofilament antiserum. *J. Neurosci.* 1982, **2**, 113–119.

50. Yen S. H., Gaskin F. and Fu S. M. Neurofibrillary tangles in senile dementia of the Alzheimer type share an antigenic determinant with intermediate filaments of the vimentin class. *Am. J. Pathol.* 1983, **113**, 373–381.

51. Goldman J. E., Yen S.-H., Chiu F.-C. and Peress N. S. Lewy bodies of Parkinson's disease contain neurofilament antigens. *Science* 1983, **221**, 1082–1084.

52. Probst A., Anderton B. H., Ulrich J., Kohler R., Kahn J. and Heitz P. U. Pick's disease: An immunocytochemical study of neuronal changes. *Acta Neuropathol.* 1983, **60**, 175–182.

53. Palo J., Haltia M., Carpenter S., Karpati G. and Mushynski W. Neurofilament subunit-related proteins in neuronal intranuclear inclusions. *Ann. Neurol.* 1984, **15**, 322–328.

54. Trojanowski J. Q., Walkenstein N. and Lee V. M.-Y. Expression of neurofilament subunits in neurons of the central and peripheral nervous system: An immunohistochemical study with monoclonal antibodies. *J. Neurosci.* 1985, in press.

55. Tremblay G. F., Lee V. M.-Y. and Trojanowski J. Q. Expression of vimentin, glial filament and neurofilament proteins in primative childhood brain tumours. *Acta Neuropathol.* 1985, in press.

26

Non-hormonal Markers in the Pituitary

P. Ordronneau and P. Petrusz

The first objective of pituitary immunocytochemistry was to demonstrate and differentiate between the various cell types according to their hormone content. It is now well established that suitable antisera will identify distinct cell populations with well-defined morphological characteristics.[1,2] Attention is beginning to be shifted to non-hormonal markers as probes to further examine the structure and functions of the various hypophyseal cells.[3-6] A good example of such a probe is the enzyme neuron-specific enolase (NSE). Originally thought to be an exclusively neuronal protein,[7] it is now known to be an important marker for neuroendocrine cells and tumours including those of the pituitary,[8-10] and gliomas.[11] Thus, it is possible to use such markers to broaden our understanding of the origin, differentiation, structure, function and abnormal behaviour of the hormone-containing cells.[12,13] In this chapter we report our initial results on the localization and distribution of five such markers in the rat pituitary: the cytoskeletal proteins actin, desmin, glial fibrillary acidic protein (GFAP) and keratin, and the glial cell protein S-100. The sources and characteristics of the antisera we used in these studies are listed in *Table* 26.1.

All immunostaining was done according to the 'double PAP' procedure described earlier[1,23] and was preceded by a 5-min treatment of the sections with trypsin.[24] Following the immunocytochemical procedure the sections were counterstained with toluidine blue.

1. ACTIN

Actin is a highly conserved protein that is the major component of microfilaments (6 nm) in both muscle and non-muscle cells.[25] Besides functioning in a

425

*Table 26.1. Characteristics of antisera**

Antiserum	Immunogen	Distribution in normal tissue
Anti-actin	48-kdal soluble actin monomer (G-actin) from chicken skeletal muscle[14]	Component of all eukaryotic cells; abundant in skeletal, cardiac and smooth muscle[15]
Anti-desmin	55-kdal protein from chicken gizzard[16]	Smooth muscle, Z-line of skeletal muscle[15]
Anti-GFAP	47-kdal protein from bovine brain[17]	Astrocytes, other glial cells[17,18]
Anti-keratin	Keratin from human stratum corneum[19]	Epithelial cells and their derivatives[19,20]
Anti-S-100	20-kdal protein from bovine brain[21]	Glial cells in rat brain, ductal epithelium in salivary glands[21,22]

*Rabbit antisera obtained from Lipshaw Co., Detroit, Michigan.

tension-generating system, it also helps control the shape of the cell and coordinate its activities.[26] Evidence is beginning to accumulate that microfilaments function to direct secretory granules to the plasma membrane in protein hormone-secreting cells[12] and to transport cholesterol to the mitochondria in steroid-producing cells.[12,27]

Our observations indicate that small amounts of actin are present in probably every cell of the adenohypophysis, resulting in a weak granular immunostaining localized usually at the periphery of the cells. In addition, a small but heterogeneous cell population in the pars distalis (PD) stains strongly for actin. Most of these cells are irregular or stellate, others are rounded or cuboidal. The dark staining in these cells is especially intense after colchicine treatment (100 µg by intracerebroventricular injection). Actin appears in the form of fine granules in the cytoplasm, often forming a ring around the periphery of the cell and a smaller inner ring surrounding the nucleus (*Fig.* 26.1). In other cells, the distribution of actin is polarized, i.e. the granules are accumulated at one end of an elongated or oval cell or in one process of a stellate cells. The cell type that stains intensely for actin has not been identified, but it seems unlikely that it represents a single class of hormone-producing cells. It is more probable that the amount and distribution of actin in pituitary cells are determined by functional parameters. Only further investigations can provide the answers to these and many other interesting questions regarding the role and significance of actin in the pituitary.

2. DESMIN

Desmin is a member of a family of proteins which polymerize to form intermediate (10 nm) filaments.[28] Filaments composed primarily of desmin are found in smooth, skeletal (Z-disc) and cardiac muscle and the protein is often used as a marker for muscle cells.[28,29]

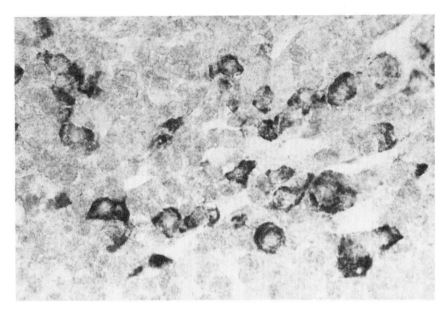

Fig. 26.1. Actin immunoreactivity in the pars distalis of a colchicine-treated rat.

Desmin is sparsely distributed in the PD of the rat pituitary: only an occasional small stellate cell is seen. Also, a thin line of positive reaction is visible along the apical surface of the cells lining the follicles of the PD. Strong staining is present in the epithelium of the cleft (the remnant of Rathke's pouch) separating the PD from the pars intermedia (PI), and in glial-like stellate cells of the PI (*Fig.* 26.2). The presence of glial cells in this region was described many decades ago (for review, *see* Wingstrand[30]) and it was assumed that these cells migrate into the PI from the neurohypophysis.[30] In view of this, it is surprising that such 'glial' cells seem to contain a protein which is generally regarded as a specific marker for muscle (mesenchymal) cells.

3. GFAP

GFAP is the predominant protein of intermediate filaments of astrocytes.[31] It has also been identified in Schwann cells,[32] pineal supportive cells[33] and the pituicytes of the neurohypophysis.[34-36]. In the human pituitary, GFAP has been reported to be localized in the cells lining Rathke's cysts and cells resembling folliculostellate cells in the PD.[36,37] In one study published on the rat pituitary, antiserum to GFAP stained neither the PD nor the PI.[34]

Surprisingly, our antiserum did detect GFAP in the rat PD in the form of fine, short, curved processes running between and around GFAP-negative cells. The origin of these processes is uncertain since GFAP-containing cell bodies were not seen in the PD. The number of these processes is relatively small in normal rats, while none can be found after colchicine treatment. However, they become more numerous in ovariectomized rats (*Fig.* 26.3). The 'sensitivity' of these structures

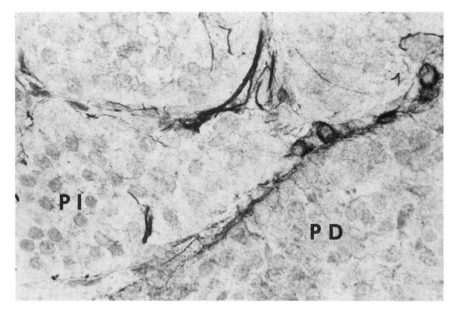

Fig. 26.2. Desmin immunoreactivity in the rat pituitary. PD, pars distalis; PI, pars intermedia.

to colchicine indicates either that GFAP contained in them is rendered non-immunoreactive by this compound or that the presence of GFAP in the PD depends on microtubule-mediated transport mechanisms. An extensive network of GFAP-positive fibrous processes (and perhaps cell bodies) was found in a spontaneous pituitary tumour from a Fischer 344 inbred male rat (*Fig.* 26.4). In the PD, our GFAP antiserum does not stain follicular cells (which stain heavily for another glial protein, S-100 — *see Fig.* 26.9). Taken together, these preliminary observations suggest that GFAP (and the glial-like cells in which it may be assumed to occur) are subject to physiological regulation in the PD of the pituitary and may play a role in the functions of this gland.

There is intense GFAP staining in some areas of the cleft. In the PI itself, there are large stellate glial-like cells which stain strongly for GFAP. They send long processes surrounding groups of parenchymal cells (*Fig.* 26.5). Other processes may end as typical glial 'feet' around blood vessels or in the septa between the nests of parenchymal cells (*Fig.* 26.6). It is not clear at this time whether these glial cells are identical with those labelled by the desmin antiserum in the same region (*see Fig.* 26.2).

4. KERATIN

The intermediate filaments of epithelial cells and cells of epithelial origin contain keratin. The keratins are a family of polypeptides with molecular weights that range from 40 to 65 kdal.[29] Despite this heterogeneity, keratins show striking similarities[28] and are immunologically closely related.[29] In the human pituitary, keratin was reported to occur in somatotrophs, lactotrophs and corticotrophs.[37] In addition, patients with Cushing's syndrome (hypercortisolaemia) have severely hyalinized corticotrophs packed with keratin intermediate filaments.[38]

We found keratin in practically all cells of the normal rat PD and PI, in highly variable amounts in the cells of the former, and more uniformly distributed in the latter (*Fig. 26.7*). Larger cells of the PD appear to contain more keratin than smaller cells. There is a characteristic subcellular distribution as well; the reaction is heaviest in the region around the nucleus and around the periphery of the cytoplasm. The cytoplasm itself seems to have a fine, uniform network of keratin fibres. All these features are dramatically enhanced after ovariectomy in the gonadotrophs, which are easily identified as enlarged, vacuolated cells (*Fig. 26.8*). Thyroidectomy cells, on the other hand, contain only moderate amounts of keratin. The cleft epithelium is strongly labelled. The cells of the PI are uniformly surrounded by granular reaction product giving a characteristic 'cobblestone' appearance (*see Fig. 26.7*). The distribution and the punctate nature of the reaction product suggest that it may represent desmosomes.[39] Some of the PI cells also have a perinuclear keratin ring.

5. S-100 PROTEIN

This soluble dimeric protein is a mixture of two component polypeptides.[40] S-100 is capable of binding Ca^{2+} [41] and it modulates protein phosphorylation by a brain-specific kinase.[42] Also, it inteferes with the assembly of microtubules.[43] There is evidence that S-100 is secreted into both the blood[44] and the cerebrospinal fluid[45] and may stimulate the release of prolactin from the pituitary.[46] Immunocytochemically, S-100 was found in astrocytes and ependymal cells of the central nervous system[47] and Schwann cells and ganglion satellite cells of the peripheral nervous system.[48] The protein is also present in the interstitial cells of the pineal,[49] stellate cells of the adrenal medulla,[48,50,51] sustentacular cells of the carotid body[52] and the pituicytes of the neruohypophysis.[53] In the adenohypophysis, S-100 was reported to occur in folliculostellate cells in the PD[53,54] and pars tuberalis[53] and in the anterior and posterior marginal cells that line the hypophyseal cleft.[53]

In our studies, S-100 has shown a remarkable and selective distribution in the pituitary. In the PD, follicular cells are heavily labelled (*Fig. 26.9*); in addition, there is a large number of irregular stellate cells which stain with our antiserum (*Fig. 26.10*). Their short processes seem to run between and around other cells. In normal rats the amount of S-100 reaction product is clearly greater in the female (*Fig. 26.11*) than in the male. This feature is even more dramatic in the pituitary of ovariectomized rats (*Fig. 26.12*). The morphology and the distribution of the S-100-positive cells are reminiscent of prolactin cells in the rat pituitary; however, the identity of the cells has not been established. The cleft epithelium also contains substantial amounts of S-100 immunoreactivity. In the PI, strong S-100 staining is present in stellate glial cells (similar to those seen with GFAP), while there is a weak staining in a few parenchymal cells as well (*Fig. 26.13*).

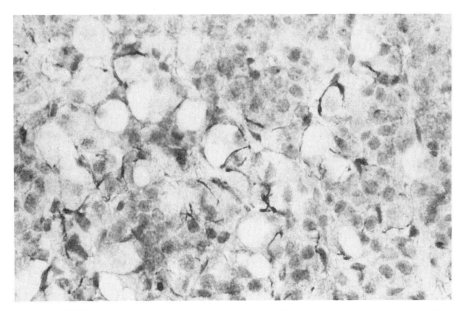

Fig. 26.3. GFAP-immunoreactive structures in the pars distalis of an ovariectomized rat. The large, vacuolated cells are the hypertrophied gonadotrophs.

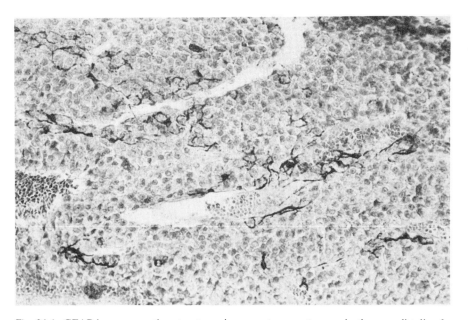

Fig. 26.4. GFAP-immunoreactive structures in a spontaneous tumour in the pars distalis of a Fischer 344 rat.

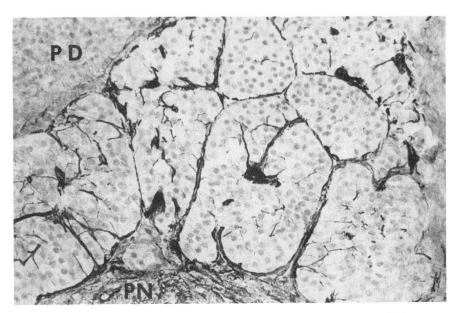

Fig. 26.5. GFAP-immunoreactive structures, presumably glial cells and processes, in the rat pars intermedia. PD, pars distalis; PN, pars nervosa.

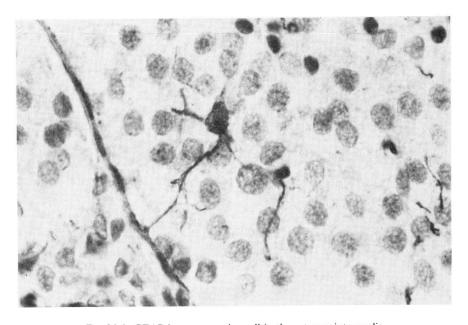

Fig. 26.6. GFAP-immunoreactive cell in the rat pars intermedia.

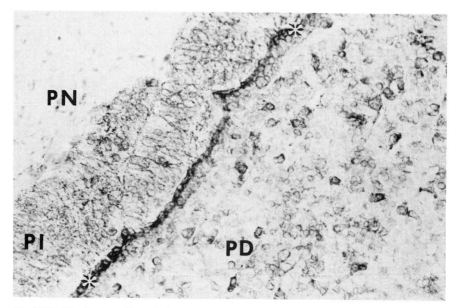

Fig. 26.7. Distribution of keratin immunoreactivity in the normal rat pituitary. PD, pars distalis; PI, pars intermedia; PN, pars nervosa;* cleft epithelium.

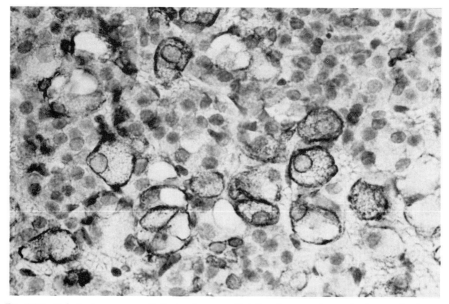

Fig. 26.8. Keratin immunoreactivity in the pars distalis of an ovariectomized rat. The large, vacuolated cells are gonadotrophs.

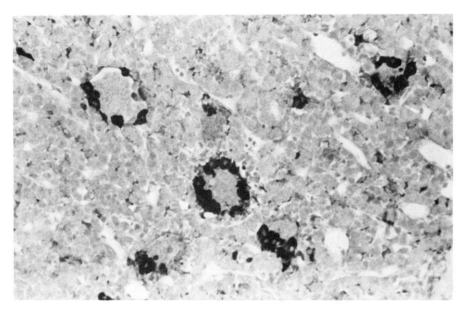

Fig. 26.9. S-100 immunoreactivity in the rat pars distalis. Note heavy staining in the follicular epithelium.

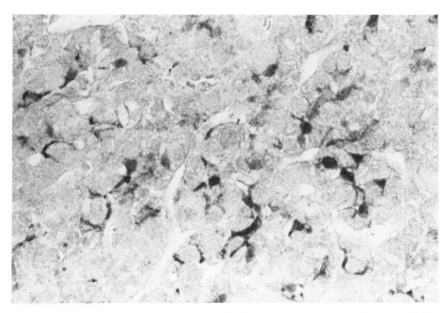

Fig. 26.10. S-100 immunoreactivity in the pars distalis of a normal male rat. Compare with *Figs.* 26.11 and 26.12.

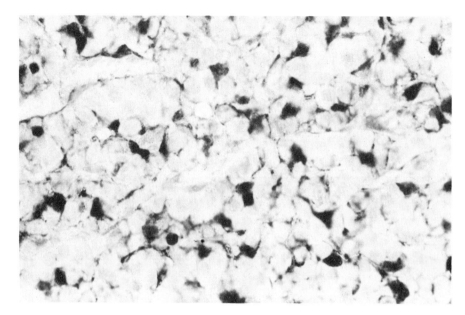

Fig. 26.11. S-100 immunoreactivity in the pars distalis of a normal female rat. Compare with *Figs.* 26.10 and 26.12.

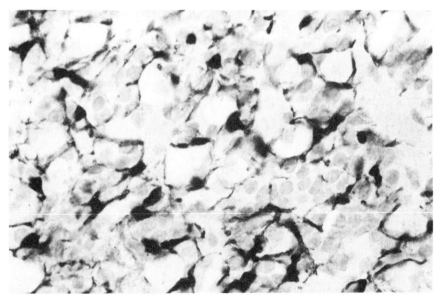

Fig. 26.12. S-100 immunoreactivity in the pars distalis of an ovariectomized rat. The large, vacuolated cells are hypertrophied gonadotrophs. Compare with *Figs* 26.10 and 26.11.

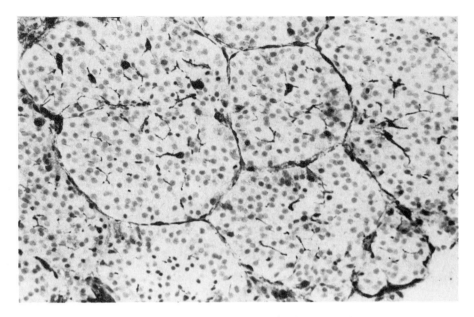

Fig. 26.13. S-100-immunoreactive structures in the rat pars intermedia.

6. CONCLUSIONS

Until recently, pituitary immunocytochemistry has been largely restricted to the localization of the known hormones of the gland. The development of hormone-specific (including monoclonal) antibodies and much improved immunocytochemical techniques and reagents have resulted in significant progress in our understanding of pituitary histophysiology and pathology. In recent years, a large number of other (non-hormonal) cell markers have become known, with excellent antibodies becoming commercially available. Since each of these markers shows a selective and specific distribution in various tissues and cell types, their presence can be expected to be correlated with specific cellular functions. Several cytoskeletal proteins and two nervous system-specific markers (GFAP and S-100) have been shown by our initial results to be excellent candidates for functional cell markers in the pituitary. The specific actions of these proteins, their precise relationship to hormone-producing cells, and their usefulness as meaningful and reliable markers in pituitary immunocytochemistry remain to be established by future research.

REFERENCES

1. Petrusz P. and Ordronneau P. Immunocytochemistry of pituitary hormones. In: Polak J. M. and Van Noorden S., eds., *Immunocytochemistry. Practical Applications in Pathology and Biology.* Bristol, Wright·PSG, 1983: 212–232.
2. Halmi N. S. The hypophysis. In: Weiss L., ed., *Histology. Cell and Tissue Biology*, 5th ed. New York, Elsevier, 1983: 1054–1078.

3. Polak J. M. and Bloom S. R. Immunocytochemistry of regulatory peptides. In: Polak J. M. and Van Noorden S., eds., *Immunocytochemistry. Practical Applications in Pathology and Biology*. Bristol, Wright·PSG, 1983: 184–211.

4. Heyderman E. Tumour markers. In: Polak J. M. and Van Noorden S., eds., *Immunocytochemistry. Practical Applications in Pathology and Biology*. Bristol, Wright·PSG, 1983: 274–294.

5. Evans D. J. Intermediate filaments in diagnostic histopathology. In: Polak J. M. and Van Noorden S., eds., *Immunocytochemistry. Practical Applications in Pathology and Biology*. Bristol, Wright·PSG, 1983: 295–301.

6. Heitz P. U. Immunocytochemistry in endocrine pathology. In: Polak J. M. and Van Noorden S., eds., *Immunocytochemistry. Practical Applications in Pathology and Biology*. Bristol, Wright·PSG, 1983: 362–371.

7. Pickel V. M., Reis D. J., Marangos P. J. and Zomzely-Neurath C. Immunocytochemical localization of nervous system specific protein (NSP-R) in rat brain. *Brain Res*. 1976, **105**, 184–187.

8. Van Noorden S., Polak J. M., Robinson M., Pearse A. G. E. and Marangos P. J. Neuron-specific enolase in the pituitary gland. *Neuroendocrinology* 1984, **38**, 309–316.

9. Springall D. R., Lackie P., Levene M. M., Marangos P. J. and Polak J. M. Immunostaining of neuron-specific enolase is a valuable aid to the cytological diagnosis of neuro-endocrine tumors of the lung. *J. Pathol*. 1984, **143**, 259–265.

10. Rode J. and Dhillon A. P. Neuron specific enolase and S100 protein as possible prognostic indicators in melanoma. *Histopathology* 1985, **8**, 1041–1052.

11. Vinores S. A., Marangos P. J., Bonnin J. M. and Rubinstein L. J. Immunoradiometric and immunohistochemical demonstration of neuron-specific enolase in experimental rat gliomas. *Cancer Res*. 1984, **44**, 2595–2599.

12. Hall P. F. The role of the cytoskeleton in hormone action. *Can. J. Biochem. Cell Biol*. 1984, **62**, 653–677.

13. Bishop A. E., Carlei F., Lee V., Trojanowski J., Marangos P. J., Dahl D. and Polak J. M. Combined immunostaining of neurofilaments, neuron specific enolase, GFAP and S-100. A possible means for assessing the morphological and functional status of the enteric nervous system. *Histochemistry* 1985, **82**, 93–97.

14. Spudich J. A. and Watt S. The regulation of rabbit skeletal muscle contraction. I. Biochemical studies of the interaction of the tropomyosin–troponin complex with actin and the proteolytic fragments of myosin. *J. Biol. Chem*. 1971, **246**, 4866–4871.

15. Albert B., Bray D., Lewis J., Raff M., Roberts K. and Watson J. D. *Molecular Biology of the Cell*. New York, Garland, 1983: 582.

16. Huiatt T. W., Robson R. M., Arakawa N. and Stromer M. Desmin from avian smooth muscle. Purification and partial characterization. *J. Biol. Chem*. 1980, **255**, 6981–6989.

17. Eng L. F. The glial fibrillary acidic protein: The major protein constituent of glial filaments. *Scand. J. Immunol*. 1982, **15**, Suppl. 9, 41–51.

18. Dahl D. and Bignami A. Immunochemical and immunofluorescence studies of the glial fibrillary acidic protein in vertebrates. *Brain Res*. 1973, **61**, 279–293.

19. Schlegel R., Banks-Schlegel S. and Pinkus G. S. Immunohistochemical localization of keratin in normal human tissues. *Lab. Invest*. 1980, **42**, 91–96.

20. Sun T.-T., Shih C. and Green H. Keratin cytoskeletons in epithelial cells of internal organs. *Proc. Natl Acad. Sci. USA* 1979, **76**, 2813–2817.

21. Moore B. W. A soluble protein characteristic of the nervous system. *Biochem. Biophys. Res. Commun*. 1965, **19**, 739–744.

22. Langley K. O., Ghandour M. S. and Gombos G. Immunohistochemistry of cell markers in the central nervous system. In: Lajtha A., ed., *Handbook of Neurochemistry, Vol. 7, Structural Elements of the Nervous System*. New York, Plenum, 1984: 545–611.

23. Ordronneau P., Lindström P. B.-M. and Petrusz P. Four unlabeled antibody bridge techniques: A comparison. *J. Histochem. Cytochem*. 1981, **29**, 1397–1404.

24. Towle A. C., Lauder J. M. and Joh T. H. Optimization of tyrosine hydroxylase immunocytochemistry in paraffin sections using pretreatment with proteolytic enzymes. *J. Histochem. Cytochem*. 1984, **32**, 766–770.

25. Pollard T. D. and Weihing R. R. Actin and myosin and cell movement. *CRC Crit. Rev. Biochem*. 1974, **2**, 1–64.

26. Pollard J. D. Cytoplasmic contractile proteins. *J. Cell Biol*. 1981, **91**, 156s–165s.

27. Osawa S., Betz G. and Hall P. F. Role of actin in the responses of adrenal cells to ACTH and cyclic AMP: inhibition by DNase I. *J. Cell Biol*. 1984, **99**, 1335–1342.

28. Geisler N. and Weber K. The amino acid sequence of chicken muscle desmin provides a common structural model for intermediate filament proteins. *EMBO J.* 1982, **7**, 7649–7656.
29. Lazarides E. Intermediate filaments: a chemically heterogeneous, developmentally regulated class of proteins. *Annu. Rev. Biochem.* 1982, **51**, 219–250.
30. Wingstrand K. G. Microscopic anatomy, nerve supply and blood supply of the pars intermedia. In: Harris G. W. and Donovan B. T., eds., *The Pituitary Gland*, Vol. 3. Berkeley, University of California. Press, 1966: 1–27.
31. Bignami A., Eng L. F., Dahl D. and Uyeda C. T. Localization of the glial fibrillary acidic protein in astrocytes by immunofluorescence. *Brain Res.* 1972, **43**, 429–435.
32. Jessen K. R., Thorpe R. and Mirsky R. Molecular identity, distribution and heterogeneity of glial fibrillary acidic protein: an immunoblotting and immunohistochemical study of Schwann cells, satellite cells, enteric glia and astrocytes. *J. Neurocytol.* 1984, **13**, 187–200.
33. Higley H. R., McNulty J. A. and Rowden G. Glial fibrillary acidic protein in pineal supportive cells: an electron microscopic study. *Brain Res.* 1984, **304**, 117–120.
34. Salm A. K., Hatton G. I. and Nilaver G. Immunoreactive glial fibrillary acidic protein in pituicytes of the rat neurohypophysis. *Brain Res.* 1982, **236**, 471–476.
35. Suess U. and Pliška V. Identification of the pituicytes as astroglial cells by indirect immunofluorescence staining for the glial fibrillary acidic protein. *Brain Res.* 1981, **221**, 27–33.
36. Velasco M. E., Roessmann U. and Gambetti P. The presence of glial fibrillary acidic protein in the human pituitary gland. *J. Neuropathol. Exp. Neurol.* 1982, **41**, 150–163.
37. Hofler H., Denk H. and Walter G. F. Immunohistochemical demonstration of cytokeratins in endocrine cells of the human pituitary gland and in pituitary adenomas. *Virchow's Arch. Abt. A Pathol. Anat.* 1984, **404**, 359–368.
38. Neumann P. E., Horoupian D. S., Goldman J. E. and Hess M. A. Cytoplasmic filaments of Crooke's hyaline change belong to the cytokeratin class. An immunocytochemical and ultrastructural study. *Am. J. Pathol.* 1984, **116**, 214–222.
39. Chatterjee P. Ultrastructure of the pars intermedia: development *in vivo* and in organ culture. In: Motta P. M., ed., *Ultrastructure of Endocrine Cells and Tissues*. Boston, Martinus Nijhoff, 1984: 39–56.
40. Isobe T., Nakajima T. and Okuyama T. Reinvestigation of extremely acidic proteins in the bovine brain. *Biochim. Biophys. Acta* 1977, **494**, 222–232.
41. Moore B. W. Chemistry and biology of the S-100 protein. *Scand. J. Immunol.* 1982, **15**, Suppl. 9, 53–74.
42. Qi D.-F. and Kuo J. F. S-100 modulates Ca^{2+} independent phosphorylation of an endogenous protein (M_r = 19K) in brain. *J. Neurochem.* 1984, **43**, 256–260.
43. Donato R. Mechanism of action of S-100 protein(s) on brain microtubule assembly. *Biochem. Biophys. Res. Commun.* 1984, **124**, 850–856.
44. Suzuki F., Kato K. and Nakajima T. Hormonal regulation of adipose S-100 release. *J. Neurochem.* 1984, **43**, 1336–1341.
45. Shashoua V. E., Hesse G. W. and Moore B. W. Proteins of the brain extracellular fluid: evidence for release of S-100 protein. *J. Neurochem.* 1984, **42**, 1536–1541.
46. Ishikawa H., Nogami H. and Shirasawa N. Novel clonal strains from adult rat anterior pituitary producing S-100 protein. *Nature* 1983, **303**, 711–713.
47. Cocchia D. Immunocytochemical localization of S-100 protein in the brain of the adult rat. *Cell Tissue Res.* 1981, **214**, 529–540.
48. Cocchia D. and Michetti F. S-100 antigen in stellate cells of the adrenal medulla and the superior cervical ganglion of the rat. *Cell Tissue Res.* 1981, **215**, 103–112.
49. Møller M., Ingild A. and Bock E. Immunohistochemical demonstration of S-100 protein and GFA protein in interstitial cells of rat pineal gland. *Brain Res.* 1979, **140**, 1–13.
50. Stefansson K., Wollman R. L. and Moore B. W. Distribution of S-100 protein outside the central nervous system. *Brain Res.* 1982, **234**, 309–317.
51. Iwanaga T. and Fujita T. Sustentacular cells in the fetal human adrenal medulla are immunoreactive with antibodies to brain S-100 protein. *Cell Tissue Res.* 1984, **236**, 733–735.
52. Kondo H., Iwanaga T. and Nakajima T. Immunocytochemical study on the localization of neuron-specific enolase and S-100 protein in the carotid body of rats. *Cell Tissue Res.* 1982, **227**, 291–295.
53. Cocchia D. and Miani N. Immunocytochemical localization of the brain-specific S-100 protein in the pituitary gland of the adult rat. *J. Neurocytol.* 1980, **9**, 771–782.
54. Nakajima T., Yamaguchi H. and Takahashi K. S-100 protein in folliculostellate cells of the rat pituitary lobe. *Brain Res.* 1980, **191**, 523–531.

27

Two-colour Immunofluorescence: Analysis of the Lymphoid System with Monoclonal Antibodies

G. Janossy, M. Bofill and
L. W. Poulter

Since the development of immunofluorescence in its direct one-step technique[1] this procedure has continuously progressed into a more and more analytical and informative method. Three years ago it was already clear that the revolution of applying monoclonal antibodies to lymphoid populations and other cell types transformed the field of immunohistology[2-4] in the area of identifying T-lymphocyte subpopulations. At that time the assessment of B-cell heterogeneity had just begun and the study of different macrophage subpopulations required complicated double staining methods. These were necessary in order to identify macrophages with large amounts of lysosomal enzymes together with those which abundantly expressed class II (HLA-DR) antigens.[4,5] During the last three years, five further steps have been made which have implications for the development of immunofluorescence techniques as applied to the analysis of the immune system.

The first three of these steps were achieved at the recent Workshop on Leucocyte Antigens in Boston[6] and in additional investigations which can be regarded as a continuation of this international comparative study. Firstly, the major B-cell-specific and B-cell-associated antigens have been defined by clustering numerous antibodies which react with the same antigens (CD19-24; *see Fig.* 27.1 and the literature[7,8]). The tissue reactivity of these antibody clusters has also been described[9] and the fetal development of some of the identifiable

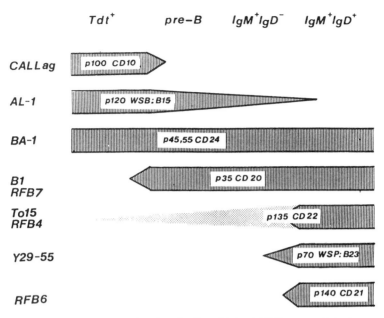

Fig. 27.1. Reactivity of B-cell monoclonal antibodies with 'early' B lineage cells in the normal bone marrow. This figure demonstrates the use of combinations of directly labelled goat anti-human Ig-FITC reagents together with indirect labelling using monoclonal antibodies and goat anti-mouse Ig-TRITC. The B-cells of IgM+ve, IgD−ve versus IgM+ve, IgD+ve isotype in the bone marrow show different phenotypical patterns. The CD numbers refer to the provisional B-cell clusters established at the Boston Workshop for Human Leucocyte Differentiation Antigens. WSB: unclustered reagent at the Boston Workshop, 1984; WSP: unclustered reagent at the Paris Workshop, 1982. ▨: Cytoplasmic antigen, absent in membrane. (From Campana et al.[8] with permission of the authors and publishers.)

B-cell subpopulations has been established.[10] Secondly, an impressive range of myeloid reagents has been described in the same way.[11] These provide a somewhat less convincingly established range of 'myeloid-associated' clusters (CD11-18). The doubts are due to the fact that, according to immunohistological investigations, these clusters may still be heterogeneous.[12] Perhaps 'cleaner' antibodies which characteristically react with certain tissue macrophage lineages are still to come (*see below*). In this review we shall describe some of these anti-macrophage reagents, as identified by the double fluorescence method. However, the ultrastructural characterization of the macrophage subsets identified with these new antibodies is still to be performed.

The third major advance is inherent in the previous two. There are now many monoclonal antibodies which react with the same antigens, and some of them express different heavy chain isotypes. The investigator is therefore in a position to select an antibody with the required isotype (heavy chain class) and combine it with another of a different isotype which reacts with a second antigen of interest. One can therefore choose antibodies of different (sub)classes for appropriate double immunofluorescent labelling. The second layers, heterologous antisera specific against the different isotypes, are commercially available in forms conjugated to different fluorochromes such as fluorescein isothiocyanate (FITC, green) and tetramethylrhodamine isothiocyanate (TRITC, red). The principle of double immunofluorescence staining by using antibodies of different isotypes in combination, together with the corresponding heavy chain specific second layers is, of course, not a new one (*see*, for example, Poulter et al.[3]). Still, during the last two years, the availability of a virtually unlimited range of monoclonal antibodies of varied Ig isotype and the corresponding second layers of exceptionally high quality have given a boost to this simple technique.

The fourth advance is a gradual conceptual development. Monoclonal antibodies are extremely specific reagents which selectively react with given antigens. Some of these antigens are lineage specific and can therefore be used, with great clarity, to identify cell types in sections. It is nevertheless not 'guaranteed' that any given antigen is expressed solely on one particular cell lineage. Furthermore, additional antigens which are related to certain cellular functions are most frequently not lineage specific. It follows that the immunopathologist wishing to enquire about the possible functional attributes of cell types in their normal microenvironments, i.e. in a tissue section, will need to use, more and more frequently, informative combinations of antibodies detecting lineage-associated antigens together with functionally related molecules. The necessity of double-marker technology, in order to gain an insight into the heterogeneity of cells with 'natural killer' function, has been demonstrated in the work of Lanier and his colleagues,[13] and parallel studies in immunohistology have also been performed.[14] The relevant point is that these double combinations can be most conveniently and precisely used by combined immunofluorescence staining (*see below*).

Finally, one should enquire about the recent developments of prefixation and paraffin embedding in relation to immunofluorescence work. This is an important issue from three points of view: (i) the safety of handling infectious material, (ii) transport of samples, fixed without the danger of deterioration and (iii) the quality of tissue preservation in comparison with conventional paraffin-embedded material. The discussion below illustrates the recent developments in these five areas.

1. IMMUNOFLUORESCENCE AND IMMUNOENZYME METHODS

Immunostaining can be optimally performed in frozen sections even following transportation in Michel's fixative.[15] For the best results, tissues should be embedded in a freezing medium such as OCT and snap-frozen in liquid nitrogen or a solid CO_2 bath. The tissue should be covered with aluminium foil or placed into a thin plastic container. After freezing, the tissue can be stored at $-70\,°C$ under OCT cover. The cryostat sections are stable for many months if they are stored in a freezer ($-20\,°C$) and are covered in order to protect them against destruction by moisture. Postfixation in pure acetone is now a generally accepted method and only a few antigens, such as the easily diffusible nuclear enzyme, terminal deoxynucleotidyl transferase (TdT), require special postfixation. In the case of staining for TdT the sections are immediately submerged into formalin within the cryostat.[16]

This short summary of tissue handling illustrates the original double principle as established by Stein and his colleagues:[17] minimal denaturation of tissue antigens and morphology almost as excellent as is observed in formalin-fixed preparations. From an immunological point of view this technique represents the standard by which the other modified methods are compared (*see below* and *Table* 27.4). Under these conditions both immunofluorescence and immunoenzyme methods can be performed with great efficiency. Indirect immunofluorescence will provide strong membrane labelling with more than 80 per cent of the monoclonal antibodies listed in *Fig.* 27.1 and *Tables* 27.1 and 27.2, and always works with antibody cocktails (*see below*). Immunofluorescence staining is very quick, a distinct advantage over immunoenzyme methods. In order to obtain morphologic- al details, haematoxylin may be used as a counterstain without quenching the fluorescence.[18] Fading is prevented by using a mounting medium containing *p*-phenylenediamine (BDH; 0·1 per cent concentration) in a 9 : 1 mixture of glycerol and phosphate-buffered saline. A mountant which retards fading and also sets permanently is available.[19]

As pointed out previously, the immunofluorescence and immunoenzyme methods are complementary rather than competitive techniques. Enzyme methods are basic in immunohistological laboratories but require the comple- mentary power of immunofluorescence for two reasons: (i) immunofluorescence is an elegantly *rapid* screening method; (ii) it is the only scientifically acceptable method for investigating *double-labelled* preparations with sufficient clarity when the populations of interest are (or might be) simultaneously labelled by two antibodies. In this case the viewing of the two labels by separate filters is an advantage, while the intermediate colours of double enzymatic reactions are difficult to judge. We will therefore discuss the techniques of *double colour immunofluorescence* below.

2. DOUBLE IMMUNOFLUORESCENCE METHODS FOR INVESTIGATING B-CELL HETEROGENEITY

The general applicability of a method is dependent upon its reliability and practicability. The reliability of double immunofluorescence techniques has been markedly increased by three factors: (*a*) it is now recognized that highly diluted monoclonal antibodies prepared from ascitic fluids rapidly deteriorate in storage, while the same antibodies prepared as culture supernatants retain their activity when stored at −30 °C for longer than 2 years;[20] (*b*) affinity purified goat anti-mouse immunoglobulin (sub)class antibodies (anti-IgM, anti-IgG$_1$, anti-IgG$_2$) are now available in FITC- and TRITC-labelled forms (e.g. Southern Biotechn. Associates, Birmingham, Ala, USA.); (*c*) it has been noticed that the penetration of reagents into the tissue sections may be restricted when larger complexes are used.[21] For example, the avidin–biotin complex method[22] is less efficient in most hands than the application of successive layers of antibodies.[23] For these reasons there is a tendency to opt for simple systems in immunofluorescence investigations (*Table* 27.1). These combinations are practical, although with weakly expressed antigens the need to use antibodies labelled with biotin or arsanilic acid is sometimes unavoidable (*Table* 27.1). However, they are not denatured by the conjugation methods.[24–26]

During the investigation of B-cell heterogeneity in man, it was necessary to employ all the practical double immunofluorescence combinations shown in *Table* 27.1, for the following reasons. Direct labelling for human Ig classes was

Table 27.1. Practical double and triple immunofluorescence combinations

Example	First layer	Second layer	Third layer	Ref.
Strong antigens				
Hetero antiserum with murine monoclonal antibody	Goat anti-human Ig* (class)	—	—	8,49,50
Murine antibodies of different (sub)class	Monoclonal antibody (μ)	Goat anti-mouse-Ig*†	—	*Fig.* 27.2, 27.3
	Monoclonal antibody (γ)	Goat anti-mouse-IgM*†	—	
		Goat anti-mouse-IgG(7S)*†	—	
Murine antibody biotinylated‡	Monoclonal antibody–biotin	Avidin*	—	51
Weak antigens				
Murine antibody	Monoclonal antibody	Goat anti-mouse Ig-biotin	Avidin*	23
Murine antibody‡	Monoclonal antibody–Ars	Rabbit anti-Ars	Goat anti-rabbit Ig*	3
hapten-labelled	Monoclonal antibody–Glu	Rabbit anti-Glu	Goat anti-rabbit Ig*	3

*Antibody is labelled with FITC or TRITC.
†Available from Southern Biotechn. Associates, Birmingham, Ala., USA.
‡If the required monoclonal antibodies have the same (sub)class, the combination used is the biotinylated together with the hapten-labelled murine antibody.
Ars = arsanilic acid.
Glu = glutamic acid.

used in combination with indirect labelling with mouse monoclonal antibodies in order to establish the different phenotypic characteristics of IgM+ve, IgD−ve and IgM+ve, IgD+ve cells in the bone marrow, and the results are demonstrated in *Fig.* 27.1.[8] These phenotypic distinctions were also found on cells in tissue sections of fetal bone marrow,[10] and in adults the peripheral B-cells of the lymphocyte corona of lymph nodes and tonsil also expressed the phenotype that was seen on the IgM+ve, IgD+ve cells depicted in *Fig.* 27.1. Interestingly, rare peripheral IgM+ve, IgD−ve cells such as B-cells in the splenic marginal zone also showed membrane CD22 and CD21 positivity indicating that IgD−ve cells in the spleen show a mature B-cell phenotype which appears to be different from the features of the IgM+ve, IgD−ve B-cells in the bone marrow. It was unnecessary to use double markers to establish the peculiar phenotype of germinal centre B-blasts because these cells occupy an anatomically distinct site. Single markers with immunoperoxidase can clearly demonstrate that these blasts are CALL antigen+ve (CD10), class II antigen+ve, CD20+ve, CD21+ve, CD22+ve and also express T10 antigen but lack BA1 (CD24).[7] On the other hand, a peculiar observation was made with double immunofluorescence combinations in fetal lymph nodes, where the primary follicles expressed unique characteristics. Their B-cells were IgM+ve, IgD+ve with 'peripheral B-cell' features and, in addition, combination staining with monoclonal antibodies to B-cells and a T-cell-associated antigen, T1 (CD5), showed them to be T1 (CD5)-positive.[10] This cell type closely resembles the malignant cells seen in centrocytic lymphoma and chronic lymphocytic leukaemia,[27,28] thus suggesting that these malignant populations may derive from B-cells which are similar to those observed in the primary nodules of fetal lymph nodes.

The investigation of conventional monoclonal antibody markers with double immunofluorescence combinations can be summarized as follows (*Table* 27.2). The majority of monoclonal antibody clusters contain individual reagents of different isotypes. These antibodies can be used in combination with (sub)class-specific goat anti-mouse Ig sera conjugated with different fluorochromes. It is also possible to make very robust B-cell 'cocktails' for particularly strong staining intensity. The following antibodies may be mixed for this purpose: IgM class: RFB7 + B2; IgG$_2$ subclass: B1 + HB5 + SCHL-1; IgG$_1$ subclass: B4 + RFB6 + RFB4. Thus selected reagents of different isotypes can be used in lymphocyte studies (i.e. T-cells with B-cells) as well as in combination with any monoclonal antibodies detecting accessory cells of the immune system, e.g. follicular reticulum cells, endothelial cells, various types of macrophage etc.

3. IMMUNOFLUORESCENCE METHODS FOR STUDYING MACROPHAGE HETEROGENEITY

The lineage relationships of the various accessory cells of the immune system are unknown and the phenotypical analysis of the infiltrated tissues containing immunologically activated cells can be confusing. During the last three years we, like other groups, have adopted a three-stage plan, based on double-staining methods, to initiate our working hypotheses in this area (*Figs.* 27.2 and 27.3).

*Table 27.2. **The cluster designation and isotype of monoclonal antibodies to human leucocytes***

	IgM†	IgG2	IgG1
T-cell antibodies			
CD2 (T11)	9–2 (1)†	OKT 11A (3)	OKT 11 (3)
pan-T	7T4 (2)	9·6 (4)	RFT 11 (5)
			MT910 (6)
CD3 (T3)	T10C5 (7)	OKT 3 (3)	UCHT1 (9)
periph.T (mitogenic)	T3/2 (2)	WT31 (8)	Leu-4 (10)
CD4 (T4)	BW264 (11)	OKT 4 (3)	Leu-3a (10)
helper	66·1 (4)	12T4 (2)	MT321 (6)
CD5 (T1)	n.a.‡	Leu-1 (10)	RFT1 (5)
pan-T + B-CLL		OKT 1 (3)	MT215 (6)
CD6 (T12)	T12 (2)		
periph.T (not mitogenic)	RFT12 (5)	12·1 (4)	MT211 (6)
	MBG6 (12)		
CD7	n.a.	3A1 (13)	MT215 (6)
pan-T, strong on blasts		WT1 (8)	
		RFT2 (5)	
CD8 (T8)	RFT8μ (5)	OKT 8 (3)	RFT8γ (5)
suppressor/cytotoxic		Leu2 (10)	MT415 (6)
Precursor cell associated			
CD10	VILA1 (14)	J-5 (2)	RFAL1 (5)
common ALL antigen	RFAL3 (5)	RFAL2 (5)	
HLA-DR	RFDR1 (5)	RFDR2 (5)	
non-polymorphic	(*Figs.* 27.2, 27.3)		
B-cell antibodies			
CD19	n.a.	n.a.	B4 (2)
pre-B + B			HD37 (15)
CD20	RFB7	B1 (2)	—
pan-B	(*Fig. 27.2b*)	2H7 (16)	
CD21	B2 (2)	HB5 (17)	RFB6 (5)
C3d receptor			(*Fig. 27.5b*)
CD22	n.a.	SHCL-1 (10)	RFB4 (5)
pan-periph. B			To15 (18)
CD23 (B blast antigen)	PL-13 (1)	n.a.	Blast-2 (18)
Unclustered pan-B	Y29/55 (19)		

*The majority of monoclonal antibody clusters as defined at the two Workshops on Leucocyte Differentiation Antigens in Paris (1982) and Boston (1984) (reacting with the same cell populations and antigens of the same molecular weight and chemical characteristics) contain individual reagents of different isotypes. These antibodies can be used in combinations with (sub)class specific goat anti-mouse IgM, -IgG2 and -IgG1 second layers conjugated with -FITC and -TRITC fluorochromes. Although the monoclonal antibodies shown include many commercially available reagents (such as Ortho, Becton-Dickinson and Coulter) the selection is essentially random and not exhaustive.
†The numbers in brackets refer to the Laboratories where the reagents are made or distributed from. (1) Naito & Dupont, New York; (2) Reinherz, Nadler, Ritz & Schlossmann (see Coulter range); (3) ORTHO, Raritan; (4) Hansen & Martin, Seattle; (5) Janossy, London; (6) Rieber, Munich; (7) Thompson, Kentucky; (8) Tax, Nijmegen; (9) Beverley, London; (10) Becton-Dickinson; (11) Kurrle, Marburg; (12) McMichael, Oxford; (13) Haynes, Durham; (14) Knapp, Vienna; (15) Dorken, Heidelberg; (16) Clark, Seattle; (17) Tedder, Birmingham, AL.; (18) Mason, Oxford; (19) Forster, Basel.
‡Not available.

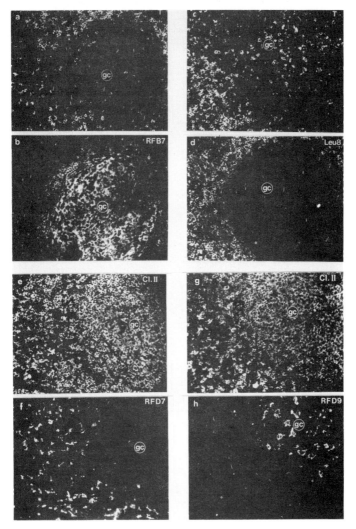

Fig. 27.2. Analysis of human palatine tonsil with double immunofluorescence staining. Lymphoid subsets were studied in (*a*)–(*d*) and macrophage subsets were analysed in (*e*)–(*h*). The double combinations were as follows: (*a–b*) T-cell cocktail of IgG1 class together with anti-pan-B (CD20) of IgM class. (*c–d*) T-cell cocktail of IgM class with Leu-8 of IgG1 class. This pair of photographs shows that Leu-8 labels the B-cells in the lymphocyte corona but is negative on T-cells as well as B-blasts of the germinal centre. (*e–f*) Anti-class II (HLA-DR, core) of IgM class (RFDR-1) in combination with an anti-tissue macrophage antibody of IgG1 class (RFD7). Note that some large class II+ve dendritic cells in the T-zone are RFD7−ve (arrows). Tingible body macrophages in the germinal centre are also RFD7−ve. (*g–h*) Anti-class II (HLA-DR, core) of IgM class (RFDR-1) in combination with an antibody reacting exclusively, in normal tissue, with tingible body macrophages (RFD9). In all these preparations the second layers were anti-IgM (μ chain) and anti-IgG1 (γ chain) specific and conjugated with TRITC and FITC, respectively. The photographs were made in pairs on the same sections with selective filters for TRITC and FITC.

Firstly, we assumed a functional heterogeneity of accessory cell types based on the proposition that macrophages with high lysosomal enzyme content (strongly

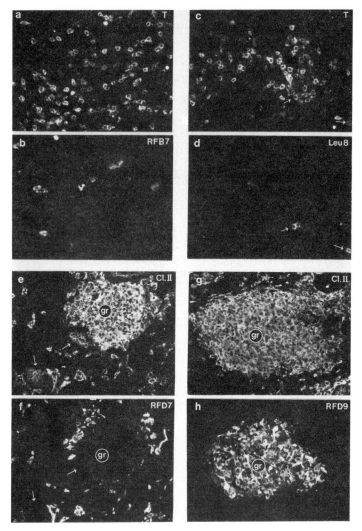

Fig. 27.3. Analysis of lymphoid and macrophage infiltrates with double fluorescence staining. Lymphoid subsets were studied in rheumatoid synovial tissue using the same combinations as shown in *Fig.* 27.2(*a–d*). Note that very few B-cells are present and the vast majority of T-cells are Leu-8+ve (except two shown by arrows). The macrophage subsets were studied in a skin granuloma from a patient with sarcoidosis using the same combinations as shown in *Fig.* 27.2(*e–h*). There are only moderate numbers of tissue macrophages that react with RFD7. These RFD7+ve cells surround the granuloma (gr; class II antigen strongly positive). Note that the vascular endothelium is class II antigen positive [arrows in (*e*)] but RFD7−ve [arrows in (*f*)]. In contrast, the activated macrophages, probably epithelioid cells, in the middle of the granuloma are strongly RFD9+ve [in (*h*)]. Inside the granuloma RFD9 appears to stain the same cells as anti-class II but the labelling with RFD9 seems to be more cytoplasmic (*h*) than labelling with anti-class-II (*g*).

acid phosphatase-positive) are likely to be scavenger cells, and that strongly class II antigen-positive cells may efficiently present foreign antigens. A combination of staining for acid-phosphatase (cytochemical) and class II (immunofluores-cence[3,5]) has indeed confirmed previous data that there were distinct populations of high class II, low acid phosphatase, e.g. interdigitating dendritic cells of the T-zone in lymph nodes and Langerhans' cells of the skin,[29,30] as opposed to numerous tissue macrophages of high acid phosphatase and low class II expression (reviewed in Poulter[31]). These separate cell types were admixed in lesions seen in histiocytosis-X.[5] It is well known that in tissues with infiltrating, immunologically activated cells, various cell types (including epithelium etc.) can express high levels of class II, and it was therefore not unexpected that in inflammatory bowel disease[32] and in granulomas, many acid phosphatase-con-taining cells were also strongly class II-positive.[33]

Secondly, as the cytochemical staining in combination with immunofluores-cence is a difficult method, monoclonal antibodies were produced against non-lymphoid tissue macrophage populations (*Table* 27.3). The class II-positive, acid phosphatase-negative populations of thymic medulla and lymph node ('interdigitating' cells) strongly react with RFD1 which also labels some B-cells (but not endothelial cells, and Langerhans' cells).[34] Tissue macrophages are, on the other hand, RFD7-positive (*Figs*. 27.2f, 27.3f), but this reagent does not label the strongly class II-positive interdigitating cells of T-zones and thymic medulla. An exciting finding is that 'conventional' bone marrow cells and blood monocytes (UCHM1+ve[35]) lack RFD7 reactivity but develop it within 5 days in culture. This finding indicates that some RFD7+ve cells in the tissues, particularly those which are double stained with UCHM1+ve, may be recently arrived derivates of blood monocytes.[34] The final, somewhat surprising, observation here is that well developed 'tingible body' macrophages of the normal germinal centres are RFD7−ve (*Fig*. 27.2f). This finding is, however, less disturbing in the light of the observations that (i) this special cell type reacts with RFD9, the third interesting antibody,[36] and (ii) the large, activated, epithelioid cells in granulomas (in sarcoidosis and leprosy) are also RFD7−ve, RFD9+ve (*Fig*. 27.3h). It appears that 'tingible body' macrophages are a unique population with some suggestive similarities to other activated epithelioid cells.[36]

Thirdly, the final stage of standardization of these reagents has been carried out in the fetus. These studies demonstrate the heterogeneity of dendritic and macrophage-like cells. The majority of these cells are RFD7+ve, class II−ve and probably represent tissue macrophages, while a significant minority show the class II+ve, RFD7−ve phenotype (precursors of 'dendritic' or interdigitating cells?). These two different cell types are found in the fetal liver, among the cells scattered in the mesenchyma and even in the yolk sac.[37] The latter class II+ve, RFD7−ve cell type also acquires positivity with RFD1 as soon as the T-cell zones in the thymic medulla and in the earliest identifiable lymph nodes develop. But the RFD9+ve macrophage population is conspicuously absent in the primary follicles of fetal lymph nodes, and appears only inside the properly developed germinal centres during later periods of life.

In conclusion, at least three types of accessory cells can be easily demonstrated with double immunofluorescence methods. Conveniently selective antibodies are now available for further investigation of the induced phenotypic changes of monocytic cell types in cultures. These studies will contribute to the definition of the relationship of these different cell types and may also help us to understand the

*Table 27.3. Summary of a few characteristic anti-macrophage reagents**

Monoclonal antibody	Class	Reactivity with circulating monocytes	Reactivity with tissues	Ref.
UCHM1	IgG1	++	Monoblasts and monocytes in bone marrow; a few monocytic cells around blood vessels in tissues	35
RFD1γ RFD1μ	IgG1 IgM	++ (small subset)	T-zone dendritic cells in lymph node and thymic medulla Strongly class II+ve dendritic cells in rheumatoid synovium and edge of granulomas	34
RFD7	IgG1	None†	Tissue macrophages throughout the body (class II−ve in the fetus)	34; *Figs. 27.2f, 27.3f*
RFD9	IgG1	None†	Tingible body macrophages in germinal centres, activated macrophages in granulomas, erythroblasts in the bone marrow	36; *Figs. 27.2h, 27.3h*
Class II: RFDR-1μ RFDR-2γ	IgM IgG2	++ + B-cells	Thymic epithelium, B cells, interdigitating, 'dendritic' cells in lymph node, Langerhans' cells and many cell types in tissues following immune activation	*Figs. 27.2e, g Figs. 27.3e, g Fig. 27.4a*

*These antibodies have been characterized with the double-marker techniques described in this chapter during fetal development and in normal as well as inflamed tissues.[34,36,37]

†These reagents show no reactivity with circulating or bone marrow monocytes, although positivity for RFD7 can be generated during 5-day culture of adherent cells.[34] These antibodies appear to be more specific for the corresponding lineages than those characterized in the Boston panel of myeloid/monocytic antibodies.[12]

cellular 'composition' of inflammatory infiltrates within infected tissues and in autoimmune disorders. These, and similar antibodies, will also be useful for analysing the ultrastructural characteristics of the respective cell types throughout development.

4. THE DILEMMA OF PHENOTYPE AND FUNCTION

There is no *a priori* reason why a single 'marker' (reactivity with a monoclonal antibody) should define the function of a cell. For example, even the first publication on OKT 4 showed that T4+ve cells included cortical thymocytes, an immature population incapable of T-cell function.[38] In spite of these limitations, the T4/T8 dichotomy of the peripheral T-lymphoid system is an important feature. Instead of discarding a well-substantiated distinction, a better approach is to define further subpopulations within the main categories with simple methods, such as double immunofluorescence labelling. The candidates for use in this further analysis of T-cell types are: monoclonal antibodies Leu-8[39], TQ1[40] and 2H4[41] which appear to distinguish genuine helper (Leu-8−ve, TQ1−ve, 2H4−ve) and suppressor-inducer cells (Leu-8+ve, TQ1+ve, 2H4+ve) within the T4+ve population.

When the T4/T8 dichotomy was described, the evidence for these populations' divergent features *in vitro*[42] was supported by the different distribution of these cell types in various normal tissues, such as lymph nodes, bone marrow and gut.[43] A similar situation is emerging with the further subset division of T4+ve cells: the Leu-8+ve, T4+ve and Leu-8−ve, T4+ve sub-subsets show different tissue localization. The former are abundant in the thymic medulla (*Fig. 27.4c–d*) and in the paracortical T zone (*Fig. 27.2d*), but are virtually absent among the T4+ve cells which reach the germinal centres (*Fig. 27.2d*). The T4+ve, Leu-8−ve cells are also seen, in abundance, within the rheumatoid synovium (*Fig. 27.3d*). These results are intriguing because they are compatible with the suggestion that in the normal medulla of thymus and in the paracortical zone of lymph nodes the induction of suppressor cells is perhaps more common than, for example, among the T-cells of the germinal centre.[44] More appropriate conclusions at this stage, however, are that (*a*) only double combinations are informative, since Leu-8+ve cells include non-lymphoid elements (*Fig. 27.4b*) and B-cells (*Fig. 27.2d*) and must therefore be studied together with T4 staining if we wish to reach even preliminary conclusions; (*b*) further antibodies such as TQ1 and 2H4 require a similar study; (*c*) from such studies, combined with double-colour immunofluorescence cell sorting and functional assays, further observations may be forthcoming which could explain the cellular defects of the immunological network in rheumatoid arthritis and systemic lupus erythematosus.

5. ATTEMPTS TO IMPROVE PREFIXATION

Hancock and his colleagues[45] have used the fixative periodate–lysine–paraformaldehyde (PLP) to preserve macrophage antigens. In studies performed on murine tissues it was demonstrated that lymphoid antigens, with the exception of

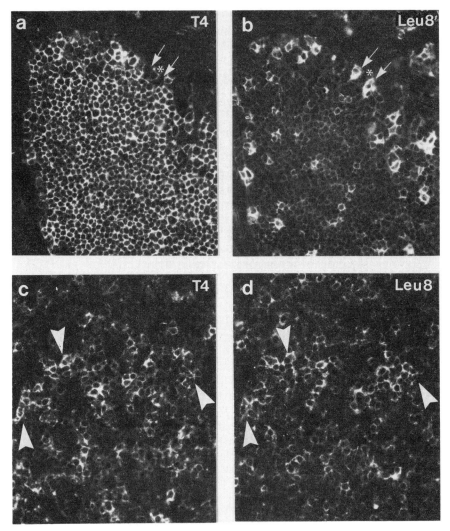

Fig. 27.4. Analysis of the human thymus with double-fluorescence staining. Double fluorescence analysis of T4+ve (CD4) and Leu-8+ve cells in the human thymic cortex (*a, b*) and medulla (*c, d*). Immature cortical thymocytes (T4+ve as well as T8+ve, not shown) are weakly Leu-8+ve (*see* bulk of cells in the middle and the solitary cell labelled with asterisk in the subcapsular region). The strongly Leu-8+ve cells are not lymphoid (arrows). In the medulla, 70–80 percent of T4+ve are Leu-8+ve (large arrows) and 20–30 per cent of T8+ve are Leu-8+ve. The double staining was obtained by selecting an anti-T4 reagent of IgM class in combination with Leu-8 of IgG class. The anti-mouse Ig second layers were anti-IgM (μ class) and anti-7S (γ chain)-specific conjugated with TRITC and FITC, respectively.

Lyt-2 antigen and the human equivalent T8 (CD8), were also preserved.[46] As it is known that only exceptional membrane antigens survive conventional formalin fixation and paraffin embedding, attempts have been made to define the range of antigens which survive the milder PLP protocol.[47] As shown in *Table* 27.4, the loss of T8 antigen was confirmed in human tissues, but a remarkable series of

Table 27.4. Comparison of labelling with monoclonal antibodies in frozen and paraffin-embedded sections of normal lymphoid (tonsil) tissue*

Antibodies	Type of tissue		
	Frozen	PLP-fixed†	NBF-fixed‡
Anti-class II:			
RFDR1 (μ)	+++	+++ (*Fig. 27.5a*)	+/−
RFDR2 (γ)	+++	+++	+/−
Dako-HLA-DR	+++	+++	+/−
2D1 (pan-leucocyte)	+++	+++	+
RFA2 (pan-leucocyte)	++	++	−
NA1/34 (CD1)*	+++	++	−
UCHT1 (CD3)	+++	+	−
Leu3 (CD4)	++	−	−
RFT1 (CD5)	+++	++	−
RFT2 (CD7)	++	++	−
RFT8 (CD8)	+++	−	−
OKT 8 (CD8)	+++	−	−
Leu-2a (CD8)	+++	−	−
RFB6 (CD21)	+++	+++ (*Fig. 27.5b*)	−
RFB4 (CD22)	+++	+++	−
Dako-pan B (CD22)	+++	++	−
BA1 (CD24)	+++	+++	−
Dako-DRC1 (foll.dendritic)	++	+++	−
Dako-anti-IgM	+++	+++	++
RFD3 (foll.dendritic)	++	−	−
RFD1 (dendritic or ID)	++	−	−
Leu M2 (Mac 120)	++	+++	−
S22 (anti-macrophage)	++	++	++
RFFVIII (factor VIII)	+++	+++	+

*Modified from Collins et al.[47] with permission of Elsevier Scientific Publishers, Amsterdam.
†Periodate–lysine–paraformaldehyde-fixed.
‡Neutral-buffered formalin-fixed.
CD numbers refer to the cluster established at the Workshop for Human Leucocyte Differentiation Antigens, Paris and Boston.

antibodies showed activity. The morphological preservation of tissue was indistinguishable from formalin-fixed tissues. This method may become important for studying infectious samples including lymph nodes from patients with HTLV-III-positive persistent generalized lymphadenopathy. A further important feature of this fixative is that autofluorescence is not generated, and that double immunofluorescence staining of paraffin-embedded sections is feasible (Fig. 27.5). It is also clear, however, that there is some, although variable, loss of antigenicity, as compared to sections of frozen tissues. In order to obtain bright fluorescence the use of triple layer labelling (*Table* 27.1), instead of double, may be required.

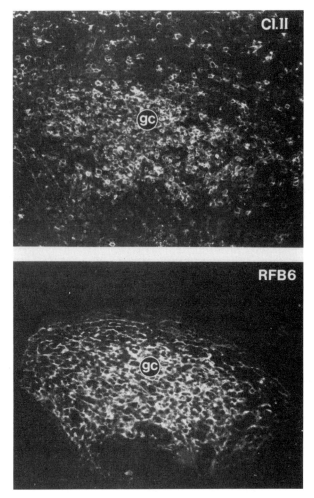

Fig. 27.5. Immunofluorescence staining of tonsil samples following fixation in periodate–lysine–paraformaldehyde (PLP) and paraffin embedding. The samples were labelled with a double combination of (*a*) anti-class II (core) of IgM class (RFDR-1) and (*b*) RFB6 antibody of IgG1 class. The latter reacts with the C3d receptor (mol. wt = 130 kdal) on B-lymphocytes and follicular dendritic cells (*Fig.* 27.1) and belongs to the CD21 cluster. Note that the method is free of artefacts due to autofluorescence, although some of the antigens (e.g. CD8) do not survive the procedure (*Table* 27.4).

6. CONCLUSIONS

Double immunofluorescence techniques are simple, rapid and reliable methods which complement cytochemical methods. Histopathologists will benefit from employing these informative techniques more frequently. It is unlikely that the final practical limits of the technique have been reached. Combinations applying monoclonal antibodies of different isotypes (*Table* 27.2) can still be improved by using the heavy chain-specific goat anti-mouse Ig reagents as bridges, followed by a third layer: a fluorochrome-labelled antibody of the same isotype as the primary reagent.[48]

Acknowledgements

Dr L. Collings has participated in some of these studies. We are indebted to Drs P. B. Beverley and N. Hogg for the gift of UCHM1 monoclonal antibody.

REFERENCES

1. Coons A. H, Creech H. J. and Jones R. N. Immunological properties of an antibody containing a fluorescent group. *Proc. Soc. Exp. Biol. Med.* 1941, **47**, 200–202.
2. Mason D. Y., Naiem M., Abdulaziz Z., Nash J. R. G., Gatter K. C. and Stein H. Immunohistological applications of monoclonal antibodies. In: McMichael A. J. and Fabre J. W., eds., *Monoclonal Antibodies in Clinical Medicine*. London, Academic Press, 1982: 585–635.
3. Poulter L. W., Chilosi M., Seymour G. J., Hobbs S. and Janossy G. Immunofluorescence membrane staining and cytochemistry, applied in combination for analysing cell interactions in situ. In: Polak J. M. and Van Noorden S., eds., *Immunocytochemistry, Practical Applications in Pathology and Biology*. Bristol, Wright·PSG, 1983: 233–248.
4. Giorno R. Technical considerations in immunohistology of lymphoid cell membrane antigens. *Surv. Synth. Pathol. Res.* 1983, **31**, 112–138.
5. Thomas J. A., Janossy G., Chilosi M., Pritchard J. and Pincott J. R. Combined immunological and histochemical analysis of skin and lymph node lesions in histiocytosis X. *J. Clin. Pathol.* 1982, **35**, 327–337.
6. Reinherz E. L. and Schlossman S. F. (eds.) Leucocyte Typing. *Second International Conference on Leucocyte Differentiation Antigens*, Boston. 1984, in press.
7. Nadler L. M., The B cell monoclonal antibodies and their differentation clusters. In: Reinherz E. L. and Schlossman S. F., eds., *Leucocyte Typing. Second International Conference on Leucocyte Differentiation Antigens*, Boston. 1984, in press.
8. Campana D., Janossy G., Bofill M., Trejdosiewicz L. K., Ma D., Hoffbrand A. V., Mason D. Y., Lebacq A.-M. and Forster H. Human B cell development. I. Phenotypic differences of B lymphocytes in the bone marrow and peripheral lymphoid tissue. *J. Immunol.* 1985, **134**, 1524–1530.
9. Mason D. Y. Immunocytochemical labelling of anti-B and anti-leukaemia antibodies. In: Reinherz E. L. and Schlossman S. F., eds., *Leucocyte Typing. Second International Conference on Leucocyte Differentiation Antigens*, Boston. 1984: in press.
10. Bofill M., Janossy G., Janossa M., Burford G. D., Seymour G. J., Wernet P. and Kelemen E. Human B cell development. II. Subpopulations in the human fetus. *J. Immunol.* 1985, **134**, 1531–1538.
11. Berstein. I. D. Analysis of the myeloid reagent clusters. In: Reinherz E. L. and Schlossman S. F., eds., *Leucocyte Typing. Second International Conference on Leucocyte Differentiation Antigens*, Boston. 1984: in press.
12. Berti E., Pandelli M. G., Parkavicini C., Cattoretti G. and Delia D. Immunohistochemical reactivity of anti-myeloid/stem cells workshop; monoclonal antibodies in thymus, lymph nodes, liver and normal skin. In: Reinherz E. L. and Schlossman S. F., eds., *Leucocyte Typing. Second International Conference on Leucocyte Differentiation Antigens*, Boston. 1984: in press.
13. Lanier L. L., Le A. M., Phillips J. H., Warner N. L. and Herzenberg L. A. Subpopulations of human natural killer cells defined by expression of the Leu-7 (HNK-1) and Leu-11 (NK-15) antigens. *J. Immunol.* 1983, **131**, 1789.
14. Pizzolo G., Semenzato G., Chilosi M., Morittu L., Warner N., Favrot M., Bofill M. and Janossy G. Distribution and heterogeneity of cells detected by HNK-1 monoclonal antibody in blood and tissues in normal, reactive and neoplastic conditions. *Clin. Exp. Immunol.* 1984, **57**, 195.
15. Giorno R. Evaluation of a tissue transport medium for immunological characterization of benign and malignant lymphoid tissues. *Am. J. Clin. Pathol.* 1982, **78**, 8–13.

16. Thomas J. A., Janossy G., Eden O. B. and Bollum F. J. Demonstration of nuclear terminal transferase deoxynucleotidyl transferase (TdT) in leukemic infiltrates of testicular tissue. *Br. J. Cancer* 1982, **45**, 709–717.
17. Stein H., Bonk A., Tolksdorf G., Lennert K., Rodt H. and Gerdes J. Immunohistologic analysis of the organization of normal lymphoid tissue and non-Hodgkin's lymphomas. *J. Histochem. Cytochem.* 1980, **28**, 746–760.
18. Chilosi M., Pizzolo G. and Vincenzi C. Haematoxylin counterstaining of immunofluorescence preparations. *J. Clin. Pathol.* 1983, **36**, 114–115.
19. Johnson G. D. and Holborow E. J. Immunofluorescence. In: Weir D. M. and Herzenberg L. A., eds., *Handbook of Experimental Immunology*, 4th ed. Oxford, Blackwell Scientific Publications, 1986: in press.
20. Ciocca D. R., Adams D. J., Bjercke R. J., Sledge G. W., Edwards D. P., Chamness G. C. and McGuire W. L. Monoclonal antibody storage conditions and concentration effects on immunohistochemical specificity. *J. Histochem. Cytochem.* 1983, **31**, 691–696.
21. Giorno R. Comparison of sensitivity of labeled avidin–biotin and avidin–biotin complex methods for in situ antigen detection. *J. Histochem. Cytochem.* 1983, **31**, 1066.
22. Hsu S. M., Raine L. and Fanger H. Use of avidin–biotin–peroxidase complex (ABC) in immunoperoxidase techniques: a comparison between ABC and unlabelled antibody (PAP) procedures. *J. Histochem. Cytochem.* 1981, **29**, 577–580.
23. Warnke R. and Levy R. Detection of T and B cell antigens with hybridoma monoclonal antibodies: a biotin–avidin–horseradish peroxidase method *J. Histochem. Cytochem.* 1980, **28**, 771–776.
24. Guesdon J. L., Ternyck T. and Avrameas S. The use of avidin–biotin interaction in immunoenzymatic techniques. *J. Histochem. Cytochem.* 1979, **27**, 1131–1139.
25. Cammisuli S. Hapten-modified antibodies specific for cell surface antigens as a tool in cellular immunology. In: Lefkovits I. and Pernis B., eds., *Immunological Methods*, Vol. 2. New York, Academic Press, 1981: 139–162.
26. Simmonds R. G., Smith W. and Marsden H. 3-Phenylazo-4-hydroxyphenyl-isothiocyanate: versatile reagent for the efficient haptenation of Ig and other carrier molecules. *J. Immunol. Methods* 1982, **54**, 23–30.
27. Martin P. J., Hansen J. A., Siadak A. W. and Nowinski R. C. Monoclonal antibodies recognizing human T lymphocytes and malignant human B lymphocytes: a comparative study. *J. Immunol.* 1981, **127**, 1920–1926.
28. Stein H., Gerdes J. and Mason D. Y. The normal and malignant germinal centre. *Clin. Haematol.* 1982, **11**, 531–559.
29. Hoefsmit Ch. M., Kamperdijk E. W. A., Hendrikes H. R., Beeten R. H. J. and Balfour B. M. Lymph node macrophages in the reticuloendothelial system. *Morphology* 1980, **1**, 1809–1834.
30. Humphrey J. H., Differentiation of function among antigen presenting cells. In *Microenvironments in Haemopoietic and Lymphoid Differentiation*. Ciba Foundation Symposium 84. London, Pitman, 1981: 302–321.
31. Poulter L. W. Antigen presenting cells in situ: their identification and involvement in immunopathology. *Clin. Exp. Immunol.* 1983, **53**, 513–520.
32. Selby W. S., Poulter L. W., Hobbs S., Jewell D. P. and Janossy G. Heterogeneity of HLA-DR positive histiocytes in human intestinal lamina propria: a combined histochemical and immunohistological analysis. *J. Clin. Pathol.* 1983, **36**, 379–384.
33. Collings L. A., Waters M. F. J. and Poulter L. W. The involvement of dendritic cells in the cutaneous lesions of tuberculoid and lepromatous leprosy. *Clin. Exp. Immunol.* 1985, in press.
34. Poulter L. W., Hobbs S., Collings L. A., Campbell D. C., Munro C. M. and Janossy G. Discrimination of human macrophages and dendritic cells using monoclonal antibodies. *Cell Immunol.* 1985, in press.
35. Linch D. C., Allen C., Beverley P. C. L., Bynoe A. G., Scott C. S. and Hogg N. Monoclonal antibodies differentiating between monocytic and non-monocytic variants of AML. *Blood* 1984, **63**, 566–573.
36. Bofill M., Poulter L. W. and Janossy G. Distribution of RFD9 positive macrophages in normal tissues. 1985, in preparation.
37. Bofill M., Janossy G., Poulter L. W. and Kelemen E. Characterization of macrophages and antigen presenting cells during human fetal development. *J. Immunol.* 1985, in press.
38. Reinherz E. L., Morimoto G., Fitzgerald K. A., Hussey R. E., Daley J. F. and Schlossman S. F. Heterogeneity of human T4 positive inducer T cells defined by a monoclonal antibody that delineates two functional subpopulations. *J. Immunol.* 1982, **128**, 463.

39. Gatenby P. A., Kotzin B. L., Kansas G. S. and Engelman E. G. Immunoglobulin secretion in human autologous mixed leukocyte reaction. Definition of a suppressor amplifier circuit using monoclonal antibodies. *J. Exp. Med.* 1982, **156**, 55.

40. Reinherz E. L., Kung P. C., Goldstein G. and Schlossman S. F. Further characterization of the human inducer T cell subset defined by monoclonal antibody. *J. Immunol.* 1979, **123**, 2894–2896.

41. Morimoto G., Letvin N. L., Distaso J. H., Aldrich W. R. and Schlossman S. F. The isolation and characterization of the human suppressor-inducer T cell subset. *J. Immunol.* 1985, **134**, 3763–3769.

42. Reinherz E. L. and Schlossman S. F. The differentiation and function of human T lymphocytes. *Cell* 1980, **19**, 821–826.

43. Janossy G., Tidman N., Selby W. S., Thomas J. A., Granger S., Kung P. C. and Goldstein G. Human T lymphocytes of inducer and suppressor type occupy different microenvironments. *Nature* 1980, **288**, 81–84.

44. Janossy G., Panayi G., Duke O., Bofill M., Poulter L. W. and Goldstein G. Rheumatoid arthritis: a disease of T lymphocyte/macrophage immunoregulation. *Lancet* 1981, **ii**, 839–842.

45. Hancock W. W., Becker G. J. and Atkins R. C. A comparison of fixatives and immunohistochemical techniques for use with monoclonal antibodies to cell surface antigens. *Am. J. Clin. Pathol.* 1982, **78**, 825–831.

46. Van Ewijk W., Coffman R. C. and Weissman I. L. Immunoelectron microscopy of cell surface antigens: a quantitative analysis of antibody binding after different fixation protocols. *Histochem. J.* 1980, **12**, 349–361.

47. Collings L. A., Poulter L. W. and Janossy G. The demonstration of cell surface antigens on T cells, B cells and accessory cells in paraffin-embedded human tissues. *J. Immunol. Methods* 1984, **75**, 227–239.

48. Falini B. and Taylor C. R. New developments in immunoperoxidase techniques and their application. *Arch. Pathol. Lab. Med.* 1983, **107**, 105–117.

49. Tedder T. F., Fearon D. T., Gartland G. L. and Cooper M. D. Expression of C3b receptors on human B cells and myelomonocytic cells but not natural killer cells. *J. Immunol.* 1983, **130**, 1668–1673.

50. Nadler L. M. and Schlossman S. F. Development of B-cells as analysed by B1, B2 and B4 monoclonal antibodies in the human fetal bone marrow and spleen. *J. Clin. Invest.* 1984, **74**, 332–341.

51. Clark B. R. and Todd C. W., Avidin as a precipitant for biotin-labeled antibody in a radioimmunoassay for carcinoembryonic antigen. *Anal. Biochem.* 1982, **121**, 257–262.

28

Immunocytochemistry in the Study and Diagnosis of Organ-specific Autoimmune Diseases

W. A. Scherbaum,   B. M. Dean and
R. Mirakian,   G. F. Bottazzo
R. Pujol-Borrell,

The detection of autoantibodies is an essential diagnostic criterion to establish the autoimmune origin of diseases previously considered of unknown aetiology. For this purpose the indirect immunofluorescence technique is still the test of choice. It is a simple and reliable method in the hands of the expert for detecting serum antibodies directed against a variety of tissue structures or cell components. This test is carried out on cryostat sections of unfixed tissue with the clear advantage of detecting autoantibodies to unaltered antigens. The antibody test requires no more than a small amount of serum from the patient and a positive result provides a valuable diagnostic tool and the basis for the study of the complex role of cellular and humoral reactions which might lead to overt disease.

Immunofluorescence also helps to identify new autoantibodies and these may be so specific that they react only with one cell type within an organ. This is the case for the pancreatic islets, the thyroid, the pituitary gland, the hypothalamus and the gut mucosa. These organs are composed of different cells which cannot be distinguished by light microscopy. For the exact characterization of single cell antibodies within these organs we have developed a four-layer double-fluorochrome immunofluorescence sandwich test (Fig. 28.1) performed on the same tissue section which enables us to determine precisely the endocrine nature of positive cells. Applying this test we were able to demonstrate specific antibodies reacting with pituitary prolactin[1] and growth hormone cells,[2]

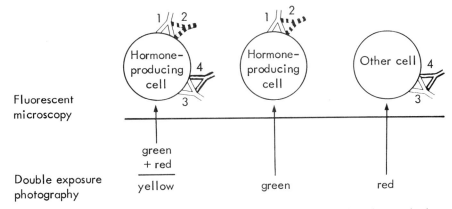

Fluorescent
microscopy

Double exposure
photography

Fig. 28.1. Diagram of double immunofluorescence technique used for characterization of autoantibodies to single cells in complex endocrine organs. Each section is sequentially stained with: 1. Anti-hormone serum (e.g. raised in rabbits), 2–6 h, depending on the affinity and avidity of the antiserum. 2. Fluoresceinated anti-rabbit Ig (e.g. raised in goats), 30–40 min. 3. Patient's serum, 2–8 h, depending on the organ. 4. Rhodaminated anti-human Ig (raised in goats), 30 min. Between every step the sections are cautiously washed in PBS (not shaken) and the reaction is finally read using a fluorescence microscope fitted with epi-illumination and an automatic camera.
Left column: Final reaction when the human autoantibody and the animal anti-hormone serum are directed against the same cells. The final yellow colour is obtained by double exposure photography of the same field.
Middle and right columns: Final results when the rabbit anti-hormone serum and the patient's serum react with different cells. Double exposure photographs show separate green and red cells. This technique may be relied on because the human antibodies are directed to intracellular membranes and not against the hormones themselves.
The filters used are: for fluorescein, Zeiss set 09 blue excitation, 450–490 with interference red barrier filter KP 560 18 × 2, and for rhodamine, set 15, green excitation H546.

antibodies to glucagon and somatostatin cells of the pancreatic islets,[3] antibodies to gastrin-,[4] secretin- and gastric inhibitory polypeptide (GIP)-,[5] somatostatin- and glucagon-producing cells of the gut,[6] and more recently antibodies to calcitonin-producing cells of the thyroid[7] and to hypothalamic vasopressin cells.[8]

The spectrum of the known organ-specific autoimmune diseases is shown in *Table* 28.1 and it is intriguing that they affect mainly the endocrine system. Their corresponding autoantibodies usually react with cytoplasmic antigens which segregate with the microsomal fraction of the target cells and appear to be integral membrane-bound receptor proteins probably involved in the intracellular transport of hormones.

Until recently it was thought that microsomal antibodies reacted only with *intracytoplasmic* membranes, but immunofluorescence studies on monolayer cultures of viable thyroid,[9] adrenocortical[10] and gastric parietal cells[11] have now shown that the internal autoantigens are also represented on the plasma membrane. This makes the microsomal antigens accessible to sensitized immunocompetent cells or antibody-dependent cytotoxic mechanisms so that a pathogenetic role of organ-specific autoantibodies is becoming increasingly acceptable. It is of interest that similar close correlations between surface immunofluorescence and cytoplasmic immunofluorescence are absent in 50 per cent of sera containing islet cell antibodies,[12] in a proportion of pernicious anaemia sera[13] and in the majority of patients with vitiligo.[14]

Table 28.1. **Spectrum of organ-specific autoimmune diseases**

Disease	Antibodies to	Frequency of antibodies (%) Patients	Controls	Recommended immunological test
Insulin-dependent diabetes mellitus (type I)	Pancreatic islets all islet cells	80 at diagnosis	0·5	IFL and CF-IFL
	Glucagon cells	Unknown	0·2	IFL
	Somatostatin cells	Unknown	0·2	IFL
Idiopathic Addison's disease	Adrenal cortex	60 at diagnosis	0·1	IFL, CF-IFL, ELISA, RIA
Premature menopause with adrenalitis	Steroid cells of gonads + placenta	80	0·01	IFL
Primary gonadal insufficiency	Sperms	Unknown	7	IFL
	Ovums	Unknown	Unknown	IFL
Hashimoto's thyroiditis	Thyroglobulin	see Refs 63, 116	Dependent on age	Passive HA
Primary myxoedema	Thyroid microsomes	64, 117	Sex (F > M)	Passive HA
Graves' thyrotox.	TSH-receptors	80, 81, 24	Test	80, 81, 24
Endocrine exophthalmos	Extra-ocular muscle membranes	118	unknown	118
Pernicious anaemia	Intrinsic factor	i.s. 56	Unknown	RIA
	Parietal cells	90	F > M 0–16	IFL
Myasthenia gravis (acquired form)	Acetylcholine receptors	45* 90†	0	alpha-Bgt assay‡
	Muscle fibrils	20–40	0·2	
Fundal gastritis (type A)	Parietal cells	60	see above	IFL
Antral gastritis (type B)	Gastrin cells	10	0·1	IFL
Primary hypopara-thyroidism	Chief cells of parathyroid gland	Very rare	Unknown	IFL
Type I diabetes in early phase	Pituitary PRL/HGH and other cells	17	0	IFL
Idiopathic central diabetes insipidus	AVP cells of the hypothalamus	33	0	IFL

*Ocular form of myasthenia gravis (type I).
†Generalized form of myasthenia gravis (types II–IV).
‡Alpha-bungarotoxin immunoprecipitation assay using human peripheral muscle as antigen.
AVP, arginine vasopressin.
CF-IFL, complement fixation using the IFL technique.
HA, haemagglutination.
HGH, human growth hormone.
IFL, indirect immunofluorescence.
i.s., in the serum.
PRL, prolactin.
RIA, radioimmunoassay.
ELISA, enzyme-linked immunosorbent assay.

1. CHOICE OF TISSUE

Human tissues are always preferable but, if they are unobtainable, monkey tissues may be employed. Rodent organs often show considerable non-specific fluorescence and when antisera are applied heterophile antibodies may bind and give non-specific results. Also, the cross-reactivity of human autoantibodies on animal tissue varies widely. The thyroid antigen–antibody system shows more restriction with regard to animal species outside the primates, whereas the parietal cell antibody cross-reacts strongly even with the 'all-purpose' gastric mucosal cells of amphibians.

Some tissue antigens are quickly autolysed after death and therefore it is advisable to use tissues obtained at operation or from kidney transplant donors. For the detection of microsomal antibodies, unfixed tissues should always be used in the first instance. Some fixatives may destroy the putative autoantigen or reveal new antigenic sites which can give false positive reactions as is now well illustrated in the use of Bouin's-fixed pancreas for the detection of islet-cell antibodies (reviewed in [15]).

Antibodies like those directed to pancreatic islets, pituitary, hypothalamic vasopressin cells and single endocrine cells of the gut are usually of a low titre and they sometimes give only a dim fluorescence compared with the strong thyroid or gastric antibody reaction. For this reason it is important to test the sera undiluted in the first place, to have an ultraviolet microscope equipped with epi-illumination and to look at the sections at high magnification.

2. AUTOANTIBODIES TO PANCREATIC ISLET CELLS

2.1. Cytoplasmic Islet Cell Antibodies

2.1.1. Islet Cell Antibodies detected with Anti-IgG Conjugate (Islet Cell Antibody–IgG)

The increased coexistence of type I (insulin-dependent) diabetes with Addison's disease,[16] thyroid autoimmune diseases,[17] pernicious anaemia and primary hypoparathyroidism[18] suggested that this type of diabetes might be of autoimmune origin. Specific autoantibodies to pancreatic islet cells were first described in some of our polyendocrine patients.[19] Lendrum et al.[20] were able to show that over 80 per cent of children with sudden-onset diabetes had islet cell antibodies at the time of diagnosis and the incidence decreased to 50 per cent within months and to 15–20 per cent after the first five years. Patients in whom these antibodies persist usually suffer from polyendocrine autoimmune diseases.[21] Islet cell antibodies are not a feature of classical type II diabetes, but when they are present in these cases we also find other organ-specific autoantibodies,[22] and these patients more often require insulin treatment later.[23]

For the islet cell antibody–IgG test it is crucial to use a fresh gland obtained from kidney transplant donors and it is essential to use blood group O tissue. Blood group antigens are secreted by the exocrine pancreas and high titres of ABH isoagglutinins in the test sera may produce disturbing immunofluorescence patterns in the exocrine acini which interfere with the reading on the islets.

Several pancreases should be collected because different specimens may possess variable amounts of autoantigens and vary in their content of the different islet cell types.[24] There are also important variations in the distribution of the different endocrine cells within the gland,[25] so that it is advisable to take blocks from the head and tail of the pancreas.

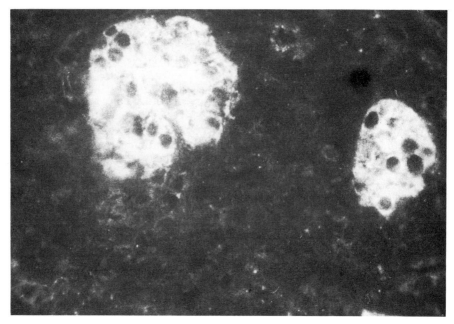

Fig. 28.2. Cryostat section of blood group O human pancreas treated with the serum of a newly diagnosed insulin-dependent diabetic patient and stained with FITC anti-human IgG conjugate. The serum contains the conventional islet-cell antibodies which stain the entire islet of Langerhans.

The antigen of islet cell antibodies is common to all types of islet cells[26] and staining of the entire islet is obtained (*Fig.* 28.2). Yet it is only the beta-cells which are selectively destroyed in type I diabetes, so cytoplasmic islet cell antibodies appear not to be responsible for the islet-cell damage, despite their great value as a serological marker.

Absorption of islet cell antibody-positive sera with excess insulin, glucagon and somatostatin does not affect the islet immunofluorescence[27] which demonstrates that these antibodies do not react with the pancreatic hormones but with the intracellular membranes in the islet cells.

Type I diabetes mellitus usually starts as an abrupt illness and it was, therefore, a surprise to discover, by regular testing of clinically unaffected relatives of diabetic children, that islet cell antibodies exist for years in some siblings who are HLA-identical or haplo-identical with the proband.[28] These siblings proved to be the most vulnerable for future diabetes. In the diabetic families, fluctuations of islet cell antibody–IgG reactions are often seen, and sometimes the antibodies appear simultaneously in more than one member including the diabetic proband.[29] The fluctuations in islet cell antibodies perhaps reflect silent intermittent infections with insulinotropic viruses and we are looking at viral antibody patterns in these individuals.

Islet cell antibody determinations in epidemiological studies have helped to clarify the role of autoimmunity in diabetic populations of different ethnic and genetic backgrounds.[30,31] However, it is now clear that this test does not help with the final diagnostic assessment of the patient. A positive or negative antibody result does not alter the natural history of the disease or the absolute need for insulin in these patients.

Various unsuccessful attempts have been made to see whether the presence of islet cell antibodies can be of value in explaining the severity of clinical diabetes, or its 'brittle' variant. Similarly, antibody determination has not helped to assess residual beta-cell function as measured by C-peptide determination.[32] Microvascular diabetic complications do not correlate with the presence of islet cell or other antibodies.[33]

A recent retrospective study on a set of identical diabetic triplets revealed the first appearance of islet cell antibody–IgG in a non-diabetic twin, 5 years before he became a diabetic; the total period of discordance was 39 years.[34] Identical twins are known to remain discordant in at least 50 per cent of cases.[35]

2.1.2. Complement-fixing Islet Cell Antibodies

In our laboratory we have developed a simple complement-fixation test which is performed by an immunofluorescence technique. In the complement-fixation–immunofluorescence test the patient's undiluted serum is applied to the tissue section for 30 min at room temperature. After washing, a drop of fresh normal human serum is added as a source of complement, and the reaction is revealed by adding fluorescein-conjugated anti-human C3 as the third layer. When this test is done on human pancreas concurrently with the standard IgG test, 50–55 per cent of islet cell antibody-positive sera fix complement.[36]

Complement-fixing islet cell antibodies are independent of the subclass composition of the islet cell antibodies and sometimes produce selective staining of portions of the islets. Some non-diabetics with polyendocrine autoimmune diseases[37] and some healthy first degree relatives of type I diabetics have islet cell antibodies in their serum and in these cases it is important to test also for complement-fixing islet cell antibodies because this subspecies seems to be more closely related to the expression of disease.[28] Fluctuations of complement-fixing islet cell antibodies have been documented in the prediabetic period.[29]

We now know that some complement-fixing islet cell antibodies are selectively directed to beta-cells,[38] but we also found sera containing complement-fixing antibodies specific for glucagon or somatostatin cells.

2.2. Pancreatic Islet Cell Surface Antibodies

The immunofluorescence test can be applied to detect autoantibodies reacting with the surface of islet cells in serum from insulin-dependent diabetic patients. These antibodies were first demonstrated with cultured insulinoma cells[39] where nearly 90 per cent of recently diagnosed young onset diabetics were positive. With ob/ob mouse islet and with dispersed rat islet cells,[40] 32 per cent of such sera were islet cell surface antibody-positive compared with 4 per cent of sera from healthy controls. Recently islet cell surface antibodies were demonstrated using human fetal cell cultures[41] and it is now clear that these antibodies are mostly restricted to type I diabetics and some of their islet cell antibody-positive first degree relatives. Islet cell surface antibodies are mostly of IgG and sometimes of IgM class.

Separate specificities exist for alpha and beta-cells.[42] Islet cell surface staining is quite often obtained in cultures with sera that give negative results on sections. This suggests an additional antigen that is expressed entirely on the plasma membrane.[43]

2.3. Pancreatic Glucagon Cell and Somatostatin Cell Antibodies

The pancreatic islets contain at least four types of endocrine cells which produce insulin, glucagon, somatostatin and human pancreatic polypeptide, respectively and can be recognized on the basis of specific antisera raised in animals[44,45] to purified hormones extracted from the pancreas or synthesized chemically.[46] Specific human antibodies to glucagon cells and somatostatin cells were identified by the double-fluorochrome immunofluorescence technique.[3] The antibodies are found in 1–4 per cent of sera from a variety of diseases and it is of interest that about half of these antibodies cross-react with the equivalent cells in the gut.

Glucagon cells and somatostatin cell antibodies are more frequent in diabetic patients, especially when there is evidence of polyendocrine autoimmunity. Yet it is difficult to establish their prevalence in the presence of islet cell antibodies unless they are of a higher titre or of different Ig class.

Autoantibodies to glucagon cells do not appear to be associated with defective alpha-cell function or with type I diabetes: of 11 positive patients studied, none had glucagon deficiency as evaluated by the arginine stimulation test.[47]

3. ANTIBODIES TO ADRENAL CORTEX

Since tuberculosis has been controlled, autoimmune destruction is the most common cause of adrenocortical insufficiency.[48] Sixty per cent of patients with newly diagnosed idiopathic adrenocortical failure have adrenal antibodies.[49] Cases with other associated autoimmune diseases such as Schmidt's or candida endocrinopathy syndromes are nearly always positive.[50] The antibodies are usually negative in patients with symptomatic Addison's disease, and in healthy controls they occur in less than 1 in 1000.[51]

These antibodies are predominantly of IgG class and the majority of positive sera react with all three layers of the adrenal cortex. Like islet cell antibodies, the titres of adrenal antibodies are usually low (1 : 1 to 1 : 64), and it has been observed that they tend to decline over the years[52] and some become negative.[53]

Adrenal antibodies are positive in 1–3 per cent of patients without overt organ-specific autoimmune diseases and in about 7 per cent of polyendocrine cases without clinical signs of adrenocortical failure. We have followed up 30 of such patients over a 3-year period. During this period, 2 patients with Graves' thyrotoxicosis developed biochemical signs of adrenocortical failure and both of them had high titres of adrenal antibodies which were also complement-fixing at the time of entry to the study.[54] Similar results were obtained by Betterle et al.[55] It is therefore advisable to perform the complement-fixation–immunofluorescence test when the outcome of an adrenal antibody-positive non-Addisonian patient is to be evaluated.

We have now developed ELISA and RIA methods for the detection of autoantibodies to adrenal cortex.[56,57] These tests are more sensitive than indirect immunofluorescence and their results are not disturbed by the presence in the serum of mitochondrial or ribosomal antibodies (*see* Chapter 29). They thus provide additional useful tools for both scientific work and clinical diagnosis of autoimmune adrenalitis.

4. ANTIBODIES TO STEROID-PRODUCING CELLS OF THE GONADS AND THE PLACENTA

Steroid-cell antibodies are a group of antibodies cross-reacting with steroid-secreting cells in the ovary, testis and placenta.[58] These antibodies can only be recognized by immunofluorescence and they have only been detected so far in sera that also react with human adrenal cortex, the only exception being sera from children with spermatic cord torsion.[59] Steroid cell antibodies have various staining patterns on corpus luteum and other ovarian structures[24] and nearly all of them stain the Leydig cells in the testis.[60]

In cases of isolated Addison's disease, steroid cell antibodies are present in 22 per cent but they are positive in over half of Addisonian patients with other endocrine disorders.[51,53] Patients with these antibodies show an increased tendency to premature menopause and ovarian failure or infertility. Steroid-cell antibodies may also appear in the serum several years before the onset of clinical symptoms and the patients in whom these antibodies are present are often polyendocrine cases. About 20 per cent of them have islet cell antibodies and are prone to sudden onset of ketotic diabetes, many have an autoimmune thyroid disease and some have overt or latent pernicious anaemia so that an extended clinical check-up is justified when the appropriate antibodies are detected.

On the other hand, steroid-cell antibodies are very rare in patients with premature menopause in the absence of Addison's disease,[61] so that separate mechanisms of ovarian failure remain to be elucidated.

5. THYROID ANTIBODIES

The detection of thyroid autoantibodies in Hashimoto's thyroiditis[62] gave the key impulse to investigate autoimmunity applied to human disease and thyroid autoimmunity is still the most extensively investigated of all organ-specific autoimmune diseases.

Although the haemagglutination test has replaced indirect immunofluorescence for routine investigation of thyroglobulin and thyroid microsomal antibodies,[63,64] the results obtained by immunofluorescence and the haemagglutination test correlate well (*Fig.* 28.3). Immunofluorescence testing may be of diagnostic value in an occasional thyrotoxic patient who may have false-negative thyroid microsomal and thyroglobulin antibody results in the haemagglutination test due to the presence of a non-specific haemagglutination inhibitor related to the blood lipoproteins.[65] When a double-antibody precipitation technique is applied, thyroid antibodies are still positive in these cases suggesting that antibody binding and immunofluorescence are not affected by these inhibitors.

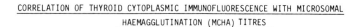

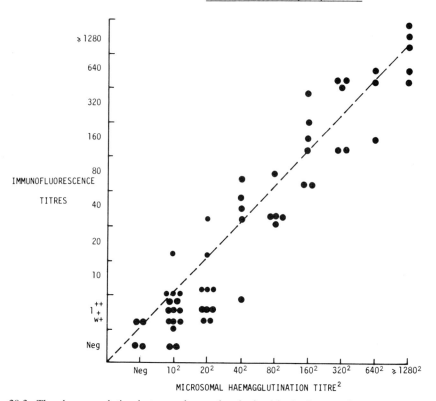

Fig. 28.3. The close correlation between the results obtained in the immunofluorescence and the haemagglutination test suggests that they detect very similar 'microsomal' antibodies. The haemagglutination test is fast and simple, it gives qualitative and semiquantitative results and can easily be combined with the determination of thyroglobulin antibodies. It is therefore the method of choice for the routine diagnosis of autoimmune thyroid disease.

Studies using indirect immunofluorescence have recently provided new insights into the immunopathological mechanisms responsible for the attack on thyroid target cells *in vivo*. Surface-reacting immunoglobulins present in the sera of autoimmune thyroid patients[66] are cytotoxic for dispersed thyroid cells in the presence of complement[67] and it was demonstrated recently that thyroid 'microsomal' autoantigens not only have a cytoplasmic localization, but they are also fully expressed on the cell surface.[9]

Thyroid microsomal antibodies are well correlated with histological[68] or biochemical[69] thyroid autoaggression, but circulating thyroid microsomal antibodies may also be present in the sera of individuals without evidence of thyroid disease.[64] We were able to demonstrate that autoreactive antigens are exclusively localized on the apical microvillar border of thyroid follicular cells facing the colloid space, where they would be relatively inaccessible to circulating antibody.[70] A preferential concentration of the cytoplasmic 'microsomal' antigens at the apical border of follicular cells was also a common observation when testing for thyroid antibodies on thyroid sections.

The polarity of antigen distribution has been shown on cultured intact thyroid follicles by Dr Hanafusa in our laboratory. He succeeded in reversing normal polarity of thyrocytes by changing the concentration of fetal calf serum from 0·5 per cent to 10 per cent in the culture medium and, by indirect immunofluorescence with human sera containing microsomal antibodies as well as by electron microscopy, he was able to demonstrate that thyroid 'microvillar' antigens were moved from the follicular to the vascular pole.[71] Spontaneous reversal of polarity was observed in six of eight glands from patients with Graves' thyrotoxicosis or toxic nodular goitre in freshly dispersed tissue. Polarity reversal may allow direct contact of the antigen with cytotoxic antibodies or sensitized immunocytes. This could initiate lesions in intact follicles and may be one of the primary events in human thyroid autoimmunity.

Another initial event in thyroid autoimmunity has recently been revealed. HLA-DR molecules are not usually expressed on resting thyroid culture cells, but incubation of the same preparations with mitogens stimulates these cells to synthesize DR molecules and express them on the cell surface.[72] This *in vitro* phenomenon is relevant to autoimmunity since we have observed that thyroid cells in frozen sections or monolayer cultures from patients with autoimmune thyroid disease express DR spontaneously.[73] This phenomenon was recently confirmed by other authors.[74]

What is the inducer of HLA-DR expression by thyroid cells *in vivo*? Interferon-gamma is known to induce DR on a variety of cells related to the immune system.[75] An extension of this work on cultures of normal thyroid cells has shown that addition of recombinant human interferon-gamma to the cultures results in a strong surface expression of DR antigens on the cells.[76] It is therefore assumed that, like macrophages and other designated cells, thyrocytes, in genetically predisposed individuals, can act as antigen-presenting cells (presenting their own surface autoantigen) under special conditions.[77] This hypothesis is also supported by the finding of local synthesis of antibodies by infiltrating lymphoid cells[78] and by the establishment of human autoreactive T-cell clones which recognize only autologous DR-positive thyrocytes.[79]

Thyroid-stimulating immunoglobulins cannot be seen by immunofluorescence owing to the small number of thyroid-stimulating hormone receptors per cell. These antibodies[80,81] as well as thyroid growth-stimulating[82] and blocking antibodies[83] are detected by other methods.

6. IMMUNOFLUORESCENCE IN MYASTHENIA GRAVIS

A screening immunofluorescence test on several tissues should be performed in every patient with acquired myasthenia gravis because a number of these patients have other associated autoimmune diseases.[84] Antibodies to striated muscle fibrils[85] can be detected by immunofluorescence in about 40 per cent of the patients, their prevalence is especially high in older myasthenic men and they are positive in nearly all thymoma cases. Antibodies to striated muscle fibrils do not occur in normal controls and may therefore be considered useful serological markers of myasthenia gravis. Yet it is now known that they do not play a pathogenetic role.[86] Antibodies to striated muscle sarcolemma are less specific for myasthenia gravis and they also occur in a proportion of patients with peri-and myocarditis or after heart surgery.

The specific autoantibodies in myasthenia gravis are directed against acetylcholine receptors where they not only block the receptors but also induce their increased degradation[87] and a complement-mediated cytotoxic effect.[88] These acetylcholine receptor antibodies can be detected by radioimmunoassay using the alpha-bungarotoxin–immunoprecipitation method of Lindstrøm[89] where about 90 per cent of patients with active generalized myasthenia gravis are positive. Recently these antibodies have also been detected by a double-fluorochrome immunofluorescence test using rat diaphragm[90] or rat tongue[91] as antigens. Yet the presence of antibodies to striated muscle fibrils in the same serum can obscure the acetylcholine receptor antibodies and the test is only positive in 21–42 per cent. The use of human tissue might increase the number of positive reactions but this has still to be tested.

7. AUTOANTIBODIES TO CELLS OF THE GUT MUCOSA

7.1. Gastric Parietal Cell Antibodies

It has been known for many years that pernicious anaemia and chronic atrophic gastritis are closely related with thyroid autoimmune diseases.[92,93] Gastric parietal cell antibodies are detected by immunofluorescence but intrinsic factor antibodies are demonstrated by radioimmunoassay. Patients with parietal cell antibodies were shown to have various degrees of atrophy of the fundal mucosa (type A gastritis), frequent achlorhydria tending to evolve towards latent or overt pernicious anaemia and a high association with endocrine autoimmune diseases.[94] Gastrin levels and gastrin cell counts were found to be increased in pernicious anaemia and subclinical fundal gastritis.[95]

7.2. Gastrin Cell Antibodies

In antral (type B) gastritis parietal cell antibodies are usually absent; there is no association with endocrine autoimmune diseases in contrast to type A gastritis, and the gastrin levels are greatly diminished.[96] About 8 per cent of these cases have antibodies to gastrin cells as characterized by the double immunofluorescence technique.[4]

For the detection of gastrin cell antibodies it is crucial to select a piece of human antrum quite devoid of parietal cells yet containing a sufficient number of gastrin cells. The patients' sera are first tested by conventional immunofluorescence on unfixed tissue and to identify the reacting cells the positive sera are retested by the four-layer double-fluorochrome immunofluorescence test (*Plates* 28, 29) applying a specific antiserum to gastrin. Gastrin cell antibodies are of IgG class, they are of low titre, they fix complement and they tend to persist over the years. Gastrin cell antibodies are rarely found in patients with pernicious anaemia or simple fundal gastritis and are not found in polyendocrine autoimmune diseases. Gastrin cell antibodies were recently confirmed using the avidin–biotin complex antibody method.[97]

Patients with gastrin cell antibodies showed a diminished response of the serum gastrin to standard protein meals, suggesting that the antibodies reflect a loss of gastrin cells in the autoimmune process. The gastrin cell antibodies may therefore be considered as markers for selective destruction of gastrin cells in cases of primary autoimmune antral gastritis.

7.3. Autoantibodies in Coeliac Disease

Our first suggestion that autoantibodies to gastric inhibitory polypeptide- and secretin-producing cells of the gut might exist was based on the finding of Besterman et al.[98] who noticed a marked reduction in the release of secretin after duodenal acidification and a diminished output of gastric inhibitory polypeptide after a standard protein meal in a proportion of patients with coeliac disease. Duodenal mucosa was chosen to test sera of patients with proven coeliac disease because this tissue is particularly rich in endocrine cells.[99] Of 173 sera tested, 16 (18 per cent) gave positive cytoplasmic immunofluorescence on single cells whereas 73 polyendocrine sera and the sera of 30 healthy blood donors were negative. Four of the positive sera reacted with gastric inhibitory polypeptide cells, 1 with secretin cells and 21 with both gastric inhibitory polypeptide and secretin cells as tested by the four-layer double-fluorochrome test.[5] Extensive metabolic studies and a better understanding of the exact physiology of the gastrointestinal hormones are required to establish the full significance of these antibodies.[100]

7.4. Antibodies to Single Endocrine Cells of the Gut Mucosa in Diabetes Mellitus

Some patients with type II (non-insulin dependent) diabetes have a poor gastric inhibitory polypeptide response to protein meals and, when the sera of such patients were investigated for antibodies to gut mucosa, 31 out of 154 sera tested had single cell antibodies. With the double-fluorochrome immunofluorescence test, 24 were shown to be directed against gastric inhibitory polypeptide cells, and only 3 sera had antibodies to other endocrine cells of the gastrointestinal tract.[6] A proportion of type I (insulin-dependent) diabetic patients have antibodies preferentially directed against gastric inhibitory polypeptide and secretin cells.

Gastric inhibitory polypeptide is known to affect the insulin secretion that follows the ingestion of glucose[101] and a defective gastric inhibitory polypeptide secretion might influence the mechanism of insulin release contributing to the pathogenesis of type II diabetes in cases where the pancreatic beta-cells still contain sufficient amounts of insulin.

7.5. Autoantibodies to Gut Epithelium in Protracted Diarrhoea

Recently, new antibodies to gut epithelium have been detected independently by Unsworth et al.[102] and our group[103] in three infants with severe diarrhoea. The two children observed by ourselves had specific complement-fixing autoantibodies reacting by immunofluorescence with the cytoplasm of human duodenum, jejunum and colon mucosal epithelium. One of our cases had severe unresponsive enteropathy with persistent gut epithelial antibodies present until death. The other had transient mucosal damage, and clinical and histological improvements were associated with antibody disappearance. An extension of these studies has revealed that more than 50 per cent of children affected by idiopathic protracted diarrhoea have similar enterocyte antibodies.[104] In cases of unexplained protracted diarrhoea mucosal damage is probably the direct consequence of an autoimmune process which can only be revealed by the antibody test.

8. AUTOANTIBODIES TO PITUITARY CELLS

8.1. Prolactin Cell Antibodies

These antibodies were first detected in polyendocrine autoimmune disease and they were finally characterized applying the double-fluorochrome test using specific anti-pituitary hormone antisera.[1] Prolactin cell antibodies have never been found in severe panhypopituitarism, but they are now known to occur in a few patients with partial pituitary defects, some of whom presented with amenorrhoea or infertility. The prolactin responses to thyrotrophin-releasing hormone in patients who possess these antibodies are either diminished or raised confirming that an autoimmune process does not always lead to destructive mechanisms.

8.2. Growth-hormone Cell Antibodies

These antibodies were first identified in a young girl with arrested growth whose mother had autoimmune adrenalitis and thyroiditis.[2] In this patient it proved possible to demonstrate a partial defect of human growth hormone secretion in response to stimulation tests. When the sera of a selected group of 220 children with growth defects were tested, 10 per cent reacted with isolated cells in the anterior pituitary.[24] The majority were directed against prolactin cells, three sera reacted with human growth hormone cells and a few remain to be identified.

8.3. ACTH Cell Antibodies

There are special difficulties in detecting ACTH cell antibodies because these cells appear to have Fc receptors on their surface and in their cytoplasm so that all normal immunoglobulins react with these cells by attachment of their Fc part giving a uniform cytoplasmic fluorescence of the oval-shaped ACTH cells.[105] This non-specific staining may be prevented by pre-incubation of the sections with a negative normal human serum.

8.4. Multiple Cell Immunofluorescence Pattern

There have been a number of reports of children with unusually tall stature presenting with type I diabetes. Boys of prepubertal age are tall at onset of diabetes and in both sexes skeletal maturity is found to be significantly advanced, suggesting a prediabetic metabolic or hormonal abnormality.[106] Thus pituitary and other endocrine defects may play a fundamental role in the early pathogenesis of type I diabetes. We have looked for pituitary antibodies in recently diagnosed diabetics and in their genetically predisposed relatives during the 'latency period'. Of the islet cell antibody-positive diabetics, 17 per cent turned out to be positive for antibodies mostly directed to more than one cell type of the pituitary. Surprisingly, these antibodies were also positive in 12 (36·4 per cent) of 33 selected islet cell antibody-positive sera from their healthy genetically predisposed relatives of whom 7 became diabetic during a 3-year follow-up period.[107] The serum of only 1 out of 48 patients with long-standing insulin-dependent diabetes was positive and pituitary antibodies were not found in 48 normal controls.

These results obtained by immunofluorescence studies suggest that in some cases temporary autoimmune processes can involve the anterior pituitary and the endocrine pancreas simultaneously. These antibodies might either represent an expression of a destructive process leading to hypophysitis or more likely they are markers for a stimulatory process at the pituitary level.[108]

9. AUTOANTIBODIES TO VASOPRESSIN-PRODUCING HYPOTHALAMIC CELLS IN IDIOPATHIC DIABETES INSIPIDUS

Diabetes insipidus is characterized by polyurea and polydipsia, and in the central form symptoms are due to a lack of arginine vasopressin (AVP) hormone. Central diabetes insipidus may be caused by a variety of underlying diseases which affect the AVP-producing cells in the supra-optic and paraventricular nuclei of the hypothalamus.[109] However, clinicians may face the diagnostic dilemma that the aetiology of spontaneous diabetes insipidus remains obscure in 30–40 per cent of the cases.[110] We can now detect autoantibodies to AVP-producing cells of the hypothalamus in 30 per cent of those cases of central diabetes insipidus which were previously considered 'idiopathic'.[8] The antibodies are demonstrated by immunofluorescence on unfixed cryostat sections of human fetal tissue obtained from therapeutic hysterotomies. Adult postmortem hypothalamus obtained several hours after death gives poor results since the cytoplasmic autoantigens accounting for the specific AVP cell staining are rapidly inactivated. Older adult donors are also unsuitable because the natural accumulation of lipofuscin granules in secretory hypothalamic cells interferes with the reading of the specific cytoplasmic immunofluorescence. Fresh baboon hypothalamus obtained from young animals gives good cross-reactivity with the human antibodies[111] and may therefore be used for clinical tests.

It was striking that in one-third of the 'idiopathic' cases, recognized autoimmune diseases coexisted with diabetes insipidus.[112] Together with the detection of AVP cell antibodies, these findings provide strong evidence for the existence of an autoimmune form of central diabetes insipidus which can now be separated from the unexplained idiopathic cases. This assumption is supported by postmortem examination of the hypothalamus in idiopathic central diabetes insipidus where gliosis of the supra-optic and paraventricular nuclei has been described in a long-standing case.[109] This feature parallels the end-stage fibrosis of the adrenals in autoimmune Addison's disease or of the thyroid in atrophic autoimmune thyroiditis.

AVP cell antibodies are not found in patients without diabetes insipidus or in normal controls and they are usually negative in symptomatic diabetes insipidus. However, about half the cases of diabetes insipidus caused by histiocytosis X are positive,[112] and the induction of hypothalamic autoimmunity may be due to the expression of HLA-class II histocompatibility molecules on infiltrating histiocytosis-X cells. The detection of AVP cell antibodies in cases of histiocytosis X may prove to be an early marker of hypothalamic infiltration.

10. FUTURE PROSPECTS

Autoantibodies have been found directed against the cytoplasm of nearly all endocrine glands, scattered endocrine cells and a number of specialized organs. Through the introduction of the double-fluorochrome immunofluorescence test and the development of specific hormone antisera it has become possible to characterize the nature of single cells reacting with the antibodies even when the form and shape of these cells is no different from the surrounding ones. Extensive metabolic and histopathological studies will be necessary to establish the clinical significance of the newly detected autoantibodies to pituitary, hypothalamic and gut cells.

Every endocrine cell is connected with the immune system through an enormous number of surface receptors, and the presence of microsomal antigens on the cell surface can now explain the pathogenetic significance of cytoplasmic antibodies. To study this directly, it will be necessary to investigate the interactions of immunoglobulins with sparse components of the cell surface which do not represent enough binding sites to show up in immunofluorescence. Far more sensitive methods are required for the detection of receptor antibodies. When the purified or synthesized antigen is available, an ELISA or RIA method can be employed. However most antibody or membrane fractions of cells are integrated in the lipid bilayer and have not been purified. To study these structures the receptors must be solubilized in detergents and then studied as soluble proteins.

The receptor protein can be labelled biosynthetically by growing the cells in the presence of amino acids or sugars that contain radioactive or heavy isotopes. The hormone can be tagged with fluorescent, radioactive, or electron-dense moieties like iron or gold, in conjunction with ultrasensitive light detectors or autoradiography and electron microscopy. These hormone preparations can be used to track the hormone or the antibody after it makes contact with the receptor on the cell.[113] For the detailed study of receptors it is also essential to apply monoclonal antibodies directed against components of the receptors.[114] The exploitation of these new methods will lead to much deeper biological insights than have been envisaged so far.

In some antibody test systems, such as parathyroid antibodies, we have not yet reached the sensitivity we would wish to achieve. In other systems, such as islet cell and thyroid microsomal antibodies, a further increase of the sensitivity of existing tests would not increase discrimination between diseased and healthy individuals who may be seropositive as well. However, autoantibodies are usually polyclonal and it may be desirable to find subspecificities that may be more closely correlated with the disease process. One helpful parameter is the complement-fixing ability of IgG islet cell or adrenal cortex antibodies, but there may be a range of additional markers that would allow better prediction of the time of clinical onset of disease which, in turn, might allow early and successful treatment of autoimmune diseases.[115]

Acknowledgements

We wish to thank Professor Deborah Doniach for helpful suggestions and advice and Professors I. M. Roitt and E. F. Pfeiffer for their constant support. Miss Queenie Jayawardena kindly typed the manuscript. Work in the Department of Immunology at the Middlesex Hospital is supported by the Medical Research Council, the Wellcome Trust Foundation, the British Diabetic Association, the Juvenile Diabetes Foundation (USA), the Joint Research Board of St Bartholomew's Hospital and Novo Research Laboratories (Copenhagen). W. A. S. is supported by a grant from the Deutsche Forschungsgemeinschaft, Sche 225/2-1.

REFERENCES

1. Bottazzo G. F., Pouplard A., Florin-Christensen A. and Doniach D. Autoantibodies to prolactin-secreting cells of human pituitary. *Lancet* 1975, **ii**, 97–101.
2. Bottazzo G. F., McIntosh C. W., Stanford W. and Preece M. Growth-hormone cell antibodies and partial growth-hormone deficiency in a girl with Turner's syndrome. *Clin. Endocrinol.* 1980, **12**, 1–9.
3. Bottazzo G. F. and Lendrum R. Separate autoantibodies to human pancreatic glucagon and somatostatin cells. *Lancet* 1976, **ii**, 873–876.
4. Vandelli C., Bottanzzo G. F., Doniach D. and Francheschi F. Autoantibodies to gastrin producing cells in antral (type B) chronic gastritis. *N. Engl. J. Med.* 1979, **300**, 1406–1410.
5. Mirakian R., Bottazzo G. F. and Doniach D. Autoantibodies to duodenal gastric-inhibitory peptide (GIP) cells and to secretin (S) cells in patients with coeliac disease, tropical sprue and maturity onset diabetes. *Clin. Exp. Immunol.* 1980, **41**, 33–42.
6. Mirakian R., Richardson C. A., Bottazzo G. F. and Doniach D. Humoral autoimmunity to gut-related endocrine cells. *Clin. Immunol. Newsletter* 1981, **2**, 161–167.
7. Scherbaum W. A. and Bottazzo G. F. Detection of autoantibodies to calcitonin-secreting cells of the thyroid gland. Which is the corresponding disease? *Acta Endocrinol. Suppl.* 1983, **103**, 256.
8. Scherbaum W. A. and Bottazzo G. F. Autoantibodies to vasopressin cells in idiopathic diabetes insipidus. Evidence for an autoimmune variant. *Lancet* 1983, **i**, 897–901.
9. Khoury E. L., Hammond L., Bottazzo G. F. and Doniach D. Presence of the organ-specific 'microsomal' antigen on the surface of human thyroid cells in culture: its involvement in complement-mediated cytotoxicity. *Clin. Exp. Immunol.* 1981, **45**, 316–328.
10. Khoury E. L., Hammond L., Bottazzo G. F. and Doniach D. Surface-reactive antibodies to human adrenal cells in Addison's disease. *Clin. Exp. Immunol.* 1981, **45**, 48–58.
11. Masala C., Smurra G., Di Prima M. A., Amendolea M. A., Celestino D. and Salsano F. Gastric parietal cell antibodies: demonstration by immunofluorescence of their reactivity with the surface of the gastric parietal cells. *Clin. Exp. Immunol.* 1980, **41**, 271–280.
12. Papadopoulos G. K. and Lernmark A. The spectrum of islet cell antibodies. In: Davies T. F., ed., *Autoimmune Endocrine Disease.* New York, John Wiley, 1983: 167–180.
13. De Azipurva H. J., Toh B. H. and Ungar B. Parietal cell surface reactive autoantibody in pernicious anaemia demonstrated by indirect membrane immunofluorescence. *Clin. Exp. Immunol.* 1983, **52**, 341–349.
14. Betterle C., Mirakian R., Doniach D., Bottazzo G. F., Riley W. and MacLaren N. K. Antibodies to melanocytes in vitiligo. *Lancet* 1984, **i**, 159.

15. Bottazzo G. F., Pujol-Borrell R. and Gale E. Etiology of diabetes: the role of autoimmune mechanisms. In: Alberti K. G. M. M. and Krall L. P., eds., *The Diabetes Annual/1.* Amsterdam, Elsevier, 1985: 16–52.

16. Nerup J. Addison's disease — clinical studies. A report of 108 cases. *Acta Endocrinol.* 1974, **76**, 127–141.

17. Solomon N., Carpenter C. C. J., Bennett J. L. and McGehee Harvey A. Schmidt's syndrome (thyroid and adrenal insufficiency) and coexistent diabetes mellitus. *Diabetes* 1965, **14**, 300–304.

18. Blizzard R. M., Chee D. and Davies W. The incidence of parathyroid and other antibodies in the sera of patients with idiopathic hypoparathyroidism. *Clin. Exp. Immunol.* 1966, **1**, 119–128.

19. Botazzo G. F., Florin-Christensen A. and Doniach D. Islet-cell antibodies in diabetes mellitus with autoimmune polyendocrine deficiencies. *Lancet* 1974, **ii**, 1279–1283.

20. Lendrum R., Walker J. G. and Gamble D. R. Islet-cell antibodies in juvenile diabetes mellitus of recent onset. *Lancet* 1975, **i**, 880–882.

21. Doniach D., and Bottazzo G. F. Polyendocrine autoimmunity. In: Franklin E. C., ed., *Clinical Immunology Update.* New York, Elsevier, 1981: 95–121.

22. Groop L., Bottazzo G. F. and Doniach D. Islet cell antibodies identify pseudo-type II diabetes in patients aged 35–75 years at diagnosis. *Diabetes* in press.

23. Gleichman H., Zorcher B., Greulich B., Gries F. A., Henrichs H. R., Bertrams J. and Kolb H. Correlation of islet cell antibodies and HLA-DR phenotypes with diabetes mellitus in adults. *Diabetologia,* 1984, **27**, 90–92.

24. Doniach D., Bottazzo G. F. and Drexhage H. A. The autoimmune endocrinopathies. In: Lachmann P. A. and Peters D. K., eds., *Clinical Aspects of Immunology,* Vol. 2. Oxford, Blackwell Scientific Publications, 1982: 903–937.

25. Orci L. Macro- and micro-domains in the endocrine pancreas. *Diabetes* 1982, **31**, 538–565.

26. Bottazzo G. F. and Doniach D. Islet-cell antibodies (ICA) in diabetes mellitus: Evidence of an autoantigen common to all cells in the islet of Langerhans. *Ric. Clin. Lab.* 1978, **8**, 29–38.

27. Bottazzo G. F., Pujol-Borrell R. and Doniach D. Humoral and cellular immunity in diabetes mellitus. *Clin. Immunol. Allergy* 1981, **1**, 139–159.

28. Gorsuch A. N., Spencer K. M., Lister J., McNally J. M., Dean B. M., Bottazzo G. F. and Cudworth A. G. The natural history of Type I (insulin-dependent) diabetes mellitus: evidence for a long pre-diabetic period. *Lancet* 1981, **ii**, 1363–1365.

29. Spencer K. M., Tarn A., Dean B. M., Lister J. and Bottazzo G. F. Family studies in Type I (insulin-dependent) diabetes: evidence of fluctuating islet-cell autoimmunity in unaffected relatives. *Lancet* 1984, **i**, 764–766.

30. Neufeld M., MacLaren N. K., Riley W. J., Lezotte D., McLaughlin J. V., Silverstein J. and Rosenbloom A. L. Islet-cell and other organ-specific antibodies in US Caucasians and Blacks with insulin-dependent diabetes mellitus. *Diabetes* 1980, **29**, 589–592.

31. Knowler W. C., Bennett P. H., Bottazzo G. F. and Doniach D. Islet-cell antibodies and diabetes mellitus in Pima Indians. *Diabetologia* 1979, **17**, 161–164.

32. Madsbad S., Bottazzo G. F., Cudworth A. G., Dean B. M., Faber O. and Binder C. Islet-cell antibodies and beta-cell function in insulin dependent diabetics. *Diabetologia* 1980, **18**, 45–49.

33. Bodansky H. J., Wolf E., Cudworth A. G., Dean B. M., Nineham L. J., Bottazzo G. F., Matthews J. A., Kurtz A. B. and Kohner E. M. Genetic and immunological factors in microvascular disease in Type I (insulin-dependent) diabetes. *Diabetes* 1982, **31**, 70–74.

34. Srikanta S., Ganda O. P., Eisenbarth G. S. and Soeldner J. S. Islet-cell antibodies and beta-cell function in monozygotic triplets and twins initially discordant for Type I diabetes mellitus. *N. Engl. J. Med.* 1983, **308**, 322–325.

35. Barnett A. H., Eff C., Leslie R. D. G. and Pyke D. A. Diabetes in identical twins. *Diabetologia* 1981, **20**, 87–93.

36. Bottazzo G. F., Dean B. M., Gorsuch A. N., Cudworth A. G. and Doniach D. Complement-fixing islet-cell antibodies in Type I diabetes: possible monitors of active beta cell damage. *Lancet* 1980, **i**, 668–672.

37. Betterle C., Zanette P., Pedini B., Presotto F., Rapp L. B., Monciotti C. M. and Rigon F. Clinical and subclinical organ-specific autoimmune manifestations in Type I (insulin-dependent) diabetic patients and their first degree relatives. *Diabetologia* 1984, **26**, 431–436.

38. Bottazzo G. F., Mirakian R., Dean B. M., McNally J. M. and Doniach D. How immunology helps to define heterogeneity in diabetes mellitus. In: Tattersall R. B. and Köbberling J. K., eds., *Genetics of Diabetes Mellitus.* London, Academic Press, 1982, 79–90.

39. MacLaren N. K., Huang S. W. and Fogh J. Antibody to cultured insulinoma cells in insulin-dependent diabetes. *Lancet* 1975, **i**, 997–1000.

40. Lernmark A., Freedman Z. R., Hofmann C., Rubenstein A. H., Steiner D. F., Jackson R. L., Winter R. J. and Traisman H. S. Islet-cell surface antibodies in juvenile diabetes mellitus. *N. Engl. J. Med* 1978, **299**, 375–380.

41. Pujol-Borrell R., Khoury E. L. and Bottazzo G. F. Islet cell surface antibodies in Type I (insulin-dependent) diabetes mellitus: use of human fetal pancreas cultures as substrate. *Diabetologia* 1982, **22**, 89–95.

42. Van De Winkel M., Smets G., Gepts W. and Pipeleers D. Islet cell surface antibodies from insulin-dependent diabetics bind specifically to pancreatic B cells. *J. Clin. Invest.* 1982, **70**, 41–49.

43. Doniach D., Bottazzo G. F. and Cudworth A. G. Etiology of Type I diabetes mellitus: heterogeneity and immunological events leading to clinical onset. *Annu. Rev. Med.* 1983, **34**, 13–20.

44. Larsson L.-I. Gastrointestinal cells producing endocrine, neurocrine and paracrine messengers. In: Creutzfeld W. ed., *Clinics in Gastroenterology*. London, Saunders, 1980: 485–516.

45. Ravazzola M. and Orci L. Glucagon and glicentin immunoreactivity are topologically segregated in the alpha granule of the human pancreatic A cell. *Nature* 1980, **284**, 66–67.

46. Polak J. M., Adrian T. E., Bryant M. G., Bloom S. R., Heitz P. U. and Pearse A. G. E. Pancreatic polypeptide in insulinomas, gastrinomas, vipomas and glucagonomas. *Lancet* 1976, **i**, 328–330.

47. Winter W. E., MacLaren N. K., Riley W. J., Unger R. H., Neufeld M. and Ozand P. T. Pancreatic alpha cell autoantibodies and glucagon response to arginine. *Diabetes* 1984, **33**, 435–437.

48. Stuart-Mason A., Maede T. W., Lee J. A.-H. and Morris J. N. Epidemiological and clinical picture of Addison's disease. *Lancet* 1968, **ii**, 744–747.

49. Scherbaum W. A. and Berg P. A. Bedeutung von Antikörpern in der Diagnostik endokrinologischer Erkrankungen. *Dtsch. Med. Wochenschr.* 1981, **106**, 308–313.

50. Doniach D., Cudworth A. G., Khoury E. L. and Bottazzo G. F. Autoimmunity and the HLA-system in endocrine diseases. In: *Recent Advances in Endocrinology and Metabolism*, 1982, **2**, 99–132.

51. Irvine W. J. Autoimmunity against steroid producing organs. In: Miescher P. A. et. al. ed., *Menarini Series on Immunopathology*. Basel, Schwabe, 1978: 35–49.

52. Wuepper K. D., Wegienka L. C. and Fudenberg H. H. Immunologic aspects of adrenocortical insufficiency. *Am. J. Med.* 1969, **46**, 206–216.

53. Sotsiou F., Bottazzo G. F. and Doniach D. Immunofluorescence studies on autoantibodies to steroid-producing cells, and to germline cells in endocrine disease and infertility. *Clin. Exp. Immunol.* 1980, **39**, 97–111.

54. Scherbaum W. A. and Berg P. A. Development of adrenocortical failure in non-Addisonian patients with antibodies to adrenal cortex. A clinical follow-up study. *Clin. Endocrinol.* 1982, **16**, 345–352.

55. Betterle C., Zanette F., Zanchetta R., Pedini B., Trevisan A., Mantero F. and Rigon F. Complement-fixing adrenal autoantibodies as a marker for predicting onset of idiopathic Addison's disease. *Lancet* 1983, **i**, 1238–1241.

56. Stechemesser E., Scherbaum W. A., Grossmann T. and Berg P. A. An ELISA method for the detection of autoantibodies to adrenal cortex. *J. Immunol. Methods* 1985, **80**, 67–76.

57. Kosowicz J., Gryczynska M. and Bottazzo G. F. Radioimmunoassay for the detection of adrenal autoantibodies. *Clin. Exp. Immunol.* in press.

58. Anderson J. R., Goudie R. B., Gray K. G. and Stuart-Smith D. A. Immunological features of idiopathic Addison's disease: an antibody to cells producing steroid hormones. *Clin. Exp. Immunol.* 1968, **3**, 107–117.

59. Zanchetta R., Mastrogiacomo I., Graziotti P., Foresta C. and Betterle C. Autoantibodies against Leydig cells in patients after spermatic cord torsion. *Clin. Exp. Immunol.* 1984, **55**, 49–57.

60. Kamp P., Platz P. and Nerup J. Steroid-cell antibody in endocrine diseases. *Acta Endocrinol.* 1974, **75**, 729–740.

61. Scherbaum W. A., Bottazzo G. F., Gehrig W. and Schindler A. E. Hypogonadism and autoimmunity in females without Addison's disease. *Acta Endocrinol. Suppl.* 253, 1983, **102**, 143.

62. Roitt I. M., Doniach D., Campbell P. N. and Hudson R. V. Autoantibodies in Hashimoto's disease (lymphadenoid goitre). *Lancet* 1956, **ii**, 820–822.

63. Doniach D. and Bottazzo G. F. Thyroid autoimmunity In: Miescher P. A. et. al., ed., *Menarini*

Series on Immunopathology. Basel, Schwabe, 1978: 22–33.
64. Scherbaum W. A., Stöckle G., Wichmann J. and Berg P. A. Immunological and clinical characterization of patients with untreated euthyroid and hypothyroid autoimmune thyroiditis. *Acta Endocrinol.* 1982, **100**, 373–381.
65. Wilkin T.J., Swanson Beck J., Hayes P. C. Potts R. C. and Young R. J. A passive haemagglutination (TRC) inhibitor in thyrotoxic serum. *Clin. Endocrinol.* 1979, **10**, 507–514.
66. Fagraeus A. and Jonsson J. Distribution of organ antigens over the surface of thyroid cells as examined by the immunofluorescence test. *Immunology* 1970, **18**, 413–416.
67. Pulvertaft R. J. V., Doniach D. and Roitt I. M. The cytotoxic factor in Hashimoto's disease and its incidence in other thyroid diseases. *Br. J. Exp. Pathol.* 1961, **42**, 496–503.
68. Yoshida H., Amino N., Yagawa K., Uemura D., Satoh M., Miyai K. and Kumahara Y. Association of serum antithyroid antibodies with lymphocytic infiltration of the thyroid gland: studies of seventy autopsied cases. *J. Clin. Endocrinol. Metab.* 1978, **46**, 859–862.
69. Hawkins B. R., Cheah P. S., Dawkins R. L., Whittingham S., Burger H. G., Patel Y., Mackay I. R. and Welborn T. A. Diagnostic significance of thyroid microsomal antibodies in randomly selected population. *Lancet* 1980, **ii**, 1057–1059.
70. Khoury E. L., Bottazzo G. F. and Roitt I. M. The thyroid 'microsomal' antibody revisited. Its paradoxical binding *in vivo* to the apical surface of the follicular epithelium. *J. Exp. Med.* 1984, **159**, 577–591.
71. Hanafusa T., Pujol-Borrell R., Chiovato L., Doniach D. and Bottazzo G. F. *In vitro* and *in vivo* reversal of thyroid epithelial polarity: its relevance for autoimmune thyroid disease. *Clin. Exp. Immunol.* 1984, **57**, 639–646.
72. Pujol-Borrell R., Hanafusa T., Chiovato L. and Bottazzo G. F. Lectin-induced expression of DR antigen on human cultured follicular thyroid cells. *Nature* 1983, **303**, 71–73.
73. Hanafusa T., Pujol-Borrell R., Chiovato L., Russell R. C. G., Doniach D. and Bottazzo G. F. Aberrant expression of HLA-DR antigen on thyrocytes in Graves' disease: relevance for autoimmunity. *Lancet*, 1983, **ii**, 1111–1115.
74. Jansson R., Karlsson A. and Forsum U. Intrathyroidal HLA-DR expression and T lymphocyte phenotypes in Graves' thyrotoxicosis, Hashimoto's thyroiditis and nodular colloid goitre. *Clin. Exp. Immunol.* 1984, **58**, 264–272.
75. Unanue E. R. Antigen-presenting function of the macrophage. *Annu. Rev. Immunol.* 1984, **2**, 395–428.
76. Todd I., Pujol-Borrell R., Hammond L., Bottazzo G. F. and Feldmann M. Interferon-gamma induces HLA-DR expression by thyroid epithelium. *Clin. Exp. Immunol.* 1985, **61**, 265–273.
77. Bottazzo G. F., Pujol-Borrell R., Hanafusa T. and Feldmann M. Hypothesis: role of aberrant HLA-DR expression and antigen presentation in the induction of endocrine autoimmunity. *Lancet* 1983, **ii**, 1115–1119.
78. MacLachlan S. M., Dickinson A. M., Malcolm A., Farndon J. R., Young E., Proctor S. J. and Rees Smith B. Thyroid autoantibody synthesis by cultures of thyroid and peripheral blood lymphocytes. I. Lymphocyte markers and response to pokeweed mitogen. *Clin. Exp. Immunol.* 1983, **52**, 45–53.
79. Londei M., Bottazzo G. F. and Feldmann M. Human T cell clones from thyroid autoimmune thyroid glands: specific recognition of autologous thyroid cells. *Science* 1985, **228**, 85–87.
80. McKenzie J. M., Zakarija M. and Sato A. Humoral immunity in Graves' disease. *Clin. Endocrinol. Metabol.* 1978, **7**, 31–45.
81. Hall R. and Rees Smith B. Role of thyroid autoantibodies in thyrotoxicosis. In: Franklin C. P. ed., *Clinical Immunology Update.* New York, Elsevier, 1979: 291–306.
82. Drexhage H. A., Bottazzo G. F., Bitensky L., Chayen J. and Doniach D. Evidence for thyroid-growth-stimulating immunoglobulins (TGI) in some goitrous thyroid diseases. *Lancet* 1980, **ii**, 287–292.
83. Drexhage H. A., Bottazzo G. F., Bitensky L., Chayen J. and Doniach D. Thyroid growth-blocking antibodies in primary myxoedema. *Nature* 1981, **289**, 594–596.
84. Oosterhuis H. J. G. H. Studies in myasthenia gravis. Part 1. A clinical study of 180 patients. *J. Neurol. Sci.* 1964, **1**, 512–546.
85. Strauss A. J. L., Seegal B. C., Hsu K. C., Burkholder P. M., Nastuk W. L. and Osserman K. E. Immunofluorescence demonstration of a muscle-binding complement-fixing serum globulin fraction in myasthenia gravis. *Proc. Soc. Exp. Biol. Med.* 1960, **105**, 184–191.
86. Vincent A. Immunology of myasthenia gravis: recent developments. *Clin. Immunol. Allergy* 1981, **1**, 161–179.
87. Drachman D. B. Myasthenia gravis. *N. Engl. J. Med.* 1978, **298**, 136–142.

88. Sahashi K., Engel A. G., Lambert E. H. and Howard F. M. Ultrastructural localization of the terminal and lytic 9th complement component (C_9) at the motor endplate in myasthenia gravis. *J. Neurol. Exp. Neuropathol.* 1980, **39**, 160–172.

89. Lindstrøm J. An assay for antibodies to human acetylcholine receptor in serum from patients with myasthenia gravis. *Clin. Immunol. Immunopathol.* 1977, **7**, 36–43.

90. Sondag-Tschroots J. R. J. M., Schulz-Raateland R. C. M., Van Walbeek H. K. and Feltkamp T. E. W. Antibodies to motor endplates demonstrated with the immunofluorescence technique. *Clin. Exp. Immunol.* 1979, **37**, 323–327.

91. Storch W. and Trautmann B. Doppelimmunofluoreszenzoptischer Nachweis von Antikörpern gegen motorische Endplatten des Skeletmuskels bei Myasthenia gravis. *Acta Histochem.* 1981, **69**, 18–22.

92. Doniach D. and Roitt I. M. An evaluation of gastric and thyroid autoimmunity in relation to hematologic disorders. *Semin. Hematol.* 1964, **1**, 313–343.

93. Doniach D., Roitt I. M. and Taylor K. B. Autoimmunity in pernicious anemia and thyroiditis: a family study. *Ann. N.Y. Acad. Sci.* 1965, **124**, 605–625.

94. Chanarin I. Gastric autoimmunity. In: Miescher P. A. et al., eds., *Menarini Series on Immunopathology*. Basel, Schwabe, 1978: 79–83.

95. Stockbrügger R., Larsson L.-I., Lundqvist G. et al. Antral gastrin cells and serum gastrin in achlorhydria. *Scand. J. Gastroenterol.* 1977, **12**, 209–213.

96. Bordi C., Gabrielli M. and Missale G. Pathological changes of endocrine cells in chronic atrophic gastritis: an ultrastructural study on peroral gastric biopsy specimens. *Arch. Pathol. Lab. Med.* 1978, **102**, 129–135.

97. Uibo R. M. and Krohn K. J. E. Demonstration of gastrin autoantibodies in antral gastritis with avidin–biotin-complex antibody technique. *Clin. Exp. Immunol.* 1984, **58**, 341–347.

98. Besterman H. S., Bloom S. R., Sarson D. L., Blackburn A. M., Johnston D. J., Patel H. R., Stewart J. S., Modigliani R., Guerin S. and Mallinson C. N. Gut hormone profile in coeliac disease. *Lancet* 1978, **i**, 785–788.

99. Bryant M. G. and Bloom S. R. Distribution of the gut hormones in the primate intestinal tract. *Gut* 1979, **20**, 653–659.

100. Jones H. W., Lendrum R., Marks J. M., Mirakian R., Bottazzo G. F., Sarson D. L. and Bloom S. R. Autoantibodies to gut hormone secreting cells as markers of peptide deficiency. *Gut* 1983, **24**, 427–431.

101. Brown J. C. and Otte S. C. GIP and the entero-insular axis. *Clin. Endocrinol. Metab.* 1979, **8**, 365–377.

102. Unsworth J., Hutchins P., Mitchell J., Phillips A., Mindocha P., Holborow J. and Walker-Smith J. Flat small intestinal mucosa and autoantibodies against the gut epithelium. *J. Pediatr. Gastroenterol. Nutrit.* 1982, **1**, 503–513.

103. Savage M. O., Mirakian R., Wozniak E. R., Jenkins H. R., Malone M., Phillips A. D., Milla J. P., Bottazzo G. F. and Harris J. T. Specific autoantibodies to gut epithelium in two infants with severe protracted diarrhea. *J. Pediatr. Gastroenterol. Nutrit.* 1985, **4**, 187–195.

104. Mirakian R., Richardson A., Milla P. J., Unsworth J., Walker-Smith J. A., Savage M. O. and Bottazzo G. F. Protracted diarrhoea of infancy: evidence in support of an autoimmune variant. *Lancet* Submitted for publication.

105. Pouplard A., Bottazzo G. F., Doniach D. and Roitt I. M. Binding of human immunoglobulins to pituitary ACTH cells. *Nature* 1976, **261**, 142–144.

106. Edelsten A. D., Hughes I. A., Oakes S., Gordon I. R. S. and Savage D. C. L. Height and skeletal maturity in children with newly-diagnosed juvenile-onset diabetes. *Arch. Dis. Child.* 1981, **56**, 40–44.

107. Mirakian R., Cudworth A. G., Bottazzo G. F., Richardson C. A. and Doniach D. Autoimmunity to anterior pituitary cells and the pathogenesis of Type I (insulin-dependent) diabetes mellitus. *Lancet* 1982, **i**, 755–759.

108. Bottazzo G. F. β-cell damage in diabetic insulitis: are we approaching the solution? *Diabetologia* 1984, **26**, 241–249.

109. Blotner H. Primary or idiopathic diabetes insipidus: a systemic disease. *Metabolism* 1958, **7**, 191–200.

110. Moses A. M. Clinical and laboratory observations in the adult with diabetes insipidus and related syndromes. In: Czernichow P. and Robinson A. G., eds., *Diabetes Insipidus in Man. Frontiers in Hormone Research*, Vol. 13. Karger, Basel, 1985: 156–175.

111. Scherbaum W. A., Bottazzo G. F., Czernichow P., Wass J. A. H. and Doniach D. Role of autoimmunity in central diabetes insipidus. In: Czernichow P. and Robinson A. G., eds.,

Diabetes Insipidus in Man. Frontiers in Hormone Research, Vol. 13. Karger, Basel, 1985: 232–239.

112. Scherbaum W. A., Wass J. A. H., Besser M., Bottazzo G. F. and Doniach D. Autoimmune central diabetes insipidus: its association with other endocrine diseases and with histiocytosis X. *Br. J. Med.* Submitted for publication.

113. Roth J. and Grunfeld C. Endocrine systems: mechanisms of disease, target cells, and receptors. In: Williams R. H., ed., *Textbook of Endocrinology*, 6th ed. London, Saunders, 1981: 15–72.

114. Kohn L. D. Monoclonal antibodies, the thyrotropin receptor and autoimmune thyroid disease. In: Labrie F. and Proulx A., eds., *Endocrinology*. Amsterdam, Excerpta Medica, 1984: 555–558.

115. Bottazzo G. F. and Doniach D. Polyendocrine autoimmunity. In: Volpe R., ed., *Autoimmunity and Endocrine Disease*. New York, M. Dekker, Inc., 1985: in press.

116. Doniach D., Bottazzo G. F. and Russell R. C. G. Goitrous autoimmune thyroiditis (Hashimoto's disease). *Clin. Endocrinol.* 1979, **8**, 63–80.

117. Scherbaum W. A., Rosenau K. O. and Seif F. J. Antikörper gegen Schilddrüsenmikrosomen und Thyreoglobulin beim Morbus Basedow und anderen Erkrankungen. *Med. Welt (Stuttg.)* 1979, **30** N.F., 1401–1406.

118. Kendall-Taylor P., Kinson A. T. and Holocombe M. A specific IgG in Graves' ophthalmopathy. Its relation to retro-orbital and thyroid autoimmunity. *Br. Med. J.* 1984, **288**, 1183–1186.

29

Spectrum and Profiles of Non-organ-specific Autoantibodies in Autoimmune Diseases

W. A. Scherbaum,   D. Doniach and
M. Blaschek,   G. F. Bottazzo
P. A. Berg

Non-organ-specific autoimmune diseases are characterized by the presence in the serum of autoantibodies reacting with distinct components of 'ubiquitous' cells. The spectrum and profiles of non-organ-specific antibodies are correlated with clinical entities which have been defined by specific marker sera. Antibody testing can thus help with the differential diagnosis and prognostic evaluation of a rheumatic disease and other 'collagen' disorders.

Early studies defined different staining patterns given by anti-nuclear antibodies, in indirect immunofluorescence on frozen sections using various normal tissues. However, the correlation with disease was weak and it proved difficult to define the staining patterns clearly by this method since the same sera could give different results when applied to various tissues.[1]

Sera containing antibodies to soluble nuclear antigens may give negative results in standard immunofluorescence tests due to a loss of antigen from the section by spontaneous solubilization in buffer solutions during rinsing. Such antibodies can still be detected on alcohol or acetone-fixed tissue or in systems such as binding tests, e.g. radioimmunoassays (RIA) or enzyme-linked immunosorbent assays (ELISA) and precitating assays like counterimmunoelectrophoresis or the Ouchterlony test. All these tests may give different results due to variable sensitivity and antibody specificity.[2] Furthermore, preparation and processing of the antigens for *in vitro* systems may lead to alteration of antibody binding.

Therefore it is advisable to apply several systems, using immunofluorescence on tissue sections for initial screening, followed by further tests, relevant to the suspected disease under investigation. The pitfalls of substrate diversity in immunofluorescence can be overcome by the introduction of cell lines with a constant set of antigens. Defined culture systems have also made it possible to look for new autoantibodies directed to the cytoskeleton, endoplasmic reticulum, Golgi apparatus and other cytoplasmic constituents which had previously been obscure. Manipulation *in vitro* of cells by detergents, drugs or virus infection can reveal further binding sites for antibodies which correlate with distinct diseases.

1. ANTI-NUCLEAR ANTIBODIES

1.1. Screening Procedures

Indirect immunofluorescence on cryostat sections of human or animal tissues is used as the initial screening test for anti-nuclear antibodies. Homogeneous, peripheral, speckled or nucleolar staining or a combination of these may be found (*Plate* 30). Nucleolar staining indicates scleroderma,[3] but all the other patterns can be observed in various diseases depending on the specificity of antigens in the test. Therefore, in the monitoring of connective tissue diseases, a battery of different antigens must be used, embracing all types of anti-nuclear antibody.

1.2. Antibodies to Extractable Nuclear Antigens

The extractable nuclear antigen (ENA) complex is a saline-soluble class of non-histone nuclear proteins, and the respective extracts from lymphoblastoid cell lines (Wil-2), thymus or spleen serve as a substrate for the detection of antibodies specific for different extractable nuclear antigens.[4,5] About 20 different specificities are now identified, some of them designated by the initials of the first patient in whom they were detected (e.g. Ro, La, Sm, Su, Jo-1, Mi-1), others named according to the biochemical nature of the ENA antigen (proliferating cell nuclear antigen, PCNA, or ribonuclear protein, RNP) and a third group denoted after their associated disease (e.g. Scl-70 for scleroderma, SS-A for Sjögren's syndrome with systemic lupus erythematosus (SLE), SS-B for primary Sjögren's syndrome, PM-1 for polymyositis, RANA for rheumatoid arthritis nuclear antigen). These defined sera serve as constant controls in the Ouchterlony test where a precipitin line of a given patient's serum identical with that of one of these markers indicates a similar clinical entity.[6]

Antibodies to extractable nuclear antigens are strongly correlated with connective tissue disease and, in their presence, speckled staining is usually detected by standard immunofluorescence testing. By studying patients known to have SS-B (=Ro) antibodies, we found evidence of subclinical sicca syndrome which was usually associated with polyarthralgia. Although detection of the SS-B specificity may antedate overt Sjögren's disease by months or years, patients with this antibody are unlikely to develop rheumatoid arthritis. We have also shown that the more commonly sought antibodies to thyroid microsomes, gastric parietal cells, and submaxillary duct tissue, are of little predictive value in connective tissue disorders.[7] SS-B antibodies are species specific and thus are preferentially detected on substrates of human origin.[8] The rheumatoid arthritis nuclear antigen specificity is only seen in continuous cultures when human diploid B-lymphocytes transformed by the Epstein–Barr virus (Wil-2) are used as an antigen.[9]

1.3. Antibodies to Deoxyribonucleic Acid

Homogeneous and peripheral nuclear immunofluorescence patterns point to DNA or histone antibodies which correlate with systemic lupus erythematosus (SLE) and pseudo-SLE, which can only be discriminated by other test systems. Antibodies to DNA are detected by the Farr ammonium sulphate assay, where radiolabelled DNA is employed and DNA antibodies in the test sera are measured after ammonium sulphate precipitation of the complex.[10] Confusion of negative, double-stranded DNA (dsDNA) with single-stranded DNA (ssDNA) is avoided, when purified dsDNA or its synthetic deoxyribose phosphate backbone is used as an antigen in RIA or ELISA.[11,12] Alternatively, a specific immunofluorescence test may be employed for the detection of anti-dsDNA when *Crithidia luciliae* is used in the system (*Plate* 31a). The kinetoplast of the haemoflagellate *Crithidia luciliae* is composed of dsDNA and its immunofluorescence reactivity with SLE sera is almost identical with the results given by RIA testing.[13]

Antibodies to dsDNA are a hallmark of SLE, whereas antibodies to ssDNA are also detected in a small proportion of normal individuals.[14] Antibodies to DNA can therefore be used as immunological markers in the diagnosis of SLE.[15] These antibodies may lead to DNA–anti-DNA immune complexes and thus be responsible for vasculitis and complex-mediated tissue damage in glomerulonephritis.[16] DNA antibodies may penetrate cells bearing Fc receptors which can cause functional disturbance of B- and T-lymphocytes, macrophages and other cells, suggesting that the imbalance of the immune system observed in SLE is partially antibody mediated.[17,18]

1.4. Antibodies to Histones

Antibodies to histones give diffuse or peripheral nuclear immunofluorescence and low titres can be detected in more than half of the sera from patients with SLE.[19] A number of drugs such as hydralazine and procainamide occasionally cause a lupus-like syndrome with the appearance in the serum of high titres of histone antibodies which may be detected by immunofluorescence using a specific histone reconstitution test.

In this test, nuclear antigens are extracted from the tissue sections by incubation with 0·1 N HCl and shaking for 30 min at room temperature. After washing with phosphate-buffered saline (PBS) for 30 min, the sections are immersed in a histone solution (now commercially available) for 30 min at room temperature, and finally rinsed with PBS. They are then ready for use as substrates in immunofluorescence.[20] The majority of the antibodies are directed against histones H2A and H2B which can be recognized by RIA.[21]

Detection of histone antibodies in the absence of antibodies to dsDNA and other non-histone nuclear antibodies strongly indicates drug-induced lupus erythematosus. In these cases, both antibodies and disease disappear following discontinuation of the drug.[22]

1.5. Antibodies to Centromeres

Antibodies to centromeres give a discrete fine-speckled pattern on interphase cells and they only reveal their characteristic centromere staining in the dividing phase (*Plate* 31*b*).[23] Therefore, cell cultures with a high turnover rate, such as the A 431 vulva carcinoma line, are preferentially used for the discrimination of centromere antibodies from others giving speckled nuclear immunofluorescence. The nature of the centromeric antigen has now been defined by immunoblotting experiments as a 19·5-kdal chromosomal non-histone protein of the kine-tochore[24,25] which is a part of the microtubule organizing centre.

The centromeric specificity of the speckled nuclear immunofluorescence is a serological marker for a benign subgroup of scleroderma with CREST syndrome (*c*alcinosis, *R*aynaud's syndrome, *e*sophageal dysmotility, *s*clerodactyly, *t*elean-giectasia).[26,27] Antibodies to centromeres are also detected in overlap cases of primary biliary cirrhosis and scleroderma[28,29] which may suggest a common aetiology that has long been suspected from the clinical correlations between the two diseases.

1.6. Antibodies to the Proliferating Cell Nuclear Antigen

The antibody to the proliferating cell nuclear antigen (PCNA) is newly described. It can only be demonstrated on proliferating cells and its detection is enhanced by mitogenic stimulation of the antigen.[30] The proliferating cell nuclear antigen is a soluble ribonucleoprotein complex and the antibody is therefore also detected by immunodiffusion tests for antibodies to extractable nuclear antigens.[31] Using continuous culture cell lines as a substrate, the morphological appearance of staining by PCNA antibodies varies with the dividing phase of the cell (*Plate* 31*c*) and can only be detected at certain stages. PCNA antibodies have been detected in 2 out of 100 patients with SLE and, in two well-studied cases, these antibodies repeatedly disappeared with glucocorticoid treatment.[32] The pathogenetic and diagnostic significance of PCNA antibodies is an exciting field for future studies.

2. NON-ORGAN-SPECIFIC ANTI-CYTOPLASMIC ANTIBODIES

Non-organ-specific cytoplasmic antibodies are directed against microsomal, lysosomal, mitochondrial, cytoskeletal or other cytoplasmic antigens and they are often associated with organ-specific autoimmune diseases (*see* Chapter 28).

2.1. Antibodies to Mitochondria

Mitochondrial antibodies are strongly associated with primary biliary cirrhosis where they appear in 96 per cent of the cases and serve as helpful serological markers in the differential diagnosis of cholestatic liver diseases[33] (*Table* 29.1). They may also be present in a subgroup of cryptogenic cirrhosis which usually affects middle-aged women, often with features of cholestasis and biopsies showing certain similarities with primary biliary cirrhosis.[34] Mitochondrial antibodies may also be positive in the sera from patients with drug-induced liver diseases,[35] pseudo-LE,[36] collagen disorders,[37] chronic active hepatitis or syphilis.[38]

Antibodies to mitochondria may be detected by immunofluorescence testing on cryostat sections where we prefer human liver and rat kidney (*Plate* 31*d*), or on liver cell lines (*Plate* 31*e*). A distinct type of mitochondrial antibody precipitates as an immune complex and may be detected in the immunodiffusion test.[39]

Mitochondrial antibodies and their clinical correlates are now better defined by the application of submitochondrial preparations in ELISA systems which we have described in detail elsewhere or by RIA.[40]

2.2. Antibodies to Ribosomes

Autoantibodies reacting with components of the cytoplasmic ribosomes have been detected in the sera of patients with systemic lupus erythematosus[41–47] and occasionally in progressive systemic sclerosis,[48] mixed connective tissue disease,[49] chronic active hepatitis[50–52] and polymyositis.[53,54]

The immunofluorescence pattern given by ribosomal antibodies depends on the distribution of ribosomes in the type of tissue tested (*Table* 29.1). In tissue sections we usually see a granular staining of the perinuclear cytoplasmic portion of the respective cells, e.g. hepatocytes, and most strongly in the exocrine pancreas.[44] Liver-specific ribosomal antibodies have also been described.[55] In cultured cells, such as human fibroblast monolayers, we can also observe staining of cytoplasmic reticular structures.

The ribosome consists of three species of RNA[56] and 70–80 identified polypeptides,[57] so that there is a great variety of possible determinants which cannot be separated by conventional immunofluorescence testing. However, this has now been achieved by electron microscopical immunocytochemistry where fluorescein was coupled to a defined amino acid of a single ribosomal protein and was then incorporated into the ribosome and identified with anti-fluorescein.[58] The reactive ribosomal protein determinants of human autoantibodies have also been described further: electrophoretic separation in polyacrylamide and blotting on nitrocellulose revealed that, in addition to antibodies against the RNA moiety or against the whole ribosome, some autoimmune sera contain antibodies against the protein moiety.[48]

2.3. Antibodies to Liver and Kidney Microsomes

Antibodies to liver and kidney microsomes (anti-LKM) were originally detected

Table 29.1. *Immunofluorescence characteristics of mitochondrial, liver/kidney microsomal and ribosomal antibodies*

Antibody	Subtype	Occurrence	Rat liver		Rat kidney			
			Hepatocytes	Glomeruli	Proximal tubules			Distal tubules
					P1	P2	P3	
Mitochondrial	M1	Secondary syphilis	++	−	+	+	+	++
	M2	Primary biliary cirrhosis	+	+	++	++	+	+++
	M3	Drug-induced pseudolupus syndrome	+	++	++	++	++	++
	M4	Mixed form chronic active hepatitis	+	−	++	++	+	+++
	M5	Collagen disorders	+	−	++	++	+	±
	M6*	Iproniazid-induced hepatitis	+++	−	+++	+ or −	−	−
Liver–kidney microsomal	LKM1	Chronic active hepatitis	+++	+ or −	+ or −	+	+++	−
	LKM2	Tienilic acid-induced hepatitis	+++	−	+++	+++	+ or −	−
Ribosomal	ARA	Collagen disorders	++	−	+	+	+	+

*M6 = Detected by ELISA test with beef heart mitochondria and found in cardiomyopathy and myocarditis.

Table 29.1 Continued

| Human stomach | | | Human pancreas | | Remarks |
| Cell type | | | | | |
Chief	parietal	Mucus	Exocrine	Endocrine	
+	+	–	+	+	Anti-cardiolipin distinct from VDRL†
+	+++	+	++ (oncocytes)	+	
+	++	++	+	+	
+	+++	+	++ (oncocytes)	+	Coexist with M2; distinguishable by complement-fixation test
+	++	–	+	+	SLE-like syndromes: autoimmune haemolytic anaemia and biological false-positive VDRL
–	– ('APUD' cells +++)	–	–	– (++ on rat islets)	
–	–	–	–	–	
–	–	–	–	–	On male mouse liver, stronger immunofluorescence in centrolobular than periportal hepatocytes
+++	–	–	+++	+	Especially SLE with nephritis

†VDRL = Venereal Disease Research Laboratory.

by Rizzetto et al.[59] and are characteristic of patients with cryptogenic chronic active heptatitis or cryptogenic cirrhosis of early onset,[59-62] but low titres of these antibodies may also be detected in drug-induced hepatitis.[63,64]

Ultrastructural studies showed that the antibodies bind to smooth and rough endoplasmic reticulum of hepatocytes.[65,66] The fluorescence pattern given by antibodies to liver and kidney microsomes can be distinguished from that of mitochondrial antibodies by their stronger staining on hepatocytes and on proximal renal tubule cells. The original antibodies, now called 'anti-LKM 1', react with the third portion of proximal renal tubules (P_3). A new liver–kidney microsome-specific antibody preferentially staining the first and second portions of proximal renal tubules (P_1 and P_2) and the centrolobular more than the periportal hepatocytes has recently been described.[67] This 'anti-LKM 2' is strongly associated with tienilic acid-induced hepatitis (*Table* 29.1).

A radioimmunoassay has also been established for the detection of anti- LKM[68] and, in this test, a low reactivity with microsomes prepared from stomach, heart, lung and skeletal muscle was detected as well as the predominant liver and kidney binding.

Liver membrane antibodies are found in all anti-LKM-containing sera, and it was shown by absorption experiments that liver cell surfaces express the LKM antigen.[69] Membrane binding has also been demonstrated for organ-specific autoantibodies to thyroid,[70] adrenal[71] and gastric parietal cell microsomes.[72]

2.4. Microsomal Antibodies associated with the Delta Agent

Carriers of the hepatitis B surface antigens (HBsAg) may be infected by a hepatotropic RNA agent called delta (δ),[73] which often causes progressive liver disease unresponsive to immunosuppressive therapy.[74] In some of the HBsAg-positive patients with infection, microsomal antibodies may be detected by immunofluorescence. These δ-associated microsomal antibodies strongly react with human hepatocytes and nephron cells and give a weak staining on thyroid follicular cells, reticular cells of adrenal cortex and pancreatic acinar cells.[75] In contrast to the antibodies to liver–kidney microsomes, this antibody reacts with human, but not with rat, tissues. The antibody was not detected in HBsAg-positive and HBsAg-negative patients without δ infections and it is therefore likely that the expression of the associated microsomal antibody is induced by persistence of the virus.

2.5. Antibodies to the Golgi Complex

Antibodies to the Golgi apparatus give a typical juxtanuclear reticular cisternal staining pattern (*Plate* 31*f*) and the characteristic appearance with perinuclear dots is achieved by pre-treatment of cells with 10 μM colchicine for 4 h.[76] The antibodies seem to be very rare; they have been detected in some patients with Sjögren's syndrome and lymphoma[77] and in only 8 out of 3600 sera from patients with various autoimmune diseases.[78]

2.6. Antibodies to the Cytoskeleton

Ultrastructural studies of eukaryotic cells have revealed the presence of complex interconnected fibre systems, the cytoskeleton, which is composed of microfilaments, microtubules and intermediate filaments.[79–81] The characterization of these cytoplasmic fibres and their associated proteins such as α-actinin, filamin, tropomyosin, fibrin and vinculin in cells and tissues has allowed the production of monoclonal antibodies and their application as reagents for immunofluorescence testing.[82,83] These monoclonals now permit the immunocytochemical distinction of different types of filaments and their presence or absence.[84–87] The intermediate filament pattern appears to be identical in tumours and their metastases, and thus provides a new tool in the exact diagnosis of metastatic tumours.[88,89]

Antibodies to the cytoskeleton are preferentially detected on established cell lines with their well-defined antigenic composition. Anti-cytoskeleton drugs such as colchicine and vinblastine serve to discriminate between the different types of intermediate filaments and tubulin. These drugs bind to tubulin and interfere with the integrity of the filament system, which in its native form cannot easily be distinguished morphologically. For pre-treatment, multispot slides are incubated with 10 mg colchicine/ml medium for 18 h, or with 10 mg vinblastine/ml medium for 30 min to 4 h. The slides are then washed with PBS pH 7·4 with gentle agitation. The cells are then fixed with methanol at $-20\,°C$ for 10 min and allowed to dry. It must be noted that formaldehyde fixation of sections is not desirable because it may alter the antigens.

Autoantibodies have now been detected to each of the cytoskeletal components and correlated with some diseases (*Table* 29.2). We have previously described three different subgroups of smooth muscle antibodies (SMA) detected on tissue sections. SMA staining on tubules (SMA-T) and on glomeruli (SMA-G) of rat kidney sections indicates actin and tubulin antibodies, respectively.[90] The nature of the SMA-V (vessels) pattern has remained unclear. Using cultured fibroblasts as an antigen (*Plate* 31g), these antibodies react with structural proteins of the cytoskeleton.[91] Smooth muscle antibodies are now known to react with a variety of cytoskeletal components and the further illumination of their respective clinical correlates is a challenge for future studies.

Cytoskeleton antibodies may be helpful in understanding the pathogenesis of diseases. Antibodies to actin (*Plate* 31h) are a characteristic finding in the autoimmune type of chronic active hepatitis,[92,93] whereas sera from patients with viral, bacterial or parasitic infections mainly contain intermediate filament or tubulin antibodies[94–96] (*Plate* 31i, j). Both, actin[97] and intermediate filaments[98,99] (*Plate* 31k, l) have been shown to be integral components of the plasma membrane and they can thus be recognized by the immune system, which, under pathological conditions, may produce the respective autoantibodies.

3. FUTURE PROSPECTS

The physician is often bewildered by identifying constellations of clinical and histological criteria, together with requesting tests and interpreting the serological findings on which the diagnosis should be based.

Autoantibodies have now been discovered that are characteristic of distinct diseases such as anti-dsDNA which indicates SLE. However, presence or absence

Table 29.2. Classification of smooth-muscle autoantibodies (SMA)

| | Staining patterns | | |
Autoantigens	Tissue sections	Cell cultures	Clinical associations
Microfilaments			
Actin		Fine linear bundles	Chronic active hepatitis
Myosin	Renal tubules		
Tropomyosin		('stress fibres')	Polymyositis
Actinin			
Microtubules			
Tubulin	Vessels	Vinblastine-induced	CREST syndrome
MTOC*		paracrystals	Viral infections
			Autoimmune thyroid diseases
Intermediate filaments			
Vimentin	Vessels	Vinblastine-induced	Viral and parasitic infections
Cytokeratin		filament 'coils and	
Desmin		whorls' of	Rheumatoid arthritis
GFA-protein†		perinuclear filaments	
Neurofilaments			

*MTOC = microtubule organizing centre (centrioles, centrosomes).
†GFA = glial fibrillary acid.

of the antibody does not reflect the broad spectrum of possible clinical courses. Therefore, we should focus our attention on the finding and clinical correlation of new antibody specifities which will be detected either by comparison with marker sera from patients as exemplified by extractable nuclear antigen testing, or by the careful study of staining patterns of sera in well-characterized cell cultures.

Further efforts are imperative concerning characterization, careful biological preparation and purification of the nuclear and cytoplasmic autoantigens. When these substances are available in a pure form, we can use them as antigens in radioimmunoassay or ELISA systems, which are more sensitive and give more quantitative results than immunofluorescence or precipitin tests.

The identification of non-organ-specific autoantibodies has been of great help in the diagnosis of connective tissue diseases and may also provide insights into genetic and cellular regulatory mechanisms and into the pathogenesis of disease.

Acknowledgements

We would like to thank Professor W. Franke, Krebsforschungsinstitut Heidelberg, and Mrs Susan J. Gardner, Glasgow, for helpful discussions. We also thank Mr Volkmar Schöllhorn, Tübingen for skilled help with cell culturing and we are indebted to Professor I. M. Roitt, London, and Professor E. F. Pfeiffer, Ulm, for their encouragement and support. W. A. Scherbaum is supported by a research grant from the Deutsche Forschungsgemeinschaft Sche 225/2-1. M. Blaschek is supported by a Scholarship from the Studienstiftung des Deutschen Volkes. P. A. Berg is supported by the Deutsche Forschungsgemeinschaft. Work in the Autoimmune Serology Laboratory at the Middlesex Hospital,

London is supported by the Medical Research Council of Great Britain, the Juvenile Diabetes Foundation and the Wellcome Trust.

Appendix

INDIRECT IMMUNOFLUORESCENCE ON CULTURED CELLS AND CRYOSTAT SECTIONS

The cells were grown in monolayers, gently fixed in methanol ($-20\,^{\circ}$C for 10 min) and incubated with the patients' sera diluted 1 : 10 for 45 min. After washing with PBS, the fluorescein (FITC) or tetramethylrhodamine isothiocyanate (TRITC)-labelled anti-human immunoglobulin conjugates were applied. After incubation for a further 45 min, the slides were washed with PBS and mounted in Moviol 4-88. Unfixed cryostat sections were treated in the same way.

The slides were read with a Leitz Dialux 22 microscope, equipped with an HBO 100 mercury source for epifluorescence and photographed with a Leitz Vario-Orthomat camera. The filter combinations for epifluorescence were for the fluorophores FITC/TRITC an excitation filter BP 450–490/546, a chromatin beam splitter FT510/580, and a barrier filter LP 520/590. A daylight colour slide film, which can be push-processed to a higher speed was used (Fujichrome, ASA 400).

The fluorescein- and rhodamine-conjugated sheep anti-human IgG reagents were supplied by Paesel, Frankfurt.

REFERENCES

1. Beutner E. H., Krasny S., Kumar V., Taylor R. and Chorzelski T. P. Prospects and problems in the definition and standardization of immunofluorescence. I. Present levels of reproducibility and disease specificity of antinuclear antibody tests. In: Beutner E. H., Nisengard R. J. and Albini B., eds., *Defined Immunofluorescence and Related Cytochemical Methods. Ann. NY Acad. Sci.* 1983, **420**, 28–54.
2. Akiziki M., Powers R. Jr. and Holman H. R. Comparative study of immunologic methods for demonstration of antibodies to soluble nuclear antigens. *Arthritis Rheum.* 1977, **20**, 693–701.
3. Bernstein R. M., Steigerwald J. C. and Tan E. M. Association of antinuclear and antinucleolar antibodies in progressive systemic sclerosis. *Clin. Exp. Immunol.* 1982, **48**, 43–51.
4. Sharp G. C., Irvin W. S. and Tan E. M. Mixed connective tissue disease, an apparently distinct rheumatic disease associated with a specific antibody to an extractable nuclear antigen (ENA). *Am. J. Med.* 1972, **52**, 148–159.
5. Sharp G. C., Irvin W. S., May C. M., Holman H. R., McDuffie F. C., Hess E. V. and Smith R. R. Association of antibodies to ribonucleoprotein and Sm antigen with mixed connective tissue disease, systemic lupus erythematosus and other rheumatic diseases. *N. Engl. J. Med.* 1976, **295**, 1149–1154.
6. Tan E. M. Autoantibodies to nuclear antigens (ANA): their immunobiology and medicine. *Adv. Immunol.* 1982, **33**, 167–240.
7. Isenberg, D. A., Hammond, L., Fisher, C., Griffiths, M., Stewart, J. and Bottazzo, G. F. Predictive value of SS-B precipitating antibodies in Sjögren's syndrome. *Br. Med. J.* 1982, **284**, 1738–1740.
8. Harmon C. E., Deng J.-S., Peebles C. L. and Tan E. M. The importance of tissue substrate in the SS-A/Ro antigen–antibody system. *Arthritis Rheum.* 1984, **27**, 166–173.

9. Catalano M. A., Carson D. A., Slovin S. F., Richman D. D. and Vaughan J. H. Antibodies to Epstein–Barr virus-determined antigens in normal subjects and in patients with seropositive rheumatoid arthritis. *Proc. Natl Acad. Sci. USA* 1979, **76**, 5825–5828.

10. Rubin R. L., Lafferty J. and Carr R. I. Re-evaluation of the ammonium sulfate assay for DNA antibody. *Arthritis Rheum.* 1978, **21**, 950–957.

11. Steinmann C. R., Deesomchok U. and Spiera H. Detection of anti-DNA antibody using synthetic antigens: characterization and clinical significance of binding of poly (deoxyadenylate–deoxythymidylate) by serum. *J. Clin. Invest.* 1976, **57**, 1330–1341.

12. Lomerfield S. D., Roberts M. W. and Booth R. J. Double-stranded DNA antibodies: A comparison of four methods of detection. *J. Clin. Pathol.* 1981, **34**, 1032–1035.

13. Aarden L. A., de Groot E. R. and Feltkamp T. E. W. Immunology of DNA. III. *Crithidia luciliae*, a simple substrate for the determination of anti-dsDNA with the immunofluorescence technique. *Ann. NY Acad. Sci.* 1975, **254**, 505–515.

14. Hughes G. R. V., Cohen S. A. and Christian C. L. Anti-DNA activity in systemic lupus erythematosus: A diagnostic and therapeutic guide. *Ann. Rheum. Dis.* 1971, **30**, 259–264.

15. Swaak A. J. G., Groenwold J., Aarden L. A. and Feltkamp T. E. W. Detection of anti-dsDNA as diagnostic tool. *Ann. Rheum. Dis.* 1981, **40**, 45–49.

16. Tan E. M., Schur P. H., Carr R. and Kunkel H. Deoxyribonucleic acid (DNA) and antibodies to DNA in the serum of patients with systemic lupus erythematosus. *J. Clin. Invest.* 1966, **45**, 1732–1740.

17. Alarcon-Segovia D., Ruiz-Arguelles A. and Fishbein E. Antibody to nuclear ribonuclear protein penetrates live human mononuclear cells through Fc receptors. *Nature*, 1978, **271**, 67–69.

18. Alarcon-Segovia D., Llorence L., Fishbein E. and Diaz-Jouanen E. Abnormalities in the content of nucleic acids of peripheral blood mononuclear cells from patients with systemic lupus erythematosus. *Arthritis Rheum.* 1982, **23**, 304–317.

19. Stollar B. D. Reactions of systemic lupus erythematosus sera with histone fractions and histone-DNA complexes. *Arthritis Rheum.* 1971, **14**, 485–492.

20. Fritzler M. J. and Tan E. M. Antibodies to histones in drug-induced and idiopathic lupus erythematosus. *J. Clin. Invest.* 1978, **62**, 560–567.

21. Rubin R. L., Joslin R. G. and Tan E. M. Specificity of anti-histone antibodies in systemic lupus erythematosus. *Arthritis Rheum.* 1982, **25**, 779–782.

22. Gioud M., Ait Kaci M. and Monier J. C. Histone antibodies in systemic lupus erythematosus. A possible diagnostic tool. *Arthritis Rheum.* 1982, **25**, 407–412.

23. Moroi Y., Peebles C., Fritzler M. J., Steigerwald J. and Tan E. M. Autoantibody to centromere (kinetochore) in scleroderma sera. *Proc. Natl Acad. Sci. USA* 1980, **77**, 1627–1631.

24. Guldner H. H., Lakomek H.-J. and Bautz F. A. Human anti-centromere sera recognise a 19.5 kD non-histone chromosomal protein from HeLa cells. *Clin. Exp. Immunol.* 1984, **58**, 13–20.

25. Cox J. V., Schenk E. A. and Olmsted J. B. Human anticentromere antibodies. Distribution, characterization of antigens and effect on microtubule organization. *Cell* 1983, **35**, 331–339.

26. Tan E. M., Prodnan G. P., Garcia J., Moroi Y., Fritzler M. J. and Peebles C. Diversity of antinuclear antibodies in progressive systemic sclerosis. *Arthritis Rheum.* 1980, **23**, 612–625.

27. Fritzler M. J., Kinsella T. D. and Garbutt E. The CREST syndrome: a distinct serologic entity with anticentromere antibodies. *Am. J. Med.* 1980, **69**, 520–526.

28. Bernstein R. M., Callender M. E., Neuberger J. M., Hughes G. R. V. and Williams R. Anticentromere antibody in primary biliary cirrhosis. *Ann. Rheum. Dis.* 1982, **41**, 612–614.

29. Makinen D., Fritzler M., Davis P. and Sherlock S. Anticentromere antibodies in primary biliary cirrhosis. *Arthritis Rheum.* 1983, **26**, 914–917.

30. Takasaki Y., Deng J. S. and Tan E. M. A nuclear antigen associated with cell proliferation and blast transformation: its distribution in synchronized cells. *J. Exp. Med.* 1981, **154**, 1899–1909.

31. Miyachi K., Fritzler M. J. and Tan E. M. Autoantibody to a nuclear antigen in proliferating cells. *J. Immunol.* 1978, **121**, 2228–2234.

32. Fritzler M. J., McCarty G. A., Ryan J. P. and Kinsella D. T. Clinical features of patients with antibodies directed against proliferating cell nuclear antigen. *Arthritis Rheum.* 1983, **26**, 140–145.

33. Berg P. A., Doniach D. and Roitt I. M. Mitochondrial antibodies in primary biliary cirrhosis. I. Location of the antigen to mitochondrial membranes. *J. Exp. Med.* 1967, **126**, 277–290.

34. Doniach D. and Walker G. Progress report. Mitochondrial antibodies (AMA). *Gut* 1974, **15**, 664–668.

35. Homberg J. C., Stelly N., Andreis J., Abuaf N., Saadoun F. and Andre J. A new antimitochondrial antibody (anti-M6) in iproniazid induced hepatitis. *Clin. Exp. Immunol.* 1982, **47**, 93–102.

36. Sayers T. J., Binder T. and Berg P. A. Heterogeneity of anti-mitochondrial antibodies: Characterization and separation of the antigen associated with the pseudo-lupus erythematosus syndrome. *Clin. Exp. Immunol.* 1979, **37**, 68–75.

37. Labro M. T., Andrieu M. C., Weber M. and Homberg J. C. A new pattern of non-organ and non-species specific anti-organelle antibody detected by immunofluorescence: The mitochondrial antibody number 5. *Clin. Exp. Immunol.* 1978, **31**, 357–366.

38. Wright D. J. M., Doniach D., Lessof M. H., Turk J. L., Grimble A. S. and Catterall R. D. New antibody in early syphilis. *Lancet* 1970, **i**, 740–744.

39. Miyachi K., Gupta C., Dickson E. R. and Tan E. M. Precipitating antibodies to mitochondrial antigens in patients with primary biliary cirrhosis. *Clin. Exp. Immunol.* 1980, **39,**, 599–606.

40. Berg P. A., Weber P., Oehring J., Lindenborn-Fotinos J. and Stechemesser E. Significance of different types of mitochondrial antibodies in primary biliary cirrhosis. In: Brunner H. and Thaler H., eds., *Festchrift for Hans Popper. Hepatology.* New York, Raven press, 1985: 231–242.

41. Sturgill, B. C. and Carpenter R. R. Antibody to ribosomes in systemic lupus erythematosus. *Arthritis Rheum.* 1965, **8**, 213–215.

42. Schur P. H., Moroz L. E. and Kunkel H. G. Precipitating antibodies to ribosomes in the serum of patients with systemic lupus erythematosus. *Immunochemistry* 1967, **4**, 447–553.

43. Lamon E. W. and Bennel J. C. Antibodies to ribosomal ribnucleic acid (rRNA) in patients with systemic lupus erythematosus (SLE). *Immunology* 1970, **19**, 439–442.

44. Homberg J. C., Rizzetto M. and Doniach D. Ribosomal antibodies detected by immunofluorescence in systemic lupus erythematosus and other collagenoses. *Clin. Exp. Immunol.* 1974, **17**, 617–628.

45. Cavanagh D. A solid-phase radioimmunoassay for the detection of antibodies to ribosomes. *Anal. Biochem.* 1977, **79**, 217–225.

46. Koffler D., Miller T. E. and Lahita R. G. Studies on the specificity and clinical correlation of antiribosomal antibodies in systemic lupus erythematosus sera. *Arthritis Rheum.* 1979, **22**, 463–470.

47. Meroni P. L., De Bartolo G., Barcellini, W., Riboldi P. S., Basile R., Betterle C. and Zanussi C. Anti-ribosomal ribonucleoprotein autoantibodies in systemic lupus erythematosus *Clin. Exp. Immunol.* 1984, **4**, 45–54.

48. Gordon J., Towbin H. and Rosenthal M. Antibodies detected against ribosomal protein determinants in the sera of patients with connective tissue diseases. *J. Rheumatol.* 1982, **9**, 247–252.

49. Miyachi, K. and Tan E. M. Antibodies reacting with ribonucleoproteins in connective tissue diseases. *Arthritis Rheum.* 1979, **22**, 87–93.

50. Koffler D., Faiferman I. and Gerber M. A. Radioimmunoassay for antibodies to cytoplasmic ribosomes in human serum. *Science* 1977, **198**, 740–743.

51. Gerber M. A., Shapiro J. M. and Smith H. Antibodies to ribosomes in chronic active hepatitis. *Gastroenterology* 1979, **76**, 139–143.

52. Richart C., Pujol-Borrell R. and Marti S. Anticuerpos anti-ribosoma (immunofluorescencia) asociados a hepatitis cronica activa. *Rev. Clin. Esp.* 1981, **161**, 259–260.

53. Goepel W., Storz H., Miethwenz D. and Banzhaf E. Subakute idiopathische Polymyositis mit seltenen antiribosomalen Antikörpern im Serum—klinischer Verlauf unter immunsuppressiver Langzeittherapie. *Z. Ges. Inn. Med.* 1983, **38**, 518–524.

54. McCrea J. D., Boyd M. W. J., McMillan S. A., Haire M., Dermott E. and Fay A. C. Polymyositis associated with high titer antibody to cytoplasmic ribosomes. *J. Rheumatol.* 1984, **11**, 396–397.

55. Storch W. Diagnostische Bedeutung von Antikörperngegen Zellorganellen (Mitochondrien, Endoplasmatisches Retikulum und Ribosomen). *Schweiz. Med. Wochenschr.* 1983, **113**, 47–59.

56. Wool I. G. and Stöffler G. In: Nomura H., Tissières A. and Lengyel P., eds., *Structure and Function of Eukaryotic Ribosomes*, Cold Spring Harbor, Cold Spring Harbor Laboratory, 1974: 417–460.

57. McConkey E. H., Bielka H. and Gordon J. Proposed uniform nomenclature for mammalian ribosomal proteins. *Mol. Gen. Genet.* 1979, **169**, 1–6.

58. Stöffler G. and Stöffler-Meilike M. Immunoelectron microscopy of ribosomes. *Annu. Rev. Biophys. Bioeng.* 1984, **13**, 303–330.

59. Rizzetto M., Swana G. and Doniach D. Microsomal antibodies in active chronic hepatitis and other disorders. *Clin. Exp. Immunol.* 1973, **15**, 331–344.
60. Homberg J. C., Micouin, C., Peltier A., Salmon Ch. and Caroli J. Un nouvel anticorps non spécifique d'organe au cours d'hépatite chronique. *Méd. Chir. Dig.* 1974, **3**, 85–92.
61. Odièvre, M., Maggiore, G. Homberg J. C., Saadoun, F. Curouce, A. M., Ivart J., Hadchonel, M. and Alagille B. Seroimmunologic classification of chronic hepatitis in 57 children. *Hepatology* 1983, **3**, 407–409.
62. Schmidt, Kienle G. J. and Wolf H. Lebererkrankungen mit mikrosomalen Antikörpern. *Z. Gastroenterol.* 1978, **8**, 512–520.
63. Smith M. G. M., Williams R., Walker G., Rizzetto M. and Doniach D. Hepatic disorders associated with liver/kidney microsomal antibodies. *Br. Med. J.* 1974, **2**, 80–84.
64. Walton B., Simpson, B. R., Strunin L., Doniach D., Perrin J. and Appleyard A. J. Unexplained hepatitis following halothane. *Br. Med. J.* 1976, **2**, 1171–1176.
65. Rizzetto M., Bianchi F. B. and Doniach D. Characterization of the microsomal antigen related to a subclass of active chronic hepatitis. *Immunology* 1974, **26**, 589–601.
66. Storch W., Cossel L. and Dargel R. The immunoelectron-microscopical demonstration of antibodies against endoplasmic reticulum (microsomes) in chronic aggressive hepatitis and liver cirrhosis. *Immunology* 1977, **32**, 941–945.
67. Homberg J. C., André C. and Abuaf N. A new anti-liver–kidney microsome antibody (anti-LKM 2) in tienilic acid-induced hepatitis. *Clin. Exp. Immunol.* 1984, **55**, 561–570.
68. Manns M., Meyer zum Büschenfelde K.-H., Slusarczyk J. and Dienes H. P. Detection of liver–kidney microsomal autoantibodies by radioimmunoassay and their relation to anti-mitochondrial antibodies in inflammatory liver diseases. *Clin. Exp. Immunol.* 1984, **57**, 600–608.
69. Lenzi M., Bianchi F. B., Cassani F. and Pisi E. Liver cell surface expression of the antigens reacting with liver–kidney microsomal antibody (LKM). *Clin. Exp. Immunol.* 1984, **55**, 36–40.
70. Khoury E. L., Hammond L., Bottazzo G. F. and Doniach D. Presence of the organ-specific 'microsomal' autoantigen on the surface of human thyroid cells in culture: its involvement in complement-mediated cytotoxicity. *Clin. Exp. Immunol.* 1981, **45**, 316–328.
71. Khoury E. L., Hammond L., Bottazzo G. F. and Doniach D. Surface-reactive antibodies to human adrenal cells in Addison's disease. *Clin. Exp. Immunol.* 1981, **45**, 48–58.
72. Masala C., Smurra G., Di Prima M. A., Amendolea M. A., Celestino D. and Salsano F. Gastric parietal cell antibodies: demonstration by immunofluorescence of their reactivity with the surface of the gastric parietal cells. *Clin. Exp. Immunol.* 1980, **41**, 271–280.
73. Rizzetto M., Hoyer B., Canese M. G., Shih J. W. K., Purcell R. H. and Gerin J. L. Delta antigen: the association of delta antigen with hepatitis B surface antigen and ribonucleic acid in the serum of delta-infected chimpanzees. *Proc. Natl Acad. Sci. USA* 1980, **77**, 6124–6128.
74. Rizzetto M., Verme G., Recchia S., Bonino F., Farci P., Aricó S., Calzia R., Picciotto A., Colombo M. and Popper H. Chronic hepatitis in carriers of hepatitis B surface antigen with intrahepatic expression of the delta (δ) antigen. An active and progressive disease unresponsive to immunosuppressive treatment. *Ann. Intern. Med.* 1983, **98**, 437–441.
75. Crivelli O., Lavarini C., Chiaberge E., Amoroso A., Farci P., Negri F. and Rizzetto M. Microsomal antibodies in chronic infection with the HBs Ag associated delta (δ) agent. *Clin. Exp. Immunol.* 1983, **54**, 232–238.
76. Ellinger A. and Pavelka M. Effect of monensin on the Golgi apparatus of absorptive cells in the small intestine of the rat. Morphological and cytochemical studies. *Cell Tissue Res.* 1984, **235**, 187–194.
77. Rodriguez J. L., Gelpi C., Thomson M., Real F. J. and Fernandez J. Anti-golgi complex autoantibodies in a patient with Sjögren's syndrome and lymphoma. *Clin. Exp. Immunol.* 1982, **49**, 579–586.
78. Fritzler M. J., Etherington J., Sokoluk C., Kinsella D. and Valencia D. W. Antibodies from patients with autoimmune disease react with a cytoplasmic antigen in the Golgi apparatus. *J. Immunol.* 1984, **132**, 2904–2908.
79. Lazarides E. and Weber K. Actin antibody: the specific visualization of actin-containing filaments in nonmuscle cells. *Proc. Natl Acad. Sci. USA* 1974, **71**, 2268–2272.
80. Weber K., Pollack R. and Bibring T. Antibody against tubulin: the specific visualization of cytoplasmic microtubules in tissue culture cells. *Proc. Natl Acad. Sci. USA* 1975, **72**, 459–463.
81. Ishikawa H., Bischoff R. and Holtzer H. Mitosis and intermediate-sized filaments in developing skeletal muscle. *J. Cell Biol.* 1968, **38**, 538–555.
82. Köhler G. and Milstein C. Continuous culture of fused cells secreting antibody of predefined specificity. *Nature* 1975, **256**, 495–497.

83. Franke W. W., Schmid E., Osborn M. and Weber K. Different intermediate-sized filaments distinguished by immunofluorescence microscopy. *Proc. Natl Acad. Sci. USA* 1978, **75**, 5034–5038.
84. Herman I. M. and Pollard T. P. Comparison of purified anti-actin and fluorescent-heavy meromyosin staining patterns in dividing cells. *J. Cell Biol.* 1979, **80**, 509–520.
85. Weingarten M. D., Guter M. M., Littman D. R. and Kirschner M. W. Properties of the depolymerization products of microtubules from mammalian brain. *Biochemistry* 1974, **13**, 5529–5537.
86. Franke W. W., Schiller D. L., Moll R., Winter S., Schmid E. and Engelbrecht I. Diversity of cytokeratins. Differentiation specific expression of cytokeratin polypeptides in epithelial cells and tissues. *J. Mol. Biol.* 1981, **153**, 933–959.
87. Huiatt T. W., Robson R. M., Avakawa N. and Stromer M. H. Desmin from avian smooth muscle. *J. Biol. Chem.* 1980, **255**, 6981–6989.
88. Osborn M. and Weber K. Biology of disease. Tumor diagnosis by intermediate filament typing: A novel tool for surgical pathology. *Lab. Invest.* 1983, **48**, 372–394.
89. Rungger-Brändle E. and Gabbiani G. The role of cytoskeletal and cytocontractile elements in pathologic processes. *Am. J. Pathol.* 1983, **110**, 361–392.
90. Bottazzo G.-F., Florin-Christensen A., Fairfax A., Swana G., Doniach D. and Gröschel-Stewart U. Classification of smooth muscle autoantibodies detected by immunofluorescence. *J. Clin. Pathol.* 1976, **29**, 403–410.
91. Kurki P., Linder E., Virtanen I. and Stenman S. Human smooth muscle autoantibodies reacting with intermediate (100 Å) filaments. *Nature* 1977, **268**, 240–241.
92. Gabbiani G., Ryan G. B., Lamelin J.-P., Vassalli P., Majno G., Bouvier C. A., Cruchaud A. and Lüscher E. F. Human smooth muscle autoantibody. Its identification as anti-actin antibody and a study of its binding to 'non-muscular' cells. *Am. J. Pathol.* 1973, **72**, 473–488.
93. Lidman K., Biberfeld G., Fagraeus A., Norberg R., Torstensson R. Utter G., Carlsson L., Luca J. and Lindberg U. Anti-actin specificity of human smooth muscle antibodies in chronic active hepatitis. *Clin. Exp. Immunol.* 1976, **24**, 266–272,
94. Mead G. M., Cowin P. and Whitehouse J. M. A. Antitubulin antibody in healthy adults and patients with infectious mononucleosis and its relationship to smooth muscle antibody. *Clin. Exp. Immunol.* 1980, **39**, 328–336.
95. Linder E., Letho V.-P., Stenman S., Lindqvist K., Bjorvatn B. and Bergquist R. Circulating antibodies to connective tissue myofibrils and dermal immunoglobulin deposits in leprosy. *Clin. Immunol. Immunopathol.* 1979, **13**, 1–8.
96. Mortazavi-Milani S. M., Badakere S. S. and Holborow E. J. Antibody to intermediate filaments of the cytoskeleton in the sera of patients with acute malaria. *Clin. Exp. Immunol.* 1984, **55**, 177–182.
97. Therien H.-M., Gruda J. and Carrier F. Interaction of filamentous actin with isolated liver plasma membranes. *Eur. J. Cell Biol.* 1984, **35**, 112–121.
98. Moll R. and Franke W. W. Intermediate filaments and their interaction with membranes. The desmosome–cytokeratin filament complex and epithelial differentiation. *Pathol. Res. Pract.* 1982, **175**, 146–161.
99. Braun J. and Unanue E. R. Surface immunoglobulin and the lymphocyte cytoskeleton. *Fed. Proc.* 1983, **42**, 2446–2451.

30

Immunocytochemistry in Cardiac and Renal Transplantation

M. L. Rose and A. Suitters

Rejection remains the main barrier to successful allogeneic organ transplantation. The detection and prevention of acute rejection is therefore of critical importance for the management of transplant patients. Experimental studies have demonstrated that during the rejection of allotransplanted heart or kidney, mononuclear cells accumulate in the rejected organ.[1] Earlier studies utilized techniques such as E-rosettes and cytochemical markers to analyse the mononuclear infiltrate, and it was established that many of the infiltrating cells of early acute renal rejection were T-lymphocytes and macrophages.[2] The advent of monospecific monoclonal antibodies against leucocyte and tissue antigens has greatly increased the potential for investigating cellular interactions during graft rejection. It is established that T-cell subpopulations (helper/inducer cells and suppressor/cytotoxic cells) recognize and interact with tissue antigens which are coded for by genes of the major histocompatibility locus (*Fig.* 30.1). The relative importance of cytotoxic T-cells and delayed hypersensitivity in eliciting graft rejection remains unresolved.[3] Use of monoclonal antibodies against T-cell subsets and the non-polymorphic determinants of class I (HLA-ABC) and class II (DP, DQ and DR) antigens on frozen sections of kidney or heart help elucidate the cellular interactions involved in rejection. Such antibodies can also be used to analyse the proportions of lymphocyte subpopulations present in peripheral blood, with the hope of obtaining a non-invasive method of detecting early rejection.

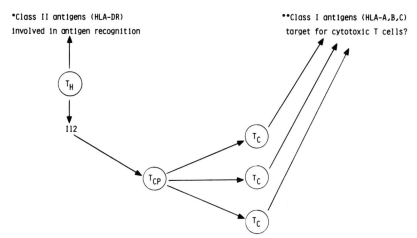

*Class II antigens (HLA-DR)
Involved in antigen recognition

**Class I antigens (HLA-A,B,C)
target for cytotoxic T cells?

* Present on macrophages, dendritic cells, endothelia, B cells, activated T cells.
  Induced in transplantation

** Present on many (but not all) nucleated cells. Found in low quantities on normal
   myocardium. Induced in transplantation.

Fig. 30.1. Interaction of T-cells and tissue antigens. Th = helper T-cell; Tcp = cytotoxic precursor
T-cell; Tc = cytotoxic T-cell; Il 2 = interleukin 2.

1. CARDIAC TRANSPLANTATION

1.1. Cardiac Biopsies

1.1.1. Nature of the Cellular Infiltrate

Transvenous endomyocardial biopsy is routinely performed on all cardiac
transplant patients. Histological assessment of cardiac biopsies is at present the
most reliable method of detecting rejection, and it is based on the presence and
appearance of a mononuclear cell infiltrate using conventional histological
criteria.[4] In order to investigate subpopulations of cells in the biopsies using
monoclonal antibodies, it is necessary to use frozen tissue. At each biopsy session
four pieces of right ventricular myocardium are removed. Three pieces are
processed for conventional histology and one piece is frozen. Pieces of tissue are
embedded in Tissue Tek on cork and snap-frozen in isopentane which is precooled
in liquid nitrogen. Cryostat sections (6 μm thick) are cut, air-dried and fixed in
acetone. The sections are covered in aluminium foil and stored at −40 °C until
use. The monoclonal antibodies used routinely in this laboratory are described in
Table 30.1. The immunoperoxidase technique used to visualize the monoclonal
antibodies is a triple layer including biotinylated goat anti-mouse IgG and IgM
(from TAGO products, TCS) and the avidin–biotin–horseradish peroxidase
complex (from Sera Labs) as recommended by Hsu et al.[8] This technique has been
found to be more sensitive than a double-sandwich technique on cardiac biopsies.

*Table 30.1. **Monoclonal antibodies used to analyse cellular interactions in cardiac biopsies***

Name	Specificity	Dilution	Source*
Leu-4	All T-cells	1/40	Becton-Dickinson
Leu-2a	Suppressor/cytotoxic T-cells	1/40	Becton-Dickinson
Leu-3a	Helper/inducer T-cells	1/40	Becton-Dickinson
HLA-DR	Non-polymorphic class II antigens	1/10	Becton-Dickinson
NFK-1	Non-polymorphic class II antigens	1/100	(5)
W6/32	Non-polymorphic class I antigens	1/2	(6)
F10-894	Common leucocyte antigen	1/80	(7)

*Reference sources are given in parentheses.

When the primary mouse monoclonal antibody is omitted, neither the secondary antibody nor the ABC complex produces non-specific staining on cardiac biopsies.

A study was performed correlating the number and phenotype of T-cells present in cardiac biopsies with the assessment of rejection in the same biopsies using conventional histological criteria.[9] It was found that biopsies showing signs of rejection contained more T-cells than those not showing signs of rejection. However, the type of immunosuppressive drug which the patient was receiving was found to affect the distribution of T-cells. Thus biopsies from patients receiving azathioprine and corticosteroids contained discrete interstitial cells (*Fig.* 30.2), whereas those from patients also taking cyclosporine contained large pools of cells many of which were found to be T-lymphocytes (*Fig.* 30.3). In patients receiving conventional immunosuppression (azathioprine and corticosteroids), rejection was associated with an influx of T-lymphocytes of suppressor/cytotoxic phenotype, this type always predominating over the helper/inducer phenotype. In contrast, biopsies from patients receiving cyclosporine sometimes contained large numbers of helper/inducer cells and there was no clear predominance of either subpopulation to denote rejection. Thus, the use of monoclonal antibodies against T-cell subsets is useful in diagnosing rejection in patients receiving conventional immunosuppression, but has not greatly helped the diagnosis of rejection in patients receiving cyclosporine.

1.1.2. Class I and Class II Antigens

Class I (HLA=ABC) antigens are highly polymorphic glycoproteins encoded by the genes of the major histocompatibility complex found on chromosome 6 in

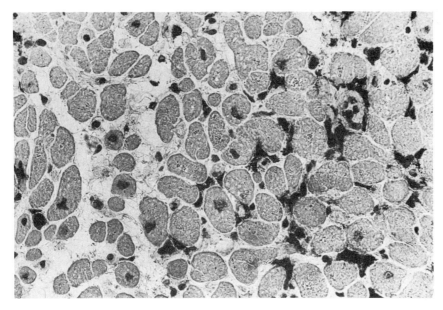

Fig. 30.2. Binding of Leu-4 (all T-cells) to cardiac biopsy showing mild rejection. Patient receiving azathioprine and corticosteroid immunosuppression. Cryostat section, 5 μm, fixed in acetone.

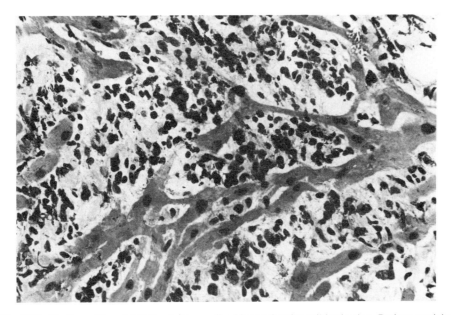

Fig. 30.3. Binding of Leu-4 (all T-cells) to cardiac biopsy showing mild rejection. Patient receiving azathioprine, corticosteroid and cyclosporine immunosuppression. Cryostat section, 5 μm, fixed in acetone.

man.[10] The distribution of class I antigens in all tissues and organs has not been well documented. Until recently, it was thought that all nucleated cells expressed class I antigen.[11] Such studies were based on the use of cell suspensions and hetero- or allo-antisera as cytotoxic reagents. The use of frozen sections, monoclonal antibodies and enzyme-linked secondary antibodies makes investigation of the distribution of antigens in fixed tissue, such as muscle, much more reliable. We have investigated the distribution of the non-polymorphic class I determinant detected by monoclonal antibody W6/32 on normal non-transplanted myocardium and have examined the effect of allotransplantation on expression of this determinant.[12] We found little or no expression of the determinant detected by W6/32 on normal myocardium, interstitial structures such as blood vessels and leucocytes being strongly positive (*Plate* 32). In contrast, 24 out of 28 biopsies from transplanted heart showed heavy staining of the myocardium (*Plate* 33). The absence or paucity of class I antigen on normal myocardium accords with a recent study[13] which also found poor binding of a monoclonal antibody against class I on myocardium. Normal skeletal muscle also lacks the antigen detected by W6/32.[14] The significance of the induction of class I antigen on transplanted heart is not clear at present. Of the biopsies showing histological signs of rejection, all contained strongly positive myocardium. However, of the biopsies not showing histological signs of rejection, 10 out of 14 also contained class I positive myocardium. It may be that class I positivity precedes rejection and is a particularly sensitive method of detecting this process.

Class II antigens are also tissue antigens encoded by genes of the major histocompatability locus. They have a more restricted distribution than class I antigens (*Fig.* 30.1). Monoclonal antibodies against class II antigens (HLA-DR and NFK-1) were found to bind to vascular endothelium in sections of normal heart and some interstitial cells. In transplanted heart there was increased binding to the blood vessels and an increase in the number of interstitial cells binding the monoclonal antibodies.[15] Myocardium was found not to bind anti-class II antibodies either on normal heart or on transplanted heart. The significance of the induction of class I and class II antigens in the transplanted heart has not yet been established. Expression of class I antigens is a necessary prerequisite for T-cell cytotoxicity against cells expressing viral[16] or allo-antigens.[17] Class II antigens are necessary for antigen recognition by helper T-cells. It seems likely that increased expression of these tissue antigens on transplanted organs could initiate or accelerate the process of graft rejection. It is known that gamma-interferon from activated T-cells can induce class I expression on human cell lines *in vitro*[18] and on cultured endothelial cells.[19] Increased expression of class II antigens on renal allografts has been found associated with rejection episodes.[20] Further studies are being carried out in this laboratory to establish the significance of class I and class II induction in transplanted heart and whether this phenomenon has any clinical diagnostic value.

1.2. Peripheral Blood

It would be extremely useful to have a non-invasive haematological assay which was predictive of rejection in heart transplant patients. A great deal of work has been devoted to studying the ratios of helper/inducer (T4) and cytotoxic/suppressor (T8) cells in the peripheral blood of transplant patients, especially renal transplant patients (*see below*). Such studies are based on the premise that, in the absence of disease, the T4 : T8 ratio in peripheral blood is relatively constant. A study of normal controls[21] has demonstrated a wide range of values for numbers of T4 and T8 cells in peripheral blood, but the values are reproducible enough to be of clinical use. A diurnal variation in the number of circulating cells has been reported.[22] It has been suggested that an increased T4 : T8 ratio is predictive of rejection[23] in renal transplant patients. A study in this laboratory was therefore undertaken.

Ten to twenty ml of heparinized venous blood were removed from patients on the same day that they received cardiac biopsy; thus the T4 : T8 ratio could be compared with the histological assessment of rejection. The mononuclear cells were separated on Ficoll–Hypaque and the T-cell subsets were identified using monoclonal antibodies of the Leu series followed by fluoresceinated goat anti-mouse immunoglobulins. Cells were visualized using fluorescence microscopy. All patients were found to have a low T4 : T8 ratio ($1 \cdot 1 \pm 0 \cdot 3$) compared to normal ($1 \cdot 71 \pm 0 \cdot 5$) but there was no significant change which correlated with the occurrence of rejection episodes. These studies are fraught with difficulties in immunosuppressed patients. They have low numbers of circulating lymphocytes and moreover up to 50 per cent of cells may be lost in the separation procedure. Treatment of an acute rejection episode with anti-thymocyte globulin produces damaged cells which have an altered buoyant density and this makes the separation of mononuclear cells from polymorphs and erythrocytes incomplete. The recent development[24] of a technique utilizing whole blood smears has been of great value in this field. Using only 2–3 drops of blood, monoclonal antibodies of the Leu series and the sensitive alkaline phosphatase–anti-alkaline phosphatase monoclonal antibody technique, it is possible to perform daily assays. It is hoped that this technique will clarify whether changes in T4 : T8 ratios are indicative of rejection.

2. RENAL TRANSPLANTATION

2.1. Renal Biopsies

2.1.1. Nature of the Cellular Infiltrate

Much of the work on renal allografts has centred on the identification of the cellular infiltrate within the graft and its relevance to the incidence of rejection.[25–28] The current methods of obtaining renal tissue are needle core and fine needle aspirate biopsies. The needle core biopsies are snap-frozen in isopentane for storage in liquid nitrogen. Sections are then cut on a cryostat in the conventional manner. Cytocentrifuge smears are prepared from the aspirate biopsies. The antibodies most commonly used to identify cells in these preparations are shown in *Table* 30.2. Immunofluorescence or immunoenzymatic techniques have been utilized to identify the primary antibody.

Most workers have reported a predominance of T-cells within the graft during rejection.[26–30] Of these T-cells the majority are positive with the antibody recognizing the cytotoxic/suppressor subset.[25–27,29,30] This predominance of the cytotoxic/suppressor subset over the helper/inducer subset is reflected in the

Table 30.2. **Monoclonal antibodies used to analyse cellular interactions in renal biopsies**

Name	Specificity	Source*
OKT 3 Leu-4 T11, T28 9·6	All T-cells	Ortho Pharmaceutical Becton-Dickinson Coulter Electronics (25)
OKT 4 Leu-3a T4	Helper/inducer T-cells	Ortho Pharmaceutical Becton-Dickinson Coulter Electronics
OKT 8 Leu-2a T8	Suppressor/cytotoxic T-cells	Ortho Pharmaceutical Becton-Dickinson Coulter Electronics
TA-1	T-cells, monocytes	(26)
F10-89-4 PHM-1	Leucocyte common antigen	(32) (25)
PHM-2 FMC-32 FMC-33	Macrophages/monocytes	(25) (25) (25)
OKM 1	Monocytes/NK cells, granulocytes	Ortho Pharmaceutical
FMC-10	Polymorph	(25)
BA-4	B-cells/PMN cells	(26)
7·2	B-cells	(25)
OKT 10	Plasma cells	Ortho Pharmaceutical
HLA-DR OK 1a-1 I2	Non-polymorphic class II antigens	Becton-Dickinson Ortho Pharmaceutical (20)

*Reference sources are given in parentheses.

T4 : T8 ratio which is lower than that seen in non-rejecting kidney.[27,31,32] Contrary to these data, some workers have shown that T-cells of the helper/inducer subset are more commonly found during moderate to severe rejection.[28]

There is some evidence to suggest a role played by leucocytes other than lymphocytes in graft rejection. McWhinnie et al.[32] found that non-lymphocytes staining with the monoclonal antibody F10-89-4 formed a larger part of the cellular infiltrate than the lymphocytes. Hancock et al.[25] also found that macrophages formed a large part of the infiltrating cells. They found that severe rejection, compared to mild or moderate rejection, was characterized by a decrease in the proportion of T-cells (expressed as a proportion of the total leucocytes) and an increase in the proportion of macrophages or polymorphs and suggested that this was probably associated with the increase in cellular destruction.

2.1.2. Class II Antigens

Recent evidence analysing the distribution and site of expression of class II antigens in normal and rejecting kidney has shown an increase in expression of these antigens on the tubular cells during rejection.[20] Using an immunoperoxidase technique, it was shown that the increase in class II expression appeared several days before deterioration of renal function on the normally class II-negative tubular cells. There was also an increase of class II expression on the monocytic cellular infiltrate. The tubular cells continued to express the class II following anti-rejection therapy and, therefore, could only be used as a possible marker of rejection following the first rejection episode.

2.2. Peripheral Blood

The initial report[23] that changes in peripheral blood T4 : T8 ratios are associated with rejection episodes stimulated many investigations into the reliability of this technique. Peripheral blood lymphocytes are obtained by separation of the blood over Ficoll–Hypaque, followed by incubation of the mononuclear cells with the relevant monoclonal antibodies (*see Table* 30.2). A fluoresceinated second antibody is used to identify the first antibody before analysis by flow-cytometry. There is now conflicting evidence as to the usefulness of this technique. Some workers have shown that an increase in the T4 : T8 ratio in the blood coincided with rejection episodes in the kidney as visualized in biopsies.[33,34] Other workers have shown a slight decrease in the T4 : T8 ratio coincided with rejection.[31] However, this decrease was not seen during all rejection episodes and was not conclusive. This discrepancy agreed with the data of other workers who have also shown that there is no correlation between the T4 : T8 ratio and events occurring in the graft.[31,32] The observations that opportunistic viral infections, commonly found in immunosuppressed transplant patients, produce inversion of the T4 : T8 ratios,[35,36] complicates these studies and makes them difficult to interpret unless viral serology is performed concomitantly with the lymphocyte marker studies.

3. CONCLUSIONS

Exciting results have been obtained using immunocytochemical methods of analysing the process of acute rejection in cardiac and renal allografts. In particular, studies on the expression of class I and class II antigens seem likely to yield interesting results which could form the basis of a method for detecting early rejection or assessing the efficiency of anti-rejection therapy. The analysis of T-cell subsets in the blood is still controversial as a method for predicting rejection and further exploration of blood using different cell markers is necessary before it can be used as a reliable aid to diagnosis of rejection.

Acknowledgements

The financial support of the British Heart Foundation is acknowledged.

REFERENCES

1. Tilney N. L., Strom T. B., Macpherson S. G. and Carpenter C. B. Surface properties and functional characteristics of infiltrating cells harvested from acutely rejecting cardiac allografts in inbred rats. *Transplantation* 1975, **20**, 323–330.
2. Von Willebrand E. and Hayry P. Composition and *in vitro* cytotoxicity of cellular infiltrate in rejecting human kidney allografts. *Cell. Immunol.* 1978, **41**, 358–372.
3. Mason D. W., Dallman M. J., Arthur R. P. and Morris P. J. Mechanisms of allograft rejection: The roles of cytotoxic T-cells and delayed hypersensitivity. *Immunol. Rev.* 1984, **77**, 167–184.
4. Billingham M. Some recent advances in cardiac pathology. *Human Pathol.* 1979, **10**, 365–370.
5. Fuggle S. V., Errasti P., Daar A. S., Fabre J. W., Ting A. and Morris P. J. Localisation of major histocompatability complex (HLA-ABC and DR) antigens in 46 kidneys. *Transplantation* 1983, **35**, 385–392.
6. Barnstable C. J., Bodmer W. F., Brown G., Galfré G., Milstein C., Williams A. F. and Ziegler A. Production of monoclonal antibodies to group A erythrocytes, HLA and other human cell surface antigens—new tools for genetic analysis. *Cell* 1978, **14**, 9–20.
7. Dalchau R., Kirkley J. and Fabre J. W. Monoclonal antibody to a human leucocyte-specific membrane glycoprotein probably homologous to the leucocyte-common (L-C) antigen of the rat. *Eur. J. Immunol.* 1980, **10**, 737–744.
8. Hsu S. M., Raine L. and Fanger M. Use of avidin–biotin–peroxidase complex (ABC)in immunoperoxidase techniques: A comparison between ABC and unlabelled antibody (PAP) procedures. *J. Histochem. Cytochem.* 1981, **29**, 577–580.
9. Rose M. L., Gracie J. A., Fraser A., Chisholm P. M. and Yacoub M. H. Use of monoclonal antibodies to quantitate T lymphocyte subpopulations in human cardiac allografts. *Transplantation* 1984, **38**, 230–234.
10. Jongsma A., Someven M., Westerwald A., Hagemeijer A. and Pearson P. Localisation of human chromosomes by studies of human Chinese hamster somatic cell hybrids. *Human Genet.* 1973, **20**, 195–200.
11. Berah M., Hors J. and Dausset J. A study of HLA antigens in human organs. *Transplantation* 1970, **9**, 185–192.
12. Rose M. L., Coles M. I., Griffin R. J., Pomerance A. and Yacoub M. H. Expression of class I and class II major histocompatibility antigens on normal and transplanted human heart. *Transplantation*, in press.
13. Daar A. S., Fuggle S. V., Fabre J. W., Ting A. and Morris P. J. The detailed distribution of HLA-ABC antigens in normal human organs. *Transplantation* 1984, **38**, 287–292.
14. Appleyard S. T., Dunn M. J., Dubowitz V. and Rose M. L. Increased expression of HLA ABC Class I antigens by muscle fibres in Duchenne muscular dystrophy, inflammatory myopathy and other neuromuscular disorders. *Lancet* 1985, **i**, 361–363.
15. Rose M. L., Coles M. I., Griffin R. J. and Yacoub M. H. Induction of class I and class II antigens during rejection of allografted human hearts. *Adv. Exp. Med. Biol.* 1985, **186**, 555–565.
16. McMichael A. J., Ting A., Zwelvink H. F. and Askonas B. A. HLA restriction of cell mediated lysis of influenza virus infected human cells. *Nature* 1977, **270**, 524–526.
17. Zinkernagel R. M. and Docherty P. C. MHC cytotoxic T cells: Studies on the biological role of polymorphic major transplantation antigens determing T cell restriction—specificity, function and responsiveness. *Adv. Immunol.* 1979, **27**, 51–177.
18. Fellows M., Kamoun M., Gresser I. and Bono R. Enhanced expression of HLA antigens and B2-microglobulin on interferon-treated lymphoid cells. *Eur. J. Immunol.* 1979, **9**, 446–449.
19. Pober J. S., Gimbrone M. A., Cotran R. S., Reiss C. S., Burakoff S. J., Fiers W. and Ault K. A. Ia expression by vascular endothelium is inducible by activated T cells and by human gamma interferon. *J. Exp. Med.* 1983, **157**, 1339–1353.
20. Hall B. M., Bishop G. A., Duggin G. G., Horvath J. S., Phillips J. and Tiller D. J. Increased expression of HLA-DR antigens on renal tubular cells in renal transplants: relevance to the rejection response. *Lancet* 1984, **ii**, 247–251.
21. Goff L. K., Habershaw J. A., Rose M. L., Gracie J. A. and Gregory W. Normal values for the different classes of venous blood mononuclear cells defined by monoclonal antibodies. *J. Clin. Pathol.* 1985, **38**, 54–59.
22. Ritchie A. W. S. and Oswald I. Circadian variation of lymphocyte subpopulations: a study with monoclonal antibodies. *Br. Med. J.* 1983, **286**, 1773–1775.
23. Cosimi A. B., Colvin R. B., Burton R. C., Rubin R. H., Goldstein G., Kung P. C., Hansen W. P., Delmonico F. L. and Russell P. S. Use of monoclonal antibodies to T-cell subsets for

immunologic monitoring and treatment in recipients of renal allografts. *N. Engl. J. Med.* 1981, **305**, 308–314.

24. Erber W. N., Pinching A. J. and Mason D. Y. Immunocytochemical detection of T and B cell populations in routine blood smears. *Lancet* 1984, **i**, 1042–1045.
25. Hancock W. W., Thomson N. M. and Atkins R. C. Composition of interstitial cellular infiltrates identified by monoclonal antibodies in renal biopsies of rejecting human renal allografts. *Transplantation* 1983, **35**, 458–463.
26. Platt J. L., LeBien T. W. and Michael A. F. Interstitial mononuclear cell populations in renal graft rejection. *J. Exp. Med.* 1982, **155**, 17–30.
27. Wood R. F. M., Bolton E. M., Thompson J. F., Morris P. J. and Mason D. Y. Characterisation of cellular infiltrates in renal allografts by fine needle aspiration: a simple technique using double labelling with monoclonal antibodies. *Transplant. Proc.* 1983, **15**, 1847–1848.
28. Hall B. M., Bishop G. A., Farnsworth A., Duggin G. G., Horvath J. S., Sheil A. G. R. and Tiller D. I. Identification of the cellular subpopulation infiltrating rejecting cadaver renal allografts. *Transplantation* 1984, **37**, 564–570.
29. Hammer C., Lard W., Stadler J., Koller C. and Brerdel W. Lymphocyte subclasses in rejecting kidney grafts detected by monoclonal antibodies. *Transplant. Proc.* 1983, **15**, 356–359.
30. Morris P. J., Carter N. P., Cullen P. R., Thompson J. F. and Wood R. F. M. Role of T-cell subset monitoring in renal allograft recipients. *N. Engl. J. Med.* 1982, **306**, 1110–1111.
31. Thompson J. F., Carter N. P., Bolton E. M., McWhinnie D. L., Wood R. F. M. and Morris P. J. The composition of the lymphocyte infiltrate in rejecting human renal allografts is not reflected by lymphocyte subpopulations in the peripheral blood. *Transplant. Proc.* 1985, **17**, 556–557.
32. McWhinnie D. L., Thompson J. F., Taylor, H. M., Chapman J. R., Bolton E. M., Wood R. F. M. and Morris P. J. Leucocyte infiltration patterns in renal allografts assessed by immunoperoxidase staining of 245 sequential biopsies. *Transplant. Proc.* 1985, **17**, 560–561.
33. Ellis T. M., Berry C. R., Mendez-Picon G., Goldman M. H., Lower R. R., Lee H. M. and Mohanakumar T. Immunological monitoring of renal allograft recipients using monoclonal antibodies to human T lymphocyte subpopulations. *Transplantation* 1982, **33**, 317–319.
34. Mazaheri R., Laupacis A., Keown P., Howson W., Sinclair N. R. and Stiller C. R. Lymphocyte subsets in the allograft recipient: correlation of helper to suppressor ratio with clinical events. *Transplant. Proc.* 1982, **14**, 676–678.
35. Schooley R. T., Hirsch M. S., Colvin R. B., Cosimi A. B., Tolkoff-Rubin N. E., McCluskey R. T., Burton R. C., Russell P. S., Herrin J. T., Delmonico F. L., Giorgi J. V., Henle W. and Rubin R. H. Association of herpes virus infections with T-lymphocyte subset alterations, glomerulopathy and opportunistic infections after renal transplantation. *N. Engl. J. Med.* 1983, **308**, 307–313.
36. Dummer J. S., Ho M., Rubin B., Griffith B. P., Hardesty R. L. and Bahnson H. T. The effect of cytomegalovirus and Epstein–Barr virus infection on T lymphocyte subsets in cardiac transplant patients on cyclosporine. *Transplantation* 1985, **38**, 433–436.

31   *Tumour Markers*

E. Heyderman

Tumour markers are tumour-derived or associated products which may be used either to diagnose or to monitor malignant disease. Most are also synthesized by non-neoplastic tissues.

In immunocytochemistry, the term is confined to those substances which are secreted and stored by the tumour or its stroma, and which may be demonstrated in tissue sections, smears, monolayer cultures, or cell suspensions by immunocytochemical techniques. Some, like most of the hormones, are present in the cytoplasm. Others, like carcinoembryonic antigen (CEA)[1] and blood group substances, are generally expressed on the cell membrane. Markers, such as CEA, when found in moderately well differentiated adenocarcinomas, tend to be mainly localized to the luminal surface of the malignant acini. In squamous cell carcinomas, the distribution is pericellular,[2] while in poorly differentiated tumours it may be cytoplasmic.[3] Markers which are not produced by the tumour itself, like acute phase proteins, but are of value as circulating markers of disease progression, are not included in this presentation.

The criteria for tumour production of these markers (*Table* 31.1) should be stringent.[4] As far as tissue markers are concerned, since endocytosis of hormones and other substances into the Golgi[5] or rough endoplasmic reticulum[6] may occur, demonstration at ultrastructural level in the Golgi apparatus or in the rough endoplasmic reticulum is insufficient evidence of secretion. In tissue sections, therefore, it is only possible to demonstrate storage. To prove synthesis, other techniques need to be employed. These include incorporation of radioactive precursors into the tumour *in vivo*, or *in vitro*, or secretion of the marker into the medium by cultured cells.[7] It may also be possible to show continued secretion of a tumour marker in a human tumour xenograft in an immune-deprived mouse,[8] or into the circulation of the mouse host.[9] The demonstration of cytoplasmic mRNA for these products by in situ hybridization

Table 31.1. Criteria for establishing the diagnosis of ectopic hormone secretion

1. Association of the neoplasm with a clinical syndrome
2. Association with inappropriately elevated plasma and/or urine levels
3. Failure of urine and/or plasma levels to respond to normal suppression
4. Exclusion of other possible causes
5. Fall in hormone levels after tumour therapy
6. Arteriovenous step-up gradient across the tumour
7. Demonstration of hormone in tumour by extraction or immunocytochemical localization
8. *In vitro* (tissue culture) hormone synthesis and release, allowing sufficient time for any entrapped hormone to be lost, measurement of the hormone in the medium or cell homogenate, or immunocytochemical localization in whole cells or in sections of the cell pellet
9. Demonstration of hormone synthesis *in vivo* by RIA of tumour extracts or of host tissue fluids of animal bearing xenograft, or by immunocytochemical localization, allowing time for entrapped hormone to be lost, and preferably after several passages
10. *In vitro* cell-free translation of tumour hormone-specific messenger RNA
11. Demonstration of specific mRNA by in situ hybridization

using either isotopically labelled or biotinylated probes is under investigation in several centres.[10–12]

1. POTENTIAL VALUE OF IMMUNOPEROXIDASE TUMOUR MARKER PROFILES

The value of tumour marker localization falls into five main categories:

a. *Determination of the site of an occult primary tumour* which presents with metastatic disease.[13–15]

b. *Differential diagnosis* of morphologically similar lesions, as in deciding whether a single cell malignant infiltrate represents lymphomatous infiltration or secondary deposits of lobular carcinoma of the breast,[13] confirming that sarcoma-like renal lesions represent sarcomatoid change in a hypernephroma,[14] or distinguishing between metastatic adenocarcinoma and pleural mesothelioma.[16–19]

c. *Functional, as well as morphological, classification* of malignant disease. This is of value, for example, in the classification of testicular tumours, where it may be difficult on morphological grounds alone to determine whether giant cells are of trophoblastic lineage, or are of undetermined tumour giant cell type, or are foreign-body type giant cells in a granulomatous response to necrosis. Here the demonstration of intracellular human chorionic gonadotrophin (HCG),[20] taken together with the morphology, is conclusive.

d. *Prognosis.* There have been conflicting reports on the prognostic significance of the presence of a variety of markers in breast carcinomas.[21–24] Our study of 44 breast carcinomas has failed to confirm any prognostic significance,[25,26] and larger studies on the presence of HCG in malignant testicular teratomas[27–29] have not confirmed the suggestion that the presence of HCG may impart a worse prognosis.[30] The best candidate for correlation with prognosis seemed to be deletion of blood group substances, especially in the bladder.[31] However, doubt has been cast on the use of fixed sections of bladder tumours for demonstration of ABO iso-antigens, and the use of these results in prognosis,[32] and a study of blood group substance A in breast cancer failed to show any adverse prognostic significance in its loss.[33]

e. *Marker prediction.* There is general agreement that only a proportion of breast carcinomas stain for CEA.[34] If this indicates failure of some breast cancers to secrete CEA, routine measurement of circulating levels in all breast cancer patients would result in failure to detect recurrence in those patients whose tumours do not secrete CEA, but who have widespread metastatic disease. Examination of a biopsy or of the mastectomy specimen could help to define this subset.

There is evidence from tissue culture studies[27] that some tumour cells secrete CEA, but have a failure of transport mechanism and are unable to excrete it. The localization of CEA in all sections of colorectal and gastric carcinomas so far studied,[25,29] while only a proportion have raised levels on metastasis, may be partly explained on this basis.

In the case of those patients with malignant testicular teratomas, who do not have raised postorchidectomy levels of HCG or of alpha-fetoprotein (AFP), demonstration of HCG or AFP in the original orchidectomy specimen would support a clinical view that there was no residual disease, since the primary tumour was capable of secretion. Where neither HCG nor AFP can be demonstrated, negative serum levels are less reliable, and patients with such tumours would be at higher risk, since biochemical determination of elevated levels is often possible long before recurrence is clinically detectable. In marker-silent patients the clinician would have to depend on physical examination, X-rays, computerized axial tomography (CAT)-scan, or ultrasound. At present all of these are poor determinants of recurrence compared with biochemical measurement of serum markers.

The demonstration of tissue markers by immunocytochemical techniques in the biopsy or resection specimen would also help to determine the most suitable antibody to use for tumour localization *in vivo* by injection of radiolabelled antibodies.[35-39]

Similarly, an indication of the most suitable marker would be of use in tumour targeting, using antibodies labelled with a variety of toxic agents,[40] such as chlorambucil,[41] diphtheria toxin[42] or cytotoxic monoclonal antibodies.[43]

However, in any histological examination of tissue, unless the whole specimen is examined by serial sectioning, there is always the possibility of sampling error. A positive result, therefore, has more credibility than a negative finding, provided that appropriate controls are carried out. Epithelial markers such as CEA tend to be present in most blocks examined from any one tumour, but other tumour products, like the alpha subunit of HCG, have a more patchy distribution and sampling error may be considerable unless multiple blocks have been examined.[44]

2. IMMUNOCYTOCHEMICAL METHODS

Localization of tumour products may be achieved by the use of antibodies labelled in a variety of ways. Fluorescein and rhodamine are used for immuno-fluorescence.[45] The most widely used enzyme labels are horseradish peroxidase[46] and alkaline phosphatase.[47] The latter may be used either alone, or together with a peroxidase-labelled antibody to demonstrate two antigens in the same tissue section.[48,49]

The immunoperoxidase technique has the advantage of an insoluble permanent reaction product, so that by using suitable counterstains, good morphological detail is retained. With immunofluorescence, although the method is simpler, requiring no inhibition of endogenous enzyme or incubation in substrate, it is difficult to assess the surrounding morphology and, with time and photographic exposure, the fluorescence fades. If the antigen under study survives fixation and processing, the immunoperoxidase technique, like other immunocytochemical techniques, may be used on formalin-fixed, paraffin-embedded tissue many years old.

Excellent results using antibodies labelled with colloidal gold have been reported.[50] The colloidal gold technique is suitable both for light and electron microscopy and there are no problems with endogenous activity. It has not, however, yet proved possible to label with gold particles of sufficient size to achieve as dense a reaction product as with peroxidase. This is no problem where the staining is intense and the distribution of the marker is well established. Where the antigen under test is in small amount, particularly on the cell surface, and the distribution is unexpected, interpretation is more difficult. However, enhancement of the gold label with silver increases the sensitivity of the reaction (*see* Chapter 5).

Antibodies labelled with radioactive isotopes may be applied to tissue sections and the preparations subjected to autoradiography[51] (and *see* Chapter 7). There are several problems with radiolabelled antibodies. Special safety facilities are necessary, non-specific scatter occurs, and the incubation time for autoradiography may be unacceptably long. However, sensitivity is increased and the method may be combined with immunocytochemistry to demonstrate two products in the same section.[52] Uptake of tritium-labelled thymidine, as well as HCG localization, has been shown in sections of metastatic testicular choriocarcinoma.[8]

Immunoperoxidase methods may employ directly labelled antibodies, double[53] or triple layers,[54] the peroxidase–anti-peroxidase (PAP)[55] or the labelled antigen method.[56] Since it is the primary specific antibody which is the most valuable and some activity is inevitably lost in conjugation, we prefer the indirect[57] to the direct method in which the first antibody is labelled with peroxidase. Experiments comparing commercially available PAP reagents and indirect peroxidase conjugates prepared in our laboratory have failed to confirm an increased sensitivity for the PAP method, and the indirect technique is shorter and less expensive. The PAP method is the most widely used, possibly because good commercial PAP reagents were available before good peroxidase conjugates. However, previous advocates of the PAP technique[58] have found the indirect method equally sensitive.[59]

Affinity-purified second, anti-species, antibodies and a periodate oxidation technique[60] are used in our laboratory. Few conjugates are required, one for each species of first antibody used. Antisera to sheep immunoglobulin cross-react with goat immunoglobulin, and rat monoclonal hybridoma antibodies may be recognized by anti-mouse peroxidase conjugates.[61] Peroxidase conjugates are stored diluted at a concentration determined by reference to a series of antisera and known positives. They are made into aliquots in azide-free 1 per cent ovalbumin in PBS pH 7·2 and stored at −20 °C.[62]

To avoid false positives due to the presence of endogenous peroxidase, the enzyme may be inhibited by a sequence of 6 per cent hydrogen peroxide to bleach any acid haematin (particularly important in haemorrhagic tumours) and commence inhibition; 2·5 per cent periodic acid, to complete the inhibition and improve sharpness of histological detail; and then 0·02 per cent potassium borohydride to reduce to basic alcohols any 'sticky' aldehyde groups produced either by the periodic acid or by fixation in aldehyde fixatives.[61] Endogenous alkaline phosphatase may be inhibited by the same schedule[63] (see Appendix for method).

At present (1985), polyclonal antisera raised in animals are still widely used. Specificity may be improved by affinity purification,[64] but with the advent of monoclonal antibodies of defined specificity,[65] affinity purification may only be required for ascites preparations. Excellent results may be achieved with monoclonal antibodies on fixed tissues,[66,67] as well as on cryostat sections.

2.1. Controls

It is always necessary to include a known positive tissue, else negative results cannot be assessed. When evaluating a new antibody, the first step is to find a positive tissue on which to determine the most suitable dilution and then, where possible, to find a positive control among the tissues under study. For example, in a study of the localization of CEA in lung tumours, it is advisable to use a CEA-positive lung tumour as a positive control as well as a colorectal carcinoma.

The only negative control we use for polyclonal antibodies is loss of activity when the specific antibody is absorbed with antigen. Our approach to other specificity controls has been fully discussed elsewhere.[49] Since many of our antisera are affinity purified we repass the purified antibody down the affinity column and use the effluent as a negative 'absorbed' control.[68] Addition of antigen to the antibody has potential problems with binding of immune complexes to Fc receptors, not destroyed in tissue sections by fixation and processing,[69] but this does not seem to be of practical importance. In the case of as yet incompletely defined reagents, it may be necessary to absorb with a partly purified preparation.[70]

For monoclonal antibodies, one is more dependent on the original characterization using immunoblotting or other biochemical techniques. Hybridoma supernatants are usually used neat, or diluted very little, and passage down an affinity column could result in sufficient dilution to cause non-specific abolition of staining. Most workers therefore rely on showing lack of 'noise' in the system, by comparing results with an irrelevant monoclonal which does not stain human tissues.

There is an inbuilt negative control in most tissue sections, since distribution of tumour markers is not uniform. If several markers are used, they may act as negative controls for each other. If the result with the 'absorbed' reagent is negative, this confirms the inhibition of endogenous peroxidase, the absence of anti-human activity in the second, anti-species, antibody, and of 'noise' in the system. Except in special circumstances, other controls such as using substrate alone, omission of the first antibody, or using so-called 'non-immune serum' are superfluous.[57]

3. TUMOUR MARKERS

Tumour markers have conventionally been divided into those which are appropriate to the tissue of origin (eutopic), such as human chorionic gonado-trophin (HCG) in gestational choriocarcinomas and gut hormones in islet cell tumours of the pancreas, and ectopic, or inappropriate, for those products, such as adrenocorticotrophic hormone (ACTH) and antidiuretic hormone (ADH) secreted by oat cell carcinomas of lung, which are not thought to be a normal product of the tissue of origin.

The concept of eutopic and ectopic tumour products is a useful one. However, it has become clear that our ideas of what is an appropriate or inappropriate product need to be revised. There is evidence of extraction of human placental lactogen (HPL) from the normal testis,[71] and of HCG[72] and human growth hormone from non-neoplastic extragonadal tissues.[73,74] The brain and gut are now considered a unified neuroendocrine system with many 'gut hormones' secreted by both.[75]

Substances secreted by tumours include hormones and other placental proteins, immunoglobulins, oncofetal antigens, membrane antigens, blood group substances and enzymes.

3.1. Placental Hormones

3.1.1. Human Chorionic Gonadotrophin

Human chorionic gonadotrophin is a glycoprotein composed of two dissimilar non-covalently linked units — alpha and beta. The alpha unit is indistinguishable immunologically from the alpha subunits of the pituitary glycoprottre hormones, luteinizing hormone (LH), follicle-stimulating hormone (FSH) and thyroid-stimulating hormone (TSH).[76] The beta subunits are different from each other and confer specificity, although probably neither the alpha nor the beta subunits are biologically active alone.

HCG is secreted by the normal placenta and is measurable in the urine early in pregnancy. Its main functions are the maintenance of the corpus luteum until the placenta takes over steroid synthesis, and Leydig cell stimulation for the masculinization of XY fetuses. Maternal levels reach a maximum at around 8 weeks, declining at 16 weeks to a low level which is maintained until the end of pregnancy.

Most, if not all, gestational, ovarian and testicular choriocarcinomas secrete HCG. It is an excellent circulating tumour marker being secreted in a stoichiometric relationship to the tumour mass. Together with adequate chemotherapy, the existence of a sensitive radioimmunoassay for HCG, which detects small volume disease long before clinical detection would be possible, has

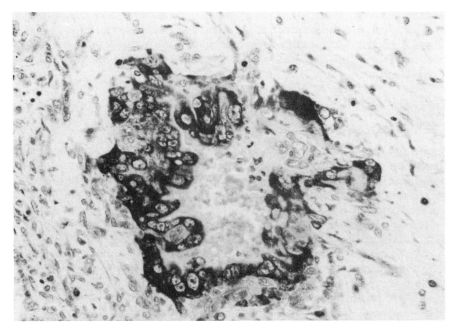

Fig. 31.1. Cluster of bizarre multinucleate syncytial giant cells form the wall of a vascular channel in a malignant testicular teratoma. They are strongly positive for human chorionic gonadotrophin.

changed a tumour with an appalling prognosis into one of the most amenable to treatment.[77] It is often easier to demonstrate small foci of choriocarcinoma in testicular or ovarian tumours, or even to make the diagnosis in atypical metastatic skin deposits[13] with an immunostain for HCG than with conventional haematoxylin and eosin preparations.[30]

One of the contributions made to tumour pathology by immunoperoxidase methods is the demonstration that giant cells in malignant testicular tumours, which have the morphology of syncytiotrophoblast cells, contain HCG[20,78] (*Plate* 34 and *Fig.* 31.1). Previously this was disputed[79] and gynaecomastia or raised levels of placental proteins in testicular tumours, not considered trophoblastic by conventional criteria, were thought to be due to the occult presence of choriocarcinoma missed by sampling error.[80] Over 50 per cent of those testicular teratomas which would be classified as non-trophoblastic by conventional criteria, and up to 17 per cent of seminomas have been found to contain HCG-positive cells.[81] Similar HCG-positive cells may be found in ovarian dysgerminomas and mixed germ-cell tumours.

Braunstein and his colleagues[82] have reported elevated levels of HCG in many kinds of extragonadal tumours. Muggia and his coworkers[83] have collected a number of such tumours, and good correlation between elevated circulating levels and the presence of giant cells containing HCG in these tumours has been shown by us.[44]

Studies on testicular tumours have employed antisera to the specific beta subunit of HCG, so excluding the demonstration of LH, FSH or TSH. Some tumours, including islet cell tumours of the pancreas, secrete the isolated alpha-glycoprotein subunit.[84,85] In an examination of 44 breast carcinomas with antisera to HCG alpha, previous reports of the presence of this substance in some of them[23,24] were confirmed, but it was not possible to demonstrate any adverse prognostic significance.[25,26]

3.1.2. Human Placental Lactogen

Human placental lactogen (HPL) is a polypeptide described by Josimovitch and Maclaren[86] and previously called human chorionic somatomammotrophin (HCS), reflecting its growth hormone- and prolactin-like properties. HPL is the normal product of the syncytiotrophoblast and levels rise progressively during pregnancy. Monitoring circulating maternal levels during pregnancy is of value as an indication of fetal distress.[87]

There are reports of elevated levels in patients with a variety of tumours, but HPL has not proved of major value as a circulating tumour marker, even in the case of choriocarcinomas of testicular or gestational origin. In a study of tissue sections of the orchidectomy specimen from 67 stage I non-trophoblastic testicular teratomas, HPL was demonstrated in 80 per cent of the tumours which contained HCG,[30] as well as in gestational choriocarcinomas. It was localized in two out of five tumours associated with elevated HPL levels — a large cell anaplastic bronchial carcinoma and a gastric adenocarcinoma with undifferentiated areas.[44]

3.2. Other Placental Proteins

3.2.1. Pregnancy-specific Glycoprotein

In a study of the wide variety of proteins which may be extracted from the human placenta, Bohn[89] isolated pregnancy-specific glycoprotein (SP1) which was shown to be a very early marker of pregnancy.[90] It has been shown in the normal placental syncytial trophoblast, as well as in choriocarcinomas and the syncytial giant cells of testicular tumours not considered trophoblastic by conventional criteria.[91,92] Horne and his colleagues[21] demonstrated the presence of SP1 in sections of breast cancers and claimed that its demonstration correlated with a worse prognosis, although others have not been able to confirm this.[25] SP1 may also be demonstrated in other tumours including colorectal carcinomas.[93,94]

3.2.2. Placental Alkaline Phosphatase

Placental alkaline phosphatase (Regan iso-enzyme) levels are raised in some patients with trophoblastic and extragonadal disease and the enzyme has been demonstrated in sections of a variety of tumours including lung and colon carcinoma.[95] Serum levels are elevated in some patients with testicular germ-cell tumours, and localization in seminomas and teratomas has been shown.[96]

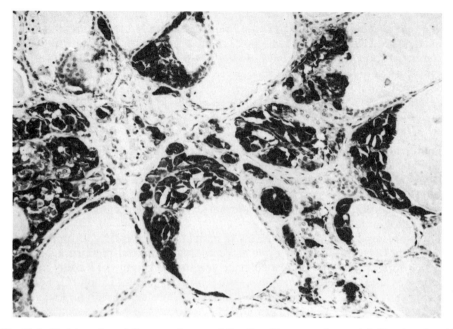

Fig. 31.2. Nodules of medullary carcinoma of the thyroid surrounding and infiltrating thyroid follicles. Immunostain for calcitonin.

3.2.3. *Other Placental Products*

Bohn and his colleagues[97] have described a number of other placental proteins which may be demonstrated in tissue sections of a variety of tumours including PP5, PP10, PP11 and PP12.[98]

3.3. Thyroid Hormones

3.3.1. *Calcitonin*

Calcitonin is the normal product of the C-cells of the thyroid and is secreted by thyroid medullary carcinomas (*see* Chapter 32). It is a polypeptide whose actions in the human are poorly understood, although it is involved in the calcium stress of pregnancy, lactation and growth.[99]

The tissue localization of calcitonin is of value in the diagnosis of medullary carcinoma of the thyroid[100] (*Fig*. 31.2) and of C-cell hyperplasia.[101] In spite of reports of extraction from breast cancers,[102] and of elevation of circulating levels of calcitonin in patients with widespread breast cancer,[103] it was not possible to demonstrate calcitonin in 44 consecutively removed breast carcinomas.[104] Calcitonin is secreted also by oat cell carcinomas,[105] and may be demonstrated in some pancreatic tumours[26] (*Fig*. 31.3).

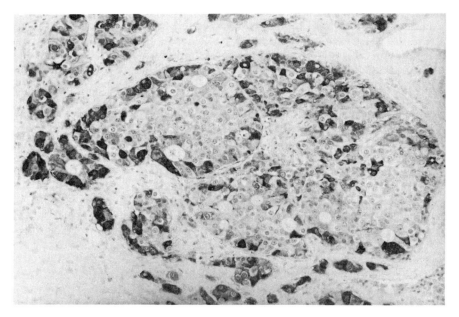

Fig. 31.3. Islet cell carcinoma of the pancreas stained for calcitonin. The marked heterogeneity of intensity of staining, so often seen in tumours, is well shown in this section.

3.4. Other Thyroid Products

3.4.1. Thyroglobulin

It is not uncommon for papillary or follicular thyroid carcinomas to present with metastatic disease. Thyroglobulin may be demonstrated in most thyroid carcinomas[106,107] (*Plate* 35) including some anaplastic tumours.[108,109] Its immunocytochemical demonstration is a useful adjunct to the diagnosis of primary undifferentiated thyroid carcinomas, and of metastatic deposits of thyroid carcinomas (*see* Chapter 32).

3.4.2. Histaminase

Histaminase is a marker of medullary carcinoma of the thyroid and of oat cell carcinomas,[110] and has been demonstrated in these tumours by the immunoperoxidase technique.[111]

3.5. Pituitary Hormones

3.5.1. Growth Hormone

Human growth hormone (HGH) is secreted by the eosinophils of the anterior pituitary. Levels are elevated in patients with eosinophil and some chromophobe adenomas, and the demonstration of HGH is of interest in confirming the presence of a tumour in hypophysectomy specimens.[112,113]

Elevation of growth hormone levels has been reported in association with a variety of lung tumours, mainly of adenocarcinomatous or of large cell type and the hormone has been extracted from 7 out of 18 lung cancers by Beck and Burger.[114] It has been suggested that growth hormone is involved in hypertrophic osteo-arthropathy,[115] though the weight of evidence is that this disease is probably not the result of elevated HGH levels.[116] Sonksen and his colleagues[117] reported one case and referred to another of a bronchial carcinoid associated with acromegaly. In their case no growth hormone could be extracted from the tumour and it was suggested that the tumour might have been secreting HGH-releasing factor. Several groups have reported synthesis of HGH-releasing factor by islet cell tumours of the pancreas.[118–120]

3.5.2. Prolactin

Prolactin is demonstrable mainly in chromophobe cells of the normal pituitary. Pituitary adenomas associated with the clinical syndrome of infertility and amenorrhoea are usually of chromophobe type,[121] but some are acidophil. Ectopic production has been reported in tumours of lung and kidney,[122,123] and testis.[124]

3.5.3. Adrenocorticotrophic Hormone (ACTH)

Since the report of Cushing's syndrome in association with the ectopic production of ACTH by a bronchial neoplasm,[125] there have been several confirmatory reports of ectopic ACTH production by lung tumours.[126,127] Its production has mainly been associated with oat cell carcinomas, but it has also been found with other cell types.[128] Gewirtz and Yalow[129] found ACTH in 14 out of 15 assorted lung tumours, but Ratcliffe and his colleagues[130] have not confirmed this high prevalence.

ACTH may readily be localized by immunocytochemical techniques in bronchial carcinoids associated with severe Cushing's disease, but only in some oat cell carcinomas associated with raised circulating levels,[26,131,132] although it has been demonstrated in other ectopic sites.[133] A possible explanation for the difficulty in the localization of ACTH in oat cell carcinomas is that, though neurosecretory granules may readily be demonstrated in carcinoids of the lung, they are much more sparse in oat cell carcinomas,[134] perhaps indicating a much lower level of storage. Immunocytochemical demonstration of ACTH in various skin carcinomas has been reported.[135]

3.5.4. Antidiuretic Hormone

Elevated circulating levels of antidiuretic hormone (ADH) may be found in some cancers, especially oat cell carcinomas of the lung,[136] and in cell lines derived from oat cell carcinomas.[137]

3.6. Steroid Hormones and their Receptors

Localization of steroid hormones in ovarian tumours has been reported,[138] and several workers have demonstrated oestrogen receptors or an oestrogen receptor-related material in tissue sections by immunofluorescence[139] or immunoperoxidase techniques,[140,141] though others have had little success in repeating the work.[142] While there is a basic biological interest in the determination of steroid receptors in breast and other oestrogen-dependent tumours, some receptor-poor tumours respond to endocrine manipulation and some receptor-rich tumours do not.[143] Anti-oestrogens and anti-adrenal agents have few undesirable side-effects, so that in some centres a trial of therapy has come to replace receptor assay. However, polyclonal and monoclonal antibodies to an oestrogen receptor-related protein, which stain fixed tissue sections have been produced.[144-146] Immunocytochemical localization with these antibodies appears to correlate well with biochemical oestrogen receptor assays (*see* Chapter 11).

3.7. Neuroregulatory Peptides (Gut Hormones)

Localization of these is important in the precise classification of gut-related tumours and is discussed fully elsewhere in this volume (*see* Chapter 20).

3.8. Oncofetal Antigens and Yolk Sac Products

So-called 'oncofetal antigens' include carcinoembryonic antigen (CEA) and alpha-fetoprotein (AFP). The terminology is based on the view that these substances are expressed by fetal tissues but not by the adult, and that in neoplasia there is gene derepression. Now that more sensitive radioimmunoassays and precise immunocytochemical tissue localization studies are available, it is clear that both these markers may be demonstrated in normal and in non-neoplastic adult tissue.

3.8.1. Carcinoembryonic Antigen

Thought by the original workers, Gold and Freedman[1] to be a specific marker for colorectal cancer, elevated levels of carcinoembryonic antigen (CEA) have been found in association with a wide variety of neoplastic and non-neoplastic lesions.[147,148] It has been localized in carcinomas of the large bowel[3] (*Fig.* 31.4), in the small bowel,[149] in tumours of the breast,[23,24,150] testis,[30] and thyroid.[111,151] While of little value as a circulating marker in the initial diagnosis of cancer, there are many reports of its use as a marker of disease progression.[152]

CEA is an incompletely defined glycoprotein. Although it was proposed that elevated serum levels of a specific isomeric species of CEA (CEA-S) might be almost entirely restricted to patients with tumours,[153] this work has not been repeated by others.

One of the problems which has beset the localization of CEA is its cross-reaction with a similar but smaller glycoprotein, normal cross-reacting antigen (NCA; CEX).[154,155] Efforts to absorb out the anti-NCA activity with spleen extract have

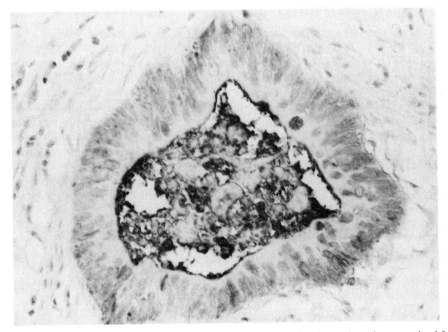

Fig. 31.4. Resin-embedded section of a moderately differentiated colorectal carcinoma stained for carcinoembryonic antigen (CEA). There is staining of the luminal membrane of the malignant acinus, and of the contained necrotic debris.

had varied success.[26,61] NCA is present in cells of the monocytic and myeloid series,[156] so that its localization in polymorphs and macrophages using antiserum to CEA unabsorbed with NCA might give rise to difficulties in interpretation. We found that the staining of these cells could be removed by absorption of affinity-purified polyclonal anti-CEA with 100 mg/ml of a lyophilized perchloric acid extract of human spleen.[61]

Investigation of a variety of monoclonal antibodies directed against CEA have indicated that most, but not all, recognize the common CEA/CEX determinant, and illustrate the variety of epitopes recognized by putative CEA antibodies.[26,157–159]

3.8.2. Alpha-fetoprotein

AFP is a fetal equivalent of albumin with levels rising progressively during fetal development. Radioimmunoassay of alpha-fetoprotein is of value in the antenatal diagnosis of neural tube defects[160] and AFP is an excellent circulating marker of malignant testicular disease.[161] Levels are raised in two-thirds of patients with metastatic disease,[162,163] and it has been localized by immunofluorescence in tumours with yolk sac morphology.[164,165] Some tumours with no evidence of yolk sac differentiation secrete AFP and though it has been claimed that AFP may be demonstrated in undifferentiated and immature somatic elements,[78,166,167] many workers experience difficulty with localization in routinely fixed and processed material, so that there is still no unanimity on what constitutes yolk sac differentiation and which types of tissue are responsible for AFP production. Fixation in cold ethanol–acetic acid results in poor histological preservation. Good results have been achieved with Bouin fixation of an AFP-producing 'seminoma' (*Plate* 36) grown as a xenograft in immune-deprived mice.[9]

AFP is much easier to demonstrate in hepatomas[168,169] and could be of value in discrimination from metastatic adenocarcinoma. Unfortunately some gastrointestinal tumours are associated with AFP secretion and some hepatomas secrete CEA.[170]

3.8.3. Fibronectin

When fibronectin[171] was demonstrated in tissue cultures, there was speculation that its demonstration would be of value in the assessment of malignant transformation. This has not proved to be the case.[172] It has, however, been shown in a mouse xenograft of a human testicular teratoma and can readily be demonstrated in tissue sections of the tumour.[26,173] It has not, unfortunately, turned out to be any more reliable than AFP as a marker of yolk sac differentiation. Fibronectin has been demonstrated in the stroma of tumours and in the cytoplasm of soft-tissue neoplasms.[174]

3.8.4. Alpha-l-antitrypsin

Alpha-l-antitrypsin is stored in the liver of homozygotes suffering from alpha-l-antitrypsin deficiency and may be demonstrated in fixed tissue sections by immunocytochemical techniques.[175] It has been localized in hepatomas and testicular tumours,[176] and is a marker of yolk sac differentiation in testicular tumours.[166] Since it is found normally in macrophages, interpretation of the results may be difficult when only isolated atypical cells are positive. An immunostain for alpha-l-antitrypsin may be used as a marker of true histiocytic lymphomas[177] (*see* Chapter 34).

3.8.5. Other Yolk Sac Products

AFP and alpha-l-antitrypsin are only two of the many substances secreted by the yolk sac.[178] Haemoglobin A and F have been demonstrated in yolk sac tumours,[179] as have a variety of other products including ferritin.[180] None have as yet proved to be clinically useful markers of yolk sac differentiation. Reports of the identification of atypical, possibly premalignant, intratubular germ cells by their ferritin content[181] await confirmation.

3.9. Immunoglobulins and Lymphocyte Markers

The localization of immunoglobulins has been of value in the differential diagnosis of reactive and malignant lymphoproliferative conditions and is discussed elsewhere in this volume, as are monoclonal antibodies to lymphocyte population subsets which are of increasing interest for the precise classification of lymphomas (*see* Chapters 34, 35).

3.10. Epithelial Membrane Antigen

Antibodies to human milk fat globule membranes (HMFG),[182] have been shown to react with normal and neoplastic epithelium in a wide variety of sites.[70,183–189] The glycoprotein(s) with which these antisera react has been termed 'epithelial membrane antigen' (EMA), since the staining in normal epithelial tissues, as well as in well-to-moderately differentiated adenocarcinomas, is mainly on the luminal or plasma membrane. The antigen has been partially purified, and shown to consist of a heterogeneous glycoprotein(s) of high molecular weight. It has been suggested that carbohydrate forms the major antigenic determinant, the principal sugars being galactose and *N*-acetylglucosamine.[190]

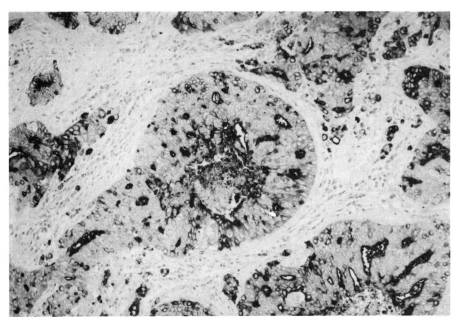

Fig. 31.5. Infiltrating ductal carcinoma of the breast stained for epithelial membrane antigen (EMA). The micro-acinar pattern of staining so characteristic of breast carcinoma is very apparent.

All the breast carcinomas (*Fig.* 31.5) in the published series, and the majority of adenocarcinomas from a variety of sites, have been positive for EMA, while sarcomas and neural tumours have been reported to be negative. The presence of EMA in a tumour is highly suggestive of epithelial derivation, while its absence virtually excludes breast origin.

Antisera to EMA have been used for the detection of bone marrow metastases from breast carcinoma,[70,183,191,192] and for identification of malignant cells in serous effusions.[193,194] In tissue sections from some patients, staining of the membrane of plasma cells has also been seen.[187,192,195] A few lymphomas, particularly of immunoblastic type, also show positivity.[195–197]

Monoclonal antibodies to human milk fat globule membranes show a similar immunohistological distribution and include E29,[67] HMFG1 and HMFG2,[198] and NCRC 11.[199] HMFG1 and HMFG2 have been used for radioimmunolocalization of a variety of tumours,[200] and it has been suggested that staining with HMFG1 or with NCRC 11 may have prognostic significance in breast cancer patients.[201,202]

3.11. Other Milk Proteins

Casein or casein-like material has been demonstrated in a variety of mammary and other carcinomas.[203,204] It is possible that the immunogen contained EMA-like material since the distribution of staining with the antisera is very similar. We have demonstrated alpha-lactalbumin in the lactating breast[205] and in some breast cancers, but as others have also found, there are problems with the specificity of the antisera available.[206]

3.12. Prostatic Tumour Markers

3.12.1. Prostatic Acid Phosphatase

Prostatic acid phosphatase is an iso-enzyme of acid phosphatase and is an excellent marker of prostatic cancer,[207] being elevated in many patients with metastatic disease. Its enzymic activity is largely destroyed by formalin fixation and processing of tissue into paraffin wax, but its antigenic determinants are retained and specific antisera may be used to demonstrate prostatic acid phosphatase in sections of fixed prostatic tumours and of their metastases, as well as in benign prostatic tissue.[208–210] Cross-reactivity with pancreatic islet cells, salivary glands, renal carcinomas and breast carcinomas with some rabbit antisera has been reported, although no such activity has been seen in 142 various non-prostatic benign and malignant lesions stained with the affinity-purified antibody used in our study of prostatic, bladder and colorectal carcinoma.[188] Antisera to acid phosphatase may be used to distinguish rectal carcinomas infiltrating the prostate, bladder tumours infiltrating the prostate and prostatic carcinoma invading the rectum, since of these only prostatic carcinomas secrete prostatic acid phosphatase. Some prostatic tumours secrete CEA but all rectal carcinomas do so.[3] The absence of CEA virtually excludes the diagnosis of colorectal carcinoma.

3.12.2. Prostate Antigen

Wang and colleagues[211] have described a circulating and tissue marker of prostatic carcinoma extracted from benign and malignant prostatic tissue and from seminal plasma. This is of lower molecular weight than prostatic acid phosphatase and is without enzymic activity. Like prostatic acid phosphatase, it may be demonstrated in fixed tissues of prostatic origin. Antisera to prostate antigen did not appear to have the cross-reactivity with other tissues found with unpurified anti-prostatic acid phosphatase, and it was thought to be more specific for tissues of prostatic origin.[212] Monoclonal antibodies have been described; some appeared to cross-react with kidney,[213] but others show greater specificity.[214]

3.13. Blood Group Substances and Factor VIII

Since the demonstration of blood group substances A and B on normal human epithelia,[215] there have been many reports on the deletion of ABO antigens in neoplasia using a mixed red cell agglutination technique.[216] These include studies on carcinomas of the gastrointestinal tract,[217] cervix,[218,219] prostate,[220] breast,[221] and bladder.[31] In most hands the mixed red cell agglutination technique is rather fickle and recent reports have employed either ordinary human blood typing antisera[222] or affinity-purified rabbit antisera raised against the synthetic oligosaccharide determinants.[223] The association between loss of blood group antigens and invasiveness seemed well established in the case of bladder cancer,[31,222] but doubt has been thrown on these studies.[32] Using rabbit antisera and an immunoperoxidase technique, we have failed to show any correlation between differentiation and prognosis in the breast or cervix.[26,33] One of the problems we have found is that normal tissues show a patchy distribution of staining so that any assessment of deletion as a marker of malignant or premalignant change is difficult.

Localization of factor VIII-related antigen has been used as a specific marker for normal and neoplastic endothelium,[224,225] although we and others[26] have found difficulty in repeating the results, particularly with tumours such as classical Kaposi's sarcoma. Others have found the lectin *Ulex europeaeus* I to be a more sensitive marker.[226,227]

3.14. Myoglobin, Myosin and Actin

The precise classification of soft tissue tumours is often difficult on morphological grounds alone. A combination of antibodies to intermediate filaments and to myoglobin, myosin and/or actin may be helpful.[228–231]

3.15. Ovarian Antigens

A variety of antigens have been extracted from human ovarian tumours.[232] Some, like CEA,[233] are found in many other tumours. Others, like CX 1,[26,234] appear to have more limited immunocytochemical specificity.

3.16. Melanoma Antigens and S-100

Melanomas may metastasize as pigmented or unpigmented (amelanotic) tumours. When they are amelanotic, diagnosis may be very difficult, especially if there is either no history of a previous malignant melanoma or a dubious one. A specific marker would be of great value. Although several melanoma markers has been described, workers have shown cross-reactivity with other tissues.[235–240] S-100 was thought to be a specific melanoma and neural marker,[241] but lack of melanoma specificity has been shown.[242] However, we have found S-100 useful in those undifferentiated 'epithelioid' tumours, with morphology compatible with that of metastatic amelanotic melanoma. Generally, anaplastic carcinomas are positive for EMA, a low-molecular-weight cytokeratin (CAM 5.2)[243] and/or CEA.[26] Where these are negative and S-100 is positive, the result is highly suggestive of metastatic melanoma.[26]

3.17. Glial Cell Markers

A variety of cell markers have been demonstrated in tumours of the central nervous system including glial fibrillary acidic protein (GFAP)[244,245] and carbonic anhydrase iso-enzyme II (CA III)[246] and have proved of value in the classification of CNS tumours.

3.18. Intermediate Filaments

There has been much recent interest in the use of antibodies to intermediate filaments to help determine the tissue of origin of tumours. These substances include a variety of cytokeratins, desmin, vimentin, glial fibrillary acidic protein and neurofilaments (see Chapters 23, 25). We have found the monoclonal antibody CAM 5.2, raised against the colon carcinoma cell line HT29[243] (Fig. 31.6) particularly useful in diagnosis.[19,26] Formalin sections of most tumours of epithelial origin are positive.

3.19. The Ca1 Marker

There have been several recent reports on a monoclonal antibody, Ca1, which the authors claimed could distinguish between malignant and non-malignant lesions.[247–249] Other groups have failed to confirm this tumour specificity, and have demonstrated positive staining in a variety of normal, non-neoplastic tissues.[250–253]

4. IMMUNOCYTOCHEMISTRY IN ROUTINE DIAGNOSIS

A battery of markers taken together with the morphology may help to elucidate a variety of histopathological problems — determination of the site of origin of an occult primary tumour which presents with metastatic deposits and no clinical symptoms or signs to indicate the tissue from which it has arisen, in differential diagnosis, functional classification of malignant disease, possibly as an indicator of prognosis, and in marker prediction for immunolocalization and tumour targeting with antibodies ligated to toxic agents.

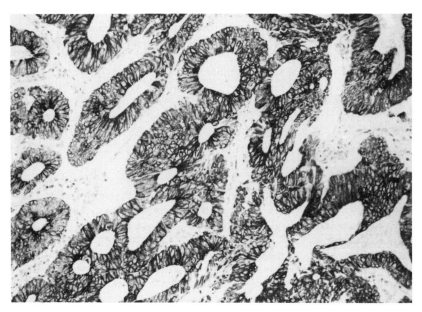

Fig. 31.6. Moderately differentiated colorectal carcinoma stained for cytokeratin (CAM 5.2). There is no luminal pattern of staining as is seen with antibodies to CEA or EMA. Although the staining appears to be pericellular, the cytoskeletal filaments lie just under the plasma membrane.

The advent of monoclonal hybridoma antibodies with their defined specificity has enabled significant progress to be made in the last few years. However, they may be directed against commonly found determinants, and since they do not represent a mixed population of antibodies, like polyclonal antisera raised in animals, unwanted specificities cannot be diluted out. They need to be selected for immunocytochemistry and where there is an interesting cross-reaction as between neural and granulocytic cells making them unsuitable for immunocytochemical distinction between metastatic neuroblastoma and normal marrow cells,[254] alternative clones have to be sought. Once established, milligramme quantities of pure immunoglobulin of defined specificity can be produced, particularly when the hybridomas are grown in mice as ascites tumours. Workers in widely separated centres are able to obtain reasonable quantities of the same reagent and compare results in a more meaningful way. At present, some inconsistencies are easily explained by the fact that antisera are obtained from different sources. Even in single centres, antibody titre and specificity of antisera raised in animals vary from animal to animal of the same species and from bleed to bleed.

There is clearly a growing role for immunocytochemistry in the pathology of malignant disease, when taken together with conventional macroscopical and microscopical examination. Enzyme labelling techniques are the most widely used, yielding a preparation which is also suitable for tissue recognition and diagnosis. Newer techniques, such as colloidal gold labelling, are of increasing interest and one looks forward to further developments in what has become an essential part of routine pathology.

Appendix

I. INDIRECT IMMUNOPEROXIDASE TECHNIQUE FOR PARAFFIN OR RESIN-EMBEDDED SECTIONS

1. Dewax through xylene and alcohols to water.
   For epoxy resin-embedded sections remove resin with saturated alcoholic sodium hydroxide (allowed to mature until dark brown — about 5 days). Wash off with absolute ethanol. Wash off with tap water.   5 min
2. Bleach acid haematin with 6 per cent hydrogen peroxide in distilled water. Wash off with tap water.   5 min
3. Inhibit endogenous peroxidase with 2·5 per cent periodic acid in distilled water. Wash off with tap water.   5 min
4. Block aldehyde groups with *fresh* 0·02 per cent potassium borohydride in distilled water. Wash off with tap water. Wash off with PBS–azide pH 7·2 cotaining 0·02 per cent sodium azide and 1 µl/ml of detergent (1 per cent BRIJ 96, Sigma). Blot dry.   2 min
5. Apply 40 µl first antibody diluted in 1 per cent ovalbumin in PBS–azide, and cover with a 40 × 33 mm coverslip. Incubate in moist chamber to prevent coverslip adhering too strongly. Wash off with PBS–azide.   1 h
6. Agitate in bath of PBS–azide containing 1µl/ml of detergent. Blot dry.   15 min
7. Apply 40 µl of peroxidase conjugate diluted in 1 per cent ovalbumin in *azide-free* PBS. Cover with 40 × 22 mm coverslip and incubate in moist chamber. Wash off with PBS–azide.   1 h
8. Agitate in PBS–azide bath containing detergent.   15 min
9. Incubate in *fresh* diaminobenzidine (DAB) (100 mg in 200 ml 0·03 per cent hydrogen peroxide in PBS). Wash in tap water.   5 min
10. Counterstain in Mayer's haemalum; blue in lithium carbonate; dehydrate, clear and mount in Ralmount or other resinous mountant.

Note: Azide should be omitted from PBS used to dilute peroxidase conjugates as it is deleterious. Its use is optional in the other steps, but it is recommended to prevent bacterial growth if the PBS is to be stored.

II. SOURCES OF IMMUNE REAGENTS USED FOR THE ILLUSTRATED PREPARATIONS

Most of the antibodies used for the illustrations in this chapter were raised in our laboratory at St Thomas' Hospital, where the affinity-purified peroxidase conjugates were also prepared.

Commercially available antibodies

EMA (monoclonal)   Dako
Cytokeratin (monoclonal)   Becton Dickinson
Affinity-purified peroxidase conju-   Amersham
gates

REFERENCES

1. Gold P. and Freedman S. O. Demonstration of tumour-specific antigens in human colonic carcinomata by immunological tolerance and absorption techniques. *J. Exp. Med.* 1965, **121**, 439–462.
2. Goldenberg D. M., Sharkey R. M. and Primus F. J. Carcinoembryonic antigen in histopathology: immunoperoxidase staining of conventional tissue sections. *J. Natl Cancer. Inst.* 1976, **57**, 11–22.
3. Heyderman E. and Neville A. M. A shorter immunoperoxidase technique for the demonstration of carcinoembryonic antigen and other cell products. *J. Clin. Pathol.* 1977, **30**, 138–140.
4. Tate H. Assessing tumour markers. *Br. J. Cancer* 1981, **44**, 643–651.
5. King A. C. and Cuatrecas P. Peptide hormone-induced receptor mobility, aggregation and internalization. *N. Engl. J. Med.* 1981, **305**, 77–88.
6. Mitra S. and Rao Ch. V. Gonadotropin and prostaglandin binding sites in rough endoplasmic reticulum and Golgi fractions of bovine corpora lutea. *Arch Biochem. Biophys.* 1978, **191**, 331–340.
7. Ellison M. L., Lamb D., Rivett J. and Neville A. M. Quantitative aspects of carcinoembryonic antigen output by a human lung carcinoma cell line. *J. Natl Cancer Inst.* 1977, **59**, 309–312.
8. Selby P. J., Heyderman E., Gibbs J. and Peckham M. J. A human testicular teratoma serially transplanted in immune deprived mice. *Br. J. Cancer.* 1979, **39**, 578–583.
9. Raghavan D., Heyderman E., Monaghan P., Gibbs J., Ruoslahti E., Peckham M. J. and Neville A. M. Hypothesis: When is a seminoma not a seminoma? *J. Clin. Pathol.* 1981, **34**, 123–128.
10. Gee C. E. and Roberts J. L. Laboratory methods. In situ hybridization histochemistry: a technique for the study of gene expression in single cells. *DNA* 1983, **2**, 157–363.
11. Coghlan J. P., Penschow P. J., Hudson P. J. and Niall H. D. Hybridization histochemistry: use of recombinant DNA for tissue localization of specific mRNA populations. *Clin. Exp. Hypertens.* 1984, **A6(1 & 2)**, 63–77.
12. Varndell I. M., Polak J. M., Sikri K. L. Minth C. D., Bloom S. R. and Dixon J. E. Visualisation of messenger RNA directing peptide synthesis by in situ hybridisation using a novel single-stranded cDNA probe. *Histochemistry* 1984, **81**, 597–901.
13. McSween A. and Heyderman E. Immunocytochemical investigation of skin metastases. In: Filipe M. I. and Lake B., eds., *Histochemistry in Pathology*. Edinburgh, Churchill Livingstone, 1983: 114–125.
14. Heyderman E. Immunocytochemistry in Cancer Diagnosis. In: Symington T., Williams A. E. and McVie J. G., eds., *The Tenth Pfizer International Symposium. Cancer Assessment and Monitoring*. Edinburgh, Churchill Livingstone, 1980: 147–171.
15. Heyderman E. The role of immunocytochemistry in tumour pathology: a review. *J. R. Soc. Med.* 1980, **73**, 655–658.
16. Wang N., Huang S. and Gold P. Absence of carcinoembryonic antigen-like material in mesothelioma: an immunohistochemical differentiation from other lung cancers. *Cancer* 1978, **74**, 438–444.
17. Whitaker D. and Shilkin K. B. Carcinoembryonic antigen in tissue diagnosis of malignant mesothelioma. *Lancet* 1981, **i**, 1369.
18. Gibbs A. R., Harach R., Wagner J. C. and Jasani B. Comparison of tumour markers in malignant mesothelioma and pulmonary adenocarcinoma. *Thorax* 1985, **40**, 91–95.
19. Heyderman E., Larkin S. E., Makin C. A. and Corrin B. Epithelial markers in pleural mesotheliomas. A comparison with primary lung carcinoma. *J. Pathol.* 1985, **145**, 81A Abstr.
20. Heyderman E. and Neville A. M. Syncytiotrophoblasts in malignant testicular tumours. *Lancet* 1976, **ii**, 103.
21. Horne C. H. W., Reid I. N. and Milne G. D. Prognostic significance of inappropriate production of pregnancy proteins by breast cancers. *Lancet* 1976, **ii**, 279–282.

22. Walker R. A. Significance of subunit HCG demonstrated in breast carcinomas by the immunoperoxidase technique. *J. Clin. Pathol.* 1978, **31**, 245–249.
23. Walker R. A. Demonstration of carcinoembryonic antigen in human breast cancer by the immunoperoxidase technique. *J. Clin. Pathol.* 1980, **33**, 356–360.
24. Shousha S., Lyssiotis T., Godfrey V. M. and Scheuer P. J. Carcinoembryonic antigen in breast cancer tissue: a useful prognostic indicator. *Br. Med. J.* 1979, **1**, 777–779.
25. Bulman A. S. The immunocytochemical identification of tumour markers in human breast cancer and their relationship to survival. *MS Thesis.* University of London, 1982.
26. Bulman A. R., Chapman D. V., Richardson T. C., Larkin S. E., Haines A. R. and Heyderman E. 1985, Unpublished data.
27. Parkinson C., Masters J. R. W., Krishnaswamy A. and Butcher D. N. HCG localisation of no prognostic value in testicular teratoma. *Lancet* 1981, **i**, 1059.
28. Heyderman E., Gibbons A. R. and Bulman A. S. Immunocytochemical identification of cells. In: Rosalki S. B., ed., *New Approaches to Laboratory Medicine. VIIIth Merz + Dade International Symposium.* Darmstdat, G-I-T Verlag Ernst Giebeler. 1981: 135–148.
29. Raghavan D., Peckham M. J., Heyderman E., Tobias J. S. and Austin D. E. Prognostic factors in clinical Stage I non-seminomatous germ-cell tumours of the testis. *Br. J. Cancer* 1982, **45**, 167–173.
30. Heyderman E. Multiple tissue markers in human malignant testicular tumours. *Scand J. Immunol.* Suppl. 8, 1978, **8**, 119–126.
31. Lange P. H., Limas C. and Fraley E. E. Tissue blood-group antigens and prognosis in low stage transitional cell carcinoma of the bladder. *J. Urol.* 1978, **119**, 52–55.
32. Thorpe S. J., Abel. P., Slavin G. L. and Feizi T. Blood group antigens in the normal and neoplastic bladder epithelium. *J. Clin. Pathol.* 1983, **36**, 873–882.
33. Neville A. M. and Heyderman E. Tumour markers in human breast cancer. In: Boelsma E. and Rumke Ph., eds., *Tumour Markers: Impact and Prospects.* Amsterdam: Elsevier/North-Holland Biomedical Press, 1979: 297–303.
34. Primus F. J., Clark C. A. and Goldenberg D. M. Immunohistochemical detection of carcinoembryonic antigen. In: DeLellis R. A., ed., *Diagnostic Immunohistochemistry.* New York, Masson Publishing USA Inc., 1981: 263–276.
35. Goldenberg D. M., DeLand F., Kim E., Bennet S., Primus F. J., van Nagell F. R., Estes N., DeSimone P. and Rayburn P. Use of radiolabeled antibodies to carcinoembryonic antigen for the detection and localisation of diverse cancers by external photoscanning. *N. Engl. J. Med.* 1978, **298**, 1384–1388.
36. Mach J. P., Carrell S., Forni M., Donath A. and Alberto P. Tumor localization of radiolabeled antibodies against carcinoembryonic antigen in patients with carcinoma. A critical evaluation. *N. Engl. J. Med.* 1980, **303**, 5–10.
37. Begent R. H. J., Searle F., Stanway G., Jewkes R. F., Jones B. E., Vernon P. and Bagshawe K. D. Radioimmunolocalisation of tumours by external scintigraphy after administration of antibody to human chorionic gonadotrophin: preliminary communication. *J. R. Soc. Med.* 1980, **73**, 624–630.
38. Halsall A. K., Fairweather D. S., Bradwell A. R., Blackburn J. C., Dykes P. W., Howell A., Reeder A. and Hine K. R. Localisation of malignant germ-cell tumours by external scanning after injection of radiolabelled anti-alpha-fetoprotein. *Br. Med. J.* 1981, **283**, 942–944.
39. Goldenberg D. M., Kim E. E. and DeLand F. H. Human chorionic gonadotropin radioantibodies in the radioimmunodetection of cancer and for disclosure of occult metastases. *Proc. Natl. Acad. Sci. USA* 1981, **78**, 7754–7758.
40. Olsnes S. Directing toxins to cancer cells. *Nature* 1981, **29**, 84.
41. Ghose T., Norvell S. T., Guclu A. and Macdonald A. S. Immunochemotherapy of human malignant melanoma with chlorambucil-carrying antibody. *Eur. J. Cancer* 1975, **11**, 321–326.
42. Thorpe P. E., Ross W. C. J., Cumber A. J., Hinson A. J., Hinson C. A., Edwards D. C. and Davies A. J. S. Yoxicity of diphtheria toxin for lymphoblastoid cells is increased by conjugation to anti-lymphocyte globulin. *Nature* 1978, **271**, 752–755.
43. Blythman H. E., Casellas P., Gros O., Jansen F. K., Paolucci F., Pau B. and Vidal H. Immunotoxins: hybrid molecules of monoclonal antibodies and a toxin subunit specifically kill tumour cells. *Nature* 1981, **290**, 145–146.
44. Heyderman E., Chapman D. V., Richardson T. C., Calvert I. and Rosen S. W. Human chorionic gonadotropin (hCG) and human placental lactogen (hPL) in extragonadal tumors. An immunoperoxidase study of ten non-germ-cell neoplasms. *Cancer* 1985, **56**, 2674–2682.

45. Coons A. H., Creech H. J. and Jones R. N. Immunological properties of an antibody containing a fluorescent group. *Proc. Soc. Exp. Biol. Med.* 1941, **47**, 200–202.
46. Graham R. C. and Karnovsky M. J. The early stages of absorption of injected horseradish peroxidase in the proximal tubules of mouse kidney: ultrastructural cytochemistry by a new technique. *J. Histochem. Cytochem.* 1966, **14**, 291–302.
47. Avrameas S. Coupling of enzymes to proteins with glutaraldehyde. Use of the conjugates for the detection of antigens and antibodies. *Immunochemistry* 1969, **6**, 43–52.
48. Mason D. Y. and Sammons R. E. Alkaline phosphatase and peroxidase for double immunoenzymatic labelling of cellular constituents. *J. Clin. Pathol.* 1978, **31**, 454–460.
49. Heyderman E., Bulman A. S. and Gibbons A. R. Immunoperoxidase techniques and controls. *J. Clin. Pathol.* 1981, **33**, 1219–1220.
50. De Mey J., Moeremans M., Geuens G., Nuydens R. and De Brabander M. High resolution light and electron microscopic localisation of tubulin with the IGS (immuno gold staining) method. *Cell Biol. Int. Rep.* 1981, **5**, 889–899.
51. Meyer-Arendt J. R. Theory and application of autoradiography. *Acta Histochem. [Suppl.] Jena* 1962, **13S**, 47–61.
52. Rooijen N. V. The application of autoradiography and other histochemical techniques to the same tissue section or smear. *J. Immunol. Method* 1981, **40**, 247–252.
53. Coons A. H. Leduc E. H. and Conolly J. M. Studies on antibody production I. A method for the histochemical demonstration of specific antibody and its application to a study of the hyperimmune rabbit. *J. Exp. Med.* 1955, **102**, 49–60.
54. Mason T., Pfifer R., Spicer S. and Dreskin R. An immunoglobulin-enzyme bridge method for localizing tissue antigens. *J. Histochem. Cytochem.* 1969, **17**, 563–569.
55. Sternberger L. A. *Immunocytochemistry*, 2nd ed. New York, John Wiley & Sons, 1979.
56. Mason D. Y. and Sammons R. E. The labeled antigen method of immunoenzymatic staining. *J. Histochem. Cytochem.* 1979, **27**, 832–840.
57. Heyderman E. Immunoperoxidase technique in histopathology: applications, methods and controls. *J. Clin. Pathol.* 1979, **32**, 971–978.
58. Burns J. Background staining and sensitivity of the unlabelled antibody–enzyme (PAP) method. Comparison with the peroxidase labelled antibody sandwich method using formalin fixed paraffin embedded material. *Histochemistry* 1975, **43**, 291–294.
59. Sinclair R. A., Burns J. and Dunnill M. S. Immunoperoxidase staining of formalin-fixed, paraffin-embedded, human renal biopsies with a comparison of the peroxidase-antiperoxidase (PAP) and indirect methods. *J. Clin. Pathol.* 1981, **34**, 859–865.
60. Nakane P. K. and Kawaio A. Peroxidase-labeled antibody: a new method of conjugation. *J. Histochem. Cytochem.* 1974, **22**, 1084–1091.
61. Heyderman E. The immunoperoxidase technique in pathology with special reference to malignant tumours of epithelial and germ cell origin. *MD Thesis.* University of London, 1985.
62. Richardson T. C., Chapman D. V. and Heyderman E. Immunoperoxidase techniques: the deleterious effect of sodium azide on the activity of peroxidase conjugates. *J. Clin. Pathol.* 1983, **36**, 411–414.
63. Bulman A. S. and Heyderman. E. Alkaline phosphatase for immunocytochemical labelling. Problems with endogenous activity. *J. Clin. Pathol.* 1982, **34**, 1349–1351.
64. Porath J., Axen R. and Ernback S. Chemical coupling of proteins to agarose. *Nature* 1967, **215**, 1491–1492.
65. Köhler G. and Milstein C. Derivation of specific antibody-producing tissue culture and tumour lines by cell fusion. *Eur. J. Immunol.* 1976, **6**, 292-295.
66. Cuello A. C., Wells C., Chaplin A. J. and Milstein C. Serotonin reactivity in carcinoid tumours demonstrated by a monoclonal antibody. *Lancet* 1982, **ii**, 771–773.
67. Heyderman E., Strudley I., Powell G., Richardson T. C., Cordell J. L. and Mason D. Y. A new monoclonal antibody to epithelial membrane antigen (EMA)-E29. A comparison of its immunocytochemical reactivity with polyclonal anti-EMA antibodies and with another monoclonal antibody, HMFG-2. *Br. J. Cancer* 1985, **52**, 355–361.
68. Heyderman E., Gibbons A. R. and Rosen S. W. Immunoperoxidase localisation of human placental lactogen: a marker for the placental origin of the giant cells in 'syncytial endometritis' of pregnancy. *J. Clin. Pathol.* 1981, **34**, 303–307.
69. McKeever P. E., Garvin A. J. and Spicer S. S. Immune complex receptors on cell surfaces. I. Ultrastructural demonstration on macrophages. *J. Histochem. Cytochem.* 1976, **24**, 948–955.
70. Heyderman E., Steele K. and Ormerod M. G. A new antigen on the epithelial membrane: its

immunoperoxidase localisation in normal and neoplastic tissues. *J. Clin. Pathol.* 1979, **32**, 35–39.

71. Payne R. A. and Ryan R. J. Human placental lactogen in the male subject. *J. Urol.* 1972, **107**, 99–103.
72. Yoshimoto Y., Wolfsen A. R. and Odell W. D. Human chorionic gonadotropin-like substance in nonendocrine tissues of normal subjects. *Science* 1977, **197**, 575–577.
73. Kaganowicz A., Farkouh N. H., Frantz A. G. and Blaustein A. U. Ectopic human growth hormone in ovaries and breast cancer. *J. Clin. Endocrinol. Metab.* 1978, **48**, 5–8.
74. Kyle C. V., Evans M. C. and Odell W. D. Growth hormone-like material in normal human tissues. *J. Clin. Endocrinol. Metab.* 1981, **53**, 1138–1144.
75. Polak J. M. and Bloom S. R. The neuroendocrine design of the gut. *Clin. Endocrinol. Metab.* 1979, **8**, 313–330.
76. Morgan F. J. and Canfield R. E. Nature of the subunits of human chorionic gonadotropin. *Endocrinology* 1971, **88**, 1045–1053.
77. Bagshawe K. D. and Begent R. H. J. Gestational trophoblastic tumours. In: Copplestone M., ed., *Gynecologic Oncology*. Edinburgh, Churchill Livingstone, 1981: 757–772.
78. Kurman R. J., Scardino P. T., McIntire K. R., Waldmann T. A. and Javadpour N. Cellular localisation of alphafetoprotein and human chorionic gonadotropin in germ cell tumors of the testis using an indirect immunoperoxidase technique. A new approach to the classification utilising tumour markers. *Cancer* 1977, **40**, 2136–2151.
79. Pugh R. C. B. (Ed.) *Pathology of the Testis*. Oxford, Blackwell, 1976.
80. Hobson B. M. The excretion of chorionic gonadotrophin by men with testicular tumours. *Acta Endocrinol.* 1965, **49**, 337–348.
81. Heyderman E. Biological markers of human teratomas and related germ cell tumors. In: Damjanov I. and Knowles B., eds., *The Human Teratomas*. Clifton, N. Jersey. The Humana Press Inc., 1983: 191–213.
82. Braunstein G., Vaitukaitis J. L., Carbone E. P. and Ross G. T. Ectopic production of human chorionic gonadotropin by neoplasms. *Ann. Intern. Med.* 1973, **78**, 39–45.
83. Muggia F. M., Rosen S. W., Weintraub B. D. and Hansen H. H. Ectopic placental proteins in nontrophoblastic tumors. Serial measurements following chemotherapy. *Cancer* 1975, **36**, 1327–1337.
84. Kahn C. R., Rosen S. W., Weintraub B. D., Fajans S. S. and Gordon P. Ectopic production of chorionic gonadotropin and its subunits by islet cell tumours. A specific marker for malignancy. *N. Engl. J. Med.* 1977, **297**, 565–569.
85. Heitz P. U., Kasper M., Klöppel G., Polak J. M. and Vaitukaitis J. L. Glycoprotein-hormone alpha chain production by pancreatic endocrine tumors: a specific marker for malignancy. Immunocytochemical analysis of tumors of 155 patients. *Cancer* 1983, **51**, 277–282.
86. Josimovich J. B. and MacLaren J. A. Presence in the human placenta and term serum of a highly lactogenic substance immunologically related to pituitary growth hormone. *Endocrinology* 1962, **71**, 209–220.
87. Nielsen P. V., Egebo., Find C. and Olsen C. E. Prognostic value of human placental lactogen (HPL) in an unselected obstetrical population. *Acta Obstet. Gynecol. Scand.* 1981, **60**, 469–474.
88. Rosen S. W. and Weintraub B. D. Ectopic placental lactogen. In: Ruddon R. W., ed., *Biological Markers of Neoplasia: Basic and Applied Aspects*. Amsterdam, Elsevier/North-Holland 1978.
89. Bohn H. Nachweis und Charakterisierung von Schwangerschafts-Proteinen in der menschlichen Placenta, sowie ihre quantitative immunologische Bestimmung im Serum schwangerer. *Frauen. Arch. Gynak.* 1971, **210**, 440–457.
90. Grudzinskas J. G., Gordon Y. B., Jeffrey D. and Chard T. Specific and sensitive determination of pregnancy-specific Bl-glycoprotein by radio-immunoassay. *Lancet* 1977, **i**, 333–335.
91. Horne C. H. W., Reid I. N. and Milne G. D. Prognostic significance of inappropriate production of pregnancy proteins by breast cancers. *Lancet* 1976, **ii**, 279–282.
92. Tatarinov Y. S., Falaleeva D. M. and Kalashrikov V. V. Human pregnancy-specific beta-globulin and its relation to chorioepithelioma. *Int. J. Cancer* 1976, **17**, 626–632.
93. Heyderman E. and Monaghan P. Immunoperoxidase reactions in resin embedded sections. *J. Invest. Cell Pathol.* 1979, **2**, 119–122.
94. Skinner J. M. and Whitehead R. Pregnancy-specific B glycoprotein (SP1) in tumours of the human gastrointestinal tract. *Br. J. Cancer* 1981, **44**, 475–478.

95. Miyayama H., Doellgast G. J., Memoli V., Gandbhir L. and Fishman W. H. Direct immunoperoxidase staining for Regan isoenzyme of alkaline phosphatase in human tumor tissues. *Cancer* 1976, **38**, 1237–1246.

96. Uchida T., Shimoda T., Miyata H., Shikata T., Iino S., Suzuki H., Oda T., Hirano K. and Sugira M. Immunoperoxidase study of alkaline phosphatase in testicular tumor. *Cancer* 1981, **48**, 1455–1462.

97. Bohn H., Inaba N. and Luben G. New placental proteins and their potential diagnostic significance as tumour markers. *Oncodev. Biol. Med.* 1981, **2**, 141–153.

98. Lee J. N., Whalstrom T., Seppala M., Salem H. T., Ouyang P. C. and Chard T. Immunohistochemical demonstration of placental protein 5 in trophoblastic tumours. *Placenta* 1982, **3**, 67–70.

99. Stevenson J. C., Hillyard C. J. and MacIntyre I. A physiological role for calcitonin: Protection of the maternal skeleton. *Lancet* 1979, **ii**, 769–770.

100. Tashjian A. H., Wolfe H. J. and Voelkel E. F. Human calcitonin: immunologic assay, cytologic localization and studies on medullary thyroid carcinoma. *Am. J. Med.* 1974, **56**, 840–849.

101. Wolfe H. J., Melvin K. E. W., Cervi-Skinner S. J., Al Saadi A. A., Juliar J. F., Jackson C. E. and Tashjian A. H. C-cell hyperplasia preceding medullary thyroid carcinoma. *N. Engl. J. Med.* 1973, **289**, 437–441.

102. Hillyard C. J., Coombes R. C., Greenberg P. B., Galante L. S. and MacIntyre I. Calcitonin in breast and lung cancer. *Clin. Endocrinol.* 1976, **5**, 1–8.

103. Coombes R. C., Easty G. C., Detre S. I., Hillyard C. J., Stevens U., Girgis S. I., Galante L. S., Heywood L., MacIntyre I. and Neville A. M. Secretion of immunoreactive calcitonin by human breast carcinomas. *Br. Med. J.* 1975, **4**, 197–199.

104. Bulman A. S. and Heyderman E. Calcitonin not a cellular tumour marker in breast cancer. *Lancet* 1983, **i**, 351.

105. Silva O. L., Becker K. L., Primack A., Doppman J. L. and Snider R. H. Increased serum calcitonin levels in bronchogenic cancer. *Chest* 1976, **69**, 495–499.

106. Lo Gerfo P., Li Volsi V., Colaccio D. and Feind C. Thyroglobulin production by thyroid cancers. *J. Surg. Res.* 1978, **24**, 1–6.

107. Burt A. and Goudie R. B. Diagnosis of primary thyroid carcinoma by immunohistological demonstration of thyroglobulin. *Histopathology* 1979, **3**, 279–286.

108. Bocker W., Dralle H., Husselmann H., Bay V. and Brassow M. Immunohistochemical analysis of thyroglobulin synthesis in thyroid carcinomas. *Virchows Arch. Abt. A. Anat. Histol.* 1980, **385**, 187–200.

109. Wilson N. W., Pambakian H., Richardson T. C. and Heyderman E. Epithelial markers in thyroid carcinomas and variety of other papillary adenocarcinomas. *J. Pathol.* 1984, **143**, 296A.

110. Baylin G. B. and Gazdar A. F. Endocrine biochemistry in the spectrum of human lung cancer: implications for the cellular origin of small cell carcinoma. In: Greco F. A., Oldhorn R. K. and Burn P. A., eds., *Small Cell Lung Cancer* New York, Grune and Stratton, 1981: 123–143.

111. Mendelsohn G. Histaminase localization in medullary thyroid carcinoma and small cell lung carcinoma in diagnostic chemistry. In: DeLellis R. A., ed., *Diagnostic Immunohistochemistry*. New York, Masson Publishing USA Inc., 1981: 299–312.

112. Ellis S. T., Beck J. S. and Currie A. R. The cellular localisation of growth hormone in the human foetal hyopohysis. *J. Pathol. Bacteriol.* 1966, **92**, 179–183.

113. Krueman A. C. N., Bots G. Th. A. M., Lindeman J. and Schaberg A. Use of immunohistochemical and morphologic methods for the identification of human growth hormone-producing pituitary adenomas. *Cancer* 1976, **38**, 1163–1170.

114. Beck C. and Burger H. G. Evidence for the presence of immunoreactive growth hormone in cancers of the lung and stomach. *Cancer* 1972, **30**, 75–79.

115. Steiner H., Dahlbach O. and Waldenstrom J. Ectopic growth hormone production and osteoarthropathy in carcinoma of the bronchus. *Lancet* 1968, **i**, 783–785.

116. Yesner R. Spectrum of lung cancer and ectopic hormones. *Pathol. Ann.* 1978 Part 1, **13**, 217–240.

117. Sonksen P. H., Ayres A. B., Braimbridge M., Corrin B., Davies D. R., Jeremiah G. M., Oaten S. W., Lowy C. and West T. E. T. Acromegaly caused by pulmonary carcinoid tumours. *Clin. Endocrinol.* 1976, **5**, 503–513.

118. Melmed S., Ezrin C., Kovacs K., Goodman R. S. and Frohman L. A. Acromegaly due to secretion of growth hormone by an ectopic pancreatic cell tumor. *N. Engl. J. Med.* 1985, **312**, 9–17.

119. Rivier J., Spiess J., Thorner M. and Vale W. Characterization of a growth hormone-releasing factor from a human pancreatic islet cell tumour. *Nature* 1982, **300**, 276–278.
120. Guillemin R., Brazeau P., Bohlen P., Esch F., Ling N. and Wehrenberg W. B. Growth hormone-releasing factor from a human pancreatic tumor that caused acromegaly. *Science* 1982, **218**, 585–587.
121. Kovacs K., Horvath E., Corenblum B., Sirek A. M. T., Penz G. and Ezrin C. Pituitary chromophobe adenomas consisting of prolactin cells. A histologic immunocytological and electron microscopic study. *Virchows Arch. Abt. A. Pathol. Anat.* 1975, **366**, 113–123.
122. Turkington R. W. Ectopic production of prolactin. *N. Engl. J. Med.* 1971, **285**, 1455–1458.
123. Rees L. H., Bloomfield G. A., Rees G. H., Cordin B., Franks L. M. and Ratcliffe J. G. Multiple hormones in a bronchial tumor. *J. Clin. Endocrinol. Metab.* 1974, **38**, 1090–1097.
124. Stepanas A. V., Samaan N. A., Schultz P. N. and Holoye P. Y. Endocrine studies in testicular tumor patients with and without gynecomastia. *Cancer* 1978, **41**, 369–376.
125. Brown H. W. A case of pluriglandular syndrome. "Diabetes of bearded women". *Lancet* 1928, **ii**, 1022–1023.
126. Liddle G. W., Givens J. R., Nicholson W. E. and Island D. P. The ectopic ACTH syndrome. *Cancer Res.* 1965, **25**, 1057–1061.
127. Rees L. H. and Ratcliffe J. G. Ectopic hormone production by non-endocrine tumours. *Clin. Endocrinol.* 1974, **3**, 263–299.
128. Imura H. Ectopic hormone syndromes. *Clin. Endocrinol. Metab.* 1980, **9**, 235–260.
129. Gewirtz G. and Yalow R. S. Ectopic ACTH production in carcinoma of the lung. *J. Clin. Invest.* 1974, **53**, 1022–1032.
130. Ratcliffe J. G., Podmore J., Stack B. H. R., Spilg W. G. S. and Gropp C. Circulating ACTH and related peptides in lung cancer. *Br. J. Cancer* 1982, **45**, 230–236.
131. Singer W., Kovacs K., Ryan N. and Horvath E. Ectopic ACTH syndrome: clinicopathological correlations. *J. Clin. Pathol.* 1978, **31**, 591–598.
132. Tsutsumi T., Osamura R. Y., Watanabe K. and Yanaihara N. Immunohistochemical studies on gastrin-releasing peptide- and adrenocorticotropic hormone-containing cells in the human lung. *Lab. Invest.* 1983, **48**, 623–632.
133. Jarett L., Lacy P. E. and Kipwis D. M. Characterization by immunofluorescence of an ACTH-like substance in non-pituitary tumors from patients with hyperadrenocorticism. *J. Clin. Endocrinol.* 1964, **24**, 543–549.
134. Bensch K. G., Corrin B., Pariente R. and Spencer H. Oat-cell carcinoma of the lung. Its origin and relationship to bronchial carcinoid. *Cancer* 1968, **22**, 1163–1172.
135. Iwasaki H., Mitsui T., Kikuchi M., Imai T. and Fukushima K. Neuroendocrine carcinoma (Trabecular carcinoma) of the skin with ectopic ACTH production. *Cancer* 1981, **48**, 753–756.
136. Schwartz W. B., Bennet W., Curelolop S. and Bartter F. C. Syndrome of renal sodium loss and hyponatraemia probably resulting from inappropriate secretion of antidiuretic hormone. *Am. J. Med.* 1957, **23**, 529–542.
137. Pettengill O. S., Faulkner C. S., Wurster-Hill D. H., Maurer L. H., Sorenson G. D., Robinson A. G. and Zimmerman E. A. Isolation and characterization of a hormone-producing cell line from human small cell anaplastic carcinoma of the lung. *J. Natl Cancer Inst.* 1977, **58**, 511–518.
138. Kurman R. J., Andrade D., Goebelsmann U. and Taylor C. R. An immunohistological study of steroid localization in Sertoli–Leydig tumors of the ovary and testis. *Cancer* 1978, **42**, 1772–1783.
139. Pertschuk L. P., Tobin E. H., Brigati D. J., Kim D. S., Bloom N. D., Gaetjens E., Berman P. J., Carter A. C. and Degenshein G. A. Immunofluorescent detection of estrogen receptors in breast cancer: comparison with dextran-coated charcoal and surcose gradient assays. *Cancer* 1978, **41**, 907–911.
140. Walker R. A., Cove D. H. and Howell A. Histological detection of oestrogen receptor in human breast carcinoma. *Lancet* 1980, **i**, 171–173.
141. Taylor C. R., Cooper C. L., Kurman R. J., Goebelsmann U. and Markland F. S. Detection of estrogen receptor in breast and endometrial carcinoma by the immunoperoxidase technique. *Cancer* 1981, **47**, 2634–2640.
142. Penney G. C. and Hawkins R. A. Histochemical detection of oestrogen receptors: a progress report. *Br. J. Cancer* 1982, **45**, 237–246.
143. Jensen E. V. Hormone dependence of breast cancer. *Cancer* 1981, **47**, 2319–2326.
144. Greene G. L., Close L. E., Fleming H., Desombre E. R. and Jensen E. V. Antibodies to estrogen receptor: immunochemical similarity of estrophilin from various mammalian species. *Proc. Natl Acad. Sci. USA* 1977, **74**, 3681–3685.

145. Greene G. L., Fitch F. W. and Jensen E. V. Monoclonal antibodies to estrophilin: probes for the study of estrogen receptors. *Proc. Natl Acad. Sci. USA* 1980, **77**, 5115–5119.
146. Coffer A. I., Lewis K. M., Brockas A. J. and King R. J. B. Monoclonal antibodies against a component related to soluble estrogen receptor. *Cancer Res.* 1985, in press.
147. Laurence D. J. R., Stevens U., Bettleheim R., Darcy D., Leese C., Turberville C., Alexander P., Johns E. W. and Neville A. M. Role of plasma carcinoembryonic antigen in the diagnosis of gastrointestinal, mammary and brochial carcinoma. *Br. Med. J.* 1972, **iii**, 605–609.
148. Neville A. M. and Laurence D. J. R. Report of the workshop on the carcinoembryonic antigen (CEA): the present position and proposals for future investigation. *Int. J. Cancer* 1974, **14**, 1–18.
149. Isaacson P. and Judd M. A. Carcinoembryonic antigen (CEA) in the normal human small intestine: a light and electron microscopic study. *Gut* 1977, **18**, 786–791.
150. Wahren B., Lidbrink E., Wallgren A., Eneroth P. and Zajicek J. Carcinoembryonic antigen and other tumor markers in tissue and serum or plasma of patients with primary mammary carcinoma. *Cancer* 1978, **42**, 1870–1878.
151. Hamada S. and Hamada S. Localisation of carcinoembryonic antigen in medullary thyroid carcinoma by immunofluorescent techniques. *Br. J. Cancer* 1977, **36**, 572–576.
152. NIH Consensus Statement (Summary) Carcinoembryonic antigen: its role as a marker in the management of cancer. *Br. Med. J.* 1981, **282**, 373–375.
153. Edgington R. S., Astarita R. W. and Plow E. F. Association of an isomeric species of carcinoembryonic antigen with neoplasia of the gastrointestinal tract. *N. Engl. J. Med.* 1975, **293**, 103–107.
154. Mach J-P. and Putztaszeri G. Carcinoembryonic antigen (CEA): demonstration of a partial identity between CEA and a normal glycoprotein. *Immunochemistry* 1972, **9**, 1031–1034.
155. Von Kleist S., Chavanel G. and Burtin B. Identification of a normal antigen that cross-reacts with the carcinoembryonic antigen. *Proc. Natl Acad. Sci. USA* 1972, **69**, 2492–2494.
156. Burtin P. and Fondaneche M-C. Characterization of the non-specific cross-reacting antigen (NCA) in human monocytes. *Clin. Immunol. Immunopathol.* 1981, **20**, 146–156.
157. Accolla R. S., Carrel S., Pham M., Heumann D. and Mach J-P. First report of the production of somatic cell hybrids secreting monoclonal antibodies specific for carcinoembryonic antigen (CEA). *Protids Biol. Fluids Proc. Colloq.* 1979, **27**, 31–35.
158. Primus F. J., Newell K. D., Blue A. and Goldenberg D. M. Immunological heterogeneity of carcinoembryonic antigen: antigenic determinants on carcinoembryonic antigen distinguished by monoclonal antibodies. *Cancer Res.* 1983, **43**, 686–692.
159. Wagener C., Petzold P., Köhler W. and Totovic V. Binding of five different monoclonal anti-CEA antibodies with different epitope specificities to various carcinoma tissues. *Int. J. Cancer* 1984, **33**, 469–475.
160. Brock D. J. H. and Sutcliffe R. G. Alpha-fetoprotein in the antenatal diagnosis of anencephaly and spina bifida. *Lancet* 1972, **ii**, 197–199.
161. Nørgaard-Pedersen B. et al. Clinical use of AFP and HCG in testicular tumours of germ cell origin. *Lancet* 1978, **ii**, 1042.
162. Grigor K. M., Detre S. I., Kohn J. and Neville A. M. Serum alpha 1-foetoprotein levels in 153 male patients with germ cell tumours. *Br. J. Cancer* 1977, **35**, 52–58.
163. Scardino P. T., Cox H. D., Waldmann T. A., McIntire R., Mittenmeyer B. and Javadpour N. The value of serum tumor markers in the staging and prognosis of germ cell tumors of the testis. *J. Urol.* 1977, **118**, 994–999.
164. Teilum G., Albrechtsen R. and Nørgaard-Pedersen B. Immunofluorescent localisation of alpha-fetoprotein synthesis in endodermal sinus tumour (yolk sac tumour). *Acta Pathol. Microbiol. Scand.* 1974, **62**, 586–588.
165. Shirai T., Itoh T., Yoshiki T., Noro T., Tomino Y. and Hayasaka T. Immunofluorescent demonstration of alpha-fetoprotein and other plasma proteins in yolk sac tumor. *Cancer* 1976, **38**, 1661–1667.
166. Beilby J. O. W., Horne C. H. W., Milne G. D. and Parkinson C. Alpha-fetoprotein, alpha-1-antitrypsin and transferrin in gonadal yolk-sac tumours. *J. Clin. Pathol.* 1979, **32**, 455–461.
167. Wagener C., Menzel B., Breuer H., Weissbach L., Tschubel K., Henkel K. and Gedigk P. Immunohistochemical localisation of alpha-fetoprotein (AFP) in germ cell tumours: Evidence for AFP production by tissues different from endodermal sinus tumour. *Oncology* 1981, **38**, 236–239.

168. Engelhardt N. V., Goussev A. I., Shipova L. J. A. and Abelev G. I. Immunofluorescent study of alpha-fetoprotein (AFP) in liver and liver tumours. 1. Technique of AFP localisation in tissue sections. *Int. J. Cancer* 1971, **7**, 198–206.

169. Palmer P. E. and Wolfe H. Immunocytochemical localization of oncodevelopmental proteins in human germ cell and hepatic tumors. *J. Histochem. Cytochem.* 1978, **26**, 523–531.

170. Thung S. N., Gerber M. A., Sarno E. and Popper H. Distribution of five antigens in hepatocellular carcinoma. *Lab. Invest.* 1979, **41**, 101–105.

171. Yamada K. M. and Olden K. Fibronectins — adhesive glycoproteins of cell surface and blood. *Nature* 1978, **275**, 179–184.

172. McDonagh J. Fibronectin. A molecular glue. *Arch. Pathol. Lab. Med.* 1981, **105**, 393–396.

173. Ruoslahti E., Jalanko H., Comings D. E., Neville A. M. and Raghavan D. Fibronectin from human germ-cell tumors resembles amniotic fluid fibronectin. *Int. J. Cancer* 1981, **27**, 763–768.

174. Stenman S. and Vaheri A. Fibronectin in human solid tumors. *Int. J. Cancer* 1981, **27**, 427–435.

175. Palmer P. E., DeLellis R. A. and Wolfe H. J. Immunochemistry of liver in α1-antitrypsin deficiency. *Am. J. Clin. Pathol.* 1974, **62**, 350–354.

176. Palmer P. E. and Wolfe H. Immunocytochemical localization of oncodevelopmental proteins in human germ cell and hepatic tumors. *J. Histochem. Cytochem.* 1978, **26**, 523–531.

177. Isaacson P., Wright D. H. and Jones D. B. Malignant lymphoma of true histiocytic (monocyte/marcophage) origin. *Cancer* 1983, **51**, 80–91.

178. Gitlin D. and Perricelli A. Synthesis of serum albumin, prealbumin, a-foetoprotein, a-l-antitry-psin and transferrin by the human yolk sac. *Nature* 1970, **288**, 995–997.

179. Albrechtsen R., Wewer U. and Wimberley P. D. Immunohistochemical demonstration of a hitherto undescribed localization of hemoglobin A and F in endodermal cells of normal human yolk sac and endodermal sinus tumor. *Acta Pathol. Microbiol. Scand. A* 1980, **88**, 175–178.

180. Wahren B. Multiple fetal antigens in germ-cell tumors. *Scand. J. Immunol. Suppl 8*, 1978, **8**, 131–136.

181. Jacobsen G. K., Jacobsen M. and Clausen P. P. Ferritin as a possible marker protein of carcinoma in-situ of the testis. *Lancet* 1980, **ii**, 533–534.

182. Ceriani R. L., Thompson K., Peterson J. A. and Abraham S. Surface differentiation antigens of human mammary epithelial cells carried on the human milk fat globule. *Proc. Natl Acad. Sci. USA* 1977, **74**, 582–586.

183. Sloane J. P., Ormerod M. G., Imrie S. F. and Coombes R. C. The use of antisera to epithelial membrane antigen in detecting micrometastases in histological sections. *Br. J. Cancer* 1980, **42**, 392–398.

184. Sloane J. P., Ormerod M. G., Carter R. L., Gusterson B. A. and Foster C. S. An immunocytochemical study of the distribution of epithelial membrane antigen in normal and disordered squamous epithelium. *Diagn. Histopathol.* 1982, **5**, 11–17.

185. Gusterson B., Lucas R. B. and Ormerod M. G. Distribution of epithelial membrane antigen in benign and malignant lesions of the salivary glands. *Virchows Arch. Abt. A. Pathol Anat.* 1982, **397**, 227–233.

186. Bamford P. N., Ormerod M. G., Sloane J. P. and Warburton M. J. An immunohistochemical study of the distribution of epithelial antigens in the uterine cervix. *Obstet, Gynecol.* 1983, **41**, 603–608.

187. Heyderman E., Graham R. M., Chapman D. V., Richardson T. C. and McKee P. H. Epithelial markers in primary skin cancer: an immunoperoxidase study of the distribution of epithelial membrane antigen (EMA) and carcinoembryonic antigen (CEA) in 65 primary skin carcinomas. *Histopathology* 1984, **8**, 423–434.

188. Heyderman E., Brown B. M. E. and Richardson T. C. Epithelial markers in prostatic, bladder and colorectal cancer; an immunoperoxidase study of EMA, CEA and prostatic acid phosphatase. *J. Clin. Pathol.* 1984, **37**, 1363–1369.

189. Graham R. M., McKee P. H., Chapman D. V., Richardson T. C., Stokoe M. R. and Heyderman E. Intracellular canaliculi in eccrine sweat glands. An immunoperoxidase study. *Br. J. Dermatol.* 1985, In press.

190. Ormerod M. G., Steele K., Westwood J. H. and Mazzini M. N. Epithelial membrane antigen: partial purification, assay and properties. *Br. J. Cancer* 1983, **48**, 533–541.

191. Gugliotta P., Botta G. and Bussolati G. Immunocytochemical detection of tumour markers in bone metastases from carcinoma of the breast. *Histochem. J.* 1981, **13**, 953–959.

192. Dearnaley D. P., Ormerod M. G., Sloane J. P., Lumley H., Imrie S., Jones M., Coombes R. C. and Neville A. M. Detection of isolated mammary carcinoma cells in marrow of patients with primary breast cancer. *J. R. Soc. Med.* 1983, **76**, 359–364.

193. To A., Coleman D. V., Dearnaley D. P., Ormerod M. G., Steele K. and Neville A. M. Use of antisera to epithelial membrane antigen for the cytodiagnosis of malignancy in serous effusions. *J. Clin. Pathol.* 1981, **34**, 1326–1332.

194. Epenetos A. A., Canti G., Taylor-Papadimitriou J., Curling M. and Bodmer W. F. Use of two epithelium-specific-monoclonal antibodies for diagnosis of malignancy in serous effusions. *Lancet* 1982, **ii**, 1004–1006.

195. Sloane J. P., Hughes F. and Ormerod J. P. An assessment of the value of epithelial membrane antigen and other epithelial markers in solving diagnostic problems in tumour histopathology. *Histochem. J.* 1983, **15**, 645–654.

196. Delsol G., Gatter K. C., Stein H., Erber W. N., Pulford K. A. F., Zinne K. and Mason D. Y. Human lymphoid cells express epithelial membrane antigen. *Lancet* 1984, **ii**, 1124–1129.

197. Heyderman E. and Macartney J. C. Epithelial membrane antigen and lymphoid cells. *Lancet* 1985, **i**, 109.

198. Burchell J., Durbin H. and Taylor-Papadimitriou J. Complexity of expression of antigenic determinants, recognized by monoclonal antibodies HMFG-1 and HMFG-2, in normal and malignant human mammary epithelial cells. *J. Immunol.* 1983, **131**, 508–513.

199. Ellis I. O., Robins R. A., Elston C. W., Blamey R. W., Ferry B. and Baldwin R. W. A monoclonal antibody, NCRC-11, raised to human breast carcinoma. 1. Production and immunohistological characterization. *Histopathology* 1984, **8**, 501–516.

200. Epenetos A. A., Britton K. E., Mather S., Shepherd J., Granowska M., Taylor-Papadimitriou J., Nimmon C. C., Durbin H., Hawkins L. R., Malpas J. S. and Bodmer W. F. Targeting of iodine-123-labeled tumour-associated monoclonal antibodies to ovarian, breast and gastrointestinal tumours. *Lancet* 1982, **ii**, 999–1004.

201. Wilkinson M. J. S., Howell A., Harris M., Taylor-Papadimitriou J. and Swindell R. The prognostic significance of two epithelial membrane antigens expressed by human mammary cells. *Int. J. Cancer* 1984, **33**, 299–304.

202. Ellis I. O., Hinton C. P., MacNay J., Elston C. W., Robins A., Owainati A. A. R. S., Blamey R. W., Baldwin R. W. and Ferry B. Immunocytochemical staining of breast carcinoma with the monoclonal antibody NCRC 11: a new prognostic indicator. *Lancet* 1985, **i**, 881–883.

203. Bussolati G., Pich A. and Alfani V. Immunofluorescence detection of casein in human mammary dysplastic and neoplastic tissues. *Virchows Arch. Abt. A Pathol Anat. His.* 1975, **365**, 15–21.

204. Pich A., Bussolati G. and Carbonara A. Immunocytochemical detection of casein and casein-like proteins in human tissues. *J. Histochem. Cytochem.* 1976, **24**, 940–947.

205. Neville A. M., Grigor K. M. and Heyderman E. Biological markers and human neoplasia. In: Woolf N. and Anthony P., eds., *Recent Advances in Histopathology*, No 10. Edinburgh, Churchill Livingstone, 1978: 3–44.

206. Lee A. K., DeLellis R. A., Rosen P. P., Herbert-Stanton T., Tallberg K., Garcia C. and Wolfe H. J. Alpha-lactalbumin as an immunohistochemical marker for metastatic breast carcinomas. *Am. J. Surg. Pathol.* 1984, **8**, 93–100.

207. Gutman A. B. The development of the acid phosphatase test for prostatic carcinoma *Bull. N.Y. Acad. Med.* 1968, **44**, 63–76.

208. Pontes J. E., Choe B., Rose N. and Pierce J. M. Indirect immunofluorescence for identification of prostatic epithelial cells. *J. Urol.* 1977, **117**, 459–462.

209. Jöbsis A. C., De Vries G. P., Anholt R. R. H. and Sanders G. T. B. Demonstration of the prostatic origin of metastases. An immunohistological method for formalin-fixed embedded tissue. *Cancer*, 1978, **41**, 1788–1793.

210. Burns J. Prostatic acid phosphatase in tissue sections revealed by the unlabelled antibody peroxidase–antiperoxidase method. *Biomedicine* 1977, **27**, 7–10.

211. Wang M. C., Papsidero L. D., Muriyama M., Valenzuela L. A., Murphy G. P. and Chu T. M. Prostate antigen: a new potential marker for prostatic cancer. *The Prostate* 1981, **2**, 89–96.

212. Nadji M., Tabei S. Z., Castra A., Chu T. M., Murphy G. P., Wang M. C. and Morales A. R. Prostate-specific antigen. An immunohistologic marker for prostatic neoplasms. *Cancer* 1981, **48**, 1229–1232.

213. Frankel A. E., Rouse R. V., Wang M. C., Chu T. M. and Herzenberg L. A. Monoclonal antibodies to a human prostate antigen. *Cancer Res.* 1982, **42**, 3714–3718.

214. Papsiderio L D., Croghan G. A., Wang M. C., Kuriyama M., Johnson E. A., Valenzuela L. A. and Chu T. M. Monoclonal antibody (F5) to human prostate antigen. *Hybridoma* 1983, **2**, 139–147.

215. Coombes R. R. A., Bedford D. and Rouillard L. M. A and B blood-group antigens on human epidermal cells. *Lancet* 1956, **i**, 461–463.

216. Davidsohn I. Early immunologic diagnosis and prognosis of carcinoma. *Am. J. Clin. Pathol.* 1972, **57**, 715–730.

217. Davidsohn I., Kovarik S. and Ling Lee C. A, B, and O substances in gastro-intestinal carcinoma. *Arch. Pathol.* 1966, **81**, 381–390.

218. Davidsohn I., Kovarik S. and Ni L. Y. Isoantigens A, B and H in benign and malignant lesions of the cervix. *Arch Pathol.* 1969, **87**, 306–314.

219. Stafl A. and Mattingley R. F. Isoantigens ABO in cervical neoplasia. *Gynecol. Oncol.* 1972, **1**, 26–35.

220. Gupta R. K., Schuster R. and Christian W. D. Loss of isoantigens A, B and H in prostate. *Am. J. Pathol.* 1973, **70**, 439–447.

221. Gupta R. K. and Schuster R. Isoantigens A, B and H in benign and malignant lesions of breast. *Am. J. Pathol.* 1973, **72**, 253–260.

222. Weinstein R. S., Coon J., Alroy J. and Davidsohn I. Tissue-associated blood group antigens in human tumors. In: DeLellis R. A., ed., *Diagnostic Immunohistochemistry*. New York, Masson Publishing USA Inc., 1981: 263–276.

223. Lemieux R. U. Human blood groups and carbohydrate chemistry. *Chem. Soc. Rev.* 1978, **7**, 423–452.

224. Mukai K., Rosai J. and Burgdorf W. H. C. Localization of factor VIII-related antigen in vascular endothelial cells using an immunoperoxidase method. *Am. J. Surg. Pathol.* 1980, **4**, 273–276.

225. Sehested M. and Hou-Jensen K. Factor VIII related antigen as an endothelial cell marker in benign and malignant conditions. *Virchows Arch. Abt. A Pathol. Anat.* 1981, **391**, 217–225.

226. Holthöfer H., Virtanen I., Kariniemi A-L., Hormia M., Linder E. and Miettenen A. *Ulex europaeus* I lectin as a marker for vascular endothelium in human tissues. *Lab. Invest,* 1982, **47**, 60–66.

227. Ordonez N. G. and Batsakis J. G. Comparison of Ulex europaeus I lectin and factor VIII-related antigen in vascular lesions. *Arch. Pathol. Lab. Med.* 1984, **108**, 129–132.

228. Brooks J. J. Immunohistochemistry of soft tissue tumors. Myoglobin as a tumor marker for rhabdomyosarcoma. *Cancer* 1982, **50**, 1757–1763.

229. Mukai K., Rosai J. and Hallaway B. Localization of myoglobin in normal and neoplastic human skeletal muscle cells using an immunoperoxidase method. *Am. J. Surg. Pathol.* 1979, **3**, 373–376.

230. Corson J. and Pinkus G. Intracellular myoglobin: a specific marker for skeletal muscle differentiation in soft tissue sarcomas. *Am. J. Pathol.* 1981, **103**, 384–389.

231. Mukai K., Schollmeyer J. and Rosai J. Immunohistochemical localization of actin: applications in surgical pathology. *Am. J. Surg. Pathol.* 1981, **5**, 91–97.

232. Raghavan D. and Neville A. M. The use of tumour markers in the management of ovarian malignancy. In: Boelsma E. and Rumke Ph., eds., *Tumour Markers: Impact and Prospects*. Amsterdam, Elsevier/North-Holland Biomedical Press, 1979: 289–296.

233. Van Nagell J. R., Donaldson E. S., Gay E. C., Sharkey P. R. and Goldenberg D. M. Carcinoembryonic antigen in ovarian epithelial cystadenocarcinomas. The prognostic value of tumor and serial plasma determinations. *Cancer*, 1978, **41**, 2335–2340.

234. Wass M., Searle F. and Bagshawe K. D. Radioimmunoassay of the normal serum glycoprotein (CX 1) in monitoring ovarian malignancy. *Eur. J. Cancer Clin. Oncol.* 1981, **17**, 1267–1273.

235. Lewis M. G. and Ikonopisov R. L. Tumour specific antibodies in human malignant melanoma. *Br. Med. J.* 1962, **3**, 547–552.

236. Werkmeister J., Edwards A., McCarthy W. and Hersey P. Prognostic significance of expression of antigens on melanoma cells. *Cancer Immunol. Immunother.* 1980, **9**, 233–240.

237. Carrel S., Schreyer M., Schmidt-Kessen and Mach J-P. Reactivity spectrum of 30 monoclonal antibodies to a panel of 28 melanoma cell lines. *Hybridoma* 1982, **1**, 387–397.

238. Garrigues T. J., Tilgen W., Hellström I., Franke W. and Hellström K. E. Detection of a human melanoma-associated antigen, p97, in histological sections of primary human melanomas. *Int. J. Cancer* 1982, **29**, 511–515.

239. Pukel C. S., Lloyd K. O., Travassos L. R., Dippold W. G., Oetiigen H. F. and Old L. J. GD3, a

prominent ganglioside of human melanoma. Detection and characterization by mouse monoclonal antibody. *J. Exp. Med.* 1982, **155**, 1133–1147.

240. Vennegoor C., Calafat J., Hageman Ph., van Buitenen F., Janssen H., Kolk A. and Rumke Ph. Biochemical characterization and cellular localization of a formalin-resistant melanoma-associated antigen reacting with monoclonal antibody NK1/C-3. *Int. J. Cancer* 1985, **35**, 297–295.

241. Hakajima T., Watanabe S., Sato Y., Kameya T. and Shimosato Y. Immunohistochemical demonstration of S100 protein in human malignant melanoma and pigmented nevi. *Gann* 1981, **72**, 335–336.

242. Kahn H. J., Marks A., Thoma H. and Baumal R. Role of antibody to S100 protein in diagnostic pathology. *Am. J. Clin. Pathol.* 1983, **79**, 341–347.

243. Makin C. A., Bobrow L. G., Bodmer W. F. Monoclonal antibody to cytokeratin for use in routine immunocytochemistry. *J. Clin. Pathol.* 1984, **37**, 975–983.

243. Eng L. F., Vanderhagen J. J., Bignami A. and Gerstl B. An acidic protein isolated from fibrous astrocytes. *Brain Res.* 1971, **28**, 351.

245. Bignami A. and Schoene W. C. Glial fibrillary acidic protein in human brain tumors. In: DeLellis R. A., ed., *Diagnostic Immunohistochemistry*. New York, Masson Publishing USA Inc., 1981: 213–227.

246. Ghandour M. S., Langley O. K., Vincendon G. and Gombos G. Double labeling immunohistochemical technique provides evidence of the specificity of glial cell markers. *J. Histochem. Cytochem.* 1979, **27**, 1624–1637.

247. Ashall F., Bramwell M. E. and Harris H. A new marker for human cancer cells. 1. The Ca antigen and the Ca1 antibody. *Lancet* 1982, **ii**, 1–6.

248. McGee J. O'D., Woods J. C., Ashall F., Bramwell M. E. and Harris H. A new marker for human cancer cells. 2. Immunohistochemical detection of the Ca antigen in human tissues with the Ca1 antibody. *Lancet* 1982, **ii**, 7–10.

249. Woods J. C., Spriggs A. I., Harris H. and McGee J. O'D. A new marker for human cancer cells in serous fluids with the Ca1 antibody. *Lancet* 1982, **ii**, 512–515.

250. Burnett R. A., Deery A. R. S., Adamson M. R., Liddle L., Thomas M. and Roberts G. H. Evaluation of Ca1 antibody in pleural biopsy material. *Lancet* 1983, **i**, 1158.

251. Krausz T. J., Van Noorden S. and Evans D. J. Experience of the Oxford tumour marker. *Lancet* 1983, **i**, 1097.

252. Pallesen G., Jepsen F. L., Hastrup J., Ipsen. A. and Hvidberg N. Experience with the Oxford tumour marker (Cal) in serous fluids. *Lancet* 1983, **i**, 1326.

253. Simpson H. W., Candlish W., Liddle C., Mcgregor F. M., Mutch F. and Tinkler B. Experience of the Oxford tumour marker. *Lancet* 1983, **i**, 1097.

254. Kemshead J. T., Bicknell D. and Greaves M. F. A monoclonal antibody detecting an antigen shared by neural and granulocytic cells. *Pediatr. Res.* 1981, (1)**15**, 1282–1286.

32

Immunocytochemistry in the Diagnosis of Thyroid Diseases

E. D. Williams

The thyroid gland, with three cell types of separate embryological origin and with characteristic polypeptide products, each giving rise to distinct major groups of tumours displaying a variety of sometimes overlapping histological appearances, would seem to be an ideal area for the diagnostic application of immunocytochemistry. This view is broadly confirmed by experience, although there are some confused areas, and the not unexpected problem of loss of antigen accompanying loss of differentiation. This chapter will first briefly discuss the typical immunocytochemical findings in the tumours derived from the follicular cells, those derived from the C cell, and those of lymphoid origin, and will then go on to consider the application of these observations in practical diagnostic problems.

The malignant follicular cell tumours are generally divided into the differentiated groups which show evidence of follicular or papillary architecture and the undifferentiated carcinomas. Those carcinomas showing a trabecular or solid packeted architecture are grouped with the differentiated tumours. One of the problems in discussing thyroid tumour nomenclature is that the words 'papillary' and 'follicular' are used to describe the pattern of differentiation present in the tumour, and also as diagnostic terms; to avoid the confusion that may arise the diagnostic terms will be capitalized.

1. FOLLICULAR, PAPILLARY AND ANAPLASTIC CARCINOMAS

The follicular cell is characterized by its ability to synthesize and iodinate thyroglobulin; it is the only cell in the body with this function, whereas iodide

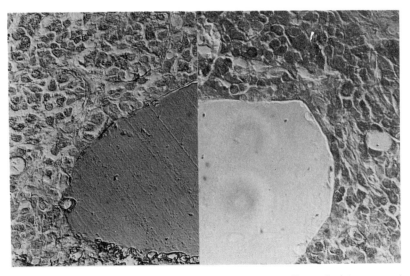

Fig. 32.1. Medullary carcinoma of the thyroid (formalin-fixed paraffin section) immunostained for bombesin (left) and calcitonin (right). The calcitonin antibody reveals moderate cytoplasmic staining in the tumour cells, and shows that the included colloid is negative. The bombesin antibody shows negative staining for the tumour cells, and positive staining for the colloid. This is due to the presence of thyroglobulin antibodies, as this molecule was used as a carrier in the generation of the anti-bombesin antibodies.

concentration is a property shared by the thyroid with a number of other tissues, including the salivary glands and the stomach. Thyroglobulin is a very large polypeptide; it is a good antigen and several excellent antibodies are available. As a large readily available protein it has unfortunately been used as a carrier for the production of commercially available antibodies to smaller polypeptides and this should be borne in mind as the possible explanation of unexpected immunocytochemical staining of a thyroid tumour when using a polyclonal antibody to a non-follicular cell peptide (*Fig.* 32.1). The use of appropriate control sections which should always include one with normal thyroid follicles should bring this problem to light if it is present. In the active normal follicular cell thyroglobulin is present in the follicular cell cytoplasm; in the inactive follicles colloid staining is often more prominent than cellular staining.

Thyroglobulin can therefore be used as a specific marker for thyroid follicular cell differentiation, and is present in the great majority of both Follicular and Papillary carcinomas. When the findings reported by five different groups[1–5] are combined together, all 87 Follicular carcinomas studied showed the presence of thyroglobulin immunoreactivity, as did 102 out of 105 Papillary carcinomas. In general the immunoreactivity for thyroglobulin is found in the cells rather than the neoplastic lumina, and is less strong in Papillary (*Fig.* 32.2) than Follicular carcinoma (*Fig.* 32.3), in keeping with the observations that Follicular carcinomas are often associated with a higher level of circulating thyroglobulin than Papillary carcinomas.[6–8] All types of Follicular carcinoma, oxyphil cell, clear cell, well and less well differentiated, showed thyroglobulin positivity. No staining for thyroglobulin was found in 144 non-thyroid tumours.

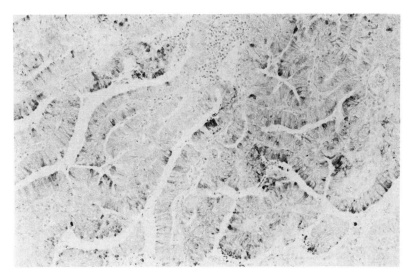

Fig. 32.2. Papillary carcinoma of the thyroid immunostained for thyroglobulin. A few positive cells are present, much of the tumour is unstained.

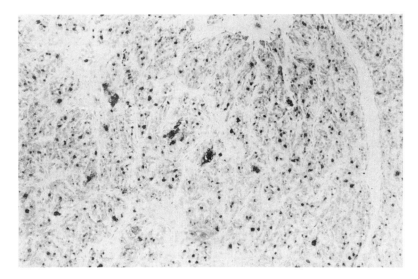

Fig. 32.3. Follicular carcinoma of the thyroid immunostained for thyroglobulin. Widespread strong focal positivity marks the lumina of microfollicles in the relatively poorly differentiated tumour.

A second marker which is not specific to thyroid epithelium, but which is often positive in Papillary carcinoma, and rarely present in Follicular carcinoma is keratin. In one study,[3] 12 out of 13 Papillary carcinomas showed strong diffuse positivity for keratin (using an antiserum to human epidermal keratin with a broad spectrum of reactivity) while none of 10 Follicular carcinomas showed

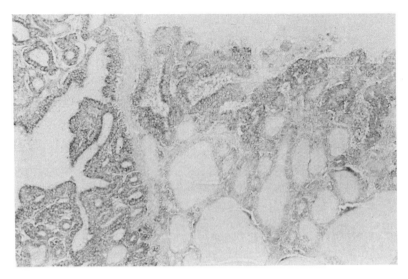

Fig. 32.4. Papillary carcinoma of thyroid immunostained for keratin to show weak cytoplasmic positivity in tumour cells to the left and at the top of the picture, and negative normal thyroid follicles lower right.

strong positivity, three showing a weak reaction. Papillary carcinomas are known to show foci of squamous metaplasia, but in the study quoted the reactivity was particularly marked in the papillary areas, the follicular areas of the tumour were largely negative for keratin with the antibody used (*Fig.* 32.4).

While the pattern of immunoreactivity of Papillary and Follicular carcinoma is relatively consistent, the reported pattern of staining in anaplastic carcinoma is much more variable. Most authors agree that strong positivity for thyroglobulin is restricted to areas also showing structural differentiation, which are interpreted as remnants of pre-existing differentiated carcinoma, as described by Ibanez et al.[9] (*Fig.* 32.5). However, while some authors find no other, or very limited, positive staining for thyroglobulin in spindle and giant cell carcinomas,[1-3] others find weak positivity in half or more of the cases studied.[4,5]

2. C CELLS AND MEDULLARY CARCINOMAS

In contrast to the follicular cell, which is derived from the thyroglossal duct, there is no doubt that C cells reach the human thyroid via the ultimobranchial body, and good evidence in birds that they are derived from the neural crest. The possibility that some C cells have a common origin with follicular cells cannot be excluded. The C cell was first clearly separated from the follicular cell by its ability to produce calcitonin; unlike thyroglobulin, however, this peptide is not confined to the thyroid, and calcitonin production has been described in various other normal tissues. Despite this it remains a good marker for C cells and their tumours. A variety of other hormonal peptides are produced by C cells—sometimes, as with somatostatin, in a minority population of C cells.[10]

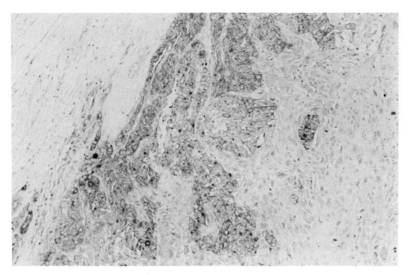

Fig. 32.5. Anaplastic carcinoma of thyroid immunostained for thyroglobulin. The diffuse spindle and giant cell component, lower right, is negative, the positive staining is present in solid islands of cells, and highlights the occasional microfollicular lumina. This pattern is typical of a poorly differentiated follicular tumour, from which in this case the anaplastic carcinoma is presumed to have arisen.

In the last few years two additional peptides produced by the calcitonin gene have been described, both of which are of value as tumour markers. Katacalcin is a flanking peptide of calcitonin, it appears to be produced in equimolar amounts with calcitonin, and has been localized to normal C cells.[11] Calcitonin gene-related peptide (CGRP) is derived from the calcitonin gene by differential RNA processing, and is present in central and peripheral nerve tissues.[12,13] It is also present in normal C cells, although in smaller amounts than is calcitonin.[14,15]

C cell tumours in man, usually referred to as medullary carcinomas, contain calcitonin (*Fig.* 32.6), almost by definition, since in its absence proof that the tumour was of C cell origin would be very difficult to obtain. They also contain katacalcin[11] and CGRP.[14,15] The distribution of katacalcin is similar to that of calcitonin, but CGRP is found in fewer cells, these tending to be large, and heavily immunoreactive, with elongated processes[15] (*Fig.* 32.7). Amyloid, commonly but not invariably found in medullary carcinoma, usually shows calcitonin but not CGRP immunoreactivity.[15] A very wide variety of additional peptides and other products have been described in medullary carcinoma; some of these are relevant to clinical problems, such as the ectopic production of ACTH leading to Cushing's syndrome in association with medullary carcinoma; others are of interest in showing the relationship between C cell tumours and other tumours of the diffuse endocrine system, for example neuron-specific enolase.[16] Another interesting product of medullary carcinoma is the enzyme histaminase; this has been shown to be present in groups of cells in medullary carcinoma by immunohistochemistry.[17] A tumour product which is of considerable diagnostic value is carcinoembryonic antigen (CEA); this membrane-bound protein is easily detected immunocytochem-

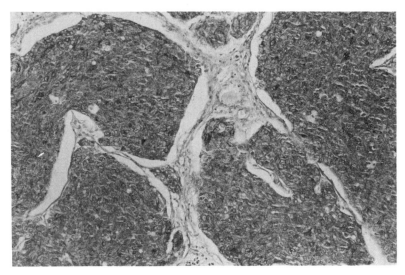

Fig. 32.6. Medullary carcinoma of the thyroid immunostained for calcitonin, showing diffuse positivity in most tumour cells.

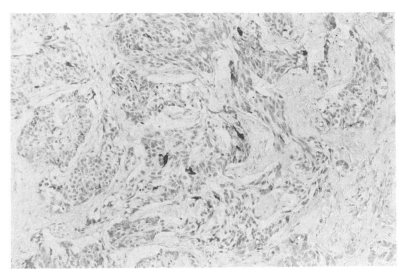

Fig. 32.7. Medullary carcinoma of the thyroid immunostained for CGRP, showing a largely negative tumour, but containing occasional large strongly immunoreactive cells. This is a common pattern of staining for CGRP in this tumour.

ically, and is present in the majority of cases of medullary carcinoma (*Fig.* 32.8). It has been detected in normal C cells by some workers but not by others; the discrepancy may reflect the different antibodies used.

It has been shown, both by immunocytochemistry and by measurement of plasma levels, that CEA may be of particular value in defining a high malignancy

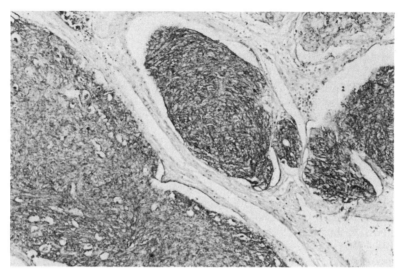

Fig. 32.8. Medullary carcinoma of the thyroid immunostained for CEA showing widespread positivity, the central area showing the cell membrane staining typically present with this antigen.

subgroup of medullary carcinoma, Mendelsohn et al.[18] showing that CEA expression may be retained, and calcitonin expression lost in cells from aggressive tumours, while several groups[19–21] have shown that plasma CEA is a useful marker of a high risk subgroup, and that a rising plasma CEA in the presence of a constant or even falling calcitonin level is an ominous sign.

The presence of small amounts of immunoreactive calcitonin has been described in some cases reported on histological grounds as anaplastic carcinoma.[22] In the absence of a strongly positive CEA, such findings should be regarded with care.

3. THYROIDITIS

Lymphoid infiltration is not found in the normal thyroid gland. It is commonly found in the thyroid of elderly females, and this focal thyroiditis must be separated from the diffuse infiltration by lymphocytes and plasma cells found in Hashimoto's thyroiditis. Studies of Hashimoto's thyroiditis have shown that the majority of plasma cells present contain IgG with κ light chain, in contrast to the very much rarer Riedel's thyroiditis, when IgA-containing plasma cells are common, and λ is the dominant light chain.[23,24]

4. LYMPHOMAS

Malignant lymphoma of the thyroid is a rare tumour of the thyroid, forming less than 10 per cent of thyroid malignancies in most series. It is typically a tumour of older females, and in about 80 per cent of cases it is associated with a severe thyroiditis of the non-neoplastic thyroid gland, often of the Hashimoto type.[25] It is almost invariably of non-Hodgkin's type, a minority of cases showing marked plasmacytoid differentiation. It behaves as a primary thyroid tumour, with spread to the cervical lymph nodes.

Immunocytochemical findings of malignant lymphomas are described in Chapter 34; in the thyroid, as might be expected, only a minority contain demonstrable immunoglobulin; studies on surface markers are difficult because of the need for fresh tissue.

5. DIAGNOSTIC USES OF IMMUNOCYTOCHEMISTRY

In the majority of cases of malignant thyroid tumours immunocytochemistry can be used in a confirmatory way, but is not really needed to establish the diagnosis. One of the commonest diagnostic problems in thyroid pathology is the differential diagnosis of a low-grade follicular carcinoma and a cellular follicular adenoma. This at present depends entirely on the identification of vascular invasion. It might be thought that immunocytochemical techniques to identify smooth muscle elastic fibres or endothelium would be of value, but smooth muscle and elastic are widely present in follicular tumour capsules. Studies to identify endothelium using antibodies to factor VIII-related antigen have been carried out, but are of little diagnostic value, although they did permit speculation that inhibition of clotting mechanisms might be important in allowing metastasis through vascular channels.[26]

There are, however, a number of well-defined situations where immunocytochemical techniques are critical in allowing an accurate diagnosis to be established. Perhaps the commonest of these is the cold nodule in the thyroid which on resection is found to be a solid amyloid negative tumour, lacking papillary or follicular differentiation, and with a differential diagnosis between a follicular tumour, a medullary carcinoma and a secondary carcinoma. The use of immunocytochemistry for thyroglobulin and calcitonin is usually adequate to allow separation of tumours of follicular or of C cell origin, although techniques for CEA may also be of value as confirmatory of a C cell origin when the calcitonin staining is weak. These three antibodies should also be used for any thyroid tumour with follicular or papillary differentiation when the cytology differs from the typical follicular tumour, or areas of solid medullary carcinoma-like tumour are present. Medullary carcinoma may show papillary or follicular differentiation[27] and immunocytochemistry is essential to establish the diagnosis in these cases.

Over the last few years a small number of cases of mixed Medullary–Follicular carcinomas have been described. These are very rare tumours and must be distinguished from medullary carcinoma with included non-neoplastic thyroid follicles. In this situation, thyroglobulin immunoreactivity may be seen in malignant C cells in the vicinity of damaged follicles, presumably due to phagocytosis of released thyroglobulin. However, production of thyroglobulin and of calcitonin has been described in a metastasis from one of these tumours,[28] so the entity of mixed Medullary–Follicular carcinoma must be accepted. Its existence calls into question the origin of all C cells from the neural crest.

Another important problem in thyroid pathology is the classification of small cell malignant tumours. For many years the entity of small cell carcinoma was recognized. More recently it has been accepted that many of the tumours previously diagnosed as small cell carcinomas were in reality malignant lymphomas of the thyroid, in which the apparent 'packeting' of the tumour was the result of survival of the pre-existing follicular reticulin network of the gland, which had been invaded but not destroyed by the lymphoma.[25] Although the distinction between the anaplastic carcinoma, with its spindle and giant cell pattern, and the malignant lymphoma with its lack of any spindle cell component, its non-cohesive pattern of growth and its cytological resemblance to lymphoma of other sites is usually straightforward, there remains a subgroup of small cell tumours whose exact nature can be difficult to ascertain from the routinely stained sections. Here immunocytochemistry is essential, and several groups have reported studies in which cases previously reported as anaplastic small cell carcinomas were reviewed using immunocytochemistry, and the diagnosis clearly established in most cases. The development of antibodies to common epithelial markers and to common leucocyte antigens is of particular value in this situation and this approach has been used by Burt et al.[29] to study 53 anaplastic small cell tumours of the thyroid; 33 were found to be positive for common leucocyte antigen, 6 were positive for epithelial membrane antigen, while 14 were negative for both markers. In a smaller study, Mambo and Irwin[30] used antibodies to light chains as B-cell markers, and an antibody to thyroxine as a follicular cell marker, with broadly similar results. The choice of an antibody to thyroxine or to thyroglobulin in this situation relies upon the epithelial cells retaining sufficient functional differentiation, and one of the various antibodies to common epithelial antigens would seem more likely to pick up a larger number of anaplastic carcinomas from this group.

Another differential diagnosis involving anaplastic carcinoma is its distinction from malignant haemangioendothelioma. This rare tumour has been described principally from areas of endemic goitre, and has been regarded by some as a variant of anaplastic carcinoma. Recent work has shown that factor VIII-related antigen, an endothelial marker, can be identified in the tumour cells, confirming its endothelial origin.[31,32] The differential diagnosis of anaplastic carcinoma from sarcoma, including malignant fibrous histiocytoma, is more difficult; for the latter tumour, histiocytic markers such as α1-antitrypsin may be of value.

Metastases to the thyroid occur occasionally, although they are rare causes of thyroid enlargement leading to surgery. The common primary tumours involved are kidney, breast, lung and melanoma;[33,34] of these, metastases from renal cell carcinoma often pose a particular problem histologically as they may closely resemble clear cell Follicular carcinoma. The use of an antibody to thyroglobulin is the most convenient way of resolving this particular differential diagnosis.

In considering metastases from primary tumours of the thyroid, the problems can be separated into two groups. Firstly, metastatic carcinoma with an indeterminate origin by conventional histology, where the site, usually cervical lymph nodes, leads to the thyroid being considered as the source of a possible primary tumour. Follicular carcinomas rarely produce metastases without a palpable primary tumour, but this can occur with Medullary carcinoma, and is common with Papillary carcinoma. Thyroglobulin, calcitonin and CEA are the antigens to search for; if all are negative the thyroid is a most unlikely primary site.

Many carcinomas may of course give rise to cervical lymph node metastases; amongst those which have caused particular problems prostatic carcinoma is worth mentioning, as it may occasionally show a papillary pattern reminiscent of a thyroid tumour. The second group of problems in dealing with metastases are those when a metastasis in any location, usually distant from the neck, is thought to be from a thyroid carcinoma because of its morphology, but proof is needed. Here obviously the antibody desired would be chosen to suit the possibility; for example thyroglobulin for a metastasis in bone with a follicular pattern, and calcitonin and CEA for a liver metastasis with a pattern suggestive of medullary carcinoma. Although rare, it should be remembered that metastases may be derived from Follicular or Papillary carcinomas in the ovary. These arise from the thyroid component of a teratoma, sometimes a so-called 'struma ovarii' in which the thyroid component is dominant. The carcinoid tumour arising in association with struma ovarii, the 'strumal carcinoid', may contain calcitonin.

Finally, this review of the value of immunocytochemistry in thyroid pathology must return to the C cell and its tumours. Medullary carcinoma can mimic a wide variety of tumours in its histological appearances—Papillary carcinoma, Follicular carcinoma, giant cell anaplastic carcinoma, 'small cell' anaplastic carcinoma, and metastasis to the thyroid. Whenever an atypical medullary carcinoma is raised as a diagnostic possibility, immunolocalization for calcitonin and CEA should be carried out.

Whenever the diagnosis of medullary carcinoma is made, the question as to whether it is sporadic or familial must also be considered. It has been shown that familial medullary carcinoma is commonly accompanied by and preceded by C-cell hyperplasia,[35] and this should be sought in all cases of medullary carcinoma, in the normal thyroid surrounding the tumour and in the opposite lobe. Differentiation of C-cell hyperplasia from islands of medullary carcinoma invading the adjacent thyroid may be difficult on a haematoxylin and eosin-stained section—as indeed is the recognition of mild degrees of C cell hyperplasia itself. For these observations immunocytochemistry is essential, and the use of antibodies to calcitonin, CEA and CGRP is recommended. Normal and hyperplastic C cells are commonly much richer in immunoreactive calcitonin than are the cells of medullary carcinoma; hyperplastic C cells commonly contain more immunoreactive CGRP than the carcinoma cells or the normal cells, and tumour cells contain more immunoreactive CEA than hyperplastic or normal cells[15] (Figs. 32.9 and 32.10). These observations may help to make the diagnosis of C-cell hyperplasia, and therefore to lead to the performance of family studies in appropriate cases.

Appendix

I. IMMUNOPEROXIDASE STAINING PROTOCOL

The formalin-fixed, paraffin-embedded thyroid sections shown in the figures were immunostained by a modified version[36,37] of the dinitrophenyl (DNP)–hapten

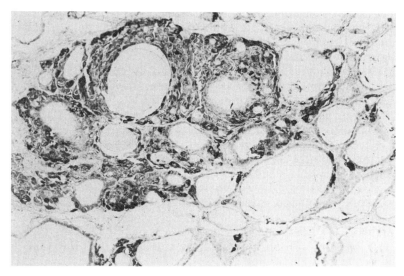

Fig. 32.9. Hyperplasia of C cells in a patient belonging to a multiple endocrine neoplasia type II family, immunostained for calcitonin.

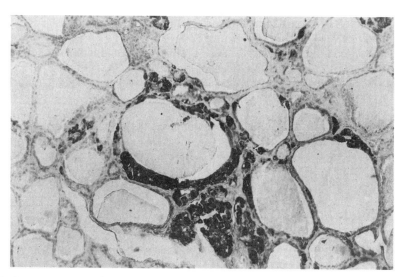

Fig. 32.10. Hyperplasia of C cells in a patient belonging to a multiple endocrine neoplasia type II family. The hyperplastic cells, unlike the normal C cells and unlike medullary carcinoma, are strongly and diffusely positive for CGRP.

sandwich staining procedure.[38,39] This is a method of high sensitivity and method specificity. The DNP group used as the hapten is linked to the primary antibody by a simple procedure. The use of a hapten such as DNP allows the use of a homologous blocking serum. A monoclonal IgM anti-DNP is used as the bridge and is linked to sequentially applied DNP-labelled peroxidase and DNP-labelled glucose oxidase. The monoclonal IgM bridge has the advantages of reagent uniformity common to all monoclonal antibodies and, in addition, the multiple binding sites of the IgM molecule allow the binding of several enzyme molecules. The concentrations of the two DNP-labelled enzymes are adjusted empirically so that an adequate number of binding sites remains free for the DNP–glucose oxidase after attachment of the DNP–peroxidase. The action of glucose oxidase on the glucose provided as a substrate results in the local generation of absolutely fresh hydrogen peroxide which acts as a substrate for the (DNP-labelled) peroxidase and ensures precise deposition of diaminobenzidine reaction product at the site of primary antibody binding with a very low background.

The staining protocol consists of the application of a series of reagents in the following manner:

1. A series of 4 changes of xylene followed by 4 changes of absolute alcohol for the dewaxing of the sections
2. Methanol (47·2 ml) and hydrogen peroxide (0·8 ml of 30 per cent H_2O_2) mixture (30 min at room temperature) for the inhibition of endogenous peroxidase
3. Phosphate-buffered saline (PBS) equilibration for 5 min
4. DNP-labelled primary antibody (diluted in non-immune homologous serum), for 16 h at 4 °C
5. IgM anti-DNP monoclonal antibody for 45 min at 20 °C
6. DNP-labelled peroxidase for 45 min at 20 °C
7. DNP–glucose oxidase for 45 min at 20 °C
8. Diaminobenzidine (0·05 per cent) plus β-D(+)-glucose (1·5 per cent) in PBS for 16 h at 20 °C in dark
9. Mayer's haemalum-based nuclear counterstaining
10. Dehydration and mounting in Permount.®

Three washes in PBS (1 min each) were included between Steps 4 to 8 to remove any excess reagent.

II. SOURCES OF REAGENTS

Antibody	Source	Dilution used
Rabbit anti-human thyroglobulin	Dako	1:1200
Rabbit anti-human CEA	Dako	1:12 000
Rabbit anti-human bombesin	Professor J. M. Polak	1:800
Rabbit anti-human calcitonin	Dr Morag Ellison	1:3200
Rabbit anti-human CGRP	Professor R. Craig	1:800
Rabbit anti-human keratin	Miles-Yeda	1:200
Mouse monoclonal IgM anti-DNP (ascites)	Sera-Lab	

REFERENCES

1. Burt A. and Goudie R. B. Diagnosis of primary thyroid carcinoma by immunohistological demonstration of thyroglobulin. *Histopathology* 1979, **3**, 279–286.

2. Bocker W., Dralle H., Husselmann H., Bay V. and Brassow M. Immunohistochemical analysis of thyroglobulin synthesis in thyroid carcinomas. *Virchows. Arch. Abt. A Pathol. Anat. Histol.* 1980, **385**, 187–200.
3. Permanetter W., Nathrath W. B. J. and Lohrs U. Immunohistochemical analysis of thyroglobulin and keratin in benign and malignant thyroid tumours. *Virchows. Arch. [Pathol. Anat.]* 1982, **398**, 221–228.
4. Albores-Saavedra J., Nadji M., Civantos F. and Morales A. Thyroglobulin in carcinoma of the thyroid. *Human Pathol.* 1983, **14**, 62–66.
5. Logmans S. C. and Jöbsis A. C. Thyroid-associated antigens in routinely embedded carcinomas. *Cancer* 1984, **54**, 274–279.
6. Schlossberg A. M., Jacobson J. C. and Ibbertson H. K. Serum thyroglobulin in the diagnosis and management of thyroid carcinoma. *Clin. Endocrinol.* 1979, **10**, 17–27.
7. Shah D. H., Dandekar S. R., Jeevanram R. K., Kumar A., Sharma S. M. and Ganatra R. D. Serum thyroglobulin in differentiated thyroid carcinoma: histological and metastatic classification. *Acta Endocrinol.* 1981, **98**, 222–226.
8. Ericsson U. B., Tegler L., Lennquist S., Christensen S. B., Stahl E. and Thorell J. I. Serum thyroglobulin in differentiated thyroid carcinoma. *Acta. Chir. Scand.* 1984, **150**(5), 367–375.
9. Ibanez M. L., Russell W. O., Albores-Saavedra J., Lampertico P., White E. C. and Clark R. L. Thyroid carcinoma—biologic behaviour and mortality. *Cancer* 1966, **19**, 1039–1052.
10. Van Noorden, S., Polak J. M. and Pearse A. G. E. Single cellular origin of somatostatin and calcitonin in the rat thyroid gland. *Histochemistry* 1977, **53**, 243–247.
11. Ali-Rachedi A., Varndell I. M., Facer P., Hillyard C. J., Craig R. K., MacIntyre I. and Polak J. M. Immunocytochemical localisation of katacalcin, a calcium-lowering hormone cleaved from the human calcitonin precursor. *J. Clin. Endocrinol. Metab.* 1983, **57**(3), 680–682.
12. Amara S. G., Jones V., Rosenfeld M. G., Ong E. S. and Evans R. M. Alternative RNA processing in calcitonin gene expression generates mRNAs encoding different polypeptide products. *Nature* 1982, **298**, 240–244.
13. Rosenfeld M. G., Mermod J.-J., Amara S. G., Swanson L. W., Sawchenko P. E., Rivier J., Vale W. W. and Evans R. M. Production of a novel neuropeptide encoded by the calcitonin gene via tissue-specific RNA processing. *Nature* 1983, **304**, 129–135.
14. Sabate M. I., Stolarsky L. S., Polak, J. M., Bloom S. R., Varndell, I. M., Ghatei M. A., Evans R. M. and Rosenfeld M. G. Regulation of neuroendocrine gene expression by alternative RNA processing. *J. Biol. Chem.* 1985, **260**, 2589–2592.
15. Williams E. D., Ponder B. J. and Craig R. CGRP in C cells and medullary carcinoma. *Clin. Endocrinol.* in press.
16. Tapia F. J., Barbosa A. J. A., Marangos P. J., Polak J. M., Bloom S. R., Dermody C. and Pearse A. G. E. Neuron-specific enolase is produced by neuroendocrine tumours. *Lancet* 1981, **i**, 808–811.
17. Mendelsohn G., Eggleston J. C., Weisburger W. R., Gann D. S. and Baylin S. B. Calcitonin and histaminase in C-cell hyperplasia and medullary thyroid carcinoma. *Am. J. Pathol.* 1978, **92**, 35–43.
18. Mendelsohn G., Wells S. A. and Baylin S. B. Relationship of tissue carcinoembryonic antigen and calcitonin to tumor virulence in medullary thyroid carcinoma. *Cancer* 1984, **54**, 657–662.
19. Rougier Ph., Calmettes C., LaPlanche A., Travagli, J. P., Lefevre M., Parmentier C., Milhaud G. and Tubiana M. The values of calcitonin and carcinoembryonic antigen in the treatment of non-familial medullary thyroid carcinoma. *Cancer* 1983, **51**, 855–862.
20. Busnardo B., Girelli M. E., Simioni N., Nacamulli D. and Busetto E. Non-parallel patterns of calcitonin and carcinoembryonic antigen levels in the follow-up of medullary thyroid carcinoma. *Cancer* 1984, **53**, 278–285.
21. Saad M. F., Fritsche H. A. and Samaan N. A. Diagnostic and prognostic values of carcinoembryonic antigen in medullary carcinoma of the thyroid. *J. Clin. Endocrinol. Metab.* 1984, **58**(5), 889–894.
22. Nieuwenhuijzen Kruseman A. C., Bosman F. T., van Bergen Henegouw J. C., Cramer-Knijnenburg G. and Brutel de la Riviere G. Medullary differentiation of anaplastic thyroid carcinoma. *Am. J. Clin. Pathol.* 1982, **77**(5), 541–547.
23. Hedinger Chr. Geographic pathology of thyroid disease. *Pathol. Res. Pract.* 1981, **171**, 285–292.
24. Harach H. R. and Williams E. D. Riedel's thyroiditis and Hashimoto's thyroiditis. *J. Pathol.* 1983, **139**, 472–473.
25. Williams E. D. Malignant lymphoma of the thyroid. In: Williams E. D., ed., *Pathology and*

Management of Thyroid Disease. Clinics in Endocrinology and Metabolism. London, W. B. Saunders Co. Ltd., 1981: 379–389.

26. Harach H. R., Jasani B. and Williams E. D. Factor VIII as a marker of endothelial cells in follicular carcinoma of the thyroid. *J. Clin. Pathol.* 1983, **36**, 1050–1054.

27. Harach H. R. and Williams E. D. Glandular (tubular and follicular) variants of medullary carcinoma of the thyroid. *Histopathology* 1983, **7**, 83–97.

28. Pfaltz M., Hedinger Chr. E. and Muhlethaler J. P. Mixed medullary and follicular carcinoma of the thyroid. *Virchows. Arch. [Pathol. Anat.]* 1983, **400**, 53–59.

29. Burt A. D., Kerr D. J., Brown I. L. and Boyle P. Lymphoid and epithelial markers in small cell anaplastic thyroid tumours. *J. Pathol.* 1984, **143**(4), 296A.

30. Mambo N. C. and Irwin S. M. Anaplastic small cell neoplasms of the thyroid. *Human Pathol.* 1984, **15**, 55–60.

31. Egloff B. The hemangioendothelioma of the thyroid. *Virchows Arch. [Pathol. Anat.]* 1983, **400**, 119–142.

32. Ruchti C., Gerber H. A. and Schaffner T. Factor VIII-related antigen in malignant hemangioendothelioma of the thyroid: additional evidence for the endothelial origin of this tumor. *Am. J. Clin. Pathol.* 1984, **82**(4), 474–480.

33. Shimaoka K., Sokal J. E. and Pickren J. Metastatic neoplasms in the thyroid gland; pathological and clinical findings. *Cancer* 1962, **15**, 557–565.

34. Czech J. M., Lichtor, T. R., Carney J. A. and van Heerden J. A. Neoplasms metastatic to the thyroid gland. *Surg. Gynecol. Obstet.* 1982, **155**, 503–505.

35. Wolfe H. J., Melvin K. E. W., Cervi-Skinner S. J., Saadi A. A., Juliar J. F., Jackson C. E. and Tashjian A. H. C-cell hyperplasia preceding medullary thyroid carcinoma. *N. Engl. J. Med.* 1973, **289**, 437–441.

36. Jasani B. and Williams E. D. Hapten Enzyme Labelling. British Patent No. 2,098,730 B, 1985.

37. Jasani B. and Williams E. D. A coupled glucose–oxidase/peroxidase immuno-enzyme system for greater accuracy in immunolabelling of tissue targets. In preparation.

38. Jasani B., Wynford-Thomas D. and Williams E. D. Use of monoclonal anti-hapten antibodies for immunolocalisation of tissue antigens. *J. Clin. Pathol.* 1981, **34**, 1000–1002.

39. Jasani B., Thomas N. D., Newman G. R. and Williams E. D. DNP–hapten sandwich (DHSS) procedure: design, sensitivity, versatility and applications. *Immunol. Commun.* 1983, **12**, 51.

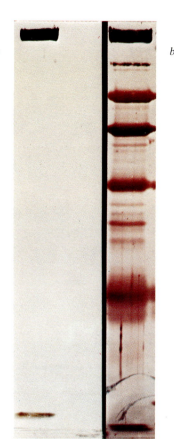

a   *b*

Plate 1. Immunodetection of myosin with the IGSS method and subsequent protein staining with AuroDye®. HMW standards (1 μg/band) were separated on a 7·5 per cent SDS-PAGE gel and electrotransferred onto nitrocellulose paper. The blot was quenched with 0·3 per cent Tween 20 in PBS. One lane was incubated with rabbit anti-myosin, 1 μg/ml in TBS + 0·3 per cent Tween 20 for 1 h, washed with TBS–Tween 20 and incubated with GAR G BL grade (Janssen Life Sciences Products), at an absorbance at 250 nm ($A_{250} = 0·05$) in PBS–Tween 20 for 4 h. After washing with PBS-Tween 20 and H_2O, the red signal was amplified with silver enhancement.[33] The result is shown in (*a*). In (*b*) another lane was treated in the same way, but subsequently stained with AuroDye®. The immunodetected protein band can now be precisely correlated with the transferred electropherogram. (Courtsey of G. Daneels and M. Moeremans, Janssen Pharmaceutica, Beerse, Belgium.)   (*Chapter 1*)

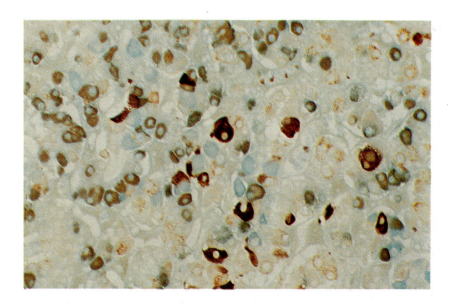

Plate 2. Human pituitary immunostained for four hormones: TSH (black), FSH (brown), ACTH (turquoise) and GH (green). Mouse monoclonal anti-TSH was used first and developed by the PAP method with DAB and cobalt chloride. This was followed by mouse monoclonal anti-FSH, rabbit anti-ACTH and mouse and rabbit antibodies to GH. Mouse antibodies were revealed by the PAP method with DAB (brown) and rabbit by an indirect β-galactosidase method (turquoise). The mixture of peroxidase and β-galactosidase labels on the GH cells results in a green colour.

(*Chapter 3*)

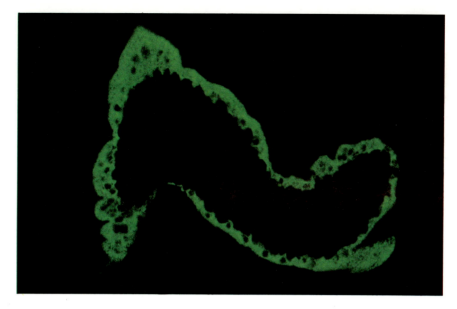

a

b

Plate 3. Immunolocalization of β subunit of chorionic gonadotrophin in human placenta. Indirect FITC-labelled avidin–biotin (*a*) and indirect immunofluorescence (*b*) methods. The photographic exposure time has been kept identical. (*Chapter 4*)

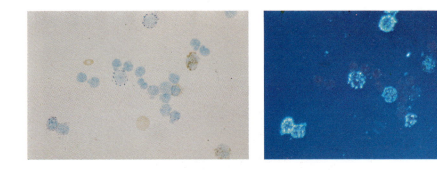

a   *b*

c

Plate 4. Human peripheral blood mononuclear leucocytes stained with OKT 4 (Ortho Diagnostic Systems) and GAM G30 (Janssen Life Sciences Products). The nuclei are counterstained with methyl green, and the preparations have been treated with diaminobenzidine and H_2O_2 for endogenous peroxidase. Photographs were taken with a Nikon Optiphot microscope equipped for epipolarization microscopy (IGS block) and a 40x Plan-Apo, N.A. 1·0 oil immersion objective. (*a*) Bright-field microscopy. OKT 4-positive lymphocytes show numerous dark granules associated with the plasmalemma. Non-lymphocytes have a granular brown deposit, corresponding to their peroxidase content. (*b*) The same field, viewed with epipolarization. The gold marker emits a strong signal but non-marked and peroxidase-containing cells are difficult to detect. (*c*) Combined epipolarization and bright-field microscopy. In this mode, the identification and counting of marked and non-marked cells, and the assessment of their enzyme activity content is greatly facilitated.

(*Chapter 5*)

Plate 5. A *Haemanthus* endosperm cell in metaphase, stained with antibodies to tubulin and the immunogold staining (IGS) method. The spindle fibres are stained red and can be observed with high resolution. Chromosomes are stained with toluidine blue. [Courtsey of A. Bajer (Eugene, Oregon, USA). From joint work with A. Lambert (Strasbourg, France) and M. De Brabander (Beerse, Belgium).] *(Chapter 5)*

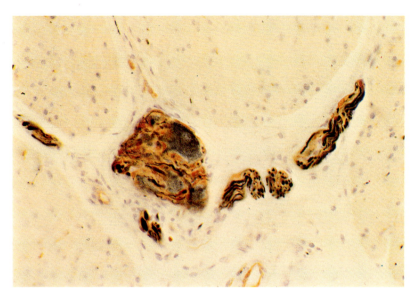

Plate 6. Ganglion cells immunoreactive with rabbit antiserum to chicken neurofilament protein triplet (black, IGSS method) and glial structures immunoreactive to rabbit antiserum to bovine S-100 (brown, PAP method) in the muscle wall of pig urinary bladder. Bouin's fluid-fixed, 5-μm wax section, counterstained with haematoxylin. (*Chapter 5*)

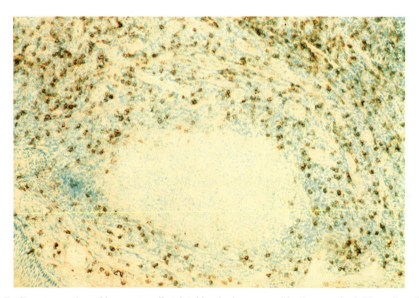

Plate 7. Cryostat section of human tonsil stained by the immunogold–silver method. The cell surface membrane of the T-cytotoxic/suppressor cells was labelled with the OKT 8 mouse monoclonal antibody and goat anti-mouse IgG antibodies adsorbed to 5-nm gold particles. Positive cells were mainly found in the paracotrical and interfollicular areas; a few only were scattered in the follicle mantles and the germinal centres. Methyl green counterstain. (Unpublished results of M. De Waele.) (*Chapter 5*)

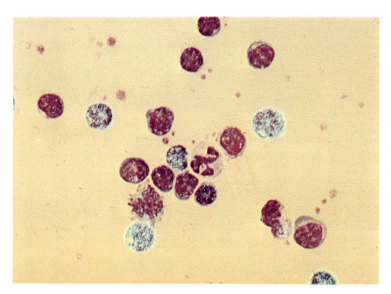

Plate 8. Human peripheral blood mononuclear leucocytes stained with OKT 4 antibody and the immunogold–silver method. The cells were counterstained with May–Grünewald/Giemsa and photographed with a Nikon Optiphot microscope using combined bright-field and epipolarization illumination. (40x plan apo, oil immersion objective N.A. 1·0.) OKT 4-positive cells are clearly visible, and cell-type recognition on the basis of conventional counterstaining is easy.   (*Chapter 5*)

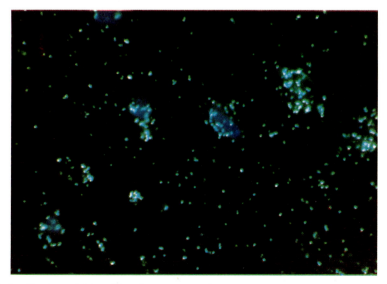

Plate 9. Co-localization of Fast Blue in cells containing POMC cDNA–mRNA hybrids as determined by in situ hybridization (in collaboration with T. O'Donahue and B. Chronwall, NICCDS, Bethesda MD). Rats were injected with 0·2 µl of a 3 per cent solution of Fast Blue in the medial preoptic nucleus. Two days later, animals were sacrificed and brain sections processed for POMC in situ cDNA–mRNA hybridization. This is a double exposure of Fast Blue-labelled arcuate cells, some showing hybridization to the [³H]POMC cDNA probe, with Fast Blue visualized by illumination at 360 nm, silver grains visualized by polarized light epiluminescence.   (*Chapter 12*)

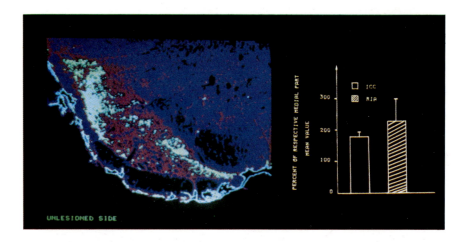

Plate 10. An absorbance map of substance P-like immunoreactivity is shown in the left part of the picture. In the right part of the picture the ratio of substance P-like immunoreactivity in the lateral to that in the medial part of the substantia nigra is shown as revealed by immunocytochemistry and by radioimmunoassay procedures (ICC versus RIA). Means ± s.e.m. are shown out of five rats. No significant differences are observed. (*Chapter 13*)

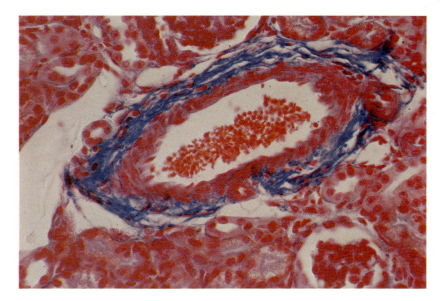

a

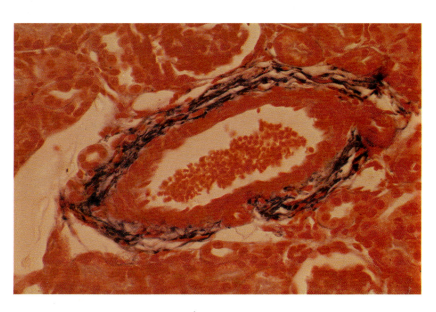

b

Plate 11. A blood vessel in the kidney cortex stained with Azan. (*a*) Photographed on colour film corrected for tungsten light, at the correct colour temperature of 3200 K. (*b*) The same field but photographed with the colour temperature reduced by 130 mireds (mired or microreciprocal degree = $10^6/T$ with T as the colour temperature in Kelvins). Note the pronounced change in colour balance. Such a change results, for example, from regulating the light *intensity* by reducing the lamp voltage. (*Chapter 14*)

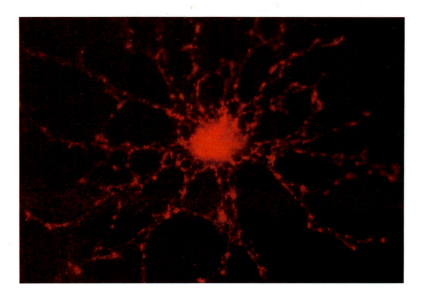

Plate 12. RAN-2 immunoreactivity on the surface of a cultured astrocyte. Rat cerebellum, 10 days in culture. (*Chapter 15*)

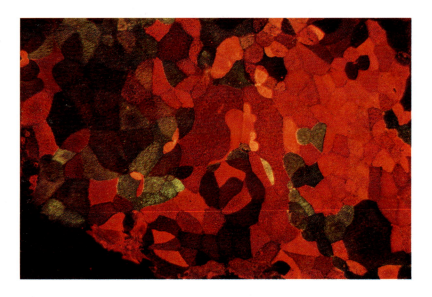

Plate 13. Double immunofluorescence on MCF7 cells. LICR-LON-M8 positive cells are green. LICR-LON-M18 positive cells are red. Cells bearing both antigens are orange. *See* Appendix, Section II for method. (Photomicrograph kindly supplied by Mr Robert Skilton, Ludwig Institute, London.) (*Chapter 16*)

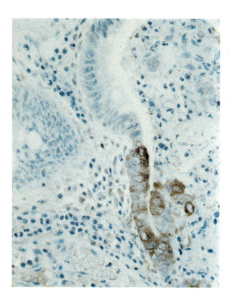

Plate 14. Focal non-specific cytoplasmic staining in a human colon carcinoma using a sheep polyclonal antiserum to a peptide in a transforming gene protein. A similar staining pattern was seen with normal non-immune rabbit serum but not with the use of an isotypic IgG. ABC immunocytochemistry, haematoxylin, formalin fixation. (*Chapter 17*)

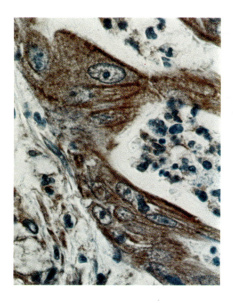

Plate 15. Prominent non-specific membrane staining of a colonic carcinoma by a sheep polyclonal antiserum to a transforming gene protein. A similar pattern of staining was seen with high concentrations of normal rabbit serum but not with an isotypic IgG. Note luminal surface staining is most prominent. ABC, haematoxylin, formalin fixation. (*Chapter 17*)

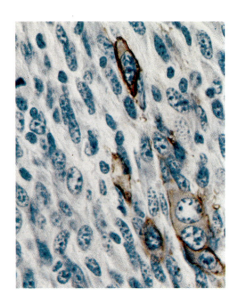

Plate 16. Prominent membrane staining in a sarcoma induced in a mouse by injected BALB/c 3T3 cells transfected by *Ras^{Ha}* oncogene, using a sheep polyclonal antiserum to a peptide in *Ras^{Ha}* p21. Note that the mitotic cell on left edge is not immunoreactive. ABC immunocytochemistry, hematoxylin. Bouin's fixation (also for *Plates* 17–22). (*Chapter 17*)

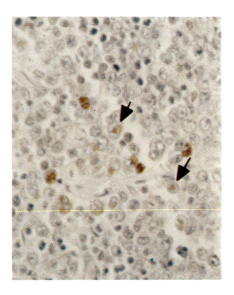

Plate 17. Lymphoblastic lymphoma in a mouse induced by *raf/myc* recombinant virus showing intranuclear staining for *myc* protein of a few tumour cell nuclei (arrows) and non-specific cytoplasmic staining of a few neutrophils and other haematopoietic cells. Intranuclear staining must be differentiated from non-specific staining reactions. ABC immunocytochemistry, haematoxylin, rabbit antibody. (*Chapter 17*)

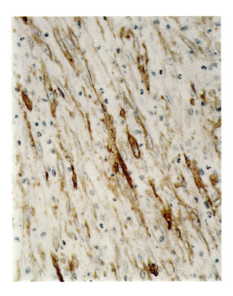

Plate 18. *Ras*^{Ha} p21 on large proportion of sarcoma cells induced by Harvey sarcoma virus/Moloney leukaemia virus. ABC, immunocytochemistry, haematoxylin, sheep antiserum to a peptide in *Ras*^{Ha} p21 (also for *Plates* 19, 20). *(Chapter 17)*

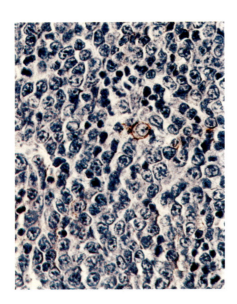

Plate 19. *Ras*^{Ha}p21 on cell membrane of a splenic erythroblast in Harvey virus-induced erythroblastosis. *(Chapter 17)*

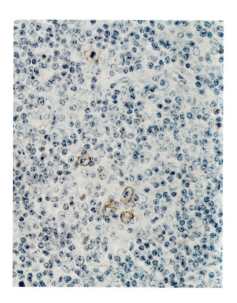

Plate 20. *Ras^{Ha}*p21 on the cell membrane of reticular cells in regional lymph node adjacent to virus-induced sarcoma. Note mitotic figure with membrane staining. (*Chapter 17*)

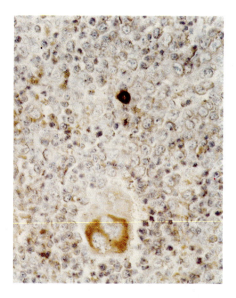

Plate 21. Rauscher leukaemia virus p30 in the cytoplasm (brown) of a megakaryocyte and *Ras^{Ha}* p21 in an erythroblast (black) from spleen of Harvey sarcoma/Moloney leukaemia virus-injected mouse. Double staining using the PAP technique and DAB (brown) for p30 using a goat antiserum and the ABC technique and DAB with cobalt chloride (black) for p21 using a sheep antiserum to a peptide in *Ras^{Ha}* p21. (*Chapter 17*)

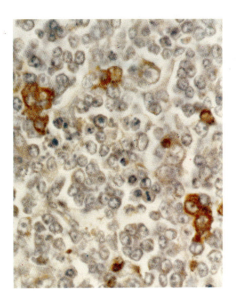

Plate 22. V-*raf* oncogene protein in lymphoblasts induced by *raf/myc* recombinant virus. A polyclonal rabbit antisera (α-SP) to a peptide in v-*raf* transforming protein was used.   (*Chapter 17*)

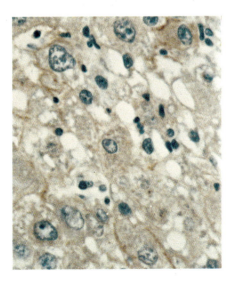

Plate 23. Immunoreactive cell membranes of chemically induced hepatocellular adenoma cells using a rabbit antiserum to SP-63, a peptide in v-*raf* transforming protein. ABC, haematoxylin.   (*Chapter 17*)

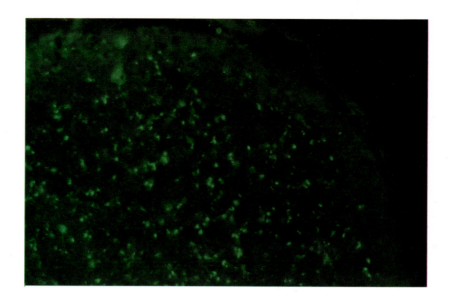

Plate 24. Gastrin/CCK-like immunoreactivity in the neural ganglion of *Ciona intestinalis* (Ascidiacea). Bouin's-fixed, paraffin section, indirect immunofluorescence. (*Chapter 19*)

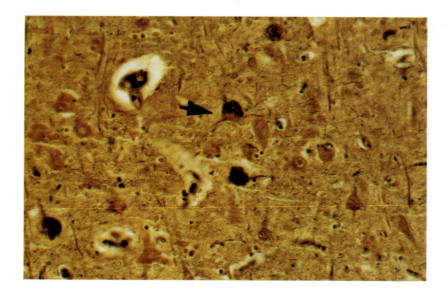

Plate 25. Temporal cortex from a case of Alzheimer's dementia. Formalin-fixed paraffin section stained with Bielchowsky's silver method and subsequently immunostained (PAP). The tangles (impregnated with silver giving a black reaction product) are seen to be in somatostatin-immunoreactive neurones (brown reaction product). (*Chapter 21*)

Plate 26. Low power photomicrograph of the caudate nucleus in Huntington's chorea showing that numerous NPY-immunoreactive neurones are still present. Formalin-fixed, 100-µm Vibratome section, immunostained by the PAP method. (*Chapter 21*)

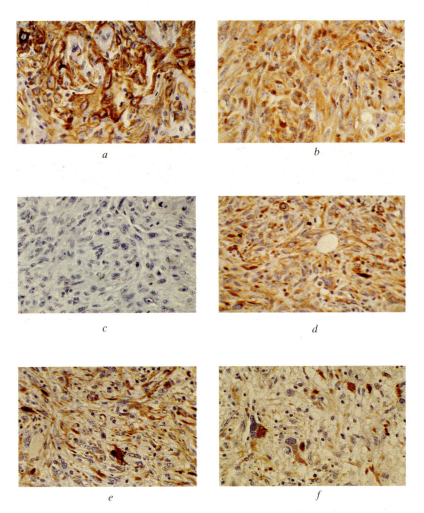

Plate 27. Immunoperoxidase staining of formalin-fixed, paraffin sections of three different tumours, initially diagnosed as malignant fibrous histiocytomas, with antiserum to human skin keratins (*a,c*), to bovine lens vimentin (*b,d,e*) and to chicken gizzard desmin (*f*). Note the strong positive reaction with the keratin antiserum in (*a*) and with the desmin antiserum in (*f*). On basis of these staining patterns two of these tumours were reclassified as mesothelioma (*a,b*) and rhabdomyosarcoma (*e,f*), respectively. Note also that all three tumours are positive for vimentin, thereby illustrating coexpression of two types of IFP in spindle cell-type mesothelioma and in rhabdomyosarcoma.

(*Chapter 23*)

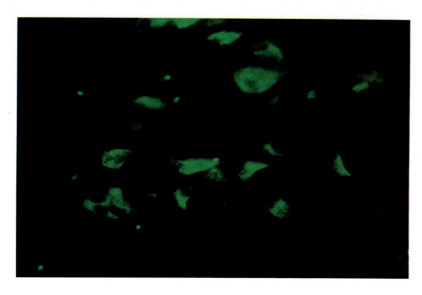

Plate 28a. Cryostat section of human antrum treated with patient's serum and FITC anti-human IgG showing bright-green cells scattered in the antral mucosa. (*Chapter 28*)

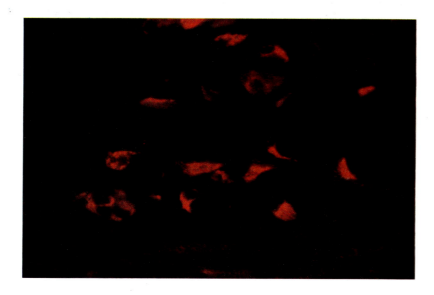

Plate 28b. Same section stained with rabbit anti-gastrin serum which was revealed with rhodaminated goat anti-rabbit antiserum. The red cells have the same distribution as in (*a*). In a double exposure photograph the green and red cells appeared yellow thus confirming the identity of the cells stained by the patient's serum and the rabbit anti-gastrin serum. (*Chapter 28*)

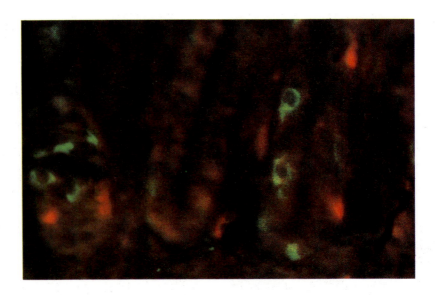

Plate 29. Double exposure photography of a section of human antrum first treated with the same patient's serum as in *Plate* 28*a* and subsequently stained with rabbit anti-somatostatin antiserum. The red somatostatin cells are clearly separate from the green cells (gastrin cells) which reacted with the patient's serum. (*Chapter 28*)

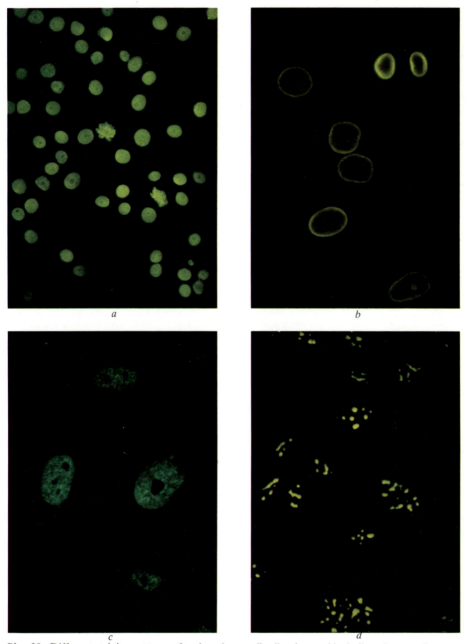

Plate 30. Different staining patterns of anti-nuclear antibodies detected by indirect immunofluorescence on tissue cultures. (*Chapter 29*)

(*a*) Diffuse, HEp-2 cells. (*Chapter 29*)
(*b*) Rim (peripheral), mouse fibroblast 3T3 cell line. (*Chapter 29*)
(*c*) Fine speckled, Alexander (liver) cell line. (*Chapter 29*)
(*d*) Nucleolar, Alexander cell line. (*Chapter 29*)

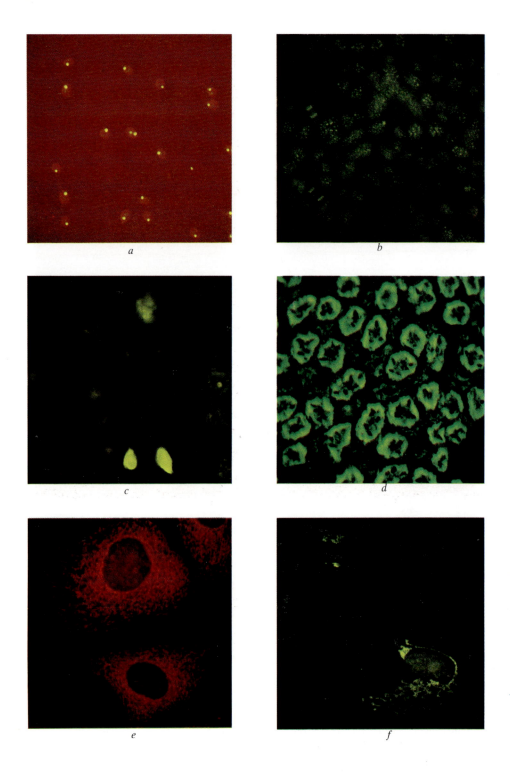

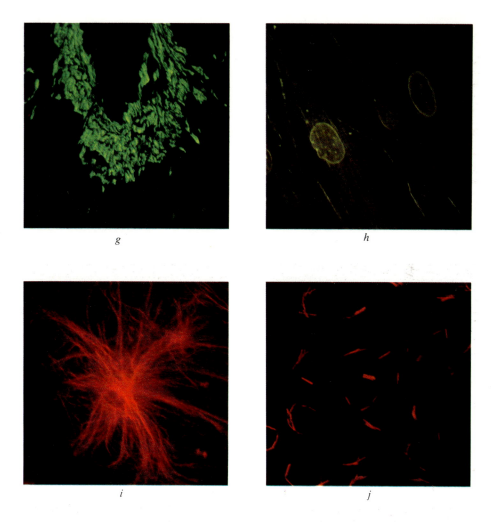

g   h

i   j

Plate 31. Immunofluorescence staining patterns of different anti-nuclear and anti-cytoplasmic antibodies.   (*Chapter 29*)

(*a*) Smear of the haemoflagellate *Crithidia luciliae* incubated with the serum of a patient with SLE. Staining of the kinetoplast indicates anti-dsDNA antibodies.   (*Chapter 29*)
(*b*) Fine-discrete speckled staining pattern of the nucleus in interphase cells of cultured epidermoid cell carcinoma indicating anti-centromeric antibodies. In mitotic cells, the immunofluorescence is polarized at the centromeric region of the chromosomes.   (*Chapter 29*)
(*c*) Antibodies to the proliferating cell nuclear antigen from the serum of patient with SLE demonstrated on the PTK-2 cell line. The staining is seen on dividing cells exclusively.   (*Chapter 29*)
(*d*) Anti-mitochondrial antibodies detected on unfixed cryostat sections of rat kidney, giving a fine-speckled pattern of tubular cytoplasm.   (*Chapter 29*)
(*e*) The same serum as *Plate* 31*c* reacted with liver cells in culture.   (*Chapter 29*)
(*f*) Golgi antibodies visualized by immunofluorescence on embryonic fibroblast cells (Wi-38). Juxtanuclear staining of the cisternae is the typical appearance.   (*Chapter 29*)
(*g*) Smooth-muscle antibodies demonstrated on unfixed cryostat section of rat stomach. (*Chapter 29*)
(*h*) Actin antibodies visualized on embryonic fibroblast cells. Densely packed microfilament bundles ('stress fibres') are apparent.   (*Chapter 29*)

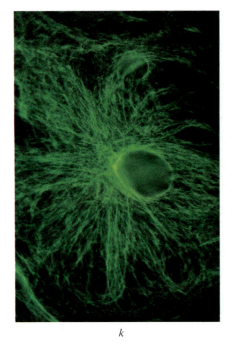

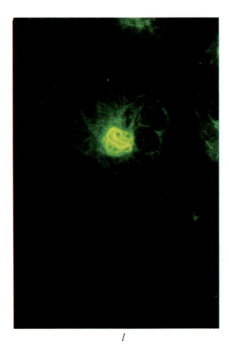

k   l

Plate 31 (contd)

(*i*) Tubulin antibodies reacting with the extensive microtubular network radiating from its organizing centre to the cell periphery. Mouse fibroblast 3T3 line   (*Chapter* 29)
(*j*) Paracristals were seen with the same serum as *Plate* 31*c* when cells were pre-treated with vinblastine.   (*Chapter* 29)
(*k*) Vimentin antibodies reacting with the typical arrays of filaments. Mouse fibroblast 3T3 line.
                                    (*Chapter* 29)
(*l*) Coils and whorls are stained with the same serum as *Plate* 31*c* when the antigen was pre-treated with vinblastine.   (*Chapter* 29)

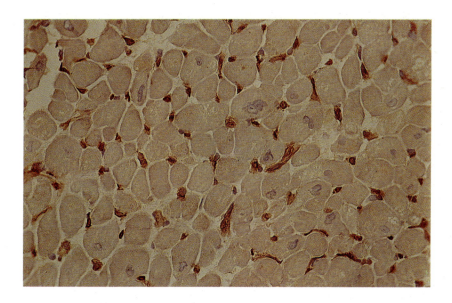

Plate 32. Binding of W6/32 monclonal antibody to normal heart. Cryostat section, 5 μm, fixed in acetone. (*Chapter 30*)

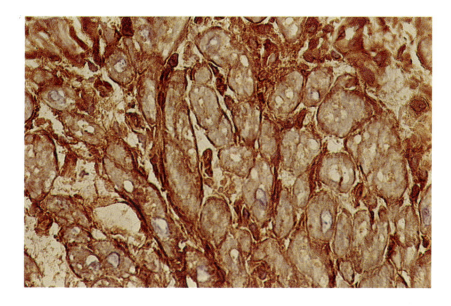

Plate 33. Binding of W6/32 monoclonal antibody to cardiac biopsy from a transplant patient showing signs of mild rejection as assessed by conventional histological criteria. Cryostat section, 5 μm, fixed in acetone. (*Chapter 30*)

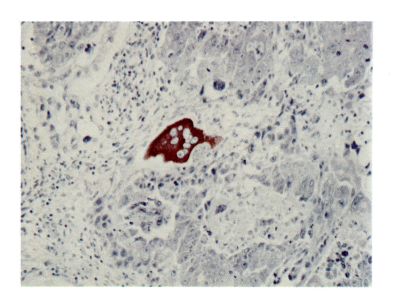

Plate 34. Single HCG-positive multinucleate cell in malignant teratoma intermediate. (*Chapter 31*)

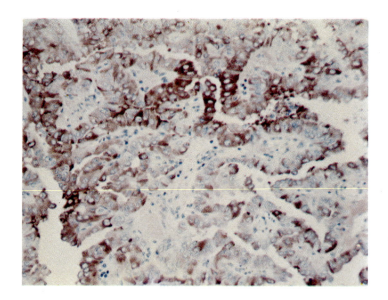

Plate 35. Papillary carcinoma of the thyroid stained for thyroglobulin. In this field the staining is mainly cytoplasmic. Elsewhere there was staining of luminal membranes. (*Chapter 31*)

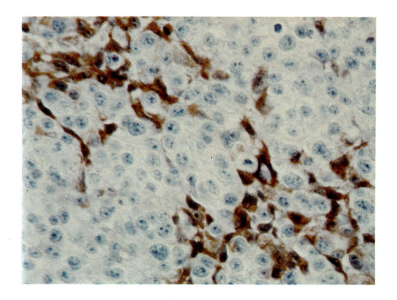

Plate 36. Metastatic testicular tumour with seminoma morphology shows strong but patchy cytoplasmic positivity for alpha-fetoprotein (AFP). (*Chapter 31*)

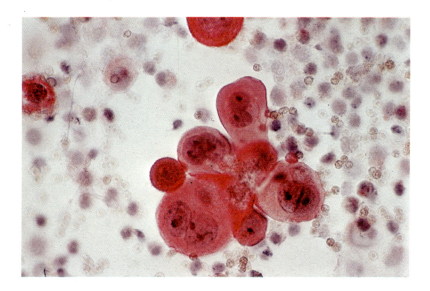

Plate 37. Malignant cells in ascitic fluid from a case of ovarian carcinoma immunostained for HMFG2. Cytospin preparation fixed for 5 min in 95 per cent ethanol. The bound mouse monoclonal antibody is visualized by an alkaline phosphatase–anti-alkaline phosphatase method with Fast Red TR as the chromogen (*see* Chapter 3 for method). (Photomicrograph by courtsey of Dr T. Krausz, Histopathology Department, Royal Postgraduate Medical School.) (*Chapter 33*)

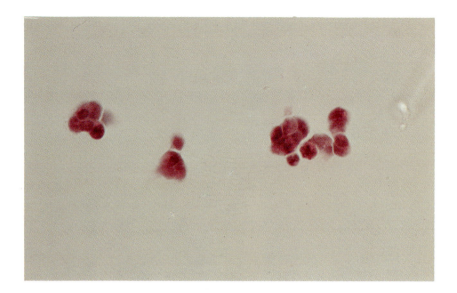

a

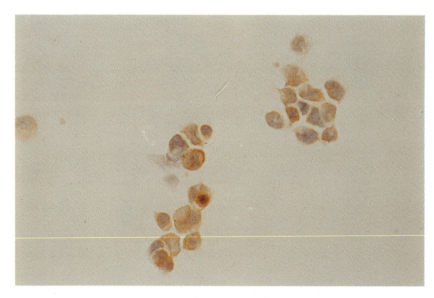

b

Plate 38. Tumour cells in smears (prepared as in Appendix, Section I) of cells in a pleural effusion from a patient with small cell carcinoma of the lung, stained by haematoxylin and eosin (*a*) and immunostained·with antiserum to neuron-specific enolase (*b*). Strong staining is seen in the small cell carcinoma cells. Smear wet-fixed in methanol and immunostained without drying after fixation, anti-NSE diluted 1:800, PAP method with haematoxylin counterstain. (*Chapter 33*)

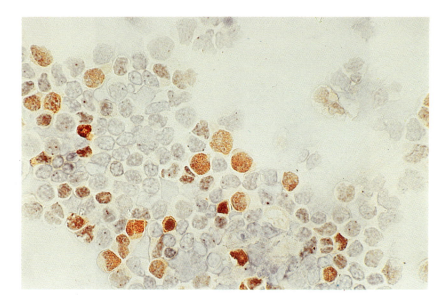

Plate 39. Pleural effusion from a case of T-acute lymphoblastic leukaemia showing nuclear staining for TdT (indirect immunoperoxidase)   (*Chapter 35*)

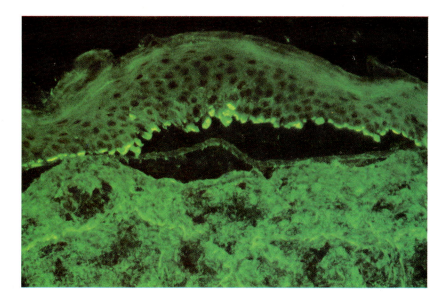

Plate 40. Junctional epidermolysis bullosa stained with bullous pemphigoid serum showing staining in the roof of the blister.   (*Chapter 36*)

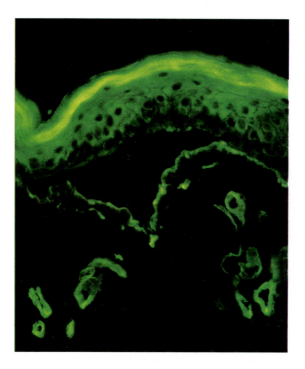

Plate 41. Junctional epidermolysis bullosa stained for type IV collagen showing staining in the floor of the blister. (*Chapter 36*)

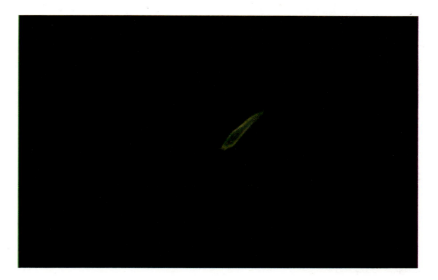

Plate 42. *Leishmania tropica major* promastigotes stained by indirect immunofluorescence with the monoclonal antibody WIC 79.3. The antibody stained the surface and the flagellum of the organism. (By kind permission of E. Liew, Wellcome Labs.). (*Chapter 38*)

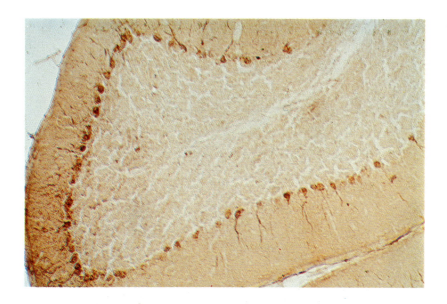

Plate 43. Frozen section of paraformaldehyde-perfused rat cereballar cortex stained with monoclonal antibody CE5 using the immunoperoxidase technique. Purkinje cell dendrites and cell bodies are stained. (Reprinted from Wood et al.[19] with permission of the editors of *Nature*, 1982, Vol. 296, No. 5852, p. 36. Copyright 1982 Macmillan Journals Ltd.)   (*Chapter 38*)

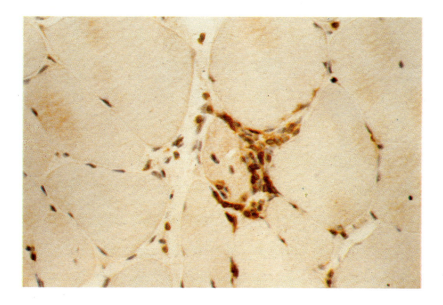

Plate 44. T-lymphocytes in a case of polymyositis shown by use of monoclonal antibody and avidin–biotin–peroxidase technique on a fixed cryostat section. Haematoxylin counterstain.
(*Chapter 39*)

33   *Immunocytochemistry in Diagnostic Cytology*

D. R. Springall

Immunostaining methods since their conception have remained largely the province of histologists. Cytopathologists (the distinction being that they examine preparations of aspirated or exfoliated whole cells rather than sections of tissue) have tended to lag behind in the development of immunocytochemical methods and their application in diagnosis. However, the advantages of cytological preparation for immunostaining, in addition to the attractive benefits of immunocytochemistry in routine diagnostic cytopathology, are now being realized. Further impetus is added by the increasing interest in automation of cervical screening. Any technique which can assist a machine to discriminate between normal and abnormal cells would be of immense value, and immunocytochemistry is a prime candidate.

In the past 30 years cytological investigation of patients with malignant or suspected malignant neoplasia has become the 'front-line' of diagnosis. The rapidly increasing use of percutaneous fine-needle aspiration techniques has enabled the cytopathologist to examine material from many organs of the body. Thus the scope of the discipline has enlarged considerably when compared with the original but still predominant approach, which involved mostly the examination of exfoliated cells. Inevitably this increased scope has also enlarged the problems in differential diagnosis of lesions from the somewhat novel sites, and there too immunocytochemistry can be of great assistance.

1. ADVANTAGES OF CYTOLOGICAL PREPARATIONS FOR IMMUNOCYTOCHEMISTRY

Cytological preparations offer advantages compared with histology for the patient, clinician and immunochemist.

a. The techniques used to obtain specimens are non-invasive, or minimally so.
b. Ample material may be obtained.
c. Cell specimens are easy to divide and many smears may be made. This permits the use of several fixatives, preparation methods and pre-treatments if necessary.
d. Fixation is rapid and controllable because material is in the form of single cells or small groups. In a tissue block material on the outside may be overfixed whilst that on the inside is underfixed.
e. Specimens may be stored unfixed in the refrigerator or as prepared smears until decisions are made as to the fixation required.
f. Time is saved since smears are easy and quick to prepare; processing and wax embedding are not required.

These advantages are sometimes partially offset by disadvantages such as the lack of tissue architecture, high background non-specific staining with some specimens, especially sputum, and occasionally poor condition of cells in fluids and sputa.

2. TYPES OF SPECIMEN

Gynaecological material, predominantly cervical scrapes, forms the major part of the workload of most cytopathology laboratories. The remainder consists of a variety of specimens including pleural, ascitic and pericardial fluids, sputum, urine, washings from gastrointestinal or respiratory tracts, cerebrospinal fluids, fine-needle aspiration biopsies, and brushings from respiratory tract or cervix. All of these may be used for immunostaining, but some more easily than others. Immunocytochemistry is particularly difficult with sputum specimens, owing to binding of antibodies by mucus. Mucolytic agents may help but they rarely remove all mucus from all specimens, and it is often easiest to resort to sectioning of blocks prepared from centrifuged sputum.

3. PREPARATION METHODS

As mentioned in Section 1, one of the principal advantages that cytology has over histology is that most cytological specimens are easily divided for different preparation methods, fixations and treatments. Thus, immunostaining of a variety of antigens which may require different fixation techniques can be accomplished with one specimen. However, since one aim must be to cause the minimum disruption of the day-to-day running of the laboratory, the preparation methods should be as simple as possible. It is also useful to examine smears microscopically before immunostaining to ensure that the required cells are actually present. For this purpose a rapid stain such as methylene blue is used and the slide is examined wet.

3.1. Smears

There are two methods of preparing cell smears on glass slides—by spreading cells manually, or by centrifuging them onto the slide using a machine such as the Shandon Cytospin (Shandon, Cheshire, UK). The former method has the advantage of simplicity and speed and does not require expensive equipment. A wire loop, an orange stick, or another glass slide ('blood film' method) are most commonly used. Cytocentrifuge preparations have the advantage of better adhesion of cells to the slide within a small regular area (6 mm diameter), thus making microscopical examination easier. The disadvantages are that they take longer to prepare and the machine must be cleaned between specimens.

The manual smearing technique may be used for all types of specimen including those containing mucus, such as sputum. For liquid specimens an initial centrifugation step to sediment the cells is necessary ($500 \times g$ for 5 min is usually adequate) following which the supernatant is drained off the cell button. If large numbers of erythrocytes are present in the cell deposit, they may be removed at this stage by density-gradient centrifugation of the resuspended deposit, using agents such as Ficoll (Pharmacia) or Lymphoprep (Nyegaard). The erythrocytes sediment through the medium leaving other cells in a layer above it, but care is necessary since aggregates of tumour cells may also be sedimented and lost. The initial centrifugation step is not usually necessary with the cytocentrifuge which has a filter paper pad interposed between slide and specimen holder to absorb the liquid.

The use of an agent such as poly(L-lysine) to ensure adhesion of cells to the slide[1] is recommended. A rapid method for smear preparation is given in the Appendix.

3.1.1. Fixation of Smears

Fixation is performed either before (wet-fixed) or after (air-dried) the smears have dried. For practical purposes, air-drying is preferable because cells are less liable to become detached during immunostaining. Furthermore, the cells flatten onto the slide making microscopical examination easier. However, antigenicity of some antigens such as neuron-specific enolase[2] may be reduced or destroyed by air-drying so caution should be exercised. For air-drying, smears should be prepared from a cell button which has been well drained so that drying occurs quickly.

Fixatives which have been used for smears are numerous. The most common are methanol, ethanol (or industrial methylated spirit), acetone and formol saline, either singly or in various combinations with or without the addition of water and acetic acid; times from 20 to 30 min are generally used. Each has its merits, and advantages are claimed for various concoctions. There is no doubt that fixation is very important for immunocytochemistry, but so is simplicity, and a balance needs to be struck between optimum fixation and avoidance of multiple and complicated procedures. The object of fixation is to preserve antigens, to bind them in situ to prevent loss by diffusion but without damaging antigenicity, and to maintain cell morphology. The cross-linking fixatives, such as formalin, are good for preservation but may reduce antigenicity. Precipitating or dehydrating agents, such as alcohol and acetone, are less damaging to antigens but may not bind them adequately. Antigens which are sensitive even to these organic fixatives may require the preparations to be fixed very briefly or at low temperatures (down to $-20\,^{\circ}$C), or even to be immunostained unfixed.

As a general rule, air-dried alcohol-fixed smears are adequate, especially when immunostaining large molecules, but for those antigens which are sensitive to air-drying or for smaller molecules, wet fixation and cross-linking fixatives may be necessary. For immunostaining of new antigens, various fixation regimes should be tried to find the simplest which works well.

3.2. Imprints

This method is useful for making cytological preparations from resected tumours or endoscopic biopsies which are to be fixed and wax embedded. They give the advantage of a rapid immunocytochemistry result, since there is no waiting for sections to be produced, and permit optimal fixation of labile antigens[2] or different fixations for multiple antigens.

To prepare imprints the biopsy is gently applied to a poly(L-lysine)-coated slide, using fine forceps, and then lifted away directly without smearing. This may be repeated with several slides which are then fixed appropriately for immunostaining.

3.3. Other Preparation Methods

Although smearing and cytocentrifuging are the methods generally used to prepare specimens, two other techniques are available. The first is filtering; liquid specimens are filtered through a porous cellulose–ester membrane leaving the cells deposited on its surface. The membrane is placed in fixative, then rinsed prior to immunostaining. The technique has no special advantages. The immunostaining procedure is more tedious, requires extended washings, and the membrane is inclined to bind proteins, thus causing high background.

The second method is to cut sections of either frozen or wax-embedded material. This technique is extremely useful for small tissue cores from fine needle aspirates which would be useless for diagnosis if squashed on a slide and stained whole. The cores should be removed from the other aspirated material before it is used to make smears. When processed and cut they provide enough sections for several slides which can be used for morphological examination and immunostaining. If the biopsy is fixed in glutaraldehyde and embedded in resin, then both electron and light microscopical examinations may be made. Unfortunately, many antigens are rendered unsuitable for immunostaining by such treatment, but the use of a highly sensitive technique, such as immunogold–silver staining, may help (*see* Appendix and Chapter 5).

To prepare cryostat sections, cells are pipetted into a gelatine capsule together with a small amount of cryostat embedding medium (e.g. OCT Compound, Lab-Tek Products) and then frozen. The capsule is sectioned from the base at right angles to its axis.

For wax embedding, the cell button is resuspended in fixative, centrifuged, rinsed in buffer, centrifuged again and then processed to wax. It is safest to embed the cell button, after fixation, either in agar or in a small celloidin bag (*see* Appendix). For these very small specimens reduced times for each step in the processing to wax should be used to avoid excessive loss of antigenicity.

3.4. Storage of Specimens

It is useful to be able to store specimens until after morphological examination has revealed the desirability of immunostaining and the antigens, and therefore the fixation methods, required. Body fluids may sometimes be stored for one or even two days in a refrigerator without too much cell degeneration, especially serous fluids (which are a good cell culture medium). However, it is preferable to store specimens on slides for two reasons—cell preservation is better and spare slides may be prepared for immunostaining at the same time as those for routine staining and discarded later if not required.

Air-dried smears may be stored by wrapping them in aluminium foil and placing in a −20 °C freezer. When required they are removed from the freezer, allowed to warm to room temperature, unwrapped and appropriately fixed. After fixation they should be rinsed in buffer and immunostained immediately, without drying.

Wet-fixed smears may be dried and stored in a refrigerator for short periods, or a freezer for longer periods. Much better results are often obtained, however, by avoiding drying and storing the slides wet in buffered saline containing 0·01 per cent sodium azide as preservative, for periods up to 7 days. Alternatively the use of fixatives containing carbowax (polyethylene glycol), as is generally used for cervical smears, prevents complete drying of the cells and seems to give adequate results.

3.5. Permeabilization

Whole fixed cells are impermeable to immunoglobulin molecules. Therefore, to immunostain intracellular antigens, pores must be created in the plasma membrane. This may be achieved by chemical or physical means. The former includes treatment with organic solvents, such as alcohol or acetone, or detergents such as Triton X-100 and Tween 20 or 80; the latter is achieved by cycles of freezing and thawing (*see also* Chapter 15). If smears are fixed in alcohol or acetone then further treatment is probably not necessary. Detergents may be used by immersing the smears in buffer containing 0·2–0·5 per cent detergent for 30 min before immunostaining, or by using the detergent buffer to dilute the primary antibody. Use of detergents may have the added advantage of reducing background staining.

Air-dried smears do not usually require permeabilization. Sufficient pores are created in the plasma membrane of the cells by distortion whilst drying.

4. IMMUNOSTAINING METHODS

Any of the available immunostaining techniques may be used for cytological preparations. Fluorescent labels, despite their choice for simplicity and speed, are not advisable for routine use due to the difficulty in assessing cell morphology. Any of the enzyme labels may be used, either conjugated to antibody (indirect, two-step method) or in the unlabelled antibody enzyme (bridge, three-step) method. Alkaline phosphatase and glucose oxidase labels have the advantage that endogenous enzyme does not need to be blocked, unlike peroxidase (*see* Chapter 3). This may be especially useful if labile surface antigens are to be stained in unfixed preparations. If peroxidase label is used, blocking and permeabilization may be achieved simultaneously by using a methanolic rather than an aqueous solution of hydrogen peroxide.

The recently developed[3] and subsequently modified[4] immunogold–silver staining technique is also proving to be of great value in cytology.[5] It has the advantage of speed and simplicity, because it is a two-step procedure, coupled with a high sensitivity permitting primary antibody dilutions around ten-fold greater than those used for the peroxidase–anti-peroxidase (PAP) technique. The high sensitivity may also be useful for enhancing the weak immunoreactivity which is often obtained when dismounted slides previously used for morphological assessment need to be immunostained.

4.1. Applications in Cytopathology

The cytopathologist tries to define whether or not a tumour or pre-cancerous lesion is present and, if it is, whether or not it is malignant, the tumour type, and the primary site if this is not known. Thus the main use of immunostaining is the detection of so-called 'tumour markers', with the possibility of using antibodies to tumour type-specific (histogenetic) or tissue-specific antigens. In addition, the detection of viruses and other infective agents is possible by these methods.

Cytopathological work may be divided into two categories: investigation of specimens of cell types from patients with symptoms, and screening of symptomless patients for signs of early disease. The latter is principally examination of cervical scrapes. The potential value of immunostaining is different for the two categories.

For the symptomatic cases the aim must be to use immunocytochemistry when a diagnostic problem arises. There is no need for any other than selected specimens to be immunostained, since the majority can be assessed on purely morphological grounds. Amongst many problems are, firstly, the location of sparse tumour cells, especially in specimens where predominant cell types are erythrocytes, inflammatory, or mesothelial cells. Secondly, the distinction of reactive mesothelial cells from adenocarcinoma cells in pleural fluids is particularly difficult when small single tumour cells showing isonucleosis are present, as may occur in 15 per cent of cases.[6] Thirdly, the classification of small undifferentiated tumour cells, such as small cell carcinoma, other small cell undifferentiated tumours and lymphoma, can be a great problem.

For the screening of specimens from symptomless patients, the value of immunocytochemistry lies in the potential ability of the technique to assist in the selection for closer scrutiny of specimens which contain abnormal cells. The ultimate aim is automation of the screening procedure in an attempt to speed up the process and reduce costs. However, the final diagnosis of each abnormal case which is selected will still rely on pathological acumen, probably with some help from immunocytochemistry.

5. IMMUNOCYTOCHEMISTRY OF NON-GYNAECOLOGICAL SPECIMENS

Serous fluids are a good source of cells for immunodiagnosis because any epithelium-derived cell in such a location is inevitably malignant. Any antiserum which can discriminate between epithelial and non-epithelial type cells is therefore a useful indicator of malignancy in pleural, peritoneal or pericardial fluids.

5.1. Carcinoembryonic Antigen

The value of immunostaining cytological specimens for carcinoembryonic antigen (CEA) was first reported by Pascal and Fenoglio.[7] They used sections of cell blocks from 112 fluids and found positive results with all adenocarcinomas but not small cell or some squamous carcinomas. These results have been subsequently confirmed by other workers.[8-12] Ghosh and coworkers[9,10] have used a monoclonal antibody to CEA which reacted with carcinoma cells in 30 of 36 malignant effusions,[10] but not with mesothelial cells. Therefore, CEA is a good discriminator of benign cells but only a fair indicator of malignancy. The former attribute is useful when the antibody is used in combination with others which are better at detecting malignant cells[9] but which may react with mesothelium, such as human milk fat globule 2 (HMFG-2) and Ca1 (see below). Thus by judicious assessment of the staining results, using the other two antibodies to pick up tumour cells missed by the anti-CEA, and ensuring no false positives occur due to HMFG-2 or Ca1 antibodies reacting with mesothelial cells (CEA negative), precision may be increased. As well as aiding in the distinction of epithelial tumours from mesothelial cells, such a panel of antibodies is also valuable in solving another problem in cytopathology: the detection of tumour cells when sparse or masked by blood cells in smears from effusions. Ghosh and colleagues have shown an increase of 20 per cent in diagnostic accuracy using this method.[9] They used acetone-fixed air-dried smears and an immuno-alkaline phosphatase technique with monoclonal primary antibodies.

Other workers have shown lower detection rates of malignant cells in fluids by anti-CEA. Walts and Said,[11] using air-dried smears of effusions and fine-needle aspirates fixed in 95 per cent ethanol, detected 50 per cent of adenocarcinomas. O'Brien and coworkers[12] obtained immunoreactivity for CEA in only 44 per cent of malignant effusions using sections of formalin-fixed wax-embedded cell blocks. This preparation technique was also used in one of the few attempts to immunostain sputum specimens.[13] Here reactivity with anti-CEA was found in all grades of squamous epithelium from normal through metaplastic and dysplastic to malignant. All small cell carcinomas were negative, but all adenocarcinomas were positive.

When using polyclonal antisera to CEA care should be taken to ensure that there are no component antibodies present which will react with non-specific cross-reacting antigens (NCA). These may be removed by pre-absorbing the antiserum with tissue powder (e.g. acetone-dried spleen) or tissue extracts[14] (*see also* Chapter 31). As NCA is present in leucocytes and myeloid cells, staining of these cells in smears suggests the presence of the unwanted antibodies.

5.2. Epithelial Membrane Antigen

This antigen has been widely assessed[6,12,15,16] for its discriminating ability in effusions. Whilst it has a detection rate for epithelial malignancies in wet-fixed (95 per cent ethanol) smears of up to 100 per cent[16] this depends on the dilution at which the antiserum is used. Reaction is also obtained with mesothelial cells.[6] This may be reduced by further diluting the antiserum, but the detection rate for malignancy then falls to 54 per cent.[16] Much reliance is placed on the grading of strong and weak reactions and ensuring precise timings for the immunostaining, as well as the need to wash the cells before preparing specimens.[6] This system is therefore technically time consuming and of limited value when used alone.

5.3. Human Milk Fat Globule Membrane Antigen

This antibody has a similar specificity to EMA, being raised by hybridization techniques to an oligosaccharide epitope of a large protein from human milk fat globule (HMFG) membrane. It gives a variably intense staining of epithelial tumour cells in the majority (98 per cent) of malignant effusions[10] (*Plate* 37), but also reacts with mesothelial cells, albeit weakly.[10] Epenetos and coworkers,[17] using ether–ethanol fixation, found no reaction with mesothelial cells. They showed that the combination of HMFG-2 and AUA 1 (an epithelial proliferation antigen) detected all of 26 cytologically malignant effusions. The HMFG-2 antibody has also been proposed by Ghosh and coauthors[9] as part of a useful panel of antisera for immunostaining effusions.

5.4. Intermediate Filaments

Antibodies to keratin,[10] desmin, vimentin and neurofilaments[18,19] are valuable in cytopathology and may help to reveal the histogenetic origin of tumours. Anti-keratins such as LE61 (cytokeratin filaments) and M73 (Mallory bodies) react with cells from most epithelial tumours, but also mesothelial cells.[10] In a study of 308 fine-needle aspiration biopsies, Droese and coworkers[18] used intermediate filament antibodies to characterize tumours in 12 cases where cytology was indecisive. Cytokeratins were revealed in all of 5 carcinomas, vimentin in 1 carcinoma, 1 sarcoma, 1 melanoma, 1 centroblastic lymphoma, 1 malignant schwannoma and a plasmacytoma, desmin in a leiomyosarcoma and neurofilaments in a phaeochromocytoma.

Antisera to glial fibrillary acid protein (GFAP), a component of normal and neoplastic astrocytes, astroblasts and some ependymoma cells, have been used by

Collins[20] for the rapid identification of tumour cells in acetone-fixed ($-20\,°C$) imprints and smears from 21 stereotactic biopsies and excised tumours of brain. They found this to be of diagnostic value. Staining was obtained in astrocytomas and glioblastomas, but not in oligodendrogliomas, meningiomas or a metastatic colonic adenocarcinoma. A different application was demonstrated by von Koskull[21] who used anti-GFAP to label glial cells in cold ($-20\,°C$) methanol-fixed cytocentrifuge preparations of fresh amniotic fluid as an aid to the prenatal diagnosis of neural-tube defects.

5.5 Ca1

This antibody detects a glycoprotein which appears to be preferentially expressed on malignant cells,[22] and has been evaluated using acetone-fixed air-dried, or methanol wet-fixed smears of cells from serous fluids by Woods and coworkers.[23] Of 25 malignant effusions, 21 were reactive with Ca1, 2 lymphomas and 2 of 6 lung carcinomas were negative, and 1 false positive was noted. They suggest that the antibody is valuable for discriminating malignant cells in effusions, since there was no reaction with mesothelial cells.[23] However, Ghosh and coworkers[10] found mesothelial cells were positive in 2 of 22 benign effusions, and whilst they found Ca1 reacted with tumour cells in 38 of 40 cytologically malignant effusions, staining was often weak and the intensity variable. Nonetheless, they suggest that Ca1 is of value as part of a panel of antibodies to assist detection of scanty malignant cells in cytologically false-negative smears.[9] Similar results were obtained by Pallesen and coworkers.[24]

5.6. Neuroendocrine Antigens

Whilst most neuroendocrine tumours are comparatively rare, there is one important exception—small cell carcinoma of the lung. This accounts for 20 per cent of lung carcinomas. Because it is highly malignant and usually responds better to radio- and chemotherapy than to surgery, early and accurate diagnosis is essential. That many of these tumours are immunohistochemically reactive with antibodies to neuron-specific enolase (NSE), a glycolytic enzyme isomer found in mammalian nerves and endocrine cells and tumours derived from them (*see* Chapter 20), was shown by Sheppard and coworkers.[25] Subsequently it was demonstrated[2] that NSE is a valuable aid to the cytological diagnosis of small cell lung carcinoma. A study of smears and cell blocks from effusions and touch imprints from endoscopic biopsies showed NSE immunoreactivity in 9 of 10 cases (*Plate* 38). The smears were wet fixed in methanol or formol saline and yielded results superior to those obtained using sections of cell buttons or biopsies.[2] All of a further 10 cases were positive for NSE (1985, unpublished results).

Yam and Winkler[26] report the use of a monoclonal antibody (534 F-8) raised against a small cell carcinoma cell line[27] to identify tumour cells in a pleural fluid from a patient with small cell lung carcinoma. The cells were stained in suspension by an immuno-alkaline phosphatase method.

Immunostaining of specific peptide hormones, as well as general neuroendocrine markers such as NSE, in cytological preparations, is sometimes useful. Thus gastrin has been detected[8] in smears from 1 aspirate and 2 touch preparations of pancreatic islet tumours. Calcitonin has been demonstrated in an imprint from 1 medullary thyroid carcinoma[8] and a neck mass in a patient with a previously resected thyroid carcinoma which subsequent histology proved to be a metastatic medullary carcinoma.[11]

Antiserum to β-human chorionic gonadotrophin has been found[8] to detect syncytiotrophoblasts in cervical smears of two pregnant women with first trimester bleeding, and malignant trophoblasts in a fine-needle aspirate of a mixed metastatic germ-cell tumour in lung.

5.7. Leucocyte Antigens and Immunoglobulins

The principal uses of immunostaining such antigens in cytological diagnosis are the confirmation of lymphoma or myeloma, either by monoclonality of κ or λ light chains[8,11] in the cell population, or by reactivity with one of the leucocyte common antigen antibodies such as 2D1.[10] The latter may also be used to confirm by negative, or exclude by positive staining a diagnosis of small cell carcinoma. Confirmation is less reliable since this implies no immunoreaction with leucocyte markers and a negative result is not strong evidence in immunocytochemistry.

5.8. Miscellaneous Antigens

Other antisera, mostly to histogenesis-associated or tissue-specific antigens, have been used. These include anti-prostatic acid phosphatase which was found to react with smears from aspiration biopsies of all of 3 histologically confirmed lymph node metastases of prostate carcinoma but not other tumours[8] and, in the study of Walts and Said,[11] to react strongly with material from another fine-needle aspirate of a lung nodule and a chylous pleural effusion. The same two groups of authors found that lysozyme was detectable immunocytochemically in histiocytes, macrophages, polymorphonuclear leucocytes and occasional mesothelial cells[8] in smears, but absent from all adenocarcinoma cells.[11]

An antibody raised against human mesothelial cell cultures has been used to immunostain pleural fluid smears, where it reacted with both benign and malignant mesothelial cells.[28]

Breast carcinoma cells were detected in all of 4 cytologically positive pleural fluids using a monoclonal antibody MRB1 raised against a low-molecular-weight glycolipid from a membrane fraction of a human mammary tumour cell line.[29] The antibody was fluorescein isothiocyanate (FITC) labelled and staining was performed in suspension. No reaction was obtained with mesothelial cells, but no data were reported for control tumours.

Szpak and coauthors[30] have used 4 monoclonal antibodies prepared using a similar immunogen—a membrane fraction of a metastatic breast carcinoma. Of these, the most useful (B72.3) was found to have good discriminating power between tumour and mesothelial cells. It reacted with 92 per cent of adenocarcinomas but not with mesothelial cells, lymphomas, or small cell or squamous cell lung carcinomas. This antibody therefore presents a similar specificity to that of HMFG-2. It is of interest that the authors used paraffin sections of formalin-fixed cell pellets and found that these gave better immunostaining results than did cytocentrifuge preparations.

6. GYNAECOLOGICAL SPECIMENS

The use of immunocytochemistry in diagnosis of abnormality in gynaecological specimens is less advanced than that in non-gynaecological specimens. There are two reasons to account for this—firstly the now common and very successful use of Papanicolaou-stained smears for cervical screening, and secondly the problem of distinguishing by immunocytochemistry normal and malignant cells of the same histogenetic type. However, the need to automate cervical screening has increased the pace of the search for antibodies to make this distinction. In addition, the implication of viruses in cervical cancer and the dramatic increase in some genital viral and bacterial infections has led to the introduction of immunocytochemical techniques for their detection.

Specimens are usually obtained by scraping the cervix with a wood or plastic Ayre's spatula. However, for sampling both the endocervix and ectocervix the cervical Biobrush is extremely useful.[31] It ensures efficient detection of lesions within the endocervical canal and, being made of soft Dicel fibres, is safe to use. Cells may be shaken off the Biobrush in buffer for use in immunocytochemistry, microdensitometry, flow cytometry or cell culture. The material remaining on the brush may be processed histologically or, for transmission or scanning electron microscopy, as a cell button after placing the brush in fixative and then dissolving the fibres in acetone. Sections may be used for immunostaining as well as morphological examination.

6.1. Cervical Cancer

Antisera tested for their ability to detect abnormal cervical cells in smears have generally been the same as those used for serous fluids. However, somewhat poorer results have been obtained, since most of these antisera will react with normal as well as abnormal cells. This reaction may be weaker with normal cells than their neoplastic counterparts, but often a strong reaction is also obtained with metaplastic cells. Therefore, although precision may be poor, the false results are at least of a 'fail safe' type, i.e. false positives. Morphological examination may then select true positives.

The most frequently used antibody is EMA. Its applicability to automation has been tested using integrating microdensitometry to quantify the expression of EMA in neoplastic cells.[32] Results obtained from 11 cases of various grades of lesions suggested that the antibody has potential for detecting abnormalities. Moncrieff and coauthors[33] have studied the detection of three antigens, EMA, CEA and pre-keratin in cervical cells, and conclude that EMA is the most suitable for detection of abnormal cells. However, a weak reaction was obtained with normal cells, even if mostly from infected cases, in 30 per cent of benign smears, and only weak staining was seen in abnormal cells in 10 per cent of abnormal smears. With the CEA and pre-keratin antibodies, a greater proportion of false positives was obtained.[33] The authors also noted that it was necessary to use smears of prefixed and washed cells in order to eliminate high non-specific staining caused by mucus and blood.

The value of transferrin receptor demonstration, using OKT 9, for detection of abnormal cervical cells has been suggested by Lloyd and coworkers.[34] They found reaction in all malignant and most severely dysplastic epithelia using unfixed cryostat sections of cervical biopsies. However, results were poorer with cytocentrifuge preparations of prefixed cells, 5 of 7 malignant and 6 of 11 severely dyskaryotic specimens were stained. The same authors had also found previously that Ca1 was a poor marker for cervical neoplasms[35] as did Krausz and coworkers.[36] Ca1 was also assessed by Jha and coauthors,[37] together with HMFG-1, HMFG-2 and two monoclonal antibodies to bladder cancer cell lines—8.30.3 and 77.1. Using paraffin sections of Bouin's fluid-fixed biopsies they found similar reactivity of all five antibodies. They were unreactive to normal squamous epithelium, and gave intense staining of metaplastic epithelium but weaker reaction or none at all with intra-epithelial or invasive neoplasia.

6.2. Flow Cytometry

The proposition of using immunostained slide preparations of cervical cells in automated screening for abnormality is somewhat optimistic; the machinery required is cumbersome and no tumour-specific antibody which will not produce false negatives has yet been found. The technique of flow cytometry has been used extensively for the study of all types of tumours, including cervical material[38,39] to provide cellular DNA measurements. Cells in suspension are passed through a small orifice. As they pass through, the volume of each cell is measured, together with other parameters, and data are stored grouped for each cell. The cell suspension may also be reacted with (usually) fluorescent probes before scanning. By using ultraviolet illumination with appropriate filters, the intensity of fluorescence in each cell may be measured. Appropriate filtration permits the simultaneous measurement in each cell of two probes, providing that their emission spectra are dissimilar. Thus, fluorescent dyes such as mithramycin or propidium iodide may be used to measure cellular content of DNA or DNA plus RNA; careful selection of a fluorescent label, such as FITC or rhodamine isothiocyanate (RITC), coupled directly or indirectly (two-step immunostaining) to an antibody allows simultaneous quantitation of the relevant antigen. If this antibody is one such as EMA which binds strongly to abnormal cells, the problem of unwanted binding to some normal cells may be overcome by the simultaneous measurement of cellular DNA. Since tumours have an aneuploid distribution of DNA, but aneuploid cells may be difficult to distinguish when diluted by many non-tumour cells in cervical exfoliative samples, the antibody may be used to dissect out the abnormal population and this may be distinguished from any stained normal cell on the basis of DNA content. Thus each cell is automatically screened for malignancy on the basis of two criteria rather than just one. Clusters of cells, which may be misread by the flow cytometer as single cells, may be excluded by volume measurement. This technique has already been used with two antibodies which have been shown to be poor discriminators of cervical malignancy, anti-EMA and anti-CEA, to detect successfully all malignant specimens in a series of 23 cases.[40]

6.3. Infection

The postulated role of some human papillomaviruses (HPV) in cervical cancer and the possible synergistic action[41] of herpes simplex virus (HSV), as well as the increasing incidence of such infections, have made their detection important. Immunocytochemistry is valuable for demonstrations of HSV in routinely fixed cervical smears[42] (*See* Appendix). It detects infection which in nearly all cases is not found by conventional cytology. It is at least as sensitive as the virus isolation technique and is also cheaper and quicker. (*See also* Chapter 38.)

Immunogold–silver staining (*Fig.* 33.1) permits the primary antibody to be used at a ten-fold higher dilution than with PAP and a Papanicolaou counterstain to be used to give enhanced demonstration of morphology, as well as allowing one to immunostain after dismounting smears which have already been used for morphological diagnosis.[5]

Immunostaining for HPV antigen detection is usually performed on biopsies (*Fig.* 33.2) rather than smears from cervical lesions, following the first report of this method.[43] However, unlike HSV, immunostaining of HPV antigens may be of somewhat limited value. In HSV infection, whilst virus-isolation methods only detect the virus during primary or secondary attacks when virus shedding occurs, there is evidence that in some cases HSV antigens may be detected immunocytochemically between attacks in the quiescent phase.[42] In other words, the immunocytochemical method detects free viral proteins rather than, or as well as, intact virus particles. This is further suggested by the cytoplasmic and perinuclear localization of staining rather than in the nucleus where the virus resides. This does not seem to be the case with HPV, where immunostaining is nuclear and only present in 50 per cent of condylomata and cervical dysplasias.[43] Such findings may either result from an inability of anti-HPV to detect viral protein other than in assembled virus particles, or lack of expression of viral protein in all cases of infection. Evidence for the latter was obtained by McCance and coworkers[44] who found immunoreaction for HPV in only half of those biopsies where HPV-6 DNA was detectable by DNA hybridization. Following infection, HPV DNA may become incorporated into the host genome and thereafter the pathognomic effects depend on the virus type—HPV-6 and 11 being associated with benign condylomata, HPV-16 and 18 being associated with malignant lesions. These HPV types cannot be distinguished by immunocytochemistry—for this it is necessary to use DNA hybridization techniques.

The detection of genital trichomonas infection in Papanicolaou-stained cervical smears is believed not to present difficulty unless the infection is very mild. However, using an antiserum to *Trichomonas vaginalis* on 100 cervical smears O'Hara and coworkers[45] found the organism to be present in 20 cases, of which only 4 were detected by routine screening. They also found smears from all of 22 culture-confirmed urine samples to have *Trichomonas* which was detectable immunocytochemically, as well as finding the organism in a renal cyst fluid where its presence was not suspected.

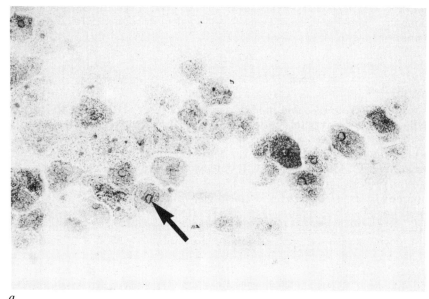

a

b

Fig. 33.1 (*a*) HSV-2 immunoreactivity in squamous epithelial cells in a routinely fixed cervical smear from a patient with genital herpes virus infection confirmed by virus isolation. The smear was immunostained after staining by the Papanicolaou method for cytological diagnosis. Another smear prepared from the same material was immunostained directly by the PAP method and showed similar results. Note the strong cytoplasmic staining and intense perinuclear immunoreactivity (arrow). Immunogold–silver method, Dako anti-HSV-2 diluted 1 : 800. (*b*) A similar smear from a patient with no herpes virus infection, immunostained in the same way as (*a*). Only very slight background staining is seen.

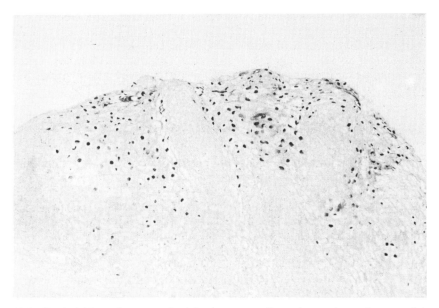

Fig. 33.2. Cervical biopsy immunostained for human papillomavirus antigens. Strong staining is seen in the nuclei of koilocytes in the upper epithelial layers. Formalin-fixed wax-embedded biopsy, 5 μm section, immunogold–silver method, Dako anti-HPV diluted 1 : 2000, eosin counterstain.

Chlamydia trachomatis is another organism which frequently causes genital infection and may cause sterility. Early detection in pregnant women is important to prevent infections of the infant at birth and possible resultant blindness. Previously, confirmation of its presence was difficult and costly since the only method for growing it involves the use of cell cultures. Woodland and coworkers[46] showed that immunoperoxidase or immunofluorescence techniques could be used to detect *Chlamydia* in smears of conjunctival scrapings or cell cultures with a high degree of sensitivity. The use of commercially available FITC-conjugated monoclonal antibodies permits very rapid diagnosis using acetone-fixed smears.

7. INTERPRETATION

Care is needed in the interpretation of immunocytochemical staining results. With cytological preparation, controls are very important and a negative control should be included for each specimen and a positive control for each primary antiserum. Negative controls may either omit the primary antiserum to ensure there is no non-specific reaction of second or third layer antibodies, or substitute the primary antiserum absorbed with homologous antigen. Positive controls are ideally a smear of known positive cells (it is worth preparing and storing many slides from typical cases for this purpose—for methods *see* Section 3.5 and Appendix) or, if unavailable, a histological section from a known positive case.

When interpreting immunostained smears, misclassification of artefactual staining as positive should be avoided. In this respect the histologist has the advantage of tissue architecture and stroma to help judge the specificity of staining. However, with experience the cytologist can also learn. General staining of cells in a smear points to poor antibodies or incorrect fixation and processing. Caution should be exercised with cases where cell staining is very weak and peripheral (unless the antigen is restricted to plasma membrane). Likewise, nuclear staining for non-nuclear antigens is suspicious even if the cytoplasm is also stained.

8. CONCLUSIONS

The value of immunocytochemistry in diagnostic cytology is being realized. Techniques and antibodies are now available to permit detection of many types of malignant cells in problematic cases of serous effusions and fine-needle aspirates. These may also be used to label tumour cells for morphological diagnosis and thus improve speed of screening and elimination of false negatives in cases where tumour cells are masked by a background of normal or inflammatory cells. Histogenesis may be determined, and this is often useful in cases where there is a differential diagnosis.

Ideally, all cytopathology laboratories should realize the value of even small scale use of immunocytochemical techniques in solving day-to-day problems.

Acknowledgements

I would like to thank Dr Max Levene, Consultant Cytopathologist, for his advice and encouragement, and my colleagues Dr Angela Russell, Richard Adams and Jenny Steel for all their help.

Appendix

I. SMEAR PREPARATION ON POLY(L-LYSINE)-COATED SLIDES

Prepare a solution of 1 mg poly(L-lysine) (molecular weight greater than 350 kdal) per ml of distilled water. The solution is stored frozen in 1-ml aliquots and thawed before use. Excess may be refrozen several times. Coated slides are best prepared just before use or at least the same day.

1. Clean the slides—wiping with a clean tissue to remove dust is usually adequate. If not, degrease in hot detergent solution and rinse in distilled water. Organic solvents may be used, but are often not effective.
2. Place a small drop of the solution near the end of the slide (slide A).
3. Take another slide (slide B), hold it at an acute angle with the end touching slide A so as to touch the drop of liquid and then push B along the surface of A as for making a blood film. An even thin layer of poly(L-lysine) solution should result which dries rapidly (within 5–8 s). If drying takes longer, the drop of poly(L-lysine) solution was too large. If the layer contracts and forms droplets, the slide is greasy [*see* step 1].

4. Mark the coated side of the slide.
5. Draw a circle about 1·5 cm in diameter in the middle of the slide (coated side) using a diamond marker. This marks the boundary of the smear and thus helps when immunostaining.
6. Apply within the circle, so as to fill it, a small quantity of cell suspension, e.g. serous fluid, or cell pellet resuspended in buffered saline. For air-dried smears a more concentrated cell suspension should be spread thinly within the circle to allow rapid drying.
7. Leave slide horizontal for about 30 s. Cells will sediment onto slide and adhere. If the slide is tipped up the liquid will drain off the cells.
8. Place slide into required fixative. Air-dried smears may be stored by wrapping in batches in aluminium foil and placing in a −20 °C freezer. When required, warm to room temperature, unwrap and fix as required.

II. WAX EMBEDDING OF CELL PELLETS

Centrifuged cell pellets are often insufficiently cohesive to permit processing to wax, especially if fixed as suspended cells prior to centrifugation. Two methods are suggested.

A. Agar Embedding

This is conveniently performed using a simple re-usable apparatus made by cutting the end with the nozzle from a 2 or 5-ml plastic syringe and glueing the 'thumb' end of the plunger to a piece of wood such that the syringe is vertical with the cut end at the top. By moving the syringe body up or down the depth of the mould is variable; it should be adjusted to a suitable depth depending on the size of the cell pellet.

1. Molten 1 per cent aqueous agar is pipetted into the mould to form a layer about 2–3 mm deep and allowed to set.
2. The cell deposit is placed centrally in the mould and covered with more molten agar, which is then left to solidify.
3. After sliding the syringe body down, the agar block may be easily removed from the top of the syringe piston for wax processing.

B. Celloidin Bag

This method is more difficult to master than agar embedding, but gives very good results.

1. Cut the bottom 1 cm from a polystyrene test tube with an internal diameter of about 5 mm (e.g. precipitin tube) to make a short tube.
2. Fill the tube with a solution of celloidin in ether–ethanol (1:1, v/v) and quickly tip out. Leave the tube to dry. A deposit of celloidin will remain as a membrane lining the tube.

3. Resuspend the cell pellet in a small volume of buffered saline and transfer to the plastic tube. Centrifuge to deposit the cells, and aspirate the supernatant carefully.
4. With a pair of fine forceps, tease the celloidin membrane gently away from the side of the tube and lift it out with the contained cells. Twist the open top around 2 or 3 times to seal the bag and then process to wax.

III. IMMUNOSTAINING OF HERPES SIMPLEX VIRUS ANTIGENS

Cervical smears wet fixed in the same way as for cytological screening may be used.[42] If using the immunogold–silver method, stained slides may be immunostained after soaking in xylene and removing the cover-slip.

PAP Method

1. Block endogenous peroxidase using 0·3 per cent (w/v) hydrogen peroxide in methanol, 30 min.
2. Wash in TBS (Tris-buffered saline: 0·9 per cent NaCl in 50 mM Tris-HCl buffer pH 7·6) containing 0·2 per cent Triton X-100 3 × 5 min.
3. Incubate with normal swine serum (NSS) diluted 1 : 20 with TBS, 10 min.
4. Tip off NSS and replace with anti-HSV2 (Dakopatts) diluted 1 : 100 in TBS containing 2 per cent NSS. Incubate 60 min.
5. Wash in TBS, 3 × 5 min.
6. Incubate with swine anti-rabbit immunoglobulins (Dakopatts) diluted 1 : 100 in TBS containing 2 per cent NSS, 30 min.
7. Wash in TBS, 3 × 5 min.
8. Incubate with rabbit PAP (Dakopatts) diluted 1 : 200 in TBS, 30 min.
9. Wash in TBS, 3 × 5 min.
10. Reveal peroxidase activity by incubating at room temperature with a freshly prepared solution of 5 mg of diaminobenzidine tetrahydrochloride in 10 ml TBS containing 5 µl of 100 vol. hydrogen peroxide solution.
11. Counterstain briefly with Harris' haematoxylin, dehydrate, clear and mount.

Immunogold–Silver Staining

The method is the same as that given in the Appendix to Chapter 5. However, the slides should be prewashed as given above for the PAP method, in TBS containing 0·2 per cent Triton X-100 for 15 min prior to treatment with Lugol's iodine solution. The anti-HSV2 is used at a dilution of 1 : 500, and the smears are counterstained by the Papanicolaou method.

Results

In HSV infection, strong immunostaining of 5–10 per cent of cervical cells in smears will be seen. The stained cells are mostly of intermediate squamous type.

REFERENCES

1. Huang Wei-Min., Gibson S. J., Facer P., Gu J. and Polak J. M. Improved section adhesion for immunocytochemistry using high molecular weight polymers of L-lysine as a slide coating. *Histochemistry* 1983, **77**, 275–279.
2. Springall D. R., Lackie P., Levene M. M., Marangos P. J. and Polak J. M. Immunostaining of neuron-specific enolase is a valuable aid to the cytological diagnosis of neuroendocrine tumours of the lung. *J. Pathol.* 1984, **143**, 259–265.
3. Holgate C. S., Jackson P., Cowen P. N. and Bird C. C. Immunogold–silver staining: a new method of immunostaining with enhanced sensitivity. *J. Histochem. Cytochem.* 1983, **31**, 938–944.
4. Springall D. R., Hacker G. W., Grimelius L. and Polak J. M. The potential of the immunogold–silver staining method for paraffin sections. *Histochemistry* 1984, **81**, 603–608.
5. Springall D. R., Tang S-K., Hacker G. W., Levene M. M., Van Noorden S. and Polak J. M. Applications in diagnostic cytology of a new sensitive staining method—immunogold–silver staining. *J. Pathol.* 1985, in press.
6. To A., Coleman D. V., Dearnaley D. P., Omerod M. G., Steele K. and Neville A. M. Use of antisera to epithelial membrane antigen for the cytodiagnosis of malignancy in serous effusions. *J. Clin. Pathol.* 1981, **34**, 1326–1332.
7. Pascal R. R. and Fenoglio C. M. Identification of carcinoembryonic antigen in cytology preparations. *Lab. Invest.* 1979, **40**, 277.
8. Nadji M. The potential value of immunoperoxidase techniques in diagnostic cytology. *Acta Cytol.* 1980, **24**, 442–447.
9. Ghosh A. K., Mason D. Y. and Spriggs A. I. Immunocytochemical staining with monoclonal antibodies in cytologically "negative" serous effusion from patients with malignant disease. *J. Clin. Pathol.* 1983, **36**, 1150–1153.
10. Ghosh A. K., Spriggs A. I., Taylor-Papadimitriou J. and Mason D. Y. Immunocytochemical staining of cells in pleural and peritoneal effusions with a panel of monoclonal antibodies. *J. Clin. Pathol.* 1983, **36**, 1154–1164.
11. Walts A. E. and Said J. W. Specific tumour markers in diagnostic cytology. Immunoperoxidase studies of carcinoembryonic antigen, lysozyme and other tissue antigens in effusions, washes and aspirates. *Acta Cytol.* 1983, **27**, 408–416.
12. O'Brien M. J., Kirkham S. E., Burke B., Ormerod M., Saravis C. A., Gottlieb L. S., Neville A. M. and Zamcheck N. CEA, ZGM and EMA localization in cells of pleural and peritoneal effusions: a preliminary study. *Invest. Cell Pathol.* 1980, **3**, 251–258.
13. Boon M. E., Lindeman J., Meeuwissen A. L. J. and Otto A. J. Carcinoembryonic antigen in sputum cytology. *Acta Cytol.* 1982, **26**, 389–394.
14. Krupey J., Gold P. and Freedman S. O. Physicochemical studies of the carcinoembryonic antigens of the human digestive system. *J. Exp. Med.* 1968, **128**, 387–398.
15. Coleman D. V., To A., Ormerod M. G. and Dearnaley D. P. Immunoperoxidase staining in tumour marker distribution studies in cytologic specimens. *Acta Cytol.* 1981, **25**, 205–206.
16. To A., Dearnaley D. P., Ormerod M. G., Canti G. and Coleman D. V. Epithelial membrane antigen. Its use in the cytodiagnosis of malignancy in serous effusion. *Am. J. Clin. Pathol.* 1982, **78**, 214–219.
17. Epenetos A. A., Canti G., Taylor-Papadimitriou J., Curling M. and Bodmer W. F. Use of two epithelium-specific monoclonal antibodies for diagnosis of malignancy in serous effusion. *Lancet* 1982, **ii**, 1004–1006.
18. Droese M., Altmannsberger M., Kehl A., Lankisch P. G., Weiss R., Weber K. and Osborn M. Ultrasound-guided percutaneous fine needle aspiration biopsy of abdominal and retroperitoneal masses. Accuracy of cytology in the diagnosis of malignancy, cytologic tumor typing and use of antibodies to intermediate filaments in selected cases. *Acta Cytol.* 1984, **28**, 368–384.
19. Ramaekers F., Haag D., Jap P. and Vooijs P. G. Immunochemical demonstration of keratin and vimentin in cytologic aspirates. *Acta Cytol.* 1984, **28**, 385–392.
20. Collins V. P. Monoclonal antibodies to glial fibrillary acidic protein in the cytologic diagnosis of brain tumours. *Acta Cytol.* 1984, **28**, 401–406.
21. von Koskull H. Rapid identification of glial cells in human amniotic fluid with indirect immunofluorescence. *Acta Cytol.* 1984, **28**, 393–400.
22. McGee J. O'D., Woods C. J., Ashall F., Bramwell M. E. and Harris H. A new marker for human cancer cells. II. Immunohistochemical detection of the Ca antigen in human tissues with the Ca 1 antibody. *Lancet* 1982, **ii**, 7–10.

23. Woods J. C., Spriggs A. I., Harris H. and McGee J. O'D. A new marker for human cancer cells. III. Immunocytochemical detection of malignant cells in serous fluids with the Ca 1 antibody. *Lancet* 1982, **ii**, 512–514.

24. Pallesen G., Jepsen F. L., Hastrup J., Ipsen A. and Hvidberg N. Experience with the Oxford tumour marker (Ca 1) in serous effusions. *Lancet* 1983, **i**, 1326.

25. Sheppard M. N., Corrin B., Bennett M. H., Marangos P. J., Bloom S. R. and Polak J. M. Immunocytochemical localization of neuron specific enolase in small cell carcinomas and carcinoid tumours of the lung. *Histopathology* 1984, **8**, 171–181.

26. Yam L. T. and Winkler C. F. Immunocytochemical diagnosis of oat-cell carcinoma in pleural effusions. *Acta Cytol.* 1984, **28**, 425–429.

27. Cuttitta F., Rosen S., Gazdar A. F. and Minna J. D. Monoclonal antibodies that demonstrate specificity for several types of human lung cancer. *Proc. Natl Acad. Sci. USA* 1981, **78**, 4591–4595.

28. Singh G., Whiteside T. L. and Dekker A. Immunodiagnosis of mesothelioma. Use of antimesothelial cell serum in an indirect immunofluorescence assay. *Cancer* 1979, **43**, 2288–2296.

29. Mariani-Constantini R., Ménard S., Clemente C., Tagliabue E., Colnaghi M. I. and Rilke F. Immunocytochemical identification of breast carcinoma cells in effusions using a monoclonal antibody. *J. Clin. Pathol.* 1982, **35**, 1037.

30. Szpak C. A., Johnston W. W., Lottich S. C., Kufe D., Thor A. and Scholm J. Patterns of reactivity of four novel monoclonal antibodies (B72.3, DF3, B1.1 and B6.2) with cells in human malignant and benign effusions. *Acta Cytol.* 1984, **28**, 356–367.

31. Levene M. M., Lynch M. J. and Wooding M. Further studies on the cells obtained by sampling the cervix with a brush, examining them by microdensitometry. In: Tressman N. J. and Wied G. L., eds., *Proceedings of the 2nd International Conference on Automation of Cancer Cytology and Cell Image Analysis.* Chicago, Illinois, Tutorials of Cytology, 1972: 63–68.

32. Sincock A. M., Middleton J. and Moncrieff D. Towards an automated procedure for the quantitative cytological screening of cervical neoplasms. *J. Clin. Pathol.* 1983, **36**, 535–538.

33. Moncrieff D., Ormerod M. G. and Coleman D. V. Tumour marker studies of cervical smears, potential for automation. *Acta Cytol.* 1984, **28**, 407–410.

34. Lloyd J. M., O'Dowd T., Driver M. and Tee D. E. H. Demonstration of an epitope of the transferrin receptor in human cervical epithelium—a potentially useful cell marker. *J. Clin. Pathol.* 1984, **37**, 131–135.

35. Lloyd J. M., O'Dowd T., Driver M. and Tee D. E. H. Immunohistochemical detection of Ca antigen in normal, dysplastic and neoplastic squamous epithelia of the human uterine cervix. *J. Clin. Pathol.* 1984, **37**, 14–20.

36. Krausz T. J., Van Noorden S. and Evans D. J. Experience of the Oxford tumour marker. *Lancet* 1983, **i**, 1097.

37. Jha R. S., Wickenden C., Anderson M. C. and Coleman D. V. Monoclonal antibodies for the histopathological diagnosis of cervical neoplasia. *Br. J. Obstet. Gynaecol.* 1984, **91**, 483–488.

38. Goerttler K. and Stöhr M. Quantitative cytology of the position region in flow sorted vaginal smears. *J. Histochem. Cytochem.* 1979, **27**, 567–572.

39. Fu Y. S., Reagan J. W. and Richart R. M. Definition of precursors. *Gynecol. Oncol.* 1981, **12**, 5220–5231.

40. Valet G., Ormerod M. G., Warnecke H. H., Benker G. and Ruhenstroth-Bauer G. Sensitive three-parameter flow–cytometric detection of abnormal cells in human cervical cancers: a pilot study. *J. Cancer Res. Clin. Oncol.* 1981, **102**, 177–184.

41. zur Hausen H. Human genital cancer: synergism between two virus infections or synergism between a virus infection and initiating events? *Lancet* 1982, **ii**, 1370–1372.

42. Adams R. L., Springall D. R., Levene M. M. and Bushell T. E. C. The immunocytochemical detection of herpes simplex virus in cervical smears—a valuable technique for routine use. *J. Pathol.* 1984, **143**, 241–247.

43. Kurman R. J., Shah K. H., Lancaster W. D. and Jenson A. B. Immunoperoxidase localization of papillomavirus antigens in cervical dysplasia and vulvar condylomas. *Am. J. Obstet. Gynecol.* 1981, **140**, 931–935.

44. McCance K. J., Walker P. G., Dyson J. L., Coleman D. V. and Singer A. Presence of human papillomavirus DNA in cervical intraepithelial neoplasia (CIN). *Br. Med. J.* 1983, **287**, 784–788.

45. O'Hara C. M., Gardner W. A. and Bennett B. D. Immunoperoxidase staining of *Trichomonas vaginalis* in cytologic material. *Acta Cytol.* 1980, **24**, 448–451.

46. Woodland R. M., El-Sheikh H., Darouger S. and Squires S. Sensitivity of immunoperoxidase and immunofluorescence staining for detecting *Chlamydia* in conjunctival scrapings and in cell cultures. *J. Clin. Pathol.* 1978, **31**, 1073–1077.

34

Immunocytochemistry of Lymphoreticular Tumours

P. G. Isaacson and D. H. Wright

Histopathological analysis of lymphoreticular neoplasms is greatly hampered by the essentially similar morphology of the different lymphoreticular cells as they appear in histological sections. It is to the credit of histopathologists that, despite this, great advances have been made in the understanding and classification of malignant lymphomas on morphological grounds alone. Patterns of cell distribution and subtle variations in nuclear size and shape have been exploited to the full, yielding useful sets of diagnostic criteria and several soundly based classifications of this group of tumours.[1,2] Nevertheless, observer error is still significant and differences in the quality and nature of fixation and staining can radically change the histological appearances of a malignant lymphoma. Recognizing this, pathologists have long sought some other means of accurately identifying lymphoma cells in order to substantiate their diagnoses and to provide a scientific basis for their classifications. Enzyme histochemistry, mainly carried out on frozen sections, was the first step in this direction. This was followed by immunological studies of dispersed cells from lymphoma tissue. The limitations of both these procedures lay in the uncertainty as to which of the numerous cell types identified were the malignant cells since, in most lymphoreticular tumours, there is a large component of benign reactive and supportive cells. Immunocytochemistry quickly became established as a precise technique for the identification of cells of the lymphoreticular system and it soon became the aim of histopathologists to apply this technique to histological sections in such a way that the morphological and immunological properties of lymphoreticular tumours could be studied simultaneously. The impact of the first immunohistochemical studies of

Table 34.1. *Antigens useful in lymphoreticular immuno-histochemistry demonstrable in paraffin or frozen sections*

Paraffin sections	Frozen sections
Immunoglobulin	Immunoglobulin
Heavy chains	Heavy chains
Light chains	Light chains
J chain	Pan T-cell antigens
Leucocyte common antigen	T-cell subset antigens
HLA-DR	Pan B-cell antigens
Granulocyte antigen	B-cell subset antigens
	Common acute lymphocytic
	leukaemia antigen (cALLA)
Lysozyme	C3B receptor
α1-Antitrypsin	Dendritic reticulum
(α1-Antichymotrypsin)	cell antigen
Vimentin	Macrophage antigens
S-100 protein	
Cytokeratin	

For references *see* text.

lymphoma[3,4] was pivotal and has been followed by an almost exponential growth of the subject. Using immunoenzyme (principally immunoperoxidase) techniques, many cellular antigens can be accurately labelled and viewed with the light microscope in either frozen or paraffin sections offering the histopathologist the setting with which he is most familiar. Variations in methods abound and each group engaged in immunohistochemistry has its favoured techniques which it will often champion as the most suitable. It is not the purpose of this chapter to compare the various techniques available. We will, however, briefly describe the techniques used by us and give our reasons for using them.

1. METHODS

A wide variety of tissue antigens have been exploited in the study of lymphoreticular tumours and we shall discuss only those to which antibodies are widely available and which we find most useful overall. The antigens fall into two main groups, the first consisting of those that are optimally demonstrated in routinely fixed paraffin-embedded tissue and the second of those demonstrable only in cryostat sections of fresh tissue (*Table* 34.1). The antibodies used may be polyclonal or monoclonal and it should be stressed that it is the antigen and not the antibody that determines the best method for its demonstration. Following the application of primary antibody to the tissue, one or other variety of immunoenzyme technique is used to demonstrate the binding sites of the primary antibody. Each technique has its adherents and in giving our preferred methods for the immunoperoxidase techniques (*see* Appendix), we are in no way assuming that other methods are of less value. When fixed and paraffin-embedded tissue is used the nature of the fixative is of great importance. Neutral-buffered formol saline (NBFS) is probably the most widely used and satisfactory fixative in histopathology, but sections from tissue fixed in NBFS require treatment with proteolytic enzymes prior to immunostaining for many antigens in order to 'unmask' reactive sites.[5] The time the section requires exposure to enzyme is directly related to the time the tissue has spent in fixative.[6] The need for proteolytic enzyme treatment becomes less and less the further the pH of formol saline diverts from neutrality and also depends to some extent on the antigen.[7] Certain antigens, such as HLA-DR,[8] cannot be demonstrated at all in NBFS-fixed tissue. If the formol saline is deliberately rendered acidic or if other fixatives such as Bouin's solution are used, proteolytic enzymes are no longer required. Staining of cryostat sections of lymphoreticular tissue is generally more simple than that of paraffin sections and the method given in the Appendix is widely used. While all antigens demonstrable in paraffin sections can also be demonstrated in cryostat sections, at present many membrane antigens defined by monoclonal antibodies can only be demonstrated in cryostat sections since the antigens appear to be destroyed or obscured by routine fixation and embedding. Gradually, however, monoclonal antibodies are being produced that recognize epitopes on these antigens that are resistant to routine fixation and processing schedules.

2. IMMUNOCYTOCHEMISTRY OF NORMAL LYMPHOID TISSUE

An important property of lymphoid tissue is the existence of antigens common to all the component cells; these are known as leucocyte common antigens (LCA). Monoclonal antibodies are available that recognize LCA in routinely fixed paraffin-embedded tissue and these are of great value in the histological separation of lymphoreticular tumours from tumours which may appear similar but which are derived from other tissues[9] (*see below*). Lymphoreticular tumours are remarkably heterogeneous and their histogenesis can only be understood with reference to the various components of normal lymphoid tissue. Thus, to appreciate the significance of immunohistochemical results obtained in cases of malignant lymphoma, it is first necessary to be familiar with the staining reactions of these normal cells. Normal lymphoid tissue (*Fig.* 34.1) is composed of B-cells, which include cells of the follicle centre, the mantle zone and plasma cells, T-cells, which are found principally in the paracortical areas, and cells of the monocyte–macrophage system and related 'reticulum' cells. The immunohistochemical characteristics of those various components are listed in *Table* 34.2 and will be briefly summarized here.

2.1. B-cells

Antigens common to B-cells can be demonstrated in cryostat sections (*Fig.* 34.2) with several monoclonal pan B-cell antibodies; antibodies defining B-cells at different stages of maturation are now becoming available. In common with several other cells, B-cells bear HLA-DR antigens on their surfaces. Plasma cells, however, are the exception and do not react with pan B-cell antibodies or bear HLA-DR antigens. B-cells are definitively identified by their production of either surface membrane Ig (SIg) or cytoplasmic Ig (CIg). The small B-lymphocytes of

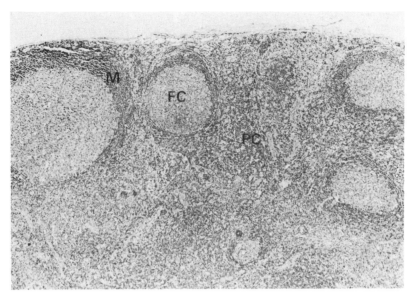

Fig. 34.1. Section of reactive lymph node showing B-cell areas consisting of follicle centres (FC) and mantle zones (M). Most T-lymphocytes are found in the paracortex (PC). (Haematoxylin and eosin.)

Table 34.2. **Immunohistochemistry of normal lymphoid tissue**

Antigen	FCC	Mantle zone cells	Plasma cells	T-cells	Macrophages	IDRC*	DRC†
CIg	±	−	+	−	−	−	−
J chain	±	−	±	−	−	−	−
LCA	+	+	+	+	+	+	?
HLA-DR	+	+	−	−	±	+	?
cALLA‡	+	−	−	−	−	−	−
Lysozyme	−	−	−	−	+	−	−
α1-AntiT (α1-AntiCT)	−	−	−	−	+	−	−
S-100	−	−	−	−	−	+	−
SIg	+	+	−	−	−	−	−
T-cell	−	−	−	+	−	±	−
B-cell	+	+	±	−	−	−	−
C3B receptor	−	±	−	−	±	−	+

*IDRC = interdigitating reticulum cells.
†DRC = dendritic reticulum cells.
‡cALLA = common acute lymphocytic leukaemia antigen.

the mantle zone of the follicle synthesize SIgM and SIgD and show polytypic membrane staining with respect to their light chain class (*Fig.* 34.3). The centrocytes and centroblasts of the follicle centre synthesize SIg but this is largely obscured in frozen sections by the large amount of extracellular Ig bound to the processes of dendritic reticulum cells. In paraffin sections of hyper-reactive follicles, CIg can be demonstrated in a population of cells which in thin (1 μm)

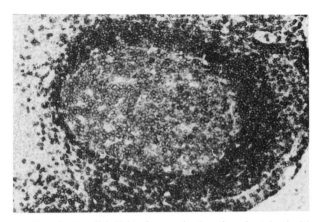

Fig. 34.2. Frozen section of a B-cell follicle of a reactive lymph node stained with monoclonal pan B-cell antibody. Follicle centre cells, mantle cells and isolated B-cells within the paracortex (at left) are staining. The immunoperoxidase methods used to stain this section and shown in subsequent figures are described in the Appendix.

sections can be clearly seen to be centrocytes and centroblasts (*Fig.* 34.4). This appears to represent a pathway of CIg synthesis distinct from that shown by plasma cells.[10,11] These CIg-synthesizing follicle centre cells also synthesize J chain[12] (*Fig.* 34.5). J or joining chain, is the molecule which links the IgA dimer or IgM pentamer and is synthesized both by cells producing these classes of CIg, and by immature CIg-synthesizing cells which may subsequently switch from the class of heavy chain they are synthesizing to another.[13] The heavy and light chains of CIg in immunoblasts and plasma cells, which are found outside the follicles, can be well demonstrated in paraffin sections (*Fig.* 34.6) and J chain can be shown in immunoblasts and immature plasma cells as well as plasma cells synthesizing dimeric IgA and IgM.

2.2. T-cells

T-cell antigens do not resist routine tissue processing. In cryostat sections they can be identified by a host of monoclonal antibodies recognizing either all T-cells (pan T) or only certain subsets (*Figs* 34.7 and 34.8). In lymphoid tissue they are found principally in the paracortex, but helper/inducer subsets are also presnt within the follicle centre. Monoclonal antibodies can usefully distinguish immature T-cells, T-inducer and T-suppressor cells.

2.3. Macrophages and 'Reticulum' Cells

Macrophages are found within the sinuses of lymphoid tissue and also in follicle centres as the so-called 'tingible body macrophages'. Lysozyme is synthesized by macrophages and can be demonstrated within them in paraffin sections. These cells can also be stained with antisera to α-1-antitrypsin (α1-antiT) and α-1-antichymotrypsin (α1-antiCT) (*Fig.* 34.9). Both these antiproteases are

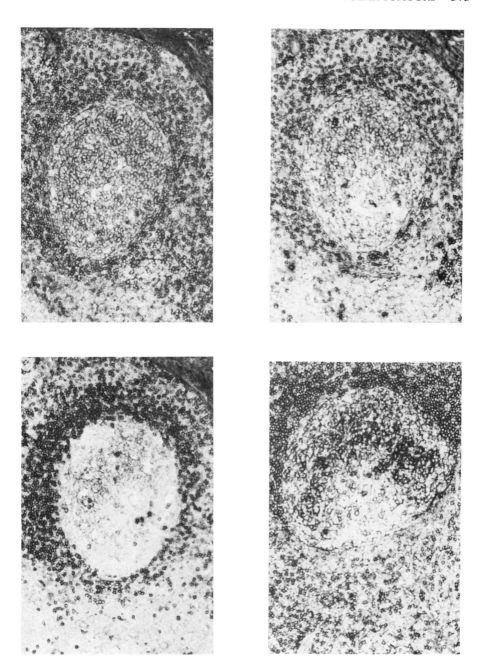

Fig. 34.3. Serial frozen sections of a B-cell follicle from a reactive lymph node stained with antibodies to κ light chain (upper left), λ light chain (upper right), IgD (lower left) and IgM (lower right).

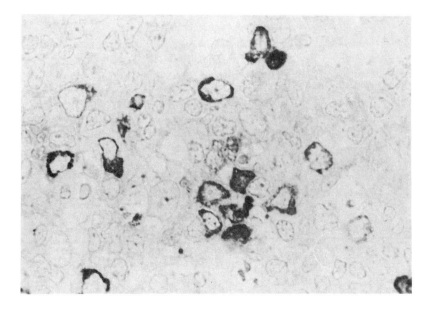

Fig. 34.4. Follicle centre of human tonsil stained for κ chain. In this 1-μm section of plastic-embedded tissue the FCC nature of the positively staining cells is clearly evident.

Fig. 34.5. Paraffin section of human tonsil stained for IgG (left) and J chain (right). A population of FCCs stains for both IgG and J chain while surrounding plasma cells are positive for IgG alone.

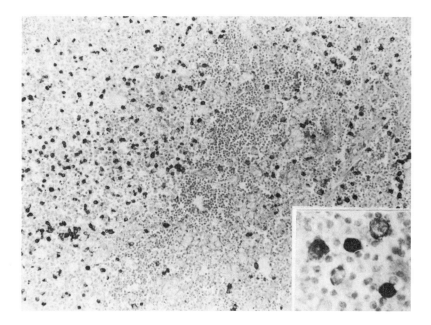

Fig. 34.6. Paraffin section of reactive lymph node stained for IgG. Many FCCs in the follicle (left) contain IgG. Positively staining cells outside the follicle (right) are a mixture of plasma cells and immunoblasts. Inset shows darkly staining plasma cells in contrast to larger immunoblasts showing a perinuclear rim of cytoplasmic staining.

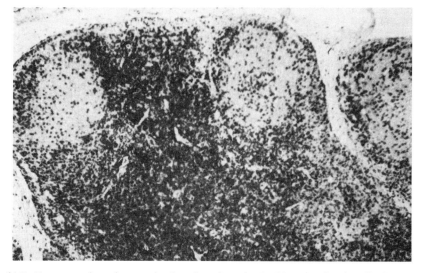

Fig. 34.7. Frozen section of a reactive lymph node stained with monoclonal antibody to T-cells (pan-T). There is staining of most of the cells in the paracortex as well as a population within follicle centres.

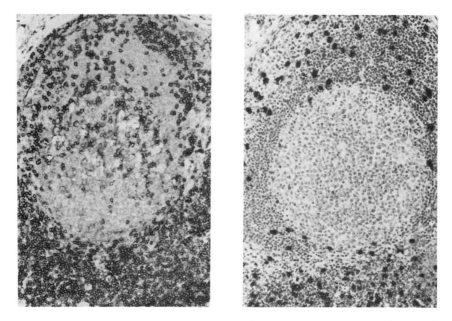

Fig. 34.8. High power of follicles shown in *Fig.* 34.7 stained with monoclonal antibody to T-helper/inducer cells (left) and T-suppressor cells (right). The majority of T-cells are of T-helper/inducer type and it is only these cells that are present within the follicle centre.

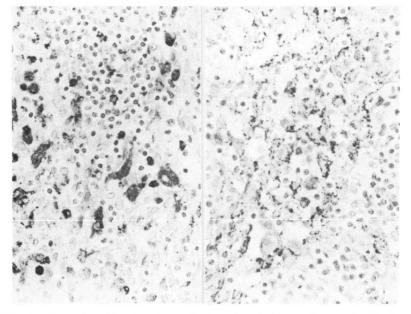

Fig. 34.9. Paraffin section of lymph node showing intrasinusoidal macrophages stained for lysozyme (left) and α1-antiT (right). There is great variation in intensity of lysozyme staining in contrast to the uniform content of α1-antiT-positive granules.

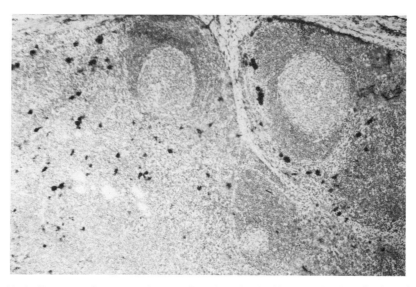

Fig. 34.10. Frozen section of reactive lymph node stained with monoclonal antibody to human thymic antigen (T6). Dark staining clusters of interdigitating reticulum cells are scattered through the paracortex.

probably synthesized by macrophages although this has only been conclusively shown for α1-antiT.[14] Monoclonal antibodies to membrane antigens of macrophages can distinguish these cells in cryostat sections but the results should be carefully interpreted since some of them may cross-react with endothelial cells. Interdigitating reticulum cells can be recognized in paraffin sections using antibodies to S-100 protein[15] and in cryostat sections they react with monoclonal antibodies to human thymic antigen[16] (*Fig.* 34.10). Dendritic reticulum cells (DRC) bear strong C3B receptors which can be shown with appropriate antibodies. Cryostat sections are preferable but C3B receptors, when strongly expressed, may be demonstrable in paraffin sections. Monoclonal antibodies specific for DRC are available for use in cryostat sections (*Fig.* 34.11).

3. MALIGNANT LYMPHOMA

The advent of monoclonal antibodies to leucocyte common antigen (LCA)[9] which will react with all lymphoid cells in either paraffin or cryostat sections has been a major advance in lymphoreticular pathology. The use of these antibodies coupled with antibodies to cytokeratin which recognize most carcinomas[19] can, at a stroke, solve the heretofore vexed question as to whether the pathologist is dealing with a lymphoproliferative condition or carcinoma. Amelanotic melanoma, when involving lymph nodes, can also be difficult to separate from lymphoproliferative disease and, here too, antibodies to LCA used in conjunction with anti-S-100 which stains most melanomas are invaluable (*Fig.* 34.12). Once other conditions have been confidently excluded, the choice of antisera and the nature of the sections (cryostat or paraffin) is based on the immunohistochemistry of normal

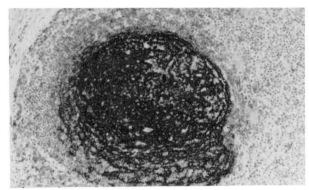

Fig. 34.11. Frozen section of a follicle of a reactive lymph node stained with the monoclonal antibody specific for dendritic reticulum cells. A dense meshwork of interdigitating reticulum cells stain strongly in the follicle centre.

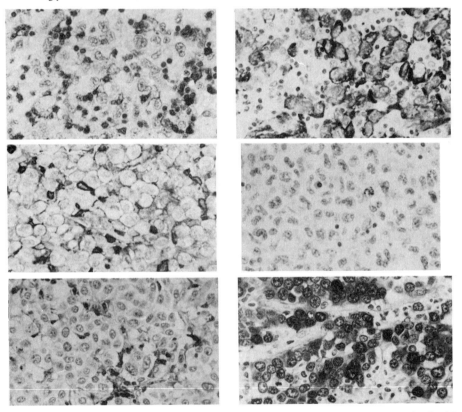

Fig. 34.12. Top left and top right: lymph node containing metastatic carcinoma stained with antibody to LCA (left) and cytokeratin (right). Large cells are negative for LCA but positive for cytokeratin while the small lymphocytes demonstrate the opposite pattern of staining. Centre left and centre right: large cell lymphoma stained for LCA (left) and cytokeratin (right). Small lymphocytes and large lymphoid cells show membrane staining of LCA but are negative for cytokeratin. Lower left and lower right: metastatic malignant melanoma stained for LCA (left) and S-100 protein (right); tumour cells are positive for S-100 protein but not for LCA while infiltrating lymphocytes demonstrate the opposite pattern of staining. This set of illustrations demonstrates the value of immunocytochemistry in the differential diagnosis of lymph node infiltrates.

Table 34.3. Causes of anomalous (polytypic) CIg staining

A. Reactive plasma cell population

B. Uptake of Ig from tissue fluid

    i. By dead or damaged cells
    ii. By macrophages
    iii. By Reed–Sternberg cells or other uncharacterized multinucleated
       giant cells

---

lymphoid tissue as outlined above. Under ideal circumstances, when fresh tissue is received, a sample can be snap-frozen in liquid nitrogen while paraffin sections are prepared from optimally fixed tissue. The histological appearances of the paraffin section will usually determine which immunohistochemical studies, if any, are most appropriate. To take extreme examples, it is pointless to study frozen sections of plasmacytoma to determine whether or not the CIg in the plasma cells is monotypic and, equally pointless, to stain paraffin sections of Burkitt's lymphoma to determine the nature of the SIg. Whichever method is chosen (and in some instances both frozen and paraffin sections should be studied), it is our practice to apply a panel of antisera and not simply those for which a positive result is expected. The reasons for this will become clearer later in this chapter but include the not infrequent finding of unexpected results, the need for continuous control of staining reactions of the different antisera, and the exclusion of spurious staining reactions which are especially common in the case of immunoglobulin.

3.1. B-cell Lymphomas

B-cells are best identified by the demonstration of synthesized immunoglobulin (Ig) on their surfaces (SIg) or in their cytoplasm (CIg). Demonstration of monotypic Ig, i.e. synthesis of only one light chain, is central to the differentiation of neoplastic and, therefore, monoclonal, B-cell proliferations from reactive B-cell hyperplasia; the latter contains cells synthesizing both light chain types.[21] The aim of the immunohistochemical study of B-cell lymphomas is, therefore, to demonstrate monotypic immunoglobulin on the surface or within the cytoplasm of tumour cells. Biclonal tumours synthesizing both κ and λ light chains have been reported but are very rare.[22,23] Despite this, numerous reports describing staining for both κ and λ light chains in paraffin sections of B-cell lymphomas have appeared in the literature.[24,25] It cannot be sufficiently stressed that, if immunohistochemical studies show both κ and λ light chains in the tumour cells, then either the B-cell proliferation is not neoplastic or the staining results are spurious. This problem of anomalous CIg staining in paraffin sections[10,26] deserves detailed discussion.

    When a polytypic CIg pattern of staining, i.e. cells staining for both κ and λ light chains, is detected in a malignant lymphoma the following causes for non-specific staining should be sought (*Table* 34.3).

1. There may be a large reactive B-cell population in the tumour. Usually the number of reactive cells containing CIg is far less than the tumour cells but this is

Table 34.4. **Immunoglobulin-containing cells in malignant lymphoma**

	Cells synthesizing Ig	Cells taking up Ig
Pattern of staining	Granular	Smooth
	Concentration in Golgi region	Concentration at periphery of cytoplasm
	Staining of perinuclear space	Diffuse, cytoplasmic staining
Immunoglobulin	Monotypic	Polytypic
J chain	+	−
Other plasma proteins	−	±

not always the case. The reactive cells are usually plasma cells, not follicle centre cells (FCCs), and stain much more intensely and uniformly than the tumour cells.

2. Polytypic immunoglobulin may be taken up from the tissue fluids either non-specifically by dead or damaged cells or possibly specifically by macrophages, including neoplastic macrophages, Reed–Sternberg cells and multinucleated cells of uncertain nature that are often found in FCC lymphomas.

While Ig in the reactive cell population does not usually lead to problems in interpretation it can be very difficult to distinguish between synthesized Ig and Ig that has been taken up from tissue fluids. Careful attention to the pattern of Ig staining will, however, permit this distinction to be made. The main features distinguishing these two types of staining are summarized in *Table* 34.4.

The earliest site of CIg synthesis is in the perinuclear space[27] and the resulting perinuclear pattern of staining is characteristically seen in many large B-cell lymphomas. As cells mature, Ig accumulates in the region of the Golgi apparatus. Accumulation of Ig in dilated profiles of rough endoplasmic reticulum gives rise to a characteristic granular staining pattern with the immunoperoxidase stain (*Fig.* 34.13). These features are highlighted in thin (1 μm) sections of either paraffin[28] or plastic-embedded[29] tissue (*Fig.* 34.14). Positive staining for J chains is a reliable indicator of Ig synthesis and the majority of CIg positive lymphomas synthesize J chain.[12]

Cells taking up Ig from tissue fluids sometimes, but by no means always, show morphological evidence of death or degeneration. The positively staining cells tend to occur in broad swathes across the section with concentration around blood vessels (*Fig.* 34.15). Other plasma proteins such as albumin or α-1-antitrypsin can also be demonstrated within these cells (*Fig.* 34.16). The quality of staining of α1-antiT taken up non-specifically in this way is quite different from synthesized α1-antiT in macrophages (*see below*). Since there is very little J chain in the plasma, cells taking up Ig do not stain for J chain. Cells of the monocyte/macrophage system, whether reactive or neoplastic, sometimes take up polytypic IgG, probably specifically. The identity of these cells will be apparent if antibodies to macrophage markers are included in the panel of antisera used in the study of lymphomas (*see below*). Reed–Sternberg cells and other multinucleated cells[10] also appear to take up IgG specifically and so stain polytypically for Ig in the same way as macrophages. When a cell has taken up Ig from its environment, the quality of the Ig staining is quite different from an Ig-synthesizing cell (*Fig.* 34.17).

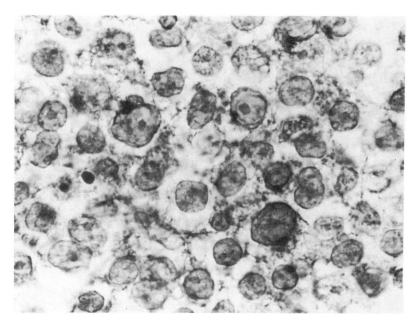

Fig. 34.13. Routine (5 μm) paraffin sections of a FCC (centroblastic) lymphoma stained for IgM. Perinuclear staining is present in most of the cells with concentration in the Golgi region evident in a few cells.

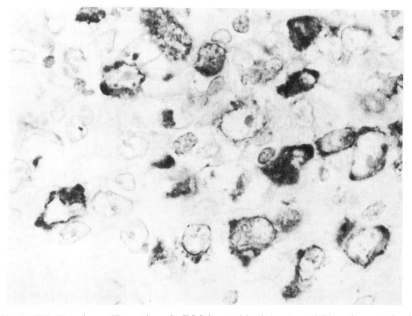

Fig. 34.14. Thin (1 μm) paraffin section of a FCC (centroblastic/centrocytic) lymphoma stained for κ chain. Staining of the perinuclear space and the granular consistency of cytoplasmic staining are well seen.

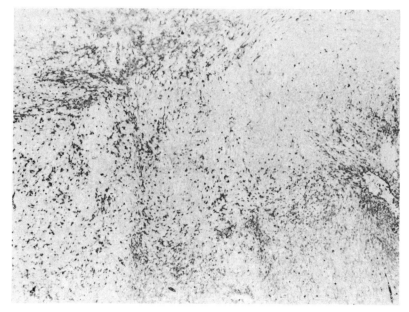

Fig. 34.15. Paraffin section of a FCC (centroblastic/centrocytic) lymphoma stained for κ chain. Swathes of positively staining cells are present with concentration around blood vessels. Stain for λ chain produced an identical picture.

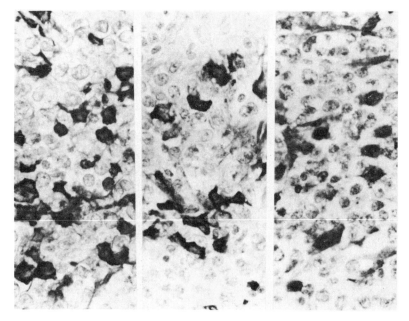

Fig. 34.16. Paraffin sections of the same tumour illustrated in *Fig.* 34.15 stained for κ chain (left), λ chain (centre) and albumin (right). Note diffuse non-granular polytypic staining of tumour cells which also stain for albumin.

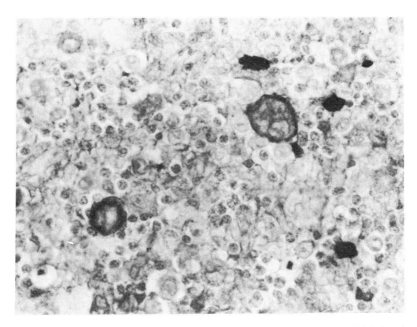

Fig. 34.17. Large multinucleated cells in a paraffin section of a FCC lymphoma (IgM κ) stained for IgG. Tumour cells are negative but there is diffuse staining of the large multinucleated cells with intensification at the periphery. Both light chains were present in these cells. Occasional reactive IgG plasma cells are also seen.

The cytoplasm is diffusely stained without the characteristic granularity of synthesized Ig and there is frequently an increased intensity of staining at the periphery of the cell.

The demonstration of SIg in cryostat sections of follicle centre cell lymphomas may be subject to similar difficulties in staining and interpretation. The interstitial tissues in neoplastic follicles may contain Ig of all classes and both light chain types and discrimination between this Ig and that on the surface of the cells can be difficult. The interstitial Ig may be extracted from the section by washing at a low pH[30] but, in our experience, this usually reduces the quality of the final stained section and we have abandoned this procedure other than in exceptional cases. The use of monoclonal antisera has reduced some non-specific staining and, in most cases, permits the identification of a monotypic neoplastic phenotype. In addition to these largely technical problems, some B-cell lymphomas synthesize little Ig and in some cases none at all.[31] In these cases, pan B-cell antibodies are useful in identifying the tumour as B-cell in nature even though monotypic Ig is not demonstrable.

In the description of the characteristic immunohistochemical staining reactions of B-cell lymphomas that follows, the need for critical interpretation of the results must always be borne in mind. Anomalous staining patterns due to uptake of Ig are especially common in paraffin sections of large cell tumours of FCC origin and immunoblastic lymphomas and, if misinterpreted, can lead to basic conceptual errors in the understanding of B-cell neoplasia.

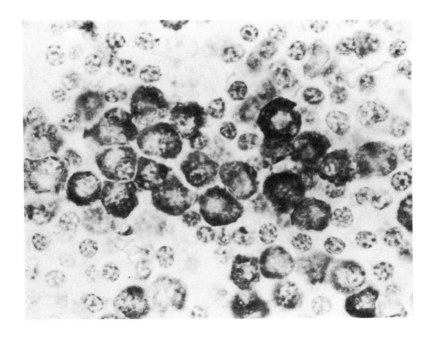

Fig. 34.18. Paraffin section of a lymphoplasmacytic lymphoma showing monotypic (κ) cytoplasmic staining of plasma cells.

3.2. Lymphocytic Lymphoma

Malignant lymphoma of lymphocytic type consists of a uniform population of small lymphocytes with occasional foci of slightly larger cells (so-called 'proliferation centres'). Most of the patients whose lymph nodes show this picture have underlying chronic lymphocytic leukaemia. The malignant cells in this condition do not usually synthesize CIg and thus are best characterized in frozen sections. They almost always synthesize IgM of single light chain type. Despite their B-cell nature, in approximately 40 per cent of cases, the cells of lymphocytic lymphoma will react with the monoclonal antibody Leu-1 which recognizes an antigen present on all T-cells.

3.3. Lymphoplasmacytic and Lymphoplasmacytoid Lymphoma

Small lymphocytic lymphoma may show plasmacytic or plasmacytoid differentiation. In this event, staining of paraffin sections will show monotypic CIg in the plasma cells or plasmacytoid cells. This Ig will be of the same class as the SIg on the small lymphocytes. The positively staining cells in lymphoplasmacytic lymphoma resemble mature plasma cells (*Fig.* 34.18), while in lymphoplasmacytoid tumours positive staining is often in the form of cytoplasmic and nuclear inclusions.

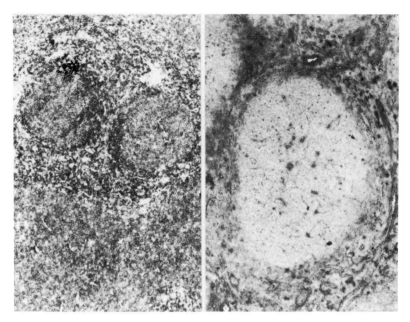

Fig. 34.19. Frozen sections of a FCC lymphoma, follicular, showing positive surface staining for κ chain (left) and absence of staining in section stained for λ chain (right).

3.4. Plasmacytoma

Plasma cell tumours, whether medullary or extramedullary, characteristically show uniform cytoplasmic staining of all the cells for a single heavy and light Ig chain. However, whereas most medullary tumours synthesize IgG or IgA, most nodal and extramedullary tumours produce IgM. In common with most other tumours synthesizing CIg plasma cell tumours stain positively for J chain regardless of the class of heavy chain being synthesized.

3.5. Follicle Centre Cell Lymphomas

Over 50 per cent of non-Hodgkin's malignant lymphomas are of follicle centre cell (FCC) origin. They may be follicular, follicular and diffuse, or diffuse in their pattern of growth. These tumours can be usefully studied for both membrane and cytoplasmic antigens and the characteristic staining patterns for each will be described separately.

Staining of frozen sections of FCC lymphomas of follicular pattern will show monotypic SIg on the tumour cells (*Fig.* 34.19). In contrast to reactive follicles, the follicles of a FCC lymphoma show poorly developed or absent mantle zones which, when present, consist of small lymphocytes bearing surface IgM and IgD of both light chain classes. While the characteristic network pattern of extracellular Ig is usually absent in neoplastic follicles, numerous dendritic reticulum cells are

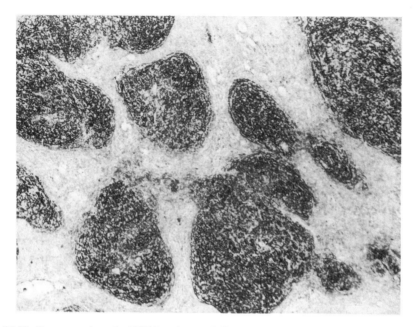

Fig. 34.20. Frozen section of a FCC lymphoma, follicular, stained with a monoclonal antibody to DRCs. There is intense 'network' staining of the malignant follicles.

present and can be demonstrated using an appropriate monoclonal antibody (*Fig.* 34.20). T-cells, which are zonally distributed in reactive benign follicles, tend to be evenly distributed in neoplastic follicles and can be shown to be predominantly of the helper/inducer subtype (*Fig.* 34.21). Vast numbers of T-cells are sometimes present between the follicles together with varying numbers of tumour cells. The FCC lymphomas of diffuse type exhibit uniform monotypic Ig surface staining of the neoplastic cells (*Fig.* 34.22). Dendritic reticulum cells can be shown in these tumours either as single cells or in small groups. T-cells, predominantly of the helper type, are distributed throughout the tumour, usually in far fewer numbers than in the follicular tumours. The cells of both follicular and diffuse FCC lymphomas stain positively for HLA class II antigens. Monotypic CIg together with J chain can be demonstrated in paraffin sections of up to two-thirds of cases of FCC lymphoma. The Ig is found predominantly in large centrocytes and centroblasts but only occasionally in small centrocytes and thus, depending on the cytology of the tumour, may be found in very few cells. It should be stressed that the monotypic CIg is found in FCCs and not in plasma cells except uncommonly when the tumour shows plasmacytic differentiation. The characteristic features of cells staining positively for CIg and the pitfalls in the interpretation of Ig staining in this group of lymphomas have already been discussed and the reader is referred to *Figs* 34.13 and 34.14.

In cryostat sections FCC lymphomas stain positively, though weakly, with anti-common acute lymphocytic leukaemia antigen (cALLA). The more recently developed B-cell differentiation antibodies may be useful in their characterization.

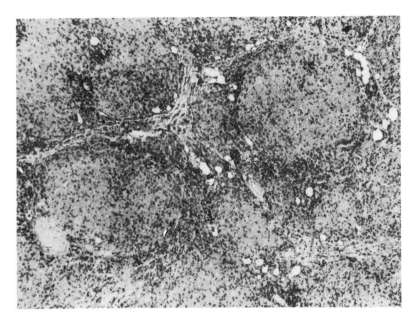

Fig. 34.21. Frozen section of a FCC lymphoma, follicular, stained with a monoclonal antibody to T-cells. Large numbers of T-cells are present, mostly between the follicles.

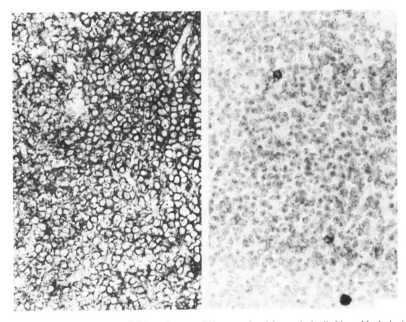

Fig. 34.22. Frozen section of a FCC lymphoma, diffuse, stained for κ chain (left) and λ chain (right). Positive surface staining for κ chain is present on all cells in contrast to only occasional cells positive for λ chain.

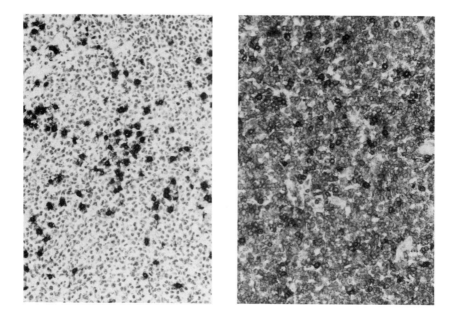

Fig. 34.23. Frozen section of malignant lymphoma, centrocytic, stained with pan T-cell antibodies UCHT1 on left and Leu-1 on right. The pan T-cell antibody UCHT1 recognizes the T3 antigen and stains only T-cells which are scattered through the tumour. Leu-1 also stains the T-cells which can be seen as darkly staining cells set against a background of centrocytes which are also stained although more weakly.

3.6. Immunoblastic Lymphoma

Large cells with central nucleoli and pyroninophilic cytoplasm, so-called 'immunoblasts', may be present in considerable numbers in FCC lymphomas. When they are the predominant cell, and especially when clear plasmacytic differentiation is present, the term 'immunoblastic lymphoma' (IBL) is justified. Immunoblasts may synthesize SIg which can be demonstrated in frozen sections and CIg synthesis is also a feature permitting the characterization of these tumours in paraffin sections.

3.7. Centrocytic Lymphoma

These tumours are composed uniformly of cells that are indistinguishable from centrocytes, but which do not share their phenotypic properties.[32] They form a distinct clinicopathological entity and have been variously called 'intermediate cell lymphoma',[33] 'mantle zone lymphoma'[34] and 'small cleaved cell lymphoma'. Immunocytochemical staining reveals strong monotypic SIg staining with IgD in addition to IgM being found in a majority of cases. Follicle-like or nodular concentrations of DRC may be shown with appropriate antibodies and the tumour cells are positive for C3B receptor. An important distinction from FCC lymphomas lies in their positive staining with the monoclonal antibody Leu-1

(*Fig.* 34.23). Unlike most FCC lymphomas they do not stain with antibodies to cALLA. It is a source of confusion that the term 'centrocyte' has been applied to these cells whereas they are behaviourally and phenotypically distinct from the centrocyte of the follicle centre.

3.8. T-cell Lymphomas

A large number of monoclonal antibodies to T-cell antigens have become available in recent years. It is unusual for these to give satisfactory staining in routinely processed tissues and either frozen sections or freeze-dried tissue is required. B-cell lymphomas frequently contain large numbers of T-cells that may out-number the neoplastic cells. Care must be exercised not to interpret these as T-cell lymphomas. In normal and reactive lymphoid tissues, helper/inducer T-cells have a ratio of approximately 4:1 to suppressor/cytotoxic T-cells. Disturbances in this ratio do not necessarily imply clonal expansion or neoplasia. Clonality of T-cell proliferations can only be determined by analysis of the T-cell receptor DNA arrangement.

Immunohistochemical studies of T-cell lymphomas have shown that no immunological phenotype is restricted to a single, morphological entity. T-cell lymphomas may be broadly divided into terminal deoxynucleotidyl transferase (TdT)-positive, thymic and pre-thymic (lymphoblastic) lymphomas and TdT-negative, post-thymic tumours (*see also* Chapter 35). T-lymphoblastic lymphomas show a range of immunophenotypes that may be related to putative differentiation subsets. T-CLL may have either a helper/inducer or a suppressor/cytotoxic phenotype. Cutaneous T-cell lymphomas[36] (Sézary's syndrome and mycosis fungoides) show a helper/inducer phenotype and usually function as helper cells. Adult T-cell lymphoma/leukaemia, both in its endemic form, and the non-endemic sporadic type, is usually of the helper/inducer phenotype, but in functional assays the cells behave as suppressor cells. Most T-zone lymphomas and T-immunoblastic lymphomas have a helper/inducer phenotype. Antibodies to B-cells, C3B receptors and dendritic reticulum cells often reveal pre-existing follicles within these tumours that have been over-run by the neoplastic T-cells (*Figs.* 34.24 and 34.25). HLA class II antigens are not expressed on normal T-cells but are exhibited by a considerable number of neoplastic T-cells. Very occasional T-cell lymphomas show granular staining for α-1-antitrypsin.

3.9. Histiocytic (Monocyte/Macrophage) Lymphoma

The diagnosis of malignant lymphoma of true histiocytic, i.e. monocyte/macrophage, origin can be very difficult on morphological grounds alone since the histological spectrum of these tumours, which include histiocytic lymphoma,[37] malignant histiocytosis,[38] and malignant histiocytosis of the intestine,[39] is very broad. Initial enthusiasm for lysozyme as the diagnostic immunohistochemical marker of this group of tumours[40] has ebbed somewhat[41,42] and, in our experience, although sometimes positive, it is an unreliable marker of malignant histiocytes. Intracytoplasmic α1-antiT, which is synthesized by macrophages, is, at present, the best marker of these tumours in paraffin sections.[14] We have found

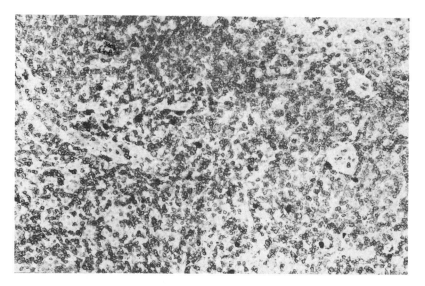

Fig. 34.24. Frozen section of a T-cell lymphoma showing staining of the majority of the cells with a pan T-cell antibody.

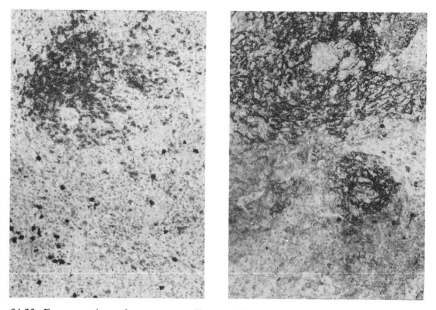

Fig. 34.25. Frozen sections of same tumour illustrated in *Fig.* 34.24 showing residual B-cell islands stained with a pan B-cell antibody at left. Clusters of dendritic reticulum cells are associated with these B-cell islands as shown when the lymphoma is stained with an antibody to DRCs as seen on right.

that α1-antiT is best demonstrated in sections of NBFS-fixed paraffin-embedded tissue following trypsin digestion. Other fixatives require a stronger concentration

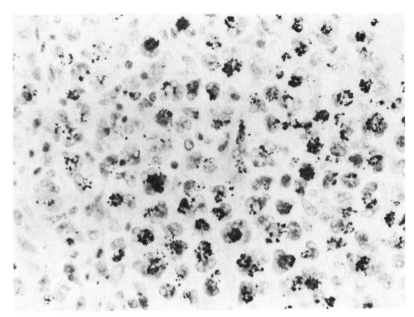

Fig. 34.26. Paraffin section from a case of malignant histiocytosis of the intestine stained for α1-antiT. Abundant positively staining granules are present in the tumour cells.

of antibody and do not result in such consistent staining. As with CIg, the nature of the positive-staining reaction is important since, along with other proteins, α1-antiT can be taken up from tissue fluids non-specifically by dead or damaged cells. The staining pattern of synthesized α1-antiT is granular often with concentration of staining in the Golgi zone. α1-AntiT which has been absorbed stains the cytoplasm diffusely and often stains the nucleus as well. The amount of α1-antiT varies considerably within the cells of any individual histiocytic tumour and from case to case. Thus cells may be stuffed with positively staining granules (*Fig.* 34.26) or only isolated granules may be present which can best be identified with an oil immersion objective. Malignant histiocytes also contain α1-antiCT which appears to mirror α1-antiT in its distribution, although the intensity of staining of the two antigens is not always the same within the same tumour. While α1-antiT is also present in neutrophils, these cells do not contain α1-antiCT and eosinophils contain neither. A proportion of histiocytic tumours show positive staining with antibody to S-100 protein. This finding is not restricted to those tumours in which the cells do not contain lysozyme and possibly reflects a common stem cell for both monocyte/macrophages and interdigitating reticulum cells (IDRC).[43,44] Malignant histiocytes sometimes stain polytypically for Ig and appear to take up IgG specifically, possibly via their Fc receptors. The quality of the staining, its polytypic nature and the absence of J chain distinguish these cells from malignant B-cells. In cryostat sections histiocytic tumours can be characterized by a number of monoclonal antibodies specifically reactive with macrophage cell membranes.[43,45] Tumour cells may react with antibodies to human thymic antigen demonstrating once more a relationship between macrophages and IDRC.

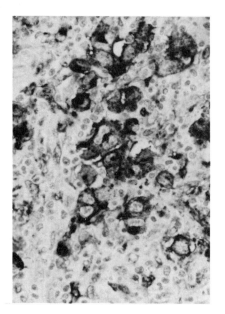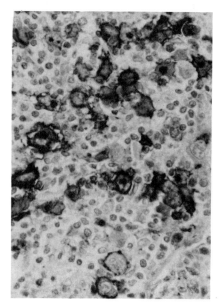

Fig. 34.27. Paraffin section from a case of Hodgkin's disease, nodular sclerosis subtype, showing strong staining of Reed–Sternberg cells with antibody to HLA-DR (left) and with anti-granulocyte antibody (right). In both there is membrane and cytoplasmic staining.

3.10. Hodgkin's Disease

Reed–Sternberg cells which are the hallmark of Hodgkin's disease can often be shown to contain polytypic Ig and sometimes α1-antiT.[46] More recently these cells have been shown to react in paraffin sections with antibodies to HLA-DR and monoclonal antibodies raised against granulocytes[47,48] (*Fig.* 34.27). The latter may be of particular diagnostic help. It is of interest that this does not apply to the Reed–Sternberg cells of lymphocyte-predominant Hodgkin's disease which may contain monotypic Ig and show other evidence of a B-cell phenotype. In cryostat sections, the monoclonal antibody Ki1 stains Reed–Sternberg cells but also a variety of other 'large cell' lymphomas.[49]

4. SUMMARY

The advent of immunohistochemical techniques has greatly increased our knowledge of the histogenesis of lymphoproliferative diseases and aided in the development of rational classifications. The increasing number of antibodies available and differences in the nature of the tissue (fixed or fresh) which can be investigated mean that careful choices must be made when studying any individual case. It cannot be overemphasized that interpretation of the results is as important as the immunohistochemical technique itself. By applying a range of antisera to any given case an appreciation of the histogenesis of most cases of malignant lymphoma is now possible.

Acknowledgements

We are grateful to Mr Keith Miller and staff for technical assistance and to Mr Brian Mepham and staff for both technical assistance and preparation of appendices.

Appendix

I. IMMUNOPEROXIDASE TECHNIQUE (PAP) FOR FIXED PARAFFIN SECTIONS

1. Deparaffinize for 10 min in two changes of xylol and take to alcohol
2. Inhibit endogenous peroxidase by treating with freshly prepared 0·5 per cent H_2O_2 in methanol for 10 min
3. Wash well in tapwater
4. Prewarm in distilled water at 37 °C for 10 min
5. Treat with 0·1 per cent trypsin in 0·1 per cent $CaCl_2$ (adjust to pH 7·8 with 0·1 N NaOH) for 15–30 min at 37 °C*
6. Rinse in cold running water 2–3 min with agitation
7. Wash in Tris-buffered saline (TBS) (*see below*) 10 min in two changes
8. Rabbit anti-human antigen (at current dilution in TBS†) 30 min
9. TBS wash three times, 10 min each
10. Swine anti-rabbit IgG (at current dilution in TBS†) 30 min
11. TBS wash three times, 10 min each
12. Peroxidase–rabbit anti-peroxidase (PAP) (at current dilution in TBS†) 30 min
13. TBS wash three times, 10 min each
14. DAB (*see below*) 10 min
15. Wash in TBS followed by a wash in running tap water, 5 min
16. Counterstain with haematoxylin, wash well 2–3 min. Differentiate in 1 per cent acid alcohol, blue by washing in tap water for at least 10 min
17. Dehydrate, clear and mount.

Tris-HCl-buffered Saline pH 7·6 (TBS) (0·05 M Tris)

Sodium chloride	80 g
Tris [tris(hydroxymethyl)methylamine]	6·05 g
1 N hydrochloric acid	38 ml
Distilled water to	10 000 ml

Check pH and adjust to pH 7·6 if necessary.

*Time may vary with batch of trypsin, fixative and/or length of fixation.
†Determined by titration.

DAB (Graham and Karnovsky)

Buffer: 0·2 M Tris (24·228 g/l) 12 ml
0·1 N HCl 19 ml
Distilled water 19 ml

50 ml pH 7·6

For use: Carefully weigh out 5 mg 3,3′-diaminobenzidine
tetrahydrochloride (possibly carcinogenic) into a glass vial
Add 10 ml Tris-HCl buffer pH 7·6
Immediately before use add 0·1 ml of freshly prepared 1 per cent H_2O_2
(0·1 ml 30 per cent H_2O_2 in 2·9 ml distilled water).

II. IMMUNOPEROXIDASE TECHNIQUE FOR CRYOSTAT SECTIONS

1. Cut cryostat sections at 6 μm
2. Air dry 30 min at room temperature (18–20 °C)
3. Store over silica gel in a closed container at −20 °C overnight or until required (up to 14 days)
4. Remove slides from container. Lay slides on bench until condensation is dry (10 min). Slides may be labelled while they are drying
5. Replace slides in racks. Fix in dry acetone for 20 min at room temperature
6. Lay slides in immunotrays and allow acetone to evaporate 10 min
7. Wash in Tris-buffered saline (TBS), two times for 2 min each.

A. Monoclonal Antibodies

8. Apply mouse anti-human antigen (at current dilution in TBS*) 30 min
9. Wash in TBS three times for 2 min each
10. Apply rabbit anti-mouse Ig, peroxidase conjugated (at current dilution in TBS*) 30 min
11. Wash in TBS three times, 2 min each
12. Apply DAB 10 min
13. Rinse in TBS followed by a wash in running tap water 5 min
14. Counterstain in haematoxylin
15. Wash in tap water. Differentiate in 1 per cent acid alcohol, blue in running tap water, 10 min
16. Dehydrate, clear and mount.

*Determined by chessboard titration.

B. Surface Immunoglobulin, e.g. IgD, IgM

8. Apply rabbit anti-human antigen (at current dilution in TBS*) 30 min
9. Wash in TBS, three times for 2 min each
10. Apply swine anti-rabbit IgG (at current dilution in TBS*) 30 min
11. Wash in TBS, three times for 2 min each
12. Apply rabbit peroxidase–anti-peroxidase (at current dilution in TBS*) 30 min
13. Wash in TBS three times for 2 min each
14. Apply DAB 10 min
15. Rinse in TBS followed by a wash in running tap water 5 min
16. Counterstain in haematoxylin
17. Wash in tap water, differentiate in 1 per cent acid alcohol, blue in running tap water, 10 min
18. Dehydrate, clear and mount.

III. SOURCES OF REAGENTS USED FOR ILLUSTRATED PREPARATIONS

Polyclonal Antibodies

Immunoglobulins (heavy and light chains)	Dako
S-100	Dako
Lysozyme	Dako
α–1-Antitrypsin	Dako
J-chain	Nordic

Monoclonal Antibodies

Pan B-cell	Dako
HLA-DR	C. Dixon (ICRF)
Leukocyte common antigen	Dako
Cytokeratin (CAM 5·2)	Becton Dickinson
Pan T-cell	Seward
T-suppressor	Seward
T-helper (Leu-3a)	Becton Dickinson
T6 thymic antigen (NA 134)	Sera Lab
Dendritic reticulum cell antigen	Nancy Hogg (ICRF)
Leu-1	Becton Dickinson
Anti-granulocytic antigen	D. Y. Mason (Oxford)

Second and Third Layer Reagents

Swine anti-rabbit immunoglobulin	Dako
Rabbit PAP	Dako
Peroxidase-conjugated rabbit anti-mouse immunoglobulin	Dako

Other Reagents

Trypsin (T8128)	Sigma
DAB (D563)	Sigma

ICRF = Imperial Cancer Research Fund, PO Box 123, Lincoln's Inn Fields, London WC2A 3PX.

REFERENCES

1. Lennert K. Follicular lymphoma. A tumour of the germinal centres. In: Akazaki K., Rappaport H., Berard C. W., Bennet J. M. and Ishikawa E., eds., *Malignant Diseases of the Hematopoietic System. GANN Monograph on Cancer Research*. No. 15. Tokyo, University of Tokyo Press, 1973: 217–231.
2. Lukes R. J. and Collins R. D. New observations in follicular lymphoma. In: Akazaki K. et al., eds., *Malignant Diseases of the Hematopoietic System. GANN Monograph on Cancer Research*. No. 15. Tokyo, University of Tokyo Press, 1973: 209–215.
3. Taylor C. R. and Burns J. The demonstration of plasma cells and other immunoglobulin-containing cells in formalin fixed paraffin-embedded tissues using peroxidase-labelled antibody. *J. Clin. Pathol.* 1974, **27**, 14–20.
4. Garvin A. J., Spicer S. S. and McKeever P. E. The cytochemical demonstration of intracellular immunoglobulin. *Am. J. Pathol.* 1976, **82**, 457–470.
5. Curran R. C. and Gregory J. The unmasking of antigens in paraffin sections of tissue by trypsin. *Experientia* 1977, **33**, 1400.
6. Mepham B. L., Frater W. and Mitchell B. S. The use of proteolytic enzymes to improve immunoglobulin staining by the PAP technique. *Histochem. J.* 1979, **11**, 345–357.
7. Curran R. C. and Gregory J. Effects of fixation and processing on immunohistochemical demonstration of immunoglobulin in paraffin sections of tonsil and bone marrow. *J. Clin. Pathol.* 1980, **33**, 1047–1057.
8. Epenetos A. A., Bobrow L. G., Adams T. E., Collins C. M., Isaacson P. G. and Bodmer W. F. A monoclonal antibody that detects HLA-DR region antigen in routinely fixed, paraffinised sectins of normal and neoplastic lymphoid tissues. *J. Clin. Pathol.* 1985, **38**, 12–17.
9. Warnke R. A., Gatter K. C., Falini B., et al. Diagnosis of human lymphoma with monoclonal antileucocyte antibodies. *N. Engl. J. Med.* 1983, **309**, 1275–1281.
10. Isaacson P. G., Wright D. H., Jones D. B., Payne S. V. and Judd M. A. The nature of the immunoglobulin-containing cells in malignant lymphoma. An immunoperoxidase study. *J. Histochem. Cytochem.* 1980, **28**, 761–770.
11. Stein H. and Tolksdorf G. Development and differentiation of the T-cell and B-cell systems: a perspective. In: van den Tweel J. G., ed., *Malignant Lymphoproliferative Diseases*. Leiden, Leiden University Press, 1980: 13–29.
12. Isaacson P. Immunochemical demonstration of J chain: a marker of B-cell malignancy. *J. Clin. Pathol.* 1979, **32**, 802–807.
13. Brandtzaeg P. Presence of J chain in human immunocytes containing various immunoglobulin classes. *Nature* 1974, **252**, 418–419.
14. Isaacson P., Jones D. B., Millward-Sadler G. H., Judd M. A. and Payne S. Alpha-1-antitrypsin in human macrophages. *J. Clin. Pathol.* 1981, **34**, 982–990.
15. Takahashi K., Yamaguchi H., Ishizeki J., Nakajima T. and Nakazato Y. Immunohistochemical

and immunoelectron microscopic localization of S100 protein in the interdigitating reticulum cells of the human lymph node. *Virchows Arch (Cell Pathol.)* 1981, **37**, 125–135.

16. McMichael A. J., Pilch J. R., Galfré G., Mason D. Y., Fabre J. W. and Milstein C. A human thymocyte antigen defined by hybrid myeloma monoclonal antibody. *Eur. J. Immunol.* 1979, **9**, 205–210.

17. Hogg N., Ross G. D., Jones D. B., Slusarenko M., Walport M. J. and Lachmann P. J. Identification of an anti-monocyte monoclonal antibody that is specific for membrane complement receptor type one (CRI). *Eur. J. Immunol.* 1984, **14**, 236–243.

18. Naiem M., Gerdes J., Abdulaziz Z., Stein H. and Mason D. Y. Production of a monoclonal antibody reactive with human dendritic reticulum cells and its use in the immunohistological analysis of lymphoid tissue. *J. Clin. Pathol.* 1982, **36**, 167–176.

19. Makin C. A., Bobrow L. G. and Bodmer W. F. Monoclonal antibody to cytokeratin for use in routine histopathology. *J. Clin. Pathol.* 1984, **37**, 975–983.

20. Cochran A. J., Wen D.-R. and Herschman H. J. Occult melanoma in lymph nodes detected by antiserum to S-100 protein. *Int. J. Cancer* 1984, **34**, 159–163.

21. Warnke R. and Levy R. Immunopathology of follicular lymphomas: a model of B-lymphocyte homing. *N. Engl. J. Med.* 1978, **298**, 481–486.

22. Sarasombath S., Mestecky J. and Skvaril F. Monoclonal IgG-κ and IgG-1λ proteins with different idiotypic determinants present in a single patient. *Clin. Exp. Immunol.* 1977, **29**, 67–74.

23. Sklar J., Cleary M. L. Thielemans K., Gralow J., Warnke R. and Levy R. Biclonal B-cell lymphoma. *N. Engl. J. Med.* 1984, **311**, 20–27.

24. Taylor C. R., Russel R. and Chandar S. An immunohistologic study of multiple myeloma and related conditions using an immunoperoxidase method. *Am. J. Clin. Pathol.* 1978, **70**, 612–622.

25. Diktor M. Immunoblastic sarcoma after thymus epithelial graft in an immunodeficient child. *Histopathology* 1980, **4**, 661–668.

26. Isaacson P. and Wright D. H. Anomalous staining patterns in immunohistologic studies of malignant lymphoma. *J. Histochem. Cytochem.* 1979, **27**, 1197–1199.

27. Avremeas S. and Leduc E. H. Detection of simultaneous antibody synthesis in plasma cells and specialised lymphocytes in rabbit lymph nodes. *J. Exp. Med.* 1970, **131**, 1137–1168.

28. Judd M. A. One micrometre paraffin sections: an aid to interpretation of immunoperoxidase staining of immunoglobulins. *J. Microsc.* 1980, **120**, 201–206.

29. Giddings J., Griffin R. L. and MacIver A. G. Demonstration of immunoproteins in araldite-embedded tissues. *J. Clin. Pathol.* 1982, **35**, 111–114.

30. Wood G. W. and Travers H. Non-Hodgkin's lymphoma: identification of the monoclonal B-lymphocyte component in the presence of polyclonal immunoglobulin. *J. Histochem. Cytochem.* 1982, **30**, 1015–1021.

31. Gregg E. O., Al-Saffar N., Jones D. B., Wright D. H., Stevenson F. K. and Smith J. L. Immunoglobulin negative follicle centre cell lymphoma. *Br. J. Cancer* 1984, **50**, 735–744.

32. Tolksdorf G., Stein H. and Lennert K. Morphological and immunological definition of a malignant lymphoma derived from germinal-centre cells with cleaved nuclei (centrocytes). *Br. J. Cancer* 1980, **41**, 168–182.

33. Weisenburger D. D., Kim H. and Rappaport H. Mantle-zone lymphoma: a follicular variant of intermediate lymphocytic lymphoma. *Cancer* 1982, **49**, 1429–1438.

34. Weisenburger D. D., Nathwani B. N., Diamond L. W., Winberg C. D. and Rappaport H. Malignant lymphoma, intermediate lymphocytic type: a clinico-pathologic study of 42 cases. *Cancer* 1981, **48**, 1415–1425.

35. Knowles II D. M. and Halper J. P. Human T-cell malignancies: correlative clinical, histopathologic, immunologic and cytochemical analysis of 23 cases. *Am. J. Pathol.* 1982, **106**, 187–203.

36. Lawrence E. C., Broder Sj., Jaffe Ej. S., Braylan R. C., Dobbins W.. O., Young Rj. C. and Waldmann T. A. Evaluation of a lymphoma with helper T-cell characteristics in Sezary's syndrome. *Blood* 1978, **52**, 481–492.

37. Isaacson P., Wright D. H. and Jones D. B. Malignant lymphoma of 'true' histiocytic (monocyte/macrophage) origin. *Cancer* 1982, **51**, 80–91.

38. Byrne G. E. and Rappaport H. Malignant histiocytes. In: Akazaki K. et al., eds., *Malignant Diseases of Hematopoietic System*, *GANN Monograph on Cancer Research*, No. 15. Tokyo, University of Tokyo Press, 1973: 145–162.

39. Isaacson P. and Wright D. H. Malabsorption and intestinal lymphomas. In: Wright R., ed., *Recent Advances in Gastrointestinal Pathology*. London, W. B. Saunders, 1980: 193–212.

40. Taylor C. R. Immunoperoxidase technique: theoretical and practical aspects. *Arch. Pathol. Lab. Med.* 1978, **102**, 113–121.
41. Risdall R. J., Sibley R. K., McKenna R. W., Brunning R. D. and Dehner L. P. Malignant histiocytosis: a light and electron microscopic and histochemical study. *Am. J. Surg. Pathol.* 1980, **4**, 439–450.
42. Mendelsohn G., Eggleston J. C. and Mann R. B. Relationship of lysozyme (muramidase) to histiocytic differentiation in malignant histiocytosis. *Cancer* 1980, **45**, 273–279.
43. Salisbury J., Ramsay A. D. and Isaacson P. G. Histiocytic lymphoma: Report of a case with an unusual phenotype. *J. Pathol.* 1985, in press.
44. Goordyal P. and Isaacson P. G. Immunocytochemistry of monocytes growing in human bone marrow culture—clue to origin of Langerhans and interdigitating reticulum cells. *J. Pathol.* 1985, in press.
45. Radzun H. J., Parwaresch M. R., Feller A. C. and Hansmann M. L. Monocyte/macrophage-specific monoclonal antibody Ki-M1 recognizes interdigitating reticulum cells. *Am. J. Pathol.* 1984, **117**, 441–450.
46. Payne S. V., Wright D. H., Jones K. J. M. and Judd M. A. The macrophage origin of Reed–Sternberg cells. An immunohistochemical study. *J. Clin. Pathol.* 1982, **35**, 159–166.
47. Stein H., Uchanska-Ziegler B., Gerdes J., Ziegler A. and Wernet P. Hodgkin and Sternberg–Reed cells contain antigens specific to late cells of granulopoiesis. *Int. J. Cancer* 1982, **29**, 283–290.
48. Hsu S-M. and Jaffe E. S. Leu M1 and peanut agglutinin stain the neoplastic cells of Hodgkin's disease. *Am. J. Clin. Pathol.* 1984, **82**, 29–32.
49. Stein H., Gerdes J., Kirchner H., Schaadt M. and Diehl V. Hodgkin and Sternberg–Reed cell antigen(s) detected by an antiserum to a cell line (L428) derived from Hodgkin's disease. *Int. J. Cancer* 1981, **28**, 425–429.

35

Cell Markers in Diagnostic Haematology

E. Matutes

The development and application of immunological techniques to the study of human leukaemias have proved to be of great value for the diagnosis of these conditions. In the past, the diagnosis of haematological malignancies was based on light microscopical examination of Romanovsky stained films in conjunction with cytochemical reactions. However, these techniques have some limitations, particularly when leukaemias arise from 'undifferentiated' blast cells. In the past decade, the development of monoclonal antibodies which specifically recognize cells from the various haemopoietic lineages has provided new insights in the characterization and classification of leukaemic cells as well as normal haemopoietic precursors. Although immunological analysis in haemopoietic disorders has to be considered in conjunction with other studies such as cytomorphology, cytochemistry, cytogenetics etc., this chapter will refer to the contribution of immunophenotyping to the precise diagnosis of human leukaemias and describe the various immunostaining approaches which are routinely used for this purpose.

1. MATERIAL AND METHODS

1.1. Specimen Collection

Peripheral blood and bone marrow samples are collected in plastic or glass tubes containing preservative-free heparin and diluted volume to volume with phosphate-buffered saline (PBS). Mononuclear cells are obtained by a density gradient centrifugation using Ficoll–Isopaque (*d*:1077) and washed three times in PBS.

599

Mononuclear cells from lymph nodes and spleen are obtained by mincing and extruding the tissue through a wire mesh and, if the red cell contamination is high, a further step of purification on Lymphoprep is performed.

1.2. Conventional Markers for Lymphoid Cells

Before the advent of monoclonal antibodies, the demonstration of receptors for sheep red blood cells (SRBC) (E-rosettes) on T-lymphocytes and surface immunoglobulins on B-cells represented the first insight in the classification of lymphoproliferative disorders. Later, it was shown that neoplastic cells from B-chronic lymphocytic leukaemia (B-CLL) had the property of binding mouse red blood cells (MRBC)—M-rosettes. These markers are still used for the diagnosis of T and B-lymphoid diseases.

1.3. Immunostaining Techniques

1.3.1. Light Microscopy

Cell surface and/or intracellular antigens can be investigated at light microscopy by two methods:

a. Direct or indirect immunofluorescence (see Appendix)
b. Indirect immunoperoxidase (see Appendix).

The detection of membrane antigens is carried out routinely in suspension on living (unfixed) cells by the immunofluorescence method although the immuno-peroxidase or other immunoenzymatic methods, e.g. using alkaline phosphatase, can also be used for this purpose. Intracellular (cytoplasmic or nuclear) antigens are assessed on previously fixed cells on cytocentrifuge slides by either immunofluorescence or immunoperoxidase. The immunoenzymatic methods offer the advantage of being a permanent record and permit the simultaneous assessment of morphological and immunological features.

1.3.2. Electron Microscopy

1.3.2.1. IMMUNOGOLD METHOD. The immunogold method is used to study the fine structure of cells labelled with a particular antibody. Cytochemical reactions, e.g. for myeloperoxidase, acid phosphatase and platelet peroxidase, can be combined with immunogold staining, providing information on the content of the enzyme and its distribution on labelled cells. This method has already proved to be useful in the identification of putative normal cell counterparts for some leukaemias and for the determination of the specificity of a monoclonal antibody (see Appendix).

2. CELL MARKERS

A selected panel of antibodies which recognize membrane markers on cells from the two main haemopoietic lineages, lymphoid and myeloid, is shown in *Tables* 35.1–35.3. Although a large battery of monoclonal antibodies is now available, the ones shown are those for which a well-defined specificity has been established. *Figs* 35.1 and 35.2 illustrate the cell reactivity with these reagents according to the normal lymphoid and myeloid differentiation pathways. Their contribution to the characterization of the haemopoietic malignancies will be described in the following section.

*Table 35.1. **Monoclonal antibodies against T-cells***

McAb	Specificity	Molecular weight of antigen (kdal)	Refs.
CD5 (T1)	Thymocytes and PB T-lymphocytes	67	13
CD3 (T3)	PB T-lymphocytes	19–29	14
CD4 (T4)	Helper–inducer T-lymphocytes	55	16
CD1 (T6)	Cortical thymocytes	45, 12	15
CD8 (T8)	Suppressor–cytotoxic T-lymphocytes	32, 33	17
CD2 (T11)	Receptor for sheep erythrocytes	50	12
OKT 17	PB T-lymphocytes and thymocytes	?	10
CD7 (3A1, WT1)	PB T-lymphocytes, thymocytes and pre-T-cells	41, 40	7, 8
BE-1, BE-2	Sézary cells	50, 78	20

CD, nomenclature proposed by the Committee on Human Leukocyte Differentiation Antigens.[51]
PB, peripheral blood.

2.1. Lymphoid Lineage

Membrane and enzyme marker studies have significantly contributed to the diagnosis and classification of lymphoproliferative disorders. The recognition of a broad spectrum of lymphoid leukaemias which correspond to proliferations of cells at the various stages of B- and T-lymphoid differentiation pathways, has been made possible by the immunological markers. This approach has permitted the recognition of two main groups of lymphoid malignancies: the B- and the T-cell disorders. The clinical entity, acute lymphoblastic leukaemia (ALL) will be considered separately because of its clinical and biological peculiarities, though lymphoblasts in this condition may show early features of B- and T-cell commitment as illustrated in *Fig.* 35.1.

2.1.1. Acute Lymphoblastic Leukaemia

In ALL, the enzyme terminal deoxynucleotidyl transferase $(TdT)^1$ has been the most relevant marker because it was consistently demonstrated by immunostain-

Table 35.2. *Monoclonal antibodies against lymphoid precursors and B-cells*

McAb	Specificity	Molecular weight of antigen (kdal)	Refs.
FMC4	HLA-DR, B-cells and monocytes	28–33	52
3C5	Early lymphoid and myeloid cells	120	35
B4	Early B-lymphoid precursors and	40,80	5
B1	B-lymphocytes	35	22
CD10 (J5)	Common-ALL antigen, BM lymphoid precursors and follicular centre cells	100	6
FMC7	Subpopulation (50%) normal PB lymphocytes	?	23
HC2	Hairy cells, activated B-cells	40, 60	26, 27
PC1	Plasma cells	?	24, 25

BM: bone marrow.
PB: peripheral blood.

Table 35.3. *Monoclonal antibodies against myeloid antigens*

Cell lineage	Monoclonal antibody	Specificity	Reference
I Granulocytic and monocytic-associated	My9	Early myeloid and monocytic precursors	33
	My7	Early myeloid precursors and granulocytes and monocytes	32
	OKM1	Granulocytes, monocytes and granulocytes and subpopulation of PB T-lymphocytes	29
I.1 Monocytic	UCHM1	Early monocytic precursors and monocytes	53
	FMC17, 32, 33	Monocytes	30
I.2 Granulocytic	3C5	Early myeloid precursors	35
	FMC10–13	Granulocytes	31
II Megakaryocytic	J15	Platelet glycoprotein IIb/IIIa } Megakaryocytic precursors and platelets	38
	C17	Platelet glycoprotein IIIa	39
	AN51	Platelet glycoprotein Ib	37
III Erythroid	LICR-LON-R10	Glycophorin A } Erythroblasts and RBCs	44
	GERO	Gerbich RBC group	47

RBC: red blood cells.

LYMPHOID   LINEAGE

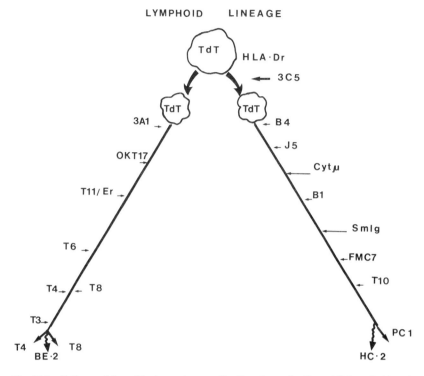

Fig. 35.1. Cell reactivity with the various antibodies along the T- and B-lymphoid pathways.

ing in blasts from all types of ALL except in the rare B- group (*Table* 35.4). Though a small proportion of acute myeloid leukaemia (AML) cases have been shown to be TdT positive,[2,3] they may correspond to malignancies in which the clonogenic cell is an early precursor cell that can differentiate along the myeloid and lymphoid pathways, so that, at any one stage, features of both pathways might be expressed simultaneously. HLA-DR-associated antigens, though not specific for lymphoblasts, are useful when considered within the ALL group, since they may help in the differential diagnosis between T-ALL (HLA-DR negative) and the remaining ALL types (HLA-DR positive).

The first feature of B-cell commitment, which is likely to be involved in most of the ALL cases, including null-ALL, is shown by the rearrangement of the immunoglobulin genes.[4] The monoclonal antibodies J5 (gp 100),[5] which demonstrate the common ALL antigen, and B4[6] are also useful markers for ALL blasts. However, they have to be considered within the context of ALL since they are expressed in other B-cell neoplasias such as follicular lymphoma. Pre-B-ALL and B-ALL are the most differentiated conditions and can be detected by the presence of cytoplasmic μ chain and surface membrane immunoglobulin, respectively.

Regarding the T-cell commitment of lymphoblasts, the monoclonal antibody 3A1,[7] or WT1[8] (CD7), detects the p40 antigen which appears to be the first expressed in the cells of this lineage. Cells from all the thymic conditions including

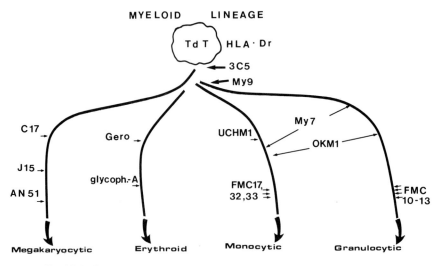

Fig. 35.2. Cell reactivity with monoclonal antibodies in the myeloid lineage.

Table 35.4. **Cell markers in acute lymphoblastic leukaemia (ALL)**

ALL-type	TdT	HLA-DR	J5	Cytμ	SmIg	3A1
Null	+	+	−	−	−	−
Common	+	+	+	−	−	−
Pre-B	+	+	+	+	−	−
B-ALL	−	+	−/+	−	+	−
Pre-T-ALL/T-ALL	+	−	−	−	−	+

Cytμ: Cytoplasmic immunoglobulin.
SmIg: Surface membrane immunoglobulin.
Boxes indicate definitive diagnostic characteristic.

the group of pre-T-ALL in which T-lymphoblasts lack the receptor for SRBC (E-rosettes) are 3A1/WT1-positive. However, since reactivity with 3A1/WT1 has been reported in a proportion of AML cases,[9] terminal transferase and myeloid markers (cytochemical and immunological) should always be used to exclude myeloid characteristics in those presumed pre-T-ALL cases.

2.1.2. T-cell Lymphoproliferative Disorders

Two major groups of T-cell disorders can be distinguished according to the stage of cell differentiation in which the leukaemic process occurs: (*a*) immature/thymic conditions, which include pre-T-ALL, T-ALL and T-lymphoblastic lymphoma (T-LbLy) and (b) mature/post-thymic proliferations which can be classified according to clinical, morphological and phenotypical criteria as T-prolymphocytic leukaemia (T-PLL), cutaneous T-cell lymphoma (CTCL), adult T-cell leukaemia–lymphoma (ATLL) and T-chronic lymphocytic leukaemia (T-CLL). The membrane phenotype for each of these conditions is shown in *Table* 35.5. TdT clearly distinguishes thymic (TdT positive) (*Plate* 39) from post-thymic (TdT negative) leukaemias. The monoclonal antibody OKT 17[10] is, overall, a consistent pan-T marker since it recognizes cells from all the types of T-cell disorders except for the group of pre-T-ALL.[11] In our experience, OKT 17 identifies a higher proportion of cases than the monoclonal antibody OKT 11 (CD2) or its equivalent, the E-rosette test,[12] even under optimal conditions.

Table 35.5. **Cell markers in T-lymphoid malignancies**

Disease type	TdT	3A1	OKT 17	ER/T11	T1	T6	T3	T4	T8
Pre-T-ALL	+	+	−	−	+/−	−	−	−	−
T-ALL	+	+	+	+	+	+/−	−	−	−
T-LbLy	+	+	+	+	+	+	+/−	+/−	+/−
T-PLL	−	+	+	+	+	−	+/−	++	−/+*
CTCL	−	−	+	+/−	+	−	+	++	
ATLL	−	−	+	+/−	+	−	+/−	++	−
T-CLL	−	−	+	+	−	−	+	−/+*	++

*A small proportion of T-PLL cases are T8+ve T4−ve or T4+ve T8+ve and a minority of T-CLL are T4+ve T8−ve.
ALL = acute lymphoblastic leukaemia.
T-LbLy = T-lymphoblastic lymphoma.
T-PLL = T-prolymphocytic leukaemia.
CTCL = cutaneous T-cell lymphoma.
ATLL = adult T-cell leukaemia-lymphoma.
T-CLL = T-chronic lymphocytic leukaemia.
Boxes indicate definitive diagnostic characteristic.

Other monoclonal antibodies regarded as pan-T reagents such as T1 (CD5)[13] or T3 (CD3)[14] are inconsistently expressed in mature or immature T-cell disorders. Although monoclonal antibody 3A1[7] is an excellent marker for the thymic proliferations including pre-T-ALL as mentioned above, post-thymic malignancies with the exception of T-PLL are frequently 3A1-negative. A combination of

TdT, OKT 17 and 3A1 allows the classification of the T-cell leukaemias according to three major stages of cell maturation: pre-T-ALL (TdT+ve, 3A1+ve, OKT 17−ve), thymic leukaemias (TdT+ve, 3A1+ve, OKT 17+ve) and post-thymic proliferations (TdT−ve, 3A1−ve, OKT 17+ve). T-PLL (TdT−ve, 3A1+ve, OKT 17+ve) occupies an intermediate position in this scheme. Of the monoclonal antibodies to T-cells that are restricted to particular stages of cell maturation, T6 (CD1), which is specific for normal cortical thymocytes,[15] is expressed in a variable proportion of thymic disorders (usually T-lymphoblastic lymphoma) but is always negative in the post-thymic leukaemias. Within the latter group, a distinct though not entirely consistent pattern of reactivity with T4 (CD4)[16] and T8 (CD8)[17] is found. As shown in *Table* 35.5, T-PLL, CTCL and ATLL generally represent neoplastic expansions of the T4+ve subset whereas T-CLL cells are frequently T8+ve.[18,19] Finally, the monoclonal antibody BE-2[20] appears to be highly specific for CTCL cells. In our experience, reactivity with BE-2 has been demonstrated in over 50 per cent of the CTCL cases tested, whereas cells from only one case (ATLL) of the other post-thymic leukaemias have been BE-2+ve.[21]

2.1.3. B-chronic Lymphoproliferative Disorders

Before the availability of monoclonal antibodies the immunological classification of the chronic B-cell disorders was based on the expression of two markers, surface membrane immunoglobulin and M-rosettes. B-CLL cells, unlike the other B-cell leukaemias, were characterized by weak expression of surface membrane immunoglobulin and high density of mouse RBC receptors. Numerous monoclonal antibodies to B-cells are now available and this has led to a better characterization of the B-cell malignancies according to the stage of differentiation of the neoplastic cells. Some of these reagents, such as B1, B4 FMC7 and PC1,[5,22–25] detect B-cell-specific antigens and others, T1 (CD5), J5 (CD10), OKT 10,[6,13] detect markers that, although expressed in cells from other lineages, are useful when considered in the context of B-cell leukaemias. A summary of the reactivity with these reagents in the chronic B-cell diseases is shown in *Table* 35.6. The selectivity of some of these reagents is exemplified by the monoclonal antibody PC1 which is expressed only in late stages of B-cell maturation as seen in Waldenström's macroglobulinaemia and multiple myeloma.[24] The membrane phenotype profile, HLA-DR−ve, OKT 10+ve, PC1+ve, is characteristic for these two conditions. Also, FMC7[23] appears to be a useful marker expressed at selected stages of cell maturation being positive in proliferations of relatively mature B-cells such as B-PLL, hairy cell leukaemia and non-Hodgkin's lymphomas, but rarely found in B-CLL except for those cases in 'prolymphocytoid' transformation. Among the non-B-cell lineage-restricted monoclonal antibodies, T1 (CD5)[13] is useful for distinguishing B-CLL (T1+ve) from other B-cell neoplasias (T1−ve). Suprisingly, this monoclonal antibody which also binds to normal T-cells,[13] is frequently negative in T-CLL. Thus, the profile T1+ve, FMC7−ve, J5−ve together with the conventional markers surface membrane immunoglobulin+ve/−ve, M-rosettes++, is typical of B-CLL. J5 (CD10) (gp100), which detects a relatively high proportion of ALL cases, is also

Table 35.6. **Cell markers in B-cell chronic lymphoid disorders**

Disease type	SmIg	M-r.	FMC7	T1	J5	T10	PC1	B1/B4/Ia
B-CLL	±	++	−/+	+	−	−	−	+
B-PLL	++	−/+	++	−/+	−	−	−	+
HCL	++	−/++	++	−	−	−	−	+
NHL*	++	+	+	−	++	+	−	+
WM and MM	−	−	−	−	−	++	++	−

*Mainly applicable to follicular (centroblastic/centrocytic) lymphoma.
B-CLL = B-chronic lymphocytic leukaemia.
B-PLL = B-prolymphocytic leukaemia.
HCL = hairy cell leukaemia.
NHL = non-Hodgkin's lymphoma.
WM = Waldenström's macroglobulinaemia.
MM = multiple myeloma.
Boxes indicate definitive diagnostic characteristic.

expressed in non-Hodgkin's lymphoma, characteristically in the follicular (centroblastic/centrocytic) type and the monoclonal antibody HC2[26] (not shown in *Table* 35.6), though not restricted to B-cells (also positive in some AMLs), appears to be a useful marker for the detection of hairy cell leukaemia.[27]

2.2. Myeloid Lineage

In contrast to the lymphoid disorders where a large number of immunological markers have been available for several years, the myeloproliferative conditions have been mainly characterized and classified on the basis of conventional cytomorphological and cytochemical methods and it has not been until recently that cell typing has become important for the diagnosis of this group of haematological diseases.

Most cases of acute myeloid leukaemias (AMLs) can be diagnosed on well prepared May–Grünwald–Giemsa-stained smears and classified according to the generally accepted criteria for the various AML subtypes from M1 to M6.[28] In some instances, the diagnosis may need confirmation by applying a few cytochemical reactions (myeloperoxidase, acid phosphatase and α-naphthyl acetate esterase). Immunophenotyping of the myeloid leukaemias is important in those cases in which the leukaemic blast cells are poorly differentiated and can not be characterized adequately by standard cytomorphology.

In the context of the granulomonocytic lineage, cell typing was imprecise until recently, since most of the early monoclonal antibodies were reactive with cells only at late stages of maturation.[29–32]. In the last three years the development of reagents, such as the monoclonal antibody My9,[33] which identify immature blast cells has demonstrated that the immunological assessment of AML can provide objective diagnostic data.[33,34] In our laboratory we have confirmed the findings reported by Griffin et al.[33] showing that My9 is a good pan-myeloid reagent, being expressed in blast cells from 87 per cent of the AML cases and 72 per cent of cases of chronic granulocytic leukaemia in blast crisis (CGL-BC). Electron microscopical immunocytochemistry by the immunogold method combined with the myeloperoxidase reaction has shown that My9 is weakly expressed in myeloblasts with little or no myeloperoxidase content and that the reactivity increases in parallel to the enzyme activity (*Fig.* 35.3). This pattern is different from that seen with another monoclonal antibody, 3C5, which reacts with myeloblasts in the same way as My9, and also recognizes early lymphoid precursors.[35] Monoclonal antibody 3C5 reacted with myeloblasts with little or no myeloperoxidase activity (*Fig.* 35.4), whereas AMLs with promyelocytic or monocytic differentiation were consistently 3C5-negative.[36] The combined use of monoclonal antibodies My9 and 3C5 appears to be helpful for the diagnosis and classification of AML in its various subtypes.

The production of monoclonal antibodies against the various glycoproteins present in the membrane of the platelets and their precursors,[37–39] has provided a major advance in the recognition of proliferating cells of the megakaryocytic lineage. The identification of such cells was previously carried out by an electron microscopical cytochemical method, the platelet peroxidase reaction.[40,41] However, the diagnosis of megakaryoblastic leukaemias, whether presenting as acute leukaemia or as the blast cell components of CGL-BC, has been improved in the past few years by the use of these monoclonal antibodies to platelets, particularly those, such as J15 or C17, that recognize the early megakarocyte precursors.[42,43] The relevance of cell typing in these conditions has been demonstrated in a study carried out in our laboratory on a series of patients with poorly differentiated leukaemias by using immunological and ultrastructural analysis. The high specificity of these reagents for the detection of megakarocyte proliferations led us to conclude that the platelet peroxidase reaction, only

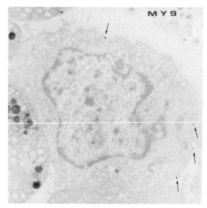

Fig. 35.3. Myeloperoxidase-negative myeloblast showing weak reactivity with My9 (arrows). Immunogold and myeloperoxidase method; no counterstain. × 7750.

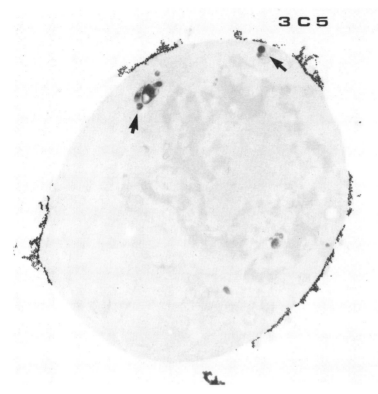

Fig. 35.4. Myeloblast reactive with monoclonal antibody 3C5 showing a few myeloperoxidase-positive granules (arrows). Immunogold and myeloperoxidase method; no counterstain. × 10000.

possible at electron microscopical level, should be reserved for rare cases in which immunophenotyping fails to demonstrate the megakarocytic nature of the blast cells.[42]

There are available a few reagents which allow identification of leukaemic blasts of the erythroid lineage. Early morphological features of erythroid differentiation, such as ropheocytosis or aggregates of ferritin particles in the cytoplasm, can be identified in blast cells by electron microscopical methods. The possibility of immunologically typing these conditions became apparent with the development of a monoclonal antibody against glycophorin A.[44] This reagent was shown to be useful for the diagnosis of some erythroleukaemias.[42,45] However, it is possible that the demonstration of erythroid involvement in haemopoietic malignancies by this monoclonal antibody is underestimated because glycophorin A is expressed at relatively late stages of normal erythroid differentiation.[46] A new monoclonal antibody, Gero, against the red blood cell group Gerbich,[47] recognizes erythroid progenitor cells at an earlier stage of differentiation than anti-glycophorin A and, thus, might be more sensitive for the detection of erythroleukaemias. These

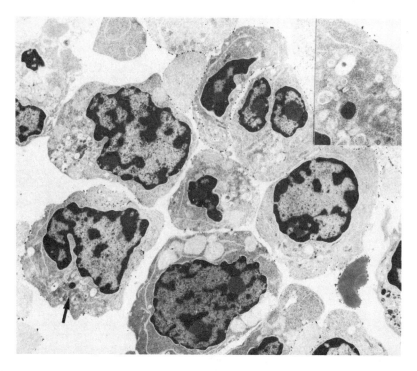

Fig. 35.5. Bone marrow from an erythroleukaemia. Cluster of erythroblasts in different stages of maturation showing reactivity with anti-glycophorin A. *Inset* (arrow)—enlargement of a siderosome. Immunogold method with uranyl acetate and lead citrate counterstain. ×5000; ×14 400.

reagents can be used in combination with morphological techniques by means of the immunogold method at electron microscopical level, as shown in *Fig.* 35.5.

Appendix

I. SURFACE MARKERS ON CELL SUSPENSIONS: IMMUNOFLUORESCENCE

A. Direct

The antibody is directly labelled with a fluorochrome, commonly fluorescein isothiocyanate (FITC). This technique is mainly used for detecting immunoglobulin heavy and light chains on the surface of lymphoid cells. Polyclonal and monoclonal antibodies to immunoglobulins are commercially available for this purpose.

1. Resuspend mononuclear cells (3×10^6) in 2 ml of acetate-buffered saline (*see below*) and incubate at 37 °C for 15 min
2. Wash twice in culture medium, e.g. RPMI-1640
3. Incubate cells in 2 ml of culture medium at 37 °C for 60 min to remove non-specifically bound cytophilic antibodies
4. Wash once and resuspend in 0·2 ml of fresh culture medium
5. Add 200 µl of FITC-conjugated anti-Ig at the optimal dilution in PBS, 0·1 M pH 7·4, and incubate at 4 °C for 30 min
6. Wash twice in culture medium
7. Resuspend the cells in one drop of PBS/glycerol (v/v), mount on a glass slide, cover and seal with nail varnish.

Acetate-buffered saline pH 5·5
   Sol. 1: Glacial acetic acid: 12 ml/l
   Sol. 2: Anhydrous sodium acetate 16·4 g/l
Mix 8·8 ml Sol. 1 with 41·2 ml Sol. 2 and make up to 200 ml with distilled water. Add 1·8 g NaCl and 0·2 g anhydrous $CaCl_2$. Aliquot and freeze at −20 °C.

B. Indirect

In this technique, the unlabelled antibody is first reacted with the cells and its binding is shown by FITC-conjugated anti-Ig. Indirect immunofluorescence is currently used to detect membrane antigens recognized by the various monoclonal antibodies.

1. Resuspend mononuclear cells (1×10^6) in 0·2 ml of PBS–azide–BSA–AB serum (*see below*)
2. Add the monoclonal antibody to give optimal dilution and incubate at 4 °C for 30 min
3. Wash twice in PBS–azide–BSA–AB serum
4. Remove supernatant and add 35 µl of FITC-conjugated goat anti-mouse Ig or F(ab)₂ fragment thereof at optimal dilution
5. Incubate at 4 °C for 30 min
6. Wash twice in PBS–azide–BSA–AB serum
7. Remove supernatant, add one drop of PBS/glycerol (v/v) and mount as for the direct method.

Set up a control replacing the monoclonal antibody by non-immune mouse serum.

PBS–azide–BSA–AB serum pH 7·4	
Phosphate-buffered saline (PBS)	100 ml
Sodium azide	0·2 g
Bovine serum albumin (BSA)	0·2 g
Pooled human AB serum (to block FC receptors on mononuclear cells)	2 ml

II. INTRACELLULAR AND SURFACE MARKERS ON CYTOCENTRIFUGE PREPARATIONS

A. Immunofluorescence (Direct or Indirect) with Polyclonal Rabbit Antibodies

The direct method is particularly used for the detection of cytoplasmic immunoglobulin in lymphoid cells, and the indirect to assess the enzyme TdT. The slides can be stored at room temperature for up to one week or kept at −20 °C in foil paper. Warm to room temperature before unwrapping. All antibodies are optimally diluted in PBS.

1. Mark the cellular area with a diamond pencil
2. Fix in methanol at 4 °C for 15 min
3. Wash in PBS for 10 min in a jar with a magnetic stirrer
4. Dry around the area of cells and add 10 μl of the fluorescein-labelled rabbit anti-human Ig (heavy or light chain) for the direct method or 10 μl of rabbit anti-TdT for the indirect method
5. Incubate in a moist chamber for 30 min at room temperature
6. Wash in PBS for 15 min
7.
   a. *Direct method:* wipe off excess PBS and mount with glycerol/PBS (v/v), cover and seal.
   b. *Indirect method:* add 10 μl of FITC-conjugated anti-rabbit Ig, incubate for 30 min in a moist chamber, wash for 15 min in PBS and mount as for the direct method.

For the indirect method, a control must be carried out omitting the first antibody, rabbit anti-TdT.

B. Immunoperoxidase Stain for Intracellular and Membrane Antigens

This method is commonly applied for the detection of cytoplasmic immunoglobulin and TdT. It also permits the analysis of surface antigens by using monoclonal antibodies; it should be noted that not all the reagents are suitable for this purpose and it is recommended that a simultaneous assessment on cell suspension and cytocentrifuge slides is carried out for each reagent in order to standardize its use.

1. Fix in acetone for 10 min at room temperature and another 10 min in chloroform
2. Air dry and wash in PBS for 15 min (jar with magnetic stirrer)
3. Wipe off excess of PBS, add 10 μl of rabbit anti-TdT and incubate for 30 min at room temperature in a moist chamber
4. Wash in PBS for 15 min, wipe off excess of PBS and add 15 μl of peroxidase-conjugated swine anti-rabbit Ig at an appropriate dilution in PBS containing 20 per cent of human AB serum
5. Incubate for 30 min in a moist chamber at room temperature
6. Wash in PBS for 15 min
7. Develop the cytochemical reaction for 5 min using 30 mg of diaminobenzidine tetrahydrochloride (DAB) in 50 ml of PBS and 20 μl of hydrogen peroxide (30 vol.)
8. Wash in PBS and distilled water and counterstain with haematoxylin.

Set up a control leaving out the first antibody or replacing it by an inappropriate rabbit Ig.

The same method can be applied to detect surface antigens by using a monoclonal antibody and a peroxidase-conjugated rabbit anti-mouse Ig as a second layer. The reaction can be enhanced by using a third layer consisting of peroxidase-conjugated swine anti-rabbit Ig.

III. IMMUNOGOLD METHOD AT ELECTRON MICROSCOPICAL LEVEL

1. Resuspend mononuclear cells (5×10^6) in 200 µl of PBS–azide–BSA–AB serum buffer (*see below*)
2. Add the monoclonal antibody at the same dilution as used for immunofluorescence and incubate for 30 min at room temperature
3. Wash three times in PBS–azide–BSA–AB buffer as below
4. Resuspend cells in 50 µl of this buffer and add 50 µl of optimally diluted goat anti-mouse IgG conjugated to colloidal gold particles, 20–40 nm diameter
5. Incubate for 60 min at room temperature
6. Wash three times in the above buffer
7. Fix the cells in 3 per cent glutaraldehyde in 0·1 M PBS pH 7·4, for 30 min at room temperature
8. Continue as for standard electron microscopy.

PBS–azide–BSA–AB serum buffer	
Phosphate-buffered saline	100 ml
Sodium azide	0·2 g
Bovine serum albumin	1 g
AB serum	2 ml
Adjust pH to 7·6 with 1 M HCl or 1 M NaOH	

Acid phosphatase and myeloperoxidase reactions can be carried out after the immunological test by applying Barka and Anderson's method[48] for acid phosphatase and Graham and Karnovsky's method[49] for myeloperoxidase. In these instances, the appropriate fixative has to be used. Platelet peroxidase reaction must be performed before the immunological reaction by following Roels' technique.[50]

IV. SOURCES OF REAGENTS

Antibody	Source
HLA-DR (OK1a)	Ortho Diagnostic Systems
CD5 (T1)	
CD3 (T3)	
CD1 (T6)	
CD4 (T4)	
CD8 (T8)	
CD2 (T11)	
OKT 10	
OKM1	

B1	Coulter Clone
B4	
CD10 (J5)	
My7	
My9	

FMC7	Sera-Lab
FMC11, FMC13	
FMC17, FMC33	

J15	Dako

CD7 (3A1/WT1)	Becton Dickinson
Monoclonal anti-human Ig (heavy and light chains)	
Polyclonal (rabbit) anti-human Ig (heavy and light chains)	Behring
Rabbit anti TdT	Supertechs
FITC-conjugated goat anti-mouse Ig	Nordic
FITC-conjugated F(ab)$_2$ fragment of goat anti-mouse Ig	Cappel
FITC-conjugated swine anti-rabbit Ig	Dako
FITC-conjugated F(ab)$_2$ fragment of goat anti-rabbit Ig	Supertechs
Peroxidase-conjugated rabbit anti-mouse Ig	Dako
Peroxidase-conjugated swine anti-rabbit Ig	Dako

Not commercially available: OKT 17, BE-2, PC1, HC2, 3C5, AN1, C17, UCHM1, anti-glycophorin A, Gero

REFERENCES

1. Bollum F. J. Terminal deoxynucleotidyl transferase as a haemopoietic cell marker. *Blood* 1979, **54**, 1203–1215.
2. Jani, P., Verbi W., Greaves M. F., Bevan D. and Bollum F. Terminal deoxynucleotidyl transferase in acute myeloid leukaemia. *Leuk. Res.* 1983, **7**, 17–29.
3. Lanham G. R., Bollum F. J., Williams D. L. and Stass S. A. Simultaneous occurrence of terminal deoxynucleotidyl transferase and myeloperoxidase in individual leukemic blasts. *Blood* 1984, **64**, 318–320.
4. Korsmeyer S. J., Arnold A., Bakhshi A., Ravetch J. V., Siebenlist U., Hieter P. A., Sharrow S. O., Le Bien T. W., Kersey J. H., Poplack D. G., Leder P. and Waldmann T. A. Immunoglobulin gene rearrangement and cell surface antigen expression in acute lymphocytic leukaemia of T cell and B cell precursor origins. *J. Clin. Invest.* 1983, **71**, 301–313.
5. Nadler L. M., Anderson K. C., Marti G., Bates M., Parke E., Daley J. F. and Schlossman S. F. B4, a human B-lymphocyte associated antigen expressed on normal, mitogen-activated and malignant B lymphocytes. *J. Immunol.* 1983, **131**, 244–250.

6. Ritz J., Pesando J. M., Notis-McConarty J., Lazarus H. and Schlossman S. F. A monoclonal antibody to human acute lymphoblastic leukaemia antigen. *Nature* 1980, **283**, 583–585.

7. Haynes B. F., Mann D. L., Hemler M. E., Schroer J. A., Shelhamer J. H., Eisenbarth G. S., Strominger J. L., Thomas Ch. A., Mostowsky H. S. and Fauci A. S. Characterization of a monoclonal antibody that defines an immunoregulatory T cell subset for immunoglobulin synthesis in humans. *Proc. Natl Acad. Sci. USA* 1980, **77**, 2914–2918.

8. Tax W. J. M., Willems H. W., Kibbelaar M. D. A., De Groot J., Capel P. J. A., De Waal R. M. W., Reekers P. and Koene R. A. P. Monoclonal antibodies against human thymocytes and T lymphocytes. In: Peeter S. H., ed., *Protides of the Biological Fluids*. Oxford, Pergamon Press, 1982: Vol. 29, 701–704.

9. Vodinelich L., Tax W., Bai Y., Pegram S., Capel P. and Greaves M. F. A monoclonal antibody (WT1) for detecting leukaemias of T-cell precursors (T-ALL). *Blood* 1983, **62**, 1108–1113.

10. Thomas Y., Rogozinsky L., Irigoyen O. H., Shen H. H., Talle M. A., Goldstein G. and Chess S. Functional analysis of human T-cell subsets defined by monoclonal antibodies. V. Suppressor cells within the activated OKT4+ population belong to a distinct subset. *J. Immunol.* 1982, **128**, 1386–1390.

11. Matutes E., Parreira A., Foa R. and Catovsky D. Monoclonal antibody OKT17 recognise most cases of T-cell malignancy. *Br. J. Haematol.* 1985, in press.

12. Van Wauwe, J., Goossens J., Decock W., Kung P. and Goldstein G. Suppression of human T cell mitogenesis and E rosette formation by the monoclonal antibody OKT11A. *Immunology* 1981, **44**, 865–871.

13. Reinherz E. L., Kung P. C., Goldstein G. and Schlossman S. F. A monoclonal antibody with selective reactivity with functionally mature thymocytes and all peripheral human T cells. *J. Immunol.* 1979, **123**, 1312–1317.

14. Kung P. C., Goldstein G., Reinherz E. L. and Schlossman S. F. Monoclonal antibodies defining distinctive human T-cell surface antigens. *Science* 1979, **206**, 347–349.

15. Reinherz E. L., Kung P. C., Goldstein G., Levey R. H. and Schlossman S. F. Discrete stages of human intrathymic differentiation: analysis of normal thymocytes and leukaemic lymphoblasts of T cell lineage. *Proc. Natl Acad. Sci. USA* 1980, **77**, 1588–1592.

16. Reinherz E. L., Kung P. C., Goldstein G. and Schlossman S. F. Separation of functional subsets of human T cells by a monoclonal antibody. *Proc. Natl Acad. Sci. USA* 1979, **76**, 4061–4065.

17. Reinherz E. L., Kung P. C., Goldstein G. and Schlossman S. F. A monoclonal antibody reactive with the human cytotoxic/suppressor T cell subset previously defined by a heteroantiserum termed TH₂. *J. Immunol.* 1980, **124**, 1301–1307.

18. Catovsky D., San Miguel J. F., Soler J., Matutes E., Melo J. V., Bourikas G. and Haynes B. F. T-cell leukaemias—immunologic and clinical aspects. *J. Exp. Clin. Cancer Res.* 1983, **2**, 229–233.

19. Catovsky D., Melo J. V. and Matutes E. Biological markers in lymphoproliferative disorders. In: Bloomfield C. D., ed., *Chronic and Acute Leukaemias in Adults 3*. The Hague, Martinus Nijhoff Publishers, 1985: 69–112.

20. Berger C. L., Morrison S., Chu A., Patterson J., Estabrook A., Takezaki Sh., Sharon J., Warburton D., Irigoyen O. and Edelson R. L. Diagnosis of cutaneous T cell lymphoma by use of monoclonal antibodies reactive with tumor-associated antigens. *J. Clin. Invest.* 1982, **70**, 1205–1215.

21. Matutes E., Brito Babapulle V. and Catovsky D. Clinical, immunological ultrastructural and cytogenetic studies in black patients with adult T cell leukemia/lymphoma. In: Miwa M., ed., *Retroviruses in Human Lymphoma/Leukemia*. Tokyo, Japan Scientific Society Press, 1985: in press.

22. Stashenko P., Nadler L. M., Hardy R. and Schlossman S. F. Characterization of a human B lymphocyte-specific antigen. *J. Immunol.* 1980, **125**, 1678–1685.

23. Brooks D. A., Beckman I. G. R., Bradley J., McNamara P. J., Thomas M. E. and Zola H. Human lymphocyte markers defined by antibodies derived from somatic cell hybrids. IV. A monoclonal antibody reacting specifically with a subpopulation of human B lymphocytes. *J. Immunol.* 1981, **126**, 1373–1377.

24. Anderson K. C., Bates M. P., Slaughenhoupt B. L., Pinkus G. S., Schlossman S. F. and Nadler L. M. Expression of human B cell associated antigens on leukemias and lymphomas: a model of human B cell differentiation. *Blood* 1984, **63**, 1424–1433.

25. Anderson K. C., Bates M. P., Slaughenhoupt B., Schlossman S. F. and Nadler L. M. A

monoclonal antibody with reactivity restricted to normal and neoplastic plasma cells. *J. Immunol.* 1984, **132**, 3172–3179.

26. Posnett D. N., Chiorazzi N. and Kunkel H. G. Monoclonal antibodies with specificity for hairy cell leukemia cells. *J. Clin. Invest.* 1982, **70**, 254–261.

27. Posnett D. N. and Marboe C. C. Differentiation antigens associated with hairy cell leukemia. *Semin. Oncol.* 1984, **11**, 413–415.

28. Bennett J., Catovsky D., Daniel M. T., Flandrin G., Galton D. A. G., Gralnick H. R. and Sultan C. Proposals for the classification of acute leukaemias. (FAB Cooperative Group). *Br. J. Haematol.* 1976, **33**, 451–458.

29. Breard J., Reinherz E. L., Kung P., Goldstein G. and Schlossman S. A monoclonal antibody reactive with human peripheral blood monocytes. *J. Immunol.* 1980, **127**, 1943–1948.

30. Brooks D. A., Zola H., McNamara P. J., Bradley J., Bradstock K. F., Hancock W. W. and Atkins R. C. Membrane antigens of human cells of the monocyte/macrophage lineage studied with monoclonal antibodies. *Pathology* 1983, **15**, 45–52.

31. Zola H., McNamara P., Thomas M., Smart I. and Bradley J. The preparation and properties of monoclonal antibodies against human granulocyte membrane antigens. *Br. J. Haematol.* 1981, **48**, 481–490.

32. Griffin J. D., Mayer R. J., Weinstein H. J., Rosenthal D. S., Coral F. S., Beveridge R. P. and Schlossman S. F. Surface marker analysis of acute myeloblastic leukemia: Identification of differentiation-associated phenotypes. *Blood* 1983, **62**, 557–563.

33. Griffin J. D., Linch D., Sabbath K., Larcom P. and Schlossman S. F. A monoclonal antibody reactive with normal and leukemic human myeloid progenitor cells. *Leuk. Res.* 1984, **8**, 521–534.

34. Chan L. C., Pegram S. M. and Greaves M. F. Contribution of immunophenotype to the classification and differential diagnosis of acute leukaemia. *Lancet* 1985, **i**, 475–479.

35. Tindle R. W., Nichols R. A. B., Chan L. C., Campana D., Catovsky D. and Birnie G. D. A novel monoclonal antibody BI-3C5 recognises myeloblasts and non-B non-T lymphoblasts in acute leukemias and CGL blast crises, and reacts with immature cells in normal bone marrow. *Leuk. Res.* 1985, **9**, 1–9.

36. Matutes E. and Catovsky D. Ultrastructural analysis of normal and leukaemia cells by the immunogold method and monoclonal antibodies. In: Neth R., Gallo R. C., Spiegelman S. and Stohlman Jr F., eds., *Modern Trends in Human Leukaemia VI. Haematology and Blood Transfusion.* New York, Grune & Stratton Inc., 1985: **29**, 174–179.

37. McMichael A. J., Rust N. A., Pilch J. R., Sochynsky R., Morton J., Mason D. Y., Ruan C., Tabelem G. and Caen J. Monoclonal antibody to human platelet glycoprotein I. I. Immunological studies. *Br. J. Haematol.* 1981, **49**, 501–509.

38. Vainchenker W., Deschamps J. F., Bastin J. M., Guichard J., Titeux M., Breton Gorius J. and McMichael A. J. Two monoclonal antiplatelet antibodies as markers of human megakaryocyte maturation: Immunofluorescent staining and platelet peroxidase detection in megakaryocyte colonies and in *in vivo* cells from normal and leukemic patients. *Blood* 1982, **59**, 514–521.

39. Tetteroo P. A. T., Lansdorp P. M., Leeksma O. C. and Kr. Von dem Borne A. E. G. Monoclonal antibodies against human platelet glycoprotein IIIa. *Br. J. Haematol.* 1983, **55**, 509–522.

40. Breton Gorius J., Reyes F., Vernant J. P., Tulliez M. and Dreyfus B. The blast crisis of chronic granulocytic leukaemia: Megakaryoblastic nature of cells revealed by the presence of platelet peroxidase—a cytochemical ultrastructural study. *Br. J. Haematol.* 1978, **39**, 295–303.

41. Bain B., Catovsky D., O'Brien M., Prentice H. G., Lawlor E., Kumaran T. O., McCann S., Matutes E. and Galton D. A. G. Megakaryoblastic leukaemia presenting as acute myelofibrosis. A study of four cases with the platelet peroxidase reaction. *Blood* 1981, **58**, 206–213.

42. San Miguel J. F., Tavares de Castro J., Matutes E., Rodriguez B., Polli N., Zola H., McMichael A. J., Bollum F. J., Thomson D. S., Goldman J. M. and Catovsky D. Characterisation of blast cells in chronic granulocytic leukaemia in transformation, acute myelofibrosis and undifferentiated leukaemia. II. Studies with monoclonal antibodies and terminal transferase. *Br. J. Haematol.* 1985, **59**, 297–309.

43. Bettelheim P., Lutz D., Majdic O., Paietta E., Haas O., Linkesch W., Neumann E., Lechner K. and Knapp W. Cell lineage heterogeneity in blast crisis of chronic myeloid leukaemia. *Br. J. Haematol.* 1985, **59**, 395–409.

44. Edwards P. A. W. Monoclonal antibodies that bind to the human erythrocyte membrane glycoproteins glycophorin A and Band 3. *Biochem. Soc. Trans.* 1980, **8**, 334.
45. Greaves M. F., Sieff C. and Edwards P. A. W. Monoclonal antiglycophorin as a probe for erythroleukemias. *Blood* 1983, **61**, 645–651.
46. Robinson J., Sieff C., Delia D., Edwards P. and Greaves M. F. Expression of cell surface HLA-Dr, HLA-ABC and glycophorin during erythroid differentiation. *Nature* 1981, **289**, 68–71.
47. Daniels G. L., Banting G. and Goodfellow P. A monoclonal antibody related to the human blood group Gerbich. *J. Immunogenet.* 1983, **10**, 103–105.
48. Barka T. and Anderson P. J. Histochemical methods for acid phosphatase using hexazonium pararosanilin as coupler. *J. Histochem. Cytochem.* 1962, **10**, 742–753.
49. Graham R. C. and Karnovsky M. J. The early stages of absorption of injected horseradish peroxidase in the proximal tubules of mouse kidney: ultrastructural cytochemistry by a new technique. *J. Histochem. Cytochem.* 1966, **14**, 291–302.
50. Roels F., Wisse E., De Brest B. and Van der Meulen J. Cytochemical discrimination between catalases and peroxidases using diaminobenzidine. *Histochemistry* 1975, **41**, 281–311.
51. Committee on Human Leukocyte Differentiation Antigens: IUIS-WHO Nomenclature Subcommittee. Differentiation human leukocyte antigens: a proposed nomenclature. *Immunology Today* 1984, **5**, 158–159.
52. Beckman I. G. R., Bradley J., Brooks D. A., Kupa A., McNamara P. J., Thomas M. E. and Zola H. Human lymphocyte markers defined by antibodies derived from somatic cell hybrids. II A hybridoma secreting antibody against an antigen expressed by human B and null lymphocytes. *Clin. Exp. Immunol.* 1980, **40**, 583–601.
53. Linch D. C., Allen C., Beverley P. C. L., Bynoe A. G., Scott C. S. and Hogg N. Monoclonal antibodies differentiating between monocytic and nonmonocytic variants of AML. *Blood* 1984, **63**, 566–573.

36 Immunocytochemistry in Dermatology

A. C. Chu

The last decade has witnessed enormous advances in our understanding of the immunological mechanisms underlying certain disease processes. The combined techniques of histology, electron microscopy and immunocytochemistry now enable us not only to diagnose various diseases with greater accuracy, but have also given us some insight into the underlying pathophysiology of these diseases which should ultimately lead to a more rational approach to therapy.

In dermatology, immunocytochemistry is of enormous diagnostic value in two fields — in the bullous dermatoses and in the cutaneous lymphoid infiltrates. In the bullous dermatoses, immunocytochemistry is still used routinely to augment clinical and histological diagnosis of the immunologically based bullous diseases. Such techniques have also demonstrated the possible mechanisms of action of tissue damage in these conditions. More recently, the greater understanding of the structure of the dermo-epidermal junction has enabled us to use immunocytochemical techniques to augment the diagnosis of even the non-immunologically based bullous dermatoses.

In the cutaneous lymphoid infiltrates the advent of monoclonal antibody production has made it possible to differentiate B- from T-cell infiltrates and benign from malignant lymphoid infiltrates. In the broader field of cellular biology, immunocytochemistry now enables us to identify a variety of important cell types in the skin, e.g. Langerhans cells and Merkel cells, which can be studied with greater ease. Monoclonal antibodies are also proving valuable in detecting tumour markers within populations of cells in the skin.

In this chapter, I will briefly describe the immunocytochemical characteristics of the common bullous dermatoses and the application of immunocytochemical techniques in problem solving in the bullous dermatoses. The use of cell markers

618

in dermatological diagnosis will also be discussed with the application of immunocytochemistry to the study of cellular interaction in the skin (*see* Appendix).

1. THE BASEMENT MEMBRANE ZONE

In the subepidermal bullous diseases, the attack on the skin is usually at the basement membrane zone. A detailed knowledge of the structure and composition of the area is thus important in understanding the pathogenic mechanisms of the bullous diseases.

The basement membrane zone is a complex structure involving the basal keratinocyte, the basement membrane itself and the upper dermis.

Ultrastructurally, it can be broadly subdivided into four parts (*Fig.* 36.1):

SCHEMATIC REPRESENTATION OF THE DERMO-EPIDERMAL JUNCTION

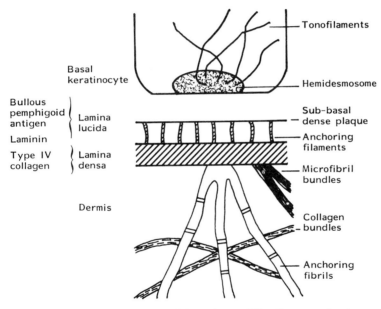

Fig. 36.1. Localization of pemphigoid, laminin and type IV collagen at the dermo-epidermal junction.

a. Basal keratinocyte plasma membrane
b. Lamina lucida
c. Lamina densa
d. Sub-basal lamina fibrous elements.

The basal keratinocyte plasma membrane is a trilaminate structure about 8-nm thick. On the cytoplasmic side are the hemidesmosomes which are thickened areas of the plasma membrane 20–40 nm thick. These act as anchoring points for the tonofilaments of the basal keratinocytes. The basal keratinocyte plasma membrane is bound on the inferior side by the lamina lucida. This is a 30-nm band which appears in electron microscopy as an amorphous pale zone. Adjacent to the hemidesmosomes and within the lamina lucida, are the electron-dense sub-basal dense plaques, within which anchoring filaments connect the plasma membrane to the lamina densa. The lamina lucida is at least in part a product of the basal keratinocyte and contains the bullous pemphigoid antigen. This is a protein of molecular weight 31 000[1] also found in the saliva and urine of normal individuals. Keratinocytes have been shown to produce the bullous pemphigoid antigen in tissue culture and, if basal keratinocytes are separated by trypsinization of epidermal sheets, the bullous pemphigoid antigen can be identified as a polar cap on the keratinocytes by the use of bullous pemphigoid antiserum. The lamina lucida is also the site of the fibrous protein laminin. Laminin is a major connective tissue protein and is widely distributed in basement membranes. In the basement membrane zone of the skin, it is restricted to the lamina lucida and is an antigenic marker of this area.[2] The basal lamina is a 30–50-nm electron-dense fibrillar and amorphous band. It is a continuous layer but reduplication may be observed due to basal cell division and upwards migration. The basal lamina contains type IV collagen, which can be used as an antigenic marker of this layer.[3]

The sub-basal lamina fibrous elements are of three types — collagen fibres, anchoring fibrils and microfibril bundles. The anchoring fibrils and microfibril bundles are attached to the basal lamina and pass through to the deeper dermis. The anchoring fibrils are composed of collagen and have a unique periodicity compared to other human collagens. The microfibril bundles are thought to be composed of elastic tissue.

2. SUBEPIDERMAL BULLOUS DISORDERS

In the different autoimmune bullous dermatoses, the target sites are different. The use of electron microscopy with immunoperoxidase techniques has done much to elucidate the events in these diseases. Also, suction blister techniques, in which blisters are raised in the skin by the application of a vaccum cup, have helped to differentiate the antigens involved in certain of the bullous diseases.

The main characteristics of the autoimmune bullous dermatoses are given in *Tables* 36.1 and 36.2. For a further account, readers are advised to read Chu et al.[4] and Chu.[5]

In bullous pemphigoid, the immunoreactants are deposited in the lamina lucida (*Fig.* 36.2). The bullous pemphigoid antigen has been identified by Diaz et al.[1] and has been shown to block bullous pemphigoid antibody binding to the basement membrane zone.

In cicatricial pemphigoid, the immunocytochemical findings are the same as those seen in bullous pemphigoid, except that the incidence of circulating anti-basement membrane zone IgG is much lower — 10–36 per cent in cicatricial pemphigoid as opposed to 60–70 per cent in bullous pemphigoid. The two diseases are, however, clinically very different, as in cicatricial pemphigoid the skin lesions recur at the same sites in the skin, and the disease has a predilection for oral and ocular mucous membranes. Recently, Fine et al.[6] have demonstrated that, although in both bullous pemphigoid and cicatricial pemphigoid, the immuno-reactants are deposited in the lamina lucida, the antigens involved are

*Table 36.1. **Immunocytochemistry in skin diseases***

Disease	Direct IMF	Indirect IMF	Pathogenesis
Bullous pemphigoid	Linear C3 and IgG at BMZ Occasionally IgM, IgA properdin, factor B (*Fig.* 36.2)	60–70% have circulating IgG against BMZ	Complement mediated
Cicatricial pemphigoid	Linear IgG and C3 at BMZ	10–36% have circulating IgG against BMZ	Complement mediated
Dermatitis herpetiformis	80% granular papillary IgA 12–18% linear IgA at BMZ (*Fig.* 36.3)	2% of granular papillary IgA and 30% of linear IgA have circulating IgA against BMZ	? complement mediated via alternative pathway
Chronic bullous disease of childhood	Linear IgA at BMZ	60% have circulating IgA against BMZ	? complement mediated via alternative pathway
Herpes gestationis	Linear C3 at BMZ 25–30% IgG at BMZ Occasionally properdin, factor B, Clq, and C4	20% have circulating IgG against BMZ 90% HG factor — complement-fixing IgG against BMZ	Complement mediated
Epidermolysis bullosa acquista	Granular IgG at BMZ	Circulating IgG against BMZ	Complement mediated
Lupus erythematosus	Granular IgM at BMZ 90% lesional skin, 0% in uninvolved skin in discoid, 60% in uninvolved skin in systemic	Antinuclear antibody	Unknown
Pemphigus	IgG in the intercellular spaces of the epidermis Occasionally IgM, Clq, C4 and C3 are found (*Fig.* 36.5)	Circulating IgG against intercellular substances	Enzyme activation

IMF, immunofluorescence.
BMZ, basement membrane zone.
HG, herpes gestationis.

anatomically distinct. In sections of blisters in bullous pemphigoid, the immunoreactants are deposited in the roof of the blister, while in cicatricial pemphigoid they are deposited in the floor of the blister. These findings suggest that different antigens are involved in the two diseases, which may explain the different clinical manifestations of the diseases.

In herpes gestationis, linear IgA dermatoses and chronic bullous diseases of childhood, the immunoreactants are found in the lamina lucida. No formal studies have as yet looked at the specific antigens in these diseases, but they are probably different from the bullous pemphigoid antigens.

In linear IgA dermatoses, the IgA may occasionally be deposited in association with the anchoring fibrils, as well as in the lamina lucida.[7]

Table 36.2. T-cell ontogeny

Stage	Antibody reactivity (in order of appearance)	
1 Prothymocyte	OKT 10 OKT 9/5E9	
2 Common thymocyte	OKT 10 OKT 11 OKT 6 or NA1/34 OKT 4 or Leu-3a OKT 5 OKT 8 or Leu-2a 3A1	
3 Mature thymocyte	*Helper* OKT 11 OKT 10 OKT 1 or Leu-1 OKT 4 or Leu-3a 3A1	*Suppressor* OKT 11 OKT 1 or Leu-1 OKT 3 OKT 5 OKT 8 or Leu-2a
Peripheral circulation	OKT 11 OKT 1 or Leu-1 OKT 3 OKT 4 or Leu-3a 3A1 (70%)	OKT 11 OKT 1 or Leu-1 OKT 3 OKT 5 OKT 8 or Leu-2a 3A1

In the granular papillary IgA dermatitis herpetiformis, the IgA is deposited in aggregations associated with microfibrillar bundles and with microfibrillar components of the elastic tissue (*Fig.* 36.3).

3. INTRA-EPIDERMAL BULLOUS DISORDERS

The major immunocytologically based intra-epidermal bullous disease is pemphigus and its four clinical variants — pemphigus vulgaris and vegetans in which the cleft is suprabasilar and pemphigus foliaceus and erythematosus in which the cleft is subcorneal (*Fig.* 36.4). Pemphigus is an autoimmune disease in which the body produces antibodies against the intercellular substance of the epidermis. This results in the breakdown of intercellular linkages and detachment of individual keratinocytes which round up and lie singly or in groups within the blister cavity — a process known as acantholysis.

The characteristic immunocytochemical finding in pemphigus is the presence of IgG in the intercellular spaces in the epidermis (*Table* 36.1, *Fig.* 36.5). The majority of patients also have demonstrable circulating IgG against the intercellular substance.

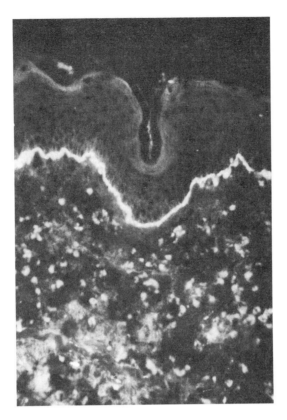

Fig. 36.2. Bullous pemphigoid: linear deposition of IgG in the basement membrane zone.

4. APPLICATION OF IMMUNOHISTOCHEMISTRY IN THE INVESTIGATION OF THE PATHOGENIC MECHANISMS OF THE BULLOUS DISORDERS

Immunocytochemistry can be used in two ways to investigate the way in which antibody deposition in the skin leads to tissue damage and blister formation. The first is by morphologically identifying other immunoreactions in the skin which will give clues as to the way in which change was mediated. The second is to demonstrate the functional activity of circulating antibodies by the indirect C3 test.

Antibody deposition in tissue generally causes damage by complement fixation. Complement can be activated by both the classical and the alternative pathways, both of which involve a different initial sequence of events and precursor proteins (*Fig*. 36.6). Using antibodies against C1, C4 and C2 these factors can be detected in the skin and, if present, would suggest that the classical pathway of complement fixation was active. Similarly, antibodies against properdin and factor B can be used and, if present, would implicate the alternative pathway.

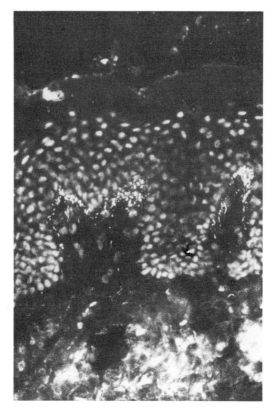

Fig. 36.3. Dermatitis herpetiformis: granular deposit of IgA in the dermal papillae.

PEMPHIGUS

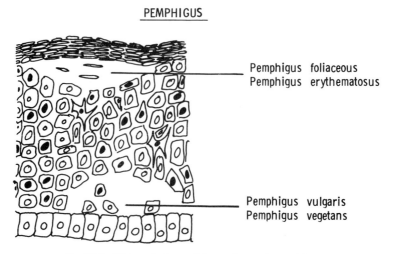

Pemphigus foliaceous
Pemphigus erythematosus

Pemphigus vulgaris
Pemphigus vegetans

Fig. 36.4. Histopathology of the variants of pemphigus.

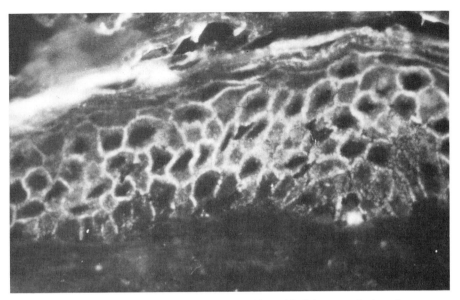

Fig. 36.5. Pemphigus vulgaris: intercellular deposition of IgG between the keratinocytes.

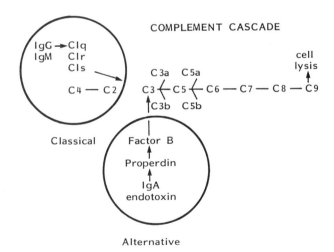

Fig. 36.6. Complement fixation by the classical and alternative pathways.

The indirect C3 technique (*Fig.* 36.7) is of enormous value in two ways — as a method of enhancement to detect small quantities of circulating complement-fixing antibody, and to investigate circulating autoantibody to see whether it is complement fixing or not. The former application has been used very successfully in the investigation of bullous pemphigoid. Using standard indirect immunofluorescence techniques, Hodge et al.[8] showed that 58 per cent of patients had

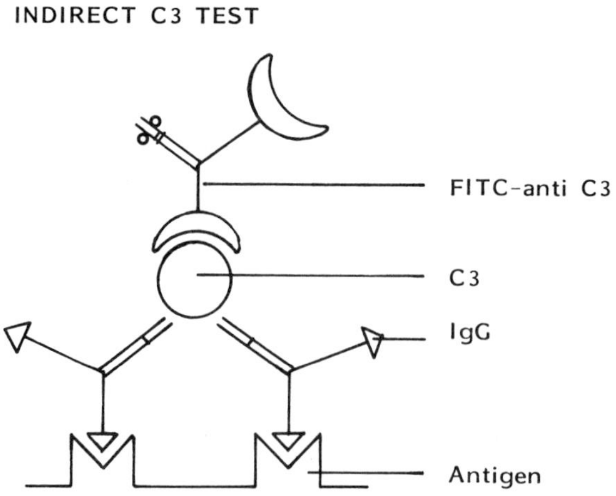

INDIRECT C3 TEST

FITC-anti C3

C3

IgG

Antigen

Fig. 36.7. Indirect C3 test.

circulating IgG against the basement membrane zone, but using the indirect C3 test, 78 per cent of the same patients had demonstrable circulating IgG.

In pemphigus, the characteristic immunocytochemical finding is of the deposition of IgG in the intercellular spaces in the epidermis. In the early stages C1q, C4 and C3 have been detected, which suggested that the acantholysis seen in this disease was mediated by complement fixation.[9]

Using the indirect C3 test, however, the circulating pemphigus antibody has been shown to be non-complement fixing *in vitro*. In 1976, Schiltz and Michel[10] demonstrated that the pemphigus antibody could induce acantholysis in organ culture in the absence of complement, and more recently, Singer et al.[11] have shown that the acantholysis can be inhibited by soya-bean trypsin inhibitor, which inhibits serine esterase, and also by β_2 macroglobulin which is a broad spectrum protein inhibitor. This suggests that the cellular change in pemphigus is mediated by enzyme activation by the antibody, which dissolves the intercellular cement and splits the desmosomes.

5. APPLICATION OF IMMUNOFLUORESCENCE TO THE INVESTIGATION AND DIAGNOSIS OF NON-IMMUNOLOGICAL BULLOUS DISORDERS

As we have already discussed, the different parts of the dermo-epidermal junction have specific antigenic markers. Antibodies against type IV collagen and laminin and high-titre bullous pemphigoid antibody can be used to localize the cleft in various non-immunological bullous dermatoses.

In the bullous diseases encompassed by the term 'epidermolysis bullosa', there are a number of clinically, genetically and prognostically different entities, united by the characteristic feature of blister formation at sites of mild trauma. The fundamental defects appear to be structural abnormalities in the dermo-epidermal junction area and, by electron microscopy, the different varieties can be identified by the structural defect present. Electron microscopy, however, is not a

EPIDERMOLYSIS BULLOSA

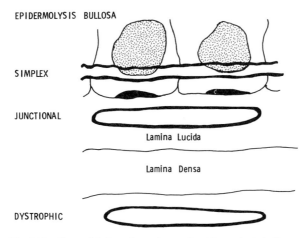

Fig. 36.8. Sites of defects in variants of epidermolysis bullosa.

technique open to all clinicians and, in situations where it is difficult or impossible to examine the tissue by electron microscopy, immunohistochemistry may provide important guides to the differentiation of the different diseases.

In epidermolysis bullosa simplex, the defect is in the basal keratinocytes and the cleft occurs through the cells (*Fig*. 36.8). This form of the disease has a good prognosis and low morbidity. In the junctional forms of epidermolysis bullosa, the defect is within the lamina lucida and clefts occur at this level. In the fetus and newborn, the junctional form of epidermolysis bullosa may be lethal, while in adults the disease has a relatively good prognosis but may be associated with severe morbidity. In the dystrophic forms of epidermolysis bullosa, the defect occurs below the basal lamina and the split is subepidermal. These forms are associated with high morbidity and are often associated with gut involvement and other disease processes.

The differentiation of these forms of epidermolysis bullosa is thus important, prognostically and also therapeutically. Using immunofluorescence techniques, the three major forms of epidermolysis bullosa can be differentiated.

In the simplex form, bullous pemphigoid antigen, type IV collagen and laminin will be found in the floor of the blister. In the junctional forms, bullous pemphigoid antigen and laminin will be found in the roof of the blister and, on occasion, in the floor of the blister and type IV collagen will be localized in the floor (*Plates* 40, 41). In the dystrophic forms, bullous pemphigoid antigen, laminin and type IV collagen will be present in the roof of the blister. This relatively easy technique may, therefore, quickly differentiate these major variants.

6. IDENTIFICATION OF T-CELLS IN CUTANEOUS TISSUE SECTIONS

6.1. Introduction

T-cells hold an important place in cutaneous pathology, and interact with the epidermis in complex and varied ways. Evidence strongly suggests that the skin is a physiological site for T-cell migration and represents a T-cell organ of the body with many similarities to the thymus.

Morphologically, the skin and thymus share certain features:

a. Both possess a keratinizing epithelium — the Hassall's corpuscle in the thymus and the epidermis in the skin
b. Both produce immunohistochemically identical keratin
c. Both possess an immigrant population of specialized macrophages — the Langerhans cells
d. A possible genetic link is seen in the nude mouse which lacks both thymus and an important skin appendage, hair.

Functionally, the skin has been shown to exert a variety of changes in selected T-cell subpopulations:

a. Human and murine epidermal cells have been shown to induce the enzyme terminal deoxyribonucleotidyl transferase in mature T-cells *in vitro*
b. Epidermal cells have been shown to induce the thymocyte antigen recognized by the mouse monoclonal antibody OKT 6, in malignant helper T-cells
c. Murine epidermal cells *in vitro* produce an interleukin-I-like substance
d. A substance immunohistochemically identical to thymopoietin has been demonstrated in basal keratinocytes of the epidermis.

In a large number of skin disorders in which lymphocytic infiltrates are present, the predominant cell type has been shown by immunohistochemistry to be the T-cell. These diseases include lichen planus, discoid lupus erythematosus, psoriasis, solar keratoses and chronic eczema. Of particular interest is the finding that the vast majority of lymphomas originating or predominantly affecting the skin are of T-cell origin.

In the ensuing section, we will describe the use of both heterologous and monoclonal antibodies in the investigation of an important group of skin malignancies, the cutaneous T-cell lymphomas.

The cutaneous T-cell lymphomas are neoplasms of T-lymphocytes which originate in or predominantly involve the skin. With the advent of immunohistochemical identification of T-cells, the incidence of this group of tumours has been shown to be much greater than previously considered, and probably approximates to that of Hodgkin's disease.[12]

The disease usually starts as a non-specific skin eruption which may simulate eczema, psoriasis or one of the other superficial dermatoses. The disease evolves through a recognized sequence of stages, the skin becoming raised into plaques and then tumid. In the early stages, histology may be non-specific and multiple biopsies are often required before the diagnosis is confirmed. Similarly, in the tumour stage, the histology may be dominated by the presence of tumour cells which may be difficult to differentiate from B-cell or histiocytic tumours.

Immunohistochemistry using antisera to T-cells thus plays an important role in the diagnosis of this condition and differentiation of cutaneous T-cell lymphomas from tumours of other cell types.

6.2 Antisera to T-cells

Both polyclonal and monoclonal antibodies against T-cells are now commercially available. Both are useful tools in the investigation of lymphoid infiltrates. As polyclonal antibodies tend to be raised in rabbits while monoclonal antibodies are normally murine, they can be successfully used together in double-labelling techniques — using a fluorescein-conjugated anti-rabbit and a rhodamine-conjugated anti-mouse immunoglobulin as the second layer in an indirect technique.

6.2.1. Polyclonal Antisera to T-cells (Anti-human T-lymphocyte Antigen)

These antisera are commercially available (see Appendix) and have been successfully used in the investigation of cutaneous cellular infiltrates. They are produced in rabbits using thymocytes as the antigen and the resulting antiserum is absorbed out on various human tissues, adherent cells and B-lymphoblastoid lines. The specificity of the antisera is controlled by microlymphocytotoxicity tests and inhibition of E-rosette formation.

Heterologous antibodies can be used in immunofluorescence or immunoperoxidase techniques and have helped to establish the T-cell nature of the infiltrate in cutaneous T-cell lymphoma.[13] They are particularly valuable in double-labelling techniques using polyclonal antisera to human T-lymphocytes to identify T-cells, and monoclonal antibodies to identify T-cell subpopulations simultaneously.

6.2.2. Monoclonal Antibodies to T-cells

The complexity of T-cell ontogeny is now becoming more and more evident with the increasing number of monoclonal antibodies now available. Commercial antibodies (see Appendix) have been successfully used to identify different subpopulations of T-cells in cell suspensions and tissue sections (Table 36.2).

OKT 9, 5E9 These monoclonal antibodies[14] are not T-cell specific but identify the transferrin receptor[15] which is present in all activated human cells. In T-cell ontogeny, bone marrow precursor cells are OKT 9-negative, but acquire this antigen as early thymocytes.

OKT 10 This monoclonal antibody[16] reacts with more than 95 per cent of thymocytes, but is generally lost by T-cells after their export to the peripheral circulation; less than 5 per cent of the circulating T-cells react with OKT 10. However, 12 per cent of circulating non-T-cell mononuclear cells and some bone marrow cells including B-cell, monocyte and myeloid precursors also express this antigen.

OKT 1, LEU-1, OKT 3, OKT 11 These monoclonal antibodies[16,17] react with pan T-cell antigens. The antibodies all react with slightly different populations of cells and different antigens. OKT 11 reacts with the E-rosette receptor of T-cells and most nearly correlates with previous methods of enumerating T-cells by E-rosette formation.

OKT 4, LEU-3a These antibodies[16,18] define the helper/inducer subpopulation of T-cells. These antigens are acquired by the common thymocyte and are lost by suppressor T-cells on maturation to the mature thymocyte. Functionally, most cells reactive with these antibodies have been found to be helper T-cells, but the OKT 4/Leu-3a-positive subpopulation does contain some cells which show functional suppression.

OKT 8, OKT 5, LEU-2a These monoclonal antibodies[18,19] define the cytotoxic/suppressor population of T-cells. In common with the helper T-cell-specific monoclonal antibodies, these antibodies are acquired by the common thymocyte and are lost, this time by the helper T-cells on maturation to the mature thymocyte.

OKT 6, NA1/34 These antibodies[19,20] react with different epitopes on the cortical thymocyte antigen HTA-1. This is a marker of the common or cortical thymocyte. In adults, no OKT 6, NA1/34 reactive circulating cells are present, but in neonates, a significant population of circulating OKT 6-reactive T-cells are present.[21]

OKT 6 and NA1/34 are not T-cell specific and have now been shown to react with a population of specialized dendritic mononuclear cells in the skin — the Langerhans cells.[22,23]

3A1 This monoclonal antibody[24] defines an antigen present on most thymocytes and most circulating T-cells. Thirty per cent of OKT 4 positive circulating T-cells are, however, 3A1 negative. This antibody is of particular relevance to the cutaneous lymphomas, as circulating Sézary cells have been shown to be 3A1-negative.

BE1 AND BE2 These monoclonal antibodies[25] were generated using leukaemic T-cells from a patient with cutaneous T-cell lymphoma (CTCL) as the immunogen. They react with different tumour-associated antigens present on the surface of CTCL cells. They are not specific for these cells as they do react with some long-term T-cell lines, EB virus-transformed B-cells and some B-cell chronic lymphocytic leukaemia. They have, however, proved very useful in differentiating CTCL from benign dermatoses by examination of both blood and tissue sections. BE2 seems to be more specific for circularing Sézary cells, while BE1 reacts with tumour cells at all stages of CTCL.[26]

OKT 1, OKT 3, OKT 6, OKT 8 and OKT 10 react with antigens which are present in high density on the T-cell cytoplasmic membranes. They are thus easily used in

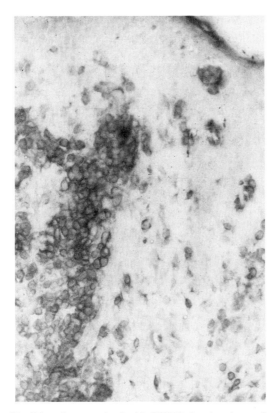

Fig. 36.9. Cutaneous T-cell lymphoma stained with OKT 3 showing the predominant T-cell nature of the infiltrate.

immunohistochemistry on tissue sections using either direct or indirect immuno-peroxidase reactions. In the indirect technique we generally employ a Dako (Dako, Denmark) peroxidase-conjugated anti-mouse antibody. OKT 4, however, is reactive to an antigen present at a much lower density and is only usable in indirect or triple-layer techniques. For use in the direct immunoperoxidase reaction, the antibodies are first conjugated to horseradish peroxidase, using a modification of the technique of Nakane and Kawaoi.[27] The direct technique has the advantage of high specificity and low background, although it is not as sensitive as indirect techniques.

T-cells are recognized by the presence of a ring of dark brown reaction product on the surface (*Fig*. 36.9). The intensity of staining varies from cell to cell, reflecting the density of the antigen on the cell's surface.

6.3. Application of Monoclonal T-cell Antibodies to the Study of Cutaneous T-cell Lymphoma

A recent study of 91 patients with cutaneous T-cell lymphoma using monoclonal antibodies showed three main patterns:[29]

a. 64 per cent of patients showed a fairly homogeneous distribution of the different T-cell subpopulations, of which 60 per cent were OKT 1+ve, 54 per cent OKT 4+ve (helper) and 8 per cent were OKT 8+ve (suppressor). Many patients showed an increase of OKT 8+ve cells up to 60 per cent at the margins of the infiltrate

b. 21 per cent of patients showed selective loss of OKT 1 antigen

c. 15 per cent of patients showed large numbers of OKT 8+ve cells (50–90 per cent), but the percentages of OKT 1+ve and OKT 4+ve cells were within the ranges seen in group 1, suggesting the presence of an immature population of T-cells which were OKT 4+ve and OKT 8+ve.

In the epidermis the majority of lymphoid cells were OKT 1+ve and OKT 4+ve, but in 26 per cent of patients OKT 8+ve cells were also identified.

The results of this study suggest that in the typical plaque stage of cutaneous T-cell lymphoma, this technique is only of limited diagnostic value, as similar results may be found in benign dermatoses. Helpful features in differentiating cutaneous T-cell lymphoma from benign disorders using these techniques are:

a. The homogeneous distribution of T-cell subpopulations in the infiltrate; in benign lymphocytic infiltrates, the lymphoid cells are often arranged in a nodular fashion

b. Selective loss of the OKT 1 antigen

c. Presence of immature cells which are OKT 4+ve and OKT 8+ve.

In tumid lesions, where a monomorphous infiltrate of tumour cells is present, these techniques are of great value in identifying the cells as T-cells.

The findings of this study have been supported by a variety of other more limited studies.[29]

More recently, studies using BE1 and BE2 have confirmed their usefulness in the tissue diagnosis of CTCL. In a recent study,[30] 20 patients with CTCL and 20 patients with benign dermatoses (pityriasis lichenoides chronica, digitate dermatosis and lymphomatoid papulosis) and 3 patients with cutaneous B-cell lymphoma were examined using BE1 and BE2. Nineteen of 20 patients with CTCL reacted with either BE1 or BE2 (*Figs.* 36.10 and 36.11), 1 of 3 patients with B-cell lymphoma reacted with BE2 and only in 1 patient with a benign dermatosis, in whom there was a clinical suspicion of transformation to CTCL from lymphomatoid papulosis, were BE2 reactive cells found in the cellular infiltrate.

Although not totally specific for CTCL, BE1 and BE2 are of obvious importance in the tissue diagnosis of this disease.

7. OKT 6, A LANGERHANS CELL MARKER

OKT 6, as has previously been described, is a mouse monoclonal antibody raised against human thymocytes and reactive with the cortical immuno-incompetent thymocytes. A chance finding, however, has made this antibody one of the most important reagents in the investigation of cutaneous disease. The antibody was found to react with dendritic Ia+ve cells in the epidermis,[22] and later studies

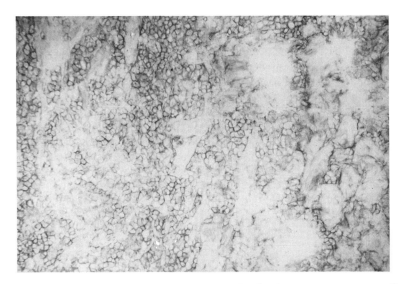

Fig. 36.10. Cutaneous T-cell lymphoma stained with BE1 showing numerous tumour cells in the dermal infiltrate.

Fig. 36.11. Cutaneous T-cell lymphoma stained with BE2 showing scattered tumour cells in both dermis and epidermis.

confirmed at the ultrastructural level that these cells were the Langerhans cells and their putative precursor cells, the indeterminate cells.

7.1. Langerhans Cell

The Langerhans cell is a specialized macrophage which is an immigrant resident in the epidermis. It was at first recognized by Paul Langerhans as a clear dendritic cell which was aureophilic. Initially it was considered to be an effete melanocyte, but is now recognized as being part of the monocyte/macrophage series of cells and as such is an important element of the cutaneous immune system. The Langerhans cell appears to be central in the development of cutaneous allergic sensitization and is the target cell in cutaneous allergenic graft rejection.

The Langerhans cell bears a variety of phenotypic characteristics which help to differentiate it from other cells in the skin. The only totally specific feature of the cell, however, is the Birbeck granule — an intracytoplasmic organelle of unknown function (*Table* 36.3).

Table 36.3. Markers of Langerhans cells

1. Aureophilic
2. Surface ATPase
3. Fc IgG receptors
4. C3 receptors
5. Ia antigen
6. OKT 6 antigen
7. S-100 antigen
8. OKT 4 antigen
9. Birbeck granule

In practical terms, OKT 6 is the most useful marker for Langerhans cells, as it can be used in cell suspension and tissue sections, and the only other cell that reacts with this monoclonal antibody is the intrathymic thymocyte.

Immunocytochemical studies have confirmed that OKT 6 is a marker of normal Langerhans cells[31] and of histiocytosis X cells which are malignant Langerhans cells. Electron microscopical immunocytochemistry has demonstrated that morphologically recognizable Langerhans cells, with Birbeck granules, and also the putative precursor cell of Langerhans cells, the indeterminate cell of the skin, react with OKT 6.[23]

Immunocytochemical studies have demonstrated the usefulness of detecting Langerhans cells in cutaneous infiltrates. Langerhans cells specifically infiltrate skin in which T-cells are present. They are thus a good marker for malignant T-cell proliferations and can be used to differentiate T-cell from B-cell tumours from which Langerhans cells are absent[32] (*Fig.* 36.12). Immunocytochemical studies have also helped in elucidating the chronological changes in certain cutaneous immunological processes, such as contact allergic dermatitis and the role of Langerhans cells in these.[33]

Acknowledgements

I wish to thank Mr B. Bhogal, St John's Hospital for Diseases of the Skin, for preparing the IMF photomicrographs. Dr A. C. Chu is a Wellcome Senior Research Fellow.

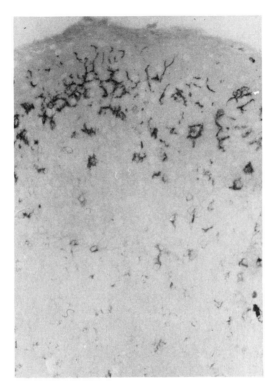

Fig. 36.12. Cutaneous T-cell lymphoma stained with OKT 6 showing numerous dendritic Langerhans cells in the epidermis and scattered round cells in the dermis.

Appendix

I. METHODS

The methods used were fully described in the first edition of this book.[4] They are essentially the following:-

A. *Direct immunofluorescence* on patient's skin with fluorescein-conjugated antibodies to immunoglobulins

B. *Indirect immunofluorescence*: using the circulating antibody in the patient's serum to bind to normal skin, the bound immunoglobulin being identified with fluorescein-conjugated antibodies to human immunoglobulins.

C. *Indirect C3 technique* to amplify the immunostain or to identify complement-fixing antibodies.

D. *Indirect immunoperoxidase* to identify T-cells and their subsets.

Air-dried or acetone-fixed cryostat sections are used.

II. SOURCES OF ANTIBODIES

Polyclonal T-cell antibodies	Institut Mérieux
Monoclonal T-cell antibodies	Coulter
	Becton Dickinson (Leu series)
	Ortho Pharmaceuticals (OKT series)
Fluorescein-conjugated rabbit anti-C3	Dako
Peroxidase-, fluorescein- and rhodamine-conjugated anti-human, rabbit and mouse immunoglobulins	Dako
Anti-collagen, type IV	The antibody used was a gift to the author

REFERENCES

1. Diaz C. A., Calvanico N. J., Tomasi T. B. and Jordan R. E. Bullous pemphigoid antigen: isolation from normal human skin. *J. Immunol.* 1977, **118**, 455–460.
2. Timpl R., Rohde H., Robey P. G., Rennard S. I., Foidart J. M. and Martin G. R. Laminin — a glycoprotein from basement membranes. *J. Biol. Chem.* 1979, **254**, 9933–9937.
3. Weber L., Kreig T., Muller P. K., Kirsch E. and Timpl R. Immunofluorescent localisation of type IV collagen and laminin in human skin and its application in junction zone pathology. *Br. J. dermatol.* 1982, **106**, 267–273.
4. Chu A. C., Bhogal B. and Black M. M. Immunocytochemistry in dermatology. In: Polak J. M. and Van Noorden S., eds., *Immunocytochemistry — Practical Applications in Pathology and Biology*. Bristol, Wrights·PSG, 1983: 302–333.
5. Chu A. C. Bullous dermatoses. In: Berry C. L., ed., *Current Topics in Pathology — Dermatopathology*, Vol. 74. Berlin, Springer-Verlag, 1985: 226–270.
6. Fine J. D., Neises G. R. and Katz S. I. Immunofluorescence and immunoelectron microscopic studies in cicatricial pemphigoid. *J. Invest. Dermatol.* 1984, **82**, 39–43.
7. Yaoita H. and Katz S. I. Immunoelectron microscopic localisation of IgA in skin of patients with dermatitis herpetiformis. *J. Invest. Dermatol.* 1976, **67**, 502–506.
8. Hodge L., Black M. M., Ramnarain N. and Bhogal B. Indirect complement immunofluorescence in the immunopathological assessment of bullous pemphigoid, cicatricial pemphigoid and herpes gestationis. *Clin. Exp. Dermatol.* 1978, **3**, 61–67.
9. Beutner E. H. and Jordan R. E. Demonstration of skin antibodies in sera of pemphigus vulgaris patients by indirect immunofluorescence staining. *Proc. Soc. Exp. Biol. Med.* 1964, **117**, 505–510.
10. Schiltz J. R. and Michel B. Production of epidermal acantholysis in normal human skin *in vitro* by the IgG fraction from pemphigus serum. *J. Invest. Dermatol.* 1976, **67**, 254–260.
11. Singer K. H., Hashimoto K. and Lazarus G. S. Antibody-induced proteinase activation: a proposed mechanism for pemphigus. In: Gigli I. N., ed., *Immunodermatology. Springer Seminars in Immunopathology*, 1981, **4**, 17–32.
12. Edelson R. L. Cutaneous T-cell lymphoma: mycosis fungoides, Sézary syndrome and other variants. *J. Am. Acad. Dermatol.* 1980, **2**, 89–106.
13. Chu A. C. and MacDonald D. M. Identification in situ of T-lymphocytes in dermal and epidermal infiltrates of mycosis fungoides. *Br. J. Dermatol.* 1979, **100**, 177–189.
14. Haynes B. F., Hemler M., Cotner T., Mann D. I., Eisenbarth G. S., Strominger H. and Fauci A. S. Characterisation of a monoclonal antibody (5E9) which defines a human cell surface antigen for cell activation. *J. Immunol.* 1981, **127**, 347–351.
15. Sutherland R., Delia D., Schneider C., Newman R., Kemshead J. and Greaves M. F. Ubiquitous cell surface glycoprotein on tumour cells in proliferation associated receptor for transferrin. *Proc. Natl Acad. Sci. USA*, 1981, **78**, 4515–4519.

16. Kung P. L., Goldstein G., Reinherz E. L. and Schlossman S. F. Monoclonal antibodies defining distinctive human T-cell surface antigens. *Science*, 1979, **206**, 347–349.

17. Wang C. Y., Good R. A., Ammirati P., Dymbort G. and Evans R. L. Identification of a p69, 71 complex expressed on T-cells showing determinants with B-type chronic lymphatic leukemia cells. *J. Exp. Med.* 1980, **151**, 1539–1544.

18. Ledbetter J. A., Evans R. L., Lipinski M., Rundles C., Good R. A. and Herzenberg L. A. Evolutionary conversion of surface molecules that distinguish T-lymphocyte helper/inducer and cytotoxic/suppressor subpopulations in mouse and man. *J. Exp. Med.* 1981, **153**, 310–323.

19. Reinherz E. L., Kung P. C., Goldstein F., Levey R. H. and Schlossman S. F. Discrete stages of human intrathymic differentiation analysis of normal thymocytes or leukemic lymphoblasts of T-cell lineage. *Proc. Natl Acad. Sci. USA*, 1980, **77**, 1588–1592.

20. McMichael A. J., Pilch J. R., Galfré G., Mason D. Y., Fabre J. W. and Milstein C. A human thymocyte antigen defined by a hybrid myeloma monoclonal antibody. *Eur. J. Immunol.* 1979, **9**, 205–210.

21. Griffiths-Chu S., Patterson J., Berger C. L., Edelson R. L. and Chu A. C. Characterisation of immature T-cell subpopulations in neonatal blood. *Blood*, 1984, **64**, 296–300.

22. Fithian E., Kung P., Goldstein G., Rubenfeld M., Fenoglio C. and Edelson R. Reactivity of Langerhans' cells with hybridoma antibody. *Proc. Natl Acad. Sci. USA*, 1981, **78**, 2541–2544.

23. Chu A. C., Eisinger M., Lee J. S., Takezaki S., Kung P. C. and Edelson R. L. Immunoelectron microscopic identification of Langerhans' cells using a new antigenic marker. *J. Invest. Dermatol.* 1982, **78**, 177–180.

24. Haynes B. F., Metzger R. S., Minna J. D. and Bunn P. A. Phenotype characterisation of cutaneous T-cell lymphoma. *N. Engl. J. Med.* 1981, **304**, 1319–1323.

25. Berger C. L., Morrison S., Chu A., Patterson J., Estabrook A., Takezaki S., Sharon J., Warburton D., Irigoyen O. and Edelson R. L. Diagnosis of cutaneous T-cell lymphoma by use of monoclonal antibodies reactive with tumour associated antigens. *J. Clin. Invest.* 1982, **70**, 1205–1215.

26. Chu A. C., Berger C. L., Edelson R. L., Spittle M., Russel C. and Smith N. Circulating T-cells in cutaneous T-cell lymphoma — a study using monoclonal antibodies against normal and malignant T-cells. *Br. J. Dermatol.* 1983, **108**, 219–220.

27. Nakane P. K. and Kawaoi A. Peroxidase-labelled antibody. A new method of conjugation. *J. Histochem. Cytochem.* 1974, **22**, 1084–1092.

28. Chu A. C., Patterson J., Berger C., Vanderheid E. and Edelson R. *In situ* study of T-cell subpopulations in cutaneous T-cell lymphoma — diagnostic criteria. *Cancer*, 1984, **54**, 2414–2422.

29. Chu A. C. The use of monoclonal antibodies in the in situ identification of T-cell subpopulations in cutaneous T-cell lymphoma. *J. Cut. Pathol.* 1983, **10**, 479–498.

30. Prendiville J., Smith N. P., Berger C. L., Edelson R. and Chu A. C. Monoclonal antibodies in the tissue diagnosis of cutaneous T-cell lymphoma. *J. Invest. Dermatol.* 1984, **82**, 560.

31. Harrist T. J., Bhan A. K., Murphy G. F., Sato S., Berman R. S., Gellis S. E., Freedman S. and Mihm M. C. Histiocytosis X. In situ characterisation of cutaneous infiltrates with monoclonal antibodies. *Am. J. Clin. Pathol.* 1983, **79**, 294–300.

32. Chu A. C., Berger C. L., Kung P. and Edelson R. L. In situ identification of Langerhans cells in the dermal infiltrate of cutaneous T-cell lymphopma. *J. Am. Acad. Dermatol.* 1982, **6**, 350–354.

33. Carr M. M., Botham P. A., Gawkrodoer D. J., McVittie E., Ross J. A., Steward I. C. and Hunter J. A. A. Early cellular reactions induced by dinitrochlorobenzene in sensitized human skin. *Br. J. Dermatol.* 1984, **110**, 637–643.

37

Immunohistology in the Diagnosis of Glomerular Disease

D. J. Evans

Immunohistology has gained a widespread acceptance in renal pathology; indeed, apart from dermatopathology, there is probably no diagnostic area in which it is so extensively used. At first sight this is rather surprising, because, with a few exceptions, the classifications of glomerular disease were largely unchanged by the introduction of the new methodology. Furthermore, in many publications stress is laid on the variability of immunohistological findings within a particular morphological subgroup, even to the extent of stating that, in membranoproliferative glomerulonephritis, 'patterns of immunofluorescence (are) too variable to group logically'.[1]

The reason for the ready acceptance of the method by morphologists is probably to be found in the experimental animal work, which had proposed the existence of two mechanisms of glomerular damage, one involving antibodies to glomerular basement membrane (GBM), and one involving the localization of circulating complexes to the glomeruli.[2] There was, therefore, a general hope that immunohistology would reveal, at least in part, the pathogenetic mechanisms in human glomerular disease. The objectives, therefore, were the recognition of antibody fixed on basement membrane or deposited as insoluble complexes, and the identification of the material known as 'fibrinoid' (from its conventional staining properties), often seen in diseased renal vessels or glomeruli. Immune complexes, by definition, contain an antigen and antibody; in addition, components of the complement cascade may be demonstrable, and these will vary with the pathway of complement fixation.[3,4]

One of the major technical problems is that the high concentration of immunoglobulins normally present in the tissues can cause substantial back-

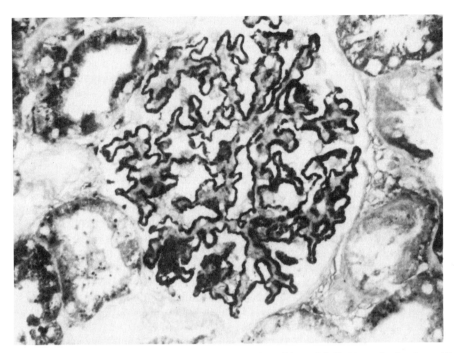

Fig. 37.1. Membranous glomerulonephrits. Granular deposition of IgG. Formalin-fixed paraffin section, digested for 17 min in 0·05 per cent protease VII (Sigma), pH 7·0 at 37 °C. PAP method with the DAB reaction product enhanced with gold chloride.

ground with conventional immunofluorescence methods, unless sections are prewashed (which is, however, bad for tissue preservation). In tissue prepared by freeze substitution, where soluble plasma components are not removed, they constitute a considerable annoyance, which may explain why the method has never achieved wide popularity. A recent innovation has been the use of tissue which has been paraffin embedded after fixation with formalin or formol mercury, where the sections are subjected to digestion, prior to staining with antibodies, with a proteolytic enzyme, such as trypsin, pronase, or other proteases of bacterial origin. This method has been reported as showing good correlation with frozen sections;[5,6] it does, however, require standardization of fixation, selection of suitable antibodies, and careful control of the enzymic digestion, but provides accurate localization of immune complexes and a permanent preparation (*Fig.* 37.1). Among the early publications, although there is general agreement on the broad outline of the findings, there are many disconcerting disagreements on details. Many of these can be attributed to technical differences, e.g. workers using indirect immunofluorescence found C1q in mesangial IgA disease much more frequently than those using direct immunofluorescence, largely because amounts present are usually very small, and often patchily distributed, requiring a sensitive method to reveal them. Some of the other differences were undoubtedly due to the inadequacies of the available sera, or to insufficient controls. The improvement in quality of antisera and the availability of monoclonal antibodies means that there is now emerging extensive and reliable information on the immunoglobulins and their subclasses[7] and accompanying components of the complement cascade in many renal diseases. It appears that, from a diagnostic point of view, patterns of distribution of immunoreactants are of more importance than details of subclass of immunoglobulin in the deposits.

Studies on the cells and membrane components in the normal and diseased glomerulus have not so far found much application in a clinical context. Contractile proteins may be demonstrated in mesangial cells,[8] and a variety of antigens (blood group antigens, factor VIII-related antigen[9,10]) in endothelial cells. Although conventional histochemical methods have been used to identify monocytes in glomeruli, some workers have preferred antibody methods.[11] In trying to resolve the problem of the composition of the crescent, antibodies to cytokeratin have been of value in demonstrating an epithelial component in addition to the monocyte-derived cells.

The composition of the GBM and the mesangial matrix has been surprisingly controversial; although there is general consensus that types IV and V collagen are present in the more electron-dense region of the peripheral membrane, findings with other antigens have differed. Madri et al.[12] were able to demonstrate fibronectin only in the mesangium, but Courtoy et al.[13] found it in addition in small quantity between epithelial cells and GBM and between endothelial cells and GBM. In an *in vivo* study Abrahamson and Caulfield[14] reported that when pure sheep anti-laminin was peroxidase labelled and given intravenously to a rat, and the animal killed within 1–2 h, reaction product, on developing the peroxidase, was present throughout the membrane. Madri et al., however, in an *in vitro* ultrastructural study, found laminin confined to the endothelial side of the GBM. Entactin, which is a sulphated glycoprotein of mol. wt 158 kdal, appears to have a similar distribution to that which Courtoy et al.[13] reported for fibronectin.[15] A view is currently being expressed, however, that all parts of the basement membrane contain all the components discussed above.[16] Published results on ultrastructural localization of the Goodpasture antigen have also given divergent results, early work suggesting that it was present throughout the membrane,[17] while later studies suggest that it is localized to the lamina rara interna.[18]

There have been several studies of alterations to these components in disease; it has been reported that type III collagen which is not normally present in glomeruli is present in the large synechiae of focal sclerosis and suggested that, in the crescentic nephritis associated with vasculitis, type III collagen is an early component, whereas in anti-GBM disease it is present later in the disease.[19] In Alport's syndrome, although the collagen appears normal, in several reports alterations to the Goodpasture antigen have been noted. We have recently reviewed the literature,[20] and found a total of 40 cases reported. Of these, 24 failed to show binding of antibodies from the serum or kidney eluates of patients with Goodpasture's syndrome, but 16 were positive. Most of the patients showing negative staining were males. Using a monoclonal antibody directed against the Goodpasture antigen,[21] we studied 10 cases of Alport's syndrome and found that staining was absent in 4 males and substantially reduced in the other 6 patients (5 males and 1 female). The reason for the variability of demonstrable antigen is not obvious; there was no clear correlation with age or severity of disease. Genetic heterogeneity of the syndrome seems the most likely explanation.[22]

Table 37.1. **Immunohistology of glomerulonephritis**

Disease	Immunohistology
Focal sclerosis	Segmental granular IgM and C3; less commonly IgG; IgA and fibrin usually absent
Membranous glomerulonephritis	Uniform granular subepithelial IgG. C3 usually similar to IgG. IgA sometimes similar to IgG. IgM, fibrin usually absent except for small patches
Crescentic glomerulonephritis	Fibrin always present in crescent—other findings vary with pathogenesis
Acute proliferative glomerulonephritis	Granular C3 as scattered deposits mainly subendothelial and subepithelial. Usually granular IgG in similar distribution. Variable IgA and IgM; little fibrin
Mesangiocapillary glomerulonephritis type I	C3 present in large subendothelial clumps, rather irregular. Immunoglobulins may be present in similar distribution; IgA is less common than IgG and IgM
Mesangiocapillary glomerulonephritis type II	C3 only. Mesangial usually but sometimes tram-track on peripheral capillary loops
Mesangial IgA disease	Mesangial deposits of IgA usually with IgG and C3; fibrin occasionally seen; IgM only in minute amounts

1. GLOMERULAR DISEASE CATEGORIES

Simply to list the immunohistological findings in the various categories of renal disease (*see Table* 37.1) does not convey adequately the way in which the technique is of value in practical problems, and it is necessary to consider the individual types of glomerular disease and their differential diagnosis.

1.1. Minimal Change Glomerulonephritis

In this condition, reported findings have varied from a complete absence of reactants, to the presence in the mesangium of granular IgM and C3. Doubt has been expressed as to whether cases with immunoreactants differ significantly from the others from a clinical or prognostic viewpoint.[23] The differential diagnosis of the light microscopical slides includes mesangial IgA disease, early membranous glomerulonephritis, and focal sclerosis. The first two of these are readily recognized by their characteristic immunohistology (q.v.).

1.2. Focal Sclerosis

The differentiation of this condition from minimal change glomerulonephritis is critically dependent on demonstrating a hyaline or sclerotic lesion; although the hyaline lesions characteristically contain IgM and C3, glomeruli normal on light microscopy are unstained, so that in this instance immunohistology is not contributory.

1.3. Membranous Glomerulonephritis

The diagnosis is readily made when the disease is at the stage in which the GBM is thickened and shows spikes on the outside on silver stains. However, it is clear that a phase exists in which there are electron-dense deposits in a subepithelial location demonstrable by electron microscopy, but no thickening of the membrane, or spike formation; at this stage the light microscopical findings resemble those of minimal change glomerulonephritis. The deposits can easily be demonstrated by immunohistology; they contain IgG (*Fig.* 37.1) and sometimes IgA and C3 as well, and the method is probably more sensitive than electron microscopy.

In advanced cases, the GBM may be grossly thickened and disorganized; ultrastructurally it shows abnormality throughout its width with spaces, in some of which there is a little residual immune deposit. The problem on light microscopy is to exclude mesangiocapillary glomerulonephritis. In most cases of this type there is little mesangial hypercellularity and the decision is not too difficult, but the immunohistology may be of assistance in showing faint granular staining for IgG and the absence of the distinctive pattern of C3 staining seen in mesangiocapillary glomerulonephritis.

1.4. Crescentic Glomerulonephritis

This is now known to be heterogeneous in its pathogenesis. It occurs not infrequently in the course of various systemic diseases (systemic lupus erythematosus, polyarteritis nodosa, Wegener's granulomatosis, Goodpasture's syndrome). Sometimes it may supervene on pre-existing nephritis of another type.[24] When it occurs in isolation, the immunohistology may show the presence of a linear deposit of antibody to GBM (*Fig.* 37.2) granular immune complexes, or deposits may be scanty or absent. The common feature in the genesis of the crescent appears to be leakage of plasma through ruptures in the glomerular capillary wall and the deposition of fibrin in Bowman's space. The groups of idiopathic crescentic glomerulonephritis first defined by immunohistology have been shown to have differing associations and, in some cases, their prognosis appears related to tissue type. In anti-GBM disease there is a significant association with HLA-DR2, and the prognosis is related to B7.[25] For immune complex disease crescentic nephritis, there is also an association with HLA-DR2 but a stronger association exists with properdin factor BfF.[26]

Fig. 37.2. Rapidly progressive glomerulonephritis due to antibodies to glomerular basement membrane. There is uniform staining for IgG in the capillary loops and in Bowman's capsule. Acetone-fixed cryostat section, direct immunofluorescence.

1.5. Proliferative Glomerulonephritis

It is perhaps surprising that any difficulty should be encountered in this category, since the presence of mesangial hypercellularity and increased polymorphs, together with a largely normal peripheral capillary loop, would seem to be distinctive. Although this is usually the case, occasionally the differential diagnosis from mesangiocapillary glomerulonephritis presents a problem, for, in this condition, mesangial proliferation and increase of polymorphs may also be features, though the polymorphs are often patchily distributed. Proliferative glomerulonephritis has a well-known 'lumpy, bumpy' distribution of C3 and IgG on immunofluoresence. Sorger et al.[27] distinguish different patterns which they call 'starry sky' and 'garland' and which they relate to the degree of proteinuria. In the later stages of the disease, the exudative changes are absent but there is residual mesangial hypercellularity, with C3 the only demonstrable reactant, this being confined to the mesangium. At this stage, differentiation from mesangio-proliferative glomerulonephritis is impossible without knowledge of the clinical course of the disease.

1.6. Mesangiocapillary Glomerulonephritis Type I

The difficulties in differential diagnosis between this (*Fig.* 37.3) and proliferative glomerulonephritis have already been considered: certain cases of lupus nephritis may be a problem as may light chain disease. The immunohistology is usually

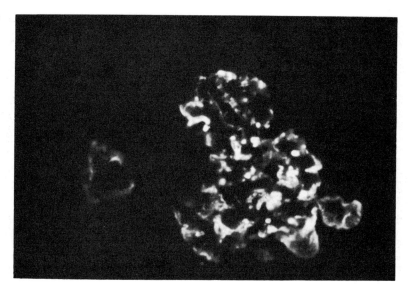

Fig. 37.3. Mesangiocapillary glomerulonephritis, type I, showing large irregular C3 deposits with a lobular outline. Acetone-fixed cryostat section, direct immunofluorescence.

characteristic, with C3 deposited along the capillary walls, outlining the lobules. Flecks of mesangial deposits may also be present. C3 may occur without immunoglobulins being present, but IgG, IgM and IgA are also found in decreasing order of frequency.

1.7. Mesangiocapillary Glomerulonephritis Type II

In this case the predominant problem is in distinguishing the changes from those of systemic lupus erythematosus. The dense deposits may occasionally be difficult to recognize on paraffin sections, though readily identifiable on semithin plastic sections. These cases usually show C3 deposition with a pattern similar to that seen in type I mesangiocapillary glomerulonephritis, though along the peripheral capillary loops staining is often weak and may show a tramline appearance.

1.8 Mesangioproliferative Glomerulonephritis

There is some doubt as to whether this constitutes a distinct entity; the appearance of mesangial proliferation as an isolated histological finding may occur during the course of systemic lupus erythematosus, in Henoch–Schoenlein purpura, mesangial IgA disease, proliferative glomerulonephritis, and subacute infective endocarditis; some cases have been observed to evolve into

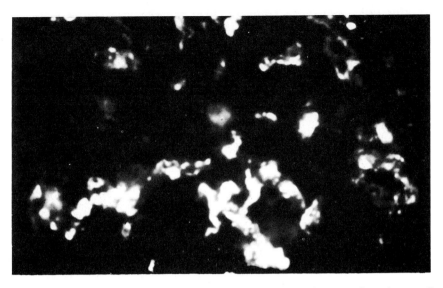

Fig. 37.4. Mesangial IgA disease showing IgA deposits confined to the mesangium. Acetone-fixed cryostat section, direct immunofluorescence.

mesangiocapillary glomerulonephritis, and in others focal sclerosis has developed.[28]

1.9. Idiopathic Focal Glomerulonephritis

1.9.1. Mesangial IgA Disease

This is perhaps the most visible effect of immunohistology on renal pathology, (*Fig.* 37.4) since here we have a disease which, since its recognition by Berger and Hinglais in 1968,[29] has been defined in terms of its immunohistology, and separated by it from other forms of focal nephritis. In the original description, stress was laid on the presence of IgA, IgG and C3 and the absence of IgM. Subsequent workers have reported that IgM and fibrin may be present in addition to IgA, IgG, and C3 and that IgA and C3 may occur alone. However, all workers are agreed that all glomeruli show deposits, even when they show no obvious lesions by light microscopy. Although in most patients the disease follows a benign course, in some cases there is progression to renal failure, and at various stages the appearances on light microscopy may be those of minimal change glomerulonephritis, of focal proliferation or of chronic glomerulonephritis.[30] The variation in the morphological and immunohistological appearance raises the serious problem of definition, which has not been adequately met in a formal sense. As it has become clear that mesangial deposits of IgA may be present diffusely in systemic lupus erythematosus, in Henoch–Schoenlein purpura and in cirrhosis, these diseases must be excluded before a diagnosis can be made.

1.9.2. Other Idiopathic Focal Glomerulonephritides

Unlike mesangial IgA disease, immunohistology here usually shows deposits localized to the lesions seen on light microscopy, usually of IgM and C3, the findings resembling those of focal sclerosis.

2. SYSTEMIC DISEASE

The problem for the pathologist here is in several separate categories.

2.1. Overt Clinical Disease

Typical examples here are of patients with vasculitic rashes and nephritis (? polyarteritis or Henoch's purpura), or with pulmonary haemorrhage and nephritis (? Goodpasture's syndrome or Wegener's granulomatosis). In both these examples, the immunohistology makes a conclusive decision with mesangial deposits of IgA, IgG and C3 revealing the Henoch's purpura and the well-known linear deposit of IgG providing the diagnosis of Goodpasture's syndrome.

2.2. Unsuspected Systemic Disease

2.2.1. Light Chain Disease

On light microscopy this may closely resemble mesangiocapillary glomerulonephritis, but immunohistology will often reveal the deposits in the membrane.

2.2.2. Systemic Lupus Erythematosus

Glomerular disease may be the presenting feature, and may precede involvement of other systems and positive laboratory tests. An extraordinary variety of morphological patterns may be present ranging from mesangial proliferation only, through focal to diffuse nephritis. Alternatively there may be a membranous morphology. In this last instance, it may be impossible to make the differentiation from idiopathic membranous glomerulonephritis, but in the other examples some immunohistological findings may be highly suggestive, such as mesangial deposits of IgG and C3, or mesangial deposits containing all the immunoglobulins and C3.

2.2.3. Subacute Infective Endocarditis

The literature on the immunohistology of this condition is sparse and rather contradictory. It is, of course, hardly surprising that with a wide variety of organisms and host responses, severity of glomerular involvement, and differing times of biopsy in relation to treatment, that the immunohistology should vary. The biggest series of biopsies studies was obtained from patients with subacute endocarditis who were mostly asymptomatic from a renal point of view;[31] these showed mainly scattered deposits of C3 and more rarely mesangial deposits of immunoglobulins; this finding may not be relevant to patients with prolonged undiagnosed infections, with symptomatic renal disease, or with persistent negative blood cultures. Patients with focal nephritis who show substantial immunoglobulin deposits of a type and distribution not seen in other forms of focal nephritis (e.g. extensive deposits of IgM and C3 on peripheral capillary loops)[31,32] should be carefully investigated; in one of our recent cases no cardiac murmurs were heard and no positive blood cultures obtained during several months as an inpatient at another hospital, but following immunohistological examination of a renal biopsy, definitive evidence of aortic valve vegetations was obtained by ultrasound. An initial symptomatic improvement on antibiotics was not sustained and the diagnosis was confirmed by the aortic valve perforations found at the time of valve replacement.

2.2.4. Amyloid

Since this diagnosis is made on the basis of tinctorial properties, the contribution of immunohistology is mainly that of differentiating secondary amyloid from the primary variety. Although it has been shown that the congophilia of secondary amyloid is sensitive to prior permanganate oxidation, whereas primary amyloid is often resistant, we have found that misleading results are quite often obtained; antisera to amyloid protein AA give far more reliable results. It is of course impossible to find an antiserum which will reliably stain immunoamyloid since this is derived from the variable part of the light chain. Amyloid P component is invariably present in renal amyloid, but it is also present in a variety of other circumstances; it appears to be a normal component of GBM in the adult, though absent in infancy, and granular deposits may be seen in diabetes and in some other forms of immune complex disease;[33] it is also present in vascular elastic.

3. CONCLUSION

Immunohistology has had an enormous impact on renal pathology, providing an increased diagnostic accuracy, and subdividing conventional categories of glomerular disease in a way which is important both clinically and scientifically.

It is to be anticipated that it will also play an important role in elucidating the nature of the antigens involved in the various types of glomerular disease.[34]

Appendix

SOURCES OF IMMUNE REAGENTS USED FOR THE ILLUSTRATED PREPARATIONS

Fluroescein—conjugated rabbit anti-human IgG, IgA, C3	Miles
Unconjugated rabbit anti-human IgG	Dako
Unconjugated swine anti-rabbit IgG	Dako
Rabbit PAP	Miles

REFERENCES

1. Burkholder P. M., Marchand A. and Krueger R. P. Mixed membranous and proliferative glomerulonephritis. *Lab. Invest.* 1970, **23**, 459–479.
2. Dixon F. G. The pathogenesis of glomerulonephritis. *Am. J. Med.* 1968, **44**, 493–498.
3. Westberg N. G., Naff G. B., Boyer J. T. and Michael A. F. Glomerular deposits of properdin in acute and chronic glomerulonephritis with hypocomplementaemia. *J. Clin. Invest.* 1971, **50**, 642–649.
4. Evans D. J., Gwyn Williams D., Peters D. K., Sissons J. G. P., Boulton Jones J. M., Ogg C. S., Cameron J. S. and Hoffbrand B. I. Glomerular deposits of properdin in Henoch–Schönlein purpura and idiopathic focal nephritis *Br. Med. J.* 1973, **3**, 326–328.
5. Eneström S. Immunohistology on formol sublimate fixed and paraplast embedded kidney tissue with comparison to formalin fixation and pre-embedding immunofluorescent staining *Stain Tech.* 1983, **58**, 259–272.
6. MacIver A. G. and Mepham B. L. Immunoperoxidase techniques in human renal biopsy. *Histopathology* 1982, **6**, 249–267.
7. Bannister K. M., Howarth G. S., Clarkson A. R. and Woodroffe A. J. Glomerular IgG subclass distribution in human glomerulonephritis. *Clin. Nephrol.* 1983, **19**, 161–165.
8. Becker C. G. Demonstration of actomyosin in mesangial cells of the renal glomerulus. *Am. J. Pathol.* 1972, **66**, 97–110.
9. Hoyer L. W., De Los Santos P. P. and Hoyer J. R. Antihaemophilic factor localisation in endothelial cells by immunofluorescence microscopy. *J. Clin. Invest.* 1973, **52**, 2737–2744.
10. Szulman A. E. The histological distribution of blood group substances A and B in man. *J. Exp. Med.* 1960, **111**, 785–799.
11. R. J. Marshall and A. G. MacIver. The monocyte/macrophage population of the normal human kidney. *J. Pathol.* 1984, **143**, 275–280.
12. Madri J. A., Roll J., Furthmayr H. and Foidart J. M. Ultrastructural localisation of fibronectin and laminin in the basement membranes of the murine kidney. *J. Cell Biol.* 1980, **86**, 682–687.
13. Courtoy P. J., Kanwar Y. S., Hynes R. O. and Farquhar M. G. Fibronectin localisation in the rat glomerulus. *J. Cell Biol.* 1980, **87**, 691–696.
14. Abrahamson D. R. and Caulfield J. P. Proteinuria and structural alterations in the rat glomerular basement membranes induced by intravenously injected antilaminin immunoglobulin G. *J. Exp. Med.* 1982, **156**, 128–145.
15. Bender P. L., Jaffe R., Carlin B. and Chung A. E. Immunolocalisation of entactin, a sulfated basement membrane component in rodent tissues and comparison with GP2. *Am. J. Pathol.* 1981, **103**, 419–426.
16. Leblond C. P., Inoue S. and Grant D. Nature of basement membranes. British Connective Tissue Society Meeting, 1985.
17. Burns J. and MacIver A. Immunoelectronmicroscopy of the glomerular basement membrane in Goodpasture's syndrome. *Rev. Eur. Étud. Clin. Biol.* 1971, **16**, 48–50.
18. Sisson S., Dysart M. K., Fish A. J. and Vernier R. L. Localisation of the Goodpasture antigen by

immunoelectronmicroscopy. *Clin. Immunol. Immunopathol.* 1982, **23**, 414–429.

19. Morel-Maroger Striker L., Killen P. D., Chi E. and Striker G. Composition of glomerulosclerosis. *Lab. Invest.* 1984, **51**, 181–192.

20. Savage C. O. S., Pusey C. D., Kershaw M., Cashman S., Harrison B., Hartley B., Turner D. R., Cameron J. S., Evans D. J. and Lockwood C. M. Variability of expression of the Goodpasture antigen in Alport's syndrome; studies with a monoclonal antibody. Submitted for publication.

21. Pressey A., Pusey C. D., Dash A., Peters D. K. and Lockwood C. M. Production of a monoclonal antibody to autoantigenic components of human glomerular basement membrane. *Clin. Exp. Immunol.* 1983, **54**, 178–184.

22. Evans S. H., Erickson R. P., Kelsch R. and Peirce J. C. Apparently changing patterns of inheritance in Alport's hereditary nephritis: genetic heterogeneity versus altered diagnostic criteria. *Clin. Genet.* 1980, **17**, 285–292.

23. Vilches A. R., Turner D. R., Cameron J. S., Ogg C. S., Chantler C. and Williams D. G. Significance of mesangial IgM deposition in minimal change nephrotic syndrome. *Lab. Invest.* 1982, **46**, 10–16.

24. Rajamaran S., Pinto J. A. and Cavallo T. Glomerulonephritis with coexistent immune deposits and antibasement membrane activity. *J. Clin. Pathol.* 1984, **37**, 176–181.

25. Rees A. J., Peters D. K., Amos N., Welsh K. I. and Batchelor J. R. The influence of HLA linked genes on the severity of anti-GBM nephritis. *Kidney Int.* 1984, **26**, 444–450.

26. Müller G. A., Gebhardt M., Kömpf J., Baldwin W. M., Ziegenhagen D. and Bohle A. Association between rapidly progressive glomerulonephritis and the properdin factor BfF and different HLA D region products. *Kidney Int.* 1984, **25**, 115–118.

27. Sorger K., Gessler U., Hübner F. K., Köhler H., Schulz W., Stuhlinger W., Thoenes G. H. and Thoenes W. Subtypes of acute postinfectious glomerulonephritis; synopsis of clinical and pathological findings. *Clin. Nephrol.* 1982, **17**, 114–118.

28. Hirzsel P., Yamase H. T., Carney W. R., Galen M. A., Graeber C. W., Johnson K. J., Kennedy T. L., Lapkin R. A., McLean R. H., Rosenworcel E. and Rowett D. A. Mesangial proliferative glomerulonephritis with IgM deposits. *Nephron* 1984, **38**, 100–108.

29. Berger J. and Hinglais N. Les dépôts intercapillaires d'IgA–IgG. *J. Urol. Néphrol.* 1968, **74**, 694–695.

30. Berger J. IgA glomerular deposits in renal disease. *Transplant Proc.* 1969, **1**, 939–944.

31. Morel-Maroger L., Sraer J. D., Herreman G. and Godeau P. Kidney in subacute endocarditis *Arch. Pathol.* 1972, **94**, 205–216.

32. Boulton-Jones J. M., Sissons J. G. P., Evans D. J. and Peters D. K. Renal lesions of subacute infective endocarditis. *Br. Med. J.* 1974, **2**, 11–14.

33. Dyck R. K., Evans D. J., Lockwood C. M., Rees A. J., Turner D. and Pepys M. Amyloid P component in human glomerular basement membrane. *Lancet* 1980, **ii**, 606–609.

34. Tomino Y., Sakai H., Endoh M., Miura M., Suga T., Kaneshige H. and Nomoto Y. Crossreactivity of eluted antibodies from renal tissues of patients with Henoch–Schonlein purpura nephritis and IgA nephropathy. *Am. J. Nephrol.* 1983, **3**, 315–318.

38 Immunocytochemistry of Micro-organisms

A. R. M. Coates

Immunocytochemistry has been widely used as a research tool in medically important microbial diseases and in the development of diagnostic tests, although there are relatively few areas where it is used in routine clinical diagnosis. Monoclonal antibodies against microbes have made a major impact on immunocytochemistry in infectious diseases because of the exquisite specificity and abundant availability of these antibodies. The main advantage of immunocytochemistry over other techniques is its ability to localize antigens in relation to the surrounding structures. This has been particularly successful in identifying surface antigens of organisms such as the sporozoites of the rodent malaria parasite *Plasmodium berghei*[1] and *Schistosoma mansoni*[2] which causes schistosomiasis in man. External antigens are thought to be important for the induction of protective immunity and may be involved in biologically significant functions such as adhesion to host cells by *Mycoplasma pneumoniae*,[3] the human respiratory pathogen.

Immunocytochemistry has also been used to locate microbial antigen in eukaryotic cells such as surface antigens of erythrocytes infected with *P. falciparum*[4,5] and *P. yoelii*,[6] which cause human and rodent malaria, respectively, and the internal nuclear and cytoplasmic antigens of the human pathogen cytomegalovirus (CMV).[7] There has also been widespread usage of immunocytochemistry in identifying microbial antigens in tissues. This has led to numerous observations of cross-reactions between human tissues and antibodies to microbes such as *Trypanosoma cruzi*,[8] the cause of Chagas' disease, and streptococcal species,[9] which may be important in the understanding of autoimmune pathology in infectious diseases. Although polyclonal antisera have been used to detect these cross-reactions in the past, monoclonal antibodies will particularly benefit this area of research.

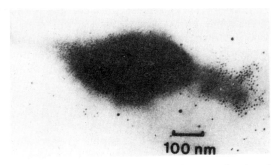

Fig. 38.1. Immunoferritin labelling of *M. pneumoniae* by the monoclonal antibody M43. Cells were grown on parlodion-carbon-coated platinum grids and were fixed with glutaraldehyde. Ferritin molecules can be seen clustered at the tip structure. (Reprinted by permission from Feldner et al.[3] in *Nature*, vol. 298, No. 5867, p. 765. Copyright 1982 Macmillan Journals Ltd.)

Another important role of immunocytochemistry in pathology has been in the development of diagnostic tests for viral[10,11] and bacterial[12,13] infections. Due to the generation of monoclonal antibodies against most of the major human pathogens, tools now exist for the emergence of a new generation of highly specific and reproducible immunological tests.

This chapter will be divided into two overlapping parts, the first of which will cover recent advances in the research uses of immunocytochemistry in infectious diseases. The second part will be devoted to its use in the development of new clinical diagnostic tests.

1. RESEARCH

1.1. Antigen Localization in the Microbe

Immunocytochemistry has been used to localize biological structures in micro-organisms. *Mycoplasma pneumoniae* has a specialized tip structure with which it can stick to animal cells. The ability of this organism to adhere to cells is thought to enhance its survival in the host. The monoclonal antibody M43[3] inhibited the binding of sheep erythrocytes to *M. pneumoniae* and the adhesin to which the monoclonal antibody is bound was identified as a protein of 160–190 kdal. The monoclonal antibody stained only one pole of the microbe. This was confirmed by immunoferritin labelling and electron microscopy (*Fig.* 38.1) which showed ferritin molecule localization predominantly at the tip structure. These findings indicate that the adhesin is located in the tip structure of the organism.

The surface antigens of parasites have been extensively studied because of their importance in inducing an immune response. One such antigen was detected on the surface of *Leishmania tropica major* promastigotes with the monoclonal antibody WIC 79.3 (E. Liew, 1984 personal communication) using indirect immunofluorescence (*Plate* 42). The monoclonal antibody stained the exterior of the organism and the flagellum but not the internal structures. Furthermore, monoclonal antibodies have been identified by screening *P. berghei* sporozoites with immunofluorescence.[1] Of these monoclonal antibodies, one bound to a stage-specific sporozoite surface antigen which had a molecular weight of 44 kdal.

Mice were passively protected against *P. berghei* sporozoite infection by the injection of the monoclonal antibody. Inoculation with the Fab fragment alone also induced protection which indicates that this antibody blocks the attachment of sporozoites to host cells.[14] Thus the antigen to which this monoclonal antibody binds may be a suitable immunogen for an experimental vaccine.

1.2. Location of Antigen in Tissues

Micro-organisms can be localized within a single mammalian cell or to certain tissues. For instance, cytomegalovirus (CMV), which is particularly troublesome in immunosuppressed patients, has a nucleus-associated antigen and a cytoplasmic antigen within the cell. Staining of the nuclear antigen with a monoclonal antibody (6E3) to CMV,[7] has been demonstrated by immunofluorescence. This monoclonal antibody precipitated a 72-kdal protein from detergent lysates of [^{35}S]methionine-labelled CMV-infected human embryonic fibroblasts. The cytoplasmic antigen was associated with another monoclonal antibody (6C5) which bound to a protein of 80 kdal. The significance of this work is that the nucleus-associated antigen appears early in infections before cytopathic changes are visible in cells and so the 6E3 monoclonal antibody might be useful in detecting early CMV disease.

Numerous monoclonal antibodies against malarial antigens on the surface of erythrocytes infected with *P. yoelii* and *P. falciparum* have been isolated by screening hybridoma supernatants with indirect immunofluorescence. Monoclonal antibodies to *P. yoelii*[6] which bind to merozoites as well as to infected red blood cells showed a protective effect against challenge infection in mice after antibody passive transfer. These antibodies blocked the penetration of parasites into mature red cells, but not into reticulocytes. Of interest as potential vaccine candidates for human malaria are monoclonal antibodies against *P. falciparum*,[4,15] three of which inhibited parasite development in an 80-h *in vitro* culture assay. The target antigen of two of the antibodies was a 41-kdal molecule and that of the third was two molecules of 56 and 96 kdal.

Microbial antigens can also be localized to certain tissues by immunocytochemistry. Of particular interest is the binding of anti-CMV polyclonal antibodies to smooth muscle cells in human carotid artery specimens which were removed from patients with atherosclerosis (*Fig.* 38.2).[16] Out of 132 patients, immunofluorescence detected 32 who had CMV antigens in the carotid artery. It is possible that these arterial cells as well as other tissues, such as lymphocytes,[17] are sites of CMV latency. Whether CMV is important in human atherosclerotic disease remains to be established.

1.3. Cross-reactions between Micro-organisms and Tissues

There have been repeated demonstrations with monoclonal antibody immunocytochemistry and other techniques of cross-reactions between sites on molecules.[18] Mechanisms which have been suggested for these cross-reactions include the hypothesis that the shared antigenic determinants have the same three-dimensional shape or, alternatively, that one antibody molecule is capable of interacting

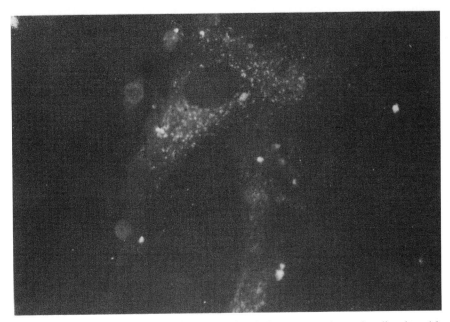

Fig. 38.2. Immunfluorescence staining of CMV antigen in human smooth muscle cells cultured from a carotid artery atherosclerotic plaque. Polyclonal antiserum against CMV was used which was raised in a guinea-pig. The CMV antigens are confined exclusively to the cytoplasm where they are situated in numerous granular structures. (Reprinted from Melnick et al.[16] with permission of the editors of the *Lancet* 1983, Vol. ii, p. 646.)

with two or more antigens of different structure. Irrespective of the underlying mechanisms, observations of cross-reactions between micro-organisms and tissues raise interesting questions about the relationship between infection and autoimmunity. These questions have been elegantly addressed by the monoclonal antibody immunocytochemical work of Wood and his colleagues[19] on *Trypanosoma cruzi* which causes severe neuronal and cardiac degeneration associated with autoantibodies against muscle, vascular structures, heart and neurones.[8] An IgM monoclonal antibody (CE5) which cross-reacted with *T. cruzi* was produced by the immunization of mice with membranes from rat dorsal root ganglia. This antibody stained rat sensory dorsal root neurones, Purkinje neurones in the cerebellar cortex (*Plate* 43), other rat neurones in the cerebral cortex, hippocampus, dentate fascia and olfactory bulb, neurones in the guinea-pig gastrointestinal submucosal plexus and the cross-striations of rat heart muscle. The antibody was cytotoxic to mammalian neurones *in vitro* in the presence of complement. This suggests that anti-parasite immunity may lead to cross-reactivity between *T. cruzi* and human tissue, and thus may be involved in the pathology of Chagas' disease.

Other immunocytochemical cross-reactions with polyclonal antibodies have been demonstrated between the heart and streptococcal species[9,20] a number of years ago. The significance of these shared antigens is unknown, although it has been suggested that they might play a role in the pathogenesis of rheumatic fever. An important consequence of these observations is that candidates for new streptococcal vaccines are tested for antigens which induce cardiac tissue-reactive antibodies. If such cross-reactive antibodies are found in immunized animals, it is unlikely that the vaccine will be used in humans. This is well demonstrated in the development of a vaccine against *Streptococcus mutans* which is a common cariogenic organism in dental plaque. Vaccination of animals with killed strains and serotypes of *S. mutans* can protect monkeys against caries.[21,22] Subsequently, Hughes and his colleagues[23] using indirect immunofluorescence, two-dimensional immunoelectrophoresis and radioimmunoassay showed that rabbits, which were immunized with *S. mutans*, produced cardiac tissue-binding antibodies. Furthermore, myocarditis could be detected in some immunized rabbits and anaphylaxis could be produced by intravenous challenge with a large dose (10 mg) of human heart tissue antigens. Crossed immunoelectrophoresis of *S. mutans* strain Ingbritt and human heart identified the ID and IF antigens to be the shared determinants between the streptococcus and the heart. Using indirect immunofluorescence, pre-vaccination rabbit antisera gave no fluorescent staining with heart tissue (*Fig. 38.3a*) in contrast to antisera against the ID antigen which bound to the sarcolemma and sarcoplasm (*Fig. 38.3b*). Antibodies to the IF antigen bound to myofibre striations (*Fig.38.3c*). Although the clinical significance of such reactions is unknown, the possibility that such dental caries vaccines which contain ID or IF antigens could induce myocarditis are likely to preclude their use in young children.

Further cross-reactions of unknown clinical significance have been observed between *Mycobacterium tuberculosis* and DNA, raising the possibility of a connection between mycobacterial infection which is extremely common[24] and autoimmune disease in subjects with a suitable immunogenetic background. Using the indirect immunoperoxidase method, an *M. tuberculosis*-specific murine monoclonal antibody TB68[25] which binds to a 14-kdal molecule (P. Baird and A. R. M. Coates et al., 1985, manuscript in preparation) has been shown to stain the nuclei of cells in ferret liver (*Fig. 38.4*) and in human lung and other organs (J. Morris, S. Van Noorden and A. R. M. Coates, 1984 unpublished observation). This is supported by the finding that TB68 binds to ssDNA and dsDNA.[26] Antibodies to DNA frequently appear in patients with tuberculosis[27] and in patients with the autoimmune disease systemic lupus erythematosus (SLE).[28] The idiotype of the DNA antibodies in SLE is predominantly the 16/6 type and may be identical to that of the DNA antibodies found in mycobacterial disease, because TB68 is also of the 16/6 idiotype.[26] Furthermore, human monoclonal DNA antibodies from patients with SLE bound to purified *M. tuberculosis* glycolipids.[26] This indicates that mycobacteria share antigens with human tissue which could be one explanation for the occurrence of autoantibodies to DNA in tuberculosis. It is not as yet known whether mycobacterial infection is important in the pathogenesis of autoimmune disease.

2. DEVELOPMENT OF DIAGNOSTIC TESTS

Immunocytochemistry has made a substantial contribution to the development of methods for the detection of microbial antigens in tissues. The reason for this is that immunocytochemistry is relatively easy to perform on a small scale, but once large-scale routine use is required, by inexperienced technicians, more objective

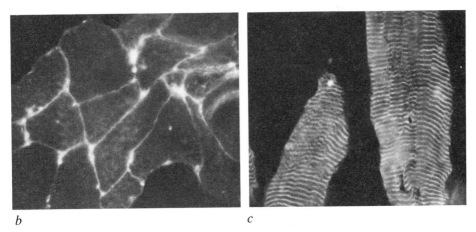

Fig. 38.3. Indirect immunofluorescence staining of human heart tissue with polyclonal rabbit antisera to *S. mutans*. Tissue was obtained at necropsy and frozen to −20 °C. Cryostat sections (4–6 mm) were cut, dried onto slides, washed with isotonic PBS and used within 24 hours. (*a*) Prevaccination antisera incubated with heart gave no staining. (*b*) Antiserum against *S. mutans* ID antigen was used to stain the sarcolemma and sarcoplasm. To raise the antibodies, rabbits were injected with ID antigen which had been isolated from immunoprecipitates in gels. (*c*) Staining of myofibre striations by antisera which had been raised by injection of rabbits with *S. mutans* IF antigen. (Reprinted by permission from Hughes et al.[23] with permission of the editors of *Infection and Immunity*, 1980, Vol. 27, p. 583.)

and less time-consuming methods such as ELISA are likely to prevail. The recent production of monoclonal antibodies against most of the medically important micro-organisms has opened the way towards new improved diagnostic tests for infectious diseases and immunocytochemistry is being used for the development of many of them.

Immunocytochemical techniques for the detection of antigens in tissues will be described in this section. Methods for measuring the level of microbial antibody in patients' serum will not be considered because they are immunochemical rather

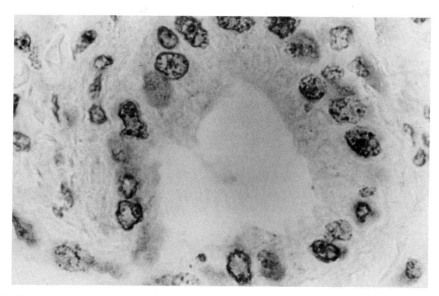

Fig. 38.4. Immunoperoxidase staining of the monoclonal *M. tuberculosis* antibody TB68 binding to paraffin-embedded, trypsinized section of ferret liver. The antibody TB68 stains nuclei in contrast to other anti-*M. tuberculosis* monoclonal antibodies which did not bind to DNA (not shown). (By courtesy of Dr J. Morris, Central Veterinary Laboratories, Weybridge, Surrey, UK.)

than immunocytochemical in nature and merely detect the level of antibody in serum with little regard for the relationship of antigen to surrounding structures.

2.1. Viral Diseases

2.1.1. Cytomegalovirus

Cytomegalovirus (CMV) infection is clinically important in immunocompromised patients[29] in whom CMV pneumonitis can be a terminal event. In addition, this virus can cause severe congenital disease. Existing methods for diagnosis of active infection are inadequate because direct microscopy is insufficiently sensitive, culture is too slow (it usually takes several weeks to yield a positive result) and a rise in the level of a patients' CMV antibody inevitably involves a delay, particularly in immunosuppressed patients. Although there is no effective treatment for CMV as yet, a prompt diagnosis is desirable to spare the patient unnecessary investigations and treatment and to develop new therapeutic regimens. Attempts have been made to diagnose active CMV infection with immunological techniques using polyclonal antibodies,[30] but the lack of large quantities of highly specific antibodies has prevented the emergence of a routine diagnostic test for this infection.

Monoclonal antibodies against CMV have been produced[7,31] and initial attempts to use them for diagnosing CMV infection have been encouraging. Goldstein and his colleagues[7] made two monoclonal antibodies of which one, 6E3, stained the early antigen which appears before the cytopathic effect in tissue culture and the other, 6C5, reacted with the late antigen. Since diagnosis is conventionally made on the appearance of cytopathic changes, a faster conclusion should be feasible with monoclonal antibodies against the CMV early antigen.[31] This has been achieved by Griffiths and his colleagues[32] with a monoclonal antibody against the CMV early antigen. They obtained positive results in an immunofluorescence test 27 h after inoculation of fibroblast cell cultures whilst the cytopathic effect did not appear until 17·5 days (mean) after inoculation. In an analysis of 385 specimens from 63 immunocompromised patients, the specificity of the test was 100 per cent and the sensitivity was 80 per cent. If it were possible to demonstrate the presence of CMV antigen directly in tissues by immunofluorescence, even faster diagnosis might be possible. Although CMV pneumonitis has been diagnosed directly by monoclonal antibody immuno-fluorescence,[7,32] insufficient numbers of patients with active disease were examined for an accurate estimate of sensitivity to be made.

2.1.2. Herpes Simplex Virus

Fast diagnosis is particularly important in genital herpes simplex virus (HSV) infection in pregnancy because acquisition of the virus by the newborn at delivery can severely damage the baby.[33] The danger to the infant can be avoided by caesarean section if the infection can be diagnosed quickly. The appearance of a cytopathic effect in HSV-infected cells in tissue culture is the most sensitive diagnostic method available, but the cytopathic effect is slow to emerge, taking up to 10 days for some specimens with a low viral titre. Using the sensitive avidin–biotin-fluorescent polyclonal antibody detection system, Nerurkar and his colleagues[10] have been able to detect HSV antigen in cell culture prior to the appearance of the cytopathic effect in cultures which had been inoculated with both low and high inocula of HSV (*Fig.* 38.5), although staining of the infected cells was delayed in cultures which had low inocula of HSV (less than 10^4 tissue culture infective does per ml). This test was as sensitive as the tissue culture cytopathic effect in specimens from 35 patients with genital herpes. Since it yielded a result in a shorter period of time it might be suitable for routine use.

Monoclonal antibodies have been developed which may speed up the diagnosis of HSV infection even further by giving a positive result directly on HSV-containing specimens without the need for tissue culture.[34] Cells were obtained from herpes lesions of 59 patients with genital, oral, ocular and mucocutaneous disease by smearing the specimens onto microscope slides. Immunofluorescence with monoclonal HSV antibodies was performed on these patients and on 43 control patients. HSV antigens were detected by the monoclonal antibody immuno-fluorescent test in 54 (88 per cent) of the 59 patients, but not in any of the 43 control specimens. Furthermore these antibodies can distinguish between HSV1 and HSV2.

2.1.3. Respiratory Syncytial Virus

For a number of years it has been known that respiratory syncytial virus (RSV) infection is the commonest cause of hospital admissions for respiratory illness in

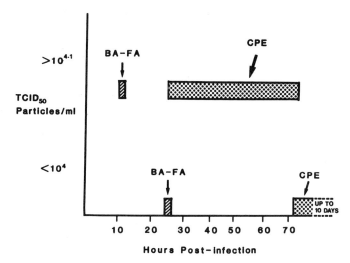

Fig. 38.5. Detection of genital HSV in 35 patients by tissue culture–biotin–avidin fluorescent antibody (BA-FA) and culture-cytopathic effects (CPE) in a human foreskin fibroblast cell line. HSV2 antibodies were raised in rabbits and linked to biotin. Cross-hatched boxes represent the appearance of staining by BA-FA and the dots in squares repesent the time of occurrence of CPE. $TCID_{50}$ = 50 per cent tissue culture infective doses per ml. (Adapted from Nerurkar et al.[10] with permission of the editors of the *Journal of Clinical Microbiology*, 1983, Vol. 17, p. 151.)

children under 5 years old.[35] The diagnosis of RSV can be achieved with equal sensitivity by immunofluorescence or tissue culture, but the immunological method has the advantage of giving a result in less than a day because it is performed directly on nasopharyngeal aspirates, whereas tissue culture may take several weeks.[11,35] In 2692 nasopharyngeal specimens from children, immuno-fluorescence and tissue culture virus isolation were in agreement in 91 per cent of the samples. Monoclonal antibodies against RSV are now available[36] which gave 93 per cent agreement between cell culture and immunofluorescence for 252 patients. Other monoclonal RSV antibodies[37] which have been used directly on nasopharyngeal aspirates include an antibody which is directed against the RSV nucleoprotein. Nasal cells of 100 infants with acute respiratory illness yielded 52 isolations of RSV by tissue culture. The monoclonal antibody immunofluores-cence test was 79 per cent sensitive and 100 per cent specific as compared with a combination of immunofluorescence with polyclonal antibody and tissue culture. It is likely that the advantages offered by the speed of the immunological tests and by the reproducible specificity of the monoclonal antibodies will lead to a reduction in the need for RSV tissue culture.

2.2. Bacterial Diseases

2.2.1. *Chlamydia trachomatis*

Sexual transmission of *Chlamydia trachomatis* now causes a large proportion of genital infections in both men and women[38] and is thought to result in permanent infertility in some women. It may also cause pneumonia, and infection of the eye

Table 38.1. Detection of Chlamydia trachomatis *from clinical specimens by immunofluorescence of smears and tissue culture*

| | No. of patients | No. of patients whose samples yielded the following results | | | | |
		Smear + Culture +	Smear + Culture −	Smear − Culture +	Total positive, %	Agreement between tests, %
Men with NGU	100	34	1	0	35	99
Men with gonorrhoea	100	21	4	2	27	94
Female contacts of men with NGU	35	9	0	0	26	100
Babies with conjunctivitis	7	4	0	0	57	100

Adapted from Thomas et al.[12] by kind permission of the Editors of the *Journal of Clinical Pathology* and the authors.
NGU = non-gonococcal urethritis.

and nasopharynx in young infants. This bacterium is an obligate intracellular parasite and so can only be grown in tissue culture. Diagnosis of *C. trachomatis* infection is routinely achieved by isolation in tissue culture and staining of the inclusions with the Giesma stain or iodine. Since diagnosis is slow (about 48 h) and relatively expensive, control of the disease has been inadequate in spite of the susceptibility of the organism to antibiotics.

Were it possible to detect chlamydial elementary bodies in smears from potentially infected sites, the diagnosis could be achieved more rapidly and tissue culture would no longer be necessary. Monoclonal antibodies have been produced which will detect *C. trachomatis* elementary bodies in clinical specimens.[39] A fluorescein-labelled monoclonal antibody was used with immunofluorescence to probe for *Chlamydia* in genital specimens from 926 patients who attended a clinic for sexually transmitted diseases.[40] Of these patients, 211 were positive with monoclonal antibody immunofluorescence, 216 yielded a positive result with antibody-stained *C. trachomatis* in tissue culture, and 207 were positive with conventional iodine staining of cultures. There was complete agreement among the three tests in 875 (94 per cent) out of 926 specimens. Of the 51 specimens in which there was disagreement, 29 were positive by direct monoclonal antibody immunofluorescence and were negative by the other tests, and 22 were negative by the direct test but positive by culture. A monoclonal antibody immunofluorescence test[12] has also been used to examine specimens from 100 men with non-gonococcal urethritis, 100 males with gonorrhoea, 35 female contacts of men with non-gonococcal urethritis and 7 babies with conjunctivitis (*Table* 38.1). The direct test on smears from these patients was compared with the conventional tissue culture method. For men with non-gonococcal urethritis, the agreement between the tests was 99 per cent, for males with gonorrhoea 94 per cent, and for the female contacts and babies there was 100 per cent agreement. The prospects for such a test being widely used seems to be hopeful, although immunofluorescence is not practical for large numbers (more than 30 per day) of specimens. Other immunological tests such as ELISA are likely to be developed for use in larger laboratories.

2.2.2. *Legionella pneumophila*

Legionnaires' disease which is caused by *Legionella pneumophila* has been found in many parts of the world and is responsible for a small proportion of the pneumonias which can occur in both previously healthy and immunocompromised persons. Clinical diagnosis of the condition is usually obtained by a four-fold or more rise in the level of *L. pneumophila* antibody which may not be reached for several weeks or by culture which can give a result in several days. Rapid diagnosis of Legionnaires' disease can be achieved by direct immunofluorescence[13] or Dieterle silver impregnation staining of clinical specimens. Although the detection of legionella antigen by immunofluorescence is in routine use it is rather insensitive. Of 19 patients who had Legionnaires' disease diagnosed by a rise in titre of antibodies to *Legionella* only 13 (68 per cent) were diagnosed by immunofluorescence antigen detection. The test is highly specific (there were no false positives amongst 22 control patients) and so is useful if it gives a positive result, because the diagnosis can be made relatively quickly in less than 24 hours.

3. CONCLUSION

Immunocytochemistry is of great value as a research tool in infectious diseases and in the development of new diagnostic tests. It is of less use in routine diagnosis due to its relatively subjective and time-consuming nature. The widespread application of monoclonal antibodies to infectious diseases will result in the emergence of many new research opportunities and the development of numerous new diagnostic tests, and so the continued importance of immunocytochemistry in this field.

Appendix

ANTIBODIES USED FOR ILLUSTRATED PREPARATIONS

Organism	Antibody	Authors/source
Mycoplasma pneumoniae	M42 (monoclonal)	Feldner et al.[3]
Leishmania tropica	WIC 79.1/.2/.3/.5 (monoclonal)	E. Liew, Wellcome Labs.

Cytomegalovirus	Guinea-pig anti-CMV	Melnick et al.[16]
Streptococcus mutans	Rabbit anti-*S. mutans*	Hughes et al.[23]
Trypanosoma cruzi	CE5 (monoclonal)	Wood et al.[19]
Mycobacterium tuber-culosis	TB68 (monoclonal)	Coates et al.[25]
Herpes simplex virus	Rabbit anti-HSV2 im-munoglobulin (Accurate Chemical and Scientific Corporation, Westbury, New York)	Nerurkar et al.[10]
Respiratory syncytial virus	Imagen anti-RSV (monoclonal)	Boots-Celltech Ltd
Chlamydia trachomatis	Micro Trak anti-*C. trachomatis* (monoclonal)	Syva UK

REFERENCES

1. Yoshida N. R. S., Nussenzweig R. S., Potocnjak P., Nussenzweig V. and Aikawa M. Hybridoma produces protective antibodies directed against the sporozoite stage of malaria parasite. *Science*, 1980, **207**, 71–73.
2. Verwarde C., Gryzch J. M., Bazin M., Capron M. and Capron A. Production of monoclonal antibodies to *Schistosoma mansonii*: preliminary studies on their biological activities. *C. R. Hebd. Séances Acad. Sci. Ser. D Sci. Nat. (Paris)* 1979, **289**, 725.
3. Feldner J., Gobel U. and Bredt W. *Mycoplasma pneumoniae* adhesion localized to tip structure by monoclonal antibody. *Nature*, 1982, **298**, 765–767.
4. Perrin L. H., Ramirez E., Er-Hsiang L. and Lambert P. H. *Plasmodium falciparum*: characterisation of defined antigens by monoclonal antibodies. *Clin. Exp. Immunol.* 1980, **41**, 91–96.
5. Langreth S. G. and Reese R. T. Antigenicity of the infected erythrocyte and merozoite surfaces in falciparum malaria. *J. Exp. Med.* 1979, **150**, 1241–1254.
6. Freeman R. R., Trejdosiewicz A. J. and Cross G. A. M. Protective monoclonal antibodies recognising stage-specific merozoite antigens of a rodent malaria parasite. *Nature*, 1980, **284**, 366–388.
7. Goldstein L. C., McDougall J., Hackman R., Meyers J. D., Thomas D. and Nowinski R. C. Monoclonal antibodies to cytomegalovirus: Rapid identification of clinical isolates and preliminary use in diagnosis of cytomegalovirus pneumonia. *Infect. Immun.* 1982, **38**, 273–281.
8. Scott M. T. and Snary D. American trypanosomiasis (Chagas' disease). In: Cohen S. and Warren K. S., eds., *Immunology of Parasite Infections*, 2nd ed. Oxford, Blackwell Scientific Publications, 1982: 261–298.
9. Kaplan M. H. Immunologic relation of streptococcal and tissue antigens. I. Properties of an antigen in certain strains of group A streptococci exhibiting an immunologic cross reaction with human heart tissue. *J. Immunol.* 1963, **90**, 595–605.
10. Nerurkar L. S., Jacob A. J., Maddens D. L. and Sever J. L. Detection of genital herpes simplex infection by a tissue-culture fluorescent antibody technique with biotin–avidin. *J. Clin. Microbiol.* 1983, **17**, 149–154.

11. McQuillin J. and Gardner P. S. Rapid diagnosis of respiratory syncytial virus infection by immunofluorescent antibody techniques. *Br. Med. J.* 1968, **1**, 602–605.

12. Thomas B. J., Evans R. T., Hawkins D. A. and Taylor-Robinson D. Sensitivity of detecting *Chlamydia trachomatis* elementary bodies in smears by use of a fluorescein labelled monoclonal antibody: comparison with conventional chlamydial isolation. *J. Clin. Pathol.* 1984, **37**, 812–816.

13. Saravolatz L. D., Russell G. and Cvitkovich, D. Direct immunofluorescence in the diagnosis of Legionnaires' disease. *Chest*, 1981, **79**, 566–570.

14. Potocnjak P., Yoshida N., Nussenweig R. S. and Nussenweig V. Monovalent fragments (Fab) of monoclonal antibodies to a sporozoite surface antigen (Pb.44) protect mice against malarial infection. *J. Exp. Med.* 1980, **151**, 1504–1513.

15. Perrin L. H., Ramirez E., Lambert P. H. and Miescher P. A. Inhibition of *P. falciparum* growth in human erythrocytes by monoclonal antibodies. *Nature*, 1981, **289**, 301–303.

16. Melnick J. L., Pressman G. R., McCollum C. H., Petrie B. L., Burek J. and Debakey M. E. Cytomegalovirus antigen within human arterial smooth muscle cells. *Lancet*, 1983, **ii**, 644–646.

17. Roche I. K., Cheung K. S., Boldogh I., Huang E. S. and Lang D. J. Cytomegalovirus: detection in human colonic and circulating mononuclear cells in association with gastrointestinal disease. *Int. J. Cancer*, 1980, **27**, 659–657.

18. Lane D. and Koprowski H. Molecular recognition and the future of monoclonal antibodies. *Nature*, 1982, **296**, 200–202.

19. Wood J. N., Hudson L., Jessell T. M. and Yamamuto M. A monoclonal antibody defining antigenic determinants on sub-populations of mammalian neurones and *Trypanosoma cruzi* parasites. *Nature* 1982, **296**, 34–38.

20. Onica D., Mihalan F., Lenkei R., Gherman M. and Tudor M. Auto-antibodies detected in rabbits hyperimmunized with group A, C and G streptococcal vaccines. *Infect. Immun.* 1977, **18**, 624–628.

21. Bowen W. H. A vaccine against dental caries. A pilot experiment with monkeys (Macaca iris). *Br. Dent. J.* 1969, **126**, 159–160.

22. Lehner T. S., Challacombe S. J. and Caldwell J. An immunological investigation into the prevention of caries in deciduous teeth of rhesus monkeys. *Arch. Oral Biol.* 1975, **20**, 305–310.

23. Hughes M., MacHardy S. M., Sheppard A. J. and Woods N. C. Evidence for an immunological relationship between *Streptococcus mutans* and human cardiac tissue. *Infect. Immun.* 1980, **27**, 576–588.

24. Edwards L. B., Edwards P. Q. and Palmer C. E. Sources of tuberculosis sensitivity in human populations. A summing up of recent epidemiological research. *Acta Tuberc. Pneumol. Scand. Suppl.* 1959, **47**, 77–82.

25. Coates A. R. M., Hewitt J., Allen B. W., Ivanyi J. and Mitchison D. A. Antigenic diversity of *Mycobacterium tuberculosis* and *Mycobacterium bovis* detected by means of monoclonal antibodies. *Lancet*, 1981, **ii**, 167–169.

26. Shoenfeld Y., Coates A. R. M., Rauch J., Lavie G., Shaul D. and Pinkhas J. Monoclonal anti-tuberculosis antibodies react with DNA and monoclonal anti-DNA autoantibodies react with *Mycobacterium tuberculosis*. Submitted for publication.

27. Deltlaff Z., Bowsyzc S. and Wiecki J. Anti-nuclear antibodies in the sera of in-patients of a pulmonary tuberculosis clinic. *Gruzhia*, 1973, **41**, 15–22.

28. Shoenfeld Y., Isenberg D. A., Rauch J., Madaio M. P., Stollar B. D. and Schwartz R. S. Idiotype cross-reactions of monoclonal human lupus autoantibodies. *J. Exp. Med.* 1983, **158**, 718–830.

29. Weller T. H. The cytomegaloviruses: ubiquitous agents with protean clinical manifestations. Parts 1 and 2. *N. Engl. J. Med.* 1971, **285**, 203–214, 267–273.

30. Weller, T. H. Clinical spectrum of cytomegalovirus infection. In: Nahmias A., Dowdle W. R. and Schinazi R. F., eds., *The Human Herpes Viruses: an Interdisciplinary Perspective*. New York, Elsevier North Holland Inc., 1981.

31. Pierera L., Hoffman M., Gallo D. and Cremer N. Monoclonal antibodies to human cytomegalovirus: three surface membrane proteins with unique immunological and electrophoretic properties specify cross-reactive determinants. *Infect. Immun.* 1982, **36**, 924–932.

32. Griffiths P. D., Stirk P. R., Ganczakowski M., Panjwani D. D., Ball M. G., Blacklock H. A. and Prentice H. G. Rapid diagnosis of cytomegalovirus infection in immunocompromised patients by detection of early antigen fluorescent foci. *Lancet*, 1984, **ii**, 1242–1244.

33. Whitley R. J., Nahmias A. J., Visintine A. M., Fleming C. L. and Alford C. A. The natural history of herpes simplex virus infection of mother and newborn. *Paediatrics*, 1980, **66**, 489–494.

34. Nowinski R. C., Tam M. K., Goldstein L. C. Strong L., Kuo C. C., Corey L., Stamm W. E., Handsfield E. H., Knapp J. S. and Holmes K. K. Monoclonal antibodies for diagnosis of infectious diseases in humans. *Science* 1983, **219**, 637–644.
35. Report to the Medical Research Council subcommittee on respiratory syncytial virus vaccines. Respiratory syncytial virus infection: admissions to hospital in industrial, urban and rural areas. *Br. Med. J.* 1978, **2**, 796–798.
36. Kim H. W., Wyatt R. G., Fernie B. F., Brandt C. D., Arrobio J. O., Jeffries B. C. and Parrott R. H. Respiratory syncytial virus detection by immunofluorescence in nasal secretions with monoclonal antibodies against selected surface and internal proteins. *J. Clin. Microbiol.* 1983, **18**, 1399–1404.
37. Bell D. M., Walsh E. E., Hruska J. F., Schnabell K. C. and Breese Hall C. Rapid detection of respiratory syncytial virus with a monoclonal antibody. *J. Clin. Microbiol.* 1983, **17**, 1099–1101.
38. Holmes K. K. The *Chlamydia* epidemic. *J. Am. Med. Assoc.* 1981, **245**, 1718–1723.
39. Stephens R. S., Tam M. R., Kuo C. C. and Nowinski R. C. Monoclonal antibodies to *Chlamydia trachomatis*: antibody specificities and antigen characterisation. *J. Immunol.* 1982, **128**, 1083–1089.
40. Tam M. R., Stamm W. E., Handsfield H. H., Stephens R., Kuo C. C., Holmes K. K., Ditzenberger K., Krieger M. and Nowinski R. C. Culture-independent diagnosis of *Chlamydia trachomatis* using monoclonal antibodies. *N. Engl. J. Med.* 1984, **310**, 1146–1150.
41. Wilson G. Miscellaneous diseases In: Wilson G. and Miles A., eds., *Topley and Wilson's Principles of Bacteriology, Virology and Immunity*, 7th ed., Vol. 3. London, Edward Arnold, 1984: 515–522.

39

Immunocytochemistry of Human Skeletal Muscle Diseases

C. A. Sewry,   M. J. Dunn and

S. T. Appleyard,   M. J. Capaldi

Immunocytochemistry is providing a new insight into the structure of skeletal muscle and is also enhancing our understanding of the pathogenesis of human neuromuscular disorders. This chapter aims to review (a) immunocytochemical methods applied to muscle and the problems associated with them and (b) immunocytochemical investigations of diseased human muscle. The application of immunocytochemistry to human pathology has been limited, but it is already proving to be of diagnostic value. More extensive use of these techniques and wider availability of suitable antibodies will undoubtedly increase our knowledge of the changes in disease states.

1. METHODOLOGY

1.1. Light Microscopy

Fixation is unnecessary for localization of many muscle antigens and we have obtained excellent results using unfixed cryostat sections. These may be stored at $-40\,°C$ without apparent loss in antigenicity. In some cases fixation, e.g. with 100 per cent acetone, is beneficial. Most standard immunocytochemical methods can be used, but biotin–avidin techniques enhance labelling and overcome some of the problems of non-specific binding, e.g. of fluorescein isothiocyanate (FITC)–IgG to the sarcolemma, connective tissue and necrotic fibres. Some conjugates of avidin from egg albumin however, bind to muscle fibre nuclei even with avidin and biotin pre-treatment. Streptococcal avidin does not have this disadvantage. For fluorescence techniques we recommend the new fluorochrome, Texas Red (Amersham), as it is brighter and more resistant to fading than FITC and rhodamine. However, lipofuscin which is stored in varying amounts in normal and diseased tissue is autofluorescent, as are some components of connective tissue, particularly at the wavelength used for fluorescein. Enzyme-label techniques do not suffer from these problems and we routinely use peroxidase conjugates, though endogenous peroxidase occurs in interstitial cells and some necrotic fibres. The use of alkaline phosphatase conjugates is not recommended because of endogenous activity in some fibres in diseased muscle.

1.2. Electron Microscopy

Where antigenicity is resistant to fixation, tissue can be labelled *en bloc* using ferritin, gold or peroxidase conjugates, but penetration is only a few micrometres and intracellular labelling does not occur. The principal problems relate to fixation, penetration and accessibility of antigenic sites. To overcome these problems, investigators have used labelling of cryostat sections prior to embedding, labelling of resin sections, manipulation of fixation conditions[1] and labelling of ultrathin frozen sections. The use of the latter for human muscle immunocytochemistry has not been reported, but we have found that the methods of Tokuyasu[2] are equally applicable to human muscle as they are to animal muscle.

2. EXTRACELLULAR MATRIX

2.1. Distribution of Collagen Types

We have used antibodies to types I, III and IV collagen to investigate the localization of collagen types in diseased human muscle.[3–5] In normal muscle, antibodies to types I and III collagen stain perimysial connective tissue. Strong staining of the endomysium is observed with antibodies to type III (*Fig.* 39.1), whereas anti-type I staining is relatively weak. Staining with antibodies to type IV collagen is confined to basement membrane of muscle fibres, major blood vessels and capillaries.

Muscle from patients with X-linked and autosomal muscular dystrophies shows increased anti-type III staining of the extensive perimysial and endomysial connective tissue (*Fig.* 39.2). A similar pattern of staining is seen with antibodies to type I collagen, except for weaker staining of endomysium. There appears to be no significant change in the distribution of type IV collagen, although the staining around small muscle fibres is occasionally more intense which may reflect thickening or folding of the basement membrane. In split and whorled fibres, antibodies to collagen types I, III and IV are associated with the abnormal membrane features. Increased endomysial staining with antibodies to type III collagen is not seen in spinal muscular atrophy, although an increase is observed in staining of the perimysium surrounding bundles of atrophied fibres. There is no alteration in the distribution of collagen type IV.

At the ultrastructural level, collagen type III is found in the reticular layer surrounding muscle fibres and in connective tissue including fibrillar collagen bundles, while type IV is in the lamina densa of the basement membrane of muscle fibres and blood vessels.

2.2. Distribution of Laminin and Fibronectin

In normal and diseased muscle (*Fig.* 39.3), staining with antibodies to laminin parallels that with antibodies to type IV collagen, whilst antibodies to fibronectin show that this protein is similarly distributed to type III collagen.[5] Electron microscopy confirms that fibronectin is found in the reticular layer and interstitial regions, whereas laminin is found external to the lamina densa of muscle fibres and blood vessels.[5]

2.3. Lectins

Lectin cytochemical analysis of carbohydrate-rich matrix components confirms the distribution of connective tissue glycoproteins established using specific antibodies.[6-8] Paljarvi and coworkers[9] found that the intensity of staining of the perimysial and endomysial connective tissue by peanut agglutinin (PNA) and wheat germ agglutinin (WGA) was increased in X-linked muscular dystrophies.

3. MEMBRANES

3.1. HLA-A,B,C Class I Antigens

Human class I (HLA-A,B,C) antigens are membrane glycoproteins involved in T-lymphocyte reactions to virus and alloantigen-bearing cells. We have shown that normal skeletal muscle expresses little or no class I antigens as detected by the monoclonal antibody W6/32, whereas all muscle fibres in inflammatory myopathies express class I antigens (*Fig.* 39.4) as do many fibres in X-linked dystrophies.[10] In autosomal dystrophies, antigen expression varies from slight to moderately strong, whereas in congenital muscular dystrophy and spinal muscular atrophy, there is little or no expression of class I antigens. In addition we have found by immunocytochemistry that normal muscle plasma membrane does not contain β2-microglobulin, a class I-associated polypeptide, whereas in polymyositis β2-microglobulin is present.

3.2. Immunoglobulin and Complement

Immunoglobulin and complement deposition occur in polymyositis, dermatomyositis, the muscular dystrophies and myasthenia gravis.[11-13] Antibodies to IgG, IgM, IgA, C1q and C3 stain the sarcolemma/basement membrane, blood vessel endothelium and some whole muscle fibres. Immunoglobulin deposition can be used to distinguish myopathic from neuropathic disorders.[12] The membrane attack complex C5b-9 (MAC) is found in necrotic fibres[11,14] and we have shown using a monoclonal antibody to C9[13] that MAC is also present on the surface of necrotic fibres and in discrete patches on the surface of non-necrotic fibres in myositic conditions (*Fig. 39.5*). Some cases of Duchenne muscular dystrophy (DMD) also appear to have some non-necrotic fibres that bind C9, suggesting early involvement of complement in fibre damage.

Studies of acquired autoimmune myasthenia gravis (MG) show immunoglobulin and C3 and C9 of the complement pathway at motor end plates.[15] This provides evidence for antibody-dependent complement-mediated injury of the postsynaptic membrane in MG.

3.3. Lectins

Peroxidase-conjugated concanavalin A (ConA)[16,17] and *Ricinus communis* I agglutin (RCA I)[1] bind strongly to the cell surface of skeletal muscle. We have demonstrated by electron microscopy[1] that RCA I is bound more intensely by muscle fibre plasma membrane than by basement membrane (*Fig. 39.6*). RCA I labelling of muscle from patients with spinal muscular atrophy is normal, but in DMD staining of the sarcolemma is weak and disrupted and, in particular, binding to plasma membrane is absent (*Fig. 39.7*). In Becker and limb–girdle dystrophies and in polymyositis RCA I binding to plasma membrane is discontinuous and patchy, although not completely absent as in DMD.[18] In addition, in some areas the basement membrane is also unstained in Duchenne and Becker dystrophies. Areas of sarcolemmal disruption are often associated with entry of RCA I–peroxidase complex into the fibre and appear to be similar to the focal defects described with ConA.[16]

4. CYTOSKELETON

4.1. Muscle 'Spectrin'

We have used a panel of monoclonal antibodies specific for human erythrocyte α- or β-spectrin subunits to investigate spectrin-like protein in human muscle.[19] Eight of 9 anti-β-spectrin antibodies are bound by muscle fibre plasma membrane. Electron microscopy confirms that the reaction product is associated with the cytoplasmic surface of the plasma membrane (*Fig. 39.8*). None of 13 anti-α-spectrin antibodies bind to muscle fibre membranes. In neuromuscular disorders, the intensity of binding of antibodies against β-spectrin determinants appears to be increased, except that regenerating fibres in dystrophic muscle show reduced binding. Fetal muscle fibres show a normal pattern of binding.

4.2. Intermediate Filaments

The intermediate filament proteins desmin (skeletin),[20] vimentin,[21] synemin,[22] and a neurofilament-associated polypeptide[23] have been localized in muscle fibres and in cultured myotubes. However, there are few studies of

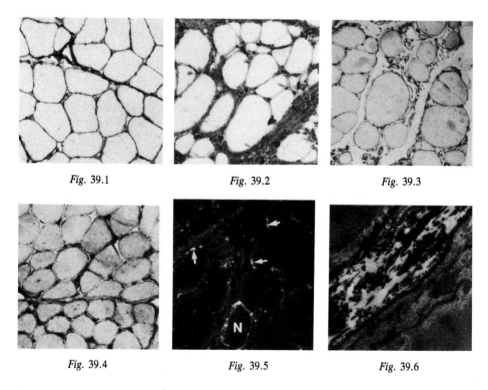

Fig. 39.1   *Fig.* 39.2   *Fig.* 39.3

Fig. 39.4   *Fig.* 39.5   *Fig.* 39.6

Fig. 39.1. Collagen type III distribution in normal muscle. Unfixed cryostat section. Avidin–biotin–peroxidase technique.

Fig. 39.2. Collagen type III distribution in Duchenne muscular dystrophy muscle. Unfixed cryostat section. Avidin–biotin–peroxidase technique. Note increased staining of perimysial and endomysial connective tissue.

Fig. 39.3. Localization of laminin in a case of Becker muscular dystrophy. Unfixed cryostat section. Avidin–biotin–peroxidase technique.

Fig. 39.4. HLA class I antigen localization in a case of dermatomyositis. Unfixed cryostat section. Avidin–biotin–peroxidase technique with haematoxylin counterstain.

Fig. 39.5. Localization of C9 in a case of polymyositis. Unfixed cryostat section. Texas Red–streptavidin technique. The periphery of a necrotic fibre (N) is completely labelled whilst some non-necrotic fibres show patches (arrows).

Fig. 39.6. Electron micrograph showing RCA-I peroxidase staining in normal muscle. Note the heavy delineation of the plasma membrane. Prefixed cryostat section, labelled then embedded.

human muscle, and only desmin has been localized.[24,25] It is present in small amounts at the periphery of the Z line and links adjacent myofibrils. Morphological evidence suggests a further linkage at the M line. The sarcolemma and the nuclear membrane may also be associated with desmin as is the case in chicken muscle.[2] In diseased muscle there are changes in the quantity and distribution of desmin[20] and it is found in abnormal structures such as cytoplasmic bodies.[26] Increased amounts of desmin occur in developing fibres and in the small, basophilic 'regenerating' fibres in some muscular dystrophies and myositic conditions. In these fibres, its distribution resembles that in fetal fibres in forming longitudinal strands before shifting to the transverse pattern of mature fibres.

Disruption of myofibrillar architecture, e.g. cores, mini-cores and ring fibres, is accompanied by changes in desmin distribution, and intermediate filaments are also involved in Z line streaming. Rod bodies in nemaline myopathy do not contain desmin, but are often surrounded by increased staining.[27]

5. MYOFIBRILLAR PROTEINS

Only a few contractile proteins have been studied immunocytochemically in man including the myosin and troponin subtypes and α-actinin. Both heavy and light chains of myosin are polymorphic and embryonic; fetal and adult forms have been isolated. Most mature muscle fibres contain only one myosin type (*Fig.* 39.9), either slow or fast, corresponding to histochemical fibre types 1 and 2, respectively.[28] A few fibres contain both slow and fast myosin and correspond to type 2C fibres. During muscle development the embryonic forms are replaced by a fetal form and later by the adult isomers though traces of fetal myosin exist up to four weeks of postnatal life.[29] Adult slow myosin can be identified in fetal myotubes as early as 15–16 weeks gestation. We have found fetal myosin in regenerating fibres in DMD (*Fig.* 39.10) and in some fibres in spinal muscular atrophy whose maturation may have been arrested. We have not found a significant quantity of fetal myosin in two neonatal cases of the X-linked form of centronuclear myopathy in which maturation arrest has been postulated.

Studies of myosin subtypes give a clearer understanding of fibre typing which is of diagnostic significance in diseased muscle. A recent study of a distal myopathy shows that, in this instance, fibre types are distinguished immunocytochemically but not by standard histochemical methods.[30]

In normal human muscle, troponin I isomers are segregated into type 1 and type 2 fibres. In muscular dystrophies and spinal muscular atrophies intermediate fibres containing both slow and fast troponin I occur.[31,32]

α-Actinin is the major protein of the Z line and has been shown to exist in large quantities in the rod-like structures (*Fig.* 39.11) that characterize nemaline myopathy.[27]

6. INTRACELLULAR STAINING

Engel and Biesecker[11] have consistently found C3, C5b-9 (MAC) and C9 in necrotic muscle fibres. Components C1q and C4 in necrotic fibres are usually not increased. We have confirmed the presence of C9 in necrotic muscle fibres in dermatomyositis, polymyositis and DMD.[13] We have found that a number of lectins also stain necrotic fibres in diseased muscle. One of these, *Bandeiraea simplicifolia* agglutinin I (BSA I lectin), also stains fibres that are not obviously necrotic and these may be at a prenecrotic stage as some also stain with anti-C9 antibody.

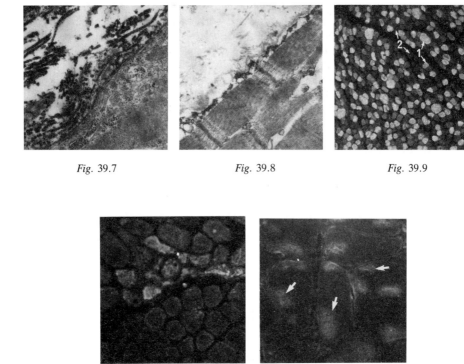

Fig. 39.7   Fig. 39.8   Fig. 39.9

Fig. 39.10   Fig. 39.11

Fig. 39.7. RCA-I peroxidase staining in Duchenne muscular dystrophy muscle. There is no staining of the muscle plasma membrane despite intense labelling of adjacent collagen. Prefixed cryostat section, labelled then embedded for electron microscopy.

Fig. 39.8. Localization of β-spectrin in normal muscle. Avidin–biotin–peroxidase procedure on an unfixed cryostat section prior to processing for electron microscopy.

Fig. 39.9. Distribution of slow myosin in a case of centronuclear myopathy. Unfixed cryostat section. Texas Red–streptavidin technique. Note segregation into two fibre types (1 and 2).

Fig. 39.10. Localization of fetal myosin in regenerating fibres in Duchenne muscular dystrophy. Note that traces are also present in some adjacent fibres. Texas Red–streptavidin technique.

Fig. 39.11. Localization of α-actinin in rods (arrows) in nemaline myopathy. Unfixed cryostat section. Indirect FITC-labelled second antibody technique.

7. INTERSTITIAL CELLS

The cellular infiltrate that occurs in inflammatory myopathies has been characterized using a panel of monoclonal antibodies specific for particular cell types.[33–35] An antibody reactive with all peripheral lymphocytes shows a significant number of these to be present (*Plate* 44) as well as some

B-lymphocytes and macrophages. Antibodies specific for T-cell subsets reveal that the T-cells are of both suppressor/cytotoxic and helper/inducer phenotypes, and some of these are activated. In addition, there are differences between endomysial and perivascular areas which may be of diagnostic significance.

Cellular infiltrates are also observed in DMD but in this case the suppressor/cytotoxic T- cell phenotype is predominant,[33] suggesting that T-cell-mediated reactions may play a role in this disease. Interstitial cells and phagocytes within necrotic fibres contain cathepsin D. This is associated with increased staining of some fibres.[36]

Acknowledgements

We acknowledge financial support from the Muscular Dystrophy Group and the Wellcome Trust.

Appendix

SUPPLIERS OF IMMUNE REAGENTS

Immune reagent	Supplier
Collagen types III and IV	Dr V. Duance (ARC, Bristol)
Laminin	Bethesda Research Laboratories Inc.
HLA class I	Dr C. Navarette and Professor H. Festenstein (London Hospital)
C9	Dr A. K. Campbell (Cardiff) and Dr P. Luzzio (Cambridge)
RCA-I peroxidase	EY Laboratories
β-Spectrin	Dr D. Shotton (Oxford)
Slow myosin	Dr R. Fitzsimons (RPMS, London)
Fetal myosin	Dr P. Hale (Sydney)
α-Actinin	Dr S. Watkins (Paris)
Leu-4	Becton Dickinson Laboratory Systems (Belgium)
Biotinylated secondary antibodies	TAGO Inc
FITC secondary antibodies	TAGO Inc
Streptavidin Texas Red	Amersham
Avidin–biotin complex	Vector Laboratories

REFERENCES

1. Capaldi M. J., Dunn M. J. Sewry C. A. and Dubowitz V. Altered binding of Ricinus communis I lectin using muscle membranes in Duchenne muscular dystrophy. *J. Neurol. Sci.* 1984, **63**, 129–142.

2. Tokuyasu K. T., Dutton A. H. and Singer S. J. Immunoelectron microscopic studies of desmin (skeletin) localization and intermediate filament organization in chicken skeletal muscle. *J. Cell Biol.* 1983, **96**, 1727–1735.

3. Duance V. C., Stephens H. R., Dunn M. J., Bailey A. J. and Dubowitz V. A role for collagen in the pathogenesis of muscular dystrophy? *Nature* 1980, **284**, 470–472.

4. Stephens H. R., Duance V. C., Dunn M. J., Bailey A. J. and Dubowitz V. Collagen types in neuromuscular diseases. *J. Neurol. Sci.* 1982, **53**, 45–62.

5. Dunn M. J., Sewry C. A., Statham, H. E., Stephens H. R. and Dubowitz V. Studies of the extracellular matrix in diseased human muscle. In: Kemp R. B. and Hinchliffe J. R., eds., *Matrices and Cell Differentiation.* New York, Alan R. Liss, 1984: 213–231.

6. Pena S. D. J., Gordon B. B., Karpati G. and Carpenter S. Lectin histochemistry of human skeletal muscle. *J. Histochem. Cytochem.* 1981, **29**, 542–546.

7. Dunn M. J. and Dubowitz V. Cytochemical studies of lectin binding by diseased human muscle. *J. Neurol. Sci.* 1982, **55**, 147–159.

8. Capaldi M. J., Dunn M. J., Sewry C. A. and Dubowitz V. Lectin binding in human skeletal muscle: a comparison of 15 different lectins. *Histochem. J.* 1985, **17**, 81–92.

9. Paljarvi L., Karjalainen K. and Kalimo H. Altered muscle saccharide pattern in X-linked muscular dystrophy. *Arch. Neurol.* 1984, **41**, 39–42.

10. Appleyard S. T., Dunn M. J., Dubowitz V. and Rose M. L. Increased expression of HLA ABC class I antigens by muscle fibres in Duchenne muscular dystrophy, inflammatory myopathy, and other neuromuscular disorders. *Lancet* 1985, **i**, 361–363.

11. Engel A. G. and Biesecker G. Complement activation in muscle fiber necrosis: Demonstration of the membrane attack complex of complement in necrotic fibers. *Ann. Neurol.* 1982, **12**, 289–296.

12. Isenberg D. A. Immunoglobulin deposition in skeletal muscle in primary muscle disease. *Q. J. Med. New Series*, 1983, **LII**, no. 207, 297–310.

13. Morgan B. P., Sewry C. A., Siddle K., Luzio J. P. and Campbell A. K. Immunolocalization of complement component C9 on necrotic and non-necrotic muscle fibres in myositis using monoclonal antibodies: a primary role of complement in autoimmune cell damage. *Immunology* 1984, **52**, 181–188.

14. Cornelio F. and Dones I. Muscle fiber degeneration and necrosis in muscular dystrophy and other muscle diseases: Cytochemical and immunocytochemical data. *Ann. Neurol.* 1984, **16**, 694–701.

15. Sahashi K., Engel A. G., Lambert E. H. and Howard F. M. Ultrastructural localisation of the terminal and lytic ninth complement component (C9) at the motor end-plate in myasthenia gravis. *J. Neuropath. Exp. Neurol.* 1980, **39**, 160–172.

16. Bonilla E., Schotland D. L. and Wakayama Y. Duchenne dystrophy: focal alterations in the distribution of Concanavalin A binding sites at the muscle cell surface. *Ann. Neurol.* 1978, **4**, 117–123.

17. Bonilla E., Schotland D. L. and Wakayama Y. Application of lectin cytochemistry to the study of human neuromuscular disease. *Muscle Nerve* 1980, **3**, 28–35.

18. Capaldi M. J., Dunn M. J., Sewry C. A. and Dubowitz V. Binding of *Ricinus communis* I lectin to the muscle cell plasma membrane in diseased muscle. *J. Neurol. Sci.* 1984, **64**, 315–324.

19. Appleyard S. T., Dunn M. J., Dubowitz V., Scott M. L., Pittman S. J. and Shotton D. M. Monoclonal antibodies detect a spectrin-like protein in normal and dystrophic human skeletal muscle. *Proc. Natl Acad. Sci. USA* 1984, **81**, 776–780.

20. Thornell L-E., Eriksson A. and Edstrom L. Intermediate filaments in human myopathies. In: Dowben R. M. and Shay J. W., eds., *Cell and Muscle Motility*, Vol. 4. New York, Plenum Publishing Corp., 1983.

21. Lazarides E. Intermediate filaments as mechanical integrators of cellular space. *Nature* 1980, **283**, 249–256.

22. Granger B. L. and Lazarides E. Synemin: a new high molecular weight protein associated with desmin and vimentin filaments in muscle. *Cell* 1980, **22**, 727–738.

23. Wang C., Asai D. J. and Lazarides E. The 68 000 dalton neurofilament-associated polypeptide is a component of non neuronal cells and skeletal myofibrils. *Proc. Natl Acad. Sci. USA* 1980, **77**, 1541–1545.

24. Thornell L-E., Edstrom L., Eriksson A., Henriksson K-G. and Angqvist K. A. The distribution of intermediate filament protein (skeletin) in normal and diseased human skeletal muscle. *J. Neurol. Sci.* 1980, **47**, 153–170.

25. Osborn M., Geisler N., Shaw G., Sharp G. A. and Weber K. Intermediate filaments. *Cold Spring Harbor Symp. Quant. Biol.* 1982, **49**, 413–429.

26. Osborn M. and Goebel H. H. The cytoplasmic bodies in a congenital myopathy can be stained with antibodies to desmin, the muscle-specific intermediate filament protein. *Acta Neuropathol.* 1983, **62**, 149–152.

27. Jockusch B. M., Veldman H., Griffiths G. W., van Oost B. A. and Jennekens F. G. I. Immunofluorescence microscopy of a myopathy: α-actinin is a major constituent of nemaline rods. *Exp. Cell Res.* 1980, **127**, 409–420.

28. Billeter R., Weber H., Lutz H., Howald H., Eppenberger H. M. and Jenny E. Myosin types in human skeletal muscle fibers. *Histochemistry* 1980, **65**, 249–259.

29. Fitzsimons R. B. and Hoh J. F. Y. Embryonic and foetal myosins in human skeletal muscle. The presence of foetal myosins in Duchenne muscular dystrophy and infantile spinal muscular atrophy. *J. Neurol. Sci.* 1981, **52**, 367–384.

30. Thornell L-E., Edstrom L., Billeter R., Butler-Browne G. S., Kjorell U. and Whalen R. G. Muscle fibre type composition in distal myopathy (Welander). An analysis with enzyme- and immuno-histochemical, gel-electrophoretic and ultrastructural techniques. *J. Neurol. Sci.* 1984, **65**, 269–292.

31. Dhoot G. K. and Pearce G. W. Changes in the distribution of fast and slow forms of troponin I in some neuromuscular disorders. *J. Neurol. Sci.* 1984, **65**, 1–15.

32. Dhoot G. K. and Pearce G. W. Transformation of fibre types in muscular dystrophies. *J. Neurol. Sci.* 1984, **65**, 17–28.

33. Arahata K. and Engel A. G. Monoclonal antibody analysis of mononuclear cells in myopathies. I. Quantitation of subsets according to diagnosis and sites of accumulation and demonstration and counts of muscle fibres invaded by T cells. *Ann. Neurol.* 1984, **16**, 193–208.

34. Engel A. G. and Arahata K. Monoclonal antibody analysis of mononuclear cells in myopathies. II. Phenotypes of autoinvasive cells in polymyositis and inclusion body myositis. *Ann Neurol.* 1984, **16**, 209–215.

35. Giorno R., Barden M. T., Kohler P. F. and Ringel S. P. Immunohistochemical characterization of the mononuclear cells infiltrating muscle of patients with inflammatory and noninflammatory myopathies. *Clin. Immunol. Immunopathol.* 1984, **30**, 405–412.

36. Whitaker J. N., Bertorini T. E. and Mendell J. R. Immunocytochemical studies of cathepsin D in human skeletal muscle. *Ann. Neurol.* 1983, **13**, 133–144.

40

Some Implications of Immunocytochemistry for Routine Diagnostic Histopathology

R. B. Goudie

Immunocytochemistry has been available as an investigative tool since the pioneering work of Coons and Kaplan with fluorescein-labelled antibodies nearly 40 years ago. Among its earliest diagnostic applications were its use as a method of detecting anti-tissue antibodies in the serum of patients suspected of having autoimmune diseases, and of demonstrating antigen-bound immunoglobulin and complement in frozen sections of kidney and skin from patients with immunological disorders of these organs. Widespread interest in the application of immunocytochemistry to the broad field of diagnostic histopathology is more recent, and has largely followed three important developments: (*a*) the introduction of enzyme labels, such as horseradish peroxidase, which allow antibodies to be used as special stains in permanent histological preparations that are counterstained in a conventional way and viewed with the light microscope; (*b*) the ready availability of good antisera against a variety of antigens of pathological importance; (*c*) the realization that, contrary to previous belief, certain immunocytochemical tests can be satisfactorily performed on formalin-fixed, paraffin-embedded tissue specimens.

Retrospective study of case material stored in departmental files and investigation of current surgical biopsies have revealed many disorders in which immunocytochemistry may be of value in the diagnosis and management of patients or in epidemiological surveys. These are now briefly reviewed, reference being made to other chapters in this volume which deal with the various topics in more detail. Some important implications of immunocytochemistry for diagnostic histopathology are also discussed. Clearly, it is desirable that all practising histopathologists should know which of their diagnostic problems immunocytochemistry may help to solve, how to preserve biopsy specimens in a suitable condition for the appropriate tests, and how or where such tests are performed.

1. USES OF IMMUNOCYTOCHEMISTRY

The special and unique function of immunocytochemistry is to demonstrate the microscopic distribution of antigen in cells and tissues, in many circumstances a matter of considerable diagnostic importance. The patterns of immunoglobulin deposition in diseases of the glomeruli are typical examples (*see* Chapter 37). So also is the localization in individual cells of surface antigens which assists in the accurate diagnosis and classification of lymphoid tumours even when they are heavily infiltrated with a mixed population of reactive (non-neoplastic) mononuclear cells (*see* Chapter 34).

Some diagnostic applications of immunocytochemistry are qualitative and simply involve demonstrating the presence of a particular substance, its histological localization within the specimen being of little importance. Thus, a metastatic carcinoma shown to contain thyroglobulin is likely to have originated in thyroid whether the thyroglobulin is found in the cells, in follicular lumina or in both. Among the many examples of the qualitative use of immunochemistry is the demonstration of tumour products such as hormones, enzymes or fetal antigens which may provide important information on the likely cell of origin of a tumour and may have a bearing on the prognosis, choice of treatment and selection of appropriate tumour markers to monitor the progress of the disease. As an alternative to immunocytochemistry, it may be possible to obtain comparable (or even more informative) results from conventional biochemical or microbiological tests on homogenates of the lesion or, in some cases, on the patient's blood. Nevertheless, the qualitative use of immunocytochemistry on a biopsy specimen may be more convenient, and it is particularly valuable in cases where the lesion has been completely excised and placed in fixative before the need for further tests is recognized.

Immunocytochemistry also has a role in detecting tissue antigens for which biochemical tests are not available, such as the markers which distinguish helper from suppressor T-cells.

It is important for clinicians to appreciate that most histopathological diagnoses can be made without the help of immunocytochemistry in spite of its great value in certain cases and in particular disorders.

Some of the circumstances in which immunocytochemical testing with readily available antibodies have been found useful in routine diagnostic histopathology are indicated in *Tables* 40.1–40.3. In the disorders of skin and kidney listed in *Table* 40.1, immunocytochemical testing demonstrates the presumed immunopathogenesis of the lesion. In *Tables* 40.2 and 40.3 various applications of the method are given as a series of specific questions. Often there is a choice of antigens which may provide the answer to each question. Sometimes a panel of antibodies is necessary to provide positive and negative evidence for a particular diagnosis. The likelihood of success in obtaining helpful information depends greatly on a number of factors considered in the following paragraphs.

*Table 40.1. **Immunological disorders of skin and kidney with diagnostic immunoglobulin and complement deposition***

Skin

Bullous pemphigoid, cicatricial pemphigoid, herpes gestationis
Chronic bullous disease of childhood
Pemphigus
Dermatitis herpetiformis
Epidermolysis bullosa acquisita
Discoid and systemic lupus erythematosus

Kidney (differential diagnoses)

Minimal change *from* early membranous glomerulonephritis and mesangial IgA disease
Mesangiocapillary *from* certain cases of proliferative or late membranous glomerulonephritis,
    systemic lupus erythematosus and light chain disease
Mesangial IgA disease *from* minimal change, focal proliferative or chronic glomerulonephritis
Categories of otherwise unexplained crescentic glomerulonephritis
Polyarteritis *from* Henoch's purpura
Goodpasture's syndrome *from* Wegener's granulomatosis

See Chapters 36 and 37.

2. FIXATION AND PROCESSING OF SPECIMENS

To obtain the best results with immunocytochemical staining, it is necessary to use whatever method of fixing and processing the specimen leads to optimum preservation and minimum diffusion or elution of the antigen under investigation. No single method achieves this purpose for all antigens, different antigens having different optimal requirements as shown in the examples in *Table* 40.4. Fortunately, methods of tissue fixation and processing other than the ones of choice may still give adequate results with many antigens, and this is increasingly so since the introduction of specially sensitive immunocytochemical techniques such as immunogold–silver.[1]

Difficulties with tissue fixation and processing should not arise if the individual who performs the biopsy works in a specialist clinic, e.g. nephrology, and knows what immunocytochemical tests are required and the preferred way of dealing with the specimen. Sometimes, however, the need for immunocytochemistry is recognized too late for suitable fixation and processing and it may be necessary to subject the patient to a second biopsy to establish the diagnosis. This problem will be less common if surgeons are encouraged to submit unfixed specimens of all lymph node biopsies and all unexplained or atypical lesions of other tissues to the histopathology laboratory as though for urgent frozen section. The pathologist should then snap-freeze several small unfixed blocks and retain them wrapped in tinfoil at $-70\,°C$ until routine paraffin histology shows whether immunocytochemistry is necessary. The snap-frozen blocks may then be used for the preparation of unfixed cryostat sections, or fixed and embedded as required (*Fig.* 40.1). In addition to their value in immunocytochemistry, samples of frozen tissue may be homogenized for biochemical or microbiological investigations should these be indicated. The importance in difficult cases of retaining unfixed frozen samples of

Table 40.2. Immunocytochemistry in tumour diagnosis

Nature of specimen	Question	Relevant antigen
Anaplastic tumour	Epithelial or lymphoid?	Cytokeratin, e.g. CAM 5.2 Epithelial membrane antigen (EMA) Leucocyte common antigen
	Melanoma?	S-100
	Glioma?	Glial fibrillary acidic protein
Spindle cell tumour	Sarcoma or carcinoma?	Cytokeratin Vimentin Desmin
Lymph node, marrow	Are micrometastases present?	CAM 5.2 EMA, CA1 CEA
Pleural biopsy	Mesothelioma or adenocarcin- oma?	CEA
Metastatic adenocarcinoma	Is the primary in thyroid or prostate?	Thyroglobulin Prostatic acid phosphatase
Endocrine tumour	Hormone production? Tumour marker?	Calcitonin Gastrin Glucagon Insulin ACTH etc.
Teratoma	Tumour marker?	α-Fetoprotein Chorionic gonadotrophin, β- chain
Adenocarcinoma	Tumour marker?	CEA

See Chapters 20, 23, 25, 31.

tissue cannot be overemphasized nor can the possible medicolegal consequences of failure to do so when a biopsy is carried out for diagnostic purposes.

An advantage of the scheme shown in *Fig.* 40.1 is that it diminishes the chance of unwittingly preparing cryostat sections of unfixed tissue infected with dangerous pathogens. In this respect the hazard of tuberculous infection is well known. Sectioning of unfixed lymph nodes from patients with T-cell leukaemia due to HTLV-1 or the lymphadenopathy of AIDS (HTLV-3) poses a new threat for the cryostat operator and other laboratory workers; this can be minimized by awareness of relevant clinical history, appropriate serological investigations prior to lymph node biopsy, and preliminary study of formalin-fixed paraffin sections by a histopathologist familiar with the microscopical appearance of these conditions. Pending further information on what precautions are appropriate, it is advisable in the meantime to trim, section and process all unfixed lymph node

Table 40.3. **Immunocytochemistry in lymphoreticular disorders**

Nature of specimen	Question	Relevant antigen
Lymphoreticular tumour	B-cells, T-cells or histiocytes?	Cytoplasmic or surface Ig
		J chain T- cell markers Muramidase α-1-Antitrypsin α-1-Antichymotrypsin
Proliferation of B-cells	Reactive or neoplastic?	Monotypic kappa or lambda
Skin	Benign dermatosis or cutaneous T-cell lymphoma	BE1, BE2
	Histiocytosis X	OKT 6

See Chapters 34, 36.

Table 40.4. **Recommended fixation and processing of tissues for immunoperoxidase staining of various antigens**

Antigen	Recommended fixative	Type of section	Chapter
Cytoplasmic immunoglobulin	Neutral buffered formol saline	Paraffin	34
Muramidase	Neutral buffered formol saline	Paraffin	34
Surface immunoglobulin	Acetone	Frozen	34
T-cell markers	Acetone	Frozen	34
Cytokeratin (mouse mono- clonal CAM 5.2)	Formol saline	Paraffin	31
Peptide hormones	Benzoquinone (solution)	Frozen	
	Benzoquinone (vapour)	Paraffin	20

biopsies in a room set aside for the purpose and equipped with a safety cabinet and dedicated cryostat and staining equipment.[2]

3. INTERPRETATION OF RESULTS

Due to the relatively recent application of immunocytochemistry to diagnostic histopathology, the continuing rapid development of techniques, the introduction of new antibodies and the specific nature of the information they provide, there are considerable problems in interpreting many results.

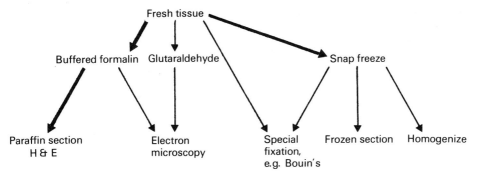

Fig. 40.1. Flow diagram for processing biopsies in which diagnostic difficulty is anticipated. Heavy arrows: recommended preliminary steps. Light arrows: optional steps.

The effects of fixation and tissue processing have already been discussed. Blocks of tissue in which positive staining is restricted to the periphery, presumably due to fixation artefact, are unsuitable for diagnostic purposes.

The choice of primary antibody and the dilution at which it is used is of great importance. Many antibodies give unwanted staining even at high dilution, e.g. antibody to epithelial membrane antigen reacts with plasma cells, and these cross-reactions must be known and taken into account. They occur not only with conventional polyclonal antibodies, but also with many monoclonals. However, the unlimited supply of each monoclonal provides a better opportunity for standardization of results and the pooling of information from various sources about problems encountered in interpretation.

The choice of immunocytochemical staining technique also has a significant effect on the results obtained. For example, immunoglobulin on the surface of B-lymphocytes cannot be reliably demonstrated in formalin-fixed paraffin sections stained by immunoperoxidase with diaminobenzidine as chromogen, whereas satisfactory results have been reported in such material with the more sensitive immunogold–silver technique.[1] Unfortunately, the increased sensitivity of the latter method may reveal previously unobserved cross-reactions which have to be recognized and considered in the interpretation of results.

Immunocytochemical staining is also subject to considerable vagaries even in experienced hands and using monoclonal antibodies and standard techniques. For this reason appropriate positive and negative controls must always be included in each batch of tests (*see* Chapter 31). Caution must be exercised in interpreting tests in which the staining is weak since this may indicate a non-specific cross-reaction even if the controls are satisfactory.

There may also be difficulties in interpretation at the histological level. These mainly arise when subtle localization of the antigen is required in a complex and heterogeneous tissue such as lymph node. When this is so, the pathologist must be familiar both with the histological features of the lesion and the immunocytochemical abnormality which is being sought. It is also necessary to recognize spurious staining such as the uptake of plasma proteins including immunoglobulin and α-1-antitrypsin by injured cells (*see* Chapter 34). The examination of control sections should prevent the mistake of confusing melanin, haemosiderin or endogenous peroxidase with the reaction products of immunoperoxidase staining.

Finally, consideration must be given to the biological and clinical significance of the results obtained. The relevance of many findings, such as the demonstration of a tissue-specific antigen in tumour cells of unknown origin, is often self evident, though the pathologist must be aware of pitfalls which may arise from antigen loss in poorly differentiated lesions, and from inappropriate expression of antigen by neoplastic cells. In the interpretation of results it is essential to understand the importance of quantitative factors and to appreciate that there is no *a priori* reason why the amount of antigen required to give a positive as opposed to a negative immunocytochemical reaction should correspond to a threshold concentration of clinical or biological significance. Well-established tests like those used in glomerulonephritis have been carefully assessed by follow-up of large numbers of cases and correlation with other findings. However, much of the information provided by immunocytochemical staining is recent and the clinical implications have still to be evaluated. It thus remains to be seen to what extent abnormal expresion of HLA antigens in cardiac and renal transplants (*see* Chapter 30) will assist in the diagnosis of graft rejection, whether immunocytochemistry will make a significant practical contribution to the diagnosis of degenerative diseases of the nervous system (*see* Chapter 25) and skeletal muscle (*see* Chapter 39) and if the expression of inappropriate antigens is related to tumour prognosis (*see* Chapter 31).

Preliminary attempts to standardize immunocytochemical staining quantitatively have been reported with an artificial step-wedge containing graded amounts of antigen.[3] This approach may also lead to the use of immunocytochemistry for quantitative investigations at present carried out, often somewhat unsatisfactorily, by assays on tissue homogenates.

4. PROVISION OF A DIAGNOSTIC IMMUNOCYTOCHEMISTRY SERVICE

It is already clear that immunocytochemistry is revolutionizing diagnostic histopathology by adding to it the new dimension of molecular biology. How can a clinically relevant cost-effective service be provided?

There are no published estimates of the proportion of cases in which immunocytochemistry is required in routine histopathology. At the Glasgow Royal Infirmary, a teaching hospital with most of the medical and surgical specialities, approximately three per cent of routine biopsies are referred by the Surgical Pathology Service to the Immunocytochemistry Laboratory, one-third to test for intermediate filaments and leucocyte common antigen, an equal number for the investigation of suspected lymphoreticular tumours and the remaining third for various other tests, mainly relating to the diagnosis of pigmented or endocrine tumours. These figures do not include specimens sent for research purposes or from other hospitals, and exclude skin and renal biopsies which are referred to other laboratories for the investigation of tissue deposits of immunoglobulin and complement.

Tests which are straightforward and commonly required, e.g. to distinguish anaplastic carcinoma from lymphoma, should be carried out in all routine histopathology laboratories. There are several advantages to performing such tests at local level. They maintain the professional interest of laboratory staff, serve as a reminder of the potential benefit of other immunocytochemical tests to the patient, avoid delay in reporting and improve opportunities for communication between clinician and pathologist. Hospitals with large departments specializing in nephrology, dermatology etc. should also provide relevant immunocytochemical tests for these specialities. Tests which are rarely required, e.g. the demonstration of hormones in endocrine tumours, or are difficult to interpret, e.g. in certain lymphoid tumours, should be performed in departments with appropriate special interests. Lists of departments willing to provide an immunocytochemistry service for others should be published. Such departments should received unstained material to carry out the tests using the methods and reagents with which they are familiar and should, if possible, return a set of stained slides with their report to the referring pathologist.

It is certain that the continuing production of new monoclonal antibodies will lead to progressive improvement in the diagnostic reagents available to the pathologist. Two related logistic problems in routine practice remain to be solved. First is the timely introduction of new reagents and other technical advances, especially when this involves discarding others which are familiar and well tried. Second is the difficulty of standardizing methods and reagents used by different laboratories, a matter of great importance if reproducible results of comparable clinical significance are to be obtained.

Undoubtedly, the manpower and reagents required for immunocytochemistry will add to the cost of histopathology, but if the development is coordinated at a national level, the increase will be justified by the resulting improvement in quality of patient care.

REFERENCES

1. Holgate C. S., Jackson P., Lauder I., Cowen P. M. and Bird C. C. Surface membrane staining of immunoglobulin in paraffin sections of non-Hodgkin's lymphomas using immunogold–silver technique. *J. Clin. Pathol.* 1983, **35**, 742–746.
2. Department of Health and Social Security. Advisory Committee on Dangerous Pathogens. *Interim Guidelines on Acquired Immune Deficiency Syndrome (AIDS).* 1985, NC(85)1.
3. Millar D. A. and Williams E. D. A step-wedge standard for the quantification of immunoperoxidase technique. *Histochem. J.* 1982, **14**, 609–620.

Index

Heavy type indicates description of technical details. Where a page number is followed by a Figure (or Table) reference in brackets the indication is to the Figure. Where a Figure is referred to without brackets the indication is to the illustration(s) within the pages quoted.

Absorption controls 36–38
Actin
  in pituitary 425–6, 427 (*Fig.* 26.1)
α-Actinin
  in skeletal muscle 669, 670 (*Fig.* 39.11)
Acute lymphoblastic leukaemia
  cell markers 601–4, *Table* 35.4
Acute myeloid leukaemia
  cell markers 602 (*Table* 35.3), 607–10, *Figs* 35.3, 35.4
Addison's disease
  autoantibodies to steroid-producing cells 463
Adherence of sections to slides 32
  poly(L-lysine) 32, **48**, 364, 562
Adjuvants in immunization 17
Adrenaline (see also Amines) 285–6, *Fig.* 18.1
Adrenocorticotrophic hormone
  ectopic production by tumours 512
Affinity purification of antibodies **6**, 38
Aging process in neurones
  morphometry and microdensitometry 220–3, *Table* 13.2
Agnathans
  regulatory peptides 317, 320 (*Fig.* 19.3)
Alkaline phosphatase as label 35
  incubating solution **47–8**
  inhibitor of endogenous enzyme 35, 48
Alkaline phosphatase—anti-alkaline phosphatase 28, 35, **47**, *Plate* 37
Alzheimer's disease 353–5
  neurofibrillary tangles 353, 355 (*Fig.* 21.1)
  neurofilaments and glial filaments 418–9, 421 (*Fig.* 25.3)
  senile plaques 353, 355 (*Fig.* 21.1)
  somatostatin and NPY 353–5
Amines
  biosynthetic pathways 285–7
  fixation for immunocytochemistry 289–91
  immunocytochemistry of dopamine, noradrenaline, adrenaline, serotonin, histamine 284–303
  sensitivity of immunostaining 301–2

Amines (*cont.*)
  specificity of immunostaining 298–800, *Tables* 18.1, 18.2
  in spinal cord 363 (*Fig.* 22.2), 365, *Table* 22.1
  in tumours 285, 302
3-Amino-9-ethylcarbazole 35
Amyloid
  in renal disease 647
Annelids
  neuropeptides 310–11
Antibodies
  affinity 17
  affinity purification **6**, 38
  to amines 288
    preparation 289
    specificity 298–800, *Tables* 18.1, 18.2
  to amine-synthesizing enzymes 287–8
  to B-cells 602 (*Table* 35.2)
  cross-reactivity 7
  to cytokeratins
    monoclonal 396–8, *Fig.* 23.3
    polyclonal 395–6
  detection in serum 5
  to glomerular basement membrane 638, 642, 643 (*Fig.* 37.2)
  to intermediate filament proteins 391
    cross-reactivity 398, 415–6
    specificity 413, 414, 415
    in tumour diagnosis 3–14, 392–8, *Tables* 23.1, 23.2, *Plate* 27
  to keratin
    for epithelial cell characterization 262–6, *Figs.* 16.1–16.5
  labelling 34
  for membrane markers on lymphoid and myeloid cells 601, *Table* 35.1, 602 (*Tables* 35.2, 35.3)
  monoclonal
    definition 15–16
  raising 13–25
  monospecific 4
  to oestrogen receptors 194–5

Antibodies (*cont.*)
  to oncogene-encoded proteins 277–9
    specificity 278–9
  polyclonal
    affinity purification 6
    immunization schedule 4
    monospecific 4
    raising 3–12
    species used for raising 4
    testing 5–11
  purification 38
  screening 5
  storage 33–4
  to T-cells 601 (*Table* 35.1), 622 (*Table* 36.2),
    629–31
  working dilution 33
Antibody
  adsorbed to colloidal gold **137–8**
  production mechanism 13–15
  specificity 5–7
  structure 13, 14 (*Fig.* 2.1)
a-1-Antichymotrypsin
  synthesized by macrophages 577
  in malignant histiocytes 591
Anti-cytoplasmic autoantibodies
  to cytoskeleton 485, 486 (*Table* 29.2)
  to Golgi complex 484, *Plate* 31
  to microsomes of liver and kidney 481–4,
    Table 29.1
    delta agent 484
  to mitochondria 481, *Plate* 31
  to ribosomes 481, 482 (*Table* 29.1)
Antidiuretic hormone
  in oat cell carcinomas 512
Antigen
  semi-quantitative analysis 219–20,
    Fig. 13.1–13.12, *Plate* 10
  immunogenicity 17
  identification
    by biochemistry 8–10
    by immunoblotting 9–10
    by immunoprecipitation 9
  purification
    by affinity chromatography 4
    by high resolution SDS-PAGE 4
    by isoelectric focusing 4
  purity
    peptides 4
    testing by silver staining 4
Antigens
  families and cross-reactivity 37, 40, 317–21,
    651, 654
Antinuclear autoantibodies 478–80, *Plates* 30,
    31
  to centromeres 480, *Plate* 31
    in CREST syndrome 480
  to DNA 479, *Plate* 31
    in systemic lupus erythematosus 479
  to extractable nuclear antigens 478–9
  to histones 479–80

Antinuclear autoantibodies (*cont.*)
  to histones (*cont.*)
    in systemic lupus erythematosus 479–80
  to proliferating cell nuclear antigen 480, *Plate* 31
  screening procedures 478
α-1-Antitrypsin
  in histiocytic lymphoma 589–91, *Fig.* 34.26
  synthesized by macrophages 576 (*Fig.* 34.9),
    577
  taken up by degenerating cells 580
  as tumour marker 515
Arthropods
  neuropeptides 312–6
Autoantibodies 456–71
  to acetylcholine receptors
    in myasthenia gravis 466
  to adrenal cortex 462–3
    detected by ELISA and RIA 463
  to cytoplasm
    in non-organ-specific autoimmune diseases
      480–5, *Plate* 31
  in diarrhoea 467–8
  to endocrine cells 456–8, *Table* 28.1
  to gastric inhibitory polypeptide cells
    in coeliac disease 467
    in diabetes mellitus 467
  to gastric parietal cells 466
  to gastrin cells 466–7, *Plates* 28, 29
  to glucagon cells 462
  to gut mucosal cells 466–8
  indirect immunofluorescence 456–7, ***Fig.* 28.1**
  in myasthenia gravis 465–6
  in non-organ-specific autoimmune diseases
    477–86
  to nuclei 478–80, *Plates* 30, 31
  to pancreatic islet cells
    cell surface 461–2
    complement fixing 461
    cytoplasmic 459–61, *Fig.* 28.2
  to pituitary
    in diabetes mellitus, Type I 468–9
    ACTH cells 468
    growth hormone cells 468
    prolactin cells 468
  to secretin cells
    in coeliac disease 467
  to smooth muscle 485, 486 (*Table* 29.2)
  to somatostatin cells 462
  to steroid-producing cells 463
  to striated muscle fibres
    in myasthenia gravis 465–6
  to thyroid 463–5, *Fig.* 28.3
Autoimmune disease
  connected with micro-organisms 653
  non-organ-specific 477–86
  organ-specific 456–71
Automatic image analysis
  clusters of transmitter-identified nerve cells
    209–11, 212 (*Table* 13.1), *Figs.* 13.1, 13.7
  of transmitter-identified neurones 205–23

Autonomic centres of spinal cord
 peptide-containing nerves 379–382, *Fig* 22.10,
  Table 22.4
Autoradiography
 in situ hybridization of oncogene mRNA 276
 in radioimmunocytochemical methods 109,
  113
Avidin 55
Avidin-biotin methods
 advantages 60, 62–3
 applications 62
 avidin-biotin-peroxidase complex 59 (*Fig.*
  4.4), 60, **67**
 bridge avidin 59–60, *Fig.* 4.3
 electron microscopical immunocytochemistry
  63, 64 (*Fig.* 4.9)
 labelled avidin 57, 58 (*Figs.* 4.2, 4.3), 59, **66–7**
 for oncogene-encoded proteins 279
 pitfalls
  endogenous biotin 64
  mast cell staining 64
 principles 55
 procedure 57–63, **65–8**
 in radioimmunocytochemistry 100, 101 (*Fig.*
  7.1), 107, **110–4**
 sensitivity 61 (*Fig* 4.6), 63 (*Fig.* 4.8)
 staining schedules **65–8**
 streptavidin-biotin 55
  for skeletal muscle immunocytochemistry
   664–5, *Plate* 44
 versatality 61, 62 (*Fig.* 4.7)
Avidin-biotin reaction 56

Background staining (see also Non-specific
  staining) 38–40
B-cell lymphomas 579–89
B-cells (see also B-lymphocytes)
 antibodies 602 (*Tables* 35.2)
 heterogeneity 439 (*Fig.* 27.1), 441–3
 lymphomas 579–89, *Figs*. 34.13–34.22
 in normal lymphoid tissue 570–2, *Fig.* 34.2
B-chronic lymphoproliferative disorders 606–10,
  Table 35.6
p-Benzoquinone fixation
 for neuropeptides 31, 248–9
 perfusion of spinal cord **383**
Biotin 56 (*Fig.* 4.1)
Biotin-avidin (see Avidin-biotin)
Biotinylation
 procedure 56–7
 spacer arm 57
Blood cells
 immunocytochemistry 599–600
 surface markers
  direct immunofluorescence **610–1**, **612**
  immunogold method for electron microscopy **613**
  indirect immunofluorescence **610–1**, **612**

Blood group substances
 as tumour markers 518
B-lymphocytes (see also B-cells)
 stimulation 13–15
Bone marrow
 immunocytochemistry 599–600
Brain cells
 development in cultures 258
Brain neuropeptides
 immunocytochemistry 349–58
 radioimmunocytochemistry 99–113
 somatostatin and NYP 349–58
  in neurodegenerative diseases 353–357
Breast carcinoma
 carcinoembryonic antigen in 513
 cultured cells 263 (*Fig.* 16.1), 265–6
  milk fat globule membrane antigen 7, 266,
   Plate 13
  immunoperoxidase for oestrogen receptors
   270–71
 epithelial membrane antigen in 516, *Fig.* 31.5
 oestrogen receptors 188–9
 peroxidase-labelled oestrogen binding 190–2,
  Figs. 11.1, 11.2
Bullous disorders
 bullous pemphigoid 620, 623 (*Fig.* 36.2)
 characteristics 620–23, *Tables* 36.1, 36.2, *Figs*
  36.2–36.5
 cicatricial pemphigoid 620
 dermatitis herpetiformis 622, 623 (*Fig.* 36.3)
 epidermolysis bullosa
  immunocytochemistry 626–7, *Fig.* 36.8,
   Plates 40, 41
 herpes gestationis 621
 immunocytochemistry 621 (*Table* 36.1),
  623–6
 intra-epidermal 622–3, 624 (*Fig.* 36.4), 625
  (*Fig.* 36.5)
 pemphigus 622, 624 (*Fig.* 36.4), 625 (*Fig.*
  36.5), 626
  indirect C3 technique 626
 sub-epidermal 620–2, 623 (*Fig.* 36.2)

Ca1
 in cytological diagnosis 555
 as tumour marker 519
Calcitonin
 in C cell hyperplasia 542, 543 (*Fig.* 32.9)
 marker for C cells 536
 in medullary carcinoma of thyroid 510,
  Fig. 31.2, 534 (*Fig.* 32.1), 536–7, 538
  (*Fig.* 32.6)
Calcitonin gene-related peptide
 in C cell hyperplasia 542, 543 (*Fig.* 32.10)
 in nerves of cardiovascular system 340
 in spinal cord 366, 371 (*Fig.* 22.5), 374,
  376–82, *Figs*. 22.8–22.10

Calcitonin gene-related peptide (*cont.*)
  in thyroid medullary carcinoma 537, 538 (*Fig.* 32.7)
Capsaicin 103, 105 (*Fig.* 7.4)
  investigating origin of nerves 366–8, *Fig.* 22.4
Carcinoembryonic antigen
  cross-reaction with normal antigen 513–4
  in cytological diagnosis 553–4
  in thyroid medullary carcinoma 538–9, *Fig.* 32.8
  monoclonal antibodies 514
  as tumour marker 513–4, *Fig.* 31.4
Carcinoma
  cytokeratins as markers 396–7, *Fig.* 23.3, 577, 578 (*Fig.* 34.12)
Cardiac biopsies
  cellular infiltrates in transplants 493–6
  immunocytochemical techniques **493**, *Table* 30.1
Cardiovascular system
  diffuse neuroendocrine system 340–41
    pathology 341
  CGRP-immunoreactive nerves 340 (*Fig.* 20.7)
  regulatory peptide distribution 340–1
Carrier protein in immunization 17
C3B receptors
  on dendritic reticulum cells 577
C cells
  hyperplasia 542, 543 (*Figs.* 32.9, 32.10)
  CGRP in 537–8, *Fig.* 32.7
  calcitonin as marker 536–7
  katacalcin in 537
Cell and tissue cultures 245–59
  antigens in developing brain cells 258
  for autoantibody localization 478
  cell density for antigen localization 246
  cell lines 261–71
    choice of myeloma 18
    double immunofluorescence for membrane carbohydrates **270**, *Plate* 13
    epidermal growth factor receptors 267–8, *Fig.* 16.8
    immunocytochemistry 261–71
    indirect immunoperoxidase **269**
    oestrogen receptors 266–7, *Fig.* 16.7, **270**
    squamous carcinoma 262–3, *Fig.* 16.2
  cell surface antigens 247 (*Fig.* 15.1), 252–4, 255 (*Fig.* 15.6), 257 (*Fig.* 15.7), *Plate* 12
    colloidal gold marking 124–6, *Fig.* 5.5
    fixation 248–9
    immunoperoxidase **269–70**
    receptors 266–8
  differentiated functions and cell type identification 262–6
  epidermal cells
    keratin expression 263–4
  intracellular antigens 248–52
    distinction from surface antigens 248–52
    fixation 140, 486

Cell and tissue cultures (*cont.*)
  intracellular antigens (*cont.*)
    immunogold labelling **139–41**
    permeabilization 140, **248–50**, *Table* 15.1
  keratinocytes 262
Central canal area of spinal cord
  peptide-containing nerves 369 (*Fig.* 22.4), 378
Central nervous system (see also Brain and Spinal cord)
  immunocytochemistry
    tissue preparation 352
  intermediate filaments and differentiation 401–9
  preservation of markers in post-mortem tissue 349
  regulatory peptide distribution 349–51, *Table* 21.1
Centrocytic lymphoma 580–9, *Fig.* 34.23
Chlamydia trachomatis 561
  detection by monoclonal antibody 658–60, *Table* 38.1
4-Chloro-1-naphthol 35
Cholecystokinin
  antibody cross-reaction with gastrin 37
  in evolution 317–21
Chromogranin
  in endocrine cells 332
    of fundus 332, *Fig.* 20.2
  marker for neuroendocrine tumours 341
Class I antigens (HLA-A, B, C)
  in cardiac biopsies 494–6
  in skeletal muscle 666, 668 (*Fig.* 39.4)
Class II antigens (HLA-DR)
  in autoimmune thyroid disease 564
  on B-cells 570
  in crescentic glomerulonephritis 642
  in leukaemia 603, 604 (*Table* 35.4)
  in macrophages 443–9, *Fig.* 27.2, *Table* 27.3
Coelenterates
  gastrin/CCK-like immunoreactivity 321
  neuropeptides 309–10, *Fig.* 19.1, 321–2
Coeliac disease
  autoantibodies 467
Co-existence of neuroactive substances (see also Co-localization)
  quantitative analysis 214–19, *Figs.* 13.8–13.11
Colchicine treatment 100
  for corticotrophic cells 200
  for origin of intrinsic spinal cord nerves 370, 372 (*Fig.* 22.6), 374, 378
Collagen
  immunocytochemistry in skeletal muscle 665–6, 668 (*Figs* 39.1, 39.2)
Colloidal gold 36
  advantages and disadvantages as marker 115–6
  antibody/gold probes for electron microscopy **137–9**
  in light microscopy 71–85

Colloidal gold (*cont.*)
  microscopical enhancement 73
    dark field illumination 73
    epipolarization 73, 74 (*Fig.* 5.1), *Plate* 4
  particle size and absorbance 117 (*Fig.* 8.1,
    Table 8.1)
  physical enhancement (silver precipitation)
    75–81
  preparation of sols 116–7, **133–7**
  protein A/gold probe **138–9**
  quality control 121–2
  red colour 71
    applications in light microscopy 71
  silver precipitation 75–81
  stabilization with carbowax 20M **139**
  streptavidin/gold probe **138–9**
Colloidal gold-labelled antibodies (see Immuno-
    gold staining)
Colloidal gold probes 115–41
  adsorption of proteins 117–8, 119 (*Table* 8.2)
  double marking 129, 130 (*Fig.* 5.9)
  electron microscopy 126–31
  influence of size on marking efficiency 124
  markers for cell surface antigens 124–6, *Fig.*
    5.5
    label fracture technique 124–5, 126 (*Fig.*
      5.6)
  minimal protecting amount of protein 119,
    120 (*Fig.* 8.2), **137**
  pre-embedding in cell monolayers 131, 132–3
    (*Fig.* 5.10), **139–41**
  preparation, purification and storage 120–1,
    137–9
  protein A compared with secondary antibody
    128–9
    quality control 121–2, 122 (*Fig.* 8.3), 123
      (*Fig.* 8.4)
  stability 122–4
  for ultrathin sections
    frozen 127–8, 129 (*Fig.* 5.8)
    resin 127, 128 (*Fig.* 5.7)
Colloidal silver 36
  combined with colloidal gold 72
Co-localization of antigens (see also Co-
    existence)
  in autoimmune diseases 456–7, *et seq.*
  intermediate filament proteins 394–5
  in lymphoid cells 441–9
  methods for multiple staining **41–4**, 79.
    for electron microscopy 154–7
  peptides in dorsal root ganglia 374–6, *Fig.*
    22.8
  pituitary hormones 44
  somatostatin and NPY in brain 352
Colour films for photomicrography 239–41
Colour microradioautography
  for double labelling with DNA probes
    275
Common acute lymphocytic leukaemia antigen
  in follicle centre cell lymphomas 586

Complement
  in diseased muscle 666, 668 (*Fig.* 39.5),
    669
  in glomerular disease 641 (*Table* 37.1), 644
    (*Fig.* 37.3)
Complement fixation
  in skin 623–6, *Fig.* 36.6
Complement fixation test
  for islet cell autoantibodies 461
Complement fixing antibody
  indirect C3 technique 625–6, *Fig.* 36.7
Controls
  absorption 36–8
  for lectin binding 176–8
  negative 36
  positive 36
  tumour markers 506–7
Counterstains
  fluorescent 40
Coverslip thickness
  importance in microscopy 230
CREST syndrome
  autoantibodies to centromeres 480
Cross-linking reagents
  to permeabilize cultured cells 248–9
Cross-reactivity 7, 37–40
  intermediate filament antibodies 398
Crustaceans
  gastrin/CCK-like immunoreactivity 319
Cryostat sections
  for amine immunocytochemistry 291–2, *Figs.*
    18.5–18.9, 18.11
  for autoantibodies 456, 486–7
  for intermediate filament immunocytochemis-
    try 398
  for lymphoid tissue 31, 45
    immunoperoxidase method **594**
  pre-fixed, for peptides in nerves 31, 364, 383
Cryoultramicrotomy 148, **158–9**
Cultured cells (see Cell and tissue cultures)
Cutaneous T-cell lymphomas 627–32 (*Fig.* 36.9),
    633 (*Figs.* 36.10, 36.11), 635 (*Fig.* 36.12)
Cytocentrifuge preparations
  immunocytochemistry **612**
Cytokeratins (see also Keratin)
  biochemical diversity 395
  in carcinoma diagnosis 577, 578 (*Fig.* 34.12)
  co-localized with other intermediate filament
    proteins 394–5
  distribution of types 395–8 (*Fig.* 23.3)
  and epithelial tumour type 395–8
  monoclonal antibodies 396–8 (*Fig.* 23.3)
    in carcinoma diagnosis 396–7 (*Fig.* 23.3)
      519, 520 (*Fig.* 31.6)
  tissue specificity and exceptions 390–2 (*Table*
    23.1)
Cytological preparations 547–64
  cell pellets
    agar for wax embedding **563**
    celloidin bag for wax embedding **563–4**

Cytological preparations (*cont.*)
  diagnostic immunocytochemistry
    gynaecological specimens 18–23, 557–61
    non-gynaecological specimens 553–7
  immunocytochemical controls 561
  Chlamydia trachomatis 561, 658–60, *Table* 38.1
  immunocytochemistry
    interpretation 561–2
    methods 551–2, **564**
  permeabilization 551
  smears **562**
  storage 551
Cytological smears
  fixation for immunocytochemistry 549–50
  preparation for immunocytochemistry 549, **562–3**
Cytology
  immunocytochemistry in diagnosis 547–64
    advantages 547–8, 552
Cytomegalovirus
  diagnosis with monoclonal antibodies 656–7
  localization by antibodies 652, 653 (*Fig.* 38.2)
Cytoskeletal proteins
  autoantibodies 485, 486 (*Table* 29.2)
  in spinal cord 361, 362 (*Fig.* 22.1)
Cytoskeleton 390

Dendritic reticulum cells
  in normal lymphoid tissue 571 (*Table* 34.1), 576, 577 (*Fig.* 34.11)
Dermatology
  immunocytochemistry 618–36, *Table* 36.1
Desmin
  co-localized with other intermediate filament proteins 394–5
  in normal and diseased skeletal muscle 667–9
  in pituitary 426–7, 428 (*Fig.* 26.2)
  tissue specificity and exceptions 390–2, *Table* 13.1
Desoxyribonucleic acid (see DNA)
Diabetes insipidus
  autoantibodies to vasopressin cells 469–70
Diabetes mellitus
  type I, insulin-dependent
    autoantibodies 459, 468–9
  type II, non-insulin-dependent
    autoantibodies to GIP cells 467
Diagnostic histopathology
  immunocytochemistry
    implications 647–81
    service 680–1
    uses 675, 676 (*Table* 40.1), 678 (*Table* 40.3)
  specimen fixation and processing 676, 678 (*Table* 40.4), 679 (*Fig.* 40.1)
Diaminobenzidine
  alternative reagents 35

Diaminobenzidine (*cont.*)
  for peroxidase development 34, 46
  and peroxidase for electron microscopical immunocytochemistry 146–7, *Fig.* 9.1
  storage **46**
Differential colour processing
  in hybridization histochemistry 201
Diffuse neuroendocrine system
  cardiovascular system 340–1
  immunocytochemical visualization 331
  in genital tract 338–9
  immunocytochemistry 328–44
  morphology
    nerves 330–1
    neoplasms 341–3
Dinitrophenyl hapten sandwich staining procedure **542–4**
Dilution of antibodies 28
Direct method 26
DNA
  labelled probes for in situ hybridization of oncogene mRNA 274–5
  cDNA:mRNA hybridization (see also Hybridization histochemistry) 199–200, *Fig.* 12.1, **202–3**
DOPA decarboxylase
  in amine synthesis 286
Dopamine (see also Amines) 284–6, *Fig.* 18.1
  distribution in mammals 284
Dopamine b-hydroxylase
  in amine synthesis 286
Dorsal horn
  peptide-containing nerves 377–8, *Table* 22.3, *Figs.* 22.1, 22.6
Dorsal root ganglia
  peptide-containing nerves 374–6, *Figs.* 22.7, 22.8
  co-localization of peptides 374–6, *Fig.* 22.8
Double direct method 42
Double immunofluorescence
  for autoantibodies 456–8, *Fig.* **28.1**
  for B-cells 441–3, *Table* 27.1
  lymphoid system 438–53
    methods 441–3, *Table* 27.1
  with monoclonal antibodies **270**, *Plate* 13
Double immunogold staining 129–30, *Fig.* 8.9, 330 (*Fig.* 20.1)
Double indirect method 42
Double labelling
  DNA probes for in situ hybridization 275
  for electron microscopical immunocytochemistry 129–30, *Fig.* 8.9, 154–7
    direct method 154, 155 (*Fig.* 9.5)
    indirect method – gold-labelled antigen detection 154
    indirect method – immunoglobulin/gold 13–4, 130, 154–7, *Figs.* 9.5–9.7
    indirect method – protein A/gold 130, 154–6
    selected surface immunolabelling 155 (*Fig.* 9.5), 156 (*Fig.* 9.6), **163–4**

Double labelling (*cont.*)
hybridization histochemistry 201
radioimmunocytochemistry combined with
immunoperoxidase 101, 104, **112**
Duchenne muscular dystrophy
complement binding 667
fetal myosin in regenerating fibres 669, 670
(*Fig.* 39.10)

Enterochromaffin-like cells
hyperplasia 335
Electron microscopical immunocytochemistry
146–62
cryoultramicrotome sections
procedure for immunolabelling **158–9**
double immunogold staining 330 (*Fig.* 20.1)
embedding media 153
fixation 150–2
glucagon cells of pancreas 330, *Fig.* 20.1
markers
colloidal gold 115–41, 148, 151 (*Fig.* 9.2)
ferritin 147–8
imposil 147–8
metals 148
peroxidase 146–7
polyethylenimine 148
radioactive 148
methods
cryoultramicrotomy 148
pre-embedding 149, **159–60**, *Fig.* 9.1
post-embedding, on-grid 149–50, **162–3**
post-embedding, semithin-thin 149
multiple staining procedures 154–7, *Figs.* 9.5,
9.6, 9.7
on-grid labelling
washing grids **157**
resin sections
etching 153
of skeletal muscle 665
tissue preparation 150–4
treatment after fixation 152–4
Elution method for multiple staining 42
Endocrine cells
in gastrointestinal tract 333
hyperplasia 335–6
Endocrine system
autoantibodies 456–7, *Table* 28.1
diffuse neuroendocrine system 329–44
thyroid diseases, 533–44
Endogenous alkaline phosphatase
blocking 35
Endogenous biotin 64
blocking 67–8
Endogenous peroxidase
blocking 35, 40, **66**, 506, **383–4, 521**
Epidermal growth factor
receptors 267–8, *Fig.* 16.8

Epi-polarization microscopy
colloidal gold 73, 74, *Fig.* 5.1, *Plate* 4
silver-coated gold particles 80, *Plate* 8
Epithelia
and cytokeratin type 395–6
Epithelial cells
characterized by antibodies to keratin 262–6,
Figs. 16.1–16.5
Epithelial membrane antigen
in cytological diagnosis 554
cervical cancer 557–8
as tumour marker 516–7, *Fig.* 31.5
Epithelial tumours
cytokeratins 395–8
Erythroleukaemia
monoclonal antibodies 609–10, *Fig.* 35.5
Evolution
pancreatic islets 321
peptide families 317–21
gastrin/cholecystokinin 317–21
thyroglobulin 321

Factor VIII
marker for endothelial tumours 518
marker for normal endothelium 518
Fading of fluorescence
prevention by mountant **441**
Fc receptors
binding of immunoglobulin 39
Ferritin
in electron microscopical immunocytochemistry 147–8
α-Fetoprotein
as tumour marker 514–5
Fibronectin
in muscle 566
as tumour marker 515
Fixation
acetone 31, 193, 440, 493
alcohol 30, 486
of amine-containing tissues 289–90
p-benzoquinone 31, **49, 383**
cell and tissue cultures 248–9
for electron microscopy **140**
cytological smears 549–50
for diagnostic histopathology 676, 678 (*Table*
40.4), 679 (*Fig.* 40.1)
for electron microscopical immunocytochemistry 150–2, 140
formalin 29, 569–70
glutaraldehyde 140, 276
for hybridization histochemistry 200, 276
for intermediate filament immunocytochemistry **398**
for lymphoid cell antigens 449–51, *Table* 27.4
lymphoreticular tumours 569–70
neutral buffered formol saline 569–70
of oncogene mRNA 275–6

Fixation (*cont.*)
  osmium tetroxide 151–2, 161
  periodate-lysine-paraformaldehyde 100, 449–51
  protease digestion 29, 178, 570
  for semethin frozen sections **90–1, 93–4**
  of spinal cord 364, **383**
  vapour 32, 152
Flow cytometry
  in cytological diagnosis 558–9
Fluorescein 34
Fluorescein-conjugated oestrogen 192
Fluorescent counterstains 40
Fluorescent labels 34
Follicle centre cell lymphomas 581 (*Figs*. 34.13, 34.14), 582 (*Figs*. 34.15, 34.16), 583 (*Fig*. 34.17), 585–7 *Figs*. 34.19–34.22
Freeze drying
  for electron microscopical immunocytochemistry 152 (*Fig*. 9.3), 153 (*Fig*. 9.4)
  for lymphocyte membrane antigens 31
  for peptides 32
Frozen sections
  for amine immunocytochemistry 291–2, *Figs*. 18.5–18.9, 18.11
  for autoantibodies 456
  for cardiac biopsies 493
  for intermediate filaments 398
  for in situ hybridization 202
  for lymphoid tissue 440, 569
  for muscle biopsies 664
  for oestrogen receptors 193
  post-fixed for neuropeptides 364, 383
  for renal biopsies 497
  semithin 89–97
  ultrathin 127–8, 129 (*Fig*. 8.8), 158–9

β-D-Galactosidase 35
Gastrin cells
  hyperplasia 335
Gastrin/cholecystokinin
  evolution 317–21
  in insects 318–9, *Fig*. 19.2
Gastrin/cholecystokinin-like immunoreactivity
  in coelenterates 309, 321
  in crustaceans 319
Gastrointestinal tract
  diffuse neuroendocrine system 333–6
  pathology 335–6
  endocrine cells 333, 334, 335–6
  peptide-containing nerves 333–4, *Fig*. 20.4, 336
    hyperplasia 336
    hypoplasia 336
  regulatory peptide distribution 333–6
Genital tract
  diffuse neuroendocrine system 338–9
  pathology 339

Genital tract (*cont.*)
  regulatory peptides 338–9
Glial fibrillary acidic protein 415
  in astrocytes 258, 402
  co-localized with vimentin 394
  distribution 415
  in CNS development 402
  in gastrointestinal tract 251, *Fig*. 15.4
  not tumour-specific 417
  in pituitary 427–8, 430 (*Figs*. 26.3, 26.4), 431 (*Figs*. 26.5, 26.6)
  tissue specificity and exceptions 390–2, *Table* 23.1
  in tumours 417–8, *Table* 25.1
Glial filaments 401–2
  in neuropathology 413–22
Glomerular basement membrane
  antibodies 638, 642, 643 (*Fig*. 37.2)
  composition 640
Glomerular disease
  crescentic glomerulonephritis 642, 643 (*Fig*. 37.2)
  immunocytochemistry 638–48, *Table* 37.1
    paraffin sections 639
  focal sclerosis 642
  mesangial IgA disease 645, *Fig*. 37.4
  membranous glomerulonephritis 639 (*Fig*. 37.1), 642
  mesangiocapillary glomerulonephritis
    Type I 643–4, *Fig*. 37.3
    Type II 644
  mesangioproliferative glomerulonephritis 644
  minimal change glomerulonephritis 641
  proliferative glomerulonephritis 643
  systemic 646–7
Glucose oxidase 35
  in hapten sandwich staining procedure **542–4**
Glutaraldehyde **140**, 276
Gold labelled antigen detection method
  for electron microscopical immunocytochemistry 154
Gold probes (see also Colloidal gold)
  preparation and use 115–41
Goodpasture's syndrome 640
Growth hormone
  as tumour marker 511–2

Haematology
  cell markers 599–614
Hapten sandwich method 35–36, **542–4**
  for electron microscopical immunocytochemistry 147
Hashimoto's thyroiditis 463
HAT selection 20, 21 (*Fig*. 2.2)
Heart transplants
  cellular infiltrate
    immunocytochemistry 493–6

Heart transplants (*cont.*)
　histocompatibility antigens, Class I and II
　　494–6, *Plates* 32, 33
　immunocytochemistry 492–9
　T-cell ratio in peripheral blood 496–7
Herpes simplex virus
　in cervical smears 559, 560 (*Fig.* 33.1) **564**
　diagnosis with monoclonal antibodies 657, 658
　　(*Fig.* 38.5)
　immunogold-silver staining 559, 560 (*Fig.*
　　33.1), **564**
　PAP method **564**
Histamine (see also Amines) 285–6, 287 (*Fig.*
　18.3)
　distribution in mammals 285
Histaminase
　in medullary thyroid carcinoma 537
　as tumour marker 511
Histiocytic lymphoma 589–91, *Fig.* 34.26
Histocompatibility antigens
　class I
　　heart transplants 494–6, *Plates* 32, 33
　class II (see also HLA-DR)
　　in rejected kidney transplants 499
HLA-DR
　in acute lymphoblastic leukaemia 603, 604
　　(*Table* 35.4)
　in autoimmune thyroid disease 564
　on B-cells 570
　in crescentic glomerulonephritis 642
　in macrophages 443–9 *Fig.* 27.2, *Table* 27.3
Hodgkin's disease
　Reed-Sternberg cells 592, *Fig.* 34.27
Human chorionic gonadotrophin
　as tumour marker 507–9, *Fig.* 31.1, *Plate* 34
Human milk fat globule membrane antigen
　in cytological diagnosis 554, *Plate* 37
　expression in cultured breast cells
　　defined by monoclonal antibody LICR-
　　　LON-M8 266, *Plate* 13
　as tumour marker 516–7
Human papillomavirus
　in cervical biopsies 559, 561 (*Fig.* 33.2)
　DNA hybridization and malignancy 559
Human placental lactogen 509
Huntingdon's chorea 356–7
　somatostatin and NPY 355 (*Fig.* 21.2), 356–7,
　　Plate 26
Hybridization histochemistry 198–203
　analysis of mRNA 198–203
　applications 200–2
　cell architecture 200
　combined with other techniques
　　immunohistochemistry 201
　　retrograde transport of fluorescent dye 201,
　　　Plate 9
　double labelling
　　differential processing 201
　　with isotopes of different energy 201
　oncogene mRNA 274–7

Hybridization histochemistry (*cont.*)
　oncogene mRNA (*cont.*)
　　conditions for reaction 275
　pro-epidermal growth factor 201
　pro-opiomelanocortin in rat brain with cDNA
　　202
　pro-opiomelanocortin systems in pituitary 199
　　(*Fig.* 12.1), 200
　quantitation of mRNA 201–2
　size of probe 200
　technique 199–200, *Fig.* 12.1, **202–3**
　tissue fixation 200
Hybridomas 15–16
　cloning 22, 23 (*Fig.* 2.3), 24 (*Fig.* 2.4)
　feeder cells 19
　fusion 19, 21 (*Fig.* 2.2)
　plating out 19
　HAT selection 20, 21 (*Fig.* 2.2)
Hydra
　neuropeptides 309–10

Imidazole
　to intensify peroxidase reaction 41
Immune response 13–5
Immunization
　adjuvants 17
　antigen immunogenicity 17
　carrier protein 17
　for monoclonal antibody production
　　route 16
　　species 16, 18
　sample schedule 4–5
Immunoblastic lymphoma 588
Immunoblot technique 9–10, *Plate* 1
Immunocytochemistry
　autoimmune diseases
　　organ-specific 456–71
　　non-organ-specific 477–86
　blood and bone marrow samples 599–600
　of brain neuropeptides 349–58
　in cardiac and renal transplantation 492–9
　of cell lines 261–71
　cell and tissue cultures 245–59
　combined with hybridization histochemistry 201
　combined with nerve tracing 373–4
　controls 36–8, 506–7
　cytological preparations 547–64
　definition 26
　dermatology 618–36, *Table* 36.1
　diagnostic service 680–1
　diffuse neuroendocrine system 328–44
　electron microscopy 146–62
　essential conditions 2
　fixation 28–30, *Fig.* 3.2, 29 (*Table* 3.1)
　glomerular disease of kidney 638–48
　haematology 599–614
　implications for diagnostic histopathology
　　674–81
　indirect immunoperoxidase technique **521**

Immunocytochemistry (*cont.*)
 intermediate filament proteins 390–99
  in CNS development 401–9
 of lymphoid tissue
  preparation 440–1, 449–51, *Table* 27.4
 lymphoreticular tumours 568–96
 methods
  alkaline phosphatase – anti-alkaline phosphatase 28, **47**
  direct 26
  dinitrophenyl hapten sandwich **542–4**
  indirect 27, **45–6**
  multiple staining 41–44
  peroxidase – anti-peroxidase 27, *Fig.* 3.1, **46**
  unlabelled antibody-enzyme 26, 27–8, *Fig.*
   3.1
 micro-organisms 650–61
 neuropeptides 328–44
 neurotransmitters in spinal cord (see also
   Spinal cord) 360–85
 oestrogen receptors 189, 194–5, 266–7
 of oncogenes 273–81
  applications 279
 oncogene-encoded proteins 277–80
  non-specific staining 278, *Plates* 14, 15, 17
 peptide evolution 308–22
 pituitary
  non-hormonal markers 426–35
 quantitative analysis 205–23
 of receptors in cell lines 266–8
 regulatory peptides 328–44
 in routine diagnosis 519–20, 675–81
 semithin frozen sections 89–97
 of skeletal muscle 664–71
 skin diseases 618–36
 spinal cord
  antibody penetration **384**
  neurotransmitters 360–85
 thyroid diseases 533–44
 tumour markers 505–22
Immunofluorescence
 direct method for cell surface markers **610–2**
 improved resolution on semithin frozen sections 91 (*Fig.* 6.1), 92
 method **45**
Immunoglobulin/gold probes
 on-grid immunolabelling procedure **160–1**
Immunoglobulins
 binding to Fc receptors 39
 classes 13, 17, 22
Immunogold-silver staining 75–81, *Figs.* 5.2–5.6,
   131, 133 (*Fig.* 8.11)
 in cell cultures 80, *Fig.* 5.6
 for cell membrane antigens 79, *Plate* 7
 compared with PAP 77
 for cytological preparations 79
 for endocrine cells 77, 79 (*Fig.* 5.5)
 epi-polarization microscopy 80, *Plate* 8
 modifications 76
 in multiple staining methods 79, *Plate* 6

Immunogold-silver staining (*cont.*)
 for nerves 77, 78 (*Figs.* 5.3, 5.4)
 potential and applications 77
 schedule **83–5**
Immunoperoxidase
 on cytocentrifuge preparations **612–3**
 indirect method **45–6**
 methods for cryostat sections **594**
 for oestrogen receptors on breast tumour cell
   line **269–70**
 PAP method for immunoglobulins **593–5**
 quantitative standardization 680–1
 for skeletal muscle components 665
 for tumour markers 505, **521**
Immunostaining
 sample schedules **44–8**
Imposil
 marker for electron microscopical immunocytochemistry 147–8
Indirect immunofluorescence **45**
 for autoantibodies 456–8, *Fig*, **28.1**
 cytoplasmic
  non-specific staining 579–83, *Tables* 34.3,
   34.4
 deposits in glomerular disease 638–48, *Figs.*
   37.1, 37.2, 37.4, *Table* 37.1
 in diseased skeletal muscle 666–7
 light chains
  monotypic in B-cell lymphoma 579
 in lymphomas
  distinction between synthesized and taken
   up 579–83, *Figs.* 34.13–34.17, *Table*
   34.17
 normal B-cell expression 570–2, *Fig.* 34.2,
   573–5 (*Figs.* 34.3–34.6)
 structure 13, 14 (*Fig.* 2.1)
 surface Ig in follicle centre cell lymphomas
  interpretation of staining 583
 in thyroiditis 539
Immunogold staining
 for cell surface antigens 72, *Plate* 4
 combined with colloidal silver 72
 combined with cytochemistry 72
  for blood cells 600, 608 (*Fig.* 35.3), 609
   (*Fig.* 35.4), 610 (*Fig.* 35.5), **613**
 for microtubules 73, *Plate* 5
 in paraffin and resin sections 72
 to define receptor mechanisms 267
 surface markers
  electron microscopy **613**
 transferrin receptors on squamous carcinoma
   268 (*Fig.* 16.9)
 on cultured cells and cryostat sections **486–7**
 endocrine cells of colon 334 (*Fig.* 20.3)
 on cultured cells
  surface or extracellular antigens **269–70**
Indirect method 26, 45–6
Insects
 gastrin/cholecystokinin-like immunoreactivity
   318–9, *Fig.* 19.2

Insects (*cont.*)
  insulin-like immunoreactivity 313
in situ hybridization (see Hybridization histo-
    chemistry)
Insulin
  similarity to prothoracotrophic hormone 313
Insulin-like immunoreactivity
  in insects 313
Interdigitating reticulum cells
  in normal lymphoid tissue 571 (*Table* 34.1),
    577, *Fig.* 34.10
Intermediate filament proteins 390–422
  antibodies
    cross-reactivity 398
  in tumour diagnosis 392–8, *Tables* 23.1, 23.2,
      Plate 27
  central nervous system differentiation 401–9
  co-localization in some tumours 394–5
  cytokeratins 390–8, 519–20
  in cytological diagnosis 554–5
  desmin
    in pituitary 426–7, 428 (*Fig.* 26.2)
    in skeletal muscle 667–9
    tissue and tumour specificity 390–2
    in tumours 392–5, *Table* 23.2, *Plate* 27
  in diffuse neuroendocrine system 332
  glial fibrillary acidic protein 390–2, *Table* 23.1
    in brain development 258
    in CNS 415–8
    in gastrointestinal tract 251, *Fig.* 15.4
    in pituitary 427–8, 430 (*Figs.* 26.3, 26.4),
        431 (*Figs.* 26.5, 26.6)
    in nervous system differentiation 401–2
    in tumours 417–8, *Table* 25.1
  homologies between classes 413, 414
  immunocytochemistry 390–9
    frozen sections **398–9**
    problems 398
    tissue preparation **398**
    in tumour diagnosis 392–8, *Tables* 23.1,
        23.2, *Plate* 27, 519
  keratin
    in pituitary 428–9, 432 (*Figs.* 26.7, 26.8)
  markers in tumour typing 392–8, *Figs.* 23.2,
      23.3, *Tables* 23.1, 23.2, *Plate* 27
  in nervous system
    differentiation 401–9
    immunocytochemical methods **420–3**
  in neuropathology
    neurofilaments and glial filaments 413–22
  in skeletal muscle 667–9
  species differences 402, 415
  specificity of antibodies 413, 414, 415
  tissue specificity 390–2, *Table* 23.1, 401, 414
  as tumour markers 519, 520 (*Fig.* 31.6)
  in tumours of the nervous system 415–8, *Table*
      25.1
  vimentin 390–5
    in tumours 392–5, *Fig.* 23.2, *Table* 23.2,
        Plate 27

Intracellular antigens
  in cell cultures 139–41, 248–52, *Figs.* 15.3–
      15.5)
Islet cell antibodies
  to cell surface 461–2
  complement fixing 461
  cytoplasmic 459–61, *Fig.* 28.2

J-chain
  B-cell lymphoma 580
  normal B-cell 571 (*Table* 34.2), 572, 574 (*Fig.*
      34.5)

Keratin (see also Cytokeratins)
  antibodies for epithelial cell characterization
      262–6, *Figs.* 16.1–16.5
  monoclonal antibody LICR-LON-1 263–4,
      Figs. 16.3–16.5
  in pituitary 428–9, 432 (*Figs.* 26.7, 26.8)
  in thyroid papillary carcinoma 535–6, *Fig.*
      32.4
  variety of proteins 263
Keratinocytes in culture 262
Kidney diseases (see also Glomerular disease)
  glomerular diseases 638–48
  systemic 646–7
Kidney transplants
  cellular infiltrate
    T4:T8 ratio 497–8
    immunocytochemistry 497–9, *Table* 30.2
Köhler illumination 229–30

Labelling of antibodies 34
Labels
  colloidal gold 71–85, 115–41
  colloidal metals 36
  for electron microscopical immunocyto-
      chemistry 146–8
  for enzymes 34–5
  fluorescent 34
  haptens 35–6
  radioactive 99–113
Laminin
  in skeletal muscle 666, 668 (*Fig.* 39.3)
Langerhans cell
  markers 632–5, *Table* 36.3
  OKT 6 as marker 632–5, *Fig.* 36.12
Lectin cytochemistry
  of skeletal muscle 666, 667, 668 (*Fig.* 39.6),
      669, 670 (*Fig.* 39.7)
Lectin histochemistry 167–86
  applications 180–5, *Figs.* 10.8–10.13
  buffer and pH 185
    carbohydrate heterogeneity revealed 180

Lectin histochemistry (*cont.*)
  carbohydrate structure 167–9
  detection of binding to tissues
    antibody method **173–4**, *Fig*. 10.4
    biotinylated lectin **175**, *Fig*. 10.5
    carbohydrate-conjugated label **175–6**, *Fig*.
      10.6
  cell markers 599–614
Leucocyte common antigen 556, 570, 577–8,
  Fig. 34.12
Levamisole
  inhibitor for endogenous alkaline phosphatase
    35
Lymphocytes
  B-lymphocytes
    heterogeneity 439 (*Fig*. 27.1), 441–3
    in normal lymphoid tissue 570–72, *Figs*.
      34.1, 34.2, *Table* 34.2, 573–5 (*Figs*.
      34.3–34.6)
  T-lymphocytes
    monoclonal antibodies 444 (*Table* 27.2),
      449, 450 (*Fig*. 27.4), 622 (*Table*
      36.2), 629–31
    in heart and kidney transplants 492–9
    in normal lymphoid tissue 571 (*Fig*. 34.1,
      Table 34.1), 572, 575 (*Fig*. 34.7), 576
      (*Fig*. 34.8)
    in polymyositis 699, *Plate* 44
    in skin diseases 627–32
Lymphocytic lymphoma 584
Lymphoid cells
  fixation 449–51, *Table* 27.4
  paraffin-embedded for immunofluorescence
    450–1, *Table* 27.4, 452 (*Fig*. 27.5)
Lymphoid leukaemia
  B-chronic disorders 606–7, *Table* 35.6
  T-cell disorders 601–6, *Table* 35.5
Lymphoid lineage in leukaemia 601, 603 (*Fig*.
  35.1)
Lymphoid system
  advances in analytical methods 438–40, *Fig*.
    27.1
  B-cell heterogeneity 439 (*Fig*. 27.1),
    441–2
  double immunofluorescence 438–53
    controls 176–8
    direct method **172–3**, *Fig*. 10.3
    methods 172–86
    enzyme treatment – non-specific
      proteolysis 178
    enzyme treatment – specific 178–80, **186**, *Fig*.
      10.7
    glycosidases 179–80, *Fig*. 10.7
    neuraminidase 179, **186**
  interpretation of binding 180
  glycoconjugate structure 167–9
  immobilized sugars for isolating lectins 168–9,
    Fig. 10.1
  neoglycoproteins for determining binding site
    168

Lectins
  definition 167
  isolation 171–2
  occurrence 169
  selective reagents for oligosaccharides 170–1,
    Fig. 10.2
  structure and nomenclature 169–70, *Table*
    10.1
  toxicity 185
Legionella pneumophila
  detection by immunocytochemistry 660
Leishmania tropica major
  surface antigen localization 651, *Plate* 42
Leucocyte common antigen
  in cytological diagnosis 556
  in lymphoma diagnosis 577, 578 (*Fig*. 34.12)
  in normal lymphoid tissue 570
Leukaemia
  macrophage heterogeneity 443–9, *Figs*. 27.2,
    27.3
  methods 441–3, *Table* 27.1
Lymphoid tissue (normal)
  B-cells 570–2, *Figs*. 34.1, 34.2, 573–5 (*Figs*.
    31.3–31.6)
  leucocyte common antigen 570
  macrophages and reticulum cells 571 (*Table*
    34.1), 572–7 *Figs*. 34.8, 34.9
  preparation for immunocytochemistry 440–1,
    449–51, *Table* 27.4
  T-cells 571 (*Fig*. 34.1, *Table* 34.1), 572, 575
    (*Fig*. 34.7), 576 (*Fig*. 34.8)
Lymphomas 577–92
  B-cell 579–89
  centrocytic 588–9, *Fig*. 34.23
  cutaneous T-cell 627–32, *Fig*. 36.9, 633 (*Figs*.
    36.10, 36.11), 635 (*Fig*. 36.12)
  follicle centre cell 581 (*Figs*. 34.13, 34.14), 582
    (*Figs*, 34.15, 34.16), 583 (*Fig*. 34.17),
    585–7
  histiocytic 589–91, *Fig*. 34.26
  Hodgkin's 592, *Fig*. 34.27
  immunoblastic 588
  lymphocytic 584
  lymphoplasmacytic and lymphoplasmacytoid
    584, *Fig*. 34.18)
  plasmacytoma 585
  T-cell 588 (*Fig*. 34.23), 589, 590 (*Figs*. 34.24,
    34.25)
    cutaneous 627–32, *Fig*. 36.9, 633 (*Figs*.
      36.10, 36.11), 635 (*Fig*. 36.12)
  thyroid 539–10
Lymphoplasmacytic lymphoma 584, *Fig*. 34.18
Lymphoplasmacytoid lymphoma 584
Lymphoreticular tumours
  immunocytochemistry 568–96
  diagnostically useful antigens 569 (*Table* 34.1)
  fixation 569–70
  paraffin sections 569–70
    protease digestion 570
Lysozyme in macrophages 572, 576 (*Fig*. 34.9)

Macrophages
  heterogeneity 443–9, *Figs.* 27.2, 27.3
  in normal lymphoid tissue 571 (*Table* 34.1),
    572–7, *Figs.* 34.8, 34.9
Malarial antigens
  monoclonal antibodies 652
Malarial parasite
  monoclonal antibodies
    screened on semithin frozen sections 90–3
Medullary carcinoma of thyroid
  calcitonin and other peptides in 510, *Fig.* 31.2,
    537–9, *Figs* 32.6–32.8
  carcinoembryonic antigen in 538–9, *Fig.* 32.8
  differential diagnosis 540–2
  histaminase in 511, 537
Medullary-follicular carcinoma
  thyroglobulin and calcitonin in 540
Megakaryoblastic leukaemias
  monoclonal antibodies to platelets 608–9
Melanoma
  antigens 518–9
  S–100 577, 578 (*Fig.* 34.12)
Messenger RNA
  analysis by hybridization histochemistry 198–
    203
Micro-organisms
  and autoimmune diseases 653
  cross-reactions with tissues 652–4, 655 (*Fig.*
    38.3), 656 (*Fig.* 38.4)
  immunocytochemistry 650–61
Microscope objectives
  achromatic 227
  apochromatic 228–9
  choice and use 226–9, *Fig.* 14.1
  fluorite 228
Molluscs
  neuropeptides 311–2, 314 (*Table* 19.1)
Monochrome emulsions for photomicrography
    234–6
Monoclonal antibodies
  bulk production 23
  to carcinoembryonic antigen 514
  class and sub-class testing 22
  to class I and II histocompatibility antigens
    in heart transplants 494–6, *Plates* 32, 33
  definition 15–16
  double immunofluorescence of B-
    lymphocytes 441–3, 444 (*Table* 27.2)
  double immunofluorescence of macrophages
    443–9, *Figs.* 27.2, 27.3, *Table* 27.3
  double labelling with colloidal gold for elec-
    tron microscopical immunocytochemis-
    try 156
  in haematology 599–610, *Tables* 35.1–35.3
  to malarial parasite 89
  38 D7 marker for peripheral nervous system
    254, 255 (*Fig.* 15.6)
  to oestrogen receptors 194–5, 266–7, *Fig.* 16.6
  to platelets 608–9
  raising 13–25

Monoclonal antibodies (*cont.*)
  screening 20
  species of origin 18
  storage 24, 441
  T-lymphocyte analysis 444 (*Table* 27.2), 449,
    450 (*Fig.* 27.4)
    in heart transplants 494, *Table* 30.1
Monoclonal antibody
  to cytokeratin 263–4, *Figs.* 16.3–16.5
  to human milk fat globule membrane
    cultured breast cells 266, *Plate* 13
  to epidermal growth factor receptors 267–8,
    Fig. 16.8
Morphometrical analysis
  of nerve cell bodies 213–4
  of nerve cell groups 206–13
  of the neuropil
    dendrites 213
    nerve terminals 214
Morphometry and microdensitometry
  aging process in neurones 220–3, *Table* 13.2
Mounting medium
  to retard fading of fluorescence 441
Multiple immunolabelling methods
  for electron microscopical immunocyto-
    chemistry 154–7
Multiple staining techniques 41–4
  double direct method 42
  double indirect method 42
  elution methods 42
  immunoenzymatic 43–4, *Plate* 2
  immunofluorescence 438–53
  immunogold 129–30, *Fig.* 8.9
  immunoperoxidase without elution 43
  'mirror image' sections 41
  radiolabelling with peroxidase **112**
  serial sections 41
Muscle (skeletal)
  immunocytochemistry 664–71
  class I (HL-A, B, C) antigens 666, 668 (*Fig.*
    39.4)
  collagen 665–6, 668 (*Figs.* 39.1, 39.2)
  desmin 667–9
  at electron microscopical level 665
  immunoglobulins and complement in
    disease 666–7, 668 (*Fig.* 39.5), 669
  intermediate filaments 667–9
  interstitial lymphocytes 670–1, *Plate* 44
  laminin and fibronectin 666, 668 (*Fig.* 39.3)
  lectin cytochemistry 666, 667, 668 (*Fig.*
    39.6), 670 (*Fig.* 39.7), 669
  myosin 669, 670 (*Fig.* 39.9)
  α- and β-spectrin 667, 670 (*Fig.* 39.8)
Myasthenia gravis
  immunocytochemistry for autoantibodies
    465–6
Mycobacterium tuberculosis antibody
  cross-reaction with DNA 654, 656 (*Fig.* 38.4)
Mycoplasma pneumoniae
  monoclonal antibody 651, *Fig.* 38.1

Myelin synthesis
  immunocytochemistry of nervous system cell
    cultures 255–8, *Fig.* 15.8
Myeloid lineage in leukaemia 607–10
Myeloma
  choice of cell line 18
  non-secretor 18
Myosin
  in normal and diseased skeletal muscle 669,
    670 (*Fig.* 39.9)

Nanometer particle video microscopy (see
    Nanovid ultramicroscopy)
Nanovid ultramicroscopy
  enhancement of gold marker 81, 82 (*Fig.* 5.7)
Neoplasms (see also Tumours)
  intermediate filaments in diagnosis 415–8,
    Table 25.1
  neuroendocrine 341–3
Nerve cell bodies
  demonstration in spinal cord 370
  morphometrical analysis 213–4
Nerve cell groups
  morphometrical analysis 206–13
Nerve fibres
  preparation of tissue 31–2, 363–5
Nerves
  peptide-containing
    in gastrointestinal tract 333–4, *Fig.* 20.4,
      336
    in spinal cord 365–85, *Table* 22.2
Nerve tracing techniques
  combined with immunocytochemistry 373–4
Nervous system
  brain
    development 258, 401–9
    peptides in neurodegenerative diseases
      349–58
    quantitative analysis of immunostained
      neurones 109, 205–23
    radioimmunocytochemistry of peptides 99–
      110
  cultures of peripheral nervous system 245–58
  diffuse neuroendocrine system 328–44
  intermediate filaments in pathology 413–22
  invertebrates 309–17
  spinal cord immunocytochemistry 360–85
Neurodegenerative diseases
  Alzheimer's disease and Huntington's chorea
    somatostatin and neuropeptide Y 349–58
  neurofilaments and glial filaments 418–9,
    Table 25.2
Neuroendocrine neoplasms 341–3
  glucagon-immunoreactive tumours 342 (*Fig.*
    20.8)
Neurofibrillary tangles 353, 355 (*Fig.* 21.1)
  neurofilaments 418–9, 421 (*Fig.* 25.3)
  somatostatin and NPY 354

Neurofilament proteins 414–5
  in Alzheimer's disease 418–9
  co-localized with cytokeratins 394
  in diffuse neuroendocrine system 382
  distribution 414, 415 (*Fig.* 25.1), 420 (*Fig.*
    25.2)
  in neuropathology 413–22
  not tumour specific 417
  in senile plaques and neurofibrillary tangles
    418–9, 421 (*Fig.* 25.3)
  in tumours 417–8, *Table* 25.1
Neurofilament protein triplet
  axon-specific antibodies 403–5, *Figs.* 24.1,
    24.2
  co-localized with vimentin 403
  in CNS development 402–3
  in Purkinje cell development 405–9, *Figs.*
    24.5, 24.6
  tissue specificity and exceptions 390–2, *Tables*
    23.1, 23.2
Neuron-specific enolase
  marker for the diffuse neuroendocrine system
    331
  marker for neuroendocrine tumours 341
  marker for small cell carcinoma of lung 341,
    342 (*Fig.* 20.9), 555–6, *Plate* 38
  in pituitary 425
Neuropathology
  neurofilaments and glial filaments 413–22
  somatostatin and NPY in Alzheimer's disease
    and Huntington's chorea 349–58
Neuropeptides
  annelids and platyhelminthes 310–1
  arthropods 312–6
  brain 349–58
  coelenterates 309–10, *Fig.* 19.1, 321–2
  diffuse neuroendocrine system 328–44
  fixation
    p-benzoquinone 31, 248–9, 383
  immunocytochemistry 99–113, 309–16, 328–
    44, 349–58, 361–85
  molluscs 311–2, 314 (*Table* 19.1)
  quantitative analysis 109, 205–23
  radioimmunocytochemistry 99–110
  spinal cord 360–85
Neuropeptide Y see NPY
NPY
  co-localization with somatostatin 352
    in Huntington's chorea 356
  in central nervous system 352
  in Huntington's chorea 356–7, *Fig.* 21.2, *Plate*
    26
Neurotransmitters
  in spinal cord 360–85
    histochemical mapping 361
    peptides as candidates 365–6
Neutral buffered formol saline fixation 569–70
Non-specific staining 38
  active groups in tissue 39
  binding of second antibody to tissue Ig 39

Non-specific staining (*cont.*)
  cross-reactivity 37, 40
  endogenous enzyme 40
  Fc-receptor binding 39
  formaldehyde-induced fluorescence 40
  ionic binding 39
  intrinsic fluorescence 40
  polytypic immunoglobulin in lymphomas 579, *Tables* 34.3, 34.4
  prevention by borohydride 506, **521**
  in hybridization histochemistry 275
Normal cross-reacting antigen
  reaction with CEA 513–4
Noradrenaline (see also Amines) 284–6, *Fig.* 18.1
  distribution in mammals 284
Numerical aperture of microscope objectives 226–7

Oat cell carcinoma
  ACTH in 512
  antidiuretic hormone in 512
  calcitonin in 510
Oestrogen binding sites and receptors
  cytochemical localization with labelled oestradiol 190–3
  histological methods 189, *Table* 11.1
    problems 193–4
  immunocytochemical localization 189–90, 194–5
  specificity of the methods 193–4
Oestrogen receptors 198
  antibodies 194–5
  assays 188–9
  cytochemistry with labelled oestradiol 190–3
  immunocytochemical localization 189–90, 194–5
  localization on breast tumour cell line 266–7, *Fig.* 16.7, **270–1**
Oncofetal antigens and yolk sac products
  as tumour markers 513–5
Oncogenes
  definition 273
  expression of *Ras*^Ha in virus-induced tumours 279–80, *Plates* 16, 18–21
  immunocytochemistry 273–81
    applications 279–80
  differential expression of *Ras* in malignant and benign lesions 279–80, *Plates* 16, 18–21
  non-specific staining 278, *Plates* 14, 15, 17
  v-*raf* and v-*myc* proteins 280, *Plates* 17, 22
  in situ hybridization of mRNA 274–7
    applications 277
    autoradiography 276
    conditions 275
    pre-treatment of cells – fixation 275
    pre-treatment of cells – proteinase digestion 275–6

Oncogenes (*cont.*)
  in situ hybridization of mRNA (*cont.*)
    probes 274–5
    sensitivity 276
On-grid procedure
  for electron microscopical immunocytochemistry 149–50, **160–1**
  immunogold method 151 (*Fig.* 9.2), **160–1**

Pancreas
  endocrine cells 336–7, *Fig.* 20.5
  glucagon-immunoreactive tumour 342 (*Fig.* 20.8)
  hyperplasia of insulin cells 336–7
  hypoplasia of somatostatin cells 337
  regulatory peptides 336–7
Pancreatic islets
  phylogeny 321
Paraffin sections
  for amine immunocytochemistry 292–4, 216 (*Fig.* 18.10)
  for brain neuropeptides 52
  of cytological specimens 563–4
  for immune complex deposits in renal biopsies 639
  immunogold staining 72–3
  immunogold-silver staining 75–6
  for intermediate filaments in neuropathology 421
  of lymphoid tissue 569–70
  for oestrogen receptors 190
Parasites
  semithin frozen sections 89–97
  surface antigen localization 651–2
Peptides
  in brain 349–58
  co-localization
    in dorsal root ganglia 374–6
    somatostatin and NPY in brain 352
  cross-reactivity
    gastrin and cholecystokinin 37–40
  in diffuse neuroendocrine system 328–44
  evolution 308–22
    gastrin/cholecystokinin 317–21
  fixation
    p-benzoquinone 31, 248–9, 383
    freeze-drying and vapour fixation 32
  as neurotransmitters 365–6
  quantitative analysis in neurons 109, 205–23
  radioimmunocytochemistry 99–110
  in spinal cord 360–85
Periodate-lysine-paraformaldehyde 100
  fixation of lymphoid cells 449–51, *Table* 27.4, 452 (*Fig.* 27.5)
Permeabilization
  of cultured cells
    chemical 248, 250, *Table* 15.1
    by detergents **140**, 249–50

Permeabilization (*cont.*)
  of cultured cells
    by freezing 250
    by protein precipitants/lipid extractants 249
  of cytological preparations 551
  of spinal cord **384**
  of tissue preparations 31, 32
Peroxidase 34
  in avidin-biotin methods 55, 57–60
Peroxidase – anti-peroxidase method 27, *Fig.*
    3.1, **45–7**
Peroxidase-conjugated oestrogen 190–93, *Figs.*
    11.1, 11.2
Peroxidase development
  incubating solution 46
  inhibition of endogenous enzyme 35, 40, **66**,
    383–4, 506, **521**
  in electron microscopical immunocytochemis-
    try 146, 147 (*Fig. 9.1*)
Phenylethanolamine-N-methyltransferase
  in adrenaline synthesis 286
Photomicrography 225–42
  exposure 241–2
  colour films 239–41
  contrast control 236–9, *Fig. 14.6*
  film size and material 23–6
  image transferred to film 230–34, *Figs. 14.3,*
    14.4
  monochrome emulsions
    contrast 235
    resolving power 235
    sensitivity to colour 236, 237 (*Fig. 14.5*)
    speed 235
Phycoerythrin 34
Pituitary hormones
  as tumour markers 511–2
Pituitary
  non-hormonal markers 425–35
    actin 425–6, 427 (*Fig. 26.1*)
    antibodies 426 (*Table 26.1*)
    desmin 426–7, 428 (*Fig. 26.2*)
    glial fibrillary acidic protein 427–8, 430
      (*Figs. 26.3, 26.4*), 431 (*Figs. 26.5,*
      26.6)
    S-100 429, 433 (*Figs. 26.9, 26.10*), 434 (*Figs.*
      26.11, 26.12), 435 (*Fig. 26.13*)
    keratin 428–9, 432 (*Figs. 26.7, 26.8*)
Placental alkaline phosphatase
  as tumour marker 509
Plasmacytoma 585
Plasmodium berghei
  sporozoite surface antigen 651–2
Plastic sections
  for amine immunocytochemistry 294 (*Fig.*
    18.8), 295–6
Platyhelminthes
  neuropeptides 310–11
Polarized light epiluminescence
  for silver grains in hybridization histochemis-
    try *Plate 9*

Polyethyleneimine
  marker for electron microscopical immuno-
    cytochemistry 148
Poly(L-lysine) 32, **48**, 364
  coating slides for cytological smears **562**
Pregnancy-specific glycoprotein
  as tumour marker 509
Pro-epidermal growth factor
  identified by hybridization histochemistry
    201
Progesterone receptors 193
Prolactin
  as tumour marker 512
Pro-opiomelanocortin
  in situ hybridization
    combined with retrograde labelling in
      hypothalamus 201, *Plate 9*
    in pituitary 199 (*Fig. 12.1*), 200, 202
Prostate antigen
  as prostatic tumour marker 517–8
Prostatic acid phosphatase
  as prostatic tumour marker 517
Protease digestion 29, **48**, 421, 425
  lectin histochemistry 178, **186**
  lymphoreticular tumours 570
  oncogene hybridization histochemistry
    276
Protein A/gold probe
  preparation **138–9**
  on-grid labelling procedure **160–61**
Protein synthesis
  identified by mRNA 198–203
Prothoracotrophic hormone
  similarity to insulin 313
Protochordates
  endocrine cells of gut 315 (*Table 19.2*), 316
  neuropeptides 315 (*Table 19.2*), 316, 320 (*Fig.*
    19.4), *Plate 24*
  thyroglobulin 321
Purkinje cells
  neurofilament protein triplet during develop-
    ment 405–9, *Figs. 24.5, 24.6*

Quantitative analysis
  by densitometry of radioimmunocytochemical
    staining 103–4, *Table 7.1*
  peptide levels in CNS
    by radioimmunocytochemistry 100
  of immunocytochemical staining of neurones
    by morphometry and microdensitometry
      205–23
  antigen content 15–6, 219–20, *Fig. 13.12,*
    Plate 10
  co-exsistence of neuroactive substances in
    nerves 214–9, *Figs. 13.8–13.11*
  occlusion method 214, 215–9, *Fig. 13.8*)
  overlap method 214–5

Radioactive ligands
  internally labelled monoclonal antibodies 99
  tritiated biotin-avidin complex 100
Radioimmunocytochemistry 99–113
  analysis of data by densitometry 103–4, *Fig.*
    7.4, 106 (*Fig. 7.5*)
  autoradiography with LKB Ultrofilm 101, 103
    (*Fig. 7.3*), **111**
  double labelling with immunoperoxidase 101,
    104–7, 108 (*Fig. 7.7*), **111**
    advantages 107–9
  for electron microscopy **113**
  immunocytochemical techniques 100, 101
    (*Fig. 7.1*), **110–33**
  peptides in CNS
    light microscopy 101, 102 (*Fig. 7.2*), 103
      (*Fig. 7.3*)
    electron microscopy 104, 106 (*Fig. 7.6*)
    semiquantitative analysis 103, 104 (*Table*
      7.1), 109
  tissue preparation
    for light microscopy 100
    for electron microscopy 100
  semiquantitative analysis
    advantages 109
    limitations 109
Radiolabels
  DNA for in situ hybridization 275
  for electron microscopical immunocyto-
    chemistry 148
  tritiated biotin-avidin 100
Receptors (see also Oestrogen receptors)
  for epidermal growth factor 267
  immunocytochemistry 266–8
  mechanism defined by immunogold staining 267
  steroid
    localization 188–95
Reed-Sternberg cells
  lectin binding 184
  specific uptake of immunoglobulins 580,
    Table 34.3
Regulatory peptides (see also Peptides)
  in agnathans 317, 320 (*Fig. 19.3*)
  in brain 349, *Table* 21.1
  in cardiovascular system 340–41
  in diffuse neuroendocrine system 328–44
  in gastrointestinal tract 333–6
  in genital tract 338–9
  in urinary system 339–40
  evolution 308–22
  immunocytochemistry 328–44
  in enteric nerves 333–4, *Fig.* 20.4, 336
  fixation and tissue preparation 329
  in nerves 330–1
  in neuroendocrine neoplasms 341–3
  in pancreas 336–7
  in protochordates 315 (*Table* 19.2), 316. 320
    (*Fig. 19.4*), *Plate* 24
  in respiratory tract 337–8
  in spinal cord 360–85

Renal biopsies
  immunocytochemical technique **497**
Renal diseases
  glomerular diseases 638–46
  systemic diseases 646–7
Resin sections
  for amine immunocytochemistry 294 (*Fig.*
    18.8), 295–6
  etching for electron microscopical immuno-
    cytochemistry 153–4
Respiratory syncytial virus
  diagnosis with monoclonal antibodies 657–8
Respiratory tract
  CGRP-immunoreactive nerves and endocrine
    cells 337–8, *Fig.* 20.6
  diffuse neuroendocrine system 337–8
    pathology 38
  regulatory peptides 337–8
  small cell carcinoma
    neuron-specific enolase 341, 342 (*Fig.* 20.1)
Reticulum cells
  in normal lymphoid tissue 572–7, 578 (*Fig.*
    34.11)
Rhodamine 34

Selected surface immunolabelling 155 (*Fig.* 9.5),
    156 (*Fig.* 9.6), **163–4**
Semithin frozen sections
  advantages for immunocytochemistry 89
  fixation **93–4**
  freezing tissue blocks **94**
  immunofluorescence
    improved resolution 91 (*Fig.* 6.1), 92
  production **93**
  sectioning **95–6**
  sucrose infusion 9
  transfer to slide and subsequent treatment
    96–7
  trimming tissue blocks **95**
Semithin-thin procedure
  for electron microscopical immunocyto-
    chemistry 149
Senile dementia (see Alzheimer's disease)
Senile plaques 353, 355 (*Fig.* 21.1)
  neurofilaments 418–9, 421 (*Fig.* 25.3)
  neuropeptides 354, 355 (*Fig.* 21.1)
Sensitivity
  increasing 41
Serotonin (see also Amines) 285, 286, 287 (*Fig.*
    18.2)
  distribution in mammals 285
Skeletal muscle (see Muscle (skeletal))
Skin
  basement membrane zone
    structure 619–20, *Fig.* 36.1
  bullous disorders 620–7
Skin diseases
  immunocytochemistry 618–36

Small cell carcinoma of lung
  cytological diagnosis
    neuron-specific enolase 555, *Plate* 38
Small cell carcinoma of thyroid 541
Somatostatin
  in Alzheimer's disease 353–5
  in CNS 352
  co-localization with NPY 352
    in Huntington's chorea 356
  in Huntington's chorea 356–7
  in neurofibrillary tangles 354, *Plate* 25
  in senile plaques 354, 355 (*Fig.* 21.1)
Spacer arm
  in biotinylation 57
Specificity 36
  of antibodies 67
α- and β-Spectrin 667, 670 (*Fig.* 39.8)
Spinal cord
  cytoskeletal proteins 361, 363 (*Fig.* 22.1)
  dissection techniques 363–4
  fixation for immunocytochemistry 364
  immunocytochemical methods 364–5, **383–4**
  neurotransmitters
    histochemical mapping 361
  investigating origin of extrinsic peptide-
      containing nerves
    capsaicin 366–8, *Fig.* 22.4
    ganglionectomy and dorsal rhizotomy 368
        (*Fig.* 22.3), 370, 371 (*Fig.* 22.5)
    peripheral nerve lesions 370, 371 (*Fig.*
        22.5)
  investigating origin of intrinsic peptide-
      containing nerves
    anterograde and retrograde tracing 373–4
    brain and brain stem lesions 373
    colchicine and vinblastine 100, 105 (*Fig.*
        7.4), 370, 372 (*Fig.* 22.6)
    cultures 374
    hemisection and transection 370–73
  peptide-containing nerves
    distribution 374–82
    origin 366–74, *Fig.* 22.3
  peptide localization 366, 367 (*Table* 22.2)
S-100 protein
  in histiocytic tumours 591
  in interdigitating reticulum cells 577
  in malignant melanoma 577, 578 (*Fig.* 34.12)
  in pituitary 429, 433 (*Figs.* 26.9, 26.10), 434
      (*Figs.* 26.11, 26.12), 435 (*Fig.* 26.13)
  as tumour marker 518–9
Squamous carcinomas
  cell lines in culture 262–3, *Fig.* 16.2
    epidermal growth factor receptors 267, *Fig.*
        16.8
    keratin expression defined by monoclonal
        antibody 263–4
    transferrin receptors 267, 268 (*Fig.* 16.9)
Steroid hormones and receptors 513
Steroid receptors (see also Oestrogen receptors)
  localization 188–95

Storage
  of antibodies 33–4
  of cytological preparations 551
  of diaminobenzidine **46**
  monoclonal antibodies 24, 441
  of peroxidase conjugates 506
Streptavidin 55
Streptavidin/gold probes
  preparation **138–9**
Streptococcus mutans antibody
  cross-reaction with heart tissue 654, 655 (*Fig.*
      38.3)
Subacute infective endocarditis
  glomerular immunocytochemistry 646–7
Substage condenser
  Köhler illumination 229–30, 231 (*Fig.* 14.2)
  in transmitted light microscopy 229–30, 231
      (*Fig.* 14.2)
Surface antigens
  on cultured cells
    indirect immunoperoxidase method **269–70**
Systemic lupus erythematosus
  autoantibodies to DNA 479
  autoantibodies to histones 479–80
  glomerular immunocytochemistry 646
Systemic renal disease 646

T-cells (see also T-lymphocytes)
  antibodies 622 (*Table* 36.2), 629–31
  in cardiac biopsies
    influence of immunosuppressive drugs 494,
        495 (*Figs.* 30.2, 30.3)
  lymphoma 588 (*Fig.* 34.23), 589, 590 (*Figs.*
      34.24, 34.25)
  lymphoproliferative disorders 604–6, *Table*
      35.5
  monoclonal antibodies 444 (*Table* 27.2), 499,
      450 (*Fig.* 27.4)
  in normal lymphoid tissue 571 (*Fig.* 34.1,
      Table 34.1), 572, 576 (*Fig.* 34.8)
  ontogeny
    antibody reactivity 622 (*Table* 36.2)
  in polymyositis 669, *Plate* 44
  T4:T8 ratio in peripheral blood
    as index of graft rejection 496–7
    in kidney transplantation 499
  T4:T8 ratio in rejected kidney graft 497–8
  in skin diseases 627–32, *Fig.* 36.9, 633 (*Figs*
      36.10, 36.11), 635 (*Fig* 36.12)
Techniques – detailed procedures
  colloidal gold
    immunogold-silver staining **83–5**
    minimum protecting amount of protein **137**
    preparation of gold sols **133–7**
    preparation of immunogold probes for
        electron microscopy **137–8**
    preparation of streptavidin- and protein
        A-gold probes **138–9**

Techniques–detailed procedures (*cont.*)
  cytological preparations
    wax embedding of cell pellets **563–4**
  electron microscopical immunocytochemistry
    cell suspensions **613**
    cultured cell monolayers **139–41**
    etching with sodium metaperiodate **161**
    on-grid procedures **160–1**
    pre-embedding PAP **159–60**
    radioimmunocytochemistry with avidin-
      tritiated biotin complex **111**
    selected surface double labelling **161–2**
    ultrathin frozen sections **158–9**
  endogenous peroxidase inhibition **383**, **521**
  fixation
    p-benzoquinone **49**, **383**
    semithin frozen sections **90–1**, **93–4**
  fluorescence
    prevention of fading by mountant **441**
  hybridization histochemistry
    POMC in rat hypothalamus **202–3**
  immunostaining schedules
    alkaline phosphatase-anti-alkaline phos-
      phatase **47–8**
    avidin-biotin methods **65–8**
      combined with radioimmunocytochemis-
        try **110–3**
    cultured cells **252–3**, **269–71**
      for autoantibodies **486–7**
    dinitrophenyl-hapten sandwich **542–4**
    immunofluorescence **45**
      in cell suspensions **610–11**
      on cytocentrifuge preparations **612**
    immunogold-silver **83–5**
    immunoperoxidase **45–46**, **521**, **594**, **612**
      for receptors on live tissue culture cells
        269–70
    multiple immunostaining
      double immunofluorescence for autoanti-
        bodies **457** (*Fig*. 28.1)
      double immunofluorescence on cultured
        cells **270**
      radioimmunocytochemistry with per-
        oxidase **112**
    peroxidase-anti-peroxidase **46–7**, **593**
      on cervical smears **564**
    radioimmunocytochemistry **110–2**
    thin frozen sections **92**
  lectin histochemistry
    buffered composition and pH **185**
    detection of binding to tissue **172–6**
    enzyme digestion **186**
    simple sugar solutions **186**
  permeabilization
    of cultured cells **248–50**
    of spinal cord section **384**
  poly(L-lysine)-coated slides **48–9**, **562–3**
  preparation of semithin frozen sections **93–7**
  raising polyclonal antisera **5**
    affinity purification **6–7**

Techniques–detailed procedures (*cont.*)
  raising polyclonal antisera (*cont.*)
    to lectins **174**
    trypsin digestion **48**
Terminal deoxynucleotidyl transferase
  fixation 440
  as marker for acute lymphoblastic leukaemia
    601–5, *Plate* 39
Texas red 34
  for skeletal muscle immunocytochemistry 665
Thyroglobulin
  carrier in antibody production
    unexpected immunostaining 534, *Fig*. 32.1
  in follicular carcinoma 533–4, *Fig*. 32.3
  in follicular cells 533–4
  in papillary carcinoma 533–4, *Fig*. 32.2
  in protochordates 321
  in thyroid anaplastic carcinoma 536, 537 (*Fig*.
    32.5)
  in thyroid carcinoma 511
Thyroid autoantigens
  localization 463–5
Thyroid carcinoma
  thyroglobulin 511, *Plate* 35
Thyroid disease
  anaplastic carcinoma
    thyroglobulin in 536, 537 (*Fig* 32.5)
  differential diagnosis by immunocytochemis-
    try 540–2
  immunocytochemistry in diagnosis 533–44
  follicular carcinoma
    thyroglobulin in 533–6, *Fig*. 32.3
  medullary carcinoma
    calcitonin, katacalcin, CGRP 510, *Fig*.
      31.2, 537–8, *Fig*. 32.6–32.8
    familial or sporadic 542
  medullary-follicular carcinoma 540
  metastasis from thyroid 511, 541–2
  metastasis to thyroid 541
  papillary carcinoma
    keratin in 535–6, *Fig*. 32.4
    thyroglobulin in 533–4, *Fig*. 32.4, *Plate* 35
  small cell malignant tumours 541
  thyroglobulin 511, 533–4
  thyroiditis
    immunoglobulins in 539
Tissue powders
  for antibody purification 38
Tissue preparation for immunocytochemistry
  28–33, *Fig*. 3.2, 676–8 *Fig*. 40.1
  adherence to slides
    sections 32–3
    smears 562–3
  for amines 289–98
  cell and tissue cultures 248–50, 269–70, 486–7
  cytological specimens 548–51, 562–4
  for electron microscopy 150–4
  freeze-drying 31, 32, 89–97, 152
  frozen sections
    for amines 291–2

Tissue preparation for immunocytochemistry (*cont.*)
  frozen sections (*cont.*)
    for autoantibodies 28
    cardiac biopsies 493
    for in-situ hybridization 199–200, 202, 275–6
    for intermediate filaments 398–9, 420–2
    lymphoid tissue 440–1, 568–70
    muscle biopsies 664
    for oestrogen receptors 193
    for radioimmunocytochemistry 100
    renal biopsies 497
    semithin 93–7
    skin biopsies 497
    ultrathin 148
  lymphoid system 440–1, 568–70
  nerve fibres 31, 363–5
  for oestrogen receptors 193
  permeabilization (see also Protease digestion) 31, 32
    cell and tissue cultures 248–50
    cytological specimens 551
    for electron microscopy 159
    for hybridization histochemistry 200, 276
    spinal cord 384
  for peptides 329, 349, 352
  renal biopsies 638–9
  sectioning media (see also Paraffin sections) 32
    epoxy resins 32, 153
    methacrylate and other acrylic resins 32, 153
  smears, suspensions and cytocentrifuge preparations, 548–51, 562–3, 599–600, 610–3
  spinal cord 363–5 ⁻
  Vibratome (vibrating knife) sections 149, 290 (*Fig.* 18.4), 296–8 *Fig.* 18.11, 364
  whole mounts 31
T-lymphocytes (see also T-cells)
  in polymyositis 699, *Plate* 44
Transferrin receptors
  immunogold staining on cultured squamous carcinoma 268 (*Fig.* 16.9)
Trichomonas
  immunocytochemical detection 559
Troponin
  in skeletal muscle 669
Trypanosoma cruzi
  cross-reaction with neural tissues 653, *Plate* 43
Trypsin digestion (see also Protease digestion) **48**, 425
Tumour cell lines
  immunocytochemistry 261–71
Tumour markers 503–22
  α-1-antitrypsin 515
  α-fetoprotein 504, 514–5
  blood group substances 518
  Ca1 519, 555

Tumour markers (*cont.*)
  calcitonin
    for medullary carcinoma of thyroid 510–11, *Figs*. 31.2, 31.3, 536–8, *Fig*. 32.6
  carcinoembryonic antigen 502, 504, 513–4 *Fig*. 31.4, 537–9, *Fig*. 32.8, 553–4
  for CNS 415–8, 519
  criteria for ectopic hormone production 502–3, *Table* 31.1
  epithelial membrane antigen 516–7, *Fig*. 31.5, 554, 557
  ectopic and eutopic 507
  factor VIII 518
  fibronectin 515
  histaminase 511
  human chorionic gonadotrophin 504, 505, 507–9, *Fig*. 31.1
  human milk fat globule membrane antigen 516–7, 554 *Plate* 37
  human placental lactogen 509
  immunocytochemical methods 505, **521**
  intermediate filaments 392–8, *Tables* 23.1, 23.2, 519, 520 (*Fig*. 31.6)
  leucocyte common antigen 556, 570, 577–8, *Fig*. 34.12
  for melanomas 518–9
  neuron-specific enolase 341, 342 (*Fig*. 20.9), 555–6, *Plate* 38
  pituitary hormones 511–2
  placental alkaline phosphatase 509
  pregnancy-specific glycoprotein 509
  prostate antigen 517–8
  prostatic acid phosphates 517
  for soft tissue tumours 518
  steroid hormones and receptors 513
  sampling error 504
  thyroglobulin 511, 533–4, *Fig*. 32. 4, *Plate* 35
  Ulex europaeus lectin 184, 518
  value of localization 503–4
Tumours
  of the diffuse neuroendocrine system 341–3
  of the lymphoreticular system 562–96
  nervous system
    intermediate filament proteins 415–8, *Table* 25.1
  of the thyroid gland 533–43
Tyrosine hydroxylase
  in amine synthesis 236
  quantitative analysis of immunostained neurones 205–23

Ulex europaeus I lectin
  marker for endothelium 184, 518
Unlabelled antibody-enzyme methods 27–8
Urinary system
  diffuse neuroendocrine system 339–40
    pathology 340
    regulatory peptides 339–40

Vertebrate peptides
  gastrin and cholecystokinin 319–21
Vibratome (vibrating knife, free-floating) sec-
    tions 100, 149, 290 (*Fig*. 18.4), 296–8,
    Fig. 18.11, 364
Vimentin
  co-localization with other intermediate fila-
    ment proteins 394–5, 414
  in embryonic neuroglia 401

Vimentin (*cont.*)
  tissue and tumour specificity and exceptions
    390–2, *Table* 23.1 *Plate* 27
Vinblastine 100, 105 (*Fig*. 7.4)
  investigating origin of spinal cord nerves 370,
    372 (*Fig*. 22.6)

Whole mount preparations
  permeabilization 31